4 T Ellard

Geological Hazards in the UK: Their Occurrence, Monitoring and Mitigation
Engineering Group Working Party Report

The Geological Society of London
Books Editorial Committee

Chief Editor
Rick Law (USA)

Society Books Editors
Jim Griffiths (UK)
Dan Le Heron (Austria)
Mads Huuse (UK)
Rob Knipe (UK)
Phil Leat (UK)
Teresa Sabato Ceraldi (UK)
Lauren Simkins (US)
Randell Stephenson (UK)
Gabor Tari (Austria)
Mark Whiteman (UK)

Society Books Advisors
Kathakali Bhattacharyya (India)
Anne-Christine Da Silva (Belgium)
Xiumian Hu (China)
Jasper Knight (South Africa)
Spencer Lucas (USA)
Dolores Pereira (Spain)
Virginia Toy (Germany)
Georg Zellmer (New Zealand)

Geological Society books refereeing procedures

The scientific and production quality of the Geological Society's books matches that of its journals. Since 1997, all book proposals are reviewed by two individual experts and the Society's Books Editorial Committee. Proposals are only accepted once any identified weaknesses are addressed.

The Geological Society of London is signed up to the Committee on Publication Ethics (COPE) and follows the highest standards of publication ethics. Once a book has been accepted, the volume editors agree to follow the Society's Code of Publication Ethics and facilitate a peer review process involving two independent reviewers. This is overseen by the Society Book Editors who ensure these standards are adhered to.

Geological Society books are timely volumes in topics of current interest. Proposals are often devised by editors around a specific theme or they may arise from meetings. Irrespective of origin, editors seek additional contributions throughout the editing process to ensure that the volume is balanced and representative of the current state of the field.

Submitting a book proposal

More information about submitting a proposal and producing a book for the Society can be found at: https://www.geolsoc.org.uk/proposals

It is recommended that reference to all or part of this book should be made in one of the following ways:

Giles, D. P. & Griffiths, J. S. (eds) 2020. *Geological Hazards in the UK: Their Occurrence, Monitoring and Mitigation – Engineering Group Working Party Report.* Geological Society, London, Engineering Geology Special Publications, **29**.

Gamble, B., Anderson, M. and Griffiths, J. S. 2020. Hazards associated with mining and mineral exploitation in Cornwall and Devon, SW England. *Geological Society, London, Engineering Geology Special Publications*, **29**, 321–368, https://doi.org/10.1144/EGSP29.13

GEOLOGICAL SOCIETY ENGINEERING GEOLOGY SPECIAL PUBLICATION NO. 29

Geological Hazards in the UK: Their Occurrence, Monitoring and Mitigation Engineering Group Working Party Report

EDITED BY

D. P. Giles
Card Geotechnics Ltd, UK

and

J. S. GRIFFITHS
University of Plymouth, UK

2020
Published by
The Geological Society
London

THE GEOLOGICAL SOCIETY OF LONDON

The Geological Society of London is a not-for-profit organisation, and a registered charity (no. 210161). Our aims are to improve knowledge and understanding of the Earth, to promote Earth science education and awareness, and to promote professional excellence and ethical standards in the work of Earth scientists, for the public good. Founded in 1807, we are the oldest geological society in the world. Today, we are a world-leading communicator of Earth science – through scholarly publishing, library and information services, cutting-edge scientific conferences, education activities and outreach to the general public. We also provide impartial scientific information and evidence to support policy-making and public debate about the challenges facing humanity. For more about the Society, please go to https://www.geolsoc.org.uk/

The Geological Society Publishing House (Bath, UK) produces the Society's international journals and books, and acts as European distributor for selected publications of the American Association of Petroleum Geologists (AAPG), the Geological Society of America (GSA), the Society for Sedimentary Geology (SEPM) and the Geologists' Association (GA). GSL Fellows may purchase these societies' publications at a discount. The Society's online bookshop is at https://www.geolsoc.org.uk/bookshop.

To find out about joining the Society and benefiting from substantial discounts on publications of GSL and other Societies go to https://www.geolsoc.org.uk/membership or contact the Fellowship Department at: The Geological Society, Burlington House, Piccadilly, London W1J 0BG: Tel. +44 (0)20 7434 9944; Fax +44 (0)20 7439 8975; E-mail: enquiries@geolsoc.org.uk.

For information about the Society's meetings, go to https://www.geolsoc.org.uk/events. To find out more about the Society's Corporate Patrons Scheme visit https://www.geolsoc.org.uk/patrons.

Proposing a book

If you are interested in proposing a book then please visit: https://www.geolsoc.org.uk/proposals

Published by The Geological Society from:
The Geological Society Publishing House, Unit 7, Brassmill Enterprise Centre, Brassmill Lane, Bath BA1 3JN, UK

The Lyell Collection: www.lyellcollection.org
Online bookshop: www.geolsoc.org.uk/bookshop
Orders: Tel. +44 (0)1225 445046, Fax +44 (0)1225 442836

The publishers make no representation, express or implied, with regard to the accuracy of the information contained in this book and cannot accept any legal responsibility for any errors or omissions that may be made.

© The Geological Society, 2020 Except as otherwise permitted under the Copyright, Designs and Patents Act, 1988, this publication may only be reproduced, stored or transmitted, in any form or by any other means, with the prior permission in writing of the publisher, or in the case of reprographic reproduction, in accordance with the terms of a licence issued by the Copyright Licensing Agency in the UK, or the Copyright Clearance Center in the USA. In particular, the Society permits the making of a single photocopy of an article from this issue (under Sections 29 and 38 of this Act) for an individual for the purposes of research or private study. Open access articles, which are published under a CC-BY licence, may be re-used without permission, but subject to acknowledgement.

Full information on the Society's permissions policy can be found at https://www.geolsoc.org.uk/permissions

British Library Cataloguing in Publication Data

A catalogue record for this book is available from the British Library.
ISBN 978-1-78620-461-5
ISSN 0267-9914

Distributors

For details of international agents and distributors see:
www.geolsoc.org.uk/agentsdistributors

Typeset by Nova Techset Private Limited, Bengaluru & Chennai, India
Printed and bound by CPI Group (UK) Ltd, Croydon CR0 4YY

Contents

Preface	xv
Acknowledgements	xvi
Dedication	xvii

Chapter 1 Introduction to Geological Hazards in the UK: Their Occurrence, Monitoring and Mitigation — 1
1.1 Introduction — 1
1.2 A history of significant geohazards in the UK — 1
    1.2.1 Gas hazards — 2
        1.2.1.1 1986 Loscoe methane gas explosion, Derbyshire — 2
        1.2.1.2 Radon hazard, Northamptonshire — 2
    1.2.2 Karst and dissolution hazard — 2
        1.2.2.1 2012 Carsington Pasture, variable rockhead, Derbyshire — 2
        1.2.2.2 Ripon dissolution subsidence, North Yorkshire — 2
    1.2.3 Landslides and slope failures — 3
        1.2.3.1 Significant inland landslides — 3
        1.2.3.2 1966 Aberfan tip failure, South Wales — 3
        1.2.3.3 2000 M25 Flint Hall Farm landslide — 3
        1.2.3.4 1979 Mam Tor landslide, Derbyshire — 3
        1.2.3.5 Coastal landslides and coastal erosion — 4
        1.2.3.6 1915 Folkestone Warren landslide, Kent — 4
        1.2.3.7 1983 Holbeck Hall landslide, Scarborough, Yorkshire — 4
    1.2.4 Periglacial legacy — 5
        1.2.4.1 1984 Carsington Dam embankment failure, Derbyshire — 5
    1.2.5 Central London, drift-filled hollows — 5
        1.2.5.1 1965 A21 Sevenoaks Bypass slope failures, Kent — 5
        1.2.5.2 1961 M6 Walton's Wood embankment failure — 5
    1.2.6 Seismic events — 5
        1.2.6.1 1884 Colchester earthquake, Essex — 5
        1.2.6.2 1931 Dogger Bank earthquake, North Sea — 5
    1.2.7 Tsunami events — 5
        1.2.7.1 1755 Lisbon earthquake-generated tsunami — 5
        1.2.7.2 c. 8150 BP Storegga submarine landslide and tsunami — 6
    1.2.8 Volcanic events — 6
        1.2.8.1 2010 Eyjafjallajökull ash fall disruption — 6
        1.2.8.2 1783–1784 Laki fissure eruption, Iceland — 6
    1.2.9 Mining hazards — 6
        1.2.9.1 2000 chalk mine collapse, Reading, Berkshire — 6
    1.2.10 Deep coal workings — 7
        1.2.10.1 1945 Ludovic Berry and Dolly the train incident, Wigan — 7
    1.2.11 Geotechnical hazards — 7
        1.2.11.1 1976 subsidence related to clay shrinkage — 7
    1.2.12 Poorly recognized geohazards — 9

1.3 Geological Society Engineering Group Working Party on Geohazards	9
1.3.1 Background	9
1.3.2 Membership	9
1.3.3 Terms of reference of the Working Party	9
1.3.4 Developing the report	9
1.3.5 Contents and structure of the report	10
1.3.6 Geological hazards: Working Party definitions and report limitations	10
1.4 Section A: tectonic hazards	12
1.4.1 Chapter 2: seismic hazard in the UK	12
1.4.2 Chapter 3: tsunami hazard with reference to the UK	14
1.5 Section B: slope stability hazards	16
1.5.1 Chapter 4: landslide and slope stability hazard in the UK	16
1.5.2 Chapter 5: debris flows	18
1.6 Section C: problematic ground and geotechnical hazards	19
1.6.1 Chapter 6: collapsible soils in the UK	19
1.6.2 Chapter 7: quick-clay behaviour in sensitive Quaternary marine clays: UK perspective	21
1.6.3 Chapter 8: swelling and shrinking soils	22
1.6.4 Chapter 9: peat hazards: compression and failure	23
1.6.5 Chapter 10: relict periglacial hazards	25
1.7 Section D: mining and subsidence hazards	26
1.7.1 Chapter 11: subsidence resulting from coal mining	26
1.7.2 Chapter 12: subsidence resulting from chalk and flint mining	26
1.7.2.1 Flint mine workings	28
1.7.2.2 Chalk mine workings	28
1.7.3 Chapter 13: hazards associated with mining and mineral exploitation in Cornwall and Devon, SW England	28
1.7.4 Chapter 14: geological hazards from salt mining and brine extraction	29
1.7.5 Chapter 15: geological hazards from carbonate dissolution	30
1.7.6 Chapter 16: geological hazards caused by gypsum and anhydrite in the UK: dissolution, subsidence, sinkholes and heave	32
1.7.7 Chapter 17: mining-induced fault reactivation in the UK	34
1.8 Section E: gas hazards	35
1.8.1 Chapter 18: radon gas hazard	35
1.8.2 Chapter 19: methane gas hazard	35
Conclusions	38
References	38
Chapter 2 Seismic hazard	**43**
2.1 Earthquakes as a geohazard	43
2.2 Distribution of earthquakes in the UK	44
2.3 Consequences of British earthquakes	47
2.4 Identifying earthquakes as a geohazard in the UK	48
2.5 Past practice of seismic hazard in the UK	50
2.6 Seismic hazard mapping in the UK	51
2.7 Earthquake monitoring in the UK	53
2.8 Remedial action	55
2.9 Limits to earthquake hazard in the UK	55

2.10 Actions to take in an earthquake	56
Glossary	56
Data sources and further reading	58
References	58

Chapter 3 Tsunami hazard with reference to the UK — 61
3.1 Introduction	61
3.2 Tsunami geohazard	61
3.3 Tsunami wave characteristics	62
3.4 Tsunami generation processes	63
3.4.1 Tsunamigenic earthquakes	63
3.4.2 Tsunamigenic landslides	63
3.4.3 Tsunamigenic volcanism	63
3.4.4 Meteotsunami	64
3.4.5 Other potential tsunami-generating mechanisms	65
3.5 UK tsunami threat	65
3.6 Notable tsunami events with a UK impact	66
3.6.1 c. 8150 BP Holocene Storegga submarine landslide and tsunami	66
3.6.2 c. 5500 BP Holocene Garth tsunami	67
3.6.3 AD 1755 Lisbon earthquake and tsunami	68
3.6.4 AD 1911 Abbot's Cliff failure, Folkestone	69
3.6.5 Other Dover Straits events	73
3.7 Tsunami management and mitigation	74
3.8 Concluding comments	77
References	77

Chapter 4 Landslide and slope stability hazard in the UK — 81
4.1 Introduction	81
4.2 Landslide types	84
4.3 The landslide inventory for Great Britain	91
4.4 The Irish landslide inventory	92
4.5 The landslide environment of the UK	94
4.5.1 Peat failures	101
4.5.2 Slope deformation: cambering and complex rock block spreads	102
4.5.3 Large rock slope failures in the Scottish Highlands	105
4.5.4 Flow slides in colliery spoil	106
4.5.5 Coastal landslides: cliff behaviour units	106
4.6 Causes of landslides	108
4.6.1 Landslides and rainfall	114
4.6.2 Anthropogenic effects	116
4.6.3 Landslide controls: the influence of geology	119
4.7 Phases of landslide activity	124
4.7.1 Repeated phases of glacial and periglacial conditions	126
4.7.2 Impact of drainage adjustments during deglaciation	127
4.7.3 Postglacial slope responses	127
4.7.4 Changing climatic conditions during the Holocene	129
4.7.5 Climatic deterioration during the Little Ice Age	129
4.7.6 Anthropogenic land-use changes	132

		4.7.7 Extreme events	134
4.8	Landslide risk		134
	4.8.1 Sources of risk		134
	4.8.2 Assessing risk		135
4.9	Landslide hazard		140
	4.9.1 Landslide hazard assessment		142
	4.9.2 Landslide investigation		142
4.10	Landslide risk management		144
	4.10.1 Avoid the risk		144
	4.10.2 Restrict or prevent access to the area at risk from landsliding		144
	4.10.3 Accept the risk		144
	4.10.4 Share the risk		144
	4.10.5 Transfer the risk through litigation to recover the costs of landslide damage		144
	4.10.6 Reduce the exposure		145
	4.10.7 Provide forewarning of potentially damaging incidents		145
	4.10.8 Incorporate specific ground movement tolerating measures into the building design		145
	4.10.9 Control the area between a landslide event and the assets at risk		145
	4.10.10 Reduce the probability of the hazard		145
4.11	The role of government in landslide management		146
	4.11.1 Provision of publicly funded coast protection works		146
	4.11.2 Control development in high-risk areas		147
	4.11.3 Control building standards		147
	4.11.4 Fund and co-ordinate the response to major events		147
	4.11.5 Protect strategic infrastructure		148
4.12	In practice: acceptable or tolerable risks?		149
4.13	Concluding remarks		151
References			152

Chapter 5 Debris flows 163
5.1 Introduction 163
5.2 Types of landslide and flow mechanisms 164
5.3 Occurrence 166
    5.3.1 A83 Glen Kinglas/Cairndow: 9 August 2004 168
    5.3.2 A9 North of Dunkeld: 11 August 2004 168
    5.3.3 A85 Glen Ogle: 18 August 2004 170
    5.3.4 A83 Rest and be Thankful: 28 October 2007 171
5.4 Hazard and risk assessment 172
5.5 Risk reduction 173
5.6 Impacts 178
5.7 Climate change 179
Conclusions 183
References 183

Chapter 6 Collapsible soils in the UK 187
6.1 What are collapsible soils? 187
6.2 Loess in the UK 188
6.3 How to recognize loessic brickearth 191
    6.3.1 Description and mineralogy 191
    6.3.2 Geotechnical properties 193

	6.3.2.1 Particle size distribution	193
	6.3.2.2 Density	193
	6.3.2.3 Plasticity	193
	6.3.2.4 Strength, consolidation and permeability of brickearth/loess	193
6.4	Non-engineered fills	196
6.5	Identifying collapsibility	196
	6.5.1 Collapse potential	196
6.6	Strategies for engineering management: avoidance, prevention and mitigation	197
6.7	Example of damage caused by collapse	198
6.8	Conclusions	200
Glossary and definitions		200
Further reading		200
References		201

Chapter 7 Quick clay behaviour in sensitive Quaternary marine clays – a UK perspective — 205

7.1 Introduction — 205
7.2 Mode of formation — 205
7.3 Geotechnical properties and behaviour — 208
7.4 Failure mechanisms — 208
7.5 The UK context — 212
7.6 Geohazard management and mitigation — 216
7.7 Conclusions — 218
References — 218

Chapter 8 Swelling and shrinking soils — 223

8.1 Introduction — 223
8.2 Properties of shrink–swell soils — 223
8.3 Costs associated with shrink–swell clay damage — 224
8.4 Formation processes — 225
8.5 Distribution — 225
8.6 Characterization of shrink–swell soils — 226
8.7 Mechanisms of shrink–swell — 230
8.8 Shrink–swell behaviour — 230
8.9 Strategies for engineering management: avoidance, prevention and mitigation — 233
8.10 Shrink–swell soils and trees — 236
8.11 Conclusions — 237
Appendix: Definitions and glossary — 238
Recommended further reading — 240
Useful web addresses — 240
References — 240

Chapter 9 Peat hazards: compression and failure — 243

9.1 Introduction and scope — 243
9.2 Engineering background: peat consolidation and compression — 245
  9.2.1 Compression of peat — 246
9.3 UK peatlands: extent and occurrence — 249
9.4 Geological hazards associated with peat compressibility — 250
  9.4.1 Subsidence of peat — 250
  9.4.2 Derrybrien landslide, wind farm construction, County Galway 2003 — 251

9.4.3 Direct loading by quarry waste, Harthope Quarry, North Pennines, UK	252
9.4.4 Failure during upland road construction, North Pennines, UK	252
9.5 Mitigation of the hazards posed by compressible peat soils	253
9.6 Conclusion	255
References	256

Chapter 10 Periglacial geohazards in the UK — 259

10.1 Introduction	259
10.2 Relict periglacial geohazards	262
10.2.1 Deep weathering	262
10.2.2 Shallow-slope movements	265
10.2.3 Cambering and superficial valley disturbances	276
10.2.4 Rockhead anomalies	280
10.2.5 Cryogenic wedges (ice-wedge pseudomorphs)	283
10.3 Subsidiary relict periglacial geohazards	284
10.3.1 The influence of periglacial climates and processes on deep-seated landslide systems	284
10.3.2 Carbonate dissolution	284
10.3.3 Buried terrains	285
10.3.4 Submerged periglacial terrains	285
10.3.5 Loess and coversand	285
10.4 Conclusions	285
References	286

Chapter 11 Coal mining subsidence in the UK — 291

11.1 Introduction	291
11.2 Subsidence characteristics	291
11.3 Overview of mining methods	291
11.3.1 Adits, drifts (inclines) and shafts	291
11.3.2 Bell pits	292
11.3.3 Room-and-pillar	292
11.3.4 Longwall mining	293
11.3.5 Subsidence associated with partial extraction of coal	293
11.3.5.1 Mine shafts and bell pits	293
11.3.5.2 Room-and-pillar workings	293
11.3.6 Subsidence associated with total extraction of coal	295
11.3.6.1 Tilt	295
11.3.6.2 Slope	296
11.3.6.3 Curvature	296
11.3.6.4 Strain	296
11.3.6.5 Horizontal displacements	296
11.3.6.6 Strain	296
11.3.6.7 Width–depth ratio	297
11.3.6.8 Angle-of-draw (limit angle)	297
11.3.6.9 Area-of-influence	297
11.3.6.10 Maximum subsidence	297
11.3.6.11 The subsidence factor	297
11.3.6.12 Dip of seam	297
11.3.6.13 Bulking	297

| | | 11.3.6.14 Time-dependent subsidence and residual subsidence | 297 |

        11.3.6.15 Multiple seams   298
   11.3.7 Subsidence and the engineering properties of soils and rocks   298
        11.3.7.1 Soils/superficial deposits   298
        11.3.7.2 Rock   299
   11.3.8 Subsidence prediction   299
        11.3.8.1 Empirical methods   300
        11.3.8.2 Analytical or theoretical   300
        11.3.8.3 Semi-empirical methods   300
        11.3.8.4 Void migration   300
11.4 Managing subsidence risks   301
   11.4.1 Desk study   302
   11.4.2 Reconnaissance (walk-over) survey   303
   11.4.3 Ground investigations   303
11.5 Mitigation and remediation   304
11.6 Summary   306
References   306

Chapter 12 Subsidence – chalk mining   311
12.1 Introduction   311
12.2 Geographical occurrence   311
12.3 Characteristics of the mine workings   313
   12.3.1 Flint mine workings   313
        12.3.1.1 Neolithic flint mines   313
        12.3.1.2 Modern flint mines   314
        12.3.1.3 Chalk mine workings   314
        12.3.1.4 Bellpits   314
        12.3.1.5 Deneholes   314
        12.3.1.6 Chalkwells   315
        12.3.1.7 Chalkangles   316
        12.3.1.8 Pillar-and-stall mines   316
12.4 Engineering management strategy   317
Appendix: Further reading   319
Websites   319
References   319

Chapter 13 Hazards associated with mining and mineral exploitation in Cornwall and Devon, SW England   321
13.1 Introduction   321
13.2 The geological model and the setting for mining-related hazards   322
   13.2.1 Geological overview   322
   13.2.2 Paleozoic rocks of the Variscan (Rhenohercynian) basement   322
        13.2.2.1 Upper Paleozoic rift basins of the Rhenohercynian passive margin   322
        13.2.2.2 Upper Paleozoic mafic and ultramafic rocks of the Lizard Complex   324
        13.2.2.3 Upper Paleozoic allochthons   325
        13.2.2.4 Lower Paleozoic (pre-rift) basement   325
   13.2.3 Regional structure   325
   13.2.4 Post-Variscan cover, magmatism, mineralization and alteration   326
   13.2.5 Superficial deposits   328
13.3 History of mining   328

13.4 Environmental legacy of mining	333
13.4.1 Underground voids and shafts	333
13.4.2 Opencast mines	336
13.4.3 Waste tips and contaminated land	337
13.4.4 Infilled or silted-up estuaries	338
13.4.5 Slurry lakes or tailings ponds	338
13.4.6 Pollution by contaminated mine water	338
13.4.7 Flooding	343
13.5 Investigating and assessing the hazards	344
13.5.1 Desk studies	344
13.5.2 Remote sensing	344
13.5.3 Geophysics	345
13.5.4 Field mapping	346
13.5.5 Ground investigations	346
13.5.6 Developing the ground model	350
13.5.7 Hazard and risk assessment	350
13.5.8 Monitoring	351
13.6 Planning, preservation, treatment and remediation	352
13.6.1 International and local planning	352
13.6.2 Preservation	352
13.6.3 Treatment and remediation through engineering works	353
13.6.4 Derelict land reclamation	355
13.6.5 Mine water contamination and remediation	355
13.6.6 Case studies of mine site treatment and remediation	356
13.6.6.1 Wheal Peevor, Redruth, Cornwall. Kerrier District Council (2003–07)	357
13.6.6.2 The National Trust	357
13.7 Conclusions	360
References	362
Chapter 14 Geological hazards from salt mining, brine extraction and natural salt dissolution in the UK	**369**
14.1 Introduction	369
14.2 Distribution of salt deposits in the Triassic and Permian rocks of the UK	370
14.3 Salt karst and natural dissolution	371
14.4 Mining and dissolution mining of salt	372
14.4.1 Natural 'wild' brine extraction	372
14.4.2 Shallow salt mining and 'bastard' brining	372
14.4.3 Modern salt mining	373
14.5 Mining of Permian salt deposits	375
14.5.1 Teesside	375
14.6 Mining of the Triassic salt deposits	377
14.6.1 Cheshire	377
14.6.2 Blackpool and Preesall	378
14.6.3 Stafford	378
14.6.4 Droitwich	379
14.6.5 Northern Ireland	380
14.7 Mitigating salt subsidence problems	380
14.7.1 Brine Subsidence Compensation Board	380
14.7.2 Salt mine stabilization	382
14.7.3 Monitoring and investigation	384

14.7.4 Planning for soluble rock geohazards		384
References		385

Chapter 15 Dissolution – carbonates 389
15.1 Introduction 389
15.2 Geographical occurrence 389
15.3 Characteristics of natural cavities formed by dissolution 392
15.4 Engineering management strategy 395
Further reading 400
Websites 400
References 400

Chapter 16 Geohazards caused by gypsum and anhydrite in the UK: including dissolution, subsidence, sinkholes and heave 403
16.1 Introduction 403
16.2 The gypsum–anhydrite transition, expansion and heave 403
16.3 The gypsum dissolution problem 404
16.4 Geology of the gypsiferous rocks 404
    16.4.1 Triassic 404
    16.4.2 Permian 404
16.5 Subsidence caused by gypsum dissolution 407
    16.5.1 Subsidence geohazards around Ripon 407
    16.5.2 Subsidence geohazards around Darlington 409
    16.5.3 Subsidence geohazards between Ripon and Doncaster 410
    16.5.4 Subsidence geohazards in the Vale of Eden 410
    16.5.5 Subsidence over Triassic gypsum 411
16.6 Ground investigation: surveying, geophysics and boreholes in gypsum areas 411
16.7 Gypsum dissolution as a hazard to civil engineering 413
16.8 Problems related to water abstraction and injection in gypsum areas 416
16.9 Planning for subsidence 417
Conclusions 419
References 419

Chapter 17 Mining-induced fault reactivation in the UK 425
17.1 Background 425
17.2 Occurrence 425
17.3 Diagnostic characteristics 426
17.4 Mitigation 427
References 430

Chapter 18 Radon gas hazard 433
18.1 Introduction 433
18.2 Other natural sources of radiation 434
    18.2.1 Gamma rays from the ground and buildings (terrestrial gamma rays) 434
    18.2.2 Cosmic rays 435
18.3 Health effects of radiation and radon 435
18.4 Radon release and migration 436
18.5 Factors affecting radon in buildings 438
18.6 Geological associations 438
    18.6.1 Granites 439
    18.6.2 Black shales 439

18.6.3	Phosphatic rocks and ironstones	444
18.6.4	Limestones and associated shales and cherts	445
18.6.5	Sands and sandstones	446
18.6.6	Ordovician–Silurian greywackes and associated rocks	446
18.6.7	Miscellaneous bedrock units	446
18.6.8	Superficial deposits	446

18.7 Measurement of radon — 446
    18.7.1 Radon testing in the home — 446
    18.7.2 Measurement of radon in soil-gas and solid materials — 447

18.8 Radon hazard mapping and site investigation — 448
    18.8.1 Radon hazard mapping based on geology and indoor radon measurements — 448
    18.8.2 Radon hazard mapping based on geology, gamma spectrometry and soil-gas radon data — 449
    18.8.3 Radon site investigation methods — 450

18.9 Strategies for management: avoidance, prevention and mitigation — 451
    18.9.1 Introduction — 451
    18.9.2 Environmental health regulations — 451
    18.9.3 Radon and the building regulations: protecting new buildings — 452
    18.9.4 Radon and workplaces — 453
    18.9.5 Radon and the planning system — 453
    18.9.6 Remedial measures — 453

18.10 Scenarios for future events — 454

References — 454

Chapter 19 Methane gas hazard — 457

19.1 The source and chemical properties of methane — 457

19.2 Guidance and best practice — 462
    19.2.1 Legislative background — 462

19.3 Developing the conceptual site model — 462
    19.3.1 Sources of methane — 462
    19.3.2 Pathways for migration — 464
    19.3.3 Potential receptors to methane — 466

19.4 Examples of methane impacts — 466

19.5 Managing risk — 468
    19.5.1 Site investigation for methane — 468
    19.5.2 UK contamination practices — 468
    19.5.3 The planning process — 468
    19.5.4 The definition of contaminated land — 469

19.6 The risk assessment process — 469
    19.6.1 Qualitative risk assessment — 470
    19.6.2 Semi-quantitative risk assessment — 471
    19.6.3 NHBC Traffic Lights — 472
    19.6.4 British Standard BS8485: 2015 — 472
    19.6.5 Quantitative risk assessment — 474
    19.6.6 Acute situation — 475

19.7 Mitigating methane risks — 475

19.8 Summary and conclusions — 477

References — 477

Index — 479

Preface

News of environmental, personal and monetary losses arising from problems with the behaviour of the ground and with surface water appear regularly worldwide. Fortunately, the UK does not suffer from the extreme effects of such geological hazards (also known as 'geohazards') such as the large earthquakes, landslides, volcanic eruptions, floods and so forth that are an important feature of the physical environment in some other countries. Nevertheless, for the UK, such phenomena remain important on a local scale.

Within the UK, the wide variety and geographical spread of geohazards are a consequence of our varied geology and geomorphology, and the potentially adverse impacts and legacies of human activity such as mining and land management. This variety presents practical problems when writing at reasonable length and in appropriate detail for the broad audience of professions who may have to deal with geohazards and their effects: engineering geologists and geomorphologists, civil engineers, planners, environmental managers, developers, government and aid organizations. The Working Party's approach has been to subdivide geohazards into four categories (geophysical, geotechnical, geochemical and georesource related) and, having explained how the geohazards occur and operate (individually or in combination), to suggest means by which they may be detected, monitored and managed.

I hope that users of the Working Party's report will agree that it has met its objective of helping geoscientists communicate an understanding of geohazards and, by doing so, to have contributed to the Geological Society's wider purpose of 'serving science, profession and society'.

David Shilston CGeol, FGS, FRSA
Past President of the Geological Society
SNC Lavalin/Atkins Fellow and Technical Director for Engineering Geology

Acknowledgements

The original Geological Society Working Party on Geological Hazards was initiated under the leadership of Prof. Mike Rosenbaum, Dr David Entwistle and Dr Alan Forster in August 2002 following informal meetings held at the British Geological Survey, Keyworth, and they are thanked for their contributions at the start of this long project. Due to many membership changes the Working Group was reformatted in 2010 with lead authors designated to facilitate and compile the identified chapters. The following Working Party members are thanked (in chapter order) for their considerable efforts in bringing this project to completion: Prof. Roger Musson (British Geological Survey), Dr Mark Lee (Ebor Geoscience), Professor Mike Winter (formerly TRL Scotland), Professor Martin Culshaw (British Geological Survey), Dr Lee Jones (British Geological Survey), Professor Jeff Warburton (Durham University), Tom Berry (Jacobs), Dr Laurance Donnelly (AHK), Dr Clive Edmonds (Peter Brett Associates), Barry Gamble (Independent consultant to UNESCO), Dr Tony Cooper (British Geological Survey), Dr Don Appleton (British Geological Survey) and Steve Wilson (EPG Ltd).

We would also like to thank the many chapter reviewers for their invaluable comments and suggestions.

Dedication

This book is dedicated to Dr Brian Hawkins

Alfred Brian Hawkins
PhD DSc FICE FIMMM FIHT CEng CGeol EurIng
(10 October 1934–22 January 2016)

Brian Hawkins in a gull, doing what he will be most remembered for.

Brian Hawkins was born on a farm in Bitton, between Bristol and Bath, in 1934. After studying at the University of Bristol, initially for a geography and subsequently a geology degree, he took a teaching qualification to fulfil his long-held ambition to teach and began work in the first comprehensive school in Bristol. Fascinated by the Quaternary, he continued his research at the University and was invited to consolidate this by undertaking a doctorate. In order to do so, he took a post in the Department of Extra-Mural Studies, organizing and teaching geology to the public. At the same time, he continued his research and began teaching in the Department of Geology and taking an active part in the recently formed Engineering Geology Group of the Geological Society. His enthusiasm for engineering geology led to his working with Bill Dearman to set up a UK branch of the International Association for Engineering Geology. A true engineering geologist, he was well respected by the engineering fraternity and became a Fellow of the Institution of Civil Engineers.

He established a strong engineering geological research unit at Bristol, embracing topics in Quaternary geology; superficial structures; slope stability in engineering soils and engineering rocks; the development of ground sulfates and the stability and remediation of mines and tunnels, particularly in the Bristol and south Cotswolds region. Brian was awarded a PhD in 1970 and DSc in 1989, being promoted to Reader in 1979.

Brian continued teaching and research in Engineering Geology at the University of Bristol in the departments of both Geology (later Earth Sciences) and Civil Engineering until his sudden death in 2016. During a university career spanning almost 50 years, he supervised over 35 PhD students and published some 120 paper and articles as well as editing books and conference proceedings.

Brian had boundless energy and enthusiasm – as will be appreciated by anyone who attended one of his lectures, or indeed worked with him on site! In the mid 1990s he took official early retirement from the University to give more time to his national and international consultancy work. He became a recognized authority on the pyrite problem beneath houses in Ireland. The practice enabled him to remain involved on a personal level, advising local authorities, public utilities, contractors and consultants on practical aspects of construction work as well as contractual issues and as an expert witness.

He was frequently invited to lecture at universities and international conferences and to be involved in field visits and has always been active in the profession. Amongst his many roles have been: Secretary (1974–76) and then Chair (1982–84) of the Engineering Group of the Geological Society; Vice President for Europe for the International Association for Engineering Geology and the Environment (IAEG) (1994–98); member of the Geotechnique Advisory Panel; Editor of the *Quarterly Journal of Engineering and Hydrogeology* (1990–93) and Editor-in-Chief of the *Bulletin of Engineering Geology and the Environment* (1998–2012).

He and co-editor Roger Cojean are cited by the IEAG as being responsible for where the journal stands today.

Brian was honoured with being the first recipient of the Marcel Arnould Medal presented at the 2014 IAEG Congress in Torino 'in recognition of people of significant repute within the IAEG and who have made a major contribution to the Association'.

Brian was a true inspiration. His delivery style for lectures and fieldtrips was the foundation for many a practical, hands-on career for countless students over the years. They are exceedingly lucky to have met such a man.

—Dr Kevin Privett and Dr Marian Trott

Chapter 1 Introduction to Geological Hazards in the UK: Their Occurrence, Monitoring and Mitigation

David Peter Giles

CGL (Card Geotechnics Ltd), 4 Godalming Business Centre, Woolsack Way, Godalming, GU7 1XW, UK

0000-0001-8016-0538

Correspondence: DavidG@cgl-uk.com

Abstract: The UK is perhaps unique globally in that it presents the full spectrum of geological time, stratigraphy and associated lithologies within its boundaries. With this wide range of geological assemblages comes a wide range of geological hazards, whether geophysical (earthquakes, effects of volcanic eruptions, tsunami, landslides), geotechnical (collapsible, compressible, liquefiable, shearing, swelling and shrinking soils), geochemical (dissolution, radon and methane gas hazards) or related to georesources (coal, chalk and other mineral extraction). An awareness of these hazards and the risks that they pose is a key requirement of the engineering geologist. This volume sets out to define and explain these geohazards, to detail their detection, monitoring and management, and to provide a basis for further research and understanding, all within a UK context.

1.1 Introduction

A geological hazard (geohazard) is the consequence of an adverse combination of geological processes and ground conditions, sometimes precipitated by anthropogenic activity. The term implies that the event is unexpected and likely to cause significant loss or harm. To understand geohazards and mitigate their effects, expertise is required in the key areas of engineering geology, hydrogeology, geotechnical engineering, risk management, communication and planning, supported by appropriate specialist knowledge of subjects such as seismology and volcanology. There is a temptation for geoscientists involved in geohazards to get too focused on the 'science' and lose sight of the purpose of the work, which is to facilitate the effective management and mitigation of the consequences of geohazards within society. The Geological Society considered that a Working Party Report would help to put the study and assessment of geohazards into the wider social context, helping the engineering geologist to better communicate the issues concerning geohazards in the UK to the client and the wider public.

1.2 A history of significant geohazards in the UK

At the risk of cultural misappropriation, people of the UK often sing of their 'green and pleasant land' in the misguided view that they are unaffected by major natural and geological hazard events that impact the rest of the world, as these all occur in far-off places that are a very long way from the shores of the UK. However, as a country, we possess the full geological spectrum of the stratigraphic column with its associated lithologies. It is hard to think of any geological assemblage that cannot be found within the British Isles with rocks dating from the Precambrian to the Quaternary along with examples of all major environments of deposition, formation and modification. The legacy of this assemblage and the associated geological hazards, whether geophysical, geotechnical, geochemical or related to georesources, are in evidence across the UK.

Impacts of more distal events can also be seen with the 2010 Icelandic Eyjafjallajökull volcanic eruption that demonstrated we are not immune from global active volcanic events; the economic impact of the ash cloud was felt through disruption to air travel. Further back in our more recent history, the 1783–1784 Icelandic Laki Fissure eruption with its toxic gas cloud saw a significant mortality crisis across the UK with as many as 23 000 British people dying from the poisoning (Grattan & Brayshay 1995). The possibility of a caldera-collapse super-volcanic event centred on the Campi Flegrei in Italy cannot be ruled out in the immediate or distant future, with a significant impact on mainland Europe, north Africa and the UK.

The UK has the potential to be harmed by the full remit of documented geological hazards ranging from earthquakes, tsunami and landslides to the significant effects of clay particles that shrink and swell with moisture.

Engineering Group Working Party (main contact for this chapter: D.P. Giles, CGL (Card Geotechnics Ltd), 4 Godalming Business Centre, Woolsack Way, Godalming, GU7 1XW, UK, DavidG@cgl-uk.com)
From: Giles, D. P. & Griffiths, J. S. (eds) 2020. *Geological Hazards in the UK: Their Occurrence, Monitoring and Mitigation – Engineering Group Working Party Report*. Geological Society, London, Engineering Geology Special Publications, **29**, 1–41,
https://doi.org/10.1144/EGSP29.1
© 2019 The Author(s). This is an Open Access article distributed under the terms of the Creative Commons Attribution License (http://creativecommons.org/licenses/by/4.0/). Published by The Geological Society of London.
For permissions: http://www.geolsoc.org.uk/permissions. Publishing disclaimer: www.geolsoc.org.uk/pub_ethics

The impact of geological hazards can be measured in terms of fatalities, landscape loss and economic impact. The tragedy of Aberfan, South Wales, exacerbated by the number of children in the overall death toll, is perhaps the most significant hazard event of modern times affecting the UK, although the impact of a future Lisbon-style earthquake and associated tsunami or Storegga-generated tsunami on the UK may well present our greatest geological hazard challenge.

In the writing of this introduction to *Geological Hazards in the UK*, expert opinion was canvassed as to which geological hazards impacting the UK in our recent geological history could be considered as the most noteworthy. The list is both specific (to individual events) and generic (to more widely impacting geohazards such as subsidence related to coal mining). The following sections (presented alphabetically) represent impacts in terms of fatalities, as well as economic and social effects.

1.2.1 Gas hazards

1.2.1.1 1986 Loscoe methane gas explosion, Derbyshire

Loscoe was the site of a landfill gas migration explosion on 24 March 1986. There were no fatalities, but one house was completely destroyed by the blast. Atmospheric pressure on the night of the explosion fell 29 mbar over a 7-hour period, drawing methane through a permeable sandstone horizon from a former landfill site (Fig. 1.1). Landfill gas collected under the ground near the house at 51 Clarke Avenue, entered the house and ignited with catastrophic effects (Williams & Aitkenhead 1991).

1.2.1.2 Radon hazard, Northamptonshire

Radon is a naturally occurring odourless, colourless radioactive gas that migrates into homes through floors and walls and is the major source of ionizing radiation exposure of the UK population. High levels of radiation have been associated with an increased incidence of lung cancer, particularly when its exposure is long term and combined with cigarette smoking. Radon is more prevalent in some areas of the country than others and Northamptonshire, with a specific Jurassic bedrock lithology, has high levels of the gas emitted into the atmosphere. Remedial action and preventative measures are necessary for house construction in these affected areas (Sutherland Sharman 1996).

1.2.2 Karst and dissolution hazard

1.2.2.1 2012 Carsington Pasture, variable rockhead, Derbyshire

Excavation of the foundations of four wind turbines at Carsington Pasture exposed buried, sediment-filled hollows in the bedrock that had formed as the result of karstification. The bedrock geology comprised dolomitized Carboniferous limestones that had been subject to lead-zinc-barite mineralization. Excavation of the foundations commenced on 8 May 2012. Difficult ground conditions were encountered that necessitated remedial engineering measures and delayed the project by 12–14 months, with consequent economic impacts (Czerewko *et al.* 2015; Raines *et al.* 2015).

1.2.2.2 Ripon dissolution subsidence, North Yorkshire

The area in and around Ripon is significantly affected by the presence of gypsum, hydrated calcium sulphate ($CaSO_4 \cdot 2H_2O$), in the local Permo-Triassic bedrock. A substantial number of sinkholes have developed in the area caused by the dissolution of the gypsum and the formation of gypsum karst. Subsequent collapse of these features (Fig. 1.2) has

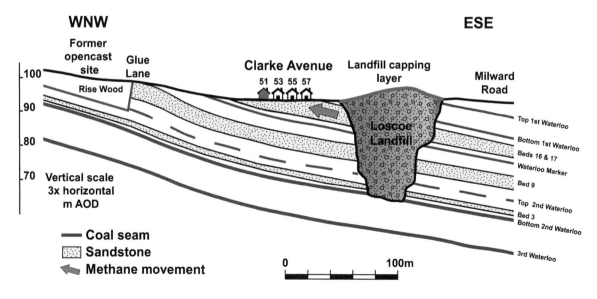

Fig. 1.1. Geological cross-section through the Loscoe landfill (Williams & Aitkenhead 1991).

Fig. 1.2. Sinkhole development near Ure Bank Terrace, Ripon, North Yorkshire (photo credit: David Giles).

led to considerable structural damage to buildings in the city (Cooper 2007).

1.2.3 Landslides and slope failures

1.2.3.1 Significant inland landslides

There are many examples of significant inland landslides that could be included here with specific events, considered as key geohazards, affecting infrastructure and other valuable assets. Such examples include: Jackfield, Shropshire 1952 (Henkel & Skempton 1955); Buildwas, Shropshire 1773 (Pennington 2008); A85 road, Glen Ogle, Lochearnhead, Stirlingshire (Winter et al. 2006); Hatfield Main Colliery 2013, South Yorkshire (BGS n.d.); Rest and be Thankful Pass, A83, Argyll and Bute 2007, 2009, 2011, 2012 (Wong & Winter 2018); Bournville and East Pentwyn, Blaina (Siddle 2000); Taren, Taff Valley (Cobb 2000); Castle Hill, Cheriton, Kent (Griffiths et al. 1995). The most noteworthy case histories are described below.

1.2.3.2 1966 Aberfan tip failure, South Wales

The catastrophic collapse of a colliery spoil tip, created on the hillslope above the village of Aberfan, occurred on 21 October 1966 (Fig. 1.3). Significantly, the tip overlaid a natural spring that fed water into the colliery spoil. The tip was further destabilized by a period of heavy rain eventually leading to a devastating mudflow, killing 116 children and 28 adults as it engulfed the local junior school and other buildings (Tribunal Appointed to Inquire into the Disaster at Aberfan on October 21st 1966).

1.2.3.3 2000 M25 Flint Hall Farm landslide

On 19 December 2000, during one of the wettest UK winters on record, a 200 m long section of the Flint Hall Farm cutting on the M25 failed and threatened to close the motorway, which carries over 120 000 vehicles a day (Fig. 1.4). Further rainfall triggered additional movements during

Fig. 1.3. Aberfan in the days immediately after the disaster, showing the extent of the spoil slip (https://en.wikipedia.org/w/index.php?curid=54882575).

January and early February 2001, further threatening the carriageway (Davies et al. 2003; Griffiths & Giles 2017).

1.2.3.4 1979 Mam Tor landslide, Derbyshire

The landslide at Mam Tor was probably initiated by the erosive steepening of valley slopes during periods of high

Fig. 1.4. Flint Hall Farm site works impacted on M25 carriageway (Griffiths & Giles 2017).

Fig. 1.5. Damage to the A625 carriageway by the Mam Tor landslide, Derbyshire (photo credit: David Giles).

rainfall and freeze–thaw action during the Devensian. The landslide was the subject of intensive investigations due to the damage that the slope instability was causing to the A625 road that traversed the landslide's displaced material (Fig. 1.5). The road has been closed since 1979 (Waltham & Dixon 2000; Griffiths & Giles 2017).

1.2.3.5 Coastal landslides and coastal erosion
The UK coastline has always been prone to erosion and major landslide events, affecting both property and land. Classic examples include the Undercliff, Isle of Wight (e.g. Moore *et al.* 2010), Happisburgh, Norfolk (e.g. Poulton *et al.* 2006) (Fig. 1.6), Black Ven, Dorset (e.g. Brunsden & Chandler 1996), Fairlight Glen, East Sussex (e.g. Moore 1986; Moore & McInnes 2011), where each location is influenced by the site-specific lithologies present.

1.2.3.6 1915 Folkestone Warren landslide, Kent
The Folkestone Warren landslide is one of the largest on the English coast and is a classic example of a deep-seated

Fig. 1.6. Erosion on the North Norfolk coast, an example from Happisburgh (photo credit: David Giles).

Fig. 1.7. Landslide at Folkestone Warren, Kent, 1914 (photo credit: Network Rail).

multiple retrogressive, compound mechanism, having translational, rotational and graben features. A major reactivation occurred throughout the complex in 1915 (Fig. 1.7), seriously disrupting the railway constructed in 1844 (British Geological Survey n.d.; Hutchinson 1969, 1988; Hutchinson *et al.* 1980; Trenter & Warren 1996; Warren & Palmer 2000).

1.2.3.7 1983 Holbeck Hall landslide, Scarborough, Yorkshire
The Holbeck Hall landslide destroyed the Holbeck Hall Hotel between the night of 3 June and 5 June 1993 (Fig. 1.8). A rotational landslide developing into a flow involving *c.* 1 Mt of glacial till cut back the 60 m high cliff by 70 m. It flowed across the beach to form a semi-circular promontory 200 m wide, projecting 135 m outwards from the foot of the

Fig. 1.8. Landslide at Holbeck Hall, Scarborough, Yorkshire, 1983 (photo credit: British Geological Survey).

Fig. 1.9. Carsington embankment under construction in 1984 (photo credit: Winter *et al.* 2017).

cliff. The likely cause of the landslide was a combination of rainfall (140 mm in the 2 months before the failure took place), issues related to slope drainage and porewater pressure build-up in the slope, all influenced by the site geology (British Geological Survey n.d.; Forster 1993; Lee 1999; Forster & Culshaw 2004).

1.2.4 Periglacial legacy

1.2.4.1 1984 Carsington Dam embankment failure, Derbyshire

An embankment dam was to be constructed 3 km south of Carsington (Fig. 1.9). An extensive ground investigation was carried out, but failed to recognize relict periglacial features that were present in the ground profile. Consequently, the embankment failed during construction in June 1984 and then had to be demolished and rebuilt to a design based on the correctly understood ground conditions. The reconstruction was successfully completed, but at a considerable cost and delay. The finished reservoir opened in 1992 (Skempton & Vaughan 2009; Martin *et al.* 2017).

1.2.5 Central London, drift-filled hollows

Engineering works carried out in central London over many decades have revealed a number of buried hollows that exhibit curious characteristics (Fig. 1.10). Some extend deep into the bedrock geology and are infilled with disturbed superficial deposits and reworked bedrock. Others are contained within the superficial deposits. The buried hollows can be up to 500 m wide and more than 60 m in depth. As the infill material often has different behavioural characteristics from the surrounding deposits, failure to identify them during an initial site investigation can prove costly. Much work has been undertaken in London to further delineate the presence of these anomalous depressions in the London Clay. Various modes of formations of these features have been proposed, including simple scour, dissolution of the underlying chalk, valley bulging, frost heave, former ice wedges or thermokarst processes, or it has even been proposed that they are former pingo remnants (Banks *et al.* 2015; Toms *et al.* 2016; Griffiths & Giles 2017).

1.2.5.1 1965 A21 Sevenoaks Bypass slope failures, Kent

The slope failures that occurred in 1965 during the construction of the Sevenoaks Bypass in Kent led to a new understanding of the behaviour and geotechnical properties of clay slopes. The failures occurred in the natural ground that had been affected by periglacial conditions during the Quaternary. Originally described as relict solifluction lobes (and now thought to be remnants of active-layer detachment slides with underlying solifluction sheets), these were reactivated during the construction works leading to considerable delays with the project and eventually a new road alignment being developed (Weeks 1969; Martin *et al.* 2017) (Fig. 1.11).

1.2.5.2 1961 M6 Walton's Wood embankment failure

The embankment failure at Walton's Wood in Staffordshire during the construction of the M6 is a seminal case study in engineering geology. Soon after the beginning of construction of an embankment, a failure occurred through the reactivation of an undetected relict landslide with movement along pre-existing shear surfaces (Fig. 1.12). The subsequent field and laboratory investigations led to major advances in the understanding of residual strength within clay slopes (Early & Skempton 1972; Griffiths & Giles 2017).

1.2.6 Seismic events

1.2.6.1 1884 Colchester earthquake, Essex

This earthquake occurred on 22 April 1884 and caused considerable damage in Colchester and surrounding villages in Essex. In terms of overall destruction (intensity), it can be considered the most destructive earthquake to have hit the UK and was estimated as a local magnitude (M_L) 4.6 event (Haining 1991).

1.2.6.2 1931 Dogger Bank earthquake, North Sea

This M_L 6.1 event, with a Modified Mercalli intensity of VI (strong) to VII (very strong), was the largest-magnitude earthquake recorded in the UK since measurements began. The epicentre in the North Sea meant that damage was significantly less than it would have been on the UK mainland (Versey 1939; Musson 2007).

1.2.7 Tsunami events

1.2.7.1 1755 Lisbon earthquake-generated tsunami

The largest historically recorded seismic event in Europe is considered to be the 1755 Lisbon earthquake, estimated as an M_s 8.5 magnitude (possibly M_w 9.0) event and between X and XI Modified Mercalli intensity scale. The impacts in the UK resulting from the earthquake and subsequent tsunami were first noted with reports of seiche (standing waves in an

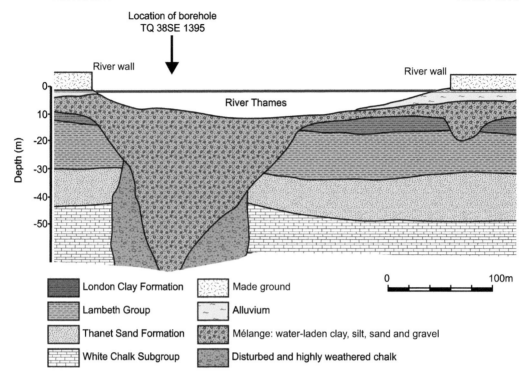

Fig. 1.10. Cross-section through a drift-filled hollow, Blackwall Tunnel, London (Griffiths & Giles 2017).

enclosed or partially enclosed body of water) in various harbours, lakes and ponds across the country. In the SW, wave trains were also reported with noticeable sea-level variations over several hours. Lisbon-related tsunami deposits have also been identified in parts of SW England (Giles 2020b).

1.2.7.2 c. 8150 BP Storegga submarine landslide and tsunami
Along the eastern and northern coasts of Scotland and at locations in NE England, sites have been investigated where a continuous layer of marine sediments can be identified. These sediments have been interpreted as tsunami deposits and have been attributed to a major submarine landslide event that displaced approximately 3500 km^3 of sediment along the mid-Norwegian margin of the North Sea (Fig. 1.13). Recent work has suggested that the occurrence of landslides with tsunami-generating potential may be more frequent, which has significant implications for the associated tsunami threat to the UK and Norwegian coasts (Giles 2020b).

1.2.8 Volcanic events

1.2.8.1 2010 Eyjafjallajökull ash fall disruption
Although relatively small for volcanic eruptions (rated 1 on the volcanic explosivity index), the 2010 eruptions of Eyjafjallajökull caused enormous disruption to air travel across western and northern Europe; around 20 countries, including Britain, closed or restricted their airspace to commercial jet traffic, affecting approximately 10 million travellers (Fig. 1.14). The restriction of UK airspace affected some 600 000 people (Gudmundsson et al. 2012).

1.2.8.2 1783–1784 Laki fissure eruption, Iceland
The 1783 Laki eruption lasted 8 months, during which time about 14 km^3 of basaltic lavas were erupted. Haze from the eruption was reported globally. An estimated 80 Mt of sulphuric acid aerosol was released by the eruption, known to be the largest air pollution incident in historic times. August temperatures in the UK in 1783 were 2.5–3°C higher than the decadal average, causing the hottest summer on record for 200 years. A bitterly cold winter followed with temperatures 2°C below average. An acid fog persisted over much of Europe, causing to an increase in sickness levels. In England, the period July 1783 to June 1784 is classified as a 'mortality crisis', with the death rate increasing by 30 000 (i.e. doubling) (Witham & Oppenheimer 2004).

1.2.9 Mining hazards

1.2.9.1 2000 chalk mine collapse, Reading, Berkshire
In January 2000, several cavities of a nineteenth century chalk mine collapsed causing major subsidence of the

Fig. 1.11. Sevenoaks bypass: (**a**) shear surfaces; and (**b**) polished shear surface (Martin *et al.* 2017).

overlying ground around the Field Road (Fig. 1.15) and Coley Road areas in Reading. Thirty homes were immediately evacuated for residents' safety, with two homes later collapsing. The mines were remediated over a 12-year period to fill the underground mine network using 1742 t of grout, costing approximately £4.3 million (Edmonds 2008; Terra Firma 2017).

1.2.10 Deep coal workings

1.2.10.1 1945 Ludovic Berry and Dolly the train incident, Wigan

There are numerous examples of hazards related to deep mines across the UK, and a notable accident associated with the coal mining industry occurred on 30 April 1945 (Fig. 1.16). *Dolly* was an engine that shunted coal wagons between the Maypole and Mains collieries in Wigan, driven by Mr Ludovic Berry. On the day of the accident, a large hole appeared in the ground under the railway lines between Abram and Platt Bridge. With the lines now unsupported they failed under the weight of the first wagons, causing them to plummet into the ground, taking the remaining wagons and Dolly with them. Ludovic, who tried to save the engine until it was too late to jump, lost his life. The hole had occurred as a result of the subsidence of a shaft sunk 60 years previously and sealed in 1932. The subsidence may have been the result of heavy rains in an area with many mine workings close by (Winstanley n.d.; K. Nicholls, pers. comm., 2019).

1.2.11 Geotechnical hazards

1.2.11.1 1976 subsidence related to clay shrinkage

Although clay shrinkage subsidence has damaged properties in the UK for hundreds of years, up until 1971 insurers did not consider it and domestic policies offered no cover. In 1971 insurance companies started to add subsidence cover to household policies, and the long hot summer of 1976 saw the first surge of subsidence claims. Many properties were affected by subsidence caused by clay shrinkage that proved

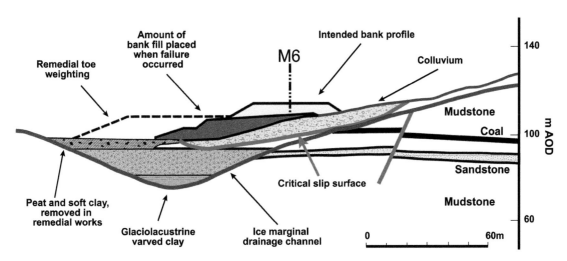

Fig. 1.12. Cross-section of the failed section of the M6 construction works, Walton's Wood, Staffordshire (Martin *et al.* 2017).

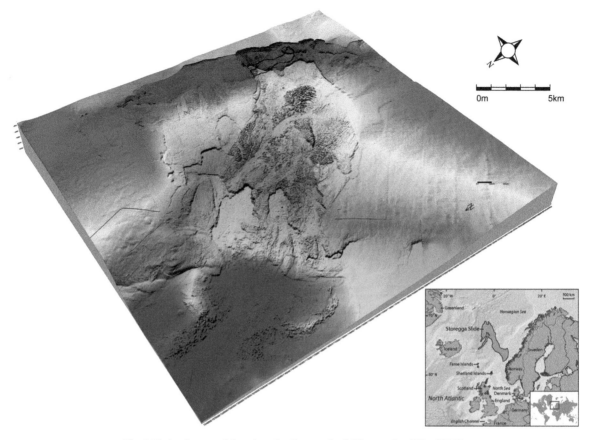

Fig. 1.13. Areal extent of the submarine Storegga landslide complex (Giles 2020b).

Fig. 1.14. Indicative map of the volcanic ash cloud (with Eyjafjallajökull volcano in red) spanning 14–25 April 2010, based on data available from the website of the London Volcanic Ash Advisory Centre (https://en.wikipedia.org/wiki/Air_travel_disruption_after_the_2010_Eyjafjallaj%C3%B6kull_eruption#/media/File:Eyjafjallaj%C3%B6kull_volcanic_ash_composite.png).

Fig. 1.15. Major subsidence through chalk mine crown hole collapse, Field Road, Reading, Berkshire (photo credit: Clive Edmonds).

Fig. 1.16. The Crooked House, Dudley, suffering from coal mining subsidence.

to be both unexpected and very expensive for the insurance industry. Further surge events occurred in 1985, 1990, 1992, 1995, 1996, 2003 and 2006. Individual surge years regularly resulted in 50 000 subsidence claims and repair bills exceeding £400 million, with over £14 billion spent during the last four decades (Giles n.d.).

1.2.12 Poorly recognized geohazards

While dramatic and dynamic geohazard events perhaps always attract media attention, there are a substantial number of more mundane static and geotechnical geohazards that have had some substantial impacts to engineering projects. Subsurface boulders (e.g. Skipper *et al.* 2005), cemented layers (e.g. Newman 2009), running sands (Newman 2009), deoxygenation (e.g. Newman *et al.* 2013), and perched water tables (e.g. Toms *et al.* 2016) can cause massive problems related to ground engineering, even when well understood and characterized.

1.3 Geological Society Engineering Group Working Party on Geohazards

1.3.1 Background

The original Geological Society Working Party on Geohazards was initiated under the leadership of Professor Mike Rosenbaum, Dr David Entwistle and Dr Alan Forster in August 2002 following informal meetings held at the British Geological Survey, Keyworth. Due to many membership changes, the Working Group was reformatted in 2010 with a view to developing a web-based resource as opposed to a hardcopy book. This initiative again stalled with the final outcome of the Working Party being a series of themed chapters compiled remotely of any formal meetings. This Engineering Geology Special Publication represents the results of this long endeavour.

1.3.2 Membership

The Working Party was developed with a UK focus, but included a global perspective in the consideration of examples of good practice and the nature of geohazard issues of a generic nature. The following principal members served as chapter lead authors in this volume: Dr David Giles (Chair & Editor; Card Geotechnics Ltd), Professor Jim Griffiths (Editor; University of Plymouth), Professor Roger Musson (British Geological Survey), Dr Mark Lee (Ebor Geoscience), Professor Mike Winter (TRL Scotland), Professor Martin Culshaw (British Geological Survey), Dr Lee Jones (British Geological Survey), Professor Jeff Warburton (Durham University), Mr Tom Berry (Jacobs), Dr Laurance Donnelly (AHK), Dr Clive Edmonds (Peter Brett Associates), Mr Barry Gamble (independent consultant to UNESCO), Dr Tony Cooper (British Geological Survey), Dr Don Appleton (British Geological Survey) and Mr Steve Wilson (EPG Ltd.).

1.3.3 Terms of reference of the Working Party

The aim of the Working Party is to help geoscientists communicate the interaction of geohazards with society.

Our objectives are to: improve awareness and understanding of geohazards, and to assist in the definition of the role of the engineering geologist in the identification, management and mitigation of hazards in the UK; improve communication between specialists, and between hazards practitioners and the wider community; consider the need for, and the form of, a strategy for the integration of geohazards studies into the planning and development process, and to define areas in which future research is needed; and summarize the current level of scientific understanding of geohazards (in terms of: types, magnitudes and frequencies; geographical locations; elements at risk in society; levels of vulnerability to the various hazards; geohazard recognition and hazard and risk evaluation; issues surrounding the dissemination of geohazard information; geohazard mitigation strategies; and future planning issues in the light of geohazards).

1.3.4 Developing the report

The proposed target audience of the Working Party are professionals who deal with geohazards and their effects, including civil engineers, planners, developers and government, as well as aid organizations. The Working Party Report will help to put the study and assessment of geohazards into the wider social context, helping the engineering geologist to better communicate the issues concerning geohazards to the client and the public. The aim is to provide the document of first choice when a geohazard occurs, able to orientate the enquirer as to 'How did this happen?', 'Where can I get help?', and 'What should I do?' This is somewhat different to the target readership of previous Working Party reports of the Engineering Group, orientated towards the specialist engineering geologist seeking a standardization of approach. The report focuses on: an outline of the nature of geohazards and their engineering consequences; a description of state-of-the-art techniques for the understanding of geohazards, and for assessing the levels of hazard and risk associated with them; a review of the range of users of geohazard information, including a consideration of strengths and weaknesses of the current position, recognizing that it is the communication of geohazards information and data that can be the most difficult part of any investigation; an account of the ways in which geohazard information is utilized within society, considering the social context and economic impact of geohazards; an examination of the potential ways in which existing and future geohazard information could/should be used and by whom; and a review of how best to communicate the information to non-geoscientists.

1.3.5 Contents and structure of the report

The Working Party Report sets out to provide an outline of the nature of the specific geohazard and its engineering consequences, in a UK context. The report provides a description of state-of-the-art techniques for the understanding of the geohazard and for assessing the levels of hazard and risk associated with it. Each section within the Special Publication sets out to summarize the character of the geohazard and considers the following topics with respect to the specific geohazard: what it is; where it might be found or occur; how to recognize it; how best to mitigate its effects; current strategies for engineering management (avoidance, prevention and mitigation); identifying actions following the occurrence of a geohazard; definitions and glossary; and data sources, essential references and further reading.

The report is structured in five sections, each addressing a variety of similarly themed hazards: Section A, Tectonic Hazards; Section B, Slope Stability Hazards; Section C, Problematic Ground and Geotechnical Hazards; Section D, Mining and Subsidence Hazards; and Section E, Gas Hazards.

1.3.6 Geological hazards: Working Party definitions and report limitations

An issue that all Geological Society working parties encounter is setting limits to the scope of their final report. In the Hot Deserts Working Party (Walker 2012), there were discussions on the definition of a 'desert' and initially whether or not cold and polar deserts should be part of the work. In the end, the Working Party decided to limit the scope to 'hot deserts' and climatic criteria were used to establish the spatial extent of these areas (Charman 2012). Within the Glacial and Periglacial Working Party (Griffiths & Martin 2017), the decision was made, after long debate (Martin *et al.* 2017), to limit the report only to the cold phases of the Quaternary, relict glacial and periglacial landforms and deposits, and specifically the conditions in the UK.

For the Geological Hazards Working Party, the spatial limit was identified from the outset as being the UK; however, when the Working Party was initiated, there was no universally agreed definition on what constituted a *geohazard*, beyond stating it was a geological source of danger. Culshaw (2018) provides the most comprehensive summary of the meaning and nature of geohazards. Quoting Nadim (2013), Culshaw (2018) defines 'hazard' as

> ... an event, phenomenon, process, situation, or activity that may potentially be harmful to the affected population and damaging to society and the environment. A hazard is characterised by its location, magnitude, geometry, frequency, or probability or occurrence and other characteristics.

Culshaw (2018) divided geohazards into three main groups: primary natural geohazards, secondary natural hazards and geohazards caused by anthropogenic activity.

Primary natural geohazards are cyclical in occurrence. They affect regions and are controlled by regional geology. They are generally unpredictable, as the geological processes are not yet well enough understood; at present, they are almost impossible to prevent. Earthquakes and volcanoes fall into this category, as do climatic conditions; when low-frequency events occur, the effects can only be dealt with through disaster mitigation plans such as evacuation, disaster response and reconstruction.

Secondary natural hazards are often triggered by the primary natural hazards; they affect sites and districts, are controlled by the local geology and are partially predictable from an understanding of geological processes. They can be controlled to some degree, and are best mitigated by land-use planning, insurance and site-specific engineering measures. Landslides and dissolution fall into this category.

Geohazards caused by anthropogenic activity include extraction of minerals and its after-effects, surface or near-surface engineering activities that go wrong, changes to surface and subsurface water conditions, and placement of waste. These geohazards will have varying degrees of geological control, but all involve anthropogenic activity.

An alternative way of classifying natural hazards that cause disasters, of which geohazards represent a subset, is to look at the causative processes. Based on this approach, CRED (2015) divided natural disasters into six categories: geophysical (earthquakes, mass movements, volcanoes); hydrological (floods, landslides, wave action); meteorological (storms, extreme temperatures, fog); climatological (drought, glacial lake outburst, wildfire); biological (animal accident, epidemic, insect infestation); and extra-terrestrial (asteroid or meteorite impact, space weather).

Under this classification, geohazards would fall under geophysical and some hydrogeological processes.

Culshaw (2018) provides a more comprehensive breakdown of geohazards (Table 1.1) based on the controlling causative process, and subdivides them into geomorphological, geotechnical, hydrological or hydrogeological, geological, marine and artificial. From this classification it is apparent that many geohazards are not relevant to the UK, which was the primary concern of this Working Party Report. However, there are some geohazards identified in Table 1.1 that are found in the UK but have not been included in this report, and this comes back to the problem of setting limits

Table 1.1. *Classification of geohazards according to Culshaw (2018)*

Process category	Nature of the geohazard
Geomorphological	Aeolian soils (loess); dissolution (karst, sinkholes etc.); erosion; desiccation; mass movement (snow avalanches, cambering, landslides, etc.); permafrost
Geotechnical	Acidity; collapsing soils; compressible soils; dispersive soils; expansive soils; quick clay; saline soils; residual soils
Hydrological or hydrogeological	Groundwater level change; floods
Geological	Earthquakes (all aspects of ground motion); fault movement; liquefaction; ground subsidence; surface rupture; tsunamis; volcanic eruptions; dome collapse; pyroclastic flows; lahars; debris flows and avalanches; lava flows; ash/tephra falls; large volcanic projectiles; volcanic gases
Marine	Coastal erosion; submarine landslides; fluid escape features (such as liquefaction); gas release (e.g. gas hydrates); scour; turbidity currents
Artificial	Acid mine drainage; artificial ground; brownfield sites; contamination; landfill; mining hazards of subsidence and collapse; pollution; unfilled, partially filled, and filled excavations and voids

to the scope of the final publication. Nevertheless, some of these warrant further discussion and explanation.

The main omission in the Working Party Report is the primary natural hazard of volcanicity, for the reason that the last active eruption in the UK took place between 60.5 and 55 Ma on the west coast of Scotland (Bell & Williamson 2002). However, there is one interesting present-day component of these eruptions; in the early Eocene deposits of East Anglia and the London Basin, there are very thin bentonite clay beds derived from chemically altered volcanic ash (King 2002). Bromhead (2013) speculated that these beds might be one of the factors controlling the occurrence of landslides in the London Clay Formation. In addition, as discussed above, the ash and gas generated by intermittent present-day volcanic activity in Iceland will continue to be a threat to air travel and air quality over the UK.

A mass movement process that is a UK geohazard but is not discussed in detail in this report is soil erosion. A 2006 report from the Parliamentary Office of Science and Technology (2006) stated that 2.2 Mt of topsoil was eroded annual in the UK, and over 17% of arable land showed signs of erosion. However, unlike countries that required terracing and other physical methods for reducing soil erosion, the main way to mitigate soil degradation in the UK is through better farming practice. Identifying appropriate changes in agricultural practice lay outside the remit of the Working Party.

One phenomenon that falls at the boundary between a geohazard and a meteorological hazard is snowfall and the potential for avalanches. The occurrence of snow is dependent on climate, and the UK is not renowned for copious amounts of snow. However, there is thriving skiing industry in Scotland, where avalanches do occur. Diggins (2018) reported that over the 10-year period from 2008/09 to 2017/18, a total of 21 people were killed by avalanches in the Scottish Highlands; over 200 avalanches occur in this area each year. However, this must be compared with the European Alps where, over the last four decades, about 100 people per year have lost their lives in avalanches (Techel *et al.* 2016). The loss of life in the UK from avalanches is similar and perhaps greater to that from landslides if the tragedy of Aberfan is excluded. It should also be noted that avalanches are not restricted to the Scottish Highlands. Indeed, the greatest loss of life in a single snow avalanche in the UK occurred in December 1836 in Lewes in East Sussex, when seven cottages were destroyed and eight people killed by the collapse of a snow corniche that had developed on a chalk cliff in the South Downs. The nineteenth century artist Thomas Henwood (Fig. 1.17) captured the event. Snow avalanches are a form of mass movement (Griffiths 2018) and, while the failure mechanisms are similar to those encountered in landslides, the techniques of investigation and mitigation are quite different and lie more in the field of snow science than engineering geology.

Another subject that crosses the boundary between geohazards, hydrogeology and meteorological hazards is flooding, whether inland from rivers or on the coast. Flooding by rivers is a natural geomorphological process, although the consequences may be exacerbated by humans who build structures in unsuitable locations, strip vegetation that would have

Fig. 1.17. The avalanche at Lewes, 1836, attributed to Thomas Henwood (Anne of Cleves House, East Sussex).

Fig. 1.18. Thames Barrier, London (photo credit: Andy Roberts).

reduced runoff, and cover areas with impermeable tarmac that increases the peak flow. On the coast, flooding by the sea is related to sea-level height, tides and waves. Coastal flooding as a result of tsunamis generated by earthquakes or submarine landslides is a phenomenon the UK does need to take into account, and this is discussed in Chapter 3 (Giles 2020b). Physical barriers to flooding are structures that require input by engineering geologists; these may be simple earthworks bunds alongside rivers or major concrete sea defences. The 1953 coastal floods in East Anglia that killed more than 300 people, caused by a storm surge in the North Sea (Orford 2005), resulted in the widespread construction of better sea defences. The most prominent of these was the Thames Barrier at Greenwich that was completed in 1984 and was designed to protect London from a similar event (Fig. 1.18); rising sea levels associated with global climate change suggest it is reaching the end of its design life. Because such events are driven by meteorological events, it was decided not to include a discussion on flooding in the Working Party Report, although it is accepted that this is a contestable viewpoint.

As demonstrated by the above discussion, deciding on what geohazards to include in any evaluation of the situation in the UK is not straightforward or without controversy. However, boundaries had to be established and, as a consequence, some topics were omitted that would have been very relevant in other countries (e.g. volcanicity in Italy). The overall aim, however, was to provide an evaluation of those geohazards that engineering geologists were most likely to encounter and have to mitigate against in UK practice.

1.4 Section A: tectonic hazards

1.4.1 Chapter 2: seismic hazard in the UK (Musson 2020)

A popular misconception among the wider public is that earthquakes do not occur in the UK; however, the UK is classified as having a low-to-moderate seismic risk with, on average, a magnitude 3.2 M_w (moment magnitude) or larger earthquake occurring once per year, and a magnitude 4.2 M_w or larger every 10 years. The latter is capable of causing non-structural damage to property. The damage caused by British earthquakes is generally not life threatening, and no one has been killed in a British earthquake (at the time of publication) since 1940. Damage is caused by shaking, not by ground rupture. Seismic hazard can be discounted for most ordinary construction in the UK, but this is not the case for high-consequence facilities such as dams, bridges and all power plants but especially nuclear power plants, where very long timescales have to be considered as a consequence of the long half-life of radioactive materials.

The diffuse spread of earthquakes across the UK means that there are hundreds of faults in the country that have been reactivated and produced (albeit minor) earthquakes. In almost all cases, however, a known, named fault cannot be shown to have been the origin of a specific earthquake. Small earthquakes have small source dimensions and require only a small fault; these are numerous, and the location of an earthquake in three dimensions is not precise beyond a few kilometres at best. There may be several potential fault sources, or the real fault source may be unmapped. In the UK, the spatial distribution of earthquakes is not uniform or random (Fig. 1.19). In Scotland, most earthquakes are concentrated on the west coast with the addition of centres of activity near the Great Glen at Inverness (earthquakes in 1816, 1890 and 1901) and a small area around Comrie, Perthshire, the site of the famous earthquake swarms principally in 1795–1801 and 1839–1846, and possibly also 1605–1622. Since 1846 small shocks have been observed at Comrie on a regular, if infrequent, basis. There has also been swarm activity in the Central Valley of Scotland by the Ochil Hills (near Stirling). This spot was active in 1736, during 1900–1916 and in 1979.

The Outer Hebrides, off the west coast of Scotland, the extreme north (including the islands of Orkney and Shetland) and most of the east of Scotland are virtually devoid of earthquakes. However, for the northwestern reaches of Scotland the absence of early written records, the small population and the recent lack of recording instruments means that there may be a data gap.

Further south in England and Wales, a similar irregularity is seen. Wales and the west of England, including the SW and NW parts of the country and the English Midlands, are much more active than the east of England. NE England seems to be very quiet; the SE has a higher rate of activity with a number of earthquakes that seem to be 'one-off' occurrences, plus a couple of important centres of activity on the south coast. It is curious that the damaging 1884 Colchester earthquake occurred in a locality of SE England that seems to have been otherwise very inactive seismically, either before or since.

Offshore, there is significant activity in the English Channel and in the North Sea off the coast of Humberside. Because only the larger events in these places are likely to be felt

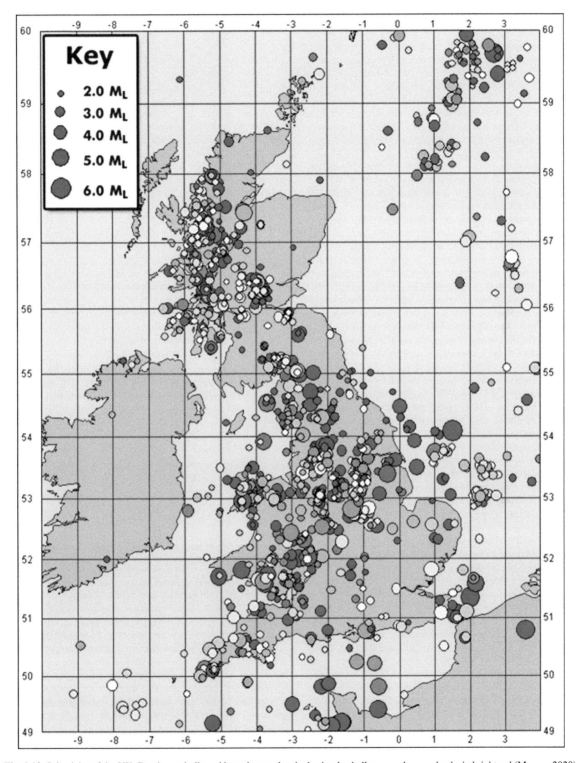

Fig. 1.19. Seismicity of the UK. Depths are indicated by colour: paler shades imply shallower; unknown depths in bright red (Musson 2020).

onshore, the catalogue is probably under-representative of the true rate of earthquake activity in these zones. The largest British earthquake for which magnitude can be estimated had an epicentre in the North Sea, off the east coast of England. This occurred on 7 June 1931 with an estimated magnitude of 5.8 M_w (moment magnitude), and the earthquake was felt over the whole of Great Britain, eastern Ireland and in all the countries bordering the North Sea. It is fortunate that this earthquake had an offshore epicentre as the damage might have been considerable otherwise; only minor damage occurred up the east coast.

Certain centres can be identified as showing typical patterns of activity. For example, the NW corner of Wales is one of the most seismically active places in the whole UK. Both large and small earthquakes, usually accompanied by many aftershocks, occur at regular intervals. In South Wales, on the other hand, although a line of major epicentres can be traced from Pembroke to Newport, only the Swansea area shows consistent recurrence. The Hereford–Shropshire area of western England adjoins South Wales, and this area has also experienced large earthquakes in 1863, 1896, 1926 and 1990; these have no common epicentre, however. In the north of England seismic activity occurs principally along the line of the Pennine Hills, which form the backbone of this part of the country. Again, it is possible to identify particular spots that have been active repeatedly.

The area of the Dover Straits is particularly significant because of the occurrence there of two of the largest British earthquakes in 1382 and 1580 (both of magnitude about 5.5 M_w). Jersey has also experienced a number of significant earthquakes, chiefly originating to the east of the island in the Cotentin peninsula area of France.

What is remarkable is the lack of correlation between this pattern and the structural geology of the UK. In the northern part of the British Isles, the geology has a strong NE–SW (Caledonian) trend and the geology of Northern Ireland is largely a SW-wards extension of the geology of Scotland. However, there is no continuity of seismicity along this trend. It is possible to draw a line roughly NNW–SSE through Scotland such that earthquakes are entirely confined to the west side of the line; yet this line has no apparent geological significance and cuts directly across the structural trend. It is clear that this pattern is persistent and not merely an artefact of recent earthquake locations; there are a number of historical sources for the east of Scotland which comment on the absence of earthquakes. The difficulty is acute in Ireland; the geological history of Ireland is very similar to that of Great Britain, and there is no clear solution to the question of why the seismicity of Ireland should be so very much lower. Scottish seismicity coincides with those areas under ice in the last phase of the last glaciation. In Scotland, stresses due to isostatic recovery after the last glaciations, with a sort of 'jostling' of different geological units in response to an overall compressive stress, were exerted from the NW in response to Atlantic widening.

The frequency of earthquakes in any region is known to be inversely related to the magnitude of the shock, according to

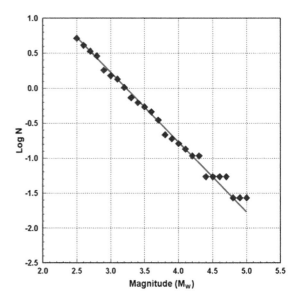

Fig. 1.20. Gutenberg–Richter relationship for UK seismicity (Musson 2020).

what is known as the Gutenberg–Richter equation. Figure 1.20 shows the application of this to the UK, using data from 1970 to 2007. The relationship represented by the red line is:

$$\log N = 3.23 - 1.00 M_w \tag{1}$$

where N is the number of earthquakes per year equal to or larger than a given M_w magnitude. The value of -1.0 coincides with the expected value from theory and practice, and generally equates to that over northern Europe.

1.4.2 Chapter 3: tsunami hazard with reference to the UK (Giles 2020*b*)

Tsunami present a significant geohazard to coastal and waterbody marginal communities worldwide. Tsunami, a Japanese word, describes a series of waves that travel across open water with exceptionally long wavelengths (up to several hundred metres in deep water) and with very high velocities (up to 950 km hr^{-1}) before shortening and slowing on arrival at a coastal zone. On reaching land, these waves can have a devastating effect on the people and infrastructure in those environments. Until relatively recently, the understanding of tsunami events and their historic catalogue had been quite poor. The 2004 Indian Ocean Boxing Day tsunami and the 2011 Tohoku event in Japan tragically brought this geohazard to the attention of the wider population and instigated a deeper investigation and research into these geological phenomena.

Tsunamis can be generated through a variety of mechanisms, including the sudden displacement of the sea floor in a seismic event as well as submarine and onshore landslides displacing a mass of water. Typically, tsunami are generated by tsunamigenic earthquakes, tsunamigenic landslides, tsunamigenic volcanism and meteotsunami.

With its 12 429 km of coastline, the UK is no less prone to the impact of tsunami as the Indian or Pacific oceans. In 2005, Defra commissioned a study precipitated by the Indian Ocean disaster to consider the potential impact on the UK from such events. This review presents those impacts together with a summary of tsunami triggers and UK case histories from the known historic catalogue. Seven potential source zones that could affect the UK (Fig. 1.21) were categorized in terms of their probability of occurrence, namely: UK coastal waters; NW European continental slope; plate boundary west of Gibraltar; Canary Islands; Mid-Atlantic Ridge; eastern North American continental slope; and the Caribbean.

Some notable tsunami events with a UK impact include the *c.* 8150 BP Holocene Storegga submarine landslide and tsunami, the *c.* 5500 BP Holocene Garth tsunami, the tsunami generated by the 1755 Lisbon earthquake, and a local

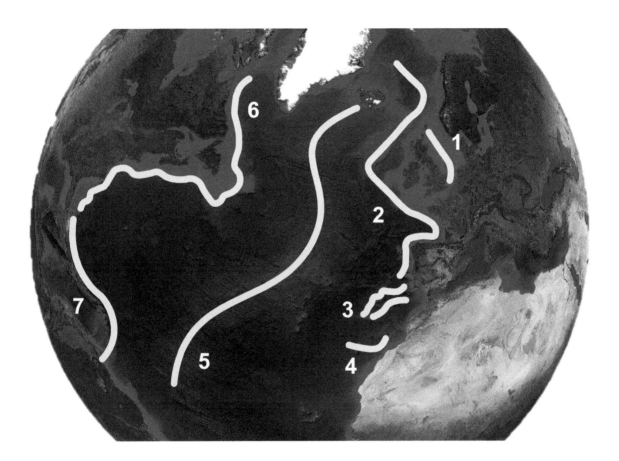

1. **UK Coastal Waters**
2. **NW European Continental Slope**
3. **Canary Islands**
4. **Plate Boundary West of Gibraltar**
5. **Mid Atlantic Ridge**
6. **Eastern North American Continental Slope**
7. **Caribbean**

Fig. 1.21. Possible tsunami source zones with a potential UK impact as considered by Defra (Giles 2020*b*).

event generated by the 1911 Abbots' Cliff failure at Folkestone, Kent.

1.5 Section B: slope stability hazards

1.5.1 Chapter 4: landslide and slope stability hazard in the UK (Lee & Giles 2020)

For many people above a certain age the word 'landslide' will evoke memories of the Aberfan disaster of 21 October 1966. A rotational failure at the front of a colliery spoil tip on the flanks of a steep-sided South Wales valley transformed into a flow slide which travelled downslope at around 10 m s^{-1} into the village. The debris ran out 605 m, building up behind the rear wall of the Pantglas Primary School, causing it to collapse inwards. The loose waste then filled up classrooms, killing 116 children and 28 teachers. The community was severely affected, with many suffering severe psychological difficulties after the disaster. However, Aberfan presents a misleading picture about the nature of landsliding in the UK. The deaths caused by this single event almost certainly exceeded the overall loss of life from all other landslide events in the UK over the last few centuries. Fatal accidents in the UK are extremely rare and tend to be the result of rockfalls or high-velocity slides on the coast, rather than in inland valleys (Fig. 1.22). For example, in July 2012 a young woman was killed by a large rockfall on the beach at Burton Bradstock, Dorset. In February 1977 a school party were studying the geology of Lulworth Cove, Dorset, when they were buried beneath a rockslide; the schoolteacher and a pupil were killed and two more pupils seriously injured, one of whom died later in hospital. In July 1979 a woman sunbathing on the beach near Durdle Door, Dorset, was killed when a 3 m overhang collapsed.

Although the incidents on the Dorset coast during the 1970s led the then Chief Inspector of Wareham police to coin the phrase 'killer cliffs', the public perception of coastal erosion is dominated by the fear that parts of the UK are being rapidly lost to the sea, raising visions of a loss of national resources to a hostile invading power (Table 1.2).

The most intense marine erosion and cliff recession rates occur on the unprotected cliffs formed of soft sedimentary rocks and glacial deposits along the south and east coasts of England, respectively. The Holderness coast, for example, has retreated by around 2 km over the last 1000 years, including at least 26 villages listed in the Doomsday survey of 1086; 75 Mm^3 of land has been eroded in the last 100 years. Rapid recession has also caused severe problems on the Suffolk coast, most famously at Dunwich where much of the former town has been lost over the last millennium. Gardner (1754) recorded that, by 1328, the port was virtually useless and 400 houses together with windmills, churches, shops and many other buildings were lost in one night in 1347. On parts of the north Norfolk coast there has been over 175 m of recession since 1885; county archives show that 21 coastal towns and villages have been lost since the eleventh century.

Today, the reality of coastal erosion is often very different, primarily because of the effectiveness of over 850 km of coastal protection measures built mainly over the last 130 years. The average annual loss of land due to cliff recession and coastal landsliding around the coast of England is probably less than 10–25 ha.

High-velocity landslide events that present a threat to people do occur inland, such as the August 2004 debris flows in the Scottish Highlands (see Chapter 5; Winter 2019). There were no fatalities, but 57 people had to be airlifted to safety by the RAF when they became trapped between debris flows on the A85 at Glen Ogle. Two people were killed in July 2012 when their car was crushed by falling debris as it emerged from the Beaminster Tunnel, Dorset, due to a landslide bringing down part of the tunnel portal. However, most inland landslides generally present only minor threats to life as movements, when they occur, usually involve only slow and minor displacements. Even when large displacements occur, the rate of movement tends to be gentle and not dramatic, as was reported graphically for the French House slide near Lympne, Kent, in 1725 where a farmhouse sank 10–15 m overnight, 'so gently that the farmer's family were ignorant of it in the morning when they rose, and only discovered it by the door-eaves, which were so jammed as not to admit the door to open' (Gostling 1756).

Nevertheless, slow-moving inland landslides can have a significant economic impact. The cumulative effects of episodes of slow movement can inflict considerable damage to buildings, services and infrastructure. Almost continuous damage from movement of the Mam Tor landslide in the High Peak of Derbyshire led to the permanent closure of the A625 Manchester–Sheffield road in 1979 and diversion of local and cross-Pennine traffic. Hutchinson (2001) describes how a power line from Dungeness Nuclear Power Station, Kent, was put out of action for over a month in the winter of 1966/67 when landslide activity on the Hythe–Lympne escarpment led to the loss of a pylon. Sustained rainfall in November 1998 led to the collapse of Greenan Road near Ballycastle, Northern Ireland, cutting off access to a farming community; as the farms quickly ran out of feed for their livestock, helicopters were used to bring in fresh supplies. There has been a history of mudslides blocking the Antrim coast road, particularly at Minnis North. For example, in a 14-month period between 1971 and 1972 there were 10 incidents when the road was blocked. Intense rainfall on the morning of Tuesday 8 November 2005 initiated a small peat slide on a hillside above the A5 London–Holyhead trunk road in the Llyn Ogwen area, Snowdonia National Park; four people were injured, a nearby construction project was delayed and A5 was blocked. In February 2013, a landslide occurred in a spoil tip at Hatfield Main Colliery and severely damaged a large section of train line along the Doncaster–Goole and Doncaster–Scunthorpe lines. The section of the line was closed for 5 months and train services in the region significantly affected.

The unforeseen presence of ancient landslides can lead to costly problems during construction. The A21 Sevenoaks

Fig. 1.22. Landslide susceptibility map of the UK (Lee & Giles 2020).

Table 1.2. *Significant landslide fatalities in the UK (British Geological Survey National Landslide Database)*

Landslide Event	Year	Fatalities	Mechanism (after Varnes 1978)	Land system
Bwlch Y Saethau pass, Snowdon, Gwenydd	2018	1*	Rock fall	Coastal
Cwmdaud, Carmarthenshire	2018	1*	Slide	River valley
Staithes, Yorkshire	2018	1	Rock fall	Coastal
Thorpeness, Suffolk	2017	1	Rock fall	Coastal
Llantwit Major, Vale of Glamorgan	2015	1	Rock fall	Coastal
Sandplace Road, Looe, Cornwall	2013	1	Slide	Coastal
Burton Bradstock, Dorset	2012	1	Rock fall	Coastal
Beaminster Tunnel, Dorset	2012	2	Slide	Hillside
Newbiggin, Northumberland	2010	1	Rock fall	Coastal
Whitehaven, Cumbria	2007	1	Debris fall	Coastal
Ben Nevis, Lochaber	2006	1	Rock fall	Upland
Nefyn, Gwynedd	2001	1	Debris flow	Coastal
Marine Drive, Gogarth, Gwynedd	1987	1	Rock fall	Coastal
Newquay, Cornwall	1986	1	Rock fall	Coastal
Durdle Door, Dorset	1979	1	Rock fall	Coastal
Lulworth Cove, Dorset	1977	3	Rock fall	Coastal
Swanage Bay, Dorset	1976	1	Rock fall	Coastal
Kimmeridge Bay, Dorset	1971	1	Rock fall	Coastal
Aberfan, South Wales	1966	144	Debris flow	Anthropogenic
Alum Bay, Isle of Wight	1959	1	Rock fall	Coastal
Boscombe, Dorset	1925	3	Rock fall	Coastal
Loch Ness, Scotland	1877	1	Rock fall	Upland
Early's Wall, Dawlish	1855	3	Rock fall	Coastal
Sonning Cutting, Reading, Berkshire	1841	9	Slide/flow	Anthropogenic
Guildford Battery, East Cliff, Dover	1810	7	Rock fall	Coastal
Pitlands Slip, Isle of Wight	1799	2	Rock fall	Coastal

*Landslides not yet confirmed; inquest currently underway.

Bypass, Kent, had to be halted in 1966 when excavation work cut through grass-covered lobes of material that proved to be the remains of a previously unidentified ancient landslide. The inadvertent removal of material from the lower portion of these landslides led to their reactivation, despite the fact that they appeared to have remained stable and stationary over the Holocene. The problems turned out to be so severe that the affected portion of the route had to be realigned. This incident and a similar landslide problem on the M6 motorway embankment at Waltons Wood provided the impetus for UK-based academic research into inland landslides.

In many instances landslide problems are less newsworthy, although they can still lead to property loss or the delay, redesign or abandonment of projects. For example, Camden Crescent in Bath is the only known asymmetric crescent in the world; half had been destroyed by the Hedgemead landslide in 1894. In 1952, a landslide occurred at the village of Jackfield, Shropshire, on the River Severn just over 2 km downstream of the Iron Bridge, destroying several houses and causing major dislocations in a railway and road. Instability problems encountered at housing developments at Bury Hill and Brierley Hill in the West Midlands, at Exwick Farm on the outskirts of Exeter, at Ewood Bridge in the Irwell Valley and at Gypsy Hill in South London are some examples of the impact of localized slope instability frequently associated with smaller-scale developments. A large landslide in 1993 at Franklands Village, West Sussex, led to the demolition of 14 flats and houses.

This chapter considers all aspects of landslide and slope stability from outlining the hazard, assessing the risk and managing the risk posed by problematic slopes.

1.5.2 Chapter 5: debris flows (Winter 2020)

Debris flows are largely fast moving and dynamic in nature; they are generally characterized by rapid movement with high proportions of either water or air acting as a lubricant for the solid material that generally comprises the bulk of their mass. Given the right circumstances, they can be highly destructive. In the UK their presence is largely, although not exclusively, restricted to mountainous areas. Indeed, the UK landslides research community has historically focused on slow-moving events that, in general, lead to economic losses such as those at Ventnor on the Isle of Wight and Folkestone Warren.

The fast-moving debris-flow events in Scotland in August 2004 and since provide a rich source of case study material; it was fortuitous that there were no major injuries to those involved in those events. However, even in the absence of serious injuries and fatalities, the socioeconomic impacts of such events may be serious. These include the severance (or delay) of access to and from relatively remote communities for: markets for goods and services; employment, health and educational opportunities; and social activities. The extent of these impacts is described by the vulnerability shadow. The work that has followed has therefore drawn on the more traditional approach to slow-moving landslides, as well as that typified by the international approach to fast-moving events that pose a real risk to life and limb.

Hillslope (or open-slope) debris flows form their own path down valley slopes as tracks or sheets, before depositing material on lower areas with lower slope gradients or where flow rates are reduced (e.g. obstructions, changes in topography; Fig. 1.23). The deposition area may contain channels and levees. The motion of such events is generally considered not to be maintained when the width exceeds five times the average depth. As the mobilized material in such events rarely persist to either the level of the slope at which transport infrastructure exists, or to the valley floor in Scotland, they are therefore of relatively little practical interest.

Channelized debris flows follow existing channel-type features, such as valleys, gullies and depressions, are often of high density, comprise 80% solids by weight, and may have a consistency equivalent to that of wet concrete; they can therefore transport boulders that are some metres in diameter.

In this chapter, the work undertaken for a hazard and risk assessment for debris flow affecting the Scottish road network is briefly referred to in terms of: a GIS-based assessment of debris-flow susceptibility; a desk-/computer-based interpretation of the susceptibility and field-based ground-truthing to determine hazard; and a desk-based exposure analysis to enable the determination of risk.

A strategic approach to landslide risk reduction allows a clear focus on that overall goal before homing in on the desired outcomes and the generic approach to achieving those outcomes. Only then are the processes that may be used to achieve those outcomes (i.e. the specific management and mitigation measures and remedial options) addressed. A top-down, rather than a bottom-up, approach is therefore targeted. Risk reduction is considered as: relatively low-cost exposure reduction (management) outcomes that allow specific measures to be extensively applied; and relatively high-cost hazard reduction (mitigation) outcomes that include measures that are targeted at specific sites.

In addition to covering the above themes, this chapter also considers the potential effects of future climate change on debris-flow hazard and risk, again using Scotland as an example.

1.6 Section C: problematic ground and geotechnical hazards

1.6.1 Chapter 6: collapsible soils in the UK (Culshaw *et al.* 2020)

Metastable soils may collapse because of the nature of their fabric. These soils have porous textures, high void ratios and low densities. They have high apparent strengths at their natural moisture content but large reductions of void ratio take place on wetting and, particularly, when on loading, because bonds between grains break down on saturation. Worldwide, there is a range of natural soils that are metastable and can collapse including: loess; residual soils derived from the weathering of acid igneous rocks and from volcanic ashes and lavas; rapidly deposited and then desiccated debris-flow materials such as some alluvial fans (e.g. in semi-arid basins); colluvium from some semi-arid areas; and cemented, high-salt-content soils such as some sabkhas. In addition,

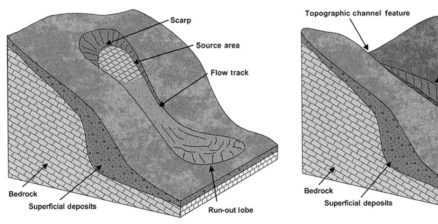

Fig. 1.23. (**a**) Hillslope and (**b**) channelized debris flow (Winter 2020).

some artificial non-engineered fills can also collapse. The main type of collapsible soil in the UK is loess, although collapsible non-engineered fills also exist. Loess in the UK can be identified from geological maps, but care is needed because it is usually mapped as 'brickearth'. This is an inappropriate term and it is suggested here that it should be replaced with the term 'loessic brickearth'. Loessic brickearth in the UK is found mainly in the SE, south and SW of England, where thicknesses greater than 1 m are found. In Great Britain, loessic deposits are mapped by the British Geological Survey mainly as 'brickearth'. Such deposits occur mainly as a discontinuous spread across southern and eastern England, notably in Essex, Kent, Sussex and Hampshire (Fig. 1.24).

Elsewhere, thicknesses are usually less than 1 m and, consequently, of limited engineering significance. There are four steps in dealing with the potential risks to engineering posed by collapsible soils: (1) identification of the presence of a potentially collapsible soil using geological and geomorphological information; (2) classification of the degree of collapsibility, including the use of indirect correlations; (3) quantification of the degree of collapsibility using laboratory and/or in situ testing; and (4) improvement of the collapsible soil using a number of engineering options.

Fig. 1.24. Surface distribution of loess/brickearth in south UK based on Soil Survey 1:250 000 scale soil maps (1983). Loess >1 m thick in black; loess >300 mm thick (and often partly mixed with subjacent deposits) shown stippled (Culshaw *et al.* 2020).

Soils that have the potential to collapse generally possess porous textures with high void ratios and relatively low densities. At their natural moisture content, these soils possess high apparent strength but are susceptible to large reductions in void ratio on wetting, especially under load. In other words, the metastable texture collapses as the bonds between the grains break down as the soil becomes saturated. As collapse is controlled both microscopically and macroscopically, both these elements need to be understood if the true nature of collapse is to be determined. The potential for soils to collapse is clearly of geotechnical significance, particularly with respect to the potential distress of foundations and services (e.g. pipelines) if not recognized and designed for. The collapse process represents a rearrangement of soil particles into a denser state of packing. Collapse on saturation usually occurs rapidly. As such, the soil passes from an underconsolidated condition to one of normal consolidation. There are two basic requirements for a soil to be collapsible: a collapsible soil is one in which the constituent parts have an open packing and which forms a metastable state that can collapse to form a closer packed, more stable structure of significantly reduced volume'; and 'in most collapsible soils the structural units will be primary, mineral particles rather than clay minerals.

The most widespread naturally collapsible soils are loess or loessic soils of aeolian origin, predominantly of silt size with uniform sorting. The majority of these soils have glacial associations in that it is believed that these silty soils were derived from continental areas where silty source material was produced by glacial action prior to aeolian transportation and deposition. There are four fundamental requirements necessary for the formation of loess: a dust source; adequate wind energy to transport the dust; a suitable depositional area or reduced wind speed; and sufficient time for its accumulation and epigenetic evolution.

These requirements are not specific to any one climatic or vegetational environment. While much loess was formed in glacial/periglacial environments, derived from the floodplains of glacial braided rivers where glacially ground silts and clays were deposited, windblown deposits can be derived in other environments, such as volcanic, tropical, desert and gypsum loesses; climatically controlled windblown deposits are referred to as trade-wind and anticyclonic.

Collapsible soils, including loess, are materials that standard soil mechanics stress–strain principles fail to adequately explain in terms of their engineering behaviour. For the ground engineering industry to avoid and mitigate the risks associated with collapse, a first significant step is to correctly identify the presence of collapsible soils. Once identified, appropriate laboratory testing procedures and, where necessary, follow-up field tests can be applied to assess collapse potential and the possible need for mitigation measures. This chapter describes the current geological, geotechnical, geochemical, mineralogical and geomorphological understanding of UK collapsible soils and may serve as a guide to aid engineering ground investigation in those areas where such natural (loessic) soils and potentially collapsible anthropogenic fills may be present. Current techniques to help mitigate the risks associated with collapse are also described. As planned expansion of the UK road and rail infrastructure progresses, it becomes ever more important that the collapse potential of poorly or non-engineered fills, including old Victorian railway embankments, is considered by ground engineers, and that the use and appropriate engineered placement of potentially metastable materials is more fully understood and designed for.

1.6.2 Chapter 7: quick-clay behaviour in sensitive Quaternary marine clays: UK perspective (Giles 2020a)

The term 'quick clay' has been used to denote the behaviour of highly sensitive Quaternary marine clays that, due to post depositional processes, have the tendency to change from a relatively stiff condition to a liquid mass when disturbed. On failure, these marine clays can rapidly mobilize into high-velocity flow slides and spreads, often completely liquefying in the process. For a clay to be defined as potentially behaving as a quick clay in terms of its geotechnical parameters, it must have a sensitivity (the ratio of undisturbed to remoulded shear strength) of greater than 30, together with a remoulded shear strength of less than 0.5 kPa. Potential quick-clay-behaving soils can be found in areas of former marine boundaries that have been uplifted through isostatic rebound after Quaternary glaciations. The presence of quick clays in the UK is unclear, but the Quaternary history of the British islands suggests that the precursor conditions for their formation could be present and should be considered when undertaking construction in the coastal zone.

Deposits prone to quick-clay behaviour develop from initially marine clays deposited from rock-flour-rich meltwater streams feeding into a nearshore marine environment (Fig. 1.25). On glacial retreat, crustal rebound (isostatic recovery) uplifts the marine sediments above current sea level, eventually exposing them to a temperate weathering environment and soil leaching by freshwater. In Norway, for example, the former syn-glacial sea level can be found up to 220 m higher than present-day sea levels.

For clay to develop 'quick' properties, the sediment must have a flocculated structure and a high void ratio. This flocculated structure would be the normal state in which fine-grained sediments formed from glacial erosion had been deposited in marine and brackish subaqueous environments. In this setting, silt- and clay-sized particles would rapidly flocculate to form these high-void ratio sediments. Generally, in freshwater sedimentary environments clay-sized particles

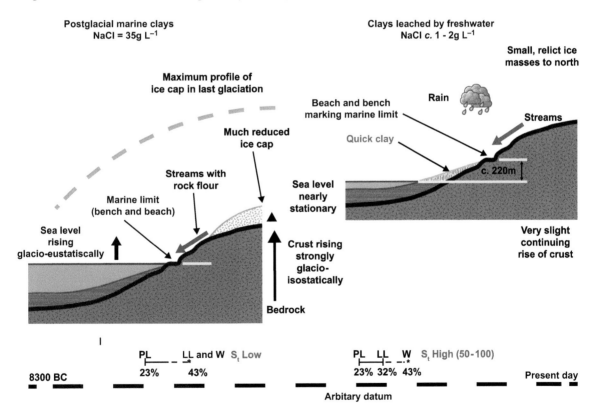

Fig. 1.25. The development of quick clays through the Holocene (Giles 2020a). PL, plastic limit; LL, liquid limit; W, natural moisture content; S_t sensitivity.

settle even more slowly than silt grains, and tend to accumulate in a dispersed structure with a parallel orientation of particles. In more saline conditions, silt and clay particles form aggregates (small flocculates) and settle together in a random pattern. This random alignment of particles (in effect a 'house of cards' structure) gives the flocculated material a higher-than-normal void space and hence potentially higher moisture content. Quick-clay sediments originally deposited in marine or brackish conditions initially had a porewater geochemistry of up to 35 g L^{-1} sodium chloride. Subsequent uplift of the strata to above sea level resulted in them being subject to temperate weathering conditions where soil leaching by freshwater occurred. This weathering created a top crust of leached material with a subsequent reduction in the strength of the former marine clays. The sodium chloride porewaters were progressively leached by rainwater and freshwater streams, reducing the salt content to around 1–2 g L^{-1}. This had the effect of generating very sensitive clay-dominated soils that exist in a metastable state. Potential quick-clay-behaving sediments can be identified by their geotechnical properties, in particular by their sensitivity, the ratio of undrained shear strength to remoulded shear strength at the same moisture content, and by their activity.

Various studies on postglacial isostatic recovery and eustatic sea-level adjustment indicate that parts of the UK coastal zone have been elevated above former sea levels. The possibility that former fine-grained marine sediments have subsequently been elevated above sea level and have been subject to weathering processes and potential porewater leaching potentially exists in these now-onshore coastal areas. The uplifted zones will have experienced the pre-conditions for quick-clay-behaving sediments to be developed. In terms of ground investigation in these areas, the geotechnical properties of any fine-grained sediments encountered need to be considered with respect to potential quick-clay behaviour, specifically with respect to the sensitivity and activity of the deposit as well as the nature of the mineral content of the soil. An awareness that these soils could be prone to rapid failure, coupled with a complete remoulding of the soil with the associated liquefaction, needs to be taken into account in the design and implementation of construction works and must form part of the hazard assessment and project risk management.

1.6.3 Chapter 8: swelling and shrinking soils (Jones 2020)

Shrink–swell soils are one of the most costly and widespread geological hazards globally, with costs estimated to run into several billion pounds annually. These soils present significant geotechnical and structural challenges to anyone wishing to build on, or in, them. Shrink–swell occurs as a result of changes in the moisture content of clay-rich soils, reflected in a change in volume of the ground through shrinking or swelling. Swelling pressures can cause heave or lifting of structures while shrinkage can cause differential settlement. This chapter aims to give the reader a basic understanding of shrink–swell soils. A review is provided on the nature and extent of shrink–swell soils, both in the UK and worldwide, discussing how they form, how they can be recognized, the mechanisms and behaviour of shrink–swell soils, and the strategies for their management (including avoidance, prevention and mitigation).

A shrink–swell soil is one that changes in volume, in response to changes in its moisture content. The extent of the volumetric change reflects the type and proportion of swelling clay in the soil. More specifically, expansive clay minerals expand by absorbing water and contract, or shrink, as they release water and dry out. Clays range in their potential to absorb water according to their different structures. For the most expansive clays, expansions of 10% are common.

In practice, the amount by which the ground shrinks and/or swells is determined by the water content in the near-surface (active) zone. Soil moisture in this zone responds to changes in the availability of atmospheric recharge and the effects of evapotranspiration. These effects usually extend to about 3 m depth, but this may be increased by the presence of tree roots. Characteristically fine-grained clay-rich soils soften, becoming sticky and heavy following recharge events such as rainfall, and commonly can absorb significant volumes of water. Conversely, as they dry, shrinking and cracking of the ground is associated with a hardening of the clay at surface. Structural changes in the soil during shrinkage, for example, alignment of clay particles, ensure that swelling and shrinkage are not fully reversible processes. For example, the cracks that form during soil shrinkage are not perfectly annealed on re-wetting. This volume increase results in a decrease in the soil density, thereby providing enhanced access by water for subsequent episodes of swelling. In geological timescales, shrinkage cracks may become infilled with sediment, thus imparting heterogeneity to the soil. Once the cracks have been infilled in this way, the soil is unable to move back, leaving a zone with a network of higher permeability infills. When supporting structures, the effects of significant changes in water content on soils with a high shrink–swell potential can be severe. In practical civil engineering applications in the UK, there are three important time-dependent situations, each with different boundary conditions, where shrink–swell processes need to be considered: (1) following a reduction in mean total stress (the most notable effects are found adjacent to cut slopes, excavations and tunnels); (2) subsurface groundwater abstraction or artificial/natural recharge under conditions of constant total stress in both unconfined and confined aquifers (regional subsidence or heave can be induced by this process); and (3) surface climatic/water balance fluctuations related to land-use change under conditions of constant total stress (the most notable effects follow the development of seasonally desiccated soils, which can cause structural damage to existing shallow foundations).

As well as effective stress changes, some deformation may be caused by biogeochemical alteration and dissolution of minerals as a result of steady-state fluid transport processes. Although surface movements and engineering problems can occur due to a loss or addition of solid material, these are not strictly shrink–swell soils. However, these processes are often combined with effective stress changes and/or fluid movements, and may therefore be difficult to separate from true shrink–swell processes that might be taking place at the same time. The main factors controlling shrink–swell susceptibility in geological formations are material composition (clay mineralogy), initial *in situ* effective stress state and stiffness of the material. Variations in the initial condition caused through processes such as original geological environment, climate, topography, land-use and weathering affect *in situ* effective stress, stiffness and hence shrink–swell susceptibility. Clays belonging to the silicate family comprise the major elements silicone, aluminium and oxygen. There are many other elements that can become incorporated into the clay mineral structure (hydrogen, sodium, calcium, magnesium and sulphur). The presence and abundance of these dissolved ions can have a large impact on the behaviour of the clay minerals. The clay minerals are defined by the ratio of silica tetrahedra to alumina, iron or magnesium octahedra.

Subsidence also occurs in superficial deposits such as alluvium, peat and laminated clays that are susceptible to consolidation settlement (e.g. in the Vale of York, east of Leeds, and in the Cheshire Basin), but these are not true shrink–swell soils.

1.6.4 Chapter 9: peat hazards: compression and failure (Warburton 2020)

Peat is a low-density, highly compressible soil that occurs at the surface or may be buried at depth. Peat is essentially an organic, non-mineral soil resulting from the decay of organic matter. In the UK, peat deposits are widespread occurring in a wide variety of upland and lowland environments covering

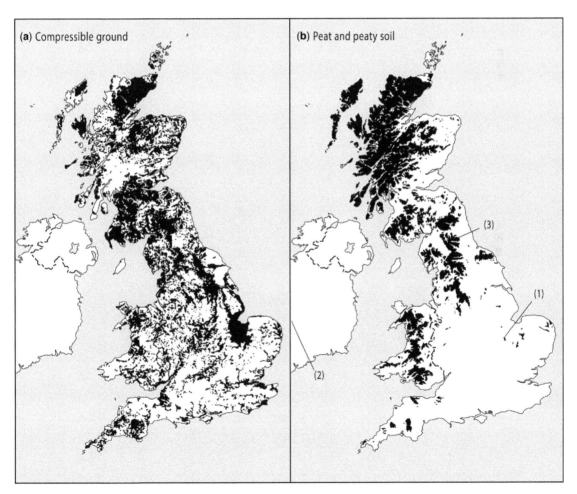

Fig. 1.26. (**a**) Compressible ground potential map and (**b**) peat and peaty soils of the UK (**Warburton 2020**). Numbers indicate key sites discussed in this volume.

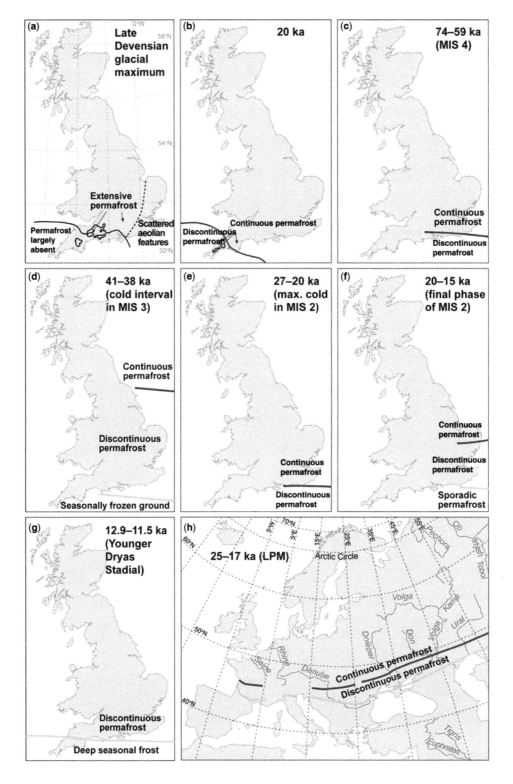

all parts of the country (Fig. 1.26). Peat accumulates wherever suitable conditions occur such as in areas of high (excess) rainfall and where ground drainage is poor leading to high water tables. In these waterlogged areas, peat develops where the rate of dry vegetative matter accumulation exceeds the rate of decay. Physiochemical and biochemical processes associated with wetland conditions ensure that the accumulating organic matter decays very slowly, safeguarding plant structures that remain partially intact for long periods of time. In the UK, temperate peat accumulates slowly at typically 0.2–1 mm a^{-1}; local rates vary depending on the topography and hydrology of the peat mire.

In the engineering community, peats and organic soils are well known for their high compressibility and long-term settlement and, in terms of engineering properties, peat is notoriously difficult to deal with. The link between the compressibility of peat, its shear strength properties and the risk of bearing capacity failure has not been explored in detail, although the mechanism has been suggested for some peat failures. Peat soils are highly organic, highly compressible and generally possess low undrained strength, and their compression and/or settlement may take a considerable amount of time to stabilize. Estimating the geotechnical properties of peat is difficult because published values are relatively few and the testing of peat using standard geotechnical tests is fraught with problems. Nevertheless, published data suggest that peat in its undisturbed state has little strength with undrained shear strength values typically varying over 5–20 kPa. These values vary with the vegetation composition of the peat (particularly fibre content) and the degree of humification, but are also affected by the method of testing. Given the high compressibility and low strength of peat, local shear failure may occur when compression and/or compaction gives rise to vertical displacements that exceed the shear strength (bearing capacity) of the soil. Shear failure may result where differential displacements of surface peat occur between the area experiencing compression (loading) and the adjacent unloaded peat. In peatlands, such sites typically include: construction embankments or waste heaps; roads and tracks; and foundations such as wind turbine bases. Although such failures are local in origin due to the sensitive nature of peat stability, under the right site conditions these may rapidly propagate to runaway failures.

In engineering practice there is a tendency to either avoid construction on these soils or, if this is not possible, remove or replace the peat material. However, in many countries, including the UK, peat extends over a substantial part of the terrestrial biosphere and peatlands are under increasing pressure for their land use. In lowland areas, particularly in the distal parts of populated deltas and estuaries, peat is common and, due to compaction, may cause land subsidence, resulting in damage to infrastructure and land inundation by the sea.

As part of its UK hazard assessment programme, the British Geological Survey has summarized key information on compressible ground as follows.

> Ground is compressible if an applied load, such as a house, causes the fluid in the pore space between its solid components to be squeezed out causing it to decrease rapidly in thickness (compress). Peat, alluvium and laminated clays are common types of deposits associated with various degrees of compressibility. The deformation of the ground is usually a one-way process that occurs during or soon after construction.

Peat soils are well known for landslide-related hazards and these have been widely reported and documented in the UK and Ireland. However, far less is known about the hazards posed by peat compression and the potential problems associated with this. The aims of this chapter are therefore to: briefly review the engineering background to peat compression; describe the occurrence of peat soils in the UK; provide examples of the compression hazards associated with these deposits; and consider some of the ways these can be mitigated.

1.6.5 Chapter 10: relict periglacial hazards (Berry 2020)

Almost all areas of the UK have experienced the effects of periglaciation and permafrost conditions during the Quaternary (Fig. 1.27) and, as such, relict periglacial geohazards can potentially be a significant technical and commercial risk for many engineering projects. The term periglacial is used here to describe areas affected by cold conditions that border, or have bordered, former Quaternary ice sheets. The term is used here to include processes as well as the resultant sediments, structures and landforms to be found in this relict environment (Ballantyne & Harris 1994; Walker 2005). Periglaciation not only affects the deposits left behind by the various phases of glaciation but also affects older geological strata that were at or near the ground surface. In contrast to present-day periglacial environments, the areal extent of former periglacial environments was much greater. Consequently, relict periglacial features are likely to have once covered the whole of the UK (Walker 2005), including offshore continental shelves of the North Sea, English Channel and Irish Sea.

The aim of this chapter is to describe the specific geological and geotechnical hazards generated from past periglacial processes and to highlight their ground-engineering-related

Fig. 1.27. Maps showing the extent of past permafrost and seasonally frozen ground in the UK (a–g) and Eurasia (h): (**a**) Late Devensian glacial maximum; (b) 20 ka; (c) 74–59 ka (Early Pleniglacial; marine isotope stage (MIS) 4); (d) 41–38 ka (Middle Pleniglacial cold interval in MIS 3); (e) 27–20 ka (maximum cold of the Late Pleniglacial, approximating the Last Glacial Maximum); (f) *c.* 20–15 ka (final phase of the Late Pleniglacial); (g) *c.* 12.9–11.5 ka (Younger Dryas Stadial); and (h) 25–17 ka (Last Permafrost Maximum). Dark blue and light blue lines in (h) indicate southern limits of the continuous and discontinuous permafrost zones, respectively, outside of mountain areas (Murton & Ballantyne 2017).

legacy in the UK. The potential impacts on engineering are considered if these relict periglacial geohazards are not identified during the investigative phase of the project. The periglacial landsystems classification proposed by Murton & Ballantyne (2017) is adopted to demonstrate its application for the assessment of ground engineering hazards within upland and lowland relict periglacial geomorphological terrains. Techniques for the early identification of the susceptibility of a site to relict periglacial geohazards are discussed, including the increasingly availability of high-quality aerial imagery such as provided by Google Earth that has proved a valuable tool in the identification of relict periglacial geohazards when considered in conjunction with the more usual sources of desk study information (such as geological, geomorphological and topographical reference material).

This chapter summarizes and builds on the landsystem approach developed by a number of authors including Higginbottom & Fookes (1971), Hutchinson (1992), Ballantyne & Harris (1994) and, most recently, the Geological Society Engineering Geology Special Publication 28, *Engineering Geology and Geomorphology of Glaciated and Periglaciated Terrains* edited by Griffiths & Martin (2017). The hierarchical classification system presented in Engineering Geology Special Publication 28 (Murton & Ballantyne 2017) categorizes periglacial processes in terms of upland and lowland terrain systems based on relative elevation. There are four landsystems defined within both upland and lowland terrains: plateaus, sediment-mantled hillslopes, rockslopes and slope–foot landsystems. Two additional landsystems described in lowland terrains only are valley and buried landsystems. The influence of past changes in sea level and its impact on the submergence of land that was previously subject to periglaciation is also considered for marine engineering.

Some periglacial processes and deposits pose a significantly increased geohazard to ground engineering projects due to their location in areas where considerable development activity occurs, such as the South English Midlands. Other periglacial geohazards may be less significant for engineering works, or are significant periglacial geohazards but located beyond the extent of frequent and high-density development, for example in more mountainous terrains in the UK.

This chapter highlights potentially the most significant periglacial geohazards in terms of their risk to civil engineering construction.

1.7 Section D: mining and subsidence hazards

1.7.1 Chapter 11: subsidence resulting from coal mining (Donnelly 2020*b*)

One of the principal geohazards associated with coal mining is subsidence. Coal was originally extracted where it cropped out, then mining became progressively deeper via shallow workings including bellpits that later developed into room-and-pillar workings. By the middle of the 1900s, coal was mined in larger open pits and underground by longwall mining methods. The mining of coal can often result in the subsidence of the ground surface. Generally, there are two main types of subsidence associated with coal mining: the generation of crown holes caused by the collapse of mine entries, mine roadway intersection and the consolidation of shallow voids; and the generation of a subsidence trough as a result of longwall mining encouraging the roof to fail to relieve the strains on the working face. This initiates round movement to migrate upwards and outwards from the seam being mined, and ultimately causes the subsidence and deformation of the ground surface. Methods are available to predict mining subsidence so that existing or proposed structures and land developments may be safeguarded. Ground investigative methods and geotechnical engineering options are also available for sites that have been or may be adversely affected by coal mining subsidence.

Many of the major cities and conurbations owe their existence and expansion to the presence of coal and associated mineral deposits (Fig. 1.28). Coal mining in the UK peaked in 1912–1915, and then experienced a wave of expansion and contraction. The last deep coal mine in the UK closed in December 2015. Coal mining has left behind a legacy of mining hazards (geohazards) that, if not properly managed and investigated, represent a risk to new construction and development. One of these hazards is subsidence. This chapter provides an overview of the occurrence, prediction and control of coal mining subsidence and is aimed at other engineering geologists, geotechnical engineers, civil engineers, planners and developers, as well as those interested in building, construction and the development of land in the abandoned (and those still active) coal mining fields of the UK.

In the context of this chapter, subsidence is considered as the ground movements that occur following the underground mining of coal, mainly the lowering of the ground surface. It should be noted, however, that the coal measures also provided other minerals, such as fireclay, ganister, ironstones, clays, shales, mudstones and sandstones for building purposes. There may be no or only incomplete records of the existence of mine workings, and these can also generate subsidence. The effects of subsidence depend on several factors: the geology, thickness and depth of the coal seam; the mining methods, and in particular the types of roof supports used; the engineering characteristics and behaviours of the strata and soils (superficial deposit); and any mitigative or engineering methods used to reduce the influence of mining subsidence. Coal mining subsidence can have serious, often dramatic and catastrophic, consequences for houses, buildings, engineered structures, underground utilities and services, and agricultural land. The inability to accurately predict the effects of ground subsidence has, in the past, resulted in the sterilization of coal mining reserves in some urban areas. This was partly associated with the expected subsidence compensation costs for damage to land, houses, roads and structures.

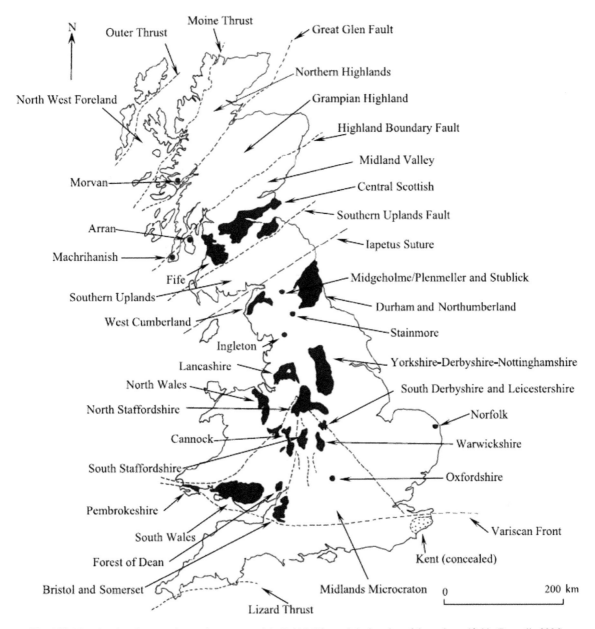

Fig. 1.28. Map showing the general tectonic structure of the British Isles and the location of the main coalfields (Donnelly 2006).

1.7.2 Chapter 12: subsidence resulting from chalk and flint mining (Edmonds 2020*b*)

Old chalk and flint mine workings occur widely across southern and eastern England. Over 3500 mines are recorded in the national Mining Cavities Database held by Peter Brett Associates LLP, and more are being discovered each year. The oldest flint mines date from the Neolithic period onwards and the oldest chalk mines from at least medieval, possibly even Roman, times. The most intensive period for mining was during the 1800s, although some mining continued into the 1900s. The size, shape and extent of the mines vary considerably, with some types being found only in particular areas. They range from crudely excavated bellpits to more extensive pillar-and-stall styles of mining (Fig. 1.29). The mines were created for a series of industrial,

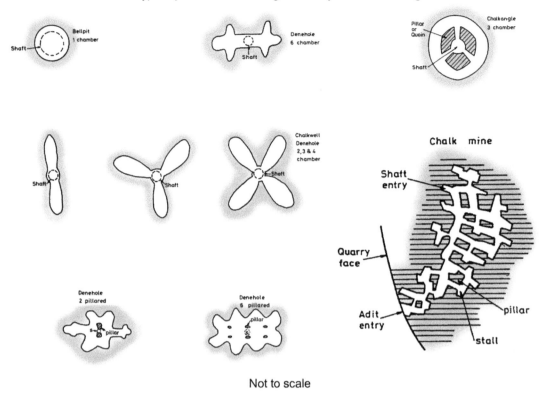

Fig. 1.29. Typical schematic plan sections through a variety of chalk mine workings (Edmonds 2020*b*).

building and agricultural purposes. Mining locations were not formally recorded, so most are discovered following collapse of the ground over poorly backfilled shafts and adits. Many of the old chalk mine workings were left open on abandonment, with just their shaft or adit entrances filled and sealed. The locations of abandoned mines are not well recorded so they pose a serious ground subsidence hazard, particularly since most of the old mines lie within 10–20 m of the ground surface. As urban development extends outwards around the historical centres of towns and cities, construction activities are revealing more mines each year as collapse of the ground occurs.

The subsidence activity, often triggered by heavy rainfall or leaking water services, poses a hazard to the built environment and people. Purpose-designed ground investigations are needed to map the mine workings and carry out follow-on ground stabilization after subsidence events. Where mine workings can be safely entered, they can sometimes be stabilized by reinforcement rather than infilling.

1.7.2.1 Flint mine workings
Flint mine workings may be referred to as ancient or modern workings. The earliest ancient workings date from the Neolithic period (*c.* 4000–2500 BC) and the later workings date from the Iron Age (from *c.* 800 BC to AD 100) or Roman period (*c.* AD 43–409). Some mines may also date from the medieval period (*c.* AD 600–1485).

1.7.2.2 Chalk mine workings
Chalk mine workings also have quite a long history, possibly from Roman times onwards. Mining styles show regional variation and both simple and more complex mine forms appear to co-exist through time, including features such as bellpits, deneholes, chalkwells, chalkangles and pillar-and-stall mines.

1.7.3 Chapter 13: hazards associated with mining and mineral exploitation in Cornwall and Devon, SW England (Gamble *et al.* 2020)

The importance of mining in the history of Cornwall is demonstrated by the county hosting the second oldest geological society in the world (established 1814), and with Cornwall and West Devon being selected in 2006 as a World Heritage Site by UNESCO for its mining landscape. The World Heritage designation was specifically related to the long history of metallic mining (mainly copper, tin and arsenic) in Cornwall and West Devon. However, while the last Cornish tin mine closed in 1998 (South Crofty), the *10th Edition of the Directory of Mines and Quarries* listed nearly 70 active mines in Cornwall and Devon that were still extracting and processing china clay, china clay waste, clay and shale (including ball clay), igneous and metamorphic rocks, sandstone, sea salt, silica sand, slate, and tungsten, and there was even a small tin streaming operation. In Cornwall, china clay alone has yielded 165 Mt of marketable clay since mining began in the mid-eighteenth century. Today, mining remains an integral part of the West Country economy, not least now because the heritage of mining is a source of revenue from tourism. Cornwall and Devon can be considered as exceptional in the UK for their long history of mining (suggested as starting in Phoenician times, i.e. *c.* 1550–300 BC), the temporal and spatial coverage of its mining infrastructure, its changing history of mineral exploitation, the range of mining-related hazards, and the nature and extent of remedial works that have been undertaken.

In this chapter, the geological basis for the mining industry in Cornwall and Devon is briefly summarized followed by a description of the history of mining and the environmental consequences. Approaches to assessing the hazards associated with mining are examined along with the varied methods of remediation, with reference made to case studies that demonstrate the various facets of the mining heritage of Cornwall and Devon.

1.7.4 Chapter 14: geological hazards from salt mining and brine extraction (Cooper 2020*b*)

In the UK rock salt (halite or sodium chloride) is present in Triassic and Permian rocks, from which it has been exploited for several millennia. Rock salt is not only a valuable industrial commodity, but also a highly soluble material responsible for natural and anthropogenic subsidence geohazards. The Triassic salt-bearing strata are widespread in the Cheshire basin area, but also common in parts of Lancashire, Worcestershire, Staffordshire and Northern Ireland (Fig. 1.30). Permian saliferous rocks are mainly present in the NE of England. This chapter considers the occurrence of salt deposits, and the way they either dissolve naturally or have been extracted by mining and anthropogenic dissolution. Subsidence problems that have arisen and continue to occur are highlighted, and methods of mitigating the problems by planning and construction/remediation techniques are considered.

Like table salt, rock salt is highly soluble and dissolves very quickly in water to make brine. This process occurs naturally in the UK, meaning that salt is not seen anywhere at outcrop. It is instead present in the subsurface, where the upper part of the sequence is dissolved, producing a buried dissolution surface (salt karst) overlain by collapsed and foundered strata. The natural dissolution processes and groundwater flow are evidenced by the presence of brine springs, many of which have been known and exploited

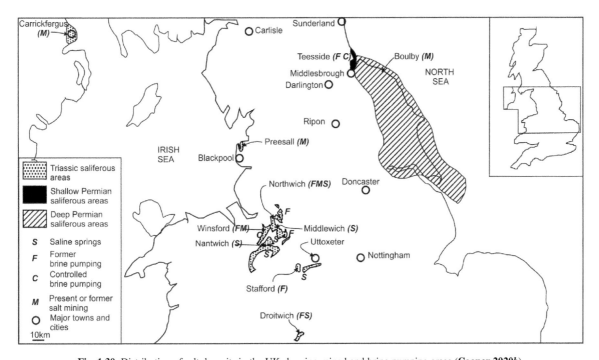

Fig. 1.30. Distribution of salt deposits in the UK showing mined and brine pumping areas (**Cooper 2020*b***).

since Roman times. Through the Middle Ages these springs were moderately exploited and gave rise to place names ending in 'wych' or 'wich'. However, it was in Victorian times that large-scale extraction both by mining and brine extraction accelerated, leading to some large and devastating instances of catastrophic subsidence. The UK is still dealing with this legacy and the effects of subsequent brine and rock salt extraction in many places, especially in parts of Cheshire, Droitwich, Stafford and Preesall. Where shallow brine extraction has occurred it mimics the natural salt karstification processes, and the results of natural and anthropogenic events can be difficult to differentiate. In Northern Ireland the salt has been mined traditionally by pillar-and-stall mining. In certain cases severe subsidence has occurred due to water ingress into the mines, causing dissolution of the pillars and catastrophic collapse.

Permian salt occurs at depth beneath coastal Yorkshire and Teesside. Here the salt deposits and the karstification processes are much deeper than in the Triassic salt, and the salt deposits are bounded up-dip by a dissolution front and collapse monocline. Salt has been won from these Permian rocks by dissolution mining, and some historical to recent subsidence due to brine extraction has occurred along the banks of the River Tees and to the NE of Middlesbrough.

Modern pillar-and-stall salt mining is deeper than old Victorian mining and located in mudstone and salt sequences that are completely dry. Modern brine extraction is controlled and restricted to deep-engineered cavities that are kept full of brine on completion, or used for other storage such as gas or waste; both methods of extraction have low or zero risks of subsidence.

1.7.5 Chapter 15: geological hazards from carbonate dissolution (Edmonds 2020*b*)

The dissolution of limestone and chalk (soluble carbonates) through geological time can lead to the creation of naturally formed cavities in the rock. The cavities can be air, water, rock or soil infilled and can occur at shallow levels within the carbonate rock surface or at deeper levels below. Depending upon the geological sequence, as the cavities break down and become unstable, they can cause overlying rock strata to settle and tilt, and the collapse of non-cemented strata and superficial deposits as voids migrate upwards to the surface. Natural cavities can be present in a stable or potentially unstable condition. The latter may be disturbed and triggered to cause ground instability by the action of percolating water, loading or vibration. The outcrops of various limestones and chalk occur widely across the UK (Fig. 1.31), posing a significant subsidence hazard to existing and new land development and people. In addition to subsidence, they can also create a variety of other problems such as slope instability or the generation of pathways for pollutants and soil gas to travel along, and impact all manner of engineering works. Knowledge of natural cavities is essential for planning, development control and the construction of safe development.

Limestone and chalk are composed of calcium carbonate that is soluble in the presence of acidic water. Rainwater combines with atmospheric and biogenic carbon dioxide to form weak carbonic acid that then dissolves the calcium carbonate. Where the water table level lies at depth within the carbonate sequence, the water can enter into the rock via joints and fissures to percolate downwards and cause dissolution. The effects of dissolution tend to be concentrated within the upper surface zone of the rock, especially where permeable overlying deposits are present. The cover deposits influence the acidity and concentration of water flows penetrating the carbonate rock surface. As solution features are formed over time, the cover deposits will tend to settle and collapse down into the enlarging features, often leading to subsidence occurring at the ground surface that can cause damage to buildings and infrastructure in urban areas.

For new construction, the challenge is to check whether solution features are present below a site and to understand the karst geohazard setting to ensure that the correct engineering solutions are put in place to permit safe development. This includes not only addressing the safe support of buildings, roads and services, but also the effects of surface water drainage disposal. Unfortunately, there are many cases where the design of development has not taken karst into account, resulting in subsidence damage. Following a subsidence event, it is essential to identify the nature and cause of movement before a suitable remedial solution to stabilize the ground can be executed. Property evacuation may be necessary for safety reasons before the remedial works can be completed and, in some instances, an economic solution might not be feasible. Blighting and dereliction of property can be an arising issue in certain circumstances.

The typical range of natural cavity forms found in limestone and chalk is shown in Figure 1.32. Where low-permeability cover deposits are present at the surface, surface water drainage will tend to collect to form streams that flow across the land surface until they meet the exposed outcrop of the limestone or chalk. At this location the water dissolves the rock surface, leading to the creation of solution-widened joints and bedding planes. As these develop, the water readily enters and flows down into the rock mass to form an underground drainage network. Over time, a depression is formed at the surface where the overland flow disappears, which is referred to as a swallow hole. In places where permeable cover deposits occur over the limestone and chalk, the water will tend to be absorbed into the surface in a diffuse manner and swallow holes are less prevalent. Given sufficient time, dissolution of joints at the surface of a limestone will tend to form linear features that extend to depth, widening upwards. Where the bare limestone surface expression of the intersecting dissolution along joints is revealed at the surface, they are known as limestone pavements. When dissolution is concentrated at the intersection of joints, a point feature may be formed centred on the intersection that becomes pipe shaped with time extending to depth. Pipe-shaped features are commonly associated with chalk

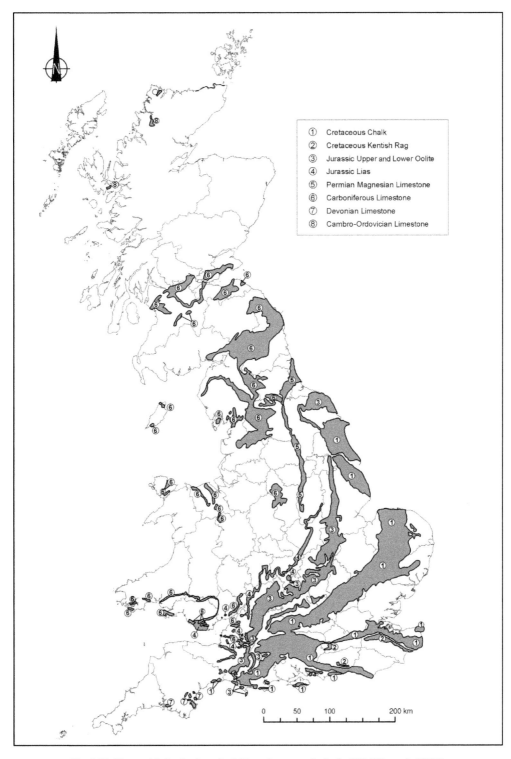

Fig. 1.31. The spatial distribution of soluble carbonate rocks in the UK (**Edmonds 2020***b*).

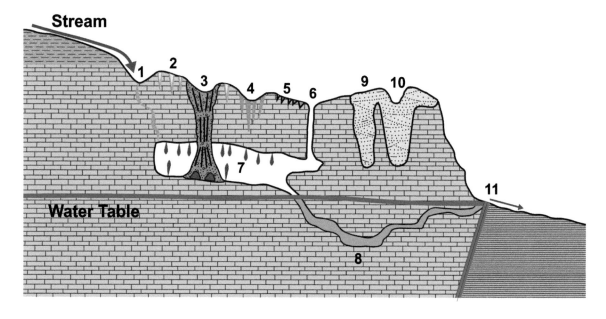

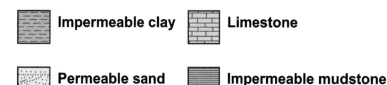

Fig. 1.32. Range of natural cavity types formed on limestone and chalk (**Edmonds 2020a**).

and referred to as solution pipes. The subsurface shape of solution pipes formed in chalk can sometimes be irregular and voided. It is common for overlying cover deposits or rocks to settle down into the enlarging cavities being formed. This leads to downwards ravelling of deposits and the development of soil arches that can suddenly collapse, causing subsidence at the surface above. The surface hollows formed are referred to as subsidence sinkholes. Where cover deposits are absent, surface hollows can be created by dissolution alone of the limestone surface, focused on the pattern of joints present or along a fault plane; these surface hollows are referred to as solution sinkholes.

1.7.6 Chapter 16: geological hazards caused by gypsum and anhydrite in the UK: dissolution, subsidence, sinkholes and heave (Cooper 2020b)

Gypsum and anhydrite are both soluble minerals that form rocks that can dissolve at the surface and underground, producing sulphate karst and causing geological hazards, especially subsidence and sinkholes. The dissolution rates of these minerals are rapid and cavities and/or caves can enlarge and collapse on a human timescale. In addition, the hydration and recrystallization of anhydrite to gypsum can cause considerable expansion and pressures capable of causing uplift and heave. Sulphate-rich water associated with the deposits can react with concrete and be problematic for construction. This chapter reviews the occurrence of these rocks in the near surface of the UK (Fig. 1.33) and looks at methods for mitigating, avoiding and planning for their associated problems.

Gypsum, hydrated calcium sulphate ($CaSO_4 \cdot 2H_2O$), is attractive as satin spar, beautiful as carved alabaster and practical as plasterboard (wallboard) and plaster. However, gypsum is highly soluble and a cause of geological hazards capable of causing severe subsidence to houses, roads, bridges and other infrastructure. Gypsum dissolves rapidly and, where this occurs underground, results in caves that evolve and quickly enlarge, commonly leading to subsidence and sometimes to catastrophic collapse. Gypsum is mostly a secondary mineral and is present in the UK mainly as fibrous gypsum (satin spar) and alabastrine gypsum (alabaster) that may include large crystals and aggregates of crystals. It occurs near the surface passing into anhydrite, the dehydrated form ($CaSO_4$) at depths below about 40–120 m, depending on the local geology and water circulation. The hydration of anhydrite to gypsum in the subsurface causes expansion and heave, problematic to engineering and hydrogeological installations such as ground source heat pumps. Furthermore, gypsum, especially in engineering fills, can react with cement causing heave. Gypsum and anhydrite

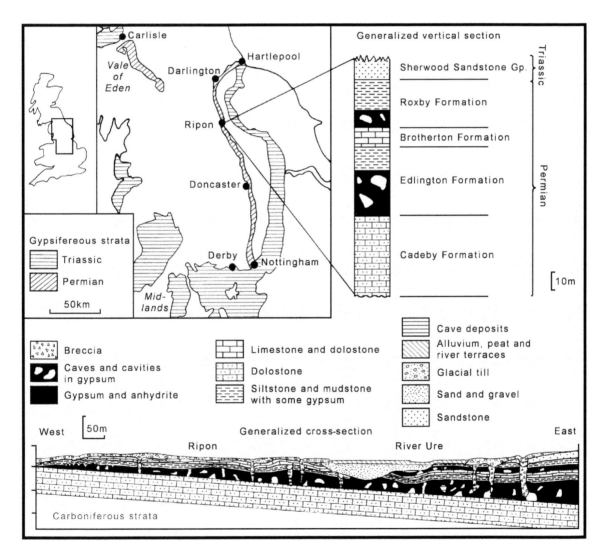

Fig. 1.33. Regional geology of the Permian and Triassic gypsiferous sequences with a cross-section from west to east through the Ripon area showing the E-dipping dolomite and gypsum sequence cut into by the glacial valley of the River Ure (**Cooper 2020a**).

are present in the Triassic strata of the Midlands and SW of the UK and in the Permian strata of the NE and NW of England (Fig. 1.30). In all these areas various geological hazards are associated with these rocks, the most visible being subsidence and sinkholes. Gypsum and anhydrite also occur to a small extent in the Jurassic of southern England, but no specific problems have been reported related to these deposits.

Gypsum dissolves more readily in flowing water; next to rivers, this can be at a rate of about 100 times faster than that seen for limestone dissolution. Under suitable groundwater flow conditions, caves in gypsum can enlarge at a rapid rate and result in large chambers. Collapse of these chambers produces breccia pipes that propagate through the overlying strata to break through at the surface and form subsidence hollows (Fig. 1.34).

1.7.7 Chapter 17: mining-induced fault reactivation in the UK (Donnelly 2020b)

Faults are susceptible to reactivation during coal mining subsidence. The effects may be the generation of a scarp along the ground surface, which may or may not be accompanied by associated ground deformation including fissuring or compression. Reactivated faults vary considerably in their occurrence, height, length and geometry. Some reactivated faults may not be recognizable along the ground surface, known only to those who have measured the ground movements or who are familiar with the associated subtle ground deformations. By comparison, other reactivated faults generate scarps up to several metres high and many kilometres long, often accompanied by widespread fissuring of the ground surface. Reactivated faults induced by mining

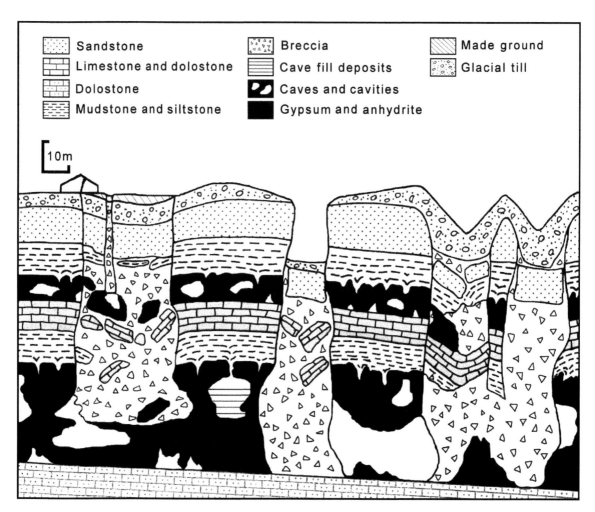

Fig. 1.34. Stylized cross-section through gypsum dissolution subsidence features in the east of the Ripon subsidence belt (Cooper 2020a).

subsidence have caused damage to roads, structures and land. The objective of this chapter is to provide a general overview of the occurrence and characteristics of fault reactivation in the UK.

Various documents and publications are available to assist with the prediction of coal mining subsidence; however, faults located in areas prone to coal mining subsidence are susceptible to reactivation, and this cannot be forecast. Reactivated faults may result in the generation of a scarp, graben, fissure or compression hump along the ground surface (Fig. 1.35).

Fault reactivation has been documented in the UK since the middle of the 1800s. However, many of the earlier theories on fault reactivation were somewhat speculative and lacked a fundamental geological appreciation of fault mechanisms. During the 1950s and later, increased mining subsidence compensation claims provided the incentive for the British coal mining industry to investigate fault reactivation. However, by the 1980s the exact mechanisms of fault reactivation still remained unclear, although numerous cases had been documented. As a result, some coal resources, particularly those located in densely populated parts of the UK, were effectively sterilized, since it was not possible to predict the ground movements and to estimate the potential compensation claims. In the 1990s, following continued cases of fault reactivation, recommendations from government (The Commission on Energy & The Environment 1981) resulted in further research to investigate fault reactivation. As with all faults, it is still not possible to predict exactly if, when and where a fault may reactivate when subjected to mining subsidence. However, this research has now enabled the factors that control fault reactivation and the different styles of ground deformation to be better understood.

1.8 Section E: gas hazards

1.8.1 Chapter 18: radon gas hazard (Appleton 2020)

Radon is a natural radioactive gas that cannot be seen, smelt or tasted by humans and can only be detected with special equipment. It is produced by the radioactive decay of radium, which in turn is derived from the radioactive decay of uranium. Uranium is found in small quantities in all soils and rocks, although the amount varies from place to place. There are three naturally occurring radon (Rn) isotopes: ^{219}Rn (actinon), ^{220}Rn (thoron) and ^{222}Rn, which is commonly called radon. ^{222}Rn (radon) is the main radon isotope of concern to people. ^{220}Rn has been recorded in houses, and about 4% of the average total radiation dose for a member of the UK population is from this source.

There are a number of different ways to quantify radon. These include (1) the radioactivity of radon gas; (2) the dose to living tissue, for example, to the lungs, from solid decay products of radon gas; and (3) the exposure caused by the presence of radon gas. The average radon concentration in houses in the UK is 20 Bq m^{-3}.

The dose equivalent indicates the potential of harm to particular human tissues by different radiations, irrespective of their type or energy. The average person in the UK receives an annual effective radiation dose, which is the sum of doses to body tissues weighted for tissue sensitivity and radiation weighting factors, of 2.8 mSv, of which about 85% is from natural sources: cosmic rays, terrestrial gamma rays, the decay products of ^{220}Rn and ^{222}Rn, and the natural radionuclides in the body ingested through food and drink. Of this natural radiation, the major proportion is from geological sources.

Mapped bedrock geology explains on average 25% of the variation of indoor radon in England and Wales, while mapped superficial geology explains, on average, an additional 2%. In the UK, relatively high concentrations of radon are associated with particular types of bedrock and unconsolidated deposits, for example, some granites, uranium-enriched phosphatic rocks and black shales, limestones, sedimentary ironstones, permeable sandstones and uraniferous metamorphic rocks. Permeable superficial deposits, especially those derived from uranium-bearing rock, may also be radon prone. Geological units associated with the highest levels of naturally occurring radon (Fig. 1.36) are: (1) granites in SW England, the Grampian and Helmsdale districts of Scotland and the Mourne Mountains in Northern Ireland; (2) Carboniferous limestones throughout the UK and some Carboniferous shales in northern England and Wales; (3) sedimentary ironstone formations in the English Midlands; (4) some Ordovician and Silurian mudstones, siltstones and greywackes in Wales, Northern Ireland and the southern uplands of Scotland; (5) Middle Old Red Sandstone of NE Scotland; and (6) Neoproterozoic psammites, semipelites and meta-limestones in the western sector of Northern Ireland.

1.8.2 Chapter 19: methane gas hazard (Wilson & Mortimer 2020)

This chapter identifies potential sources, and the key chemical properties, of methane. Guidance is provided on deriving a conceptual site model for methane, utilizing various lines of evidence to inform a robust, scientific, reasoned and logical assessment of associated gas risk. Discussion is provided regarding the legislative context of permanent gas risk assessment for methane, including via qualitative, semi-quantitative and detailed quantitative (including finite element modelling) techniques. Strategies for mitigating risks associated with methane are also outlined, together with the legal context for consideration of methane both in relation to the planning regime and under Part 2A of the Environmental Protection Act 1990.

Methane (historically known as 'marsh gas') was discovered in 1776 by Alessandro Volta, who collected gas bubbles from disturbed sediments on Lake Maggiore. Methane is the most abundant organic compound in the Earth's atmosphere. Its occurrences in the Earth's crust are predominantly of biogenic origin (i.e. it is formed by bacterial decomposition of

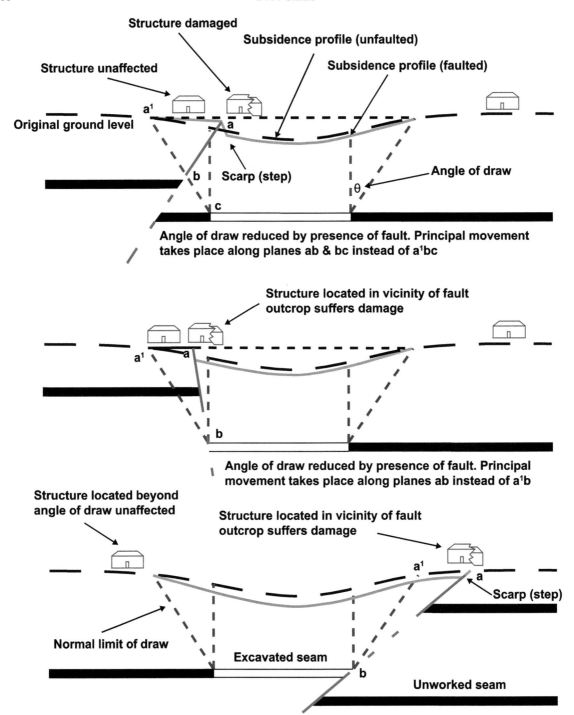

Fig. 1.35. The influence of faults on mining subsidence and the angle-of-draw (Donnelly 2020*a*).

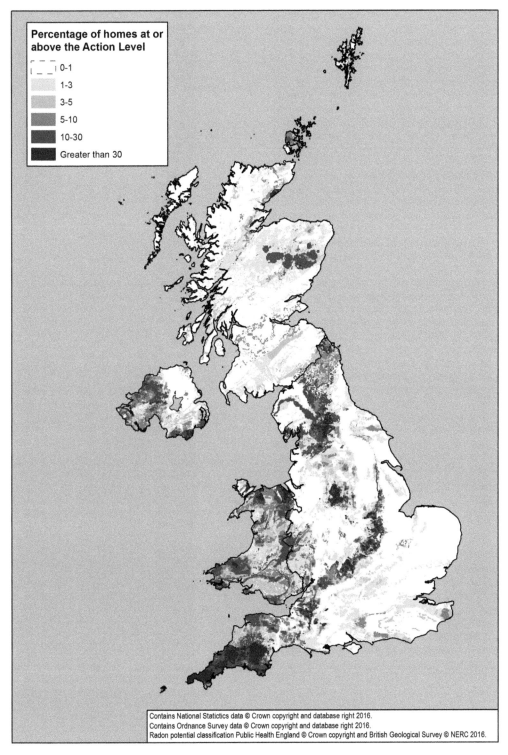

Fig. 1.36. Radon potential map of England and Wales (Appleton 2020).

organic matter). Methane can also be formed by decomposition of organic matter as a result of geothermal heat and/or pressure, when it is known as thermogenic gas. Such gas is generated at great depth, but the methane can migrate to the surface along faults or other features and accumulate in near-surface rocks. Abiogenic methane is thought to be formed by chemical reactions, for example during cooling of magma or serpentinization of ultramafic rocks. Methane has been detected in many shallow drift deposits in UK soils where there is no apparent external source such as landfill. It is thought that the methane occurs from disturbances caused by installing monitoring wells, and the oxidation of small volumes of organic material in the soils to produce carbon dioxide that is subsequently reduced by methanogens. This low-level source of methane is not known to pose a hazard to developments. Methane is ubiquitous in the subsurface environment and is present in soils and rocks below many parts of the UK and other countries. Methane is often present at elevated concentrations in uncontrolled and engineered fill materials in the unsaturated zone, especially where the soil is wet, and an anaerobic zone exists below the groundwater.

The greatest hazard posed by methane is that it is flammable and/or explosive. If an explosive mix occurs in a building, tunnel or mine, for example, there is a risk of explosion. Generally, the lower explosive limit (LEL) of methane is 5% by volume in air and the upper explosive limit (UEL) is 15% by volume in air. The explosive limits of methane will change as the oxygen concentration reduces. When carbon dioxide reaches 25% concentration, or nitrogen reduces to 36%, methane is not flammable. In addition to the explosive and/or flammable hazard posed by methane, at concentrations in excess of 33% by volume it may also act as an asphyxiant by displacing oxygen. Typically, physiological effects are observed when oxygen concentrations fall below 18% by volume. Displacement of oxygen at the root ball can also result in phytotoxic effects to plants and vegetation.

Conclusions

The UK is perhaps unique globally in that it presents almost the full spectrum of geological time, stratigraphy and associated lithologies within its boundaries. With this wide range of geological assemblages comes the full range of geological hazards, whether geophysical, geotechnical, geochemical or related to georesources. An awareness of these hazards and the risks that they pose is a key requirement of the engineering geologist. This volume has set out to define and explain these key hazards, to detail their detection, monitoring and management, and to provide a basis for further research and understanding.

Acknowledgments Professor Jim Griffiths is thanked for his helpful and constructive comments on this chapter. The lead authors of the Geological Society Engineering Group Working Party on UK Geohazards are thanked for their significant contributions to this work, namely; Dr Roger Musson, Dr Mark Lee, Professor Mike Winter, Professor Martin Culshaw, Dr Lee Jones, Professor Jeff Warburton, Mr Tom Berry, Dr Laurance Donnelly, Dr Clive Edmonds, Dr Tony Cooper, Mr Mike Gamble, Dr Don Appleton and Mr Steve Wilson. Mr David Shilston, past President of the Geological Society and President at the relaunch of the Working Party, is thanked for his Preface. Members of the LinkedIn Engineering Geology Community are thanked for their suggestions for the list of the most significant UK geohazards events. Dr Catherine Pennington from the British Geological Survey is thanked for her useful comments on aspects of landslides in the UK and for providing data from the National Database on UK landslides and their impact.

Funding This research received no specific grant from any funding agency in the public, commercial, or not-for-profit sectors.

References

APPLETON, D. 2020. Radon gas hazard. *In*: GILES, D.P. & GRIFFITHS, J.S. (eds) *Geological Hazards in the UK: Their Occurrence, Monitoring and Mitigation – Engineering Group Working Party Report*. Geological Society, London, Engineering Geology Special Publications, **29**, 433–456, https://doi.org/10.1144/EGSP29.18

BALLANTYNE, C.K. & HARRIS, C. 1994. *The Periglaciation of Great Britain*. Cambridge University Press, Cambridge.

BANKS, V.J., BRICKER, S.H., ROYSE, K.R. & COLLINS, P.E.F. 2015. Anomalous buried hollows in London: development of a hazard susceptibility map. *Quarterly Journal of Engineering Geology and Hydrogeology*, **48**, 55–70.

BELL, B.R. & WILLIAMSON, I.T. 2002. Tertiary igneous activity. Chapter 14. *In*: TREWIN, N.H. (ed.) *The Geology of Scotland*. 4th edn. The Geological Society, London, 371–407.

BERRY, T. 2020. Relict periglacial hazards. *In*: GILES, D.P. & GRIFFITHS, J.S. (eds) *Geological Hazards in the UK: Their Occurrence, Monitoring and Mitigation – Engineering Group Working Party Report*. Geological Society, London, Engineering Geology Special Publications, **29**, 259–289, https://doi.org/10.1144/EGSP29.10

BRITISH GEOLOGICAL SURVEY n.d. Landslide case studies. Retrieved from https://www.bgs.ac.U.K./research/engineeringGeology/shallowGeohazardsAndRisks/landslides/caseStudies.html

BROMHEAD, E. 2013. The Twelfth Glossop Lecture. Reflections on the residual strength of clay soils, with special reference to bedding-controlled landslides. *Quarterly Journal of Engineering Geology and Hydrogeology*, **46**, 132–155, https://doi.org/10.1144/qjegh2012-078

BRUNSDEN, D. & CHANDLER, J.H. 1996. Development of an episodic landform change model based upon the Black Ven mudslide, 1946–1995. *Advances in Hillslope Processes*, **2**, 869–896.

CHARMAN, J. 2012. Introduction. *In*: WALKER, M.J. (ed.) 2012. *Hot Deserts: Engineering Geology and Geomorphology*. Geological Society, London, Engineering Geology Special Publication, **25**, 1–6, https://doi.org/10.1144/EGSP25

COBB, A.E. 2000. Taren landslide: road construction across a landslide. *In*: SIDDLE, H.J., BROMHEAD, E.N. & BASSETT, M.G. (eds). *Landslides and landslide management in South Wales*. Amgueddfeydd ac Orielau Cenedlaethol Cymru, **18**.

COOPER, A. 2020a. Geological hazards caused by gypsum and anhydrite in the UK: dissolution, subsidence, sinkholes and heave. *In*: GILES, D.P. & GRIFFITHS, J.S. (eds) *Geological Hazards in the UK: Their Occurrence, Monitoring and Mitigation – Engineering

Group *Working Party Report*. Geological Society, London, Engineering Geology Special Publications, **29**, 403–423, https://doi.org/10.1144/EGSP29.16

COOPER, A. 2020b. Geological hazards from salt mining and brine extraction. *In*: GILES, D.P. & GRIFFITHS, J.S. (eds) *Geological Hazards in the UK: Their Occurrence, Monitoring and Mitigation – Engineering Group Working Party Report*. Geological Society, London, Engineering Geology Special Publications, **29**, 369–387, https://doi.org/10.1144/EGSP29.14

COOPER, A.H. 2007. Gypsum dissolution geohazards at Ripon, North Yorkshire, U.K. *Engineering Geology for Tomorrow's Cities, Conference Proceedings IAEG*. Nottingham, IAEG.

COMMISSION ON ENERGY & THE ENVIRONMENT 1981. *Coal and the Environment*. Her Majesty's Stationery Office, London.

CRED 2015. *The Human Cost of Natural Disaster – a Global Perspective*. United Nations Centre for Research on the Epidemiology of Disaster, Brussels, Belgium.

CULSHAW, M.G. 2018. Geohazards. *In*: BOBROWSKY, P.T. & MARKER, B. (eds) *Encyclopedia of Engineering Geology*. Springer International, Cham, Switzerland, 381–389.

CULSHAW, M.G., NORTHMORE, K.J., JEFFERSON, I., ASSADI-LANGROUDI, A. & BELL, F.G. 2020. Collapsible soils in the UK. *In*: GILES, D.P. & GRIFFITHS, J.S. (eds) *Geological Hazards in the UK: Their Occurrence, Monitoring and Mitigation – Engineering Group Working Party Report*. Geological Society, London, Engineering Geology Special Publications, **29**, 187–203, https://doi.org/10.1144/EGSP29.6

CZEREWKO, M.A., BASTEKIN, A., TUNNICLIFFE, J. & O'ROURKE, R. 2015. Assessment of ground conditions for development of wind turbine ground support in an area of challenging ground at Carsington Pasture, Derbyshire, U.K. *Geotechnical Engineering for Infrastructure and Development*, 2367–2372.

DAVIES, J.P., LOVERIDGE, F.A., PERRY, J., PATTERSON, D. & CARDER, D. 2003. Stabilisation of a landslide on the M25 highway London's main artery. In: *12th Pan-American Conference on Soil Mechanics and Geotechnical Engineering*. Boston, USA, 22–25 June 2003.

DEFRA 2005. The threat posed by tsunami to the U.K. Study commissioned by Defra Flood Management. *Produced by British Geological Survey, Proudman Oceanographic Laboratory Met Office and HR Wallingford*.

DIGGINS, M. 2018. Scottish Avalanche Information Service Winter Report 2017/18, https://www.sais.gov.uk/wp-content/uploads/2018/11/SAIS-Winter-Report-2017-18.pdf

DONNELLY, L.J. 2006. A review of coal mining induced fault reactivation in Great Britain. *Quarterly Journal of Engineering Geology and Hydrogeology*, **39**, 5–50.

DONNELLY, L. 2020a. Mining induced fault reactivation in the UK. *In*: GILES, D.P. & GRIFFITHS, J.S. (eds) *Geological Hazards in the UK: Their Occurrence, Monitoring and Mitigation – Engineering Group Working Party Report*. Geological Society, London, Engineering Geology Special Publications, **29**, 425–432, https://doi.org/10.1144/EGSP29.17

DONNELLY, L. 2020b. Subsidence resulting from coal mining. *In*: GILES, D.P. & GRIFFITHS, J.S. (eds) *Geological Hazards in the UK: Their Occurrence, Monitoring and Mitigation – Engineering Group Working Party Report*. Geological Society, London, Engineering Geology Special Publications, **29**, 291–309, https://doi.org/10.1144/EGSP29.11

EARLY, K.R. & SKEMPTON, A.W. 1972. Investigations of the landslide at Walton's Wood, Staffordshire. *Quarterly Journal of Engineering Geology and Hydrogeology*, **5**, 19–41, https://doi.org/10.1144/GSL.QJEG.1972.005.01.04

EDMONDS, C. 2020a. Geological hazards from carbonate dissolution. *In*: GILES, D.P. & GRIFFITHS, J.S. (eds) *Geological Hazards in the UK: Their Occurrence, Monitoring and Mitigation – Engineering Group Working Party Report*. Geological Society, London, Engineering Geology Special Publications, **29**, 389–401, https://doi.org/10.1144/EGSP29.15

EDMONDS, C. 2020b. Subsidence resulting from chalk and flint mining. *In*: GILES, D.P. & GRIFFITHS, J.S. (eds) *Geological Hazards in the UK: Their Occurrence, Monitoring and Mitigation – Engineering Group Working Party Report*. Geological Society, London, Engineering Geology Special Publications, **29**, 311–319, https://doi.org/10.1144/EGSP29.12

EDMONDS, C.N. 2008. Karst and mining geohazards with particular reference to the Chalk outcrop, England. *Quarterly Journal of Engineering Geology and Hydrogeology*, **41**, 261–278, https://doi.org/10.1144/1470-9236/07-206

FORSTER, A. 1993. Scarborough Landslip. *Geoscientist*, **3**, 2–3 and cover photograph.

FORSTER, A. & CULSHAW, M. 2004. Feature: implications of climate change for hazardous ground conditions in the UK. *Geology Today*, **20**, 61–66, https://doi.org/10.1111/j.1365-2451.2004.00442.x

GAMBLE, B., ANDERSON, M. & GRIFFITHS, J.S. 2020. Hazards associated with mining and mineral exploitation in Cornwall and Devon, SW England. *In*: GILES, D.P. & GRIFFITHS, J.S. (eds) *Geological Hazards in the UK: Their Occurrence, Monitoring and Mitigation – Engineering Group Working Party Report*. Geological Society, London, Engineering Geology Special Publications, **29**, 321–367, https://doi.org/10.1144/EGSP29.13

GARDNER, T. 1754. An historical account of Dunwich, … Blithburgh, … Southwold,… with remarks on some places contiguous thereto. Re-printed by Gale ECCO, Print Editions (30 May 2010).

GILES, C. n.d. Subsidence, the forgotten peril. http://www.bv-solutions.co.uk/we-are-bvs/news/blog-subsidence-the-forgotten-peril/

GILES, D.P. 2020a. Quick clay behaviour in sensitive Quaternary marine clays: a UK perspective. *In*: GILES, D.P. & GRIFFITHS, J.S. (eds) *Geological Hazards in the UK: Their Occurrence, Monitoring and Mitigation – Engineering Group Working Party Report*. Geological Society, London, Engineering Geology Special Publications, **29**, 205–221, https://doi.org/10.1144/EGSP29.7

GILES, D.P. 2020b. Tsunami hazard with reference to the UK. *In*: GILES, D.P. & GRIFFITHS, J.S. (eds) *Geological Hazards in the UK: Their Occurrence, Monitoring and Mitigation – Engineering Group Working Party Report*. Geological Society, London, Engineering Geology Special Publications, **29**, 61–80, https://doi.org/10.1144/EGSP29.3

GOSTLING, REV. W. 1756. Land sunk in Kent. *Letter to Gentleman's Magazine*, **26**, 160.

GRATTAN, J.P. & BRAYSHAY, M.B. 1995. An amazing and portentous summer: environmental and social responses in Britain to the 1783 eruption of an Iceland Volcano. *The Geographical Journal*, **161**, 125–134, https://doi.org/10.2307/3059970

GRIFFITHS, J.S. 2018. Mass movement. *In*: BOBROWSKY, P.T. & MARKER, B. (eds) *Encyclopedia of Engineering Geology*. Springer International, Cham, Switzerland, 597–604.

GRIFFITHS, J.S. & GILES, D.P. 2017. Conclusions and illustrative case studies. *In*: GRIFFITHS, J.S. & MARTIN, C.J. (eds) *Engineering Geology and Geomorphology of Glaciated and Periglaciated Terrains – Engineering Group Working Party Report*. Geological Society, London, Engineering Geology Special Publication, **28**, 891–936, https://doi.org/10.1144/EGSP28.9

GRIFFITHS, J.S. & MARTIN, C.J. (eds) 2017. *Engineering Geology and Geomorphology of Glaciated and Periglaciated Terrains*. Geological Society, London, Engineering Geology Special Publication, **28**, https://doi.org/10.1144/EGSP28

GRIFFITHS, J.S., BRUNSDEN, D., LEE, E.M. & JONES, D.K.C. 1995. Geomorphological investigations for the Channel Tunnel terminal and portal. *Geographical Journal*, **161**, 275–284, https://doi.org/10.2307/3059832

GUDMUNDSSON, M.T., THORDARSON, T., HÖSKULDSSON, Á., LARSEN, G., BJÖRNSSON, H., PRATA, F.J. & HAYWARD, C.L. 2012. Ash generation and distribution from the April-May 2010 eruption of Eyjafjallajökull, Iceland. *Nature, Scientific Reports*, **2**, 572, https://doi.org/10.1038/srep00572

HAINING, P. 1991. *The Great English Earthquake* (p. 224). Robert Hale Ltd.

HENKEL, D.J. & SKEMPTON, A.W. 1955. A landslide at Jackfield, Shropshire, in a heavily over-consolidated clay. *Géotechnique*, **5**, 131–137, https://doi.org/10.1680/geot.1955.5.2.131

HIGGINBOTTOM, I.E. & FOOKES, P.G. 1971. Engineering aspects of periglacial features in Britain. *Quarterly Journal of Engineering Geology and Hydrogeology*, **3**, 85–117, https://doi.org/10.1144/GSL.QJEG.1970.003.02.02

HUTCHINSON, J.N. 1969. A reconsideration of the coastal landslides at Folkestone Warren, Kent. *Géotechnique*, **19**, 6–38, https://doi.org/10.1680/geot.1969.19.1.6

HUTCHINSON, J.N. 1988. General report: morphological and geotechnical parameters of landslides in relation to geology and hydrogeology. BONNARD, C. (ed.) *Proceedings, Fifth International Symposium on Landslides*. Rotterdam, Balkema, **1**, 3–35.

HUTCHINSON, J.N. 1992. Engineering in relict periglacial and extraglacial areas in Britain. *In*: GRAY, J.M. (ed.) *Applications of Quaternary Research*. Quaternary Research Association, Cambridge, Quaternary Proceedings, **2**, 49–65.

HUTCHINSON, J.N. 2001. Reading the ground: morphology and geology in site appraisal. *Quarterly Journal of Engineering Geology and Hydrogeology*, **34**, 7–50, https://doi.org/10.1144/qjegh.34.1.7

HUTCHINSON, J.N., BROMHEAD, E.N. & LUPINI, J.F. 1980. Additional observations on the Folkestone Warren landslides. *Quarterly Journal of Engineering Geology*, **13**, 1–31, https://doi.org/10.1144/GSL.QJEG.1980.013.01.01

JONES, L. 2020. Swelling and shrinking soils. *In*: GILES, D.P. & GRIFFITHS, J.S. (eds) *Geological Hazards in the UK: Their Occurrence, Monitoring and Mitigation – Engineering Group Working Party Report*. Geological Society, London, Engineering Geology Special Publications, **29**, 223–242, https://doi.org/10.1144/EGSP29.8

KING, C. 2002. Palaeogene and Neogene: uplift and cooling climate. *In*: BRENCHLEY, P.J. & RAWSON, P.F. (eds) *The Geology of England and Wales*. 2nd edn. The Geological Society, London, 395–428.

LEE, E.M. 1999. Coastal planning and management: the impact of the 1993 Holbeck Hall landslide, Scarborough. *East Midlands Geographer*, **21** pt 2 1998 & v.22 pt1, 78–91.

LEE, E.M. & GILES, D.P. 2020. Landslide and slope stability hazard in the UK. *In*: GILES, D.P. & GRIFFITHS, J.S. (eds) *Geological Hazards in the UK: Their Occurrence, Monitoring and Mitigation – Engineering Group Working Party Report*. Geological Society, London, Engineering Geology Special Publications, **29**, 81–162, https://doi.org/10.1144/EGSP29.4

MARTIN, C.J., MORLEY, A.L. & GRIFFITHS, J.S. 2017. Introduction to engineering geology and geomorphology of glaciated and periglaciated terrains. *In*: GRIFFITHS, J.S. & MARTIN, C.J. (eds) 2017. *Engineering Geology and Geomorphology of Glaciated and Periglaciated Terrains*. Geological Society, London, Engineering Geology Special Publications, **28**, 1–30.

MOORE, R. 1986. *The Fairlight Landslips: the Location, Form, and Behaviour of Coastal Landslips with Respect to toe Erosion*. King's College, London, Occasional Paper No. 27.

MOORE, R. & MCINNES, R.G. 2011. *Cliff Instability and Erosion Management in Great Britain: a Good Practice Guide*. Published by Halcrow Group Ltd, Retrieved from https://www.researchgate.net/publication/323550427_Cliff_Instability_and_Erosion_Management_in_Great_Britain_A_Good_Practice_Guide

MOORE, R., CAREY, J.M. & MCINNES, R.G. 2010. Landslide behaviour and climate change: predictable consequences for the Ventnor Undercliff, Isle of Wight. *Quarterly Journal of Engineering Geology and Hydrogeology*, **43**, 447–460, https://doi.org/10.1144/1470-9236/08-086

MURTON, J.B. & BALLANTYNE, C.K. 2017. Periglacial and permafrost ground models for Great Britain. *In*: GRIFFITHS, J.S. & MARTIN, C.J. (eds). *Engineering Geology and Geomorphology of Glaciated and Periglaciated Terrains*. Geological Society, London, Engineering Group Working Party Report, **28**, 501–597, https://doi.org/10.1144/EGSP28.5

MUSSON, R.M.W. 2007. British earthquakes. *Proceedings of the Geologists' Association*, **118**, 305–337, https://doi.org/10.1016/S0016-7878(07)80001-0

MUSSON, R.M.M. 2020. Seismic hazard in the UK. *In*: GILES, D.P. & GRIFFITHS, J.S. (eds) *Geological Hazards in the UK: Their Occurrence, Monitoring and Mitigation – Engineering Group Working Party Report*. Geological Society, London, Engineering Geology Special Publications, **29**, 43–60, https://doi.org/10.1144/EGSP29.2

NADIM, F. 2013. Hazard. *In*: BOBROWSKY, P.T. (ed.) *Encyclopedia of Natural Hazards*. 425–426.

NEWMAN, T. 2009. The impact of adverse geological conditions on the design and construction of the Thames Water Ring Main in Greater London, UK. *Quarterly Journal of Engineering Geology and Hydrogeology*, **42**, 5–20, https://doi.org/10.1144/1470-9236/08-035

NEWMAN, T.G., GHAIL, R.C. & SKIPPER, J.A. 2013. Deoxygenated gas occurrences in the Lambeth Group of central London, UK. *Quarterly Journal of Engineering Geology and Hydrogeology*, **46**, 167–77, https://doi.org/10.1144/qjegh2012-013

ORFORD, J. 2005. Coastal environments. *In*: FOOKES, P.G., LEE, E.M. & MILLIGAN, G. (eds) *Geomorphology for Engineers*. Whittles, Caithness, 576–602.

PARLIAMENTARY OFFICE OF SCIENCE AND TECHNOLOGY 2006. UK Soil Degradation. Postnote. Retrieved from https://www.parliament.uk/documents/post/postpn265.pdf [accessed 14th Jan 2019]

PENNINGTON, C.V.L. 2008. *A Geological Assessment of the Landslides in the Ironbridge Gorge, Shropshire*. British Geological Survey, Keyworth, http://nora.nerc.ac.uk/id/eprint/508100 [accessed 14th Jan 2019]

POULTON, C.V., LEE, J., HOBBS, P., JONES, L. & HALL, M. 2006. Preliminary investigation into monitoring coastal erosion using terrestrial laser scanning: case study at Happisburgh, Norfolk. *Bulletin of the Geological Society of Norfolk*, 45–64.

RAINES, M.G., BANKS, V.J., CHAMBERS, J.E., COLLINS, P.E., JONES, P.F., MORGAN, D.J. & ROYSE, K. 2015. The application of passive seismic techniques to the detection of buried hollows.

SIDDLE, H.J. 2000. East Pentwyn landslide, Blaina. *In*: SIDDLE, H.J., BROMHEAD, E.N. & BASSET, M.G. (eds) *Landslides and Landslide Management in South Wales*, National Museums & Galleries of Wales, 77–80, https://doi.org/10.1046/j.1365-2451.2002.03439.x

SKEMPTON, A.W. & VAUGHAN, P.R. 2009. The failure of Carsington dam. *In*: VAUGHAN, P.R. (ed.) *Selected Papers on Geotechnical Engineering*. Thomas Telford Publishing, 257–279.

SKIPPER, J., FOLLETT, B., MENKITI, C.O., LONG, M. & CLARK-HUGHES, J. 2005. The engineering geology and characterization of Dublin Boulder Clay. *Quarterly Journal of Engineering Geology and Hydrogeology*, **38**, 171–187, https://doi.org/10.1144/1470-9236/04-038

SUTHERLAND SHARMAN, D.G. 1996. Radon–in Northamptonshire? *Geology Today*, **12**, 63–67, https://doi.org/10.1046/j.1365-2451.1996.00009.x

TECHEL, F., JARRY, F. *ET AL*. 2016. Avalanche fatalities in the European Alps: long-term trends and statistics. *Geographica Helvetica*, **71**, 147–159, https://doi.org/10.5194/gh-71-147-2016

TERRA FIRMA 2017. Chalk Mine Collapse at Field Road, Reading, Berkshire. Retrieved from https://www.terrafirmasearch.co.uk/whatliesbeneath/2017/7/18/chalk-mine-collapse-at-field-road-reading-berkshire

TOMS, E., MASON, P.J. & GHAIL, R.C. 2016. Drift-filled hollows in Battersea: investigation of the structure and geology along the route of the Northern Line Extension, London. *Quarterly Journal of Engineering Geology and Hydrogeology*, **49**, 147–153, https://doi.org/10.1144/qjegh2015-086

TRENTER, N.A. & WARREN, C.D. 1996. Further investigations at the Folkestone Warren landslide. *Geotechnique*, **46**, 589–620.

TRIBUNAL APPOINTED TO INQUIRE INTO THE DISASTER AT ABERFAN ON OCTOBER 21ST 1966. *Report of the tribunal appointed to inquire into the disaster at Aberfan on October 21st, 1966*. HM Stationery Office, London.

VARNES, D.J. 1978. Slope movement types and processes. *In*: SCHUSTER, R.L. & KRIZEK, R.J. (eds) *Landslides, Analysis and Control*. Special Report 176: Transportation research board, National Academy of Sciences, Washington, D.C., 11–33.

VERSEY, H.C. 1939. The North Sea earthquake of 1931 June 7. *Geophysical Supplements to the Monthly Notices of the Royal Astronomical Society*, **4**, 416–423.

WALKER, J. 2005. Periglacial form and processes. *In*: FOOKES, P.G., LEE, E.M. & MILLIGAN, G. (eds) *Geomorphology for Engineers*, Whittles Publishing, Caithness.

WALKER, M.J. (ed.) 2012. *Hot Deserts: Engineering Geology and Geomorphology*. Geological Society, London, Engineering Geology Special Publication, **25**, https://doi.org/10.1144/EGSP25

WALTHAM, A.C. & DIXON, N. 2000. Movement of the Mam Tor landslide, Derbyshire, UK. *Quarterly Journal of Engineering Geology and Hydrogeology*, **33**, 105–123, https://doi.org/10.1144/qjegh.33.2.105

WARREN, C. D., PALMER, M. J., BROMHEAD, E., DIXON, N. & IBSEN, M. L. 2000. Observations on the nature of landslipped strata, Folkestone Warren, United Kingdom. *In*: *Landslides in Research, Theory and Practice, Proceedings of 8th International Symposium on Landslides*, **3**, 1551–1556.

WARBURTON, J. 2020. Peat hazards: compression and failure. *In*: GILES, D.P. & GRIFFITHS, J.S. (eds) *Geological Hazards in the UK: Their Occurrence, Monitoring and Mitigation – Engineering Group Working Party Report*. Geological Society, London, Engineering Geology Special Publications, **29**, 243–257, https://doi.org/10.1144/EGSP29.9

WEEKS, A.G. 1969. The stability of natural slopes in southeast England as affected by periglacial activity. *Quarterly Journal of Engineering Geology and Hydrogeology*, **2**, 49–61, https://doi.org/10.1144/GSL.QJEG.1969.002.01.04

WILLIAMS, G.M. & AITKENHEAD, N. 1991. Lessons from Loscoe: the uncontrolled migration of landfill gas. *Quarterly Journal of Engineering Geology and Hydrogeology*, **24**, 191–207, https://doi.org/10.1144/GSL.QJEG.1991.024.02.03

WILSON, S. & MORTIMER, S. 2020. Methane gas hazard. *In*: GILES, D.P. & GRIFFITHS, J.S. (eds) *Geological Hazards in the UK: Their Occurrence, Monitoring and Mitigation – Engineering Group Working Party Report*. Geological Society, London, Engineering Geology Special Publications, **29**, 457–478, https://doi.org/10.1144/EGSP29.19

WINSTANLEY, I. n.d. The Coal Mining History Resource Centre. Retrieved from http://www.healeyhero.co.uk/rescue/Collection/ian/2004/page_30.htm

WINTER, M. 2020. Debris flows. *In*: GILES, D.P. & GRIFFITHS, J.S. (eds) *Geological Hazards in the UK: Their Occurrence, Monitoring and Mitigation – Engineering Group Working Party Report*. Geological Society, London, Engineering Geology Special Publications, **29**, 163–185, https://doi.org/10.1144/EGSP29.5

WINTER, M.G., HEALD, A.P., PARSONS, J.A., SHACKMAN, L. & MACGREGOR, F. 2006. Photographic feature: Scottish debris flow events of August 2004. *Quarterly Journal of Engineering Geology and Hydrogeology*, Feb 2006; **39**, 73–78, https://doi.org/10.1144/1470-9236/05-049

WINTER, M.G., TROUGHTON, V., BAYLISS, R., GOLIGHTLY, C., SPASIC-GRIL, L., HOBBS, P.R.N. & PRIVETT, K.D. 2017. Design and construction considerations. *In*: GRIFFITHS, J.S. & MARTIN, C.J. (eds) *Engineering Geology and Geomorphology of Glaciated and Periglaciated Terrains – Engineering Group Working Party Report*. Geological Society, London, Engineering Geology Special Publications, **28**, 831–890.

WITHAM, C.S. & OPPENHEIMER, C. 2004. Mortality in England during the 1783–4 Laki Craters eruption. *Bulletin of Volcanology*, **67**, 15–26, https://doi.org/10.1007/s00445-004-0357-7

WONG, J.C.F. & WINTER, M.G. 2018. The Quantitative Assessment of Debris Flow Risk to Road Users on the Scottish Trunk Road Network: A83 Rest and be Thankful (No. PPR798).

Chapter 2   Seismic hazard

R. M. W. Musson

Lyell Centre, Currie, Edinburgh EH14 4BA, UK

rmwm@bgs.ac.uk

Abstract: It is often thought that earthquakes do not occur in the UK; however, the seismicity of the UK is usually classified as low-to-moderate. On average, a magnitude 3.2 M_w moment magnitude or larger earthquake occurs once per year, and 4.2 M_w or larger every 10 years. The latter is capable of causing non-structural damage to property. The damage caused by British earthquakes is generally not life-threatening, and no-one has been killed in a British earthquake (at the time of writing, May 2013) since 1940. Damage is caused by shaking, not by ground rupture, so the discovery of a fault surface trace at a construction site is not something to be worried about as far as seismic hazard is concerned. For most ordinary construction in the UK, earthquake hazard can be safely discounted; this is not the case with high-consequence facilities such as dams, bridges and nuclear power plants.

2.1 Earthquakes as a geohazard

An earthquake is a hazardous phenomenon that occurs when crustal strain builds up in such a way that its local concentration exceeds the restraining force of friction in the rocks, leading to sudden displacement, usually along a pre-existing plane of weakness known as a fault. This results in the release of wave energy that radiates out from the fault break. Several different types of wave are created, which travel at different speeds. The fastest are the P waves (or primary waves), which consist of a back-and-forward compressional motion in the direction of travel, similar to sound waves. These are followed by the S waves (secondary, or shear), which move transverse to the direction of propagation. Finally, in large earthquakes surface waves are produced; these consist of a rolling motion like sea waves and which travel, as the name suggests, only over the Earth's surface. All these manifest themselves to the observer as shaking.

The prime concern when dealing with earthquakes is the damage that is done to human structures as a result of earthquake shaking. Ordinary buildings are designed to stand up under their own weight, that is, resist vertical forces. An earthquake applies a lateral force that a building may not be able to cope with. Damage due to earthquake shaking is considered to be the primary hazard from earthquakes, but there are also secondary perils.

Firstly, if the fault that breaks in an earthquake ruptures all the way to the surface and causes displacement at the surface along the line of the fault, this can rip apart anything that is built across the fault line. A spectacular example of this was the failure of the Shih-Kang concrete gravity dam on the Ta-Chia River, Taiwan, during the Chi Chi earthquake of 1999.

Secondly, under strong shaking, soft sandy surface deposits turn to quicksand, and water and sand may fountain up. This is called liquefaction. Buildings built on liquefying soils may sink and tilt over.

Thirdly, earthquake shaking may cause various different types of slope failure, ranging from landslides on steep slopes (burying buildings at the bottom of the slope and causing buildings at the top to tip over) to slumping and mass movement (damaging to buildings built on the slope).

Fourthly, certain very large earthquakes can cause sudden upwards displacement of the sea bed. This displaces a large volume of water, creating a tsunami; this is a strong wave that sweeps onshore, destroying everything in its path, made vivid in the public imagination after the devastating tsunamis that struck the Indian Ocean in 2004 and eastern Japan in 2011. Tsunamis can also occur when an earthquake triggers a large underwater landslide, as happened off the coast of Newfoundland in 1929.

There are other possible dangerous secondary consequences of earthquakes, such as the starting of serious fires, but these four above are the principal geological hazards.

The phrase 'seismic hazard' is generally taken to refer to the probability, explicitly or implicitly expressed, of earthquake shaking occurring at a given location over a given time frame. It can also refer in a general way to secondary hazards, but these would more usually be referred to as rupture hazard, tsunami hazard, etc. 'Seismic risk' is specifically the probability that actual harm will be caused. An area may have high hazard but low risk, either because structures are very resistant to earthquakes and won't be damaged, or because there are no structures to be damaged.

Engineering Group Working Party (main contact for this chapter: R. M. W. Musson, Lyell Centre, Currie, Edinburgh EH14 4BA, UK, rmwm@bgs.ac.uk)
From: GILES, D. P. & GRIFFITHS, J. S. (eds) 2020. *Geological Hazards in the UK: Their Occurrence, Monitoring and Mitigation – Engineering Group Working Party Report.* Geological Society, London, Engineering Geology Special Publications, **29**, 43–60, https://doi.org/10.1144/EGSP29.2
© 2020 The Author(s). Published by The Geological Society of London. All rights reserved.
For permissions: http://www.geolsoc.org.uk/permissions. Publishing disclaimer: www.geolsoc.org.uk/pub_ethics

To express seismic hazard quantitatively, a way of quantifying earthquake ground motion is needed. There are various physical measures available of which the most common is the peak ground acceleration (PGA), but ground velocity and displacement can also be used. Increasingly, engineers use spectral acceleration – the ground acceleration produced by waves of a specific period of vibration – since some buildings are more vulnerable to short-period vibration and others are more vulnerable to long-period vibration. Accelerations are often given as a decimal fraction of the acceleration produced by gravity, written as g. Buildings are particularly vulnerable to shaking close to the natural period of the building (think of it as an inverted pendulum fixed at the base), so short buildings are more susceptible to damage to short-period vibration, and tall buildings to long-period vibration (from large, and often distant, earthquakes). This phenomenon is called resonance.

Another possible measure is the intensity of earthquake shaking, which is a classification of strength of shaking at a location in terms of its effects, in a manner analogous to the Beaufort Wind Scale. The scale in use in the UK is the European Macroseismic Scale (EMS); the word 'macroseismic' is used for visible earthquake effects and their study (Grünthal 1998). Shaking strong enough to rattle windows is intensity 4, knock over ornaments is 5, knock down chimney pots is 6, and so on. This is too unspecific to be useful for engineering purposes, but is useful for planners, insurers and the public.

It is important not to confuse intensity with magnitude. Whereas intensity is a measure of the strength of shaking at a place, and will therefore vary from place to place for the same earthquake, magnitude is a single value per quake, roughly related to the energy released in the fault rupture. Before the introduction of magnitude, maximum intensity or epicentral intensity (usually but not always the same) was used as a rough measure of earthquake 'size', but it is not a good one. Aside from the fact that intensity is subjectively assessed from a consideration of reported effects at each location, if the epicentre is in an uninhabited area or offshore, the intensity value from the nearest settlement may be a poor indicator of what the maximum intensity might have been.

Magnitude, on the other hand, is generally calculated from reading the maximum amplitude on instrumental earthquake recordings (seismograms) and adjusting for the distance between the earthquake location and the recording station. Typically, magnitude will be calculated from as many different seismic recording stations as possible and the values obtained averaged to obtain the final value. This is why magnitude values released immediately after an earthquake are frequently revised in the next 24 hours; more stations have been used in the computation of the average.

A number of different methods for calculating magnitude have been proposed in the past, of which three such magnitude scales have seen significant use in the UK: local magnitude (M_L), surface wave magnitude (M_s) and moment magnitude (M_w). These are further explained in the Glossary. For historical reasons it is necessary to refer to all three in this paper; for any earthquake in the UK, M_s and M_w values are expected to be roughly equivalent, and the M_L value is expected to be slightly larger by around 0.2 or 0.3 units. Note that the phrase 'Richter Scale' is a journalistic term only.

For historical earthquakes that predate the introduction of seismometers, magnitude can be estimated with reasonable accuracy from the size of the felt area, reconstructed from contemporary written accounts. There is a good correlation between magnitude and the log of the felt area and, because of the logarithmic nature of the relationship, the felt area need only be approximately known to obtain a fairly reliable estimate of the magnitude. Even for early twentieth century earthquakes, the macroseismic magnitude may be more dependable than the instrumental magnitude.

2.2 Distribution of earthquakes in the UK

The global distribution of earthquakes is closely linked to plate tectonics, since the main cause of strain in the crust is tectonic movement. The planet's major earthquake zones are therefore all along the plate boundaries, especially where two plates collide, as for instance the Nazca Plate and the South American Plate along the west coast of South America. The UK is far away from the nearest plate boundaries, which are in the mid-Atlantic to the north and west, and Italy to the south and east. The situation of the UK is therefore what is known as 'intraplate', also referred to as a stable continental region or SCR (Johnston et al. 1994).

Earthquakes are far less frequent in intraplate regions, and their size is much more limited. Nevertheless, they still occur. The whole of the Earth's crust is subject to stress away from plate boundaries and very old faults, relics of ancient phases of mountain-building millions of years ago, can still slip if local stress concentrations exceed their strength.

A major difference exists in the relationship between earthquakes and faults in intraplate and plate boundary areas. In mobile regions such as Anatolia, the collision between the Arabian and Eurasian plates is creating continuing, measurable deformation as crustal blocks continue to slide past one another. This deformation is controlled by major faults such as the North and East Anatolian faults, which will continue to produce major earthquakes for the foreseeable future. Active deformation of this kind is not taking place in an intraplate area. The fact that an ancient fault may be reactivated and slip, producing an earthquake, does not mean that it must necessarily continue to slip in the future (e.g. Stein 2007). The concept of an 'active fault' is therefore unhelpful in the UK (Musson 2005).

The diffuse spread of earthquakes across the country means that there are hundreds of faults in the UK that have been reactivated and produced (albeit minor) earthquakes. In almost all cases, however, clear cases where a known, named fault can be shown to have been the origin of a specific earthquake cannot be identified. Small earthquakes have small source dimensions and require only a small fault.

These are numerous, and the location of an earthquake in three dimensions is not precise beyond a few kilometres at best; there may be several potential fault sources or the real fault source may be unmapped.

That said, the spatial distribution of earthquakes is not uniform or random (Fig. 2.1). The following description is based on Musson (1994, 2007). Most earthquakes in Scotland are concentrated on the west coast with the addition of centres of activity near the Great Glen at Inverness (earthquakes in 1816, 1890 and 1901) and a small area around Comrie, Perthshire, the site of the famous earthquake swarms, principally 1795–1801 and 1839–1846 and possibly also 1605–1622. Since 1846 small shocks have been observed at Comrie on a regular, if infrequent, basis. There has also been swarm activity in the Central Valley of Scotland by the Ochil Hills (near Stirling). This spot was active in 1736, during 1900–1916 and in 1979.

The Outer Hebrides off the west coast of Scotland, the extreme north (including the islands of Orkney and Shetland) and most of the east of Scotland are virtually devoid of earthquakes. For NW Scotland the absence of early written records, the small population and the recent lack of recording instruments means that there may be a data gap.

A similar irregularity is seen further south in England and Wales. Wales and the west of England, including the SW and NW parts of the country and the English Midlands, are much more active than the east of England. The NE of England seems to be very quiet; the SE has a higher rate of activity with a number of apparent 'one-off' earthquakes, plus a couple of important centres of activity on the south coast. It is curious that the damaging 1884 Colchester earthquake occurred in a locality of the SE of England that seems to have been otherwise totally inactive seismically, either before or since.

Offshore, there is significant activity in the English Channel and in the North Sea off the coast of Humberside. Because only the larger events in these places are likely to be felt onshore, the catalogue is probably under-representative of the true rate of earthquake activity in these zones. The largest British earthquake for which magnitude can be estimated had an epicentre in the North Sea off the east coast of England was and occurred on 7 June 1931. Felt over the whole of Great Britain, eastern Ireland and in all the countries bordering the North Sea, its magnitude was estimated as around 5.8 M_w (moment magnitude). It was fortunate that this earthquake had an offshore epicentre, otherwise the damage might have been considerable. As it was, only minor damage occurred up the east coast.

Certain centres can be identified as showing typical patterns of activity. For example, the NW corner of Wales is one of the most seismically active places in the whole of the UK. Both large and small earthquakes, usually accompanied by many aftershocks, occur at regular intervals. On the other hand, in South Wales, although a line of major epicentres can be traced from Pembroke to Newport, only the Swansea area shows consistent recurrence. The Hereford–Shropshire area of western England adjoins South Wales, and this area has also produced large earthquakes in 1863, 1896, 1926 and 1990, but none of these share a common epicentre. In the north of England seismic activity occurs principally along the line of the Pennine Hills, which form the backbone of this part of the country. Again, it is possible to identify particular spots which have been active repeatedly.

The area of the Dover Straits is particularly significant because of the occurrence there of two of the largest British earthquakes in 1382 and 1580 (both of magnitude $c.$ 5.5 M_w). Jersey has also experienced a number of significant earthquakes, chiefly originating to the east of the island in the Cotentin peninsula area of France.

What is remarkable is the lack of correlation between this pattern and the structural geology of the UK. In the northern part of the British Isles, the geology has a strong NE–SW (Caledonian) trend and the geology of Northern Ireland is to a large extent a SW-wards extension of the geology of Scotland. However, there is no continuity of seismicity along this trend. It is possible to draw a line roughly NNW–SSE through Scotland such that earthquakes are entirely confined to the west side of the line; this line has no apparent geological significance, however, and cuts directly across the structural trend. It is clear that this pattern is persistent and not merely an artefact of recent earthquake locations; there are a number of historical sources for the east of Scotland which comment on the absence of earthquakes. The difficulty is acute in Ireland; the geological history of Ireland is very similar to that of Great Britain, and there is no clear solution to the question of why the seismicity of Ireland should be so very much lower.

Some different hypotheses that attempt to explain the distribution of seismicity in the UK are summarized in Musson (2007). It is remarkable, for instance, how Scottish seismicity coincides with those areas under ice in the last phase of the last glaciation, as shown by Dawson (1992) and discussed by Musson (1996); however, there are earlier references as far back as the 1930s to deglaciation as a possible cause of North Sea earthquakes (Kolderup 1930; Versey 1939). A recent paper by Baptie (2010) finds evidence from focal mechanisms that, at least in Scotland, stresses due to isostatic recovery after the last glaciation are significant in influencing seismicity.

Chadwick *et al.* (1996) suggest there may be a sort of 'jostling' of different geological units in response to an overall compressive stress exerted from the NW in response to Atlantic widening. This idea has been utilized in the development of recent seismic hazard models (Musson & Sargeant 2007).

The frequency of earthquakes in any region is known to be inversely related to the magnitude of the shock, according to what is known as the Gutenberg–Richter equation (Gutenberg & Richter 1936). Figure 2.2 shows the application of this to the UK, using data from 1970 to 2007. The relationship expressed by the red line is

$$\log N = 3.23 - 1.00 \, M_w, \qquad (2.1)$$

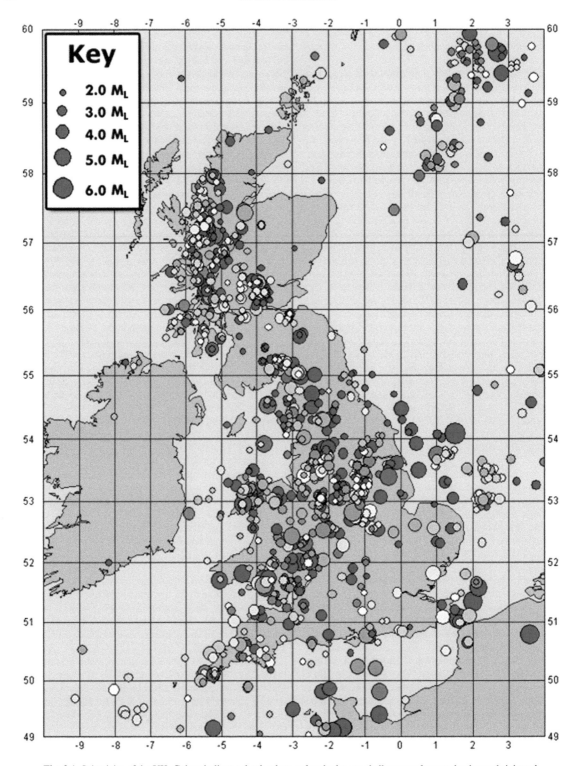

Fig. 2.1. Seismicity of the UK. Colour indicates depth where paler shades are shallower; unknown depths are bright red.

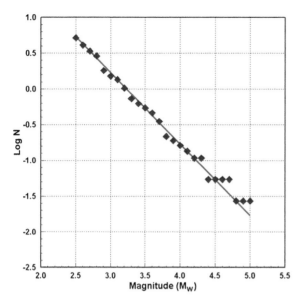

Fig. 2.2. Magnitude frequency plot for the UK.

where N is the number of earthquakes per year equal to or larger than a given M_w magnitude (Musson & Sargeant 2007). The value of -1.0 coincides with the expected value from theory and practice, and equates to that over northern Europe generally (Musson 2011).

2.3 Consequences of British earthquakes

Given the lack of major earthquakes in Britain in historical times, the impact of past earthquakes in terms of damage and death has not been great, although questions can be raised about some medieval earthquakes that are insufficiently well documented. The earthquake of 11 September 1275, for example, is described by an Oxfordshire source as having thrown down churches and houses and killed people across England, while the 11 April 1185 event caused heavy damage to Lincoln cathedral and other buildings (Musson 2008a).

The more typical consequence of the stronger British earthquakes has tended to be a diffuse distribution of minor damage, chiefly to interior plaster and elderly chimney stacks. If there are concentrations of damage, these may be more related to the distribution of vulnerable structures rather than an indication of the epicentre. On the day of the 17 July 1984 Lleyn Peninsula earthquake, the epicentre of which was in NW Wales, initial reports prior to the location of the event suggested the epicentre might have been near Liverpool, which is where most of the damage was, to the chimneys of vulnerable terraced houses in poor condition, and built on estuarine deposits which tended to amplify the shaking.

Structural damage, or damage to engineered buildings, is extremely rare. There are three cases of damage to dams: Earl's Burn reservoir near Stirling in 1839 (Musson 1991), Blackbrook reservoir near Loughborough in 1957 (Lees 1957), and Glendevon, Ochil Hills, in 1979. There is no published description of the latter, which consisted of 'displaced pegs, coping stones and ... damage to the Meter House at Lower Glendevon ...' according to a letter preserved in BGS archives. Neither this nor the damage at Blackbrook were cases of structural damage imperilling the dam itself. In the case of the Earl's Burn embankment dam, the dam failed some hours after the earthquake; the role of earthquake shaking in this failure is speculative.

Apart from the general smallness of British earthquakes, there are other reasons for this lack of significant damage. Firstly, depth is also important. The larger British earthquakes tend to occur at 15 km focal depth or deeper (Musson 2007), which means that seismic energy is partly dissipated by the time it reaches the surface (Fig. 2.3). The most damaging British earthquakes have also been those that were exceptionally shallow, such as the earthquakes of 1865 (Rampside) and 1884 (Colchester); the former event was sufficiently small to be virtually imperceptible 10 km from the village of Rampside, near Barrow-in-Furness, yet Rampside itself was wrecked (Musson et al. 1990; Musson 1998).

The second factor is that windstorms are much more damaging a phenomenon in Britain than earthquakes. Traditional British buildings therefore tend to have an innate resistance to lateral forces, which they need in order to stand up to British weather. Any that lack this are likely to be removed from the scene by the next winter storm before they are tested by any earthquake.

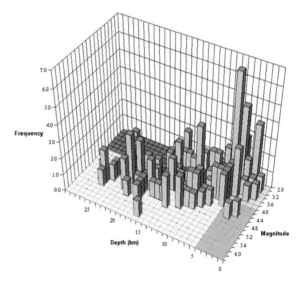

Fig. 2.3. Magnitude–depth distribution for British earthquakes (M_L magnitudes).

Secondary hazards have also not proved very damaging in the past. There is no recorded instance of surface displacement from earthquake faulting in the UK in historical times (with good reason, as can be seen from Fig. 2.3). Given that the width of an earthquake rupture for an earthquake of magnitude c. 5 M_w is likely to be of the order of a few kilometres, if it occurs at a depth of more than 5 km it will not intersect the free surface.

There are instances of falling rocks in hilly districts as a result of British earthquakes, especially the movement of scree slopes, but none that have ever caused damage (Davison 1904); movements in loose ground (cracks appearing) have also been rare. Two cases in Essex in 1884 (Meldola & White 1885) and Inverness in 1901 (Davison 1902) are the only that are well documented.

As for tsunamis, the nearest credible source is offshore from Portugal. This was the source of the tsunami from the 1755 Lisbon earthquake, which was strongly perceptible on the south coast of England but without damage (Kerridge 2005). Although massive landslides on the continental slope north of the North Sea caused a significant tsunami in prehistoric times (Dawson *et al.* 1993), there is limited potential for underwater landslides to be a source of tsunami hazard in the near future. The Storegga slides, as they are known, involved the failure of thousands of cubic kilometres of sediments. However, these were accumulated glacial sediments and having once slumped cannot slump again until replenished, perhaps by some future glaciation.

It is also probably not the case that the UK is at risk from a massive tsunami caused by the flank collapse of a volcano in the Azores. Tsunamis caused by landslides, either submarine or subaerial, can be very severe at short ranges but tend to attenuate rapidly (Musson 2012*a*). The largest Storegga slide tsunami had large run-ups in NE Scotland, but run-up heights decrease rapidly towards the south and there is no trace of the tsunami having crossed the Atlantic (D. Long, pers. comm.). Likewise, the 1929 Grand Banks (Newfoundland) tsunami was locally devastating, but barely registered on tide gauges on the eastern side of the Atlantic (Fine *et al.* 2005).

Fatalities in British earthquakes have also been few and far between; they are considered in detail in Musson (2003). The exact total depends on mortality criteria; according to one method of counting deaths the total death toll since 1550 is at most 11, and under the strictest method it is two. The most recent cases occurred in North Wales in 1940, one from heart failure and another woman who fell downstairs trying to leave her house in the dark.

A study of earthquake risk in the UK by Arup (1993) attempted to calculate an expected fatality rate from earthquakes in the UK. They concluded that there was a 1% chance per year of an earthquake causing between 1 and 500 deaths, and a 0.1% chance per year of an earthquake causing 100 to 4000 deaths. This may seem rather high if converted to a mean annual rate and compared with the historical record, but the population of the UK is much higher today than it has been historically, and average (mean) rates of fatalities are meaningless in a case where there is a very high probability of no deaths and a very small probability of a high-fatality event.

2.4 Identifying earthquakes as a geohazard in the UK

Unlike some other geohazards, there is no visible indication that a location in Britain may be susceptible to earthquake shaking. There is no such thing as 'earthquake country', at least, not so far as variations in hazard in the UK are concerned. In particular, the presence or absence of a fault trace, major or minor, on a UK site, is not indication of earthquake hazard. As discussed above, the discovery of a geological fault on a site normally has no bearing on the likelihood of seismicity. What is seen at the surface may not bear any relation to what is happening 5–15 km down, which is where earthquakes in the UK typically originate. The chances of earthquake rupture along any given fault are so tiny as to be only of concern when the consequences of rupture are extremely high, for example, for nuclear facilities. The discovery of a fault trace on a building site for a planned supermarket is therefore of no consequence as regard to earthquakes in the UK.

It is also important to realize that, as earthquakes are relatively infrequent in the UK, the potential of a site to be affected by earthquakes may not be apparent from a short time sample; in fact, the whole of written history may be too short a period to understand the earthquake potential of a site. It is not unknown for engineers at a planning stage in a project to request the last 10 years of earthquake data for the local area. However, this is completely insufficient and no sound conclusions can be drawn from it. A far more sophisticated approach needs to be taken to quantify hazard, as discussed below.

One conclusion that could be drawn from site investigation is whether a location is likely to suffer from amplified shaking as a result of soft soil conditions. A site on very soft sediments, or naturally subject to settlement, is likely to be shaken more than a nearby site on stronger ground. Damage distributions often reflect this. Site investigation can also identify potentially liquefiable soils, although the likelihood of a British earthquake causing liquefaction is very low indeed. Liquefaction is normally only associated with large earthquakes with sufficiently long durations of strong shaking (Papadopoulos & Lefkopoulos 1993). One exceptional case in the UK has been identified, where a small but shallow earthquake with strong shaking hit saturated tidal sand flats in Morecambe Bay in 1865, producing sand fountains (Musson 1998).

Since earthquake potential for intraplate regions cannot be identified from ground investigation, it is necessary to assess it by other means. Examination of a seismicity map such as Figure 2.1 will give a first impression if a site lies in a more earthquake-prone part of the country or not, although there is still the possibility of an earthquake occurring in an area previously seismically quiet during the historical period.

Earthquake hazard for a site is normally assessed probabilistically, in terms of the earthquake shaking (according to some measure, most often PGA) expected with some annual frequency. This can either be expressed as an annual probability (e.g. 10^{-4} chance per year) or as a return period (e.g. 10 000 years). It is important to understand that a return period is simply the reciprocal of the annual probability. To say that the PGA with 10 000-year return period for a site is 0.2 g does not mean that such a PGA occurs each 10 000 years; it says nothing about what might happen thousands of years in the future. It means only that the probability of a PGA of 0.2 g next year is 1 in 10 000. This has to be stressed as it is the subject of frequent misunderstandings.

In order to take account of the possibility of earthquakes that have not been seen in the historical catalogue, the assessment of seismic hazard for a site is based on two considerations: seismic source characterization (SSC) and ground motion characterization (GMC). The SSC part of any project involves building a numerical model of the sources of the seismicity that could affect a site, that is, where earthquakes occur and how frequently. Sources can either be defined as individual faults, or areas where it is believed there is a population of similar faults, such that an earthquake is equally likely to occur anywhere within that area. Because of the absence of clearly discernible active tectonic deformation in the UK, individual faults are seldom used as sources, and usually only tentatively if at all. More common is a zone model, where the region in which earthquakes of relevance to the site might occur is divided up into a series of 'source zones'. These will typically be defined from a consideration of geology, geophysics, tectonics and seismicity, and the construction of such models is a highly specialized task, requiring a degree of expert judgement.

The GMC part is concerned with estimating the expected strength of shaking from any earthquake that might occur, as a function of magnitude, distance, style of faulting and site conditions. Equally important is the aleatory variability around the expected value, since the same magnitude–distance combination has been observed to yield a range of results that follows a log-normal distribution (Bommer & Abrahamson 2006). Twenty years ago it was a reasonable assumption that ground motion could be modelled with a functional form similar to that of:

$$\ln Y = a + bM + c \ln R + dR \quad (2.2)$$

where Y is the ground motion measure (e.g. PGA), M is magnitude, R is distance, and a, b, c and d are constants. The coefficient a is a scaling constant, b represents magnitude scaling, c is geometric spreading and d is anelastic absorption. This equation is now known to be far too simplistic (e.g. Cotton et al. 2006) and some recent models (e.g. Abrahamson & Silva 2008) are extremely complex in functional form, in an attempt to handle the complexity of generation and transmission of ground motion from fault to site.

The term attenuation used to be commonly applied to GMC studies, but this word has fallen from favour due to a somewhat pedantic insistence that attenuation is strictly speaking the decrease of ground motion with distance from the earthquake source. In the first case, ground motion in reality does not decrease monotonically with distance due to complex variability over even short distances. In the second case, models such as Equation (2.2) yield an expected absolute value and not a decay with respect to some peak. The recent literature on the subject tends to use the term 'ground motion prediction equation' (or GMPE); this is not much of an improvement because (1) the term 'prediction' has specific meaning in seismology (predictions are phrased in terms of time and place for future events), and (2) modern models are too complex to be described simply as 'equations'. At least older terms such as 'attenuation laws' are infrequently used today, completely inappropriate since models are never laws. A more appropriate term is 'ground motion model' since today the basic equation will be accompanied by tables of coefficients, limits of applicability and other guidelines and explicatory apparatus, as with the Abrahamson & Silva (2008) model already referred to.

Ideally, knowledge of what ground motions are likely from a large earthquake in Britain, reflecting British conditions, is sought. The problem is that there are next to no UK strong motion records from which an appropriate empirical model could be derived. Estimation of expected ground motions in Britain is therefore very uncertain (as is common in intraplate areas in general). The seismologist is driven to other strategies, which can be to (1) borrow an empirical model from another region of similar tectonic character; (b) take an empirical model from another a region and attempt to modify it for known differences between the origin region of the model and the target region; or (c) construct an artificial ground motion model using stochastic methods. In any case, it is likely (indeed, is now standard practice) that several models will be used to try and represent the overall uncertainty as to how ground motions from future British earthquakes will behave. In the past, the most common solution has been to use empirical models from elsewhere in the world, or in the case of Principia (1982) and Woo (1988) to construct empirical models from selected overseas data. Host-to-target conversions are a relatively new idea not yet applied in the UK. Two synthetic models calibrated to the UK have been derived using stochastic techniques: Winter et al. (1996) and Rietbrock et al. (2013).

It is also worth mentioning the discrimination between earthquakes and other geohazards. Although apparently obvious, it is not unknown for householders to wishfully attribute some damage to property (generally related to subsidence) to a real or imagined earthquake.

Earthquake prediction should also be mentioned, since it is often assumed by non-seismologists that prediction is an effective way of mitigating the consequences of earthquakes. In the 1960s it was commonly assumed that earthquake prediction would be routine by the end of the century; not only was this not achieved, but it is now commonly assumed by seismologists that routine earthquake prediction is not possible. If the problems of earthquake prediction cannot be solved in highly seismic areas where data are abundant, it is futile to look for earthquake prediction in a low-seismicity area such as Britain.

2.5 Past practice of seismic hazard in the UK

Because the seismic hazard in the UK is low in world terms, for most civil construction it can be discounted as an acceptable risk. For high-consequence facilities, especially nuclear, even the low probability of significant earthquake shaking needs to be taken into consideration, and there is a strong regulatory environment to oversee this. This was not always the case; the widespread misapprehension that Britain is a country without earthquakes extended to the UK engineering community until the 1970s, despite the fact that the first commercial nuclear power plant in the UK commenced operation in 1956. Around the mid-1970s, awareness dawned that earthquakes still occur even in a low-seismicity country like the UK, and can be sufficiently large (albeit rarely) to be significant for high-consequence structures. The first structure in the UK that was built with potential earthquake hazard in mind was probably the Kessock Bridge near Inverness, where the impetus was the questionable belief that the Great Glen Fault, which the bridge spans, is still an active strike-slip feature. (Current research suggests the timing of the last reactivation of the fault is possibly Late Oligocene; see Le Breton et al. 2013.) At that date (1976) the last study of historical earthquakes in the UK was that of Davison (1924). The report on local seismicity compiled for the Kessock Bridge was therefore the first in over 50 years (Browitt et al. 1976). This was followed by a second report by Burton & Browitt (1976), in which selected strong motion records from California were proposed as suitable analogues for what might be expected at Inverness in a worst case.

At about the same time, Roy Lilwall of the Institute of Geological Sciences (IGS, now British Geological Survey, BGS) was working on an attempt to prepare a numerate version of Davison's (1924) catalogue, updated to the present day with data largely from Tillotson (1974). This catalogue was the basis of Lilwall's (1976) hazard map of Great Britain, the first quantitative assessment of seismic hazard in the UK. Lilwall's catalogue never progressed beyond a working file and, as it did not refer back to original sources, incorporated many errors due to Davison's (1924) faulty compilation.

It was now very apparent that British seismicity needed to be addressed by the nuclear industry, and a lead in this was taken by the then Central Electricity Generating Board (CEGB) and the Nuclear Installations Inspectorate (NII, now Office of Nuclear Regulation, ONR), part of the Health and Safety Executive (HSE), to whom by law regulatory oversight is given. As a result, a large amount of investigation into UK seismicity was undertaken, principally by two consultancies (Principia 1982; Soil Mechanics 1982), Imperial College London (Ambraseys & Melville 1983) and BGS (Burton et al. 1984). In addition, funding was secured to expand the UK seismic monitoring network, which in the 1970s amounted only to a handful of stations in the Scottish Midlands (LOWNET), hardly suitable for detecting earthquakes in the south of England. Interest in seismicity was also taken by the offshore oil and gas industries, in connection with hydrocarbon development in the North Sea.

One interesting fact about the development of seismic hazard analysis in the UK during this early period is that there was never any interest in the quasi-deterministic method popular in Europe, based on maximum observed intensity (e.g. IAEA 1991). From the outset, hazard was conceived in probabilistic terms, either from a Cornell-like approach (Cornell 1968) or from extreme-value methods (which is what Lilwall (1976) employed). The only other method used for hazard calculation has been a stochastic approach introduced by BGS in the mid-1990s, and yielding identical results to the Cornell method given the same input (Musson 2012b).

The earliest hazard calculation specifically intended to be relevant to the nuclear industry was made by Irving (1982), who used a single uniform source covering the whole country to come up with a 'typical' UK hazard value of 0.25 g peak ground acceleration (PGA) for a 10 000-year return period, a result that was later to be much used for mental 'anchoring' (i.e. harbouring a preconceived idea of what sort of value might be correct).

Early hazard software used for site-specific analysis was EQRISK (McGuire 1976), used by the consultancy Principia Mechanica, and in BGS an extreme-value hazard programme developed by Kostas Makropoulos (1978) was employed. When the use of extreme-value methods was discontinued, SEISRISK III (Bender & Perkins 1987) was used for a while in BGS, later replaced by in-house software for stochastic simulation hazard estimation. Early examples of studies include hazard assessment for a nuclear site using EQRISK by Woo (1983) and hazard assessment for a hydrocarbon facility using extreme-value statistics by Burton et al. (1981).

During the 1980s and 1990s, the majority of site-specific hazard studies for nuclear power plants were undertaken by a group of consultants led by David Mallard of the CEGB, and including staff from Principia Mechanica and Soil Mechanics. The Seismic Hazard Working Party (SHWP), as they were called, developed a consistent set of working practices for probabilistic seismic hazard analysis (PSHA) based around a rewritten version of EQRISK called PRISK, developed by Gordon Woo, which included the ability to implement a logic tree structure for handling epistemic uncertainty. A typical example is SHWP (1987).

The regulatory framework for nuclear facilities in the UK requires a formal and detailed safety case to be submitted for nuclear power plant sites before permission to construct or operate can be granted; such permissions are granted under the site license issued to the site owner/operator by ONR. The requirement for the site licensee (normally the operator) is to demonstrate that hazard at the site has been adequately characterized, and ONR has published guidance on what this means in its Safety Assessment Principles (SAP). The benchmark by which seismic hazard is measured is typically the PGA (with an associated response spectrum), conservatively defined with an annual probability of being exceeded of 10^{-4}.

The seismic hazard 'case' then normally forms a supporting reference to the wider safety case justifying nuclear safety of the activities for which permission is being sought. For a major seismic hazard study that would be expected to support permission to operate a nuclear power plant, ONR will assess the technical adequacy of the case using the SAP as a guide, and historically has taken advantage of independent expertise in the Earth sciences community to assist in this task.

For nuclear sites without operating nuclear reactors (for instance, some medical facilities), a more pragmatic approach can be adopted subject to the overriding principle that risks generally, and those from seismic hazard specifically, are as low as reasonably practicable (ALARP).

In addition to PSHA studies for those nuclear power plant sites constructed in the 1980s and 1990s, retrospective studies were carried out for those sites that had been built previously to 1980, with a view to demonstrating safety.

After 1995, there was an end to new nuclear construction in the UK, and after a 2001 BGS assessment of seismic hazard for the Wylfa site in North Wales (opened 1971) no further major nuclear power plant seismic hazard assessments were completed. However, a government policy review in 2006 announced the resumption of nuclear power plant construction, and since then there has been extensive work on site selection (a mixture of previous locations and new sites). ONR recently commissioned a report on capable faulting in the UK as a general set of guidelines, and founded a permanent panel of reviewers to provide guidance on future hazard assessments. PSHA studies are currently in progress for the first new build sites to be considered, which will doubtless reflect advances in PSHA methodology since the last century. Although the Senior Seismic Hazard Advisory Committee (SSHAC) system developed in the USA (Budnitz et al. 1997; USNRC 2012) has not been officially adopted in a UK context, SSHAC terminology is now practically unavoidable in PSHA discourse (Musson 2012c).

2.6 Seismic hazard mapping in the UK

Even if seismic hazard in the UK is a not a serious issue for most civil engineering, it is useful to know how it varies spatially as it is clearly not uniform, as can be seen in Figure 2.1. The earliest attempt to map seismic activity specifically in the UK, and hence indirectly the hazard, was made by O'Reilly (1884). The (perhaps better known) map of Montessus de Ballore (1896) does show a division of the British Isles into distinct zones. However, at this early date, hazard could only be expressed in terms of simple distinctions between levels of frequency of earthquake occurrence, usually subjectively assessed, and not as ground motion, which is how seismic hazard is expressed today.

The first true hazard map (in the modern, probabilistic, understanding of the word) for Great Britain was therefore that of Lilwall (1976), which expressed hazard in terms of intensity with a 200-year return period, using an extreme-value technique based on the work of Milne & Davenport (1969). A subsequent study by Arup (1993) used true PSHA to calculate hazard at selected points in the UK, but these points were too few to be contoured. The first contour maps of hazard on the UK territory produced using PSHA were therefore those of Musson & Winter (1996), prepared for the then Department of Trade and Industry (DTI). A map for the UK territorial waters was completed even earlier in around 1993, but not published until 1997 (Musson et al. 1997); this project was a collaboration between Arup and BGS for the Offshore Division of the Health and Safety Executive.

The UK was also included in two major international seismic hazard mapping projects, the Global Seismic Hazard Assessment Programme (GSHAP) and Seismotectonic and Seismic Hazard Assessment of the Mediterranean Basin (SESAME) (Grünthal et al. 1996; Jiménez et al. 2001). Both these studies used a common source model for the UK, which was derived from a simplified version of the Musson & Winter (1996) model. The results are in conformity with those of Musson & Winter (1996), although different ground motion models were used.

An updated hazard map for the UK was published by Jackson et al. (2004), but only for intensity. This is described in Musson (2004).

A specific zoning map for the UK was produced for a report on dam safety for the UK, and has since been widely circulated (Halcrow 1990). This map assesses hazard in a completely subjective way into high, medium and low classes, which are to be understood as entirely relative terms. Despite its informal nature, it proved to be a reasonable depiction of relative hazard levels when compared with later quantitative maps.

The first UK national seismic hazard maps to be issued specifically in connection with an earthquake building code are those of Musson & Sargeant (2007). They were based on a new seismic source model, elaborated from that of Musson (2004) and based rather heavily on neotectonic considerations described by Chadwick et al. (1996). The work also took advantage of what were then very recent advances in the modelling of strong ground motion. The underlying earthquake catalogue was that of Musson (1994) with minor unpublished revisions and extended up to the beginning of June 2007.

The results of the study were expressed as peak horizontal bedrock acceleration values (PGA), and mapped at two return periods of 475 and 2475 years. The latter map is reproduced here as Figure 2.4.

The source characterization model developed for Musson & Sargeant (2007) was further developed in the context of the European FP-7 project Seismic Hazard Harmonisation in Europe (SHARE), and was reconciled with the national

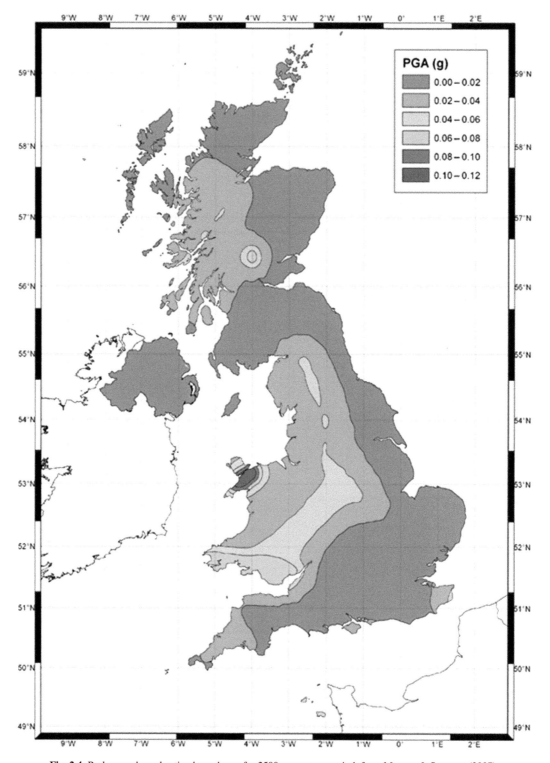

Fig. 2.4. Peak ground acceleration hazard map for 2500 year return period, from Musson & Sargeant (2007).

source models for surrounding nations: France, Belgium, the Netherlands and Norway. Hazard for NW Europe could therefore be calculated on the basis of a source zone model that had the agreement of national representatives of all the territories involved (Arvidsson et al. 2011). Zones extending out as far as the mid-Atlantic ridge were adapted from Musson et al. (1997).

It is worth reiterating here that different principles apply to seismic hazard maps for large areas and to studies for engineering design purposes for specific sites (Page & Basham 1985; Musson & Henni 2001). As an example, consider the case of assessing seismic hazard for a site near Colchester in Essex. What is to be done about the 1884 Colchester earthquake? There is no known geological factor that ties this event to a unique source feature. From the point of view of seismic hazard in the east of England, a repeat of the 1884 earthquake could occur in any geologically similar location with equal likelihood. The normal solution would be to let the earthquake occur in the future anywhere in a broad zone, in which case the hazard at Colchester would be no higher than anywhere else in eastern England.

However, if concerned with the hazard to a proposed structure specifically in Colchester, it would be irresponsible not to consider the possibility that the 1884 earthquake originated on an unknown fault with properties such that it might repeat the same earthquake in the future. In this case a local seismic source would be proposed, and this would elevate the calculated hazard well above the value plotted in the hazard map.

The same procedure cannot really be applied when constructing a hazard map, unless this procedure was to be repeated for all strong earthquakes. This runs the risk that the hazard map simply recapitulates the historical seismicity, although some compromise solutions have been attempted including Musson & Winter (1996) and discussed further in Musson (1997).

2.7 Earthquake monitoring in the UK

The use of instruments to monitor earthquakes has a long history in the UK, and the investigations of earthquakes near Comrie, Perthshire, in the early 1840s featured the first ever attempt to deploy a network of earthquake-recording devices (Musson 1993, 2013).

The very start of specific UK-oriented earthquake monitoring using modern seismometers capable of detecting small-magnitude earthquakes was in 1967. At this date work began on the Lowlands Network (LOWNET), a network of short-period vertical component seismometers (Willmore Mk II) deployed at seven outstations in Central Scotland and a central three-component set in Edinburgh (Crampin et al. 1970). This network was formally declared open in January 1969, and was operated by the then Institute of Geological Sciences, now the BGS. This network proved very capable of detecting small earthquakes in Central Scotland (this was a period when a large number of mining-induced earthquakes were observed in the Midlothian, Fife and Clackmannanshire coalfields), but its detection capacity for events in England and Wales was limited. Annual catalogues from LOWNET up to 1978 are given by Burton & Neilson (1980).

In response to a variety of factors, the network gradually expanded after its original foundation. The first expansion south came in 1976 when the Department of the Environment sponsored new stations near Stoke-on-Trent, Leeds, Leicester and Hereford. The occurrence of significant felt earthquakes in the Kintail area in 1974/75 resulted in extra stations also being deployed in the Kyle of Lochalsh area. Further additions to the network were made in an ad hoc manner in response to particular demands: stations in Shetland, NE Scotland and East Anglia in connection with hydrocarbon developments in the North Sea, and a dense local network in Cornwall and Devon to monitor the Hot Dry Rock geothermal project (Browitt 1991).

In 1989 the network was put on a new footing with the establishment of a Customer Group to sponsor the BGS Seismic Monitoring and Information Service. This Customer Group includes members from the government, nuclear, hydrocarbon and water sectors, among others. Under this project the network was gradually expanded in order to provide coverage over the whole country. This development, which combined specific local objectives with a recognition of the need to provide uniform coverage of the country, is detailed in a series of annual reports; the most recent at the time of writing is Baptie (2012).

The network reached a peak density early in the 2000s, with just over 140 stations. However, most of these were single high-frequency, high-gain Willmore Mk III instruments, and these tended to saturate in the event of even a moderately large earthquake. In recent years, there has been a programme of replacing the old Willmores with modern broadband instruments, with the intention that fewer stations will actually produce better data. The network at the time of writing is depicted in Figure 2.5. The current configuration has just over 40 broadband stations, and allows the detection and location of any earthquake of magnitude M_L 2.5 anywhere in the UK, even in bad noise conditions. For many parts of the country detection capability is even better than this, giving data completeness down to M_L 1.5 or lower. The equivalent M_w values are around 2.4 and 1.6 according to the quadratic conversion from Grünthal et al. (2009).

The net result of the seismic monitoring activity of BGS since 1969 is that the seismicity of the UK in the last 40 years is known in great detail, even down to quite small magnitudes. Although the seismicity itself is low in world terms, the dataset available for study is quite rich because it extends to low magnitudes. It is interesting to note that the spatial pattern of minor earthquake activity from recent instrumental data, even if only 5–10 years of data are considered, very closely mirrors the long-term spatial pattern. The most significant deviation is that little or no recent seismicity in South Wales west of the Rhondda has been detected, although this area has been notable for strong earthquakes in the past.

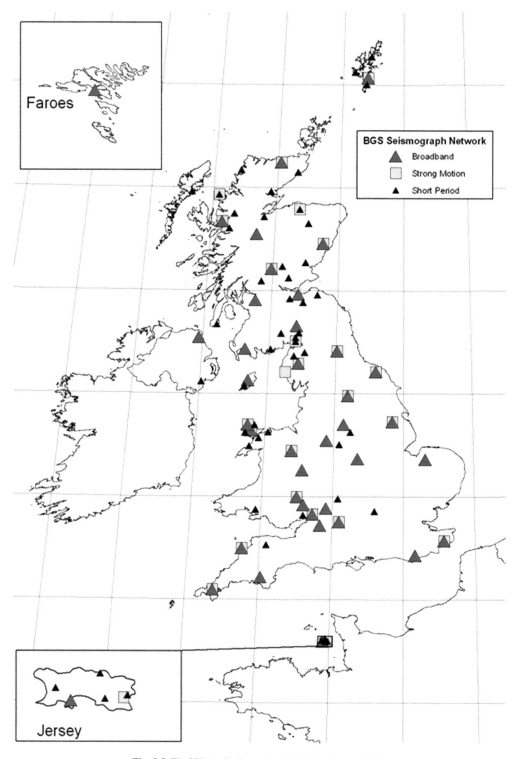

Fig. 2.5. The UK monitoring network in 2012 (Baptie 2012).

2.8 Remedial action

On those occasions when antiseismic design is justified in the UK, the actions needed are a matter for the specialist earthquake engineer and are outside the scope of this volume. However, it is worth remarking that in recent British earthquakes that have caused damage (the most notable example being the 28 April 2007 Folkestone earthquake), the predominant type of damage has been the fall of parts of old (Victorian or Edwardian) chimney stacks in poor repair. This causes secondary damage as stones from the chimneys either cascade onto the roof, damaging slates as they fall or, in the worst cases, actually fall through the roof into the interior of the house. There was a conspicuous absence of damage to modern buildings. It is therefore sensible for owners of property of this age to ensure that chimney stacks are maintained in good repair. The same can be said of architectural ornaments, such as might be found on churches or municipal buildings. These are naturally vulnerable to earthquake shaking and, again, may cause secondary damage if dislodged.

2.9 Limits to earthquake hazard in the UK

For obvious reasons it would be useful to know what might be the upper limits on earthquakes that could be experienced in the UK, but this is an intractable problem and only speculative answers are found.

At the lower end of estimates, Ambraseys & Jackson (1985) suggested that British earthquakes do not exceed 5.5 M_s (c. 5.5 M_w), given that nothing larger has ever been observed in the UK historical record. On the other hand, in the course of the major European project SHARE, Meletti et al. (2011) considered that it was impossible to completely rule out a 7.5 M_w event occurring anywhere in northern Europe, including the British Isles.

It is striking that all the larger UK earthquakes have been offshore, either in the North Sea, the Dover Straits or in the region of the Channel Islands. The largest UK earthquake onshore was the 19 July 1984 Lleyn Peninsula event (5.4 M_L, a little over 5 M_w). The tendency for the larger British earthquakes all to be offshore is statistically significant; if the larger events occurred randomly over the area of the UK and surrounding waters, it is improbable that none would be onshore (Musson 2007).

The contrary argument is that experience elsewhere shows that strong intraplate earthquakes can have extremely long recurrence intervals, and that no extrapolation from the rather short (in geological terms) historical catalogue can be validly made (e.g. Crone et al. 2003). A number of studies have been made of the difficulty of anticipating large intraplate earthquakes, starting with a ground-breaking paper by Sykes (1978) and a major study into the seismicity of stable continental regions (SCRs) by Johnston et al. (1994). Both these studies concluded that large SCR events were more likely to occur in certain tectonic settings, particularly failed rifts (the case for New Madrid, Missouri) and passive margins.

The latter setting is compatible with what may have been the largest UK earthquake in historical times on 19 September 1508. The documentation for this event is very slender, and condenses to the facts that the earthquake was felt throughout Scotland and England, was alarming and was felt particularly in churches. No damage is recorded. It is argued by Musson (2008a, b) that, given historical and geographical constraints, the information available is consistent with what might be expected from an earthquake similar to the 1929 Grand Banks (Newfoundland) earthquake (7.0 M_w) if it occurred on the continental slope NW of Scotland (e.g. on the Geikie Escarpment). The 1929 isoseismals, transferred to such an epicentre, would indicate damage only in areas with virtually no written records. The nearest cultural centres (Glasgow, Stirling) would experience intensity of around 5 EMS: alarming, but not damaging. Long-period motion from a large earthquake would explain why the tallest buildings of the period (churches) were more affected. An alternative explanation, that the earthquake was a Viking Graben event not larger than 5.5 M_w, also fits the facts except for the perceptibility in churches (Fig. 2.6).

Whether the conjecture that the 1508 earthquake was around 7 M_w in magnitude is correct or not, it is still conceivable that such an earthquake could occur anywhere on the

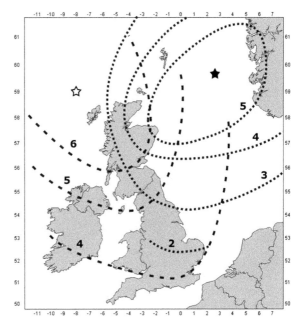

Fig. 2.6. Two possible reconstructions of the 19 September 1508 earthquake from Musson (2008b): open star, possible location on Geikie escarpment, with conjectural isoseismals (dashed lines); solid star, historical epicentre of the 1927 Viking Graben event with actual isoseismals (dashed lines).

passive margin and shake the entire British Isles. The actual probability of this is very low, however.

It is likely that 7 M_w is around the upper limit of what could be expected as the extreme event in the UK. The largest historical earthquake in northern Europe was the 1356 Basel (Switzerland) earthquake, the magnitude of which has been assessed as 6.6 M_w (SED 2009). Great earthquakes (magnitude 8 M_w or larger) are simply a physical impossibility for the UK, as there are no fault sources large enough that could credibly host such large earthquakes.

Looking instead at the possibility of strong shaking, this is to some extent decoupled from considerations of earthquake magnitude, perhaps surprisingly. Moderate or even small earthquakes can still produce high levels of ground motion in a localized area if they are shallow, or simply from aleatory variability in ground motion (e.g. due to the complexity of earthquake wave transmission, certain spots may experience a focusing effect). To some degree the difference between a large earthquake and a moderate earthquake is that strong shaking from the former is distributed over a larger area, rather than being necessarily stronger. An extreme case is the Eugowra (Australia) earthquake of 1994. Although only 4.1 M_L (c. 3.7 M_w) in magnitude, the recorded PGA from this earthquake was 0.97 g (Gibson et al. 1995). This extremely high value was not accompanied by significant damage, as the peak value occurred as a high-frequency spike in the record with little associated energy. The highest recorded PGA for a British earthquake was 0.1 g from the 2007 Folkestone earthquake, but given that strong-motion instruments have not been running for a long time in the UK and are still fairly few in number, it is certain that higher PGA values have been generated in past earthquakes in the UK. Scott (1977) estimated a PGA of around 0.2 g for the 1884 Colchester earthquake, but this is based on macroseismic observations and may not be very reliable.

It may seem strange that the value of 0.1 g recorded at Folkestone is so much higher than the 1 in 2475 annual probability hazard value for Folkestone shown in Figure 2.4, but this is the nature of probability; the chance of observing 0.1 g somewhere in the UK in any year is much higher than the probability of observing it at one specific place. As a simple analogy, imagine rolling ten dice of different colours. The probability of rolling 6 on the red die is 1 in 6 (17%), but the chance of rolling 6 on some dice is 94%. Whichever die it was, the chance of throwing 6 on that die was still 1 in 6.

The highest assessed intensity in a British earthquake is 8 EMS (heavy damage, many ordinary brick buildings would suffer large and extensive cracks in walls with fall of complete chimneys). This has been observed twice in the last 400 years, in both cases in earthquakes that were shallow in focus rather than large. These were the Barrow-in-Furness earthquake of 1865 and the Colchester earthquake of 1884. The 1865 event wrecked the small village of Rampside, but was barely felt 10 km away.

Intensity 8 EMS may have been reached in some early earthquakes, but the descriptions are not adequate to assess the damage accurately. The earthquake of 15 April 1185 reportedly caused the partial collapse of Lincoln cathedral as well as many houses. The condition of the building at the time of the shock is not known. Nevertheless, given the vulnerability of the historical fabric of many of Britain's cathedrals, the possibility of earthquake damage in the future, with loss of cultural heritage, should not be dismissed.

2.10 Actions to take in an earthquake

Many sets of guidance notes have been published regarding what to do in an earthquake, generally with respect to large damaging earthquakes of the sort that might be encountered overseas but are very unlikely in the UK. These do not always give a consistent message; for instance, is it best to hide under a table or not? This is often recommended, but it is possible that the table may simply collapse and provide no benefit. Similarly, there is really no hard and fast rule about exiting a building during an earthquake. Sometimes this can be done safely, and is the safest course; at other times it is the most dangerous thing to do.

In Britain, because of the nature of typical damage patterns in moderate earthquakes, running outside is probably a bad idea as there is a good chance of being hit by falling stones from the roof. No one in recorded history in the UK has been killed by being crushed inside a collapsed building, although there was a case in the Market Rasen earthquake of 2008 of someone suffering injury from part of a chimney falling through the roof into the house. In addition, it is not unknown for people attempting to run during an earthquake to fall and suffer injury that way. It is important to be careful of the possibility that heavy objects may fall from shelves and cause injury. If outside, stay away from buildings or steep slopes.

After an earthquake in the UK, information is quickly available from the British Geological Survey at http://quakes.bgs.ac.uk. In cases of damaged domestic property, other than in the case of very minor damage (e.g. slight cracks in plaster), the building should not be used until it has been cleared by the emergency services. Managers of large buildings or infrastructure in the worst affected areas should arrange for safety checks to be made as soon as possible after the earthquake.

Glossary

Aftershock: A small earthquake occurring shortly after a larger earthquake in roughly the same location. A series of such events can be thought of as a settling-down process as the rocks accommodate to their new post-earthquake position.

Aleatory variability: Uncertainty arising from randomness in a natural system. For instance, the probability distribution of earthquake depths in a region can be known, but not the depth of the next earthquake.

Attenuation: Decay of earthquake ground motion with distance.

Catalogue: A chronological listing of earthquakes. Early catalogues were purely descriptive, that is, they gave the date of each earthquake and a textual description. Modern catalogues are usually numerate, that is, earthquakes are listed as a set of numerical parameters describing location, size, etc.

Depth: The vertical distance between the hypocentre and the Earth's surface.

Epicentre: The point on the Earth's surface above the hypocentre.

Epistemic uncertainty: Uncertainty arising from a lack of knowledge, for instance, whether a given fault is potentially seismogenic or not.

Event: A loose term used to represent an earthquake or any similar occurrence.

Failed rift: An area where at one point in geological history a continent started to break apart, but the break-up ceased.

Fault: A discontinuity in the rocks of the Earth's crust, where the crust has broken at some time in the past due to tectonic forces.

Fault style: The type of displacement that has occurred or is expected to occur on a particular fault. Strike-slip or transform faulting involves purely horizontal displacement. Normal faulting involves vertical movement under extensional forces. Reverse or thrust faulting involves movement under compressional forces. Oblique faulting is a mixture of vertical and horizontal displacement.

Felt area: The area of the Earth's surface over which the effects of an earthquake were perceptible.

Focal mechanism: The determination of the style of faulting that occurred in an earthquake (see fault style).

Forecast: Similar to a prediction, but expressed in vaguer terms (e.g. 'There is likely to be a great earthquake on the such-and-such fault in the next 20 years.')

Foreshock: A small earthquake immediately preceding a larger one. While this could conceivably give warning, there is no way to identify a foreshock as being different from any normal small earthquake except with hindsight.

Frequency: The number of peaks per second in a seismic wave. Frequency is in the inverse of period.

GMPE: Ground motion prediction equation. See 'ground motion model'.

Ground motion model: A set of equations and coefficients for estimating the likely ground motion that would be generated by an earthquake as a function of magnitude, distance fault style and site conditions. The estimated value will be the median of a log-normal distribution.

Hypocentre: The point at depth at which an earthquake initiates.

Intensity: A measure of the strength of earthquake shaking at a given place, made by applying a classification of the observed effects. The classification system is known as an intensity scale, of which the most recent (used in this paper) is the European Macroseismic Scale, which has a scale from 1 (not felt) to 12 (total devastation).

Intraplate: Referring to locations far away from plate boundaries.

Isoseismal: A line bounding the area within which the intensity from a particular earthquake was consistently equal to or higher than a given value.

Macroseismic: Pertaining to the observed (felt) effects of earthquakes.

Magnitude: Used as a measure of the 'size' of an earthquake, magnitude is an analogue of the amount of energy released, on a logarithmic scale proceeding in steps of 30 (more or less). Three versions of magnitude have been used in the UK: M_L, M_s and M_w (see separate entries). For any given British earthquake, the M_L value will be slightly higher than the M_w value, and the M_s and M_w value will be about the same.

M_L: Abbreviation for local magnitude scale, the original definition of magnitude as formulated by Richter. This is often used for British earthquakes as it is easy to calculate and is suitable for small events. However, it becomes unusable much above M_L 6, so is not suitable for international use in highly seismic regions.

M_s: Abbreviation for surface magnitude scale, calculated from measurements of surface waves on a seismogram. Only usable for earthquakes between about M_s 4 and 8. It was popular in the 1980s, but now practically obsolete.

M_w: Abbreviation for moment magnitude scale, based on the seismic moment of the earthquake rupture. Now the preferred scale in seismology, it is usable for earthquakes of all sizes.

Passive margin: The boundary between a continental land mass and oceanic crust that is part of the same plate, for example, the edge of the continental shelf in the Atlantic west of Ireland.

PGA: Peak ground acceleration, measured in g, cm/s^2, or m/s^2.

Period: The interval between peaks in seconds in a seismic wave. Period is the inverse of frequency.

Plate: Part of the Earth's crust, capable of moving cohesively with respect to other parts. Typically, these are the size of continents. Smaller plates are often referred to as microplates.

Prediction: A statement that an earthquake will occur in the future, giving date, location and magnitude (with some margins). It is sometimes stated that prediction should also give the probability of the earthquake occurring by chance (if the margins are very wide the prediction may be trivial), but there is no agreement as to how such probability should be calculated.

Recurrence interval: The average expected time between earthquakes of a given magnitude.

Return period: The inverse of the annual probability of some value of ground motion. Not to be confused with recurrence interval.

Rupture: The process of a fault breaking during an earthquake.

Seiche: Waves in water (typically a lake, river or harbour) caused directly by earthquake shaking. Not to be confused with a tsunami.

Seismic hazard: The probability that some specified degree of earthquake shaking will occur in a given period of time.

Seismic risk: The probability that some specified degree of earthquake damage will occur in a given period of time. Generally, risk = hazard × vulnerability.

Seismicity: The occurrence of earthquakes.

Seismogenic: Capable of producing earthquakes.

Seismogram: An instrumental, graphic record of an earthquake made by a seismometer.

Seismograph: A device for recording earthquakes that records the time of occurrence. In a strict sense, the name implies that time is recorded but not the exact ground movement, but it is common to use the term interchangeably with 'seismometer'.

Seismometer: A device for recording earthquakes that records the time of occurrence and also enables the exact movement of the ground to be determined.

Seismoscope: A device that indicates in some way that an earthquake has occurred, but without recording the time.

Sigma (δ): A parameter used for the log-normal scatter around the median in a ground motion model. It is the measure of the aleatory variability of ground motion, and contributes significantly to hazard estimates.

Slip: The displacement observed during an earthquake measured in the plane of the fault.

Tectonics: The large-scale movement and deformation of the crustal plates that make up the outermost layer of the planet, in response to convection movements in the mantle. Also plate tectonics.

Teleseism: A large, distant earthquake, or the recording of one.

Tremor: Journalistic term for a small earthquake. Also 'earth tremor'. Technically speaking, an earthquake is still an earthquake no matter how small, so this term is not used in scientific discourse except in the case of volcanic tremor, which is a period of continuous vibration that may precede a volcanic eruption.

Tsunami: A large wave in the sea or a lake caused by massive displacement of water, either from an earthquake violently displacing the sea bed, a landslide falling into the water, a submarine landslide or a meteorite impact.

Vulnerability: Specifically, the probability that a structure will be damaged given a certain strength of earthquake shaking; loosely, the propensity of a building to be damaged by earthquake shaking.

Data sources and further reading

Earthquake data for the UK, both up to the minute and historical, can be obtained from the BGS web pages at http://quakes.bgs.ac.uk. For European data see http://www.emsc-csem.org, and for world data see http://earthquake.usgs.gov/regional/neic/.

For an introduction to seismology for the general reader, *The Million Death Earthquake* by Roger Musson (Palgrave Macmillan, 2012) covers most aspects. There is a lack of approachable works on seismic hazard. *Earthquake Hazard Analysis* by Leon Reiter (Columbia University Press, 1990) is now very dated, but still worth reading.

This chapter was composed in 2013 and does not represent advances in knowledge since that date.

References

ABRAHAMSON, N.A. & SILVA, W.J. 2008. Summary of the Abrahamson & Silva NGA ground-motion relations. *Earthquake Spectra*, **24**, 67–98.

AMBRASEYS, N.N. & MELVILLE, C.P. 1983. The seismicity of the British Isles and the North Sea. Report. SERC Marine Technology Centre, London.

AMBRASEYS, N.N. & JACKSON, D.D. 1985. Long-term seismicity in Britain. *Earthquake Engineering in Britain*, Thomas Telford, London, 49–66.

ARUP 1993. *Earthquake Hazard and Risk in the UK*. Department of Environment, London.

ARVIDSSON, R., GRÜNTHAL, G. & SHARE WORKING GROUP ON THE SEISMIC SOURCE ZONE MODEL 2011. Compilation of existing regional and national seismic source zones. SHARE Project Report, Deliverable D3.1.

BAPTIE, B. 2010. State of stress in the UK from observations of local seismicity. *Tectonophysics*, **482**, 150–159, https://doi.org/10.1016/j.tecto.2009.10.006

BAPTIE, B. 2012. *UK Earthquake Monitoring 2011/2012: Twenty-Third Annual Report*. British Geological Survey, Edinburgh.

BENDER, B.K. & PERKINS, D.M. 1987. *SEISRISK III: a Computer Program for Seismic Hazard Estimation*. Bulletin. USGS, Golden.

BOMMER, J.J. & ABRAHAMSON, N. 2006. Why do modern probabilistic seismic hazard analyses often lead to increased hazard estimates? *Bulletin of the Seismological Society of America*, **96**, 1967–1977, https://doi.org/10.1785/0120060043

BROWITT, C.W.A. 1991. UK earthquake monitoring 1989/90: BGS Seismic Monitoring and Information Service. British Geological Survey Technical Report **WL/90/13**, Edinburgh.

BROWITT, C.W.A., BURTON, P.W. & LIDSTER, R. 1976. *Seismicity of the Inverness region*. Global Seismology Unit Report. Institute of Geological Sciences, Edinburgh.

BUDNITZ, R.J., APOSTOLAKIS, G., BOORE, D.M., CLUFT, L.S., COPPERSMITH, K.J., CORNELL, C.A. & MORRIS, P.A. 1997. Recommendations for probabilistic seismic hazard analysis: guidance on uncertainty and use of experts. Report. US Nuclear Regulatory Commission, Washington DC.

BURTON, P.W. & BROWITT, C.W.A. 1976. Comparable accelerograph records for the Inverness region. Global Seismology Unit Report No **76**, Institute of Geological Sciences, Edinburgh.

BURTON, P.W. & NEILSON, G. 1980. *Annual Catalogues of British Earthquakes Recorded on LOWNET (1967–1978)*. Institute of Geological Sciences Seismological Bulletin No 7. HMSO, London.

BURTON, P.W., MCGONIGLE, R. & NEILSON, G. 1981. Seismicity and seismic risk evaluation, Nigg Bay and St. Fergus. Global Seismology Unit Report **145**. Institute of Geological Sciences, Edinburgh.

BURTON, P.W., MUSSON, R.M.W. & NEILSON, G. 1984. Studies of historical British earthquakes. Global Seismology Report **284**. British Geological Survey, Edinburgh.

CHADWICK, R.A., PHARAOH, T.C., WILLIAMSON, J.P. & MUSSON, R.M.W. 1996. Seismotectonics of the UK. Technical Report **WA/96/3C**. British Geological Survey, Keyworth.

CORNELL, C.A. 1968. Engineering seismic risk analysis. *Bulletin of the Seismological Society of America*, **58**, 1583–1606.

COTTON, F., SCHERBAUM, F., BOMMER, J.J. & BUNGUM, H. 2006. Criteria for selecting and adjusting ground-motion models for specific target regions: application to central Europe and rock sites. *Journal of Seismology*, **10**, 137–156, https://doi.org/10.1007/s10950-005-9006-7

CRAMPIN, S., JACOB, A.W.B., MILLER, A. & NEILSON, G. 1970. The LOWNET radio-linked seismometer network in Scotland. *Geophysical Journal of the Royal Astronomical Society*, **21**, 207–216, https://doi.org/10.1111/j.1365-246X.1970.tb01776.x

CRONE, A.J., DE MARTINI, P., MACHETTE, M.N., OKUMURA, K. & PRESCOTT, J.R. 2003. Paleoseismicity of two historically quiescent faults in Australia: implications for fault behavior in stable continental regions. *Bulletin of the Seismological Society of America*, **93**, 1913–1934, https://doi.org/10.1785/0120000094

DAVISON, C. 1902. The Inverness Earthquake of September 18th, 1901, and its accessory shocks. *Quarterly Journal of the Geological Society*, **58**, 377–398, https://doi.org/10.1144/GSL.JGS.1902.058.01-04.27

DAVISON, C. 1904. The Carnarvon earthquake of June 19th 1903. *Quarterly Journal of the Geological Society*, **60**, 233–242, https://doi.org/10.1144/GSL.JGS.1904.060.01-04.19

DAVISON, C. 1924. *A History of British Earthquakes*. Cambridge University Press, Cambridge.

DAWSON, A.G. 1992. *Ice Age Earth*. Routledge, London.

DAWSON, A.G., LONG, D., SMITH, D.E., SHI, S. & FOSTER, I.D.L. 1993. Tsunamis in the Norwegian Sea and North Sea caused by the Storegga submarine landslides. *In*: TINTI, S. (ed.) *Tsunamis in the World*. Kluwer Academic Publishers, Dordrecht, 31–42.

FINE, I.V., RABINOVICH, A.B., BORNHOLD, B.D., THOMSON, R.E. & KULIKOV, E.A. 2005. The Grand Banks landslide-generated tsunami of November 18, 1929: preliminary analysis and numerical modelling. *Marine Geology*, **215**, 45–57, https://doi.org/10.1016/j.margeo.2004.11.007

GIBSON, G., WESSON, V. & JONES, T. 1995. Strong motion from shallow intraplate earthquakes. *In: Proceedings of the Pacific Conference on Earthquake Engineering*, Melbourne, Australia, 20–22 November 1995, 185–193.

GRÜNTHAL, G. (ed.) 1998. *European Macroseismic Scale 1998, vol 15. Cahiers du Centre Européen de Géodynamique et de Seismologie*. Conseil de l'Europe, Luxembourg.

GRÜNTHAL, G., BOSSE, C. ET AL. 1996. Joint seismic hazard assessment for the central and western part of GSHAP Region 3 (Central and Northwest Europe). *In*: THORKELSSON, B. (ed.) *Seismology in Europe*. Icelandic Met Office, Reykjavik, 339–342.

GRÜNTHAL, G., WAHLSTRÖM, R. & STROMEYER, D. 2009. The unified catalogue of earthquakes in central, northern, and northwestern Europe (CENEC) – updated and expanded to the last millennium. *Journal of Seismology*, **13**, 517–541, https://doi.org/10.1007/s10950-008-9144-9

GUTENBERG, B. & RICHTER, C.F. 1936. Magnitude and energy of earthquakes. *Science*, **83**, 183–185, https://doi.org/10.1126/science.83.2147.183

HALCROW 1990. *Research Contract of Seismic Risk to UK Dams: Guidance Document for Engineers: Concrete and Masonry Dams*. Halcrow, Swindon.

IAEA 1991. Earthquakes and associated topics in relation to nuclear power plant siting. International Atomic Energy Agency Report **50-SG-S1**. Vienna.

IRVING, J. 1982. Earthquake hazard. Report. **CEGB/GDCD Report C/JI/SD/152.0/R019**, Barnwood.

JACKSON, P.D., GUNN, D.A. & LONG, D. 2004. Predicting variability in the stability of slope sediments due to earthquake ground motion in the AFEN area of the western UK continental shelf. *Marine Geology*, **213**, 363–378, https://doi.org/10.1016/j.margeo.2004.10.014

JIMÉNEZ, M.J., GIARDINI, D., GRÜNTHAL, G. & SESAME WORKING GROUP 2001. Unified seismic hazard modeling throughout the Mediterranean region. *Bolletino di Geofisica Teorica ed Applicata*, **42**, 3–18.

JOHNSTON, A.C., COPPERSMITH, K.J., KANTER, L.R. & CORNELL, C.A. 1994. The earthquakes of stable continental regions. EPRI Report. Electric Power Research Institute, Palo Alto.

KERRIDGE, D.J. (ed.) 2005. *The Threat Posed by Tsunami to the UK*. Study Commissioned by Defra Flood Management. HMSO, London.

KOLDERUP, C.F. 1930. Jordskjælv I Norge 1926–1930. *Bergens Museum Aarbok*, **6**.

LE BRETON, E., COBBOLD, P. & ZANELLA, A. 2013. Cenozoic reactivation of the Great Glen Fault, Scotland: additional evidence and possible causes. *Journal of the Geological Society*, **170**, 403–415, https://doi.org/10.1144/jgs2012-067

LEES, G. 1957. The East Midlands earthquake of February 1957. *East Midlands Geographer*, **7**, 52–55.

LILWALL, R.C. 1976. *Seismicity and Seismic Hazard in Britain*. IGS Seismological Bulletin. Institute of Geological Sciences, UK.

MAKROPOULOS, K.C. 1978. *The Statistics of Large Earthquake Magnitude and an Evaluation of Greek Seismicity*. Edinburgh University, Edinburgh.

MCGUIRE, R.K. 1976. FORTRAN computer program for seismic risk analysis. USGS Open File Report.

MELDOLA, R. & WHITE, W. 1885. Report on the East Anglian earthquake. Essex Field Club Special Memoirs.

MELETTI, C., D'AMICO, V. & MARTINELLI, F. 2011. Homogeneous determination of maximum magnitude. *SHARE Project Report*, Deliverable D3.3.

MILNE, W.G. & DAVENPORT, A.G. 1969. Distribution of earthquake risk in Canada *Bulletin of the Seismological Society of America*, **59**, 729–754.

MONTESSUS DE BALLORE M.F. 1896. Seismic phenomena of the British Empire. *Quarterly Journal of the Geological Society*, **52**, 651–668, https://doi.org/10.1144/GSL.JGS.1896.052.01-04.41

MUSSON, R.M.W. 1991. The Earl's Burn dam burst of 1839: an earthquake triggered dam failure in the UK? *Dams & Reservoirs*, **1**, 20–23.

MUSSON, R.M.W. 1993. Comrie: a historical Scottish earthquake swarm and its place in the history of seismology. *Terra Nova*, **5**, 477–480, https://doi.org/10.1111/j.1365-3121.1993.tb00288.x

MUSSON, R.M.W. 1994. A catalogue of British earthquakes. Global Seismology Report **WL/94/04**. British Geological Survey, Edinburgh.

MUSSON, R.M.W. 1996. The seismicity of the British Isles. *Annali di Geofisica*, **39**, 463–469.

MUSSON, R.M.W. 1997. Seismic hazard studies in the UK: source specification problems of intraplate seismicity. *Natural Hazards*, **15**, 105–119, https://doi.org/10.1023/A:1007970907854

MUSSON, R.M.W. 1998. The Barrow-in-Furness earthquake of 15 February 1865: liquefaction from a very small magnitude event. *Pure and Applied Geophysics*, **152**, 733–745, https://doi.org/10.1007/s000240050174

MUSSON, R.M.W. 2003. Fatalities in British earthquakes. *Astronomy and Geophysics*, **44**, 14–16.

MUSSON, R.M.W. 2004. An intensity hazard map for the UK. *In*: *ESC XXIX General Assembly*, Potsdam, 2004. Abstracts, p. 155.

MUSSON, R.M.W. 2005. Faulting and hazard in low seismicity areas. *In*: OECD (ed.) *Stability and Buffering Capacity of the Geosphere for Long-Term Isolation of Radioactive Waste*. Nuclear Energy Agency, Paris, 61–66.

MUSSON, R.M.W. 2007. British earthquakes. *Proceedings of the Geologists' Association*, **118**, 305–337, https://doi.org/10.1016/S0016-7878(07)80001-0

MUSSON, R.M.W. 2008a. The seismicity of the British Isles to 1600. British Geological Survey Technical Report **OR/08/049**, Edinburgh.

MUSSON, R.M.W. 2008b. The case for large (M > 7) earthquakes felt in the UK in historical times. *In*: FRÉCHET, J., MEGHRAOUI, M. & STUCCHI, M. (eds) *Historical Seismology: From the Archive to the Waveform*. Springer, Dordrecht, 187–208.

MUSSON, R.M.W. 2011. Assessment of activity rates for seismic source zones. SHARE Project Deliverable 3.7a, Zurich

MUSSON, R.M.W. 2012a. *The Million Death Earthquake*. Palgrave Macmillan, New York.

MUSSON, R.M.W. 2012b. PSHA validated by quasi observational means. *Seismological Research Letters*, **83**, 130–134, https://doi.org/10.1785/gssrl.83.1.130

MUSSON, R.M.W. 2012c. A introduction to SSHAC. *SECED Newsletter*, **23**, 1–4.

MUSSON, R.M.W. 2013. A history of British seismology. *Bulletin of Earthquake Engineering*, **11**, 715–861, https://doi.org/10.1007/s10518-013-9444-5

MUSSON, R.M.W. & HENNI, P.H.O. 2001. Methodological considerations of probabilistic seismic hazard mapping. *Soil Dynamics and Earthquake Engineering*, **21**, 385–403, https://doi.org/10.1016/S0267-7261(01)00020-3

MUSSON, R.M.W. & SARGEANT, S.L. 2007. Eurocode 8 seismic hazard zoning maps for the UK. British Geological Survey Technical Report, Edinburgh.

MUSSON, R.M.W. & WINTER, P.W. 1996. Seismic hazard of the UK. AEA Technology Report **AEA/CS/16422000/ZJ745/005**, Risley.

MUSSON, R.M.W., NEILSON, G. & BURTON, P.W. 1990. *Macroseismic Reports on Historical British Earthquakes XIV: 22 April 1884 Colchester*. British Geological Survey, Edinburgh.

MUSSON, R.M.W., LONG, D., PAPPIN, J.W., LUBKOWSKI, Z.A. & BOOTH, E. 1997. UK Continental Shelf Seismic Hazard. Offshore Technology Report **OTH 93 416**. Health and Safety Executive, Norwich.

O'REILLY, J.P. 1884. Catalogue of the earthquakes having occurred in Great Britain and Ireland during historical times. *Transactions of the Royal Irish Academy*, **28**, 285–316.

PAGE, R.A. & BASHAM, P.W. 1985. Earthquake hazards in the offshore environment. USGS Bulletin 1630.

PAPADOPOULOS, G.A. & LEFKOPOULOS, G. 1993. Magnitude-distance relations for liquefaction in soil from earthquakes. *Bulletin of the Seismological Society of America*, **83**, 925–938.

PRINCIPIA MECHANICA LTD 1982. *British Earthquakes*. Principia Mechanica Ltd, Cambridge.

RIETBROCK, A, STRASSER, F.O. & EDWARDS, B. 2013. A stochastic earthquake ground-motion prediction model for the United Kingdom. *Bulletin of the Seismological Society of America*, **103**, 57–77, https://doi.org/10.1785/0120110231

SCHWEIZERISCHER ERDBEBENDIENST (SED) 2009. Erdbebenkatalog ECOS-09. http://hitseddb.ethz.ch:8080/ecos09/index.html

SCOTT, R.F. 1977. The Essex earthquake of 1884. *Earthquake Engineering and Structural Dynamics*, **5**, 145–155, https://doi.org/10.1002/eqe.4290050204

SEISMIC HAZARD WORKING PARTY 1987. *Report on Seismic Hazard Assessment: Hinkley Point*. CEGB, Barnwood.

SOIL MECHANICS LTD 1982. *Reassessment of UK Seismicity Data*. Report by Soil Mechanics Ltd., Bracknell.

STEIN, S. 2007. Approaches to continental intraplate earthquake issues. *In*: STEIN, S. & MAZZOTTI, S. (eds) *Continental Intraplate Earthquakes: Science, Hazard, and Policy Issues*. Geological Society of America, Boulder CO, 1–16.

SYKES, L.R. 1978. Intraplate seismicity, reactivation of preexisting zones of weakness, alkaline magmatism, and other tectonism postdating continental fragmentation. *Reviews of Geophysics and Space Physics*, **16**, 621–688, https://doi.org/10.1029/RG016i004p00621

TILLOTSON, E. 1974. Earthquakes, explosions, and the deep underground structure of the United Kingdom. *Journal of Earth Sciences*, **8**, 353–364.

USNRC 2012 *Practical Implementation Guidelines for SSHAC Level 3 and 4 Hazard Studies*. NUREG. US Nuclear Regulatory Commission, Washington DC.

VERSEY, H.C. 1939. The North Sea earthquake of 1931, June 7. *Monthly Notices of the Royal Astronomical Society Geophysical Supplement*, **4**, 416–423.

WINTER, P.W., MARROW, P.C. & MUSSON, R.M.W. 1996. A stochastic ground motion model for UK earthquakes. AEA Technology GNSR(DTI)/P(96)275 Milestone ECS 0263, AEA/16423530/R003.

WOO, G. 1983. Seismic hazard assessment for the Sellafield site in West Cumbria. Principia Mechanica Report No **175/83**, London.

WOO, G. 1988. UK uniform risk spectra. Hinkley Point 'C' Project Management Board Report No. **HPC-IP-096013**.

Chapter 3 Tsunami hazard with reference to the UK

David Peter Giles

Card Geotechnics Ltd, 4 Godalming Business Centre, Woolsack Way, Godalming, Surrey GU7 1XW, UK

Correspondence: davidg@cgl-uk.com

Abstract: Tsunami present a significant geohazard to coastal and water-body marginal communities worldwide. Tsunami, a Japanese word, describes a series of waves that, once generated, travel across open water with exceptionally long wavelengths and with very high velocities before shortening and slowing on arrival at a coastal zone. Upon reaching land, these waves can have a devastating effect on the people and infrastructure in those environments. With over 12 000 km of coastline, the British Isles is vulnerable to the tsunami hazard. A significant number of potential tsunami source areas are present around the entire landmass, from plate tectonic boundaries off the Iberian Peninsula to the major submarine landslides in the northern North Sea to more localized coastal cliff instability which again has the potential to generate a tsunami. Tsunami can be generated through a variety of mechanisms including the sudden displacement of the sea floor in a seismic event as well as submarine and onshore landslides displacing a mass of water. This review presents those impacts together with a summary of tsunami triggers and UK case histories from the known historic catalogue. Currently, apart from some very sensitive installations, there is very little in the UK in the way of tsunami management and mitigation strategies. A situation that should be urgently addressed both on a local and national level.

3.1 Introduction

Tsunami present a significant geohazard to coastal and water-body marginal communities worldwide. Tsunami, a Japanese word, describes a series of waves that, once generated, travel across open water with exceptionally long wavelengths (up to several hundred metres in deep water) and with very high velocities (up to 950 km h^{-1}) before shortening and slowing on arrival at a coastal zone (Dawson *et al.* 2004; Long 2017). On reaching land, these waves can have a devastating effect on the people and infrastructure in those environments. Until relatively recently the understanding of tsunami events and their historic catalogue had been quite poor. The 2004 Indian Ocean Boxing Day tsunami (Fig. 3.1) and the 2011 Tohoku event in Japan tragically brought this geohazard to the attention of the wider population and instigated a deeper investigation and research into this geological phenomenon.

Tsunamis can be generated through a variety of mechanisms, including the sudden displacement of the sea floor in a seismic event as well as submarine and onshore landslides displacing a mass of water. The UK, with its 12 429 km of coastline, is no less prone to the impact of tsunami as the Indian or Pacific oceans. In 2005 Defra commissioned a study precipitated by the Indian Ocean disaster to consider the potential impact on the UK from such events (Defra 2005; Richardson *et al.* 2006). This review presents those impacts, together with a summary of tsunami triggers and UK case histories from the known historic catalogue.

3.2 Tsunami geohazard

A tsunami is a series of (usually) large ocean waves generated by impulses from geophysical events such as earthquakes, landslides and volcanic eruptions on the ocean bottom or along the coastline. A displaced column of water generates waves, which in deep water have a long wavelength, low amplitude and high velocity. As these waves approach the shallower waters of a coastline the wave velocity decreases and the wavelength shortens, resulting in the wave amplitude increasing significantly. The consequence of these waves arriving at the shoreline is the inundation of low-lying areas causing flooding with a significant threat to life and damage to infrastructure. Figure 3.2 details the terminology utilized to characterize a tsunami arriving at a coastal environment. The inundation distance is defined as the horizontal distance inland that a tsunami penetrates, measured perpendicular to the coastline. The run-up elevation is the difference between the elevation of maximum tsunami penetration and the sea level at the time of the tsunami. Definitions of wave height are calculated as the difference between the elevation of the highest local water mark and the elevation of the sea level at the time of the tsunami (International Oceanographic Commission 2008).

Engineering Group Working Party (main contact for this chapter: D. P. Giles, CGL, 4 Godalming Business Centre, Woolsack Way, Godalming, Surrey GU7 1XW, UK, davidg@cgl-uk.com
From: GILES, D. P. & GRIFFITHS, J. S. (eds) 2020. *Geological Hazards in the UK: Their Occurrence, Monitoring and Mitigation – Engineering Group Working Party Report*. Geological Society, London. Engineering Geology Special Publications, **29**, 61–80,
https://doi.org/10.1144/EGSP29.3
© 2020 The Author(s). Published by The Geological Society of London. All rights reserved.
For permissions: http://www.geolsoc.org.uk/permissions. Publishing disclaimer: www.geolsoc.org.uk/pub_ethics

Fig. 3.1. Indian Ocean Tsunami, 2004. Aftermath in the coastal zone around Aceh, Indonesia, 2005 (Photo: AusAID, https://commons.wikimedia.org/wiki/File:Tsunami_2004_aftermath._Aceh,Indonesia,_2005._Photo-_AusAID_(10730863873).jpg, https://creativecommons.org/licenses/by/2.0/legalcode.

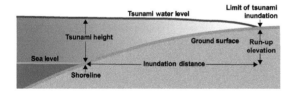

Fig. 3.2. Tsunami nomenclature (adapted from International Oceanographic Commission 2008).

3.3 Tsunami wave characteristics

In the event of an undersea disturbance, either through fault rupture, landslide or seafloor gas release, for example, a large body of water can undergo a vertical displacement. Gravity will act to counter this upwards force which will result in a wave propagating away from the original disturbance (Power & Leonard 2013). These waves are capable of travelling long distances with little energy dissipation and have long periods and wavelengths. Another key tsunami characteristic is that they involve the motion of the entire water column from the seabed to the sea surface. Shorter-period waves such as from wind action normally only involve the first tens of metres of the water column. This is significant in the total energy that can be transferred through a tsunami. The propagation velocity of a tsunami wave is a function of water depth. Truong (2012) provides a more detailed account of the experimental relationships. In shallow water the velocity of the wave can be defined:

$$V_s = \sqrt{gD}, \qquad (3.1)$$

where V_s is wave velocity (m s^{-1}), g is acceleration due to gravity (m s^{-2}) and D is water depth (m), providing that the

TSUNAMI HAZARD

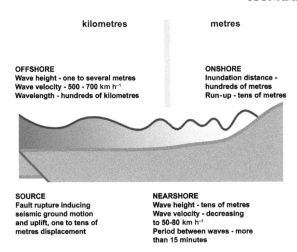

Fig. 3.3. Tsunami wave characteristics in offshore, nearshore and onshore environments.

wavelength is long compared with the water depth. For example, for a water depth of 4000 m and a gravitational acceleration field of 9.8 m s^{-2}, the wave velocity will be in the order of 200 m s^{-1} or 720 km h^{-1} (Truong 2012).

Wave amplitude in shallow or deep water can also be related to wave velocity by the equation:

$$\frac{A_s}{A_d} = \sqrt{\frac{V_d}{V_s}} \quad (3.2)$$

where A_s is amplitude of the wave in shallow water (m), A_d is amplitude of the wave in deep water (m), V_d is velocity of the wave in deep water (m s^{-1}) and V_s is velocity of the wave in shallow water (m s^{-1}).

Figure 3.3 compares the wave properties offshore and onshore for the varying water depths.

As the tsunami reaches the shallower nearshore waters, the wave velocity reduces and wave amplitude increases. In deep water the waves will be barely noticeable due to their very long wavelengths. The waves will start to bunch together in the nearshore, a process termed shoaling (Power & Leonard 2013), with the front of the wave train slowing first; the bunching results in an associated increase in wave height. As the velocity of a tsunami is related to water depth, the underlying bathymetric profile will control the wave propagation and have a significant influence.

3.4 Tsunami generation processes

Tsunamis can be generated via a variety of geophysical and other processes, sometimes operating in tandem, for example an initial seismic event triggering a submarine landslide generating a tsunami. There is a growing consensus that a submarine landslide contributed to the 2011 Tohoku tsunami as the earthquake faulting alone cannot fully explain the Tohoku tsunami run-up or waveforms (Tappin *et al.* 2014). The 1883 eruption of Krakatoa generated another substantial tsunami (Latter 1981), and there is much concern over other unstable volcanoes adjacent to the coast and their potential for flank collapse generating another tsunami, for example, the Cumbre Vieja volcano at La Palma, Canary Islands (Ward & Day 2001). Tsunami are typically generated by the mechanisms described in the following sections.

3.4.1 Tsunamigenic earthquakes

Offshore earthquakes that rupture the seafloor with a vertical displacement can displace the water column and generate a tsunami. These are most likely to occur in areas of active tectonic plate motion, usually subduction (Fig. 3.4). Up to 24 million people worldwide live in coastal communities and can be considered as being vulnerable from such events, particularly around the Pacific Rim, Hawaii and along the many SE Asian subducting plate boundaries.

3.4.2 Tsunamigenic landslides

Tsunami can be generated either by submarine (Fig. 3.5) or coastal landslides (Fig. 3.6) disturbing or entering a water body. Underwater tsunamigenic landslides can be generated by a seismic event, either simultaneously or with a time delay after the earthquake (Fig. 3.7). There are important distinctions in terms of tsunami wave arrival time after such seismic events (Løvholt *et al.* 2015). Submarine slides can range from quite modest events and volume, involving tens to hundreds of cubic metres, to giant underwater movements of considerable surface area and volume, involving 3500 km^3 in the case of the Storegga landslide. These landslides have the potential to be long run-out events, which can generate tsunami with a significant regional impact. Their potential occurrence is not uniformly distributed over the world's oceans, but they tend to occur where there are steep submarine slopes and thicker sediment piles, and where loads exerted by the local environment are high. Areas of deltaic or glacial sediments are particularly prone to such submarine slope failures. Such landslide-generated tsunami can also occur in lakes, fjords, inland seas, estuaries, deltas and undersea canyons (Locat & Lee 2009). The highest-ever recorded tsunami occurred after the 1958 Lituya Bay, Alaska, rock slide, which generated a tsunami of *c.* 162 m in height and destroyed forest up to maximum run-up height of 524 m (Heller & Hager 2014).

3.4.3 Tsunamigenic volcanism

Areas of volcanic activity adjacent to or underneath water bodies have the potential to generate tsunami waves through flank instability (Fig. 3.8), underwater eruptions, pyroclastic

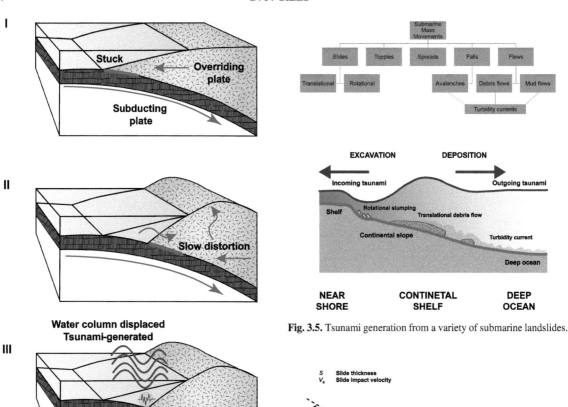

Fig. 3.5. Tsunami generation from a variety of submarine landslides.

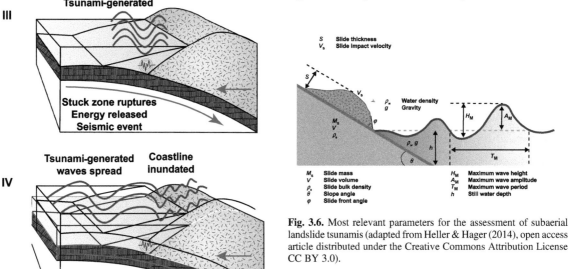

Fig. 3.6. Most relevant parameters for the assessment of subaerial landslide tsunamis (adapted from Heller & Hager (2014), open access article distributed under the Creative Commons Attribution License CC BY 3.0).

Fig. 3.4. Tsunami generation through seismic events occurring at a typical subduction zone.

flows entering water, caldera collapse, explosion-generated shock waves or volcano-tectonic earthquakes (Begét 2000; Paris 2015). This form of tsunami event tends to produce shorter wavelengths but with a greater areal dispersion with limited far-field effects (Dawson *et al.* 2004). Table 3.1 lists some examples of where tsunami have been recorded, probably generated through volcanic-related mechanisms.

3.4.4 Meteotsunami

Tsunami waves, as opposed to wind produced waves, have been recorded generated from high latitude atmospheric forcing (Šepić *et al.* 2015, 2018). Long-period ocean oscillations are generated by intense small-scale air pressure disturbances and can have far-field damaging effects. Long (2017) catalogued several UK instances of tsunami-like effects which

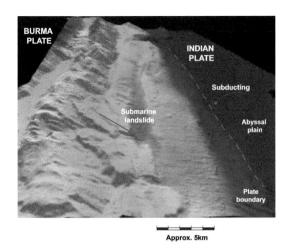

Fig. 3.7. Submarine landslide and debris visible along the Burma Plate Margin after the 2004 Indian Ocean earthquake (adapted from NOC, Southampton).

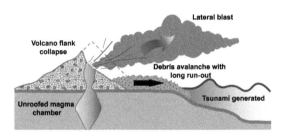

Fig. 3.8. Tsunami generated through volcanic flank collapse and associated debris avalanche and lateral blast.

Table 3.1. *Some examples of tsunamigenic volcanism (from Paris 2015)*

Date	Volcano source	Mechanism
1329	Etna, Sicily	Volcano-tectonic earthquake
1827	Avachinsky, Kamchatka, Russia	Volcano-tectonic earthquake
1857	Umboi, Papua New Guinea	Volcano-tectonic earthquake
1914	Sakurajima, Japan	Volcano-tectonic earthquake
1916	Stromboli, Italy	Volcano-tectonic earthquake
1792	Mount Mayuyama, Japan	Debris avalanche
1997, 2003	Soufrière Hills, Montserrat	Pyroclastic flows
1994	Rabaul, Papua New Guinea	Pyroclastic flows
1952	Myojin-sho, Japan	Submarine eruptions

can be attributed to certain meteorological conditions such as cyclones, hurricanes and thunderstorms, together with atmospheric pressure jumps and atmospheric gravity waves. These waves can have similar properties to geophysical tsunamis. There has been considerable debate as to whether the Bristol Channel floods of 1607 were a possible meteotsunami event (Bryant & Haslett 2002; Horsburgh & Horritt 2006). Eyewitness accounts reported a sudden flooding event under fair-weather conditions following a low-pressure storm event.

3.4.5 Other potential tsunami-generating mechanisms

Other underwater seafloor mechanisms have the potential to generate tsunami. The submarine release of gas hydrates such as methane (clathrates) contained within ocean floor sediments has the ability to displace a water column and generate a wave (Fig. 3.9). This can also have the effects of causing local submarine landslides, which again has wave-generation potential (Talling *et al.* 2014).

Asteroid impact or even large meteorites can also generate tsunami. These are extremely rare events, but have the capability of generating large wave amplitudes that are controlled by the diameter of the impacting asteroid (Dawson *et al.* 2004).

3.5 UK tsunami threat

Following the M_w 9.1 Indian Ocean earthquake and consequent devastating tsunami of 26 December 2004, the UK became reawakened to the threat posed by tsunami geohazards. Several studies were rapidly commissioned, complimented by academic research, to consider the global threat of tsunamis with specific reference to the potential impact on the UK (Dawson *et al.* 2004; Defra 2005, 2006; Richardson *et al.* 2006; Horsburgh *et al.* 2008). Key tsunami source areas were identified from where potential UK damaging events could be generated. Table 3.2 details these possible sources of tsunamis affecting the UK, their possible nature and potential impact (Defra 2005, 2006). Areas of the North Sea, Celtic Sea and offshore of Lisbon, Portugal and La Palma, Spain in the Canary Islands were investigated and modelled. The initial 2005 study was further enhanced to consider specific stretches of the UK coastline at risk from both a Lisbon-type event, which occurred after the Lisbon 1755 earthquake, and a nearshore North Sea event (Defra 2006; Richardson *et al.* 2006). Seven potential source zones (Fig. 3.10) were categorized in terms of their probability of occurrence, namely: UK coastal waters, NW European continental slope, plate boundary west of Gibraltar, Canary Islands, Mid-Atlantic Ridge, eastern North American continental slope, and the Caribbean. Table 3.3 details some notable UK wave-related events which may be tsunami (after Long & Wilson 2007; Long 2015, 2017).

Fig. 3.9. The submarine release of gas hydrates such as methane (clathrates) contained within ocean floor sediments has the ability to displace a water column and generate a wave. Image shows a crater formed by an underground gas well blowout in 1965 demonstrating the significant energy release from such an event (open access article is distributed under Creative Commons Attribution-Non Commercial-No Derivatives License 4.0 CC BY-NC-ND).

3.6 Notable tsunami events with a UK impact

3.6.1 *c.* 8150 BP Holocene Storegga submarine landslide and tsunami

Along the eastern and northern coasts of Scotland and at locations in the NE of England, sites have been investigated where a continuous layer of marine sediments can be identified (Table 3.3; Long 2015, 2017). These sediments have been interpreted as tsunami deposits and have been attributed to a major submarine landslide event that displaced approximately 3500 km^3 of sediment along the mid-Norwegian margin of the North Sea. This submarine landslide complex, termed the Holocene Storegga Slide (Fig. 3.11), has been extensively studied both in terms of potential tsunami propagation as well as the site of the Ormen Lange gas field (Dawson *et al.* 1988; Haflidason *et al.* 2004; Smith *et al.* 2004; Bondevik *et al.* 2005*a*; Bryn *et al.* 2005). The orthodox view is that this was a tsunamigenic landslide which has been dated as approximately 8150 calendar years BP (7250 ± 250 ^{14}C BP) from the offshore deposits, which equates to the 8.2 ka Holocene Cold Event (Haflidason *et al.* 2005). Other research has suggested a significant role in the slope failures of the submarine release of gas hydrates such as methane (clathrates) contributing to the tsunami, with the suggestion that gas-hydrate dissociation contributed to slope instability (Bugge *et al.* 1988; Mienert *et al.* 1998; Bünz *et al.* 2003). Given the volume of gas hydrates present in this area, the generating mechanism was probably a combination of the various effects, for example, gas-hydrate dissociation and earthquakes. Studies on Storegga have indicated that the *c.* 8150 BP event was only one of several failures that had occurred throughout the Pleistocene and Holocene, with one event approximately every

Table 3.2. *Possible sources of tsunamis affecting the UK, their possible nature and potential impact (after Defra 2005)*

Source	Nature and potential impact
UK coastal waters	Largest likely event: 6.5 M_S
	Probability of tsunamigenic event: very low
	Type of tsunami: local
	Likelihood of tsunami reaching UK if occurring: very high
	Likely coasts affected: east England and east Scotland
NW Europe continental slope	Largest likely event: 7+ M_S?
	Probability of tsunamigenic event: very low
	Type of tsunami: local
	Likelihood of tsunami reaching UK if occurring: very high
	Likely coasts affected: NW Scotland, north coast of Northern Ireland, Bristol Channel, Cornwall, south and west Wales
Plate boundary area west of Gibraltar	Largest likely event: 8.5 M_S
	Probability of tsunamigenic event: high
	Type of tsunami: transoceanic
	Likelihood of tsunami reaching UK if occurring: high
	Likely coasts affected: Cornwall, north Devon, south Wales
Canary Islands	Largest likely event: landslide (collapse of volcanic edifice)
	Probability of tsunamigenic event: low
	Type of tsunami: local or transoceanic, depending on nature of collapse
	Likelihood of tsunami reaching UK if occurring: moderate
	Likely coasts affected: Cornwall, north and south Devon
Mid-Atlantic Ridge	Largest likely event: 6.5 M_S
	Probability of tsunamigenic event: very low
	Type of tsunami: local
	Likelihood of tsunami reaching UK if occurring: very low
	Likely coasts affected: west Scotland, SW England
Eastern North America continental slope	Largest likely event: 7.5 M_S
	Probability of tsunamigenic event: very low
	Type of tsunami: local
	Likelihood of tsunami reaching UK if occurring: very low
	Likely coasts affected: west Scotland, SW England
Caribbean	Largest likely event: 8.3 M_S
	Probability of tsunamigenic event: high
	Type of tsunami: probably local
	Likelihood of tsunami reaching UK if occurring: very low
	Likely coasts affected: SW England

100 ka (Bryn *et al.* 2003; Solheim *et al.* 2005). More recent work (Watts *et al.* 2016) has suggested that the occurrence of landslides may be more frequent, which has significant implications for the associated tsunami threat to the UK and Norwegian coasts. Storegga-related field research has suggested that run-up heights decrease from north to south with the highest recorded up to *c.* 20 m or above in Shetland. Storegga-associated tsunami deposits have also been recorded in the Faroe Islands at heights of 14 m above present-day sea level, at 3–6 m in NE Scotland and at 3–13 m along the Norwegian coast (Bondevik *et al.* 2005*a*). The event would have had a significant impact on Doggerland in the North Sea, which would have been catastrophically flooded by the ensuing tsunami (Weninger *et al.* 2008). With the onset of high-resolution bathymetric profiling, numerous offshore submarine landslides and landslide complexes have been identified that may pose a UK tsunami threat (Long & Holmes 2001; Evans *et al.* 2005). Table 3.4 lists the major submarine landslide complexes in the North and Norwegian seas, and Figure 3.12 depicts their indicative location (see references for exact geographies).

3.6.2 c. 5500 BP Holocene Garth tsunami

From field evidence on Shetland, other significant North Sea tsunami events have been postulated (Bondevik *et al.* 2005*b*). Sites around Garth Loch and Loch of Benston (Fig. 3.13) have yielded potential tsunami deposits up to 0.65 m thick. A tsunami event with a run-up of greater than 10 m above sea level has been suggested, with no source other than the North and Norwegian seas being identified. This has implications for the developing wind farm installations and hydrocarbon platforms in this area, as well as the potential onshore damaging effects for repeat events.

1. UK Coastal Waters
2. NW European Continental Slope
3. Canary Islands
4. Plate Boundary West of Gibraltar
5. Mid Atlantic Ridge
6. Eastern North American Continental Slope
7. Caribbean

Fig. 3.10. Possible tsunami source zones with a potential UK impact as considered by Defra (adapted from Defra 2005, 2006).

3.6.3 AD 1755 Lisbon earthquake and tsunami

The largest historically recorded seismic event in Europe is considered to be the 1755 Lisbon Earthquake. Estimated to be a magnitude M_s 8.5 (possibly M_w 9.0) event and measuring X–XI on the Modified Mercalli Intensity Scale, the epicentre was in the vicinity of the Azores–Gibraltar boundary of the African and Eurasian plates around the Azores–Gibraltar Fault Zone (Baptista *et al.* 1998; Chester 2001; Muir-Wood & Mignan 2009).

The earthquake struck at 09:50 local time, with a tsunami arriving onshore at Lisbon some 20 minutes later. The impacts in the UK resulting from the earthquake and subsequent tsunami were first noted at 11:00 with reports of seiches (standing waves in an enclosed or partially enclosed body of water) in various harbours, lakes and ponds across the country (Long 2017). In the SW, wave trains were also reported with noticeable sea-level variations over several hours. Lisbon-related tsunami deposits have also been identified (Foster *et al.* 1993; Dawson *et al.* 2000; Banerjee *et al.* 2001). Tsunami travel times from Lisbon to Penzance (315 minutes) and Lisbon to Plymouth (390 minutes) have been

Table 3.3. *Some notable UK wave-related events which may be tsunami (after Long & Wilson 2007; Long 2015, 2017)*

Date	Location of recorded evidence	Comment	Further reading
c. 8150 BP	Eastern and northern coasts of Scotland	Tsunami event	Smith *et al.* (2004)
c. 5500 BP	Shetland	Uncertain tsunami event	Bondevik *et al.* (2005b)
c. 1500 BP	Shetland	Uncertain tsunami event	Bondevik *et al.* (2005b)
1014	North Wales	Unlikely tsunami event	Haslett & Bryant (2007)
11 November 1099	St Michaels's Mount, Cornwall	Unlikely tsunami event	Anglo-Saxon Chronicle (n.d.)
6 April 1580	English Channel	Uncertain tsunami event	Haslett & Bryant (2007)
30 January 1607	Bristol Channel	Unlikely tsunami event	Haslett & Bryant (2007)
2 July 1749	Milford Haven, Pembrokeshire	Unlikely tsunami event	Scots Magazine (1749)
1 November 1755	UK-wide	Tsunami event	Foster *et al.* (1991, 1993)
31 May 1759	Lyme Regis, Dorset	Unlikely tsunami event	Dawson *et al.* (2000)
31 March 1761	SW Cornwall	Tsunami event	Borlase 1762
9 or 10 August 1802	SW England	Uncertain tsunami event	Dawson *et al.* (2000)
31 May 1811	Plymouth, Devon	Unlikely tsunami event	Dawson *et al.* (2000)
13 September 1821	Plymouth, Devon, Truro, Cornwall	Uncertain tsunami event	Anon. (1821)
5 July 1843	Penzance, Cornwall and Plymouth, Devon	Unlikely tsunami event	Dawson *et al.* (2000)
23 May 1847	SW England	Unlikely tsunami event	Musson (1989)
5 June 1858	English Channel	Unlikely tsunami event	Newig & Kelletat (2011)
4 October 1859	North Cornwall and north Devon	Unlikely tsunami event	Dawson *et al.* (2000)
29 Sept 1869	Scilly and west Cornwall	Unlikely tsunami event	Dawson *et al.* (2000)
28 August 1883	Devonport and Portland, Devon	Tsunami event	Berninghausen (1968)
22 April 1884	Essex	Unlikely tsunami event	Haslett & Bryant (2008)
18 August 1892	Pembrokeshire	Unlikely tsunami event	Haslett & Bryant (2009)
31 December 1911	Folkestone, Kent	Tsunami event	Anon. (1912a, b)
24 January 1927	Eastern Scotland	Unlikely tsunami event	Tyrrell (1932)
25 November 1941	West Cornwall	Tsunami event	Dawson *et al.* (2000)
1 February 1953	Norfolk	Unlikely tsunami event	GTDB (n.d.)
23 May 1960	West Cornwall	Tsunami event	Van Dorn (1987)
28 February 1969	Scilly and west Cornwall	Tsunami event	Dawson *et al.* (2000)
26 May 1975	Scilly and west Cornwall	Tsunami event	Dawson *et al.* (2000)
27 December 2004	English Channel and Milford Haven	Tsunami event	Woodworth *et al.* (2005)
28 May 2008	Peterhead	Tsunami event	Glimsdal *et al.* (2010)

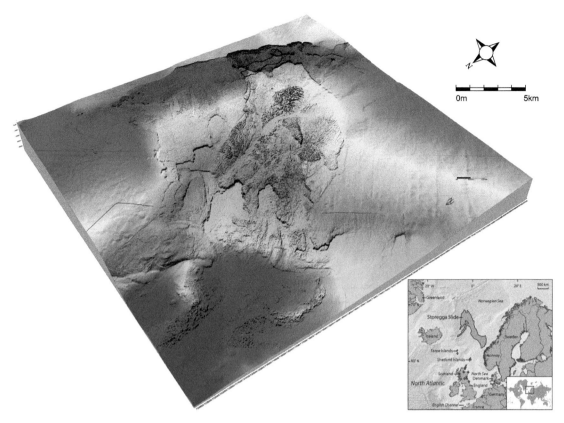

Fig. 3.11. Areal extent of the submarine Storegga landslide complex (adapted from http://www.bgs.ac.uk/discoveringGeology/hazards/land slides/sea.html and http://extras.springer.com/2005/978-3-540-24122-5/data).

projected based on contemporaneous Cornish observations (Baptista *et al.* 2003). Wave height was estimated as a maximum of 2.1 m at Penzance.

In 2005 Defra reviewed the Lisbon event for possible future tsunami impact scenarios (Defra 2005, 2006). Due to the complexity of the fault zones surrounding the plate boundary, predicting precise source locations and developing subsequent models is challenging (Fig. 3.14). In the Defra scenarios, four different source models were constructed based on the structural and fault regimes: (1) Gorringe Bank model, a single large thrust fault source; (2) West Iberian margin, a thrust fault system; (3) composite source, involving triggered fault rupture on two separate systems; and (4) Cadiz subduction model, a seismic event on a single subduction fault.

The Defra models (Defra 2005, 2006) suggest that slightly lower wave heights would be expected for such a 1755 Lisbon-style event. The research concluded that the Cornish coast would be at most risk from a repeat event of the 1755 magnitude and intensity, with wave heights typically in the range 1–2 m, although local amplification could potentially increase this to around 4 m. Travel times for the tsunami wave front would be in the region of 6–7 hours, sufficient time for management strategies to be implemented (assuming that they were in place and deliverable).

3.6.4 AD 1911 Abbot's Cliff failure, Folkestone

On 31 December 1911, a significant slope failure occurred at Abbot's Cliff to the east of Folkestone, Kent. An estimated 500 km^3 of chalk failed, principally as a rockfall (Hutchinson 2002; Long 2015, 2017). The majority of the chalk escarpment along this section of coast is protected by a large apron from the Folkestone Warren landslide complex (Fig. 3.15), but at Abbot's Cliff the chalk is exposed to coastal erosion and undercutting. The landslide debris from this failure propagated a tsunami which was recorded in the adjacent Folkestone Harbour. Such failures in these chalk cliffs are a relatively common occurrence (Williams *et al.* 2004). Figure 3.16 shows a more recent failure adjacent to Dover Harbour, where rock collapsed into the sea between Langdon Cliffs and South Foreland Lighthouse in 2012. These can be high-volume high-velocity events. There are many other examples of high rock cliffs that are prone to coastal erosion

Table 3.4. *Major submarine landslide complexes in the North and Norwegian seas (after Evans et al. 2005)*

Submarine landslide	Location	References
Trænadjupet Slide	North Vøring Margin	Laberg & Vorren (2000); Laberg *et al.* (2002)
Nyk Slide	North Vøring Margin	Lindberg *et al.* (2004)
Sklinnadjupet Slide	Vøring Margin	Rise *et al.* (2006)
Helland Hansen Slide	SE of Sklinnadjupet Slide	Leynaud *et al.* (2004)
Storegga Slide Complex	South of Vøring Margin	Bryn *et al.* (2005)
Tampen Slide	SE Nordic Sea Margin	Gafeira *et al.* (2010)
North Faroe Slide Complex	North Faroes Margin	Evans *et al.* (2005)
Miller Slide	Northern slope of Faroe–Shetland Channel	Long & Bone (1990)
Afen Slide	Faroe–Shetland Channel	Wilson *et al.* (2004)
GEM Raft	Western slope of Faroe–Shetland Channel	Long *et al.* (2003)
NE Sula Sgeir Slide	NE corner of Rockall Trough	Baltzer *et al.* (1998)
SW Sula Sgeir Slide	NE corner of Rockall Trough	Baltzer *et al.* (1998)
Geikie Slide	Geikie Escarpment	Strachan & Evans (1991)
Rockall Bank Mass Flow	Western margin of Rockall Trough	Elliott *et al.* (2010)
Peach Slide Complex	Northern slope Barra–Donegal Fan	Owen *et al.* (2010); Holmes *et al.* (1998)
Foreland Slide Complex	Southern slope Barra–Donegal Fan	Holmes (2003)
Porcupine Bank Mass Flow	SW flank of Porcupine Bank	Shannon *et al.* (2001)

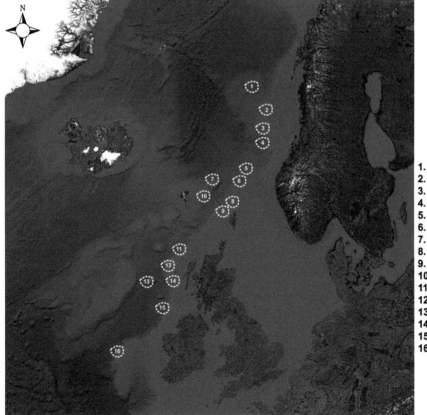

1. Trænadjupet Slide
2. Nyk Slide
3. Sklinnadjupet Slide
4. Helland Hansen Slide
5. Storegga Slide Complex
6. Tampen Slide
7. North Faroes Slide Complex
8. Miller Slide
9. Afen Slide
10. GEM Raft
11. Sula Sgeir Slides
12. Geikie Slide
13. Rockall Bank Mass Flow
14. Peach Slide Complex
15. Foreland Slide Complex
16. Porcupine Bank Mass Flow

Fig. 3.12. Indicative location of major submarine landslide complexes in the North and Norwegian seas.

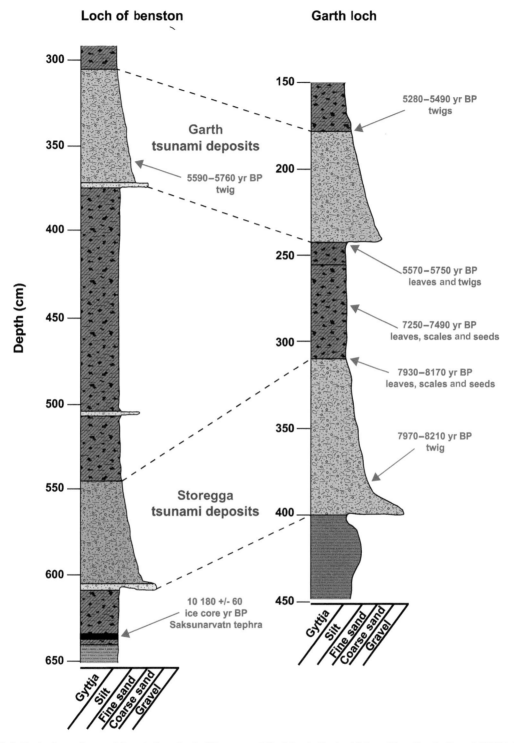

Fig. 3.13. Indicative logs of tsunami deposits from Loch of Benston and Garth Loch, Shetland (adapted from Bondevik *et al.* 2005*b*). Carbon 14 dates from retrieved organic matter. Gyttja, a sediment rich in organic matter deposited at the bottom of a eutrophic lake.

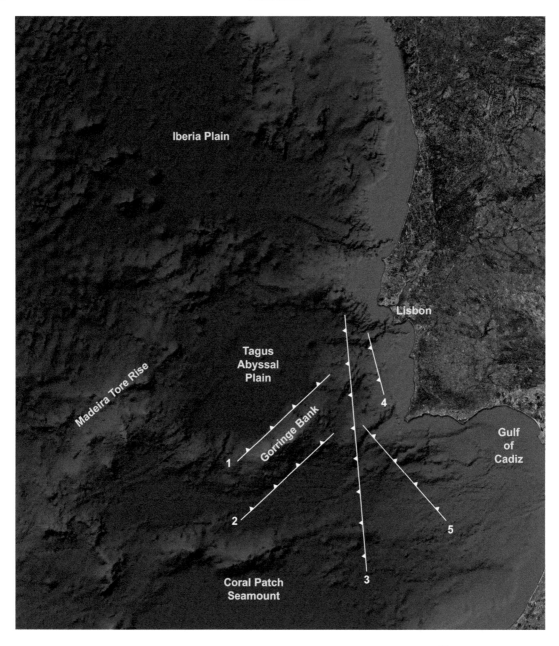

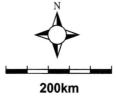

1. Johnston (1996)
2. Baptista *et al.* (1998) Model 1
3. Baptista *et al.* (1998) Model 2
4. Baptista *et al.* (1998) Model 3
5. Baptista *et al.* (1998) Model 3

Fig. 3.14. Indicative locations of potential source zones and fault models for a Lisbon-style earthquake generated tsunami (adapted from Defra 2005, 2006; Richardson *et al.* 2006).

Fig. 3.15. Slope instability at Folkestone Warren, Kent (photo credit: David Giles).

Fig. 3.16. Rockfall adjacent to Dover Harbour where chalk collapsed into the sea between Langdon Cliffs and South Foreland Lighthouse in 2012 (photo credit: David Giles).

and potential catastrophic failure all along the UK shoreline, all with the potential to generate a tsunami.

3.6.5 Other Dover Straits events

The Dover Straits is an area of known and recorded earthquake activity with seismic events generated along the Kent–Artois Shear Zone beneath the English Channel (Musson 1996, 2004; García-Moreno *et al.* 2015). The seismic background and risk in this area was extensively assessed for the Channel Tunnel given the local structural complexity (Varley 1996). Two of the largest recorded British earthquakes occurred in the area of the Dover Straits in 1382 and 1580, both estimated to be of local magnitude M_L 5.8 (Musson 1996). The fault activity in the epicentral area of the 1580 Dover Strait earthquake is associated with a major Variscan tectonic structure that traverses the Channel (García-Moreno *et al.* 2015). The faulting in this zone can be considered as being active, although seismic events are infrequent. The 1580 event had contemporary accounts of ground shaking of 6–7 minutes, possibly longer (Melville 1981). A historical review of tsunami in Britain since AD 1000 (Haslett & Bryant 2008) reported several potential tsunami events that may be associated with this fault activity. This level of event certainly has the potential to trigger coastal landslides such as the 1911 Abott's Cliff failure. Tsunami were certainly reported either from fault displacement or landsliding (Haslett & Bryant 2008). Relatively low-magnitude earthquakes occurring close to the coast have the potential to generate tsunami in nearshore UK coastal waters.

Fig. 3.17. (a) Deep Ocean Assessment of Tsunami (DART) Easy to Deploy (ETD). (b) C-Stat-2 Mobile Buoy System (https://nctr.pmel.noaa.gov/Dart/ https://www.asvglobal.com/product/c-stat/).

3.7 Tsunami management and mitigation

Across the globe tsunami warning systems are primarily designed to deal with earthquake-centric tsunami (Bernard & Titov 2015). These technology-based warning systems utilize estimation of earthquake location and magnitude, wave detection from buoys (Fig. 3.17) or satellite-based observation, and numerical simulation of wave propagation and potential inundation areas, all linked to communication systems and networks to issue timely alarms (Fig. 3.18).

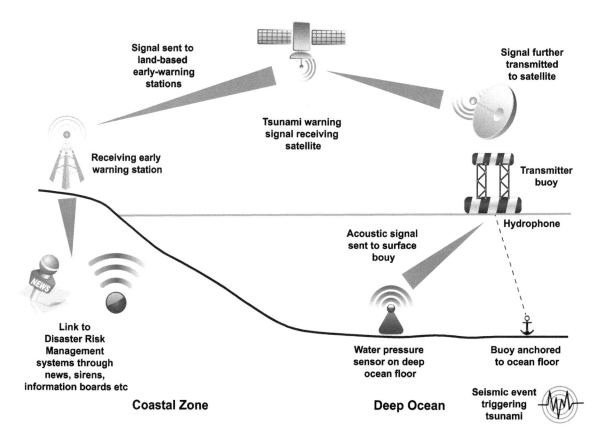

Fig. 3.18. The Pacific Ocean model for tsunami wave detection and reporting.

Fig. 3.19. Examples of tsunami-resistant engineered structures and escape buildings built to withstand the tsunami; their reinforced pillars are strengthened and mounted on a concrete base (adapted from Wong 2012).

Fig. 3.20. Examples of tsunami walls and barriers (https://aaj.tv/2018/03/seven-years-after-tsunami-japanese-live-uneasily-with-seawalls/ https://www.wired.com/story/photo-gallery-japan-seawalls/).

Fig. 3.21. Examples of tsunami hazard communication and awareness signage.

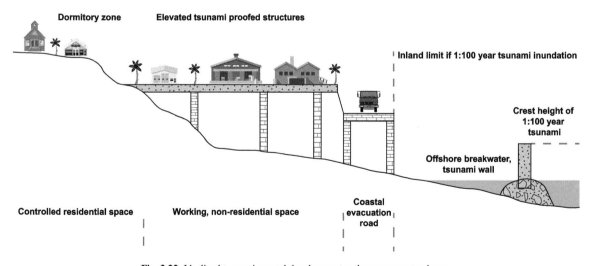

Fig. 3.22. Idealized tsunami coastal development and management scheme.

As with any geological hazard it is important in risk mitigation and management to delineate the hazard, usually through mapping, and to establish the risk posed in terms of likelihood and possible consequences. Many stretches of the UK coastline in the SW, English Channel, and north and east Scotland are prone to the impacts of tsunami. Some areas have a significantly increased vulnerability; for example, coastal cities such as Portsmouth and Aberdeen, sensitive infrastructure such as the Devonport Naval Base, Hinckley Point, Hunterston and Sizewell nuclear power stations, or ecologically and environmentally sensitive areas such as around The Wash or Caithness.

Tsunami hazard assessment can be based on specific scenario modelling with the management and mitigation relating specifically to the consequences of such events. Globally mitigation has taken the form of engineered solutions such as tsunami-resistant buildings and shelters (Fig. 3.19), tsunami walls and barriers (Fig. 3.20) as well as disaster planning with controlled development zones (Fig. 3.21) and detailed evacuation routes. The simplest mitigation schemes involve raising public awareness, signage, visual and audible warning systems, as well as practised evacuation procedures (Fig. 3.22). All of these management and mitigation solutions are difficult to implement in a UK context when these events are extremely rare and their magnitude usually quite low, although a 2 m wave from a Lisbon-type event would certainly have an impact along the UK coastal zone. One of the few positives to emerge from the 2004 Indian Ocean event was a greater UK awareness and, perhaps for the first time, knowledge of tsunami, their physical behaviour and their devastating impact. Education and awareness still remains the most effective risk management tool. Ribeiro *et al.* (2009) considered a proposal for tsunami early warning systems in the Gulf of Cadiz that would alert the UK to a potential tsunami along the SW coasts. This early warning system would be of little use unless there were formal and practised procedures in place to potentially evacuate the previously defined inundation zones that could be at risk from such an event. Risk-informed tsunami warnings (Woo 2017) still have some way to go in the UK, given the overriding lack of general awareness of this geohazard and its potential UK impact.

3.8 Concluding comments

With over 12 000 km of coastline, the British Isles is vulnerable to the hazard presented by tsunami. A significant number of potential tsunami source areas are present around the entire landmass, from plate tectonic boundaries off the Iberian Peninsula to the major submarine landslides in the northern North Sea to more localized coastal cliff instability, which again has the potential to generate a tsunami.

Up until the 2004 Boxing Day Indian Ocean Tsunami, there was very little awareness in the UK of this geological phenomena and the devastating impacts that it could present.

Although it is extremely unlikely that the UK will experience an event of this magnitude, much smaller events would still have a significant impact on the UK coastal areas, communities and infrastructure.

Currently, apart from some very sensitive installations, there is very little in the UK in the way of tsunami management and mitigation strategies; this situation should be urgently addressed, at both local and national levels.

References

ANGLO-SAXON CHRONICLE. n.d. http://www.britannia.com/history/docs/asintro2.html

ANON. 1821. The Windsor and Eton Express, Sunday 16 September 1821, page 3.

ANON. 1912*a*. The London Daily News, Tuesday 2nd January, 1912.

ANON. 1912*b*. The Folkestone, Hythe, Sandgate and Cheriton Herald, Saturday 6th January 1912.

BALTZER, A., HOLMES, R. & EVANS, D. 1998. Debris flows on the Sula Sgeir Fan, NW of Scotland. *In*: STOKER, M.S. EVANS, D. & CRAMP, A. (eds) *Geological Processes on Continental Margins: Sedimentation, Mass-Wasting and Stability*. Geological Society, London, Special Publications, **129**, 105–115, https://doi.org/10.1144/GSL.SP.1998.129.01.07

BANERJEE, D., MURRAY, A.S. & FOSTER, I.D.L. 2001. Scilly Isles, UK: optical dating of a possible tsunami deposit from the 1755 Lisbon earthquake. *Quaternary Science Reviews*, **20**, 715–718, https://doi.org/10.1016/S0277-3791(00)00042-1

BAPTISTA, M.A., HEITOR, S., MIRANDA, J.M., MIRANDA, P. & VICTOR, L.M. 1998. The 1755 Lisbon tsunami; evaluation of the tsunami parameters. *Journal of Geodynamics*, **25**, 143–157, https://doi.org/10.1016/S0264-3707(97)00019-7

BAPTISTA, M.A., MIRANDA, J.M., CHIERICI, F. & ZITELLINI, N. 2003. New study of the 1755 earthquake source based on multi-channel seismic survey data and tsunami modeling. *Natural Hazards and Earth System Science*, **3**, 333–340, https://doi.org/10.5194/nhess-3-333-2003

BEGÉT, J.E. 2000. Volcanic tsunami. *In*: SIGURDSSON, H., HOUGHTON, B., MCNUTT, S., RYMER, H. & STIX, J. (eds) *Encyclopedia of Volcanoes*. Elsevier, Amsterdam, 1005–1013.

BERNARD, E. & TITOV, V. 2015. Evolution of tsunami warning systems and products. *Philosophical Transactions of the Royal Society A*, **373**, 20140371, https://doi.org/10.1098/rsta.2014.0371

BERNINGHAUSEN, W.H. 1968. Tsunamis and seismic seiches reported from western North and South Atlantic and the coastal waters of north western Europe. Naval Oceanographic Office, Informal **Rep. No. 68–85**, Washington, D.C **20396**.

BONDEVIK, S., LØVHOLT, F., HARBITZ, C., MANGERUD, J., DAWSON, A. & SVENDSEN, J.I. 2005*a*. The Storegga Slide tsunami – comparing field observations with numerical simulations. *In*: SOLHEIM, A., BRYN, P., BERG, K., SEJRUP, H.P. & MIENERT, J. (eds) *Ormen Lange–an Integrated Study for Safe Field Development in the Storegga Submarine Area*. Elsevier, Amsterdam, 195–208.

BONDEVIK, S., MANGERUD, J., DAWSON, S., DAWSON, A. & LOHNE, O. 2005*b*. Evidence for three North Sea tsunamis at the Shetland Islands between 8000 and 1500 years ago. *Quaternary Science Reviews*, **24**, 1757–1775, https://doi.org/10.1016/j.quascirev.2004.10.018

BORLASE, W. 1762. Some Account of the extraordinary agitation of the waters in Mount's-Bay, and other places, on the 31st of

March 1761: in a Letter for the Reverend Dr. C Lyttelton. *Philosophical Transactions of the Royal Society*, **52**, 418–431.
BRYANT, E.A. & HASLETT, S.K. 2002. Was the AD 1607 coastal flooding event in the Severn Estuary and Bristol Channel (UK) due to a tsunami? *Archaeology in the Severn Estuary*, **13**, 163–167.
BRYN, P., SOLHEIM, A., BERG, K., LIEN, R., FORSBERG, C.F., HAFLIDASON, H. & RISE, L. 2003. The Storegga Slide complex; repeated large scale sliding in response to climatic cyclicity. *In*: LAMARCHE, G., MOUNTJOY, J. ET AL. (eds) *Submarine Mass Movements and Their Consequences*. Springer, Dordrecht, 215–222.
BRYN, P., BERG, K., FORSBERG, C.F., SOLHEIM, A. & KVALSTAD, T.J. 2005. Explaining the Storegga slide. *Marine and Petroleum Geology*, **22**, 11–19, https://doi.org/10.1016/j.marpetgeo.2004.12.003
BUGGE, T., BELDERSON, R.H. & KENYON, N.H. 1988. The Storegga slide. *Philosophical Transactions of the Royal Society A*, **325**, 357–388, https://doi.org/10.1098/rsta.1988.0055
BÜNZ, S., MIENERT, J. & BERNDT, C. 2003. Geological controls on the Storegga gas-hydrate system of the mid-Norwegian continental margin. *Earth and Planetary Science Letters*, **209**, 291–307, https://doi.org/10.1016/S0012-821X(03)00097-9
CHESTER, D.K. 2001. The 1755 Lisbon earthquake. *Progress in Physical Geography*, **25**, 363–383, https://doi.org/10.1177/030913330102500304
DAWSON, A.G., LONG, D. & SMITH, D.E. 1988. The Storegga slides: evidence from eastern Scotland for a possible tsunami. *Marine Geology*, **82**, 271–276, https://doi.org/10.1016/0025-3227(88)90146-6
DAWSON, A.G., MUSSON, R.M.W., FOSTER, I.D.L. & BRUNSDEN, D. 2000. Abnormal historic sea-surface fluctuations, SW England. *Marine Geology*, **170**, 59–68, https://doi.org/10.1016/S0025-3227(00)00065-7
DAWSON, A.G., LOCKETT, P. & SHI, S. 2004. Tsunami hazards in Europe. *Environment International*, **30**, 577–585, https://doi.org/10.1016/j.envint.2003.10.005
DEFRA 2005. The threat posed by tsunami to the UK. Study commissioned by Defra Flood Management. Produced by British Geological Survey, Proudman Oceanographic Laboratory, Met Office and HR Wallingford.
DEFRA 2006. Tsunamis – assessing the hazard for the UK and Irish coasts. Study commissioned by the Defra Flood Management Division, the Health and Safety Executive and the Geological Survey of Ireland. Produced by British Geological Survey, Proudman Oceanographic Laboratory, Met Office and HR Wallingford.
ELLIOTT, G.M., SHANNON, P.M., HAUGHTON, P.D. & ØVREBØ, L.K. 2010. The Rockall Bank Mass Flow: collapse of a moated contourite drift onlapping the eastern flank of Rockall Bank, west of Ireland. *Marine and Petroleum Geology*, **27**, 92–107, https://doi.org/10.1016/j.marpetgeo.2009.07.006
EVANS, D., HARRISON, Z., SHANNON, P.M., LABERG, J.S., NIELSEN, T., AYERS, S. & LONG, D. 2005. Palaeoslides and other mass failures of Pliocene to Pleistocene age along the Atlantic continental margin of NW Europe. *Marine and Petroleum Geology*, **22**, 1131–1148, https://doi.org/10.1016/j.marpetgeo.2005.01.010
FOSTER, I.D.L., ALBON, A.J. ET AL. 1991. High energy coastal sedimentary deposits: an evaluation of depositional processes in southwest England. *Earth Surface Processes and Landforms*, **16**, 341–356. https://doi.org/10.1002/esp.3290160407
FOSTER, I.D.L., DAWSON, A.G., DAWSON, S., LEES, J. & MANSFIELD, L. 1993. Tsunami sedimentation sequences in the Scilly Isles, south west England. *Science of Tsunami Hazards*, **11**, 35–46.

GAFEIRA, J., LONG, D., SCRUTTON, R. & EVANS, D. 2010. 3D seismic evidence of internal structure within Tampen Slide deposits on the North Sea Fan: are chaotic deposits that chaotic? *Journal of the Geological Society*, **167**, 605–616, https://doi.org/10.1144/0016-76492009-047
GARCÍA-MORENO, D., VERBEECK, K., CAMELBEECK, T., DE BATIST, M., OGGIONI, F., ZURITA HURTADO, O. & TRENTESAUX, A. 2015. Fault activity in the epicentral area of the 1580 Dover Strait (Pas-de-Calais) earthquake (northwestern Europe). *Geophysical Journal International*, **201**, 528–542, https://doi.org/10.1093/gji/ggv041
GLIMSDAL, S., HARBITZ, C.B. & LØVHOLT, F. 2010. The May 28th 2008 Faeroe tsunami. *In*: HARBITZ, C.B. (ed.) *EU Transfer project – ICG Contributions*. ICG **Report 2010-10-1**, NGI Document 20071129-00-66-R.
GTDB n.d. Global Tsunami Database, https://doi.org/10.7289/V5PN93H7
HAFLIDASON, H., SEJRUP, H.P., NYGÅRD, A., MIENERT, J., BRYN, P., LIEN, R. & MASSON, D. 2004. The Storegga Slide: architecture, geometry and slide development. *Marine Geology*, **213**, 201–234, https://doi.org/10.1016/j.margeo.2004.10.007
HAFLIDASON, H., LIEN, R., SEJRUP, H.P., FORSBERG, C.F. & BRYN, P. 2005. The dating and morphometry of the Storegga Slide. *Marine and Petroleum Geology*, **22**, 123–136, https://doi.org/10.1016/j.marpetgeo.2004.10.008
HASLETT, S.K. & BRYANT, E.A. 2007. Evidence for historic coastal high energy wave impact (tsunami?) in North Wales, United Kingdom. *Atlantic Geology*, **43**, 137–147, https://doi.org/10.4138/4215
HASLETT, S.K. & BRYANT, E.A. 2008. Historic tsunami in Britain since AD 1000: a review. *Natural Hazards and Earth System Sciences*, **8**, 587–601, https://doi.org/10.5194/nhess-8-587-2008
HASLETT, S.K. & BRYANT, E.A. 2009. Meteorological tsunamis in southern Britain: an historical review. *Geographical Review*, **99**, 146–163.
HELLER, V. & HAGER, W.H. 2014. A universal parameter to predict subaerial landslide tsunamis? *Journal of Marine Science and Engineering*, **2**, 400–412, https://doi.org/10.3390/jmse2020400
HOLMES, R. 2003. Holocene shelf-margin submarine landslides, Donegal Fan, eastern Rockall Trough. *In*: MINIERT, J. & WEAVER, P. (eds) *European Margin Sediment Dynamics*. Springer, Berlin, Heidelberg, 179–182.
HOLMES, R., LONG, D. & DODD, L.R. 1998. Large-scale debrites and submarine landslides on the Barra Fan, west of Britain. *In*: STOKER, M.S., EVANS, D. & CRAMP, A. (eds) *Geological Processes on Continental Margins: Sedimentation, Mass-Wasting and Stability*. Geological Society, London, Special Publications, **129**, 67–79, https://doi.org/10.1144/GSL.SP.1998.129.01.05
HORSBURGH, K. & HORRITT, M. 2006. The Bristol Channel floods of 1607–reconstruction and analysis. *Weather*, **61**, 272–277, https://doi.org/10.1256/wea.133.05
HORSBURGH, K.J., WILSON, C., BAPTIE, B.J., COOPER, A., CRESSWELL, D., MUSSON, R.M.W. & SARGEANT, S.L. 2008. Impact of a Lisbon-type tsunami on the UK coastline and the implications for tsunami propagation over broad continental shelves. *Journal of Geophysical Research: Oceans*, **113**, https://doi.org/10.1029/2007JC004425
HUTCHINSON, J.N. 2002. Chalk flows from the coastal cliffs of northwest Europe. *In*: EVANS, S.G. & DE GRAFF, J.V. (eds). *Catastrophic Landslides: Effects, Occurrence, and Mechanisms. Reviews in Engineering Geology XV*. Geological Society of America, 257–302.

International Oceanographic Commission 2008. *Tsunami Glossary*. UNESCO, Paris. IOC Technical Series, 85.

Laberg, J.S. & Vorren, T.O. 2000. The Trænadjupet Slide, offshore Norway – morphology, evacuation and triggering mechanisms. *Marine Geology*, **171**, 95–114, https://doi.org/10.1016/S0025-3227(00)00112-2

Laberg, J., Vorren, T.O., Mienert, J., Bryn, P. & Lien, R. 2002. The Trænadjupet Slide: a large slope failure affecting the continental margin of Norway 4,000 years ago. *Geo-Marine Letters*, **22**, 19–24, https://doi.org/10.1007/s00367-002-0092-z

Latter, J.H. 1981. Tsunamis of volcanic origin: summary of causes, with particular reference to Krakatoa, 1883. *Bulletin Volcanologique*, **44**, 467–490, https://doi.org/10.1007/BF02600578

Leynaud, D., Mienert, J. & Nadim, F. 2004. Slope stability assessment of the Helland Hansen area offshore the mid-Norwegian margin. *Marine Geology*, **213**, 457–480, https://doi.org/10.1016/j.margeo.2004.10.019

Lindberg, B., Laberg, J.S. & Vorren, T.O. 2004. The Nyk Slide – morphology, progression, and age of a partly buried submarine slide offshore northern Norway. *Marine Geology*, **213**, 277–289, https://doi.org/10.1016/j.margeo.2004.10.010

Locat, J. & Lee, H. 2009. Submarine mass movements and their consequences: an overview. In: Sassa, K. & Canuti, P. (eds) *Landslides – Disaster Risk Reduction*. Springer, Berlin Heidelberg, 115–142.

Long, D. 2015. A catalogue of tsunamis reported in the UK. British Geological Survey Internal Report, **IR/15/043**. 63

Long, D. 2017. Cataloguing tsunami events in the UK. In: Scourse, E.M., Chapman, N.A., Tappin, D.R. & Wallis, S.R. (eds) *Tsunamis: Geology, Hazards and Risks*. Geological Society, London, Special Publications, **456**, 143–165, https://doi.org/10.1144/SP456.10

Long, D. & Bone, B.D. 1990. Sediment instability on the continental slope of northwest Europe. In: *Proceedings of Oceanology International 90*, Brighton, UK, https://trove.nla.gov.au/version/39089701

Long, D. & Holmes, R. 2001. Submarine landslides and tsunami threat to Scotland. In: *Proceedings of the International Tsunami Symposium (ITS)*, University of Washington, Seattle, 7–9.

Long, D. & Wilson, C.K. 2007. *A catalogue of tsunamis in the UK*. British Geological Survey Commissioned Report, **CR/07/077**.

Long, D., Stevenson, A.G., Wilson, C.K. & Bulat, J. 2003. Slope failures in the Faroe–Shetland Channel. In: Lamarche, G., Mountjoy, J. et al. (eds) *Submarine Mass Movements and Their Consequences*. Springer, Dordrecht, 281–289.

Løvholt, F., Pedersen, G., Harbitz, C.B., Glimsdal, S. & Kim, J. 2015. On the characteristics of landslide tsunamis. *Philosophical Transactions of the Royal Society A*, **373**, 20140376, https://doi.org/10.1098/rsta.2014.0376

Melville, C. 1981. The historical seismicity of England. *Disasters*, **5**, 369–376, https://doi.org/10.1111/j.1467-7717.1981.tb01109.x

Mienert, J., Posewang, J. & Baumann, M. 1998. Gas hydrates along the northeastern Atlantic margin: possible hydrate-bound margin instabilities and possible release of methane. In: Henriet, J.-P. & Mienert, J. (eds) *Gas Hydrates: Relevance to World Margin Stability and Climate Change*. Geological Society, London, Special Publications, **137**, 275–291, https://doi.org/10.1144/GSL.SP.1998.137.01.22

Muir-Wood, R. & Mignan, A. 2009. A phenomenological reconstruction of the Mw 9 November 1st 1755 earthquake source. In: Mendes-Victor, L., Ribeiro, A. & Azevedo, J. (eds) *The 1755 Lisbon Earthquake: Revisited*. Springer, Dordrecht, 121–146.

Musson, R.M.W. 1989. *Seismicity of Cornwall and Devon*. British Geological Survey, Global Seismology Report. **WL/89/11**.

Musson, R.M.W. 1996. The seismicity of the British Isles. *Annals of Geophysics*, **39**, https://doi.org/10.4401/ag-3982

Musson, R.M.W. 2004. Design earthquakes in the UK. *Bulletin of Earthquake Engineering*, **2**, 101–112, https://doi.org/10.1023/B:BEEE.0000039047.77494.c7

Newig, J. & Kelletat, D. 2011. The North Sea Tsunami of June 5, 1858. *Journal of Coastal Research*, **27**, 931–941, https://doi.org/10.2112/JCOASTRES-D-10-00098.1

Owen, M., Day, S., Long, D. & Maslin, M. 2010. Investigations on the Peach 4 Debrite, a Late Pleistocene mass movement on the northwest British continental margin. In: Lamarche, G., Mountjoy, J. et al. (eds) *Submarine Mass Movements and Their Consequences*. Springer, Dordrecht, 301–311.

Paris, R. 2015. Source mechanisms of volcanic tsunamis. *Philosophical Transactions of the Royal Society A*, **373**, 20140380, https://doi.org/10.1098/rsta.2014.0380

Power, W. & Leonard, G.S. 2013. Tsunami. In: Bobrowsky, P.T. (ed.) *Encyclopedia of Natural Hazards*. Springer, Dordrecht, 1036–1046.

Ribeiro, A., Mendes-Victor, L.A., Matias, L., Terrinha, P., Cabral, J. & Zitellini, N. 2009. The 1755 Lisbon earthquake: a review and the proposal for a tsunami early warning system in the Gulf of Cadiz. In: Mendes-Victor, L., Ribeiro, A. & Azevedo, J. (eds) *The 1755 Lisbon Earthquake: Revisited* Springer, Dordrecht, 411–423.

Richardson, S., Musson, R. & Horsburgh, K. 2006. Tsunamis – assessing the hazard for the UK and Irish coast. In: *Proceedings of 41st Defra Flood and Coastal Management Conference*, 4–6 July 2006, York, UK.

Rise, L., Ottesen, D., Longva, O., Solheim, A., Andersen, E.S. & Ayers, S. 2006. The Sklinnadjupet slide and its relation to the Elsterian glaciation on the mid-Norwegian margin. *Marine and Petroleum Geology*, **23**, 569–583, https://doi.org/10.1016/j.marpetgeo.2006.05.005

Šepić, J., Vilibić, I., Rabinovich, A.B. & Monserrat, S. 2015. Widespread tsunami-like waves of 23-27 June in the Mediterranean and Black Seas generated by high-altitude atmospheric forcing. *Scientific Reports*, **5**, 11682, https://doi.org/10.1038/srep11682

Šepić, J., Vilibić, I., Rabinovich, A. & Tinti, S. 2018. Meteotsunami ("Marrobbio") of 25–26 June 2014 on the South western Coast of Sicily, Italy. *Pure and Applied Geophysics*, **175**, 1573–1593, https://doi.org/10.1007/s00024-018-1827-8

Scots Magazine 1749. *Mutinous Sailors Execute, Riots, High Tides*, **11**, 348, https://hdl.handle.net/2027/hvd.32044092547207

Shannon, P.M., O'Reilly, B.M., Readman, P.W., Jacob, A.W.B. & Kenyon, N. 2001. Slope failure features on the margins of the Rockall Trough. In: Shannon, P.M., Haughton, P.D.W. & Corcoran, D.V. (eds) *The Petroleum Exploration of Ireland's Offshore Basins*. Geological Society, London, Special Publications, **188**, 455–464, https://doi.org/10.1144/GSL.SP.2001.188.01.28

Smith, D.E., Shi, S. et al. 2004. The Holocene Storegga slide tsunami in the United Kingdom. *Quaternary Science Reviews*, **23**, 2291–2321. https://doi.org/10.1016/j.quascirev.2004.04.001

Solheim, A., Berg, K., Forsberg, C.F. & Bryn, P. 2005. The Storegga Slide complex: repetitive large scale sliding with similar cause and development. *Marine and Petroleum Geology*, **22**, 97–107, https://doi.org/10.1016/j.marpetgeo.2004.10.013

Strachan, P. & Evans, D. 1991. A local deep-water sediment failure on the NW slope of the UK. *Scottish Journal of Geology*, **27**, 107–111, https://doi.org/10.1144/sjg27020107

TALLING, P.J., CLARE, M.L., URLAUB, M., POPE, E., HUNT, J.E. & WATT, S.F. 2014. Large submarine landslides on continental slopes: geohazards, methane release, and climate change. *Oceanography*, **27**, 32–45, https://doi.org/10.5670/oceanog.2014.38

TAPPIN, D.R., GRILLI, S.T., HARRIS, J.C., GELLER, R.J., MASTERLARK, T., KIRBY, J.T. & MAI, P.M. 2014. Did a submarine landslide contribute to the 2011 Tohoku tsunami? *Marine Geology*, **357**, 344–361, https://doi.org/10.1016/j.margeo.2014.09.043

TYRRELL, G. 1932. Recent Scottish earthquakes. *Transactions of the Geological Society of Glasgow*, **19**, 1–41, https://doi.org/10.1144/transglas.19.1.1

VAN DORN, W.G. 1987. Tide gage response to tsunamis. Part II: other oceans and smaller seas. *Journal of Physical Oceanography*, **17**, 1507–1516, https://doi.org/10.1175/1520-0485(1987)017<1507:TGRTTP>2.0.CO;2

VARLEY, P.M. 1996. Seismic risk assessment and analysis. *In*: HARRIS, C.S. (ed.) *Engineering Geology of the Channel Tunnel*. Thomas Telford Publishing, 194–216.

WARD, S.N. & DAY, S. 2001. Cumbre Vieja volcano – potential collapse and tsunami at La Palma, Canary Islands. *Geophysical Research Letters*, **28**, 3397–3400, https://doi.org/10.1029/2001GL013110

WATTS, M., TALLING, P., HUNT, J., XUAN, C. & VAN PEER, T. 2016. A new date for a large pre-Holocene Storegga Slide. *In*: *EGU General Assembly Conference Abstracts*, 2016, April, **18**, 17055.

WENINGER, B., SCHULTING, R., BRADTMÖLLER, M., CLARE, L., COLLARD, M., EDINBOROUGH, K. & WAGNER, B. 2008. The catastrophic final flooding of Doggerland by the Storegga Slide tsunami. *Documenta Praehistorica*, **35**, 1–24, https://doi.org/10.4312/dp.35.1

WILLIAMS, R.B.G., ROBINSON, D.A., DORNBUSCH, U., FOOTE, Y.L.M., MOSES, C.A. & SADDLETON, P.R. 2004. A sturzstrom-like cliff fall on the Chalk coast of Sussex, UK. *In*: MORTIMORE, R.N. & DUPERRET, A. (eds) *Coastal Chalk Cliff Instability*. Geological Society, London, Engineering Geology Special Publications, **20**, 89–97, https://doi.org/10.1144/GSL.ENG.2004.020.01.06

WILSON, C.K., LONG, D. & BULAT, J. 2004. The morphology, setting and processes of the Afen Slide. *Marine Geology*, **213**, 149–167, https://doi.org/10.1016/j.margeo.2004.10.005

WONG, P.P. 2012. Impacts, recovery and resilience of Thai tourist coasts to the 2004 Indian Ocean Tsunami. *In*: TERRY, J.P. & GOFF, J. (eds) *Natural Hazards in the Asia–Pacific Region: Recent Advances and Emerging Concepts*. Geological Society, London, Special Publications, **361**, 127–138, https://doi.org/10.1144/SP361.11

WOO, G. 2017. Risk-informed tsunami warnings. *In*: SCOURSE, E.M., CHAPMAN, N.A., TAPPIN, D.R. & WALLIS, S.R. (eds) *Tsunamis: Geology, Hazards and Risks*. Geological Society, London, Special Publications, **456**, 191–197, https://doi.org/10.1144/SP456.3

WOODWORTH, P.L., BLACKMAN, D.L. ET AL. 2005. Evidence for the Indonesian tsunami in British tidal records. *Weather*, **60**, 263–267, https://doi.org/10.1256/wea.59.05

TRUONG, H.V.P. 2012. Wave-propagation velocity, tsunami speed, amplitudes, dynamic water-attenuation factors. *In*: *Proceedings of the 15th World Conference on Earthquake Engineering*, 2012, September, **15WCEE**, 24–28. http://www.iitk.ac.in/nicee/wcee/fifteenth_conf_purtgal/

Chapter 4 Landslide and slope stability hazard in the UK

Edward Mark Lee[1]* & David Peter Giles[2]

[1]Ebor Geoscience Limited, 15 Whernside Avenue, York, YO31 0QB, UK
[2]CGL (Card Geotechnics Ltd), 4 Godalming Business Centre, Woolsack Way, Godalming, GU7 1XW, UK
DPG, 0000-0001-8016-0538

*Correspondence: marklee626@btinternet.com

Abstract: With its rich lithological variation, upland, lowland and coastal settings, and past climatic changes, the UK presents a wide variety of landslide features that can pose significant hazards to people, construction and infrastructure, or simply add to landscape character and conservation value of an area. This chapter describes and defines the nature and extent of this landsliding; the causes, effects and geological controls on failure; and their mitigation and stabilization. A risk-based approach to landslide management is outlined with qualitative and semi-quantitative methodologies described. Numerous case studies are presented exemplifying landslide and slope stability hazards in the UK.

4.1 Introduction

For many people above a certain age, the word 'landslide' will evoke memories of the Aberfan disaster of 21 October 1966. A rotational failure at the front of a colliery spoil tip on the flanks of a steep-sided South Wales valley transformed into a flow slide, which travelled downslope at around 10 m s^{-1} into the village (Bishop *et al.* 1969; Miller 1974). The debris ran out 605 m, building up behind the rear wall of the Pantglas Primary School and causing it to collapse inwards. The loose waste then filled up classrooms, killing children and teachers (Penman 2000). Overall, 116 children and 28 adults were killed by the landslide; the community was severely affected, with many suffering severe psychological difficulties after the disaster (Lacey 1972; Morgan *et al.* 2003; Cabinet Office 2011).

However, Aberfan presents a misleading picture about the nature of landsliding in the UK (Jones & Lee 1994). The deaths caused by this single event almost certainly exceeded the overall loss of life from all other landslide events in the UK over the last few centuries. Fatal accidents in the UK are extremely rare and tend to be the result of rockfalls or high-velocity slides on the coast, rather than in inland valleys. For example, in July 2012 a young woman was killed by a large rockfall on the beach at Burton Bradstock, Dorset (Fig. 4.1). In February 1977 a school party were studying the geology of Lulworth Cove, Dorset, when they were buried beneath a rockslide; the schoolteacher and a pupil were killed and two more pupils seriously injured, one of whom died later in hospital. In July 1979 a woman sunbathing on the beach near Durdle Door, Dorset, was killed when a 3 m overhang collapsed.

Although the incidents on the Dorset coast during the 1970s led the then Chief Inspector of Wareham police to coin the phrase 'killer cliffs' (Jones & Lee 1994), the public perception of coastal erosion is dominated by the fear that parts of the UK are being rapidly lost to the sea, raising visions of a loss of national resources to a hostile invading power. As John Gummer (2000), former Minister at the Ministry of Agriculture, Fisheries and Food (now the Department for Food, Environment and Rural Affairs) wrote:

> Of course it has happened before. It's just that the last time our shores were successfully invaded was 1066. Now the east coast is being crossed again, and more effectively than by Norman or Dane. East Anglia is threatened as far inland as Bedfordshire. If erosion goes on at its present rate, my own constituency of Suffolk Coastal will simply continue falling into the sea.

The most intense marine erosion and cliff recession rates occur on the unprotected cliffs formed of soft sedimentary rocks and glacial deposits along the south and east coasts of England, respectively (Table 4.1). The Holderness coast, for example, has retreated by around 2 km over the last 1000 years, including at least 26 villages listed in the Domesday survey of 1086; 75 Mm^3 of land has been eroded in the last 100 years (Valentin 1954; Pethick 1996). Rapid recession has also caused severe problems on the Suffolk coast, most famously at Dunwich where much of the former city has been lost over the last millennium (e.g. Bacon & Bacon

Fig. 4.1. Bridport Sands (Lower Jurassic) East Cliff in the foreground and Burton Cliff, Burton Bradstock in the background, West Bay (Bridport Harbour) (photo credit: David Giles).

1988). Gardner (1754) records that by 1328 the port was virtually useless and that 400 houses together with windmills, churches, shops and many other buildings were lost in one night in 1347. On parts of the north Norfolk coast there has been over 175 m of recession since 1885 (Clayton 1980, 1989); county archives show that 21 coastal towns and villages have been lost since the eleventh century (Fig. 4.2).

Today, the reality of coastal erosion is often very different, primarily because of the effectiveness of over 850 km of coastal protection measures built mainly over the last 130 years (Lee & Clark 2002). The average annual loss of land due to cliff recession and coastal landsliding around the coast of England is probably less than 10–25 ha (Cosgrove et al. 1997; Lee 2001b; Lee & Clark 2002).

High-velocity landslide events that present a threat to people do occur inland, such as the August 2004 debris flows in the Scottish Highlands (Winter et al. 2005, 2006, 2010; see Winter 2019). There were no fatalities, but 57 people had to be airlifted to safety by the RAF when they became trapped between debris flows on the A85 at Glen Ogle (Fig. 4.3). Two people were killed in July 2012 when their car was crushed by falling debris as it emerged from the Beaminster Tunnel, Dorset, due to a landslide bringing down part of the tunnel portal (Morris 2012).

However, most inland landslides generally present only minor threats to life as movements, when they occur, usually involve only slow, minor displacements. Even when large displacements occur, the rate of movement tends to be gentle and not dramatic, as was reported graphically for the French House slide near Lympne, Kent, in 1725 where a farmhouse sank 10–15 m overnight, 'so gently that the farmer's family were ignorant of it in the morning when they rose, and only discovered it by the door-eaves, which were so jammed as not to admit the door to open' (Gostling 1756).

Nevertheless, slow-moving inland landslides can have a significant economic impact. The cumulative effects of episodes of slow movement can inflict considerable damage to buildings, services and infrastructure. Almost continuous damage from movement of the Mam Tor landslide (Fig. 4.4) in the High Peak of Derbyshire led to the permanent closure of the A625 Manchester–Sheffield road in 1979 and diversion of local and cross-Pennine traffic (e.g. Skempton et al. 1989; Waltham & Dixon 2000; Dixon & Brook 2007; Rutter & Green 2011). Hutchinson (2001) describes how a power line from Dungeness Nuclear Power Station, Kent was put out of action for over a month in the winter of 1966/67 when landslide activity on the Hythe–Lympne escarpment led to the loss of a pylon (Fig. 4.5). Sustained rainfall in November 1998 led to the collapse of Greenan Road near Ballycastle, Northern Ireland, cutting off access to a farming community; as the farms quickly ran out of feed for their livestock, helicopters were used to bring in fresh supplies (Cummings et al. 2000). There has been a history of mudslides blocking the Antrim coast road, particularly at Minnis North (Craig 1981). For example, in a 14-month period between 1971 and 1972 there were ten incidents when the road was blocked (Hutchinson et al. 1974). Intense rainfall on the morning of Tuesday 8 November 2005 initiated a small peat slide on a hillside above the A5 London–Holyhead trunk road in the Llyn Ogwen area, Snowdonia National Park; four people were injured, a nearby construction project was delayed and A5 was blocked (Nichol et al. 2007). In February 2013 a landslide occurred in a spoil tip at Hatfield Main Colliery and severely damaged a large section of train line along the Doncaster–Goole and Doncaster–Scunthorpe lines (Fig. 4.6). The section of the line was closed for 5 months and train services in the region significantly affected.

The unforeseen presence of ancient landslides can lead to costly problems during construction. The A21 Sevenoaks Bypass, Kent had to be halted in 1966 when excavation work cut through grass-covered lobes of material that proved to be the remains of a previously unidentified ancient landslide (Skempton & Weeks 1976). The inadvertent removal of material from the lower portion of these landslides led to their reactivation, despite the fact that they appeared to have remained stable and stationary over the Holocene. The

Table 4.1. *A selection of coastal cliff recession rates around England (compiled by Lee & Clark 2002)*

Site	Average recession rate (m a^{-1})*	Period	Source
Blue Anchor Bay, Somerset	0.2		Williams et al. (1991)
Downderry, Cornwall	0.11	1845–1966	Sims & Ternan (1988)
St Marys Bay, Torbay	1.03	1946–1975	Derbyshire et al. (1975)
Bindon, East Devon	0.1	1904–1958	Pitts (1983)
Charton Bay, East Devon	0.25	1905–1958	Pitts (1983)
Black Ven	3.14	1958–1988	Chandler (1989); Bray (1996)
Stonebarrow, Dorset	0.5	1887–1964	Brunsden & Jones (1980); Bray (1996)
West Bay (W), Dorset	0.37	1887–1962	Bray (1996)
West Bay (E), Dorset	0.03	1902–1962	Bray (1996)
Purbeck, Dorset	0.3	1882–1962	May & Heaps (1985)
White Nothe, Dorset	0.22	1882–1962	May (1971)
Barton-on-Sea, Hampshire	1.9	1950–1980	Barton & Coles (1984)
Highcliffe, Hampshire	0.27	1931–1975	University of Strathclyde (1991)
Undercliff, Isle of Wight	0.05		Hutchinson (1991b)
Blackgang, Isle of Wight	5		Clark et al. (1995)
Chale Cliff, Isle of Wight	0.41	1861–1980	Hutchinson et al. (1981)
Shanklin, Isle of Wight	0.68	1907–1981	Clark et al. (1991)
Seven Sisters, Sussex	0.51	1873–1962	May (1971)
Fairlight Glen, Sussex	1.43	1955–1983	Robinson & Williams (1984)
Beachy Head, Sussex	0.9		May & Heaps (1985)
Warden Point, Kent	1.5	1865–1963	Hutchinson (1973)
Studd Hill, Kent	1.5	1872–1898	So (1967)
Beltinge, Kent	0.83	1936–1966	Hutchinson (1970)
North Foreland, Kent	0.19	1878–1962	May (1971)
Walton-on-Naze, Essex	0.52	1922–1955	Hutchinson (1973)
Covehithe, Suffolk	5.1	1925–1950	Steers (1951)
Southwold, Suffolk	3.3	1925–1950	Steers (1951)
Pakefield, Suffolk	0.9	1926–1950	Steers (1951)
Dunwich, Suffolk	1.6	1589–1783	So (1967)
Runton, Norfolk	0.8	1880–1950	Cambers (1976)
Trimmingham, Norfolk	1.4	1966–1985	University of Strathclyde (1991)
Cromer–Mundesley, Norfolk	4.2–5.7	1838–1861	Mathews (1934)
Marl Buff–Kirby Hill, Norfolk	1.1	1885–1927	Hutchinson (1976)
Hornsea–Withernsea, Holderness	1.8	1852–1990	Pethick (1996)
Withernsea–Kilnsea, Holderness	1.75	1852–1952	Valentin (1954)
Flamborough Head, North Yorks	0.3		Mathews (1934)
Robin Hoods Bay, North Yorks	0.31	1892–1960	Agar (1960)
Saltwick Nab, North Yorks	0.04	1892–1960	Agar (1960)
Whitby (W), North Yorks	0.5		Clark & Guest (1991)
Whitby (E), North Yorks	0.19	1892–1960	Agar (1960)
Runswick Bay, North Yorks	0.27		Rozier & Reeves (1979)
Port Mulgrave, North Yorks	1.12	1892–1960	Agar (1960)
Crimdon–Blackhall, Durham	0.2–0.3		Rendel Geotechnics (1995)

*The National Coastal Erosion Risk Mapping (NCERM; England and Wales) results are available through the Environment Agency's 'What's in Your Backyard' website (http://www.environment-agency.gov.uk/homeandleisure/37793.aspx). The project provides coastal erosion projections for 20, 50 and 100-year periods, taking into account the impact of climate change and relative sea-level rise.

problems turned out to be so severe that the affected portion of the route had to be realigned. This incident and a similar landslide problem on the M6 motorway embankment (Fig. 4.7) at Waltons Wood (Early & Skempton 1972) provided the impetus for UK-based academic research into inland landslides (e.g. Skempton 1976).

In many instances landslide problems are less newsworthy, although they can still lead to property loss or the delay, redesign or abandonment of projects. For example, Camden Crescent in Bath is the only known asymmetric crescent in the world; half had been destroyed by the Hedgemead landslide in 1894 (Hawkins 1977, 1996). In 1952, a landslide occurred at the village of Jackfield, Shropshire (Fig. 4.8) on the River Severn, just over 2 km downstream of the Iron Bridge, destroying several houses and causing major dislocations in a railway and road (Henkel & Skempton 1954).

Fig. 4.2. Houses fatally damaged through cliff erosion and subsequently demolished, Happisburgh, North Norfolk (photo credit: David Giles).

Instability problems encountered at housing developments at Bury Hill (Hutchinson *et al.* 1973) and Brierley Hill (Thompson 1991) in the West Midlands, at Exwick Farm on the outskirts of Exeter (Crofts & Berle 1972), at Ewood Bridge in the Irwell Valley (Douglas 1985) and at Gypsy Hill in South London (Allison *et al.* 1991) are some examples of the impact of localized slope instability frequently associated with smaller-scale developments. A large landslide at Franklands Village, West Sussex in 1993 led to the demolition of 14 flats and houses (Stevenson 1994; Walbancke *et al.* 2000).

4.2 Landslide types

The term 'landslide' can be very confusing for the non-specialist. This is partly because it is used to embrace all forms of mass movement, including flows and falls, while at the same time used to convey the particular mechanism of sliding. In addition, the term conveys no scale significance. It can be used to describe extremely small, shallow soil failures (as along river or stream banks), the widespread 'solifluction' sheets of sediment (slow downslope-moving saturated soil or rock debris formed during periods of periglacial conditions) that mantle many slopes in southern England, and extensive deeper-seated landslides that occur on many escarpments and valley side-slopes or form coastal 'undercliffs'. As a result, the term simply describes a gravity-driven surface process and does not convey any indication of scale or hazard.

The term landslide also fails to make the important distinction between the occurrence of new landslides (so-called first-time failures) and the renewed movement of old (pre-existing) landslides (i.e. slides) that have remained in the landscape ('reactivation'; Hutchinson 1988, 1992, 1995). The importance of this distinction is that once a landslide has occurred it can be made to move again under conditions that the slope, prior to failure, could have resisted. Thus reactivations can be triggered much more readily than first-time failures. For example, they may be associated with lower rainfall/groundwater level thresholds than first-time failures in the same materials (e.g. Jones & Lee 1994). Many pre-existing landslides exhibit a number of movement episodes, separated by long or short periods of relative quiescence.

Fig. 4.3. View of the northern A85 Glen Ogle debris flow 2 days after the event, showing the sharp bend in the channel just above road level (photo credit: Mike Winter).

The term 'slope failure' is also ambiguous in its meaning. 'Failure' is often used to describe the single most significant movement episode in the known or anticipated history of a landslide, which usually involves the first formation of a fully developed rupture surface as a displacement or strain discontinuity (e.g. Hungr *et al.* 2014). The development of a landslide can therefore involve pre-failure deformations, failure itself and post-failure displacements. However, in other contexts, 'slope failure' is simply used as a synonym for 'landslide'.

Five distinct landslide mechanisms are recognized, which can be further subdivided as shown in Figure 4.9 (Cruden & Varnes 1996; Hungr *et al.* 2014).

(1) *Falling*: involving the detachment of soil or rock from a steep face or cliff, along a surface on which little or no shear displacement occurs. The material then descends through the air by falling, before impact with the ground results in disintegration with fragments subsequently bouncing and rolling some distance. This type of failure is widespread along the many rocky cliff faces in Upland Britain and around the coast (Fig. 4.10).

(2) *Toppling*: involving the forwards rotation out of a slope of a mass of soil or rock about a point or axis below the centre of gravity of the displaced mass. The subsequent evolution of the displaced mass is similar to that of a fall. The spatial distribution is similar to falls (Fig. 4.11).

(3) *Spreading or lateral spreading*: the extension of a cohesive soil or rock mass over a lower deformable layer, combined with a general subsidence of the upper fractured mass into the softer underlying material. The surface of rupture is not a surface of intense shear and so is often not well-defined. Spreading may also result from liquefaction or the flow and extrusion of a softer underlying layer of material. This is not a common failure mechanism in the UK (Fig. 4.12).

(4) *Flowing*: the turbulent movement of a fluidized mass over a rigid bed, with either water or air as the pore fluid (e.g. like wet concrete or running dry sand; see Hungr *et al.* 2001). There is a gradation from flows to slides depending on water content and mobility (Fig. 4.13a, b). Debris flows are common in upland Britain and are discussed in Chapter 5 in this volume.

(5) *Sliding*: the downslope movement of a soil or rock mass as a coherent body on surfaces of rupture or on zones of intense shear strain. Slides are characterized by the presence of a clearly defined shear surface at the contact between the moving mass and the underlying soil or rock. This is the most widespread of all landslide mechanisms in the UK, occurring in upland Britain, the South Wales valleys, along many escarpments and on hillslopes in southern England (Figs 4.14 & 4.15).

Composite and complex movements occur where two or more types of displacement occur within the same landslide area (Fig. 4.16).

Many highly stressed rock masses forming mountain slopes show signs of large-scale gravitational deformation (slope deformation) involving bulging, trenches, benches and scarps.

Numerous landslide classifications have been developed, most based on shape (morphology), the materials involved and some aspect of the principal mechanisms. The most comprehensive scheme in the English-speaking world is that of Hutchinson (1988). Simpler alternatives are the USA-focused scheme of Varnes (1978, with modifications proposed in Cruden & Varnes 1996; Hungr *et al.* 2014; Table 4.2) and the system developed by the EPOCH project for use in Europe (Dikau *et al.* 1996; Table 4.3). However, the overwhelming importance of local site conditions means that the landslide features observed on the ground can be very difficult to classify.

Pre-existing landslides are generally recognized in the field or from aerial photography of their surface forms (morphology). However, the process can at times be subjective

Fig. 4.4. The Mam Tor Landslide, Derbyshire showing the Mam Tor Beds overlying the Bowland Shale Formation (photo credit: David Giles).

and highly dependent on the experience of the personnel involved. This is because many landslides in the UK are ancient and, over time, have become degraded and partially removed by erosion, obscured by agricultural and forestry practice or built on. It is therefore possible to define a continuum of landslide features, the end-members of which are either highly recognizable or very poorly defined landslides. The latter have 'text book' features such as fresh scars, widespread tension cracks, surface ponding, tilted trees, and non-vegetated to partially vegetated debris lobes. These are very easy to identify, even by non-specialists, especially when they have caused damage to property. However, very poorly defined landslides have indistinct surface forms and are difficult to interpret. These features can be extremely difficult to recognize with confidence. Indeed, there can be disagreement between experts over the nature and origin of some suspected landslide features. It is not unknown for features such as glacial deposits, abandoned quarries, old mine workings and waste dumps, or agricultural terraces (strip lynchettes) to be mistaken for ancient landslides.

The reverse is also true, with some landslides being misinterpreted as being the product of other surface processes. For example, Davies *et al.* (2013) suggest that, in glaciated areas, many large-scale slope features have tended to be explained in terms of ice action rather than landsliding. They describe how the recent use of geomorphological mapping and ground penetrating radar has led to the reinterpretation of Threlkeld Knotts in the Lake District as a major postglacial landslide rather than a glacially modified microgranite intrusion, as had been previously thought (e.g. Boardman 1982).

The overwhelming majority of landslides in the UK lie somewhere between these two end-members, largely because of their antiquity and the widespread impact of humans on the landscape. It should be stressed that surface evidence is not a completely reliable approach to landslide classification, especially where slides are present; different slide types can generate similar surface forms. The subsurface geometry can only be confirmed through boreholes, trial pits/trenches or geophysics.

Fig. 4.5. Rotational, translation and flow failures in the Hythe and Atherfield clay formations, The Roughs, Hythe, Kent (photo credit: David Giles).

Fig. 4.6. Stainforth–Hatfield Colliery spoil tip landslide, February 2013 (photo credit: Network Rail Media Center).

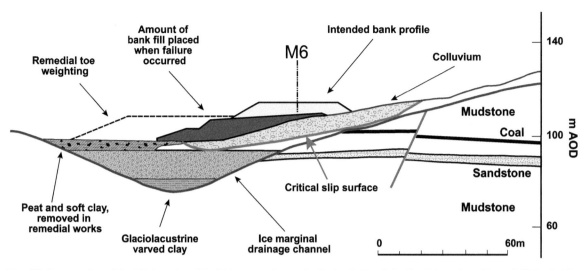

Fig. 4.7. Cross-section of the failed section of the M6 construction works, Walton's Wood, Staffordshire (adapted from Griffiths & Giles 2017).

Fig. 4.8. Remediation of the Jackfield Landslide, Ironbridge, Telford (photo credit: Telford District Council).

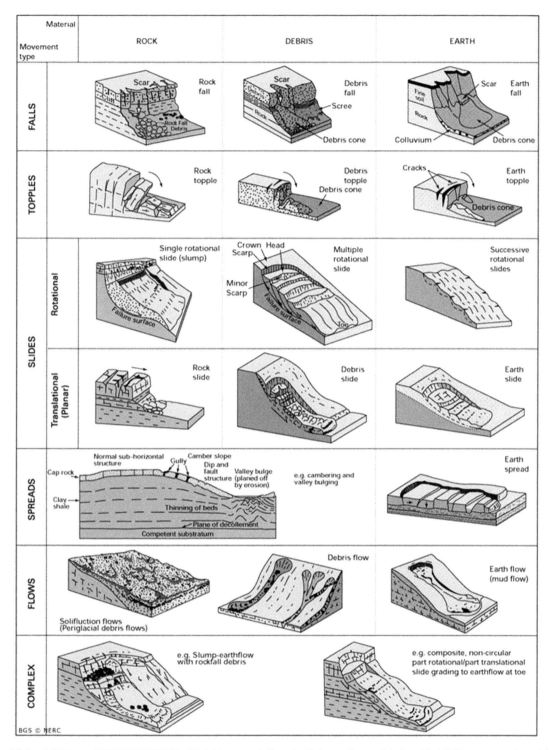

Fig. 4.9. Landslide types (BGS © NERC 2014. All rights reserved. Contains Ordnance Survey data (coastline) © Crown copyright and database right 2014).

Fig. 4.10. Example of a rockfall failure, East Pentwyn landslide, South Wales (photo credit: David Giles).

Fig. 4.11. Example of a rock topple failure from more columnar blocks developed in a microgabbro near Staffin, Isle of Skye (photo credit: David Giles).

Fig. 4.12. Example of a spreading failure in sensitive quick clays, St Jude, Quebec, Canada (photo credit: MTQ).

4.3 The landslide inventory for Great Britain

In the 1980s, growing awareness of the impact of landslide problems on land use and development led the Department of the Environment (DoE) to commission a review to consolidate knowledge on the nature, occurrence and significance of landsliding in Great Britain (England, Scotland and Wales; Geomorphological Services Limited 1986–87). Among the many objectives was the requirement to establish the geographical distribution of landslides and identify any associations with particular geological sequences.

A landslide database (the DoE Database) was established through a review of all reasonably accessible information held in the public domain regarding landslides, including British Geological Survey (BGS) maps and technical reports (e.g. Conway *et al.* 1980; Smith 1984). Here, landsliding was defined as natural gravity-driven mass movement forms (including falls to flows and slides), but excluding creep and solifluction (considered virtually ubiquitous across much of southern England) and failures of man-made slopes. By the time the review was completed in 1991 a total of 8835 landslides had been identified, of which 15% were coastal (e.g. Jones & Lee 1994). The database is heavily biased towards 'slides' (>70% of features that had been classified in the original sources; Table 4.4), with falls and flows almost certainly underreported.

The product of a pre-GIS age, the database was accompanied by a folio of hand-drawn 1:250 000-scale county/administrative region maps; many landslides were shown simply as 'dots' with no distinction between features of different sizes. The study was effective in providing a strategic overview of the nature and extent of landsliding across Great Britain, but it was not without its limitations (e.g. Jones & Lee 1994; Jones 1999; Foster *et al.* 2008).

It is important to stress that the DoE study identified only a sample of the actual landslides in existence. Subsequent experience has confirmed that the survey did not access many reports of landslides published in obscure journals and local newspapers, or held within the confidential files of consulting engineers, local authorities and even some national organizations (e.g. Lee *et al.* 2000). In addition, there must be numerous other landslides that have not yet been recorded because they exist in remote areas, are concealed by woodland, are relatively small-scale or have yet to be recognized as landslides. This is well illustrated by the detailed landslide-mapping programme in the Rhondda

Fig. 4.13. (a) Recent mud flows in the Gault Clay Formation, St Catherine's Point, Isle of Wight. (b) Sand flows in glaciofluvial sands, West Runton, Norfolk (photo credit: David Giles).

valleys (Halcrow 1988), which resulted in an increase in the number of recorded landslides from 102 to 346. Similarly, a mapping programme on the Torbay coastline raised the number of known landslides from 4 to 304 (Geomorphological Services Limited 1988). Jones & Lee (1994) concluded that the pattern of recorded landslides was dominated by the results of a small number of detailed studies set against a background of varying ignorance; in some areas, 'the harder you look the more examples you find'.

Since 1995, the British Geological Survey (BGS) has developed a national landslide database incorporating the DoE Database and other literature sources (e.g. Foster et al. 2008, 2012; Pennington et al. 2009b). The BGS database also includes the results of specific field mapping campaigns in a number of areas, including Bath (Hobbs & Jenkins 2008), Chesterfield (Pennington et al. 2009a), Derby (Jenkins & Booth 2008), Lincoln (Booth & Jenkins 2009), Market Rasen (Jenkins et al. 2010), the North York Moors (Jenkins et al. 2005), the Wellington district of Somerset (Freeborough et al. 2005) and York (Foster et al. 2007). The BGS also operate a Landslide Response Team that visits and assesses significant landslides when they occur (http://www.bgs.ac.uk/landslides/visits.html). The database is freely available via the BGS onshore GeoIndex, under the subject heading 'hazards' (http://www.bgs.ac.uk/geoindex).

By the end of 2013, this database contained information on over 16 500 landslides in Great Britain, including over 200 incidents that occurred during the extremely wet period between April 2012 and January 2013 (Fig. 4.17; Pennington & Harrison 2013; Pennington et al. 2014); 2012 was the second wettest year in the UK national record dating back to 1910 (see http://www.metoffice.gov.uk). When compared with Italy (a similar size and population) where around 500 000 landslides have been recorded (Trigila & Iadanza 2008), this figure emphasizes the relatively benign nature of the UK environment.

The summary distribution map (Fig. 4.18) revealed a spatially variable distribution with marked concentrations of reported inland landslides in (1) the Scottish Highlands; (2) the North Yorkshire Moors; (3) the Pennines, especially the Northumbrian Fells, the Northern Pennines, the Forest of Bowland, the Central Pennines and the High Peak; (4) the East Midlands Plateau; (5) the Ironbridge Gorge; (6) the Vale of Clwyd and Denbighshire Moors; (7) the South Wales Coalfield; (8) the Cotswolds, including Bredon Hill and other outliers; (9) the area around Bath; (10) Exmoor; (11) the slopes bounding the East Devon Plateau; (12) the Weald, especially the Upper and Lower Greensand escarpment, the Central Weald and the Lower Chalk escarpment at Folkestone; and (13) the London Basin.

The BGS have incorporated the landslide information within a GIS-based geohazard susceptibility model, the new GeoSure Insurance Product dataset (e.g. Foster et al. 2008, 2011; Lee et al. 2014). The level of susceptibility is communicated differently to different audiences, either as a number range, a rating A–E or as 'low to nil' to 'significant' (Fig. 4.19). GeoSure is produced at 1:50 000 scale and can be integrated to show the spatial distribution of landslide susceptibility in relation to low-rise buildings and infrastructure. It is estimated that 350 000 households in the UK, representing 1% of all housing stock, are in areas considered to have 'significant' landslide susceptibility (Gibson et al. 2012). The dataset is typically accessed directly through a GIS, operated by a licensee or through automatically generated reports, for instance as part of the Home Information Pack (Foster et al. 2008).

4.4 The Irish landslide inventory

In 2004, the Irish Landslide Working Group compiled a database of landslide events in Eire and Northern Ireland from available published sources (Creighton 2006). The landslide information is stored within a GIS also containing topographic and geological map datasets. The database includes 117 landslide events, 100 in Eire and 17 in Northern Ireland.

Fig. 4.14. Example of a translational rockslide, Chilton Chine, Isle of Wight (photo credit: David Giles).

Fig. 4.15. Example of a rotational mudslide, Chale, Isle of Wight (photo credit: David Giles).

Fig. 4.16. The landslide complex at St Catherine's, Isle of Wight with rockfalls, mud slides (rotational and translational) and mud flows all occurring (photo credit: IoWCC).

This is recognized to be a considerable underestimate of the actual distribution of landslides, being more a reflection of limited research and study of landslides in Ireland (as was the case for the DoE Database for Great Britain).

The landslides tend to be associated with the blanket bog areas (e.g. Sollas *et al.* 1897; Alexander *et al.* 1985) and the edge of the basalt escarpment in Co. Antrim (e.g. Carney 1974; Forster 1998). The majority of events involved peat failure (63), while some (31) were composed of coarse debris.

4.5 The landslide environment of the UK

The national landslide database has revealed a stark contrast between the landslide environments on the coast and inland. While coastal landsliding is a present-day process, the overwhelming majority of inland landslides are ancient features, believed to have been inherited from the Pleistocene period (Jones & Lee 1994). Dating of large ancient landslides is extremely problematic (see Johnson 1987); however, it could be expected that many of the large inland failures of southern England are of considerable antiquity because of what appears to be a very slow destruction rate by erosion processes. Destruction is a function of the connectivity between the landslide site and downslope sediment transport pathways along which material can be removed. For example, along the major landslide escarpments of SE England and the Midlands, such as the Lower Greensand escarpment in the Weald and the Cotswolds, there appears to be virtually no active removal of material from the landslide toe or debris apron by rivers or major streams. Many of these large landslides (i.e. major rock slope failures, deep-seated rotational and compound slides, and landslide complexes) could be very persistent forms, that is, once created they would remain in the landscape for considerable periods of time. First-time failures of these sites may have occurred in the distant past (possibly 10^3–10^5 years ago), followed by numerous phases of reactivation and less frequent retrogression by failure of the rear cliff.

It is suspected that the often repeated change from glacial and periglacial episodes to more temperate conditions during the Quaternary led to pulses of landsliding on hillslopes, valley sides and escarpments. Jones & Lee (1994) estimated that over 90% of inland landslides were pre-existing features. Away from the coast and river banks, first-time failures are

Table 4.2. *Adaptations of the Varnes (1978) classification of landslides; top from Cruden & Varnes (1996), bottom from Hungr et al. (2014), where material options are shown in italics*

Type of movement	Bedrock	Engineering soils	
		Predominantly coarse	*Predominantly fine*
Fall	Rockfall	Debris fall	Earth fall
Topple	Rock topple	Debris topple	Earth topple
Slide	Rockslide	Debris slide	Earth slide
Spread	Rock spread	Debris spread	Earth spread
Flow	Rock flow	Debris flow	Earth flow
Complex	Combination of two or more principal types of movement		

Type of movement	Rock	Soils
Fall	*Rock/ice* fall	*Boulder/debris/silt* fall
Topple	Rock block topple Rock flexural topple	*Gravel/sand/silt* topple
Slide	Rock rotational slide Rock planar slide Rock wedge slide Rock compound slide Rock irregular slide	*Clay/silt* rotational slide *Clay/silt* planar slide *Gravel/sand/debris* slide *Clay/silt* compound slide
Spread	Rock slope spread	*Sand/silt* liquefaction spread Sensitive clay spread
Flow	*Rock/ice* avalanche	*Sand/silt/debris* dry flow *Sand/silt/debris* flow slide *Sensitive clay* flow slide Debris flow Mud flow Debris flood Debris avalanche Earth flow Peat flow
Slope deformation	Mountain slope deformation Rock slope deformation	Soil slope deformation Soil creep Solifluction

Table 4.3. *Classification of landslide types proposed by the EPOCH project (Dikau et al. 1996)*

Type	Material		
	Rock	*Debris*	*Soil*
Fall	Rockfall	Debris fall	Soil fall
Topple	Rock topple	Debris topple	Soil topple
Slide (rotational)	Single (slump) Multiple Successive	Single Multiple Successive	Single Multiple Successive
Slide (translational; non-rotational)	Block slide	Block slide	Slab slide
Planar	Rockslide	Debris slide	Mudslide
Lateral spreading	Rock spreading	Debris spread	Soil (debris) spreading
Flow	Rock flow (sackung)	Debris flow	Soil flow
Complex (with run-out or change of behaviour downslope; note that nearly all forms develop complex behaviour)	e.g. Rock avalanche	e.g. Flow slide	e.g. Slump-earthflow

Note: A compound landslide is one that consists of more than one type, for example a rotational–translational slide. This should be distinguished from a complex slide where more than one form of failure develops into a second form of movement, that is, a change in behaviour downslope by the same material.

Table 4.4. *The relative significance of landslide types recorded within the DoE database (after Jones & Lee 1994)*

Landslide type	Number	Percentage of the DoE database
Unspecified	5214	59.0
Falls and topples	449	5.1
Rotational slides	994	11.3
Translational slides	887	10.0
Complex slides	777	8.8
Flows	370	4.2
Cambered/foundered strata	144	1.6

rare and generally confined to shallow debris or peat slides or flows on steep upland slopes, and failures of man-made slopes. There are, however, important differences between the antiquity of landslide features in glaciated northern Britain and unglaciated southern England. In addition to causing slope steepening, stripping bedrock and depositing an extensive mantle of superficial materials, the ice sheets and glaciers would have removed or obscured the evidence of earlier phases of landsliding.

Considered from a global perspective, however, landslides in the UK are somewhat unremarkable for their size and frequency (Cooper 2007) as the scale of failure is limited by the relatively low relief compared with mountainous regions of the world. Inland, most contemporary landslide events are less than 1–2 ha in area, rarely exceeding 10 ha (Table 4.5). The largest event in recent centuries was the 'eruption' of Solway Moss in November 1771 (Walker 1772; McEwen & Withers 1989). An estimated 520 ha of 'very deep and tender moss' flowed onto the adjacent plain towards the River Esk, covering 320 ha of farmland with up to 4.5 m of peat. No one was killed, but 35 families were affected, with the loss of most of their corn and some cattle. Larger events have been reported in Ireland. A bog burst in February 1883 affected around 1600 ha in County Roscommon (Forsyth 1883). An even larger burst (>5 Mm^3) occurred in the Knocknageedha bog, County Kerry in 1896 (Sollas *et al.* 1897; McCahon *et al.* 1987).

As discussed in Section 4.6, the vast majority of recorded landslides are believed to be associated with climatic conditions that developed in periglacial and immediate postglacial conditions (Fig. 4.20). New landslides on previously unfailed slopes are relatively rare events (e.g. Pennington *et al.* 2014). Major rockfall avalanches and other long-run-out landslides do not appear to have been a feature of the landslide environment, with the exception of the Beinn Alligin rock avalanche (a possible 'sturzstrom') in the Scottish Highlands (Fig. 4.21). This event occurred around 4 ka ago, and involved the failure of 8.8 million metric tonnes from the mountain crest which cascaded onto the floor of the corrie below

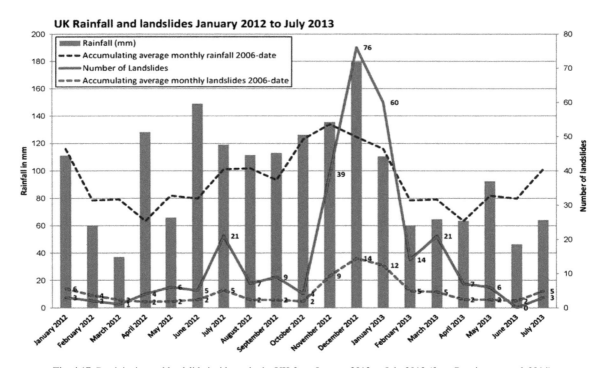

Fig. 4.17. Precipitation and landslide incidence in the UK from January 2012 to July 2013 (from Pennington *et al.* 2014).

Fig. 4.18. The National Landslide Database: distribution of landslides in Great Britain (BGS © NERC 2014. All rights reserved).

Despite the unspectacular nature of many landslides, there has been a long tradition of landslide research in the UK universities and the BGS focused primarily on improving the understanding of landslide controlling factors, triggering events, mechanisms and displacement patterns. This research has had a key role in the development of global knowledge of mass movements, their mechanisms and their countermeasures (e.g. Skempton 1964; Skempton & Hutchinson 1969; Chandler 1970, 1971, 1976; Hutchinson & Bhandari 1971; Hutchinson 1973, 1977, 1984, 1988; Hutchinson *et al.* 1980; Dixon & Bromhead 2002). This work has also been influential in the way in which landslide problems are investigated and managed by landowners, developers, local authorities and other stakeholders. In recent years, significant advances have been made in subject areas as diverse as the understanding of peat slides (e.g. Mills 2002; Dykes & Warburton 2007, 2008), seismic triggering of postglacial landslides (e.g. Jarman 2006, 2009; Ballantyne 2012, 2013) and the use of probabilistic models for predicting cliff recession rates (e.g. Lee *et al.* 2001, 2002; Walkden & Hall 2005; Walkden & Dixon 2006).

All of the six landslide mechanisms described by Hungr *et al.* (2014; see Table 4.2) can be found in the UK. A detailed discussion of the landslide types recorded in Great Britain can be found in Jones & Lee (1994) and will not be repeated here (see Winter 2020 Chapter 5 for debris flows). However, in the decades since that document was published, there has been significant research into a number of landslide features including peat failures, cambering and complex rock block spreads, large rock slope failures in the Scottish Highlands, flow slides in colliery spoil and coastal cliff behaviour units. A brief summary of these landslide features is provided in the following sections. Before doing so, it is important to stress that landsliding is a natural process and should not be viewed as being wholly detrimental to society.

In other circumstances landslides can be of importance to Earth science research and training; some landslides have been designated Sites of Special Scientific Interest (SSSI) for their geomorphological importance. The Joint Nature Conservation Committee (JNCC) has identified 33 Geological Conservation Review (GCR) sites considered to be of importance to the international Earth scientist community, demonstrating the presence of exceptional (classic, rare or atypical) features, and national importance for features that are representative of geological events or processes that are fundamental to understanding the geological and/or geomorphological history of Great Britain (Cooper 2007). These include the world-famous sites at Alport Castles and Mam Tor (Derbyshire), the Axmouth–Lyme Regis Undercliff, Black Ven (Dorset), Folkestone Warren and Warden Point (Kent), and the Trotternish Escarpment (Skye) (Fig. 4.22).

Landslides also create unique landscapes and habitats, as in the Landslip Nature Reserve on the east Devon coast. On the coast, cliffs are shaped by and dependent on landslide processes. These cliffs are among the nation's greatest assets, with many safeguarded by the protection afforded by their inclusion in National Parks and Areas of Outstanding Natural

(Ballantyne 2003). The exceptionally large mass and height of fall provided sufficient energy to cause the rock debris to move as a flow that surged up the opposite side of the corrie then moved along the corrie floor over a distance of 1.25 km.

Secondary hazards such as rock avalanches and rockslides blocking narrow, steep-sided valleys and forming landslide dams (Schuster 1986), or running-out into lakes and triggering tsunami, are not features of the contemporary UK landslide environment, although they are common in many mountainous parts of the world. Such events, however, may have occurred in the past. For example, after the last glaciation 50 Mm^3 of rock collapsed to form a landslide dam at Graig Goch, mid Wales, impounding the 1.6-km-long Tal-y-llyn lake (Hutchinson & Millar 2001).

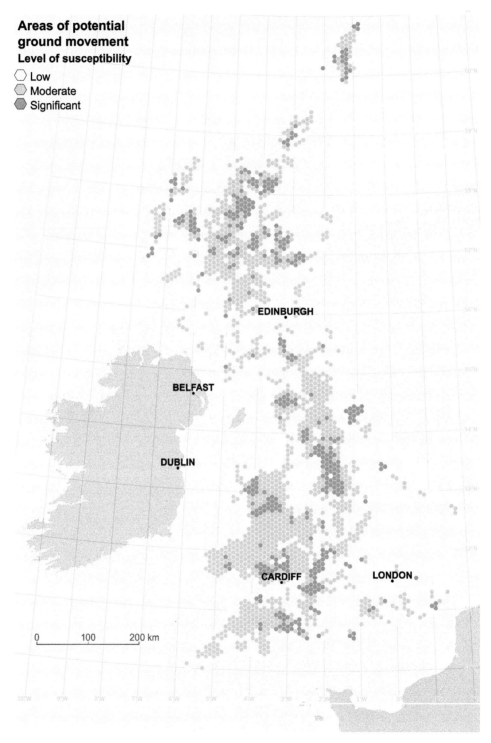

Fig. 4.19. Landslide susceptibility map of Great Britain, from the BGS GeoSure Model (BGS © NERC 2014. All rights reserved. Contains Ordnance Survey data (coastline) © Crown copyright and database right 2014).

Table 4.5. *High-magnitude/low-frequency inland landslide events in the UK over the last 300 years*

Date	Location	Area (ha)	Source
February 1745	Pilling Moss	20	Richmond (1744/5) Gemsege (1756)
March 1755	Whitestone Cliff (Thirsk)	15	Wesley (1755)
Christmas 1755	Toys Hill (near Sevenoaks, Kent)	1.1	Granticola (1756)
March 1774	Hawkley (Selbourne)	20	White (1789)
November 1771	Solway Moss	520	Walker (1772)
1773	Buildwas (Ironbridge)	7	Harral (1824)
1869	Troedrhiwfuwch, Rhymney Valley	c. 15	Bentley (2000)
1872	Clay Bank, North Yorkshire	14	Marsay (2010)
1893	Bournville, Ebbw Fach	c. 10	Bentley & Siddle (2000a)
March 1905	New Tredegar, Rhymney Valley	<10	Bentley & Siddle (2000b)
March 1930	New Tredegar, Rhymney Valley	10–15	Bentley & Siddle (2000b)
1952	Jackfield, Ironbridge Gorge	4	Henkel & Skempton (1954)
1954	East Pentwyn, Ebbw Fach	8	Siddle (2000a)
December 1965*	Mam Tor, Derbyshire	20–30	Cooper & Jarman (2007)
1989	Blaencwm, Rhondaa Fawr	c. 4	Siddle (2000b)
1994	St Dogmaels, Pembrokeshire	30	Maddison (2000)

*Mam Tor has been affected by active ground movement throughout the twentieth century (e.g. Skempton *et al.* 1989; Waltham & Dixon 2000).

Beauty or through their status as parts of heritage coasts. Eroding cliffs can also be of considerable environmental significance for their biological, Earth science and landscape value. The main benefits of cliff recession (e.g. Lee 2000) can be described as follows.

(1) *Creating and maintaining the landforms which support important habitats*. Numerous threatened species are found in such settings, such as hoary stock (*Matthiola incana*) found only on eroding chalk cliffs. Maritime grasslands occur on many cliffs and slopes, often comprising a maritime form of red fescue (*Festuca rubra*), thrift (*Armenia maritima*), sea plantain (*Plantago maritima*) and sea carrot (*Daucus carota ssp. gummifer*). Soft cliffs provide important breeding sites for sand martins (*Riparia riparia*) and are particularly important for invertebrates such as the ground beetle *Cincindela germanica*, the weevil *Baris analis* and the Glanville fritillary butterfly (*Melitaea cinxia*). Seepages, springs and pools provide habitats for many species of solitary bees and wasp, the craneflies *Gonomyia bradleyi* and the water beetle *Sphaerius acaroides*. Many hard rock cliffs provide prime breeding grounds for seabirds, with cliffs from Flamborough Head north to Dunnet Head, Cape Wrath to Land's End and the Northern and Western Isles containing the bulk of Europe's seabird population. Over 20% of the world's population of razorbills nest around the Great Britain coast.

(2) *Stimulating change within cliffs through promoting instability and ensuring that habitats evolve through natural successions, rather than remaining static.* Many active landslides support a range of vegetation from pioneer communities on freshly exposed faces through grassland communities to scrub and woodland. Wet flush vegetation occurs in areas of seepage.

(3) *Providing important geological exposures, including international reference localities for vast periods of geological time*, such as the Bartonian Stratotype between Highcliffe and Milford Cliff in Hampshire. Cliffs also provide opportunities for geological and geomorphological teaching and research.

(4) *Supplying sediment to the coastal zone and, hence, maintaining other coastal landforms* such as beaches, sand dunes, mudflats and saltmarshes. These landforms absorb wave and tidal energy arriving at the coast and can form important components of flood defence or coast protection solutions, either alone or where they front embankments or seawalls. Continued cliff recession can therefore be important in managing flood or erosion risks elsewhere on the coast.

Finally, it is worth mentioning the minor role that the landslide events at Whitestone Cliff, North Yorkshire (March 1755) and Buildwas, Shropshire (May 1773) played in the late eighteenth-century arguments about the causes of natural phenomena such as earthquakes and landslides (e.g. Cooper 1997). The Methodist preacher John Wesley used the 'home grown' example of a 'convulsion of nature' at Whitestone Cliff in his sermon on the philosophical meaning of the great Lisbon earthquake of November 1755 which resulted in between 10 000 and 100 000 deaths (e.g. Pereira 2006). This was the first modern natural disaster and led people to question the notion that such events were providential Acts of God. It was hard to imagine any offense by man that could justify such massive and non-discriminating punishment, or to defend the infliction of such loss merely to awaken

Fig. 4.20. Relict periglacial active layer detachment slides at Sevenoaks, Kent (photo credit: David Giles).

Fig. 4.21. Beinn Alligin rock avalanche. Photo taken from Na Rathanan looking to Tom na Gruagaich on Beinn Alligin (Torridon, NW Highlands, Scotland) 4 ka 1.25 km rock avalanche (photo credit: BGS Scotland).

Fig. 4.22. The Quiraing landslide complex in Tertiary lava flows, Trotternish Escarpment, Isle of Skye (photo credit: David Giles).

humanity to our sin (Maddox & Maddox 2008). Others tried to distance God from direct responsibility for the event by emphasizing that God works through secondary causes in nature, and suggesting that these secondary causes occasionally go astray.

John Wesley's response to the suggestion that earthquakes be seen as mere 'acts of nature' was published in December 1755 as Serious Thoughts occasioned by the Late Earthquake at Lisbon. He argued that if we set aside God's providential action in nature we have no reason to hope for God's protection in any specific setting. Making reference to the Whitestone Cliff landslide, he wrote:

> How then can we account for it? If not by natural causes, only by supernatural.... I believe it was by God (by himself or his Angel) who arose to shake terribly the Earth: that he purposely chose such a Place...that he wrought in such a manner, that many might see it and fear.... It must likewise for many years, maugre all the Art of Man, be a visible Monument of his Power: all that ground being so incumbered with Stones and Rocks, that it cannot be either ploughed or grazed. Nor can it well serve to any other use than to tell all Beholders, who can stand before his great God (Wesley 1755).

John Fletcher, a key interpreter of Wesleyan theology in the eighteenth century, used the 1773 Buildwas landslide that blocked the River Severn as another example of providential action:

> But whatever the second or natural cause of our phenomenon was, it is certain that the first or moral cause of it is twofold; on our part aggravated sin; and on God's part warning justice (Fletcher 1774).

The notion of landslides or earthquakes as being Acts of God declined in popularity throughout the late eighteenth and nineteenth centuries, as progressively fewer Christians accepted explanations that involved divine retribution (Chester & Duncan 2009). However, it is worth noting that, following the 2004 Indian Ocean earthquake and tsunami disaster, a fierce debate raged in the religious press over the fact that a retributive theodicy was still being proposed by some Christians to explain this event and its impact (e.g. Bradford 2005). Indeed, the Anglican Dean of Sydney, Phillip Jensen, provoked a storm of protests among religious leaders of all faiths when he characterized the tsunami as 'a warning of God's judgment'.

4.5.1 Peat failures

Mills (2002) has suggested that over 80% of the world's peat failures have been reported in the British Isles. These range in magnitude from a few hundred cubic metres of peat to 6–10 Mm^3 (e.g. the Solway Moss 'eruption' of 1771). Peat failure can cause substantial fish-kills, and loss of other aquatic life can result if peat debris reaches watercourses (McCahon et al. 1987; Wilson et al. 1996). Occasionally, peat failures can block roads (as occurred at Channerwick, Shetland, in September 2003; Moore et al. 2006; Dykes & Warburton 2008) and damage property (Evans & Warburton 2007) (Fig. 4.23a, b).

Dykes & Warburton (2007) proposed a peat failure classification scheme that is based on the type of peat that failed and the mechanism of outflow or shearing: failure of the peat matrix at the peat/mineral soil substrate interface or in the underlying substrate (Table 4.6). Peat slides correspond in appearance and mechanism to translational landslides, and tend to occur in shallow peat ($\leq$ 2.0 m) on steeper slopes (5–15°). Bog bursts correspond in appearance and mechanism to spreading failures and tend to occur in deeper peat (>1.5 m) on shallow slopes (2–10°) where deeper peat deposits are more likely to be found (Mills 2002). All the mechanisms appear to be closely linked to hillslope hydrology and the physical properties of peat (Warburton et al. 2004; Boylan et al. 2008; Winter et al. 2009; Winter 2019).

In recent years, there has been concern about the impact of onshore wind farms on peat stability. This has led the Scottish Executive (2006) to publish guidance on best practice for identifying, mitigating and managing peat slide hazards and their associated risks. Wind farm applications are required to be supported by a peat stability risk assessment.

4.5.2 Slope deformation: cambering and complex rock block spreads

The term 'cambering' is used to describe the large-scale gravity-driven slope deformation processes that occurred during periods of intense periglacial conditions where valleys cut through rigid caprocks into weaker clay layers below (e.g. Hutchinson 1988; Parks 1991; Fig. 4.24). Evidence of cambering is widespread in the Cotswolds, especially around Bath (e.g. Chandler et al. 1976; Goudie & Parker 1996; Hobbs & Jenkins 2008) and in the Hambleton Hills of North Yorkshire (e.g. Cooper et al. 1976; Cooper 1980, 2007). It is suspected that the resulting large-scale structures prepared the ground for later landsliding (Jones & Lee 1994).

The term generally encompasses a variety of features: bending downslope of competent strata; infilled fissures or 'gulls' (Fig. 4.25); mass movement caves or 'windypits' comprising large parallel-sided passages between displaced, wedged blocks, as reported by Cooper (1981, 1983a, b, 2007); complex patterns of large-scale 'ridge and trough topography', as in the Windrush Valley in the Cotswolds (Briggs & Courtney 1972); dip and fault structures, as originally observed in Northamptonshire, where large areas of the Northampton Ironstone have been affected by cambering (Hollingworth et al. 1944); and valley bulging, involved deformed thrust structures in valley floors as in the Frome Valley, near Stroud (Ackerman & Cave 1967).

Cambered structures are typically associated with strong caprocks (e.g. sandstone) overlying potentially ductile mudstone and/or shale beds (e.g. Poulsom 1995; Brunsden et al. 1996). Cambering is believed to be the product of phases of intense freeze–thaw activity under periglacial conditions (e.g. Parks 1991), involving downslope creep and/or extrusion (ductile failure) of the softened plastic clay bedrock followed by brittle failure of the caprock to form tension cracks (gulls and caves) and displaced blocks. Ductile failure is promoted by the thawing of ground ice contained within the slope, causing increased pore pressures.

Recent research has suggested that similar slope deformation features, described as 'complex rock block spreads', occur at the crest of the deeply incised valleys that dissect the moorland plateaux in South Wales and the Pennines (Donnelly et al. 2000a, b, 2002; Donnelly 2008) (Fig. 4.26). These features include: open ground fissures ('gulls') that are oblique or sub-parallel to the valley sides, often in en echelon arrays, conjugate sets, or in box-work and saw-toothed networks (e.g. at the Daren Ddu landslide, South Wales); single or clusters of major fissures and scarps parallel to the slope crests (e.g. west of the Darren Goch landslide, South Wales); single fissures or fault scarps that trend obliquely towards the slope crests and across escarpments and hillslope faces (at Allderman's Hill, Derbyshire a south-facing fault scarp is 4 m high and extends for 0.7 km in length); multiple graben that run parallel or oblique to the slope crests (e.g. at Gillot Hey, Derbyshire; Donnelly et al. 2002); and sets of regularly spaced obsequent or antislope scarps on steep slope faces (e.g. at Craig y Ddelw in the Rhondda Fawr Valley). Holmes & Jarvis (1985) reported up to 10-m-high upslope-facing scarps on the flanks of Ben Attow in the Scottish Highlands; these had been classified by Jones & Lee (1994) as 'sackung' failures (i.e. rock flows). Antiscarps have also been described on the flanks of Kirkfell, Wadale (Wilson 2005).

These grabens and extensive faults scarps are believed to be indicative of horizontal extension of the massive Carboniferous sandstone rocks along the plateaux margins. Although the causes of these features remains uncertain, it is believed that they are related to stress relief following the retreat of valley glaciers during the Anglian and Devensian glaciations (480–420 ka BP and 71–10 ka BP, respectively) and subsequent river incision (Donnelly et al. 2002). These processes would have reduced support of the sandstones, causing dilation of joints and fissures and promoting cambering and failure of the strata at the plateaux margins. In addition to stress relief, the melting of ground ice is likely to have resulted in increased pore pressures within the underlying mudstone and/or shale beds. This may have contributed to the initiation of major landslides, plastic deformation and 'squeezing' of the beds. These conditions would have promoted horizontal extension (spreading) and cambering of the sandstones at the plateaux margins.

Cambered structures are believed to be inactive in the UK, the process being associated with periglacial conditions and not purely a function of stress relief (e.g. Parks

Fig. 4.23. (a, b) Peat slides, Ben Gorm, County Mayo, Ireland (photo credit: Emily Farrugia).

Table 4.6. *Peat failure classification (Dykes & Warburton 2007)*

Peat failure type	Definition	Peat deposit	Indicative volume (m^3)	Example	Source
Bog burst	Failure of a raised bog involving break-out and evacuation of (semi) liquid basal peat	Raised bog	10^5–10^7	Solway Moss 'eruption' of 1771 (6–10 Mm^3)	Walker (1772)
Bogflow	Failure of a blanket bog involving the break-out and evacuation of semi-liquid highly humified basal peat from a clearly defined source area	Blanket bog	18 000–375 000	Glendun bogflow, Antrim, November 1963 (0.375 M m^3)	Colhoun et al. (1965)
Bog slide	Failure of a blanket bog involving sliding of intact peat on a shearing surface within the basal peat	Blanket bog	1000–75 000	Meenagharvy bog slides, January 1945, County Donegal, Ireland	Bishop & Mitchell (1946)
Peat slide	Failure of a blanket bog involving sliding of intact peat on a shearing surface at the interface between the peat and the mineral substrate material or immediately adjacent to the underlying substrate	Blanket bog	500–50 000	Channerwick peat slides, Shetland, September 2003	Moore et al. (2006)
Peaty-debris slide	Shallow translational failure of a hillslope with a mantle of blanket peat in which failure occurs by shearing wholly within the mineral substrate and at a depth below the interface with the base of the peat such that the peat is only a secondary influence on the failure	Blanket bog	500–50 000	Prod Hill, North Yorkshire, June 2005	Dykes & Warburton (2007)
Peat flow	Failure of any other type of peat deposit (fen, transitional mire, basin bog, etc.) by any mechanism, including flow failure in any type of peat caused by head loading	Fen, transitional mire, basin bog	4000–5 × 10^6	Derrybrien windfarm, County Galway, Ireland, October 2003 (0.45 Mm^3)	Dykes & Warburton (2007)

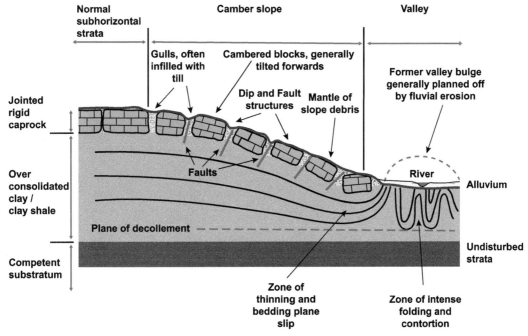

Fig. 4.24. Schematic section showing cambering and valley bulging (adapted from Griffiths & Giles 2017).

1991; Hutchinson 1991a). However, it is suspected that mining subsidence may have reactivated some slope deformation features in the South Wales Valleys where some fault scarps have offset small gullies both laterally and vertically (Donnelly *et al.* 2000a, b).

Fig. 4.25. Gull formed through cambering in the Jurassic Combe Down Oolite Member, an oolitic limestone, overlying the weak mudstones of the Fuller's Earth Formation (photo credit: David Giles).

4.5.3 Large rock slope failures in the Scottish Highlands

Systematic aerial photography and field surveys of the Scottish Highlands have resulted in the identification of 550 rock slope failures (Jarman 2006, 2007, 2009), including rockslides and deep-seated rock slope deformations (rock creep, sagging and rebound; see Hutchinson 1988). The dataset includes 147 failures greater than 0.25 km^2 in area, of which 54 can be considered rockslides and 92 potential deep-seated slope deformations. However, many rock slope failures exhibit characteristics of more than one category, so classification of individual features can be problematic (Ballantyne 2013). Similar features have been identified in the English Lake District (e.g. Wilson 2003, 2005; Clark & Wilson 2004; Wilson & Smith 2006); Davies *et al.* (2013) report that 69 rock slope failures have been recognized, including 1.7 km^2 failure at Robinson–Hindscarth (Wilson *et al.* 2004). Stead & Wolter (2015) provide a critical review of rock slope failure mechanisms and discuss the importance of structural geological controls. Figure 4.27 depicts a classification of failure modes related to structure, specifically large rock glide, rough translational slide, planar translational slide, toe-buckling translational slide, biplanar compound slide, curved compound slide, toppling failure and irregular compound slide.

It is believed that rock slope failure activity was related to conditions following deglaciation ('paraglacial' features) and

Fig. 4.26. Fissures on the moorland plateau on Mynydd James, South Wales, can occur as isolated features or in sets and may reach several tens of metres long and a few metres wide (photo credit: David Giles).

that the process is of little geohazard significance under current conditions (e.g. Jarman 2006). Indeed, it is unclear what environmental changes might be required to reactivate the numerous failures which have undergone only limited post-failure development, and currently occupy what appear to be marginally stable positions on the mountain valley sides.

4.5.4 Flow slides in colliery spoil

Flow slides develop as a result of the metastable collapse of loose, predominantly cohesionless material, resulting in the temporary transfer of part of the total stress onto the pore fluids. This causes a sudden rise in pore pressures and loss in strength. The failed material has a semi-fluid character and can flow at extremely rapid rates of up to 111 m s^{-1} (Hutchinson 1988; Hutchinson *et al.* 2000).

The industrial development of the South Wales Coalfield was accompanied by the largely uncontrolled disposal of various forms of mine waste in heaps (tips) on hillslopes near the mines. The Aberfan disaster of October 1966 highlighted the inherent dangers in this practice (Fig. 4.28).

However, the Aberfan event was not the only flow slide to occur in the coalfield. Between 1898 and 1966 there were a total of 18 events, mainly from tips in the central and eastern parts of the coalfield (Table 4.7; Siddle *et al.* 1996, 2000*b*). Most failures involved fresh or recently tipped spoil, highlighting the importance of incremental loading in the initiation of the slides, and the looseness of the spoil in promoting significant run-out. The largest run-out event was at Abergorchi colliery in 1931, 35 years before the Aberfan disaster. The flow slide travelled around 610 m from the initial tip failure, entering the colliery yard and filling the boiler house. This left barely enough steam to raise the 700 miners who were underground at the time. A major disaster was narrowly averted (Siddle *et al.* 1996). In December 1939 a flow slide from the Cilfynydd colliery tip travelled around 435 m, blocking the main Cardiff–Merthyr road for 5 days and damming the River Taff.

4.5.5 Coastal landslides: cliff behaviour units

Over the last decade or so, the concept of a cliff behaviour unit (CBU) has proved to be an important tool for managing

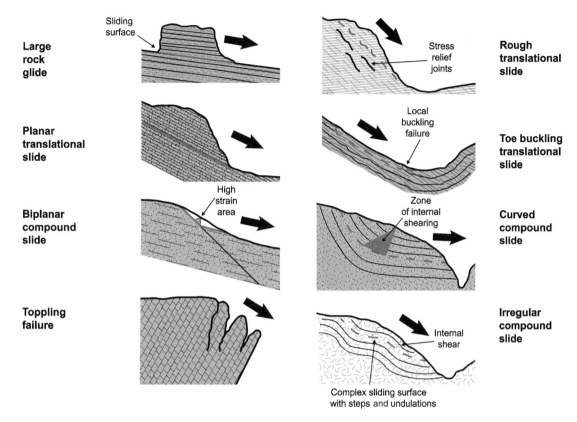

Fig. 4.27. Classification of failure modes related to structures (adapted from Stead & Wolter 2015).

coastal cliffs (e.g. Moore *et al.* 1998, 2010). These units are defined as spanning the nearshore to the cliff top and are coupled to adjacent CBUs within the framework provided by littoral and/or sediment cells. In this context, it is useful to view cliffs as open sediment transport systems characterized by inputs, throughputs and outputs of material, that is, cascading systems (e.g. Brunsden & Lee 2004). A range of types of cliff system can be recognized on the basis of the throughput and storage of sediment within the system (Fig. 4.29).

(1) *Simple cliff systems in hard or soft rock or superficial materials*: these comprise a single sequence of sediment inputs (from falls or slides) and outputs, with limited storage. A distinction can be made between cliffs prone to falls and topples and those shaped by simple landslides. The former is characterized by limited storage of sediment within the cliff system, with material from the cliff top and face reaching the foreshore in a single event. Examples include 'soft' unconsolidated sands and gravels, (e.g. the Suffolk coast), 'harder' soft rock cliffs (e.g. the chalk cliffs of East Sussex; Fig. 4.30) or cliffs developed in highly jointed or faulted rocks (e.g. the Lias Group cliffs of Glamorgan).

(2) *Simple landslide systems*: these comprise a single sequence of inputs and outputs with variable amounts of storage within the failed mass. Debris from the cliff may only reach the foreshore after a sequence of events involving landslide reactivation. Examples include rotational failures on the London Clay Formation cliffs of north Kent, 'soft' glacial till failures (e.g. the Holderness coast) and small mudslides on the north Norfolk (Fig. 4.31) and east Dorset coasts.

(3) *Composite systems*: these comprise a partly coupled sequence of contrasting simple subsystems. The output from one system may not necessarily form an input for the next (e.g. where material from the upper unit falls directly onto the foreshore). Examples include the Durham cliffs, which comprise mudslide systems developed in till over limestone, cliffs that are prone to rockfalls, and the cliffs at Flamborough Head where tills overlie near-vertical chalk cliffs (Fig. 4.32).

(4) *Complex systems*: these are strongly linked sequences of subsystems, each with their own inputs and outputs of sediment. The output from one subsystem forms the input for the next. Such systems are often characterized by a high level of adjustment between process and

Fig. 4.28. The colliery spoil tip failure at Aberfan, South Wales (photo Credit: Paul Maliphant/BGS).

form, with complex feedback mechanisms. Examples include landslide complexes with high rates of throughput and removal of sediment, such as the Naish Farm to Barton-on-Sea cliffs of Christchurch Bay (Fig. 4.33), the west Dorset cliffs and cliffs affected by seepage such as Chale Cliff, Isle of Wight (Fig. 4.34).

(5) *Relict systems* comprise sequences of pre-existing landslide units that are being gradually reactivated and exhumed by the progressive retreat of the current seacliff, for example, parts of the Isle of Wight Undercliff, the Axmouth Undercliff, East Cliff (Lyme Regis; Fig. 4.35) and the 'slope-over-wall' cliffs of SW England.

The concept of the cliff behaviour unit emphasizes the linkage between cliff and foreshore processes. The foreshore (i.e. the intertidal area) is an essential part of the CBU, controlling the scale, duration and frequency of interactions between the cliff system and wave attack. It is an important link between the CBU and the broader littoral cell (coastal process unit). The beach has an important role in determining the rate and timing of landslide events (e.g. Lee 2008). It is also important to appreciate the linkages with other landforms within a littoral cell, with many CBUs acting as sediment sources for beaches, dunes, saltmarsh and mudflats on the neighbouring coastline.

4.6 Causes of landslides

The ultimate cause of all landsliding is the downwards pull of gravity. The stress imposed by gravity is resisted by the strength of the materials forming the slope. A stable slope is one where the resisting forces are greater than the destabilizing forces and, therefore, can be considered to have a margin of stability. By contrast, a slope at the point of failure has no margin of stability, as the resisting and destabilizing forces are approximately equal. The quantitative comparison of these opposing forces gives rise to a ratio known as the Factor of Safety, F:

$$\text{Factor of Safety } F = \frac{\text{Resisting forces}}{\text{Destabilizing forces}} = \frac{\text{Shear strength}}{\text{Shear stress}}. \quad (4.1)$$

The Factor of Safety of a slope at the point of failure is assumed to be 1.0. On slopes of similar materials, progressively higher values of F represent increasingly stable situations with greater margins of stability (Fig. 4.36).

A measure of the maximum shear stress that a material can withstand before failure occurs is given by the shear strength (Fig. 4.37). In 1776 Coulomb, a French engineer, recognized that shear strength was a function of the interaction between mineral particles, notably the friction (the result of the compressive forces that hold particles together, $\tan \varphi$) and cohesion (produced by interparticle chemical bonding or cementation, c) and the normal stress σ (imposed at right angles to the slope):

$$s = c + \sigma (\tan \varphi). \quad (4.2)$$

However, the presence of water within the pore spaces or fissures can reduce the shear strength. This is because it exerts its own pressure, which serves to reduce the amount of particle on particle contact and hence the frictional component of shear strength. The frictional resistance therefore depends on the difference between the applied total normal stress and the porewater pressure. The part of the normal stress that is effective in generating shear resistance is known as the effective stress.

The effect of pore pressure (u) results in a modification to the Coulomb equation:

$$s = c' + (\sigma - u)(\tan \varphi') \quad (4.3)$$

in which c' and φ' are modified parameters with respect to effective stress.

The amount of shear strength required to resist the destabilizing forces is termed the mobilized shear strength, which is

Table 4.7. *South Wales coalfield: the chronology of flow slide failures at colliery spoil tips (from Siddle et al. 1996). ROM, run of mine waste; WD, washery discard; BA, boiler ash; T, tailings*

Tip	Date of flow slide	Colliery established	Period of tipping on failed tip	Tipping activity at date of failure	Type of spoil	Run-out (m)	Run-out slope (°)
National	1898	1881	?	?	ROM	170	21
Pentre	1909	1864	1875–1908	Disused 1 year	ROM	?	18
Craig–Duffryn	1910	c. 1855	c. 1898–1914	Active	ROM	60	<15
Maerdy	1911	1875	1878–1935	Active	ROM	270	11
Cefn Glas	1925	?	c. 1920–1950	Active?	ROM	190	22
Bedwellty	1926	?	c. 1900–1926	Active	ROM	250	13
Rhondda Main	1928	1912	1912–1924	Disused 4 years	ROM, WD	140	23
Abergorchi	1931	Early 1870s	pre-1914	?	ROM	610	8
Fforchaman	1935	Pre-1869	1913–1965	Active	ROM	100	10
Cilfynydd	1939	?	1910–1963	Disused 2 months	ROM, WD, BA	495	<17
Glenrhomnda	1943	c. 1910	c. 1920–1966	Active	ROM	280	11
Aberfan	1944	1869	1933–1944	Active	ROM	200	12.5
Nantewlaeth	1947–1960?	1913	c. 1937–1960	Active?	ROM, WD	120	8
Fernhill	1960	?	?	Active	ROM, WD	?	15
Aberfan	1963	1869	1958–1966	Active	ROM, T	230	10.5
Mynydd Corrwg Fechan	1963	1905	1958–1970	Active	ROM, WD, T	235	17
Parc	1965	1865	1959–1965	Active	ROM, WD, T	140	<23
Aberfan	1966	1869	1958–1966	Active	ROM, T	605	<12

SIMPLE CLIFFS

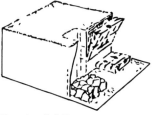

Topples & falls

Rotational landslide

Mudslide

COMPOSITE CLIFFS

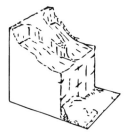

Rotational landslide in glaciogenic deposits over hard rock

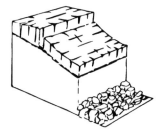

Block slide in hard rock over a thin clay layer

COMPLEX CLIFFS

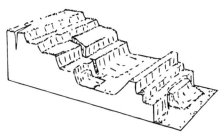

Deep seated landslide with failure at more than one level

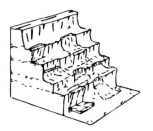

Seepage erosion cliff, alternating sand & clay

RELICT CLIFFS

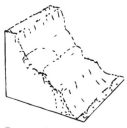

Dormant

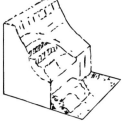

Reactivated

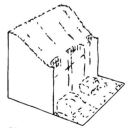

Slope-over-wall

Fig. 4.29. Types of coastal cliff (adapted from Lee & Clark 2002).

Fig. 4.30. Highly jointed chalk cliffs in the Seaford Chalk Formation, Birling Gap, East Sussex (photo Credit: David Giles).

generally less than the total shear strength available. As destabilizing forces increase, for example, due to the oversteepening of a slope by basal erosion, deformations occur as shear strength is mobilized. At the point when failure occurs, the shear strength is fully mobilized along the failure surface. The shear strength of a material depends upon both the nature of the material itself and the presence of water in the fissures and pores (porewater pressure).

A slope is only as strong as its weakest horizon, often a clay. Clays present in the Gault Clay and London Clay formations are known as brittle materials because, once they have been subject to more than the maximum stress they can withstand and have failed, further displacements are possible at lower levels of stress; the shear strength of the clay decreases from a peak value to a lower residual value. For a non-brittle material, the shear strength will increase to an ultimate value and will then remain constant.

Reactivation occurs when part or all of a previous landslide mass is involved in new movements along pre-existing shear surfaces where the materials are at residual strength (e.g. Hutchinson 1987, 1988). Reactivation can occur when the initial failed mass remains confined along part of the original shear surface. As the shear strength of the material along the shear surface is at or close to a residual value, landslide movement can be initiated by smaller triggering events than those needed to initiate first-time failures. Such failures are generally slow moving with relatively limited displacements associated with each phase of movement.

The stability of a slope can be viewed in terms of its ability to withstand potential changes: stable, when the margin of stability is sufficiently high to withstand all transient forces in the short to medium term (i.e. hundreds of years), excluding excessive alteration by human activity; marginally stable or metastable, where the balance of forces is such that the slope will fail at some time in the future in response to transient forces attaining a certain magnitude; or actively unstable slopes, where transient forces produce continuous or intermittent movement.

When a slope fails, the displaced material moves to a new position so that equilibrium can be re-established between the

Fig. 4.31. Example of a mud flow in glaciogenic deposits, Overstrand, Norfolk (photo credit: David Giles).

destabilizing forces and the strength of the material. Landsliding therefore helps change a slope from a less stable to a more stable state. No subsequent movement will occur unless the slope is subject to processes that, once again, affect the balance of opposing forces. In many inland settings landslides can remain inactive or dormant for thousands of years. However, on the coast and along active river cliffs, erosion continues to remove material from the cliff foot, reducing the margin of stability and promoting further recession.

It possible to recognize four categories of causal factors that contribute to landsliding as follows.

(1) *Controlling factors* influence the overall susceptibility of an area to landsliding and control the types of event that occur should a slope fail (e.g. ground materials, geological structure and topography). Of particular importance is the distribution of past landslides and the pre-existing shear surfaces of non-landslide origin (e.g. flexural shearing during the folding of interbedded sequences of hard rocks and clay-rich strata, and periglacial solifluction).

(2) *Preparatory factors* work to make the slope increasingly susceptible to failure without actually initiating landsliding (e.g. the long-term effect of erosion at the base of a slope, weathering or the prolonged periods of heavy rainfall).

(3) *Triggering factors* actually initiate landslide events (e.g. rainstorm events).

(4) *Sustaining factors* continue to promote the landslide events (e.g. continual rainfall).

Few mass movements can be ascribed to a single causal factor. However, rainfall is a factor in most contemporary inland landslide activity, acting either as a preparatory or triggering factor (or both).

On the coast, the frequency and magnitude aspects of cliff recession can be very complex and involve episodic mass movement events. These are associated with the periodic failure of cliffs in response to preparatory factors, such as

Fig. 4.32. Glaciogenic tills capping vertical chalk cliffs, Sewerby, near Flamborough Head, Yorkshire (photo credit: David Giles).

Fig. 4.33. Coastal slope failures in the Becton Sand and Chama Sand Formations, Barton on Sea, Hampshire (photo credit: David Giles).

Fig. 4.34. The complex landslide system at Blackgang, Isle of Wight. Upper Greensand Formation overlying Gault and Sandrock formations of the Lower Cretaceous (photo credit: David Giles).

weathering, and triggering factors, such as large storms or periods of heavy rainfall (Lee & Clark 2002; Brunsden & Lee 2004). For example, at Black Ven, Dorset there is an estimated 50–60 year cycle of major activity associated with gradual increases in slope angle and landslide events associated with the occurrence of wet periods (Chandler & Brunsden 1995; Brunsden & Chandler 1996).

4.6.1 Landslides and rainfall

It has long been recognized that contemporary landslide activity in the UK can be associated with periods of heavy rainfall or intense storms; Jones & Lee (1994) reported that 'Changes in water content (of the ground materials) can quickly affect slope stability and have been responsible for triggering or reactivating more landslides in Great Britain than any other factor'.

As the porewater pressures increase, so the effective shear strength available to resist the destabilizing forces declines. A common landslide scenario is for a transient event (e.g. an intense storm) to trigger landslide activity after there had been a gradual decline in stability (e.g. a prolonged wet period); in other words, rainfall and groundwater can have both a long-term and short-term influence on slopes. Clark *et al.* (1995) were able to demonstrate that the extensive 1994 landslide movements at Blackgang, within the Isle of Wight Undercliff, were linked to almost-continuous, and at times intense, rainfall over the previous month (Fig. 4.38). A similar pattern was revealed through the analysis of landslide activity in SW England and South Wales between March 2006 and August 2013 (Pennington *et al.* 2014). They demonstrated that the probability of at least one landslide varied over time with the changing antecedent precipitation conditions (Fig. 4.39).

Waltham & Dixon (2000) established a relationship between landslide movements and rainfall at Mam Tor, Derbyshire. Increased movements occur when winter rainfall (November–February) exceeds 210 mm (i.e. 50% above the mean winter rainfall) during a winter that follows a 6-month period of more than 750 mm (i.e. the mean rainfall for the period August–December, plus the wetter of the adjacent July or January). These combined 1-month and 6-month thresholds correlate with landslide movements for 89 out of 96 years (1903–1998). However, there are years when no

Fig. 4.35. Progressive retreat of the sea cliff in the Charmouth Mudstone Formation, East Cliff, Lyme Regis, Dorset (photo credit: David Giles).

movements occurred when rainfall exceeded the thresholds, suggesting a complex response to rainfall.

The variable temporal and spatial response to rainfall events ensures that it is not always possible to define a clear basis for forecasting landsliding activity (e.g. Dikau & Schrott 1999). An important factor in understanding this complexity is that different types of landslides may be related to different climatic threshold conditions.

Characteristic settings can be categorized as either: shallow translation slides, peat and debris flows; or deep-seated landslide.

Shallow translational slides, peat slides and debris flows in steep catchments are often associated with high-intensity rainstorms. Landslides tend to be triggered within minutes or hours of the event. For example, surface run-off supplies water to debris masses which have accumulated within and adjacent to stream channels. This increases the pore pressures within the debris, initiating a debris flow. Heavy rainfall was the trigger for a bog flow in Co. Antrim in November 1963 that surged down the Glendun River, destroying a farmhouse and damaging farm equipment (Colhoun *et al.* 1965). Dykes & Kirk (2000) describe the occurrence of a multiple peat slide

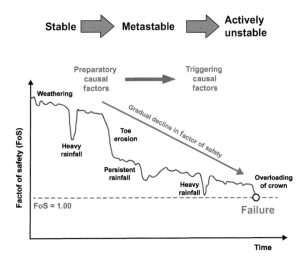

Fig. 4.36. Progressive changes in slope Factor of Safety over time, leading to ultimate failure: timeline from stable to metastable to actively unstable conditions (adapted from Popescu 2002).

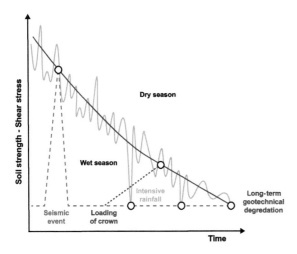

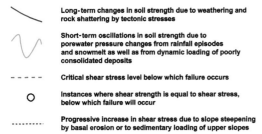

Fig. 4.37. The relationship of slope stress conditions and slope strength over time (adapted from Julian & Anthony 1996).

event on Cuilcagh Mountain, Northern Ireland, in response to extreme intense rainfall in October 1998; the total area affected by sliding was 30 500 m^2.

Multiple peat slides and debris flows occurred in the Shetland Islands on 19 September 2003 during a period of intense rainfall (Moore *et al.* 2006 suggest up to 30 mm h^{-1}). These slides caused temporary closure of the A970 between Cunningsburgh and Levenwick on the south mainland of Shetland. The largest failures initiated above the A970 and resulted in high-velocity run-out of peat debris and floodwater onto the road which became impassable. Several cars narrowly missed being hit by the peat slides and debris flows.

Around 66 landslides occurred in SW England and South Wales during the very wet period between November 2012 and January 2013. The majority of these events were small failures of man-made slopes such as road and railway embankments and cuttings. Pennington *et al.* (2014) demonstrated that these failures were closely correlated with the antecedent precipitation over 1–7 day periods.

Deep-seated landslides are generally associated with prolonged heavy rainfall. Positive pore pressures along a shear surface, induced by a rising groundwater table, often trigger this type of failure. As it is the relative porewater pressure (the ratio between pore pressure and the total normal stress on the shear surface) which determines stability, deeper landslides need larger absolute amounts of water for triggering conditions than shallower slides. In general, the deeper the slide the longer the period of antecedent heavy rainfall needed to initiate failure. The period may vary from several days (e.g. Reid 1994) to many months (e.g. Lee *et al.* 1998 for the Isle of Wight Undercliff). In many areas, it may be that pattern of wet years that appears to control the occurrence of landslides (e.g. Bromhead *et al.* 1998 at the Roughs, Kent (Fig. 4.40); Brunsden & Lee (2004) at Lyme Regis, Dorset). The time span of antecedent rainfall conditions for analysing triggering events depends on landslide size (i.e. depth), type and the geological geomorphological characteristics of the site (Van Asch *et al.* 1999).

4.6.2 Anthropogenic effects

Anthropogenic activities are often important factors in landslide activity in the UK; Jones and Lee (1994) reported that 'instability problems are not Acts of God; unpredictable, entirely natural events .... Man's role in initiating or reactivating many slope problems should not be underestimated'.

For example, the provision of and subsequent neglect of land drainage systems has had a significant impact on some slopes in southern England (e.g. Bromhead & Ibsen 1997). Blocked drainage pipes (e.g. as a result of crushing by heavy farm machinery) cause porewater pressures to build up, promoting the progressive spread of shallow landsliding upslope from the blockage. This process has been suggested for development of a series of shallow slides in Wadhurst Clay Formation near Robertsbridge, Kent (Bromhead *et al.* 2000).

One of the most serious effects is often the artificial recharge of the groundwater table, and hence increased porewater pressures. Table 4.8 provides an indication of the water usage per person in a typical home and highlights how consumption has risen by around 50% over the last 20 years. Disposal of this water in urban areas generally involves sophisticated collection and treatment systems. However, isolated and rural areas rely on onsite biological treatment using septic tanks and cesspits. Surplus water is allowed to soak into the ground. As Watson & Bromhead (2000) point out, the discharge of water consumption for an average home corresponds to a local increase in effective rainfall from temperate to tropical levels. The resulting increase in groundwater levels can lead to the generation of active instability, especially on unstable coastal cliffs as reported by Watson & Bromhead (2000) at Warden Point, Isle of Sheppey, Kent, in the 1970s (Fig. 4.41).

Water supply and drainage pipes can be particularly vulnerable to ground movement and may leak for considerable periods before repairs are undertaken. Following the 1988 landslide problems at Luccombe, Isle of Wight, it was concluded that poor maintenance of the water supply and drainage systems were the key factors in triggering the problems (Lee & Moore 1989; Moore *et al.* 1991). Water pipe leakage

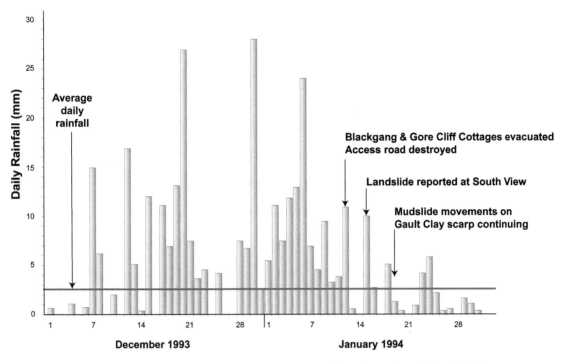

Fig. 4.38. Blackgang, Isle of Wight: rainfall and landslide activity (adapted from Clark *et al.* 1995).

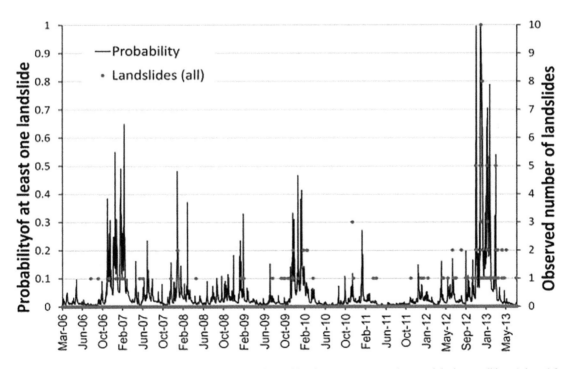

Fig. 4.39. The probability of at least one landslide on a given date, taking into account antecedent precipitation conditions (adapted from Pennington *et al.* 2014).

Fig. 4.40. Oblique aerial photograph of the Roughs landslide area near Hythe, Kent, taken in summer 1996 (photo credit: Eddie Bromhead).

was shown to be 4350 L/day (29% of the total supply), whereas the onsite soakaway drainage contributed a further 15 000 L/day.

Housing development can also affect slope stability through inappropriate excavations to create level plots for buildings, as occurred during regrading prior to house building along Marine Parade, Lyme Regis, Dorset in February 1962 (Hutchinson 1962; Lee 1992). Around 20 000 m³ of material was removed from the slope, which had been frequently affected by shallow landsliding. A few days after the earth-moving operation had finished a large, deep-seated slide developed which moved several metres in a few minutes.

Mineral extraction, especially underground mining, has been a contributory cause to many recent failures. Many of the current instability problems in South Wales and in the Ironbridge Gorge appear to be associated with ancient landslides

Table 4.8. *Typical UK domestic water consumption (from Tebbutt 1998)*

Use	1978 (L per person per day)	1992 (L per person per day)	1998 (L per person per day)
Toilet flushing	32	35	50
Drinking, cooking and dishwashing	33	45	38
Baths and showers	17	25	35
Washing machines	12	15	20
Garden watering	1	5	7
Car washing	1	2	2
Total	**96**	**127**	**142**

that have been reactivated by housing development or industrial activity (Halcrow 1988; High-Point Rendel 2005*b*). For example, a detailed review of damaging landslide activity in the Rhondda valleys, South Wales, revealed 82 events over the last 100 years including: the dumping of colliery spoil at Pentre, Rhondda Cynon Taf, Wales, reactivated an old translational slide in 1916, destroying a billiard hall and several houses (Fig. 4.42); the East Pentwyn and Bournville landslides at Blaina, where in 1954 the advancing toes reached within 10 m of a row of 12 cottages (Fig. 4.43a, b) that were later evacuated and demolished (the landslide reactivation is believed to have been associated with the collapse of shallow mine workings; Halcrow 1989); and a landslide at Penrhiwgwynt, Porth, in 1965, caused by disposal of waste from water tanks onto an old landslide, threatened properties.

In the South Wales coalfield mining subsidence may have led to the reactivation of prominent faults that appear to control the distribution of some major landslides, for example, the Darren Goch landslide in Garw valley, and the East Pentwyn and Bournville landslides (Fig. 4.43a, b) in the Ebbw Fach valley (Donnelly *et al.* 2000*a*, *b*). Coal mining is believed to have contributed to the development of a large (2–3 ha) compound landslide at Bolsover, east Derbyshire, between 1991 and 1993. The movements resulted in severe damage to 27 properties, 10 of which had to be demolished, and around £1 M worth of losses (Cobb *et al.* 2000). This slide involved the retrogression and widening of an earlier slide that had occurred in 1923. Following detailed analysis, it was concluded that subsidence of abandoned longwall workings (the last workings had been in 1984) had led to a substantial reduction in horizontal stresses at the landslide toe, initiating strain-softening and progressive failure of the strata.

Contemporary landslide activity within the Ironbridge Gorge, Shropshire (e.g. the Buildwas and Jackfield landslides, see Table 4.5), is likely to be related, at least in part, to the effects of mining of coal and clay for brick and tile manufacture (High-Point Rendel 2005*a*, *b*). The legacy of the industrial revolution has been the loading of the slopes with tile and mine waste and subsidence into the voids created by mining. In addition, mining would also have affected the natural hydrogeology of the area, providing preferential pathways for water flow and enabling the possible build-up of pressure at mined faces and faults. The problem of groundwater is likely to have been made worse as the pumping of mine water ceased with the abandonment of the mines.

Coastal instability can also be exacerbated by the disruption of sediment transport, as occurred at Folkestone Warren, Kent, following harbour construction (Hutchinson 1969; Hutchinson *et al.* 1980). The most recent phase of landslide activity was probably a consequence of the expansion of the Folkestone Harbour facilities over the years 1810–1905. In particular, the extension of the main pier resulted in the disruption of wave- and current-induced littoral drift of sand and shingle eastwards through the area, leading to the build-up of trapped material west of the pier and beach shrinkage through undernourishment to the east, at the foot of the Warren. Reduced beach volume would have led to increased wave erosion of the toes of pre-existing rotated blocks that slid seawards, removing support from the base of the cliffs.

4.6.3 Landslide controls: the influence of geology

An important feature of the pattern of landsliding in the UK is the strong association between hillslope instability and the underlying materials (e.g. Jones & Lee 1994; Cooper 2007). In general, the shear strength of a material decreases as the clay content rises. Clay slopes are therefore particularly prone to landsliding. Slides also occur frequently on slopes developed in a combination of impermeable fissured clays overlain by massive, well-jointed caprocks of limestone or sandstone. Jones & Lee (1994) identified a number of geological settings associated with particular types of landsliding, referred to as Groups A–F and described in the following.

Group A includes stiff fissured clays and mudrocks where low shear strengths and a high susceptibility to weathering and stain-softening has led to large numbers of failures on oversteepened slopes. The most common forms of landsliding on these materials include single and successive rotational slides, debris slides and mudslides. The most landslide-prone materials in this category are the argillaceous lithologies within the London Clay Formation (Fig. 4.44), the Gault Formation and Lias Group, and other overconsolidated clays of southern and eastern England.

Group B includes well-jointed, faulted, cleaved and foliated hard rocks in which the pattern of discontinuities provides potential failure surfaces or weak zones within the rock mass. Rockfalls, topples, sagging failures and rockslides are the dominant modes of failure of these rocks, which include horizons such as the Upper Dalradian quartz-mica-schist and Moinean mica-schist in the Scottish Highlands, Permian basal breccias and Devonian limestones in Torbay, the Carboniferous limestones of Wales (Fig. 4.45) and northern England, and the Chalk along the south coast of England.

Group C describes the occurrence of sequences of lithologically variable rock types that create potentially unstable conditions. For example, many areas of known landsliding are associated with the presence of thick horizons of impermeable fissured clays or mudrocks overlain by a massive, but well-jointed, permeable caprock of sandstone, limestone or volcanic rocks. Multiple rotational slides and compound failures are the dominant forms of landsliding associated with this setting. Classic examples of areas with landsliding promoted by these unstable combinations of rocks include lithologies from the Upper Greensand Formation and Chalk Group overlying the clays of the Gault Formation along the Isle of Wight Undercliff and Folkestone Warren (Fig. 4.46), the Carboniferous Coal Measures of South Wales, the Inferior Oolite Group and underlying Lias Group clays in the Cotswolds, and the Millstone Grit Group of the Pennines. In Northern Ireland, the combination of basalts, chalk and Liassic shales give rise to spectacular multiple rotational slides on the margins of the Antrim plateau.

Group D includes rocks which weather to produce sandy regoliths with high infiltration rates, susceptible to

Fig. 4.41. Major coastal landslides in the London Clay Formation cliffs, Warden Point, Isle of Sheppey, Kent, in 2011 (photo credit: Eddie Bromhead).

debris-flow or debris-slide activity, especially during intense rainstorms. In Scotland, debris flows are more frequent on rocks which weather to a coarse-grained matrix, such as granites and sandstones, rather than those such as schist which yield cohesive clay-rich soils (see Winter 2019; Fig. 4.47).

Group E describes the occurrence of weak superficial deposits, especially glaciogenic tills and 'head' deposits. A combination of relatively small rotational failures and mudslides are the most frequent types of landsliding, notably in parts of South Wales (Fig. 4.48), the Pennines and the Vale of Eden.

Group F are thick blanket peats that mantle upland slopes in Britain and Ireland (Fig. 4.49). These materials are prone to peat failures, especially after heavy rainfall or during snow melt, when the surface layers become too weak to retain the semi-liquid mass of peat below (e.g. Warburton *et al.* 2004).

The association between materials and landslide activity is particularly pronounced on the coast. Here, a combination of the geology, exposure to wave attack and associated landslide types give rise to a variety of characteristic cliff forms (Hutchinson 1984; Jones & Lee 1994; Lee & Clark 2002; Brunsden & Lee 2004).

(1) *Cliffs developed in weak superficial deposits*: the east coast of England from Flamborough Head to Essex and parts of west Wales and the Cumbrian coast are largely developed in thick sequences of glacial till interbedded with sands and gravels. The sea can rapidly erode these deposits; for example, on parts of the glacial till cliffs along the north Norfolk coast there has been over 175 m of recession since 1885 (Hutchinson 1976; Clayton 1980, 1989; Poulton *et al.* 2006). McGreal (1977, 1979) recorded annual recession rates of between 0.2–0.8 m on the glacial till cliffs on the Mourne coast of Northern Ireland (Fig. 4.50).

(2) *Cliffs developed in weak superficial deposits overlying jointed rock*: much of the NE coast, from Durham to Flamborough Head, is developed in glaciogenic till overlying Jurassic sedimentary rocks (Fig. 4.51). Cliff recession generally involves the relatively slow retreat of the rock cliff through falls and cave collapses, and shallow mudslide activity and surface erosion of the tills above. However, in certain places, these cliffs can be prone to major dramatic landslides; the Holbeck Hall failure of June 1993 in Scarborough, North Yorkshire was the most recent example (e.g. Lee 1999).

Much of the coast around the SW peninsula comprises near-vertical hard rock cliffs capped by thin periglacial head deposits. This combination gives rise to the

Fig. 4.42. Pentre, Rhondda Cynon Taf, colliery spoil tips being removed in the 1980s (photo credit: Peter Brabham).

characteristic 'slope-over-wall' cliffs of this area, with a steep upper cliff section developed under periglacial conditions and a lower vertical sea cliff fashioned by contemporary wave action.

(3) *Cliffs developed in stiff clay*: stiff clays are particularly prone to landsliding with classic examples occurring along the shore of the Thames estuary in parts of Essex and Kent, where cliffs up to 40 m high developed in the London Clay Formation have repeatedly failed in response to marine erosion (e.g. Hutchinson & Gostelow 1976; Dixon & Bromhead 1991). This results in average retreat rates of up to 2 m a^{-1} (Hutchinson 1973). The Liassic shales and till cliffs of the Antrim coast are famous for the mudslide activity, notably at Minnis North (e.g. Prior *et al.* 1968, 1971; Prior & Stevens 1972; Hutchinson *et al.* 1974; Craig 1979).

(4) *Cliffs developed in weak sandy strata*: along the south coast of England, cliffs developed in Palaeogene sands and gravels occur at Newhaven, west of Lee-on-the-Solent, the Isle of Wight (Fig. 4.52) and at Bournemouth. These materials are prone to rapid erosion, mainly through frequent small-scale slumps, seepage erosion, cliff falls and surface erosion by water (e.g. Hutchinson 1980).

(5) *Cliffs developed in sequences of stiff clays and weak sandy strata*: this geological setting can give rise to some of the most dramatic forms of cliff recession. There are major landslide complexes on the north coast of the Isle of Wight, especially at Bouldnor, and at Fairlight Glen on the Sussex Coast (Fig. 4.53). At Barton-on-Sea in Christchurch Bay landslides extend for 5 km on 30-m-high cliffs developed in the clays and sands of the Barton Clay Formation overlain by Plateau Gravel. The classic landslide areas of the west Dorset coast – Black Ven, Fairy Dell and Golden Cap – are developed in Lias clays overlain by relatively weak Foxmould Member of the Upper Greensand Formation and head deposits. This setting can also give rise to cliffs prone to seepage erosion (Hutchinson 1982), as at Chale, Isle of Wight, and the eastern parts of Christchurch Bay.

(6) *Cliffs developed in stiff clay with a hard cap-rock*: the largest coastal landslides occur in situations where a thick clay stratum is overlain by a rigid caprock of sandstone or limestone, or sandwiched between two such layers. Among the most dramatic examples are Folkestone Warren, Kent, where the high Chalk cliffs have failed on the underlying Gault Clay, the Isle of Portland (Fig. 4.54) and the Isle of Wight Undercliff. The Landslip Nature Reserve, on the East Devon coast, is another such area and is the site of the 1839 Bindon landslide (Pitts & Brunsden 1987). Prominent rotational slides have developed in basalts and the underlying Liassic shales in White Park Bay on the Antrim coast.

(7) *Cliffs developed in bedded, jointed weak rock*: the steep, jointed chalk cliffs of Kent, Sussex, Isle of Wight and Dorset (Fig. 4.55) are prone to frequent rockfalls, weathering and relatively high rates of erosion. Less

Fig. 4.43. Landslides at (**a**) East Pentwyn, Bournville, in 1954 and (**b**) Rhondda Cynon Taf, Wales, in 2017 (photo credit: Peter Brabham).

Fig. 4.44. Shear surface in a stiff fissured clay, London Clay Formation, Warden Point, Isle of Sheppey, Kent. *In situ* mass below hammer, displaced mass above (photo credit: David Giles).

Fig. 4.45. Rock topples and rockfalls in heavily jointed Carboniferous strata, Ebbw Fach, South Wales (photo credit: Peter Brabham).

Fig. 4.46. Folkestone Warren, Kent, landslide complex with White Chalk Subgroup overlying the Gault Formation (photo credit: David Giles).

commonly, large falls occur on a number of coasts including the Triassic sandstone cliffs of Sidmouth, Devon, and the Liassic limestone cliffs of south Glamorgan. In some settings, sequences of sandstone, mudrocks and limestones can give rise to composite cliff profiles, because of the differences in erodibility between the rock types. Examples include the Wadhurst sandstones and overlying clays on the Sussex coast, east of Fairlight, and the variable sequences of Jurassic shales and sandstones on the North Yorkshire coast between Robin Hoods Bay and Staithes.

4.7 Phases of landslide activity

Stabilised or fossil landslides are perhaps too often considered as a finite condition which occurs in a short time interval, but in many instances, landsliding results from slope material deterioration over long periods of time when the original stresses were initiated by valley bulging, cambering or topple flexuring. These deformation structures were often created at times which anticipated the landslip movement by several thousand years, but they had a profound influence upon the way in which the subsequent event took place. (Johnson 1987).

It is widely recognized that the UK landscape rarely reflects any one climate or period of geomorphological change; rather, it is usually a palimpsest of superimposed histories (i.e. like a surface which has been written on many times after previous inscriptions have only been partially erased; e.g. Twidale 1985). It therefore follows that a landscape can comprise a mosaic of landslide features of different ages and origins, some of which may be of considerable antiquity. For example, the South Wales Coalfield is a landscape dominated by glacial- and post-glacial-related landslides upon which are superimposed the recent landslides caused, directly or indirectly, by mining practices.

Different landslides may represent the response to different combinations of causal factors operating over different timescales. The distribution of recorded landslides in

Fig. 4.47. The semipelite metamorphic bedrock of the Ben Lui Schist Formation, A85 Glen Ogle, Scotland (photo credit: Mike Winter).

Fig. 4.48. Small rotational failures and subsequent flows in glaciogenic tills, Cwmtillery, South Wales (photo credit: David Giles).

Fig. 4.49. Peat slide, Ben Gorm, County Mayo, Ireland (photo credit: Emily Farrugia).

Britain (Jones & Lee 1994) may therefore reflect a variety of factors, as described in the following sections.

4.7.1 Repeated phases of glacial and periglacial conditions

The occurrence of repeated phases of glacial and periglacial conditions during the Pleistocene has left a legacy of numerous major inland landslides and solifluction sheets (e.g. during the Devensian glaciation). Pollen analysis of lakebed sediments indicates that the Whitestone Cliff landslide in the North York Moors occurred during 15–10 ka BP (Blackham *et al.* 1981). Chandler (1976) proposed a Devensian date for slides in the Gwash Valley, Leicestershire on the evidence of landslide debris having overrun an Ipswichian age terrace.

Although southern Britain remained free of glacial ice, it would have experienced harsh periglacial conditions with deeply frozen ground and limited vegetation, interspersed with warmer periods when the ground thawed. The thaw would have released larger volumes of meltwater than could be accommodated in the available pore spaces. With drainage impeded by permanently frozen ground below, the water would have been unable to flow away fast enough, with the result that excess porewater pressures (in excess of hydrostatic) would have been generated. When the pore pressures equalled the normal stress, the coefficient of friction would be zero and the available strength equivalent to the undrained shear strength (C_u). Instability would have occurred even on very gentle slopes, and often on angles as low as 3° for Gault Clay (7° is necessary for landsliding in present-day conditions; Weeks 1969). At Sevenoaks, Kent, it was estimated that that sliding could have occurred on slopes as gentle as 1.5–2.0° (Skempton & Weeks 1976).

Sustained periglacial conditions are also considered to be associated with the development of rockfall talus slopes in upland regions. Retreat of the Late Devensian ice sheets would have exposed numerous glacially steepened, unstable rockwalls that subsequently became the site of accelerated rockfall activity (Fig. 4.56). Ballantyne & Harris (1994) recognize two broad groups of talus slope: (1) those that are the result of the rapid disintegration of rockwalls immediately after glaciation; these slopes would have accumulated very rapidly in the Late glacial period (at two orders of magnitude greater than current rates of rockfall activity; Ballantyne & Harris 1994) and have since experienced only limited growth; and (2) other slopes may be the product of periglacial conditions at the end of the Loch Lomond Stadial (LLS) and may still be actively accumulating, albeit at restricted rates

Fig. 4.50. Rapidly eroding cliffs in glaciogenic deposits, Sidestrand, North Norfolk (photo credit: David Giles).

(estimates of rockwall retreat rates range from $c.$ 0.01 mm a^{-1} to $c.$ 0.06 mm a^{-1}; Ballantyne & Eckford 1984; Stuart 1984).

4.7.2 Impact of drainage adjustments during deglaciation

An example of the impact of drainage network adjustments during deglaciation is the extensive instability along the Cotswolds escarpment, believed to be the product of the intensive basal erosion caused by the drainage of an Anglian phase (480–428 ka BP) pro-glacial lake in Leicestershire (Lake Harrison) along the Avon valley into the Severn (e.g. Bowen *et al.* 1986). Ironbridge Gorge was fashioned by the overspill of a glacial lake (Lake Lapworth) developed in the Cheshire Basin during the Wolstonian and Devensian glaciations. This lake overspilled southwards through the gorge, initiated intense fluvial erosion and created extensive areas of unstable slopes (e.g. Henkel & Skempton 1954). Maddison *et al.* (2000) have suggested that rapid drainage of Lake Teifi in West Wales led to undercutting of the lakebed sediments that had infilled pre-existing valleys, triggering the development of multiple translational failures. Reactivation of one such slide in 1994 at St Dogmaels, Pembrokeshire, damaged 25 properties (e.g. Hobbs *et al.* 1994; Maddison 2000).

4.7.3 Postglacial slope responses

Postglacial slope responses, including Late Glacial and early Holocene rock slope failures, are associated with the long-term impacts of glaciation and subsequent deglaciation. Repeated glacial loading and unloading is believed to promote stress relief within the rock masses, reducing them to a state of conditional stability whereby failure may be triggered by factors such as postglacial stress release (the development of tensile stresses in a rock mass as a result of glacial unloading), the removal of supporting ice during deglaciation ('debuttressing'), thaw of permafrost ice in rock joints,

Fig. 4.51. Glaciogenic deposits overlying mudstones of the Ampthill Clay and Kimmeridge Clay formations, Filey Bay, Yorkshire (photo credit: David Giles).

seismic activity, build-up of joint-water pressures or progressive rock slope weakening by incremental gravitational slope deformation (Ballantyne 2013).

Deglaciation of the Scottish Highlands occurred in two distinct phases: (1) the shrinkage of the last British–Irish Ice Sheet during the Last Glacial Maximum (LGM) of around 26–21 ka covered all of Scotland; cosmogenic dating has shown that summits in NW Scotland were emerging from the ice sheet by 16–15 ka (Fabel *et al.* 2012); and (2) retreat of the LLS (12.9–11.7 ka) glaciers, which occurred during the period 12.5–11.5 ka (Ballantyne 2012).

Recent surface exposure dating of the run-out debris from six major rock slope failures outside the LLS limits has indicated failure dates of between 16.9 and 6 ka (Table 4.9; Ballantyne *et al.* 2009; Ballantyne & Stone 2009, 2013). However, the 11 dated rock slope failures within the LLS limits produced exposure ages of around 11.9–1.5 ka (Ballantyne & Stone 2013).

Ballantyne & Stone (2013) showed that the rockslides fall into two groups: those that occurred during or within a few centuries of local deglaciation (6 cases) and those where failure was delayed and occurred at least a millennium after deglaciation (11 cases). The results indicate that failure was relatively frequent immediately following deglaciation, but also that major rock slope failures continued to occur throughout the Holocene (Fig. 4.57). Rock slope failure activity peaks shortly after $t = 0$ followed by gradual decline until $t \approx 5.5$ ka, followed by two secondary peaks at $t \approx 7.5$ ka and $t \approx 9.9$ ka.

Ballantyne (2013) also suggests that numerous late glacial failures would have occurred within LSS limits, during the interval between ice-sheet deglaciation (*c.* 15–14 ka) and the final retreat of glacier ice (12.5–11.5 ka) at the end of the LLS, but have not been recorded because the run-out debris was removed by LLS glaciers.

The rapid response failures are probably associated with the development of near-surface fractures in rock slopes as a result of paraglacial stress release that results when glacially induced confining stresses are removed (Ballantyne & Stone 2013; Ballantyne *et al.* 2014). However, the delayed response failures are believed to be the result of high-magnitude seismic activity associated with differential isostatic uplift following glacial retreat (e.g. Ballantyne 1991, 1997). In the western Highlands, for example, Davenport *et al.*

Fig. 4.52. Rapid erosion and cliff failure in Palaeogene sands, Alum Bay, Isle of Wight (photo credit: David Giles).

(1989) demonstrated that a combination of isostatic uplift and tectonic stress resulted in appreciable seismic activity throughout the Holocene, producing earthquakes of magnitude 6.5–7.0 at the end of the LLS, diminishing to magnitude 5.0–6.0 events by $c.$ 3 ka BP.

4.7.4 Changing climatic conditions during the Holocene

An example of the changing climatic conditions during the Holocene is the variability of soil hydrology, suggested by Brooks *et al.* (1993) as being an important factor in Holocene slope evolution. It was demonstrated using physically based models that soils at an early stage of development (e.g. podsols and brown earths) are only unstable under high-intensity rainfall events, whereas more mature soils (e.g. sol lessivés) are vulnerable to failure under a wider range of conditions. Indeed, there is a steady decline in stability over time as soils become finer-grained with more pronounced horizonation. The translocation of materials from the surface horizons to the underlying B horizon results in higher moisture storage, together with a reduction in the Factor of Safety F and a range of lower angles of limiting stability (Brooks 1997).

4.7.5 Climatic deterioration during the Little Ice Age

Historical records studied by Grove (1988) and others have revealed that the climatic deterioration associated with the onset of the Little Ice Age ($c.$ AD 1300–1850) led to an increased frequency of avalanches, rockfalls, landslides and bog bursts in the Alps and Scandinavia. A review by Hutchinson (1965) reached the conclusion that there had been increased landslide activity in southern Britain during the period 1550–1850. Landslides that occurred during this period include the 1575 Wonder Landslide at Marcle (Hall & Griffiths 1999), the 1773 Birches slide at Buildwas in the Ironbridge Gorge, the 1774 Hawkley slide near Selborne, Hampshire, the 1790 Beacon Hill landslide at Bath and the 1792 failure of the NE slopes of the Isle of Portland (Hutchins 1803) (Fig. 4.58).

Fig. 4.53. Coastal instability at Fairlight Glen, Hastings, East Sussex coast (photo credit: David Giles).

Fig. 4.54. Competent limestone and sandstone cliffs of the Portland Sand and Lulworth formations overlying the weaker mudstones of the Kimmeridge Clay Formation, West Weare, Isle of Portland, Dorset (photo credit: David Giles).

Fig. 4.55. Major failure in 2013 occurring in the jointed weak chalk of the Seaford, Newhaven and Culver Chalk formations, St Oswalds Bay, Dorset (photo credit: David Giles).

Records for the eighteenth century indicate cold and damp conditions with lengthy accumulations of snow. For example, Gilbert White (1789) wrote of the Hawkley failure:

> The months of January and February, in the year 1774, were remarkable for great melting snows and vast gluts of rain. The beginning of March also went in the same tenor; when on the night between the 8th and the 9th of that month, a considerable part of the great woody hanger at Hawkley was torn from its place'.

Similarly, writing of the 1792 Isle of Portland failure, Hutchins (1803) stated 'the season had been very wet ... the week preceding these there had been some strong gales of wind'.

Jones (2001) notes that evidence for Little Ice Age failures is well preserved on the Jurassic outcrops of the Midlands where Chandler (1970) identified periodic mudsliding and slips post-1660 AD in the Gwash Valley, Rutland; Butler (1983) concluded that there had been major localized rotational failures and mudsliding along the Cotswolds escarpment over the period 1500–1820. Clear evidence is to be found on the flanks of Ebrington Hill, Warwickshire, where mudslide lobes can be seen to override medieval ridge-and-furrow agricultural patterning and to disrupt enclosure field boundaries established in 1810. However, as both ridge-and-furrow and 1810 enclosures also pass across pre-existing lobes, the question remains as to just how significant the mid-millennium climatic deterioration really was in terms of the generation of slope instability, since the features identified could represent the final stages of postglacial slope evolution or the product of extreme rainfall events which occur irrespective of the prevailing climatic regime.

The Little Ice Age has also been associated with an increase in the reported incidence of major coastal landslides in eastern and southern England (Lee 2001a), including:

- the 1682 landslide at Runswick, North Yorkshire, when the whole village slipped into the sea (Young & Bird 1822);
- the 1737 landslide at The Spa in Scarborough's South Bay (Schofield 1787);
- the major failure in 1780 which destroyed the main road into Robin Hoods Bay, North Yorkshire, and two rows of cottages (Lee & Pethick 2003);
- the great landslide at the Haggerlythe, Whitby, on Christmas Eve 1787 which resulted in the destruction of 5 houses and led to 196 families being made destitute (Anon 1788);

Fig. 4.56. Talus slopes in Ordovician rhyolites, Penmaenbach, North Wales (photo credit: David Giles).

- the landslide of 1792 on the northwest of the Isle of Portland which involved more than a mile of cliff and is believed to have been one of the largest coastal landslides to have occurred in historical times (Hutchins 1803);
- the major reactivation of parts of the Isle of Wight Undercliff, at Gore Cliff in 1799 and in the landslip in 1810 and 1818 (e.g. Hutchinson 1991b);
- the 1829 landslide at Kettleness, North Yorkshire, when the whole village slid into the sea with the inhabitants having to be rescued by alum boasts lying offshore (e.g. Jones & Lee 1994; Lee & Pethick 2003);
- the great landslides on the North Norfolk coast near Overstrand of 1825 and 1832 (Hutchinson 1976); and
- the famous Bindon landslide, east of Lyme Regis of Christmas Eve 1839 (Conybeare et al. 1840) and the Whitlands landslide of 1840 (Pitts 1983).

Similar coastal landslide events have occurred since this period, such as the Holbeck Hall landslide of 1993 in Scarborough (Lee 1999), but they have been much rarer.

4.7.6 Anthropogenic land-use changes

Landslide activity has also changed in response to human-induced vegetation and land-use changes, particularly over the last millennium (Higgett & Lee 2001; Jones 2001). Innes (1983, 1989) applied lichenometry to 780 debris-flow deposits in the Scottish Highlands. His interpretation suggests that there were few events prior to 1700 and that the majority of debris flows occurred in the last 250 years, coinciding with land-use change in the Highlands, particularly a sharp increase in sheep grazing densities in the nineteenthth century (see Winter 2019). It is notable that the timing of Scottish debris flows post-dates the main climatic deterioration of the Little Ice Age.

Reports of peat slides, bog flows and bursts are widespread in the uplands of the British Isles (Crisp et al. 1964; Mills 2002; Dykes & Warburton 2008). Sollas et al. (1897) provide a record of peat slides in Ireland since around 1697, including the 1896 bog burst at Knocknageehan, Co. Kerry which involved 5 Mm^3 of peat. The clustering of peat slides

Table 4.9. *Dated major rock slope failure in the Scottish Highlands (from Ballantyne & Stone 2013; Ballantyne et al. 2014). LLS, Loch Lomond Stadial*

Location	Site	Landslide	Geology	Weighted mean age	Uncertainty (years)
Outside the LLS limits	Strath Nethy, Cairngorms	Multiple translational slides seated on sheeting joints; spread of boulders across valley floor	Granite	16 933	598
	Lairig Ghru, Cairngorms	Boulder deposit c.300 m wide and c.300 m long deposited on valley floor by rockwall failure	Granite	16 234	1469
	Baosbheinn, Torridon	Rockslide (0.5–0.6 Mt); run-out debris forms an arcuate run-out ridge overlapping a moraine	Sandstone	14 017	491
	Coire Beanaidh, Cairngorms	Possible rock-slide-sourced 'rock glacier' with transverse ridges crossing the run-out zone	Granite	13 354	1215
	Coire Etchachan, Cairngorms	Full-slope translational rockslide seated on sheeting joints; lobate, ridged run-out tongue	Granite	12 758	518
	The Storr (Skye)	Retrogressive block glide of basalt caprock over shale; shattered basalt pinnacles	Basalt	6089	488
Within the LSS limits	Carn Ghluasaid, Glen Cluanie	Translational rockslide; main run-out tongue of bouldery debris 350–400 m long	Schist	11 955	466
	Coire nan Arr, Applecross	Large boulders sourced by a major rockfall transported 2–3 km downvalley by glacier ice	Sandstone	11 658	473
	Maol Cheann-dearg, Torridon	Rock avalanche with glacially transported run-out boulders extending 500 m downvalley	Quartzite	11 543	373
	Beinn an Lochain, Argyll	Rockslide with minor toppling failure; run-out debris forms boulder tongue c. 500 m long	Schist	11 037	563
	Druim nan Uadhag, Dessary	Complex (sliding plus toppling) failure with arrested debris run-out of very large boulders	Schist	9798	1250
	Carn nan Gillean, Sutherland	Arrested translational rockslide; lobate debris tongue descends to slope foot	Granulite	7338	536
	Coire Ban, Glen Lyon	Rockslide with a run-out tongue of bouldery debris 250–300 m long	Schist	4638	464
	Beinn Alligin, Torridon	Rock avalanche (c.9.0 Mt) with run-out of large boulders extending 1.25 km downvalley	Sandstone	4115	202
	Hell's Glen, Argyll	Translational rockslide; run-out zone of boulders is part of a larger arrested rock slope failure	Schist	3670	397
	Mullach Coire a'Chuir, Argyll	Translational rockslide/topple; part of larger rock slope failure; downslope boulder run-out over c.250 m.	Schist	1534	165
	Coire Gabhail, Glencoe	Complex failure (c. 0.6 Mt) that accumulated as a massive talus cone	Ignimbrite	1682	220

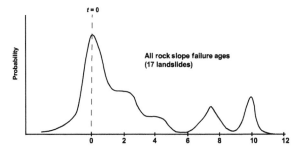

Fig. 4.57. Probability density distributions (PDDs) of rock slope failure ages, calculated as time elapsed since deglaciation (adapted from Ballantyne & Stone 2013).

in recent centuries may reflect peat digging coupled with periods of increased storminess.

4.7.7 Extreme events

Rare and extreme events, such as large first-time failures (e.g. the 1993 Holbeck Hall landslide, Scarborough, North Yorkshire; Lee 1999) affect landslide activity. Over 60 m of the cliff (developed in Jurassic sandstones, mudstones and siltstones, and mantled by up to 30 m of glacial till) was lost overnight, leaving the hotel in a very dangerous position as cracks began to develop in the building. A further 35 m of cliff collapsed over the next 3 days undermining the hotel which had to be abandoned; it was later demolished (Clark & Guest 1994; Clements 1994).

Harvey (1986) evaluated the geomorphological impact of a 100-year event in the Howgill Fells, Cumbria. Peak rainfall intensities were sufficient to trigger shallow slides on slopes, to reactivate stabilized gully slopes and to entrench gully beds. A convectional storm over the Forest of Bowland in 1967 also triggered shallow slope failures (Newson & Bathurst 1990), many of which remain unvegetated. Heavy rainfall was the trigger for a bog flow in Co. Antrim in November 1963 that surged down the Glendun River, destroying a farmhouse and damaging farm equipment (Colhoun *et al.* 1965). Dykes & Kirk (2000) describe the occurrence of a multiple peat slide event on Cuilcagh Mountain, Northern Ireland, in response to extreme intense rainfall in October 1998; the total area affected by sliding was 30 500 m^2.

Coastal cliffs can be very sensitive to the impact of large storm events. For example, the cliffs at Bacton, Norfolk, were cut back by as much as 30 m during the 1953 storm surge (Grove 1953). On the Covehithe cliffs, Suffolk, up to 27 m of land was lost during the same storm surge (Williams 1956). Cambers (1976) notes how a smaller surge in November 1971 led to significant landsliding on the Norfolk coast, with 11% of the total cliff volume lost in 1971 occurring in that one day.

4.8 Landslide risk

4.8.1 Sources of risk

Although the UK is considered to be a low-risk environment compared with other parts of the world (e.g. Gibson *et al.* 2012), significant threats to people and property can arise in a variety of different circumstances, described as follows.

Known landslides within a proposed development area or infrastructure corridor introduce a risk. For example, the

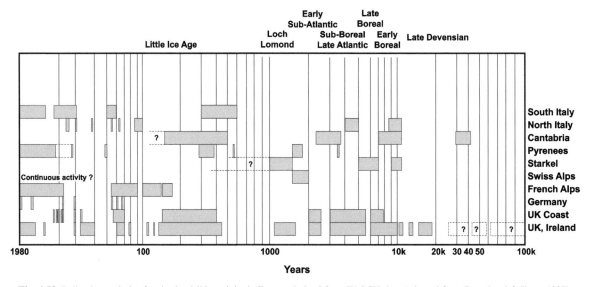

Fig. 4.58. Indicative periods of major landslide activity in Europe derived from EPOCH data (adapted from Bromhead & Ibsen 1997).

Channel Tunnel terminal and portal on the UK side is located immediately below the Lower Chalk escarpment (the North Downs), which is mantled by a series of large, ancient pre-existing landslides. These landslide features were known of prior to the start of the project (e.g. Aarons et al. 1977), meaning that their presence could be taken into account in the investigations to support layout and design (e.g. Duggleby et al. 1991; Griffiths et al. 1995).

A previously unknown landslide could be reactivated during the construction or operational phase of a project, or during the lifetime of a development. For example, the A4061 in South Wales was opened in 1929. Carriageway damage became apparent in 1954 when cracks 5 cm wide and 100 m long were observed near Treherbert. Subsequent investigations revealed that a section of the road had been constructed across the head of an ancient rock slope failure and mudslide complex (Chambers 2000). Stabilization measures were installed, including ground anchors and soil nail, allowing the road to be maintained on its original alignment.

An unexpected landslide incident can threaten people or property. For example, in 2001 the run-out from a failure of the talus slope at the foot of the Shanklin cliffs, Isle of Wight, entered the rear of the Shanklin Beach Hotel. A guest trapped by the landslide debris had to remove his trousers and unscrew a false leg to struggle free. In the winter of 2000/01 there was a series of cliff collapses behind Brighton Marina, including a major collapse in April 2001 that led to the temporary closure of the Asda supermarket at the foot of Black Rock (Lee & Jones 2014). An elderly woman was killed and her husband injured in January 2001 when their car was carried onto the beach by landslides off the 40-m-high sand and clay cliffs near Nefyn, North Wales (Statham 2002). All of these incidents led to landslide hazard or risk investigations.

Man-made slopes (e.g. cuttings and embankments), especially along infrastructure routes, can fail unexpectedly. Delays caused by rail tracks being blocked by landslides is an all-too-familiar story for many commuters. For example, in May 1992 a large debris flow from a chalk cutting blocked the Metropolitan Line between Chalfont Latimer and Chorleywood. This incident disrupted services for 33 hours and lead to £400 000 worth of mitigation measures. This led to the development of a qualitative risk assessment for the line (Phipps & McGinnity 2001).

An area of landsliding can unexpectedly encroach into a development, often leading to the abandonment of property. At Knipe Point, near Scarborough, a number of cliff-top houses have been demolished since 2008 as the backscar of a pre-existing coastal landslide has retreated through a series of small failures (e.g. Fish et al. 2006). A landslide investigation and qualitative risk assessment were commissioned by Scarborough Borough Council (Halcrow Group Limited 2009).

Ongoing coastal cliff recession or episodic landslide activity can increase risk. Perhaps the most dramatic incident in recent memory was the loss in June 1993 of the Holbeck Hall Hotel, which had been built on the coastal cliff top in South Bay, Scarborough (Clark & Guest 1994; Clements 1994). Emergency coastal protection measures were constructed and a risk assessment was undertaken to identify threats elsewhere along the Scarborough coast.

Coastal protection structures can unexpectedly fail or degrade; the prevention of marine erosion does not eliminate the potential for coastal slope failure because of the importance of porewater pressures in promoting instability (Lee & Clark 2002). While slope degradation behind defences generally involves relatively small and minor events, large-scale dramatic events do occur and can result in considerable loss of land. For example, a landslide occurred at Overstrand Norfolk (Fig. 4.59) where around 100 m of cliff-top land was lost during a 3-year period between 1990 and 1993. The slope toe had been protected by wooden breastwork defences (Frew & Guest 1997). A rock revetment was constructed at the cliff foot to protect cliff-top properties (Fig. 4.60).

4.8.2 Assessing risk

Responding to landslide problems has always involved some rudimentary form of risk assessment, although it may not have been recognized as such (e.g. Fell & Hartford 1997). However, many stakeholders are becoming increasingly aware of the need to introduce more rigorous and systematic procedures to formalize the evaluation process and thereby enhance the openness, objectivity and consistency of such judgements (e.g. Lee & Jones 2004, 2014). Adopting a risk-based approach to assessing landslides provides a valuable framework for prioritizing decision-making, especially where resources are limited.

Formal risk assessment also provides a means of comparing, for example, landslide risks with other project risks (e.g. fire, explosion or equipment failure). To achieve this there needs to be a common language between all parties. In the UK and other parts of the world this is provided by the Royal Society (1992) and the various publications of the Health and Safety Executive (e.g. HSE 2001, 2006).

Risk is the potential for adverse consequences, loss, harm or detriment (Royal Society 1992). The focus is on negative impacts of events or circumstances (i.e. threats).

Hazards are situations that in particular circumstances could lead to harm (Royal Society 1992), and can be viewed as any event presenting the possibility of danger (HSE 2001). The International Society for Soil Mechanics and Geotechnical Engineering Technical Committee on Risk Assessment and Management define hazard as the probability that a particular danger (threat) occurs within a given period of time (ISSMGE 2004).

Adverse consequences might include: accidents; loss of life; damage to property, services and infrastructure; environmental impacts; and associated financial losses.

Risks include both economic and environmental losses and/or damage (economic and environmental risks, respectively), and the potential for loss of life or injury (individual or societal risk). Risk is expressed as the product of the

Fig. 4.59. Effects of coastal landsliding, Overstrand, North Norfolk, in 2006 (photo credit: David Giles).

probability of a hazard (e.g. a landslide event) and its adverse consequences. For a single landslide event scenario,

Risk = Probability(landslide event) × Consequences. (4.4)

There are numerous versions of this simple equation in the literature (e.g. Cruden & Fell 1997; Bonnard *et al.* 2004; Lee & Jones 2004, 2014; Bromhead 2005; Glade *et al.* 2005; Hungr *et al.* 2005). All are based around the following considerations:

- the probability of an event that could cause damage (i.e. an event of sufficient magnitude or intensity to cause harm) occurs in the 'danger zone';
- the probability that this event impacts or hits the elements at risk (i.e. the probability of the asset being in the 'danger zone'; spatial probability);
- the exposure of the asset to the event (i.e. whether the elements at risk are fixed in space, such as a building, or are passing through the 'danger zone', such as traffic moving through a road cutting), in other words, the probability of being in the danger zone when the event occurs (temporal probability);
- the vulnerability of the asset to the event (i.e. the level of potential damage, or degree of loss, of a particular asset subjected to a landslide of a given intensity) expressed on a scale of 0 (no loss) to 1 (total loss; Fell 1994); and
- the cost or value of the elements at risk.

The risk assessment process is often portrayed as a 'bow-tie' diagram (Fig. 4.61; e.g. HSE 2006) with the two wings representing hazard assessment and consequence assessment. Hazard assessment is directed towards estimating the probability of the full range of credible threats that have the potential to cause adverse consequences, that is, their occurrence initiates a potential accident and/or damage scenario. Consequence assessment involves the identification and quantification of the full range of adverse consequences arising from the accident scenarios, including the estimation of the probabilities of these consequences. (Further details can be found in Fell *et al.* 2005; Lee & Jones 2014.)

Fig. 4.60. Landslide toe protection with larvikite rock revetment, Overstrand, North Norfolk (photo credit: David Giles).

At the centre of the bow-tie is the 'incident' or 'state' that initiates the various consequence scenarios that might develop. These so-called 'top events' can also be referred to as 'initiating events' by risk specialists. Common examples of 'top events' include: a landslide hitting a building, causing structural damage and subsequent collapse (this can lead to injuries or fatalities if the building was occupied at the time); a landslide hitting a person or vehicle travelling through, or stationary (albeit temporarily) within, the impact zone (this may result directly in injuries or fatalities); or landslide debris being deposited along a transport route, forming an obstacle for traffic (e.g. rockfall debris on a railway track leading to a derailment if a train fails to stop in time and collides with it, or a vehicle swerving to avoid landslide debris on the road and crashing into a tree or oncoming traffic).

Bow-tie analysis is useful because it emphasizes that the focus of hazard assessment should not just be on the annual probability of a landslide event occurring on a hillslope or cliff, but also on whether this event reaches the danger zone where assets might be present and whether the event is of sufficient intensity to cause harm.

The various components of the risk equation can be organized into a simple conditional probability model that can be

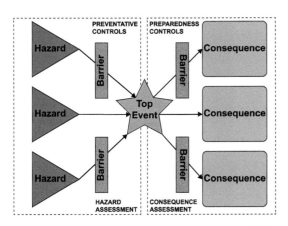

Fig. 4.61. The risk 'bow-tie' (adapted from Lee & Jones 2014).

Table 4.10. *Economic risk: example of a simple risk model (adapted from Lee & Clark 2002)*

A conference centre has been constructed at the foot of a potentially unstable 5-km-long escarpment slope. Major, high-intensity, landslide events have occurred at similar locations along the escarpment, with a historical frequency of four events in 400 years. Typically, events have been around 500 m wide.
If the event occurs it will cause severe structural damage to the building, leading to the loss of the asset.
Direct losses: estate agents have estimated the value of the conference centre to be £5 million.
Indirect losses: the conference centre is important for the local economy; the loss of the asset has been estimated to result in £1 million lost income.

P (event)	From records, it is estimated that, on average, four landslide events (500 m wide) occur every 400 years. The annual probability of a landslide is estimated to be 0.01 (4/400).	1.0×10^{-2}
P (wrong place)	The proportion of the 5000-m-long escarpment 'danger zone' that is affected by a 500-m-wide landslide is 500/5000	1.0×10^{-1}
P (wrong time)	Time in the 500-m-long 'danger zone' as a fraction of a year; the conference centre is a stationary asset.	1.0
P (hit)	P (wrong place) × P (wrong time)	1.0×10^{-1}
P (damage)	The vulnerability of the building is assumed to be 1.0.	1.0
Economic risk	P (event) × P (hit|event) × P (damage|hit)	$1.0 \times 10^{-3} \times$ £5 M

used in most situations (Tables 4.10 and 4.11; Lee & Jones 2014):

$$\text{Risk} = P(\text{event}) \times P(\text{hit}|\text{event}) \times P(\text{damage}|\text{hit}) \times C \quad (4.5)$$

where $P(\text{event})$ is a measure of the expected likelihood of a landslide event, usually expressed as an annual probability, and $P(\text{hit}|\text{event})$ is the probability of a 'hit' given that the landslide event occurs. This can involve two components: the probability of being in the wrong place (spatial probability); or $P(\text{wrong place})$ at the wrong time (temporal probability), or $P(\text{wrong time})$. For a rockfall impacting a cliff-foot footpath used by a single pedestrian, $P(\text{damage}|\text{hit})$ is the probability of damage (i.e. 'vulnerability') given that a hit has occurred; it is a measure of the chance (0–1) that a component would be damaged by the geohazard event. C is the value of the consequences expected to arise as a result of the event. When the focus is on the risk to individuals or groups (individual and societal risk), C is ignored and risk is expressed as the probability of a fatality or injury:

$$\text{Risk} = P(\text{event}) \times P(\text{hit}|\text{event}) \times P(\text{fatality}|\text{hit}) \quad (4.6)$$

where $P(\text{fatality}|\text{hit})$ is the vulnerability of a person to the landslide impact.

$P(\text{hit}|\text{event})$ can be defined as:

$$P(\text{hit}|\text{event}) = P(\text{wrong place}) \times P(\text{wrong time}) \quad (4.7)$$

where $P(\text{wrong place})$ is the size of fall and/or length of the danger zone, and $P(\text{wrong time})$ is the time in the danger zone, usually expressed as a fraction of a year.

For assets located on a pre-existing landslide or in the source area of a first-time failure, the probability of being

Table 4.11. *Individual risk: example of a simple risk model (adapted from Lee & Jones 2014)*

An individual makes a single journey, walking along a 500-m-long seaside promenade (the 'danger zone') at 2.5 km h^{-1}. The promenade is exposed to occasional rockfall events (1 m wide) from high cliffs that have occurred, on average, at a frequency of 1 every 10 years. These events occur randomly along the cliff line.
If the person is hit by the rockfall, then he/she will suffer a fatal injury.

P (event)	From records it is estimated that, on average, 1 rockfall event (1 m wide) occurs every 10 years. The annual probability of a 1-m rockfall is estimated to be 0.1 (1/10).	1.0×10^{-1}
P (wrong place)	The proportion of the 500-m-long 'danger zone' that is affected by a 1-m rockfall is 1/500	2.0×10^{-3}
P (wrong time)	Time in the 500-m-long 'danger zone', as a fraction of a year. Walking at 2500 m h^{-1}, the individual spends 500/2500 hours in the danger zone = 0.2 hours Expressed in years, this becomes 0.2/(365 × 24)	2.28×10^{-5}
P (hit)	P (wrong place) × P (wrong time)	4.57×10^{-8}
P (fatality)	The vulnerability of a person hit by a 1 m rockfall is assumed to be 1.0	1.0
Individual risk	P (event) × P (hit|event) × P (fatality|hit)	4.57×10^{-10}

Table 4.12. *Urban landslide reactivation: relationship between winter rainfall events and landslide damage (from Lee & Jones 2014)*

Reactivation scenario	Winter rainfall event (return period)	Property damage and destruction (£ million)	Traffic disruption (£ million)
Continuous creep	1 in 1	0.25	0.1
Significant localized ground movement	1 in 10	0.5	0.2
Major localized ground movement	1 in 50	1	0.5
Major widespread reactivation	1 in 100	5	1
Extensive reactivation	1 in 500	10	5

in the wrong place, P(wrong place), when reactivation or movement occurs is 1.0.

The temporal probability P(wrong time) is 1.0 for stationary assets, as they are present in the danger zone for 100% of the time.

The total risk in an area is the sum of the risk (R) that arises from each different landslide event (scenario) that may occur:

$$\text{Total risk} = \sum R_s \text{ (landslide scenarios 1},\ldots, n). \quad (4.8)$$

If the different scenarios involve different processes (e.g. a rockfall, debris flow or slide reactivation), then the risk associated with each scenario can simply be added to yield a total risk value, that is,

$$\text{Total risk} = P(\text{rockfall}) \times \text{losses})$$
$$+ (P(\text{debris flow}) \times \text{losses})$$
$$+ (P(\text{recatvation}) \times \text{losses}). \quad (4.9)$$

However, there are many situations where the scenarios are part of the same magnitude–frequency distribution, in which there are numerous potential combinations of event size and probability, each with a slightly different loss. Simply adding together the annual risk associated with each scenario would underestimate the total risk.

As an example, a small town has grown within an area of deep-seated landsliding. Throughout its history it has been subjected to very slow ground movements, with less frequent episodes of more active movement. The landslide hazard is associated with the reactivation of the landslides and varies from almost-continuous deep-seated movement in the order of mm a^{-1} to infrequent, short periods of significant ground movement that can result in widespread surface cracking and heave. Reactivation events are triggered by high groundwater levels that coincide with wet winters (as determined from historical analysis). The wetter the winter, the greater the resulting ground movement; minor events are regular occurrences, while large events are infrequent. It is therefore assumed that there is a continuous range of discrete reactivation events (Scenarios 1–5), with each magnitude of event associated with a different return period (i.e. probability) of winter rainfall.

The losses associated with reactivation can be identified as a combination of repairable damage and property destruction, together with traffic disruption, with the severity of losses related to the size of the reactivation event; hypothetical values are shown in Table 4.12.

From this information it is possible to construct a landslide reactivation damage curve that relates losses to the probability of the scenario (Fig. 4.62). As any of the scenarios could occur in a given year, depending on the winter rainfall, it is necessary to calculate the average annual damage in order to take account of every possible combination of scenario probability and loss (Table 4.13). In Figure 4.62 the average annual risk (£0.68 M) equates to the area below the damage curve, calculated as a series of slices between event probabilities. This approach has been used to estimate landslide risk in the Ventnor Undercliff, Isle of Wight (Lee & Moore 2007) and is identical to that used to calculate flood risk (e.g. DEFRA 1999), and can be used to estimate the total annual risk associated with events such as rockfalls and debris flows that have a defined magnitude–frequency distribution.

It should be stressed that landslide risk assessment presents major challenges to the geoscience community, mainly

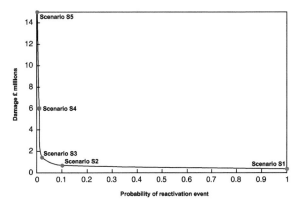

Fig. 4.62. Urban landslide reactivation example: damage curve (adapted from Lee & Jones 2014).

Table 4.13. *Urban landslide reactivation: annual risk calculation (see Fig. 4.11 for the landslide damage curve; from Lee & Jones 2014)*

Damage category	Return period (years) and annual probability of reactivation event: damage (£ million)				
	1 1.0	10 0.1	50 0.02	100 0.01	500 0.002
Residential property	0.25	0.5	1	5	10
Industrial property	0	0	0	0	0
Indirect losses (e.g. tourism)	0	0	0	0	0
Traffic disruption	0.1	0.2	0.5	1	5
Emergency services	0	0	0	0	0
Other	0	0	0	0	0
Total damage	0.35	0.7	1.5	6	15
Area (damage × frequency)		0.47	0.09	0.04	0.08
Average annual risk (area beneath curve): £ million					0.68

Notes: Area (damage × frequency) for between the 1 in 1 (Scenario 1) and 1 in 10 (Scenario 2) year events is calculated as: Area (damage × frequency) = (Probability of Scenario 1 − Probability of Scenario 2) × (Total P Damage 2 + Total Damage 1)/2; Average annual risk = $\sum$ Area (damage × frequency).

because it requires the numerical expression of the chance of future landsliding:

> The difficulty lies in different perceptions about how precise and reliable this numerical expression has to be in order to be of value to the users of a risk assessment. To some it should be an objective statement of truth, derived from observable facts, supported by comprehensive historical data without which the estimated probabilities are guesses. Others take a more pragmatic view that this ideal is simply not achievable. Indeed, historical landslide records are just not available in most areas. Many projects take the view that risk management decisions must be made even if we don't know the 'true' probability. In this context the numerical expression of chance should be a best estimate judgement, based on the available knowledge. (Lee 2009)

4.9 Landslide hazard

The destructive intensity of a landslide is related to kinetic parameters, such as velocity, acceleration, total and differential displacement, to dimensions (e.g. depth and volume (mass) of the moving material, depth of deposits after the movement ceases) and to material characteristics (Léone *et al.* 1996; Hungr 1997). The movement velocity is a key factor in landslide destructiveness (Table 4.14). An important limit appears to be at *c.* 5 m s^{-1}, approximately the speed of a person running away from a slide. In the UK, most 'slides' involve extremely slow to slow ground movements, whereas flows and falls may be more rapid.

In the UK, landslides can present a threat to people and property in a variety of ways (Fig. 4.63) in the form of either: (1) the loss of productive land through the retreat of a cliff or the formation of a scar; (2) movement of the landslide mass, expressed as a function of the static or dynamic loads exerted on those assets that are affected (Léone *et al.* 1996); or (3) boulder impacts, which are mainly a feature of cliff falls although they can be associated with other types of major failures on steep slopes.

In the first case, this can lead to the undermining and collapse of cliff-top property, infrastructure and services. Examples of property being lost as a result of coastal cliff recession are actually quite rare, probably no more than one per year on average, and often the buildings have been declared unsafe by the local authority and demolished before they fall over the cliff. Examples of loads in the second case include: horizontal displacement and deformation; vertical displacement and deformation; lateral pressure; pressure from the impact of moving debris; and accumulation or loading from run-out, either instantaneous or progressive due to the build-up of debris over several events.

Different parts of a landslide mass come under tension, compression, tilt, contra-tilt or heave according to their location and the stresses involved. As a result, assets impacted by the same landslide may suffer damage from a number of different loads, depending on their location relative to the landslide. Burial is a characteristic feature of flows, although it can occur below most types of large failure on steep slopes.

Most slides in the UK tend to move at extremely slow to moderate speeds. However, some landslides, such as debris flows and debris slides, can travel at high speed beyond the source area and present a threat significant some distance away from the original failure site (see Winter 2019). The travel distance of landslide debris is generally estimated in terms of a travel angle (defined as the slope of a line joining the tip of the debris to the crest of the main landslide source; also termed the reach angle; Fig. 4.64). The key factors that influence the travel distance include (Wong & Ho 1996): slope characteristics, including slope height, slope gradient,

Table 4.14. *Velocity classes for landslides (after Cruden & Varnes 1996)*

Velocity class	Description	Velocity (mm s^{-1})	Typical velocity	Nature of impact
7	Extremely rapid	5×10^3	5 m s^{-1}	Catastrophe of major violence; exposed buildings totally destroyed and population killed by impact of displaced material or by disaggregation of the displaced mass
6	Very rapid	5×10^1	3 m min^{-1}	Some lives lost because the landslide velocity is too great to permit all persons to escape; major destruction
5	Rapid	5×10^{-1}	1.8 m h^{-1}	Escape and evacuation possible; structure, possessions and equipment destroyed by the displaced mass
4	Moderate	5×10^{-3}	13 m/month	Insensitive structures can be maintained if they are located a short distance in front of the toe of the displaced mass; structures located on the displaced mass are extensively damaged
3	Slow	5×10^{-5}	1.6 m a^{-1}	Roads and insensitive structures can be maintained with frequent and heavy maintenance work, if the movement does not last too long and if differential movements at the margins of the landslide are distributed across a wide zone
2	Very slow	5×10^{-7}	16 mm a^{-1}	Some permanent structures undamaged or, if they are cracked by the movement, they can be repaired
1	Extremely slow			No damage to structures built, construction possible with precautions

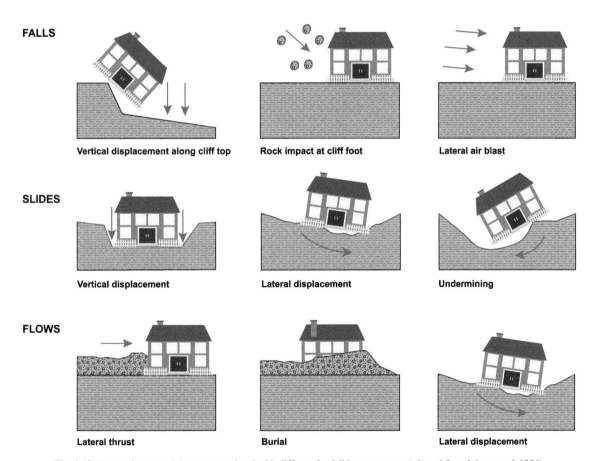

Fig. 4.63. Types of structural damage associated with different landslide movements (adapted from Léone *et al.* 1996).

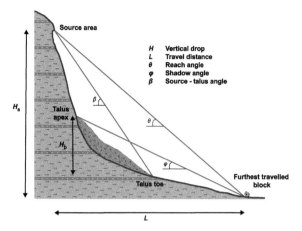

Fig. 4.64. Travel/run-out distance terminology (adapted from Hungr et al. 2005).

and the slope materials; failure mechanism, including collapse of metastable or loose soil structures leading to the generation of excess porewater pressures during failure, the degree of disintegration of the failed mass, the fluidity of the debris, the nature of the debris movement (e.g. sliding, rolling, bouncing, viscous flow) and the characteristics of the ground surface over which the debris travels (e.g. the response to loading, surface roughness, etc.); and the condition of the downhill slope, including the gradient of the deposition area, the existence of irregularities and obstructions or the presence of well-defined channels.

A variety of regression equations have been developed that describe the relationship between the tangent of the travel/reach angle (tan α; H/L) and the landslide volume (V), generally in the form:

$$\log(\tan \alpha) = A + B \log(V), \quad (4.10)$$

where A and B are constants.

Values for A and B for different landslide types and flow paths are presented in Table 4.15 (from Corominas 1996; Hunter & Fell 2003). However, Hungr et al. (2005) have stressed that care is needed when using these equations for predicting travel distance, as there can be significant scatter around the mean values and many landslides will travel beyond the calculated distance. A range of analytical (numerical) methods are available to assess landslide travel distance, including sliding block models, two-dimensional models that consider a typical profile of the slide, and three-dimensional models that analyse the flow of a landslide over irregular terrain (Hungr et al. 2005).

4.9.1 Landslide hazard assessment

Hazard assessment is a key stage in the risk assessment process and generally requires the specialist input from a combination of ground engineering disciplines, including engineering geomorphology, engineering geology and geotechnical engineering. However, there is no standard approach that can be used in all situations; landslides typically present a variety of problems ranging from urgent threats to the public, constraints to the safe development of an area, to threats to the long-term viability of a project. As a result, there can be many reasons for a client to commission an investigation and no fixed scope of work. However, the objectives of these investigations generally include one or more of: characterizing landslide behaviour as part of an assessment of landslide hazard and risk to people and property; developing a ground model to support the appraisal of landslide management options, especially the technical feasibility of stabilization measures; and monitoring slope conditions or landslide movements as part of a management strategy and providing early warning of developing problems.

4.9.2 Landslide investigation

Landslide investigation requires an understanding of the slope processes and the relationship of those processes to geomorphology, geology, hydrogeology, climate and existing development. From this understanding it should be possible to:

Table 4.15. *Regression equation parameters for predicting landslide travel distance (from Corominas 1996; Hunter & Fell 2003)*

Landslide type	Flowpath	A	B	R^2	Author
Translational slides	Obstructed	−0.13	−0.06	0.76	Corominas (1996)
	Unobstructed	−0.14	−0.08	0.8	
Debris flows	Obstructed	−0.05	−0.11	0.85	
	Unobstructed	−0.03	−0.10	0.87	
Earth flows	Unobstructed	−0.22	−0.14	0.91	
All	Unconfined	0.77	0.09	0.71	Hunter & Fell (2003)
	Partly confined	0.69	0.11	0.52	
	Confined	0.54	0.27	0.85	

- assess the physical extent of each potential landslide being considered, including the location, areal extent and volume involved;
- classify the types of existing or potential landsliding, the physical characteristics of the materials involved, and the slide mechanisms (an area or a site may be affected by more than one type of landslide hazard; e.g. deep-seated landslides on the site, and rockfall and debris flow from above the site);
- locate and define the three-dimensional shape of any shear surfaces and identify the engineering properties and groundwater conditions of the landslide mass and the surrounding ground;
- estimate the displacement history of pre-existing landslides, or the magnitude–frequency relationship for first-time failures (e.g. rockfalls, debris slides and debris flows);
- assess the likely triggering events and potential for future movement; and
- estimate the resulting anticipated displacement and landslide intensity.

On the coast, the focus is often slightly different. Key areas of investigation include the determination of: the type of recession event (i.e. landslide type, nature of surface and seepage erosion); the magnitude and frequency of recession events, including the potential for rare, large recession events; the causes of recession (e.g. marine erosion and internal slope factors); the significance of short- or long-term beach profile changes or foreshore lowering in the recession process; and the relationship between the cliff and the littoral cell, including the sediment budget.

Overseas, landslide hazard mapping of large areas is often carried out at the local government level for planning urban development and by state or federal governments for regional land-use planning or disaster management planning (e.g. Cascini et al. 2005; Fell et al. 2008a, b). This high-level strategic approach is rarely used in the UK, where the focus has traditionally been on the investigation of individual sites. There are exceptions, of course, including the landslide hazard maps produced to support land-use planning in Rhondda Borough, South Wales (Halcrow 1988), the Isle of Wight Council (Lee & Moore 1991), and the landslide distribution map of the City of Bradford (Waters et al. 1996); however, these were government-funded demonstration projects rather than studies commissioned directly by local authorities.

There are many books that describe the range of techniques available for the site-specific investigation of landslides, including Bromhead (1986, 1992), Clark et al. (1996), Turner & Schuster (1996), Lee & Clark (2002), Cornforth (2005) and Clague & Stead (2012). These techniques range from the standard ground investigation techniques (e.g. desk study, geomorphological mapping, boreholes and trial pits) that will be familiar to most practitioners to state-of-the art remote monitoring and dating methods that can be provided by specialist companies. However, the main challenge is designing and undertaking an effective investigation within the constraints imposed by time (e.g. project schedule or urgency of the landslide threat), resources (e.g. the budget allocated for investigation or the availability of equipment) and site access (e.g. the problems associated with very steep slopes, dense vegetation or very wet ground).

Although it would be misleading to suggest that there is a standard landslide investigation procedure, many investigations involve the following stages:

(1) desk study review of available information, including assessment of historical cliff recession rates;
(2) detailed geomorphological mapping of pre-existing landslide surface forms;
(3) subsurface investigation through the use of boreholes, trial pits and, possibly, the use of geophysics;
(4) laboratory testing of sampled materials;
(5) stability analysis to determine the Factor of Safety of the landslide or unstable slope;
(6) prediction of cliff recession rates (see the discussions in Lee et al. 2001, 2002; Lee & Clark 2002; Lee 2005, 2011); and
(7) site instrumentation and monitoring.

Investigations should be directed towards the development of hazard models. The hazard models should classify the different types of threat and quantify their future frequency and magnitude. This can involve defining the different mechanisms and scales of failure, each with a corresponding frequency and impact potential. Consideration must be given to hazards originating offsite or beyond the boundaries of the area under consideration, as well as within the immediate site or area, as it is possible for landslides both upslope and downslope to affect a site or location. The effects of proposed developments should also be considered, as these anthropogenic changes have the potential to alter the nature, scale, frequency and significance of future landslide hazards.

Each situation will be different because of the variation in client requirements or the purpose of study and the uniqueness of individual sites or areas. It follows that hazard models need to be individually designed to reflect site conditions and cannot be provided 'off-the-shelf'. However, hazard models should aim to answer the following questions.

- *What could happen*? What is the nature and scale of the landslide events that might occur in the foreseeable future? Often an important issue will be the way in which hazards develop, from incubation, via the occurrence of a triggering or initiating event, to the slope response and all possible outcomes. For coastal cliffs, the focus should be on defining the retrogression potential, that is, the size and style of the range of recession events that may occur.
- *Where could it happen*? The hazard model will need to provide a spatial framework for describing variations in hazard across a site or area. In many instances this framework will be provided by a geological map, although a geomorphological map is preferable.
- *Why might such events happen*? What are the circumstances associated with particular landslide events?

- *When might events happen*? What are the timescales within which particular events are expected to occur? On the coast, this is termed the recurrence interval, that is, the timing and sequence of recession events.

A hazard model, no matter how good it is, is not a risk assessment; it is merely an essential component of a risk assessment (Lee & Jones 2004, 2014).

4.10 Landslide risk management

In the UK, landowners or occupiers have a 'measured duty of care' to reduce or remove hazards to their neighbours (Goldman v Hargrave 1967 AC 645; see also Holbeck Hall Hotel Ltd v Scarborough B C 1995 ORB 561). Property owners also have duties of care imposed by the Health and Safety at Work Act 1974 and the Occupiers Liability Acts 1957 and 1984. Under health and safety legislation, operators of utilities and infrastructure such as roads, pipelines and railways have a duty to protect employees and the public from any potential harm caused by their operations. The Construction (Design and Management) Regulations (2007) are relevant to safe construction in landslide areas.

In general, a range of options are available to property owners, developers and operators for managing landslide risks (e.g. Clark *et al.* 1996).

4.10.1 Avoid the risk

Modify a project layout or alignment to avoid areas at risk from landsliding. For example, Birch (2001) describes how gas transmission pipeline routes were modified to avoid landslides in South Wales. An alternative strategy would be to maintain the proposed alignment and tunnel beneath a landslide or construct an overbridge (e.g. Bromhead 2005), a solution that was used on the A467 in South Wales at the Trinant Hall landslide (Jones 2000).

4.10.2 Restrict or prevent access to the area at risk from landsliding

Avoid the risk or limit the exposure, for example, the closure of a public footpath that runs along the top of a cliff prone to rockfalls. For example, following the rockfall incident in 2012 that caused the death of a 22-year-old woman at Burton Bradstock, Dorset, 16 miles of coastal cliff-top path was temporarily closed between Lyme Regis and West Bexington while safety assessments are carried out.

4.10.3 Accept the risk

If the perceived risks are low, the benefits of living in a landslide area (e.g. attractive scenery) may exceed the costs of constructing mitigation measures. Property in landslide areas is often cheaper than equivalent property in other areas; the savings in property cost can be balanced against the probability of losses or repair costs.

In other circumstances, simple repairs or removal of debris may be the appropriate response to a minor landslide incident. For example, the Highways Authority has a duty to remove soil that obstructs the highway and may recover expenses reasonably incurred from the person responsible for the obstruction (the Highways Act 1980 Section 150). Under S.151 of the Act, the authority may serve a notice on a landowner requiring him to undertake, within 28 days, such work as will prevent soil being washed onto the road. In 1990, for example, the Isle of Wight County Council served notice to 35 farmers or landowners under S.151 of the Highways Act (1980) when intense rainfall caused extensive erosion of agricultural land and flooding; the highway clearance costs were estimated to be about £25 000 (Clark *et al.* 1996).

4.10.4 Share the risk

Insurance is the equitable transfer of the risk of a loss, from one party to another in exchange for payment. It is a form of risk management primarily used to hedge against the risk of a contingent, uncertain loss. The transaction involves the insured assuming a guaranteed and known relatively small loss in the form of payment (the premium) to the insurer in exchange for the insurer's promise to compensate (indemnify) the insured in the case of a financial (personal) loss. In Britain, insurance related to landsliding is almost without exception taken out as part of the normal insurance for buildings. Although buildings insurance is not compulsory in the UK, it is generally required by mortgage providers.

According to Edwards (1988), the typical wording of a buildings policy is: 'damage to buildings caused by subsidence and/or heave of the site in which the buildings stand and/or landslip'. However, typical exclusions include damage due to coastal or river erosion.

The BGS have developed a landslide susceptibility model (the new GeoSure Insurance Product) to provide information to the insurance industry (e.g. Foster *et al.* 2008; Lee *et al.* 2014). The product provides the potential insurance risk due to natural ground movement and is licensed to a range of insurers.

4.10.5 Transfer the risk through litigation to recover the costs of landslide damage

Common law can be used for dealing with those issues not covered by specific legislation and involves liability for naturally occurring conditions and the questions of natural rights, negligence (duty of care) and right to protection from nuisance (i.e. where a landowner's natural rights are contravened).

Leakey v National Trust for Places of Historic Interest or National Beauty (1978) 2 WLR 774, in conjunction with the subsequent ruling by the Court of Appeal (Leakey v The National Trust (1980) 1 QB 485), provides a clear statement to date on the landowner's responsibility for a natural hazard. The case concerned a slope failure in a mound located on National Trust land called Burrow Mump. Natural erosion of Burrow Mump over a number of years had led to soil and rubble falling from the mound onto land owned by the plaintiffs and threatening their houses. The plaintiffs accordingly brought an action in nuisance calling for an abatement of the nuisance and for damages. In 1978 the court decided in favour of the plaintiffs, but the defendants chose to appeal against the decision.

The 1980 appeal by the National Trust was dismissed because the court felt that an occupier of land owned a general duty of care to a neighbouring occupier in relation to a hazard occurring on his land whether such a hazard was natural or man-made. This is a fundamentally important decision as far as landslides and landslide hazards are concerned, not least because it arises from a case of slope failure. The general duty referred to the judgement was held to be: 'to take such steps as were reasonable in all the circumstances to prevent or minimise the risk of injury or damage to the neighbour or his property of which the occupier knew or ought to have known'.

In a review of the Holbeck Hall Hotel Ltd v Scarborough Borough Council case, the solicitors Dibb, Lupton & Alsop (1997) identified a number of issues that go well beyond the dispute between the Council and the hotel owners. Of particular importance is that the judgement implies a duty of care between neighbouring landowners in respect of an entirely natural loss of support. It may follow that if a landowner is aware of any natural or man-made ground hazard on his land, this would make him liable in negligence for any subsequent damages.

In the subsequent appeal by Scarborough BC over the Holbeck judgement, it was stated that although the landowner owed a measured duty of care, that duty depended on foreseeability. In this instance, the Council had not foreseen the magnitude of the risk and would not have done so without expert evidence derived from a geological survey. The duty might extend only to warning the owner of the dominant land of the foreseen risk and did not necessarily require expensive preventative works. Furthermore, it would be unfair and unreasonable to find liability in such circumstances where the danger had been equally apparent to the dominant owner.

4.10.6 Reduce the exposure

By abandoning the area or moving assets at risk out of the area affected by landsliding (i.e. reducing the exposure), risk can be avoided. At Knipe Point near Scarborough the Council recommended that houses at risk from the ongoing landsliding should be demolished. The owners would be eligible to benefit from a share of £1 million Pathfinder funding from DEFRA. The money could be used to buy suitable land, enabling those homeowners to build elsewhere.

4.10.7 Provide forewarning of potentially damaging incidents

Exposure to the risk could be avoided by warning systems enabling the emergency services to alert people at risk from imminent events, giving them time to move to safety, and make advance preparations to lessen the impact of the event. For example, West Dorset Council has established a landslide warning and emergency response procedure to inform the residents and businesses in Lyme Regis of imminent threats from ground movement (Cole & Davis 2002).

4.10.8 Incorporate specific ground movement tolerating measures into the building design

Vulnerability to landsliding could be reduced by incorporating specific ground movement tolerance into the building design (e.g. Clark *et al.* 1996). In many unstable areas, most properties and modern buildings are unsuited to accommodate the ground movement that could occur at the site. The most widely used foundation and building types (traditional strip footings and non-framed structures) are particularly vulnerable to ground movement. However, a great deal can be done to limit the effects of ground movement in building designs. The most important is the adoption of raft-type foundations which can 'float' over the movement.

4.10.9 Control the area between a landslide event and the assets at risk

Risk can be reduced or avoided by reducing the spatial probability of the hazard by introducing a preventative barrier. A variety of methods are available to prevent the hazard associated with landslide run-out or rockfall debris being realized. These include rock-face meshing, catch fences, rock catching ditches and rock shelters (e.g. Bromhead 2005). Roads and railways constructed on piered viaducts may allow landslides to pass beneath the structure with minimal damage.

4.10.10 Reduce the probability of the hazard

Remedial measures can be implemented to reduce the likelihood of damaging landslide incidents (i.e. reducing the probability of the hazard); this option is widely regarded as the 'engineering solution' to landslides. While it may be possible to prevent further movement of small landslides, for large complex landslides there will usually be a degree of uncertainty about the level of risk reduction that has been achieved.

Excellent reviews of the available methods can be found in Hutchinson (1977, 1984) and Bromhead (1986, 1992, and 2005) and include the following (Fig. 4.65): removal of the landslide mass by digging it out and replacing with stable material; reducing porewater pressures in the slope by surface and subsurface drainage; reducing destabilizing forces by removing landslide material, particularly from the upper part of the slide; increasing stabilizing forces by adding weight to the toe of an unstable area or by increasing the shear strength along the failure surface; supporting unstable areas by the construction of retaining structures; or preventing the basal erosion of landslides by toe protection structures.

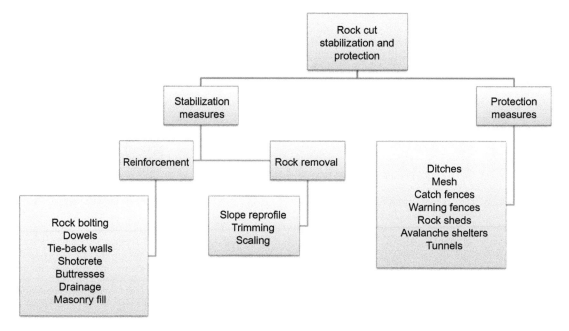

Fig. 4.65. Rock-cut stabilization and protection measures.

Effective stabilization of high-risk landslides can generally only be achieved following a detailed ground investigation and stability analysis.

In recent years a series of practical guides on landslide management have been produced, largely based on experiences in Wales (Siddle *et al.* 2000a; Nichol *et al.* 2002), the Ventnor Undercliff (McInnes 2000, 2007) and elsewhere on the UK coast (McInnes 2009; McInnes & Moore 2011). These documents provide an excellent introduction to the challenges and experiences associated with landslide management (for a more international view, see Moore & McInnes 2014).

4.11 The role of government in landslide management

Although the ultimate responsibility for managing landslide risks in the UK rests with individual property owners, the state has gradually acquired an important role in addressing a number of specific problems, discussed in the following sections.

4.11.1 Provision of publicly funded coast protection works

Publicly funded coast protection works have been provided to prevent erosion or encroachment by the sea. The Coast Protection Act 1949 provides coast protection authorities (i.e. maritime district councils or unitary authorities) with permissive (not mandatory) powers to carry out such coast protection works, whether within or outside their areas, as may be needed for the protection of any land in their area.

Grant-aid is made available to operating authorities by the government department with the policy responsibility for administering the Act. In England, this is the Department for Environment, Food and Rural Affairs (DEFRA); the National Assembly for Wales and the Scottish Executive have similar responsibilities.

A key feature of the powers is that they are permissive rather than mandatory; the authorities are not obliged to carry out works. This clearly limits the role of the State to only providing defences that are deemed to be in the national interest. However, this subtle distinction can cause considerable public misunderstanding and frustration. It also recognizes that complete protection is impossible:

> ...a balance has to be struck between costs and benefits to the nation as a whole. For example, to attempt to protect every inch of coastline from change would not only be uneconomic but would work against the dynamic processes which determine the coastline and could have adverse effect on defences elsewhere and on the natural environment. (MAFF 1993)

Over the last 100 years or so some 860 km of coast protection works have been constructed to prevent coastal erosion (MAFF 1994); this figure probably includes some low-lying areas prone to erosion. Shoreline management plans set out the strategic framework for managing erosion and flood risks along region-wide stretches of coastline (DEFRA 2006). For further details see Lee & Clark (2002), McInnes & Moore (2011).

4.11.2 Control development in high-risk areas

Another role of government is to control development in areas at risk and minimize the impact of new development on risks experienced elsewhere through the land-use planning system. The planning system (originally established in 1947), as currently defined by the Town and Country Planning Act 1990 and the Town and Country Planning (Scotland) Act 1997, aims to regulate the development and use of land in the public interest. Development is defined as: 'the carrying out of building, engineering, mining or other operations, in, on, over or under land, or the making of any material change in the use of any building or other land' according to the 1990 Act S.55.

In England and Wales, guidance covering unstable land (including landslides) was given in PPG14 (Department of the Environment/Welsh Office 1990; Department for Transport, Local Government & the Regions 2002) which set out the responsibility of the developer to determine whether the land is suitable for the proposed purpose: 'It is important that the stability of the ground is considered at all stages of the planning process. It therefore needs to be given due consideration in development plans as well as in decisions on individual planning applications' (Department of the Environment 1990).

In March 2012, PPG14 was replaced by the National Planning Policy Framework (Department for Communities & Local Government 2012) that states that: 'To prevent unacceptable risks from pollution and land instability, planning policies and decisions should ensure that new development is appropriate for its location' (Para 120).

Planning policies and decisions should also ensure that the site is suitable for its new use, taking account of ground conditions, land instability (including from natural hazards or former activities such as mining), pollution arising from previous uses, and any proposals for mitigation (including land remediation or impacts on the natural environment arising from that remediation).

Scottish Planning Policy (Scottish Government 2010) provides broad guidance to developers and planners under which environmental problems including unstable land should be taken into consideration. Specific guidance in Scotland is also given for peat slides affecting wind farm construction (Scottish Executive 2006).

Peat stability assessments are usually required as part of the environmental planning process for Section 36 electricity-generating applications in Scotland (under the Electricity Act 1989). The Scottish Government checks these assessments against its own published guidance (Scottish Executive 2006; Scottish Government 2011). The Countryside Council for Wales (now Natural Resources Wales) advocate peat stability assessments in all cases where developments involve construction of infrastructure on peat (CCW 2010), while in England, the Department of Energy and Climate Change (DECC) advise that, where relevant, applicants may be instructed to provide information on landslide risk related to development work (DECC 2011).

Local planning authorities can minimize risks from landslides by ensuring that development is suitable or preventing the development of land where slope instability poses major problems. Development plans can be used to set out broad policies that establish a framework for restricting development in landslide areas. Where development is proposed on unstable land, the planning authority can ensure that landslide issues are properly addressed by development proposals, especially the possible adverse effects of slope instability on the development and the possible adverse effects of the development on the stability of adjoining land. For further details see Clark *et al.* (1996).

4.11.3 Control building standards

Building regulations are made by the government to secure 'the health, safety, welfare and convenience of persons in and about the building'. The 2004 regulations state that:

> The building shall be constructed so that ground movement caused by (a) swelling, shrinkage or freezing of the subsoil; or (b) landslip or subsidence (other than subsidence arising from shrinkage), in so far as the risk can be reasonably foreseen, will not impair the stability of any part of the building (Anon 2004).

Building regulations apply to building work in general, controlled services and fittings, and material change of use. The local authority building control department has to ensure that building work complies with the regulations. If the work fails to comply, the developer may be required to alter or remove it.

Lutas (2002) describes the approach adopted in the Isle of Wight Undercliff, highlighting the importance of close co-operation between planning departments and building control in evaluating development and building plans.

4.11.4 Fund and co-ordinate the response to major events

Peace-time emergency planning is not a statutory duty for local authorities, although they would take a lead role in a major landslide incident. Under the Local Government Act 1972, a local authority has the permissive power to: incur expenditure that in their opinion is in the interests of their area or its inhabitants (S.137); incur such expenditure as they consider necessary in an emergency or disaster involving destruction of or danger to life or property, or where there are reasonable grounds for preventing such an event (S.138(1)); or to make grants or loans to other people or bodies in an emergency or disaster (S.138(1)).

The involvement of Scarborough Borough Council after the Holbeck Hall landslide of June 1993 is, perhaps, typical of the concern of many local authorities to ensure that the problems associated with major landslide events are managed in the interests of the community (Clements 1994).

In England, rehabilitation costs may be eased by disaster funds or grant payments made to the local authority under a 'Bellwin' scheme established under the Local Government and Housing Act 1989 S.155, by which the Government makes available financial assistance for up to 85% of local authority expenditure incurred above a threshold level (DoE 1993). Such funding would be for: providing relief or carrying out immediate works to safeguard life or property, or preventing suffering or severe inconvenience; an incident specified in the scheme; or for works completed before a specified deadline (usually within 2 months of the incident).

A Bellwin scheme may be activated, at the discretion of the Secretary of State, where an emergency or disaster involving destruction of or danger to life or property occurs and, as a result, one or more local authorities incur expenditure on, or in connection with the taking of immediate action to safeguard life or property, or to prevent suffering. Scotland, Wales and Northern Ireland run similar schemes.

The Bellwin scheme has traditionally been seen as a response to incidents in which bad weather has caused threats to life and property beyond all previous local experience, and where an undue financial burden would otherwise fall on the local authority (Stanford 2014). The scheme is only intended to cover uninsurable risk, and there is no automatic entitlement to financial assistance.

In exceptional cases, ministers may be able to announce the approval of a scheme shortly after an incident. However, most local events would only be eligible for financial assistance after detailed information on the local authorities expenditure had been forwarded to the department. The Bellwin scheme was invoked to provide assistance to South Wight Borough Council after the 1993–1994 landslide problems at Blackgang and Castlehaven, Isle of Wight (Clark *et al.* 1996) (Fig. 4.66).

4.11.5 Protect strategic infrastructure

The Government has identified the need to improve the security and resilience of the infrastructure most critical to keeping

Fig. 4.66. (a–c) Major coastal landsliding in the Gault Formation, Castlehaven, Isle of Wight, causing significant building and structural damage to properties on and above the landslide (photo credit: David Giles).

the country running against attack, damage or destruction. International terrorism, cyber-attacks, major accidents and natural hazards (e.g. floods and landslides) were identified as among the most serious risks to the UK's national security interests.

The Infrastructure Resilience programme, led by the Civil Contingencies Secretariat, was established in March 2011 to enable public and private sector organizations to build the resilience of their infrastructure, supply and distribution systems to disruption from all risks (hazards and threats) as set out in the National Risk Assessment (Cabinet Office 2012). A guide ('Keeping the country running: natural hazards and infrastructure') has been developed to support infrastructure owners and operators, emergency responders, industry groups, regulators and government departments to work together to improve the resilience of critical infrastructure and essential services.

The Natural Hazards Partnership (NHP) was established in 2011 in order to bring together expertise from across public sector agencies (including the BGS) with the aim of drawing upon scientific advice in the preparation, response and review of natural hazards (Meteorological Office 2013; Hemingway & Gunawan 2018). The NHP focuses on hazards that may disrupt the normal activities of communities or damage environmental services, including: landslides, floods, drought, extreme temperatures, space weather, volcanic ash, earthquakes, wildfire, snow, ice, fog and air quality (British Geological Survey 2013).

The BGS issues a daily landslide hazard warning directly to the NHP and its stakeholders. This hazard warning is primarily based on recent and historic landslide information in the national landslide database. These events are used in conjunction with forecasted and antecedent rainfall, BGS landslide susceptibility maps (GeoSure) and expert judgement. The potential for national landslide hazard for the upcoming 24-hour period is highlighted using a traffic-light-plus system (green, yellow, amber and red).

4.12 In practice: acceptable or tolerable risks?

How landslide risks are managed depends to a large extent on the attitudes, responsibilities and resources available to the affected parties, as this provides the context for what can and cannot be achieved (e.g. Palm 1990; Lee 2002). Landslide management in the UK often involves the planning of public expenditure to protect property and increase social welfare by reducing land instability losses. As only a minority of the tax-paying community (i.e. the nation) is affected, the use of public funds can be seen as extending the property life and safeguarding investments.

Allocation of public resources for landslide management is therefore influenced by the need to find an acceptable balance between investments in a wide range of competing public services. Economic evaluation (i.e. benefit–cost analysis) provides a mechanism for comparing the benefits of landslide management with the costs incurred, in order to determine: whether, and by how much, the benefits exceed the costs; the strategy that is expected to deliver the greatest economic return, that is, the most efficient use of resources; and the anticipated 'loss' to be incurred if it is decided to proceed with an 'uneconomic' strategy.

When considering risk to people (i.e. individual and societal risk), a cornerstone of risk management in hazardous industries is the concept that there is a degree of risk that is tolerable and that this can be defined through the use of risk acceptance criteria (HSE 1988, 1992; Fig. 4.67). Above a certain threshold, the risks might be considered intolerable or unacceptable; below another much lower threshold, the risk might be considered to be so small that it is broadly acceptable. The zone between the unacceptable and broadly acceptable regions is the tolerable region, where the level of risks are typical of the risks posed by activities that people are prepared to tolerate in order to secure benefits (HSE 2001).

In the tolerable region, it is generally expected that the level of risk should be reduced to a level which is as low as reasonably practicable (the so-called ALARP principle; HSE 2001). In England, the definition of reasonably practicable has been established by case law:

> Reasonably practicable is a narrower term than 'physically possible' and seems to me to imply that a computation must be made by the owner in which the quantum of risk is placed on one scale and the sacrifice involved in the measures necessary for averting the risk (whether in money, time or trouble) is placed in the other, and that, if it be shown that there is a gross disproportion between them – the risk being insignificant in relation to the sacrifice – the defendants discharge the onus on them. (Judge Asquith, Edwards v National Coal Board, All England Law Reports Vol. 1, 747 (1949))

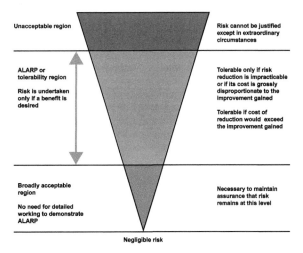

Fig. 4.67. Risk tolerability and the ALARP concept (adapted from HSE 1992).

This case established that a balance must be made between the level of risk and the sacrifice (e.g. time or money) involved in averting the risk (HSE 1999). Risk is tolerable only if risk reduction is impracticable or if its cost is grossly disproportionate to the improvement gained.

Residual risk is tolerable only if further risk reduction is impracticable or requires action that is gross disproportionate in time, trouble and effort to the reduction in risk achieved.

The ALARP principle forms the basis of the approach used by the UK Health and Safety Executive in its regulation of the major hazardous industries, such as the nuclear, chemical and offshore oil and gas industries. The concept implies that if the risk is unacceptable it must be avoided or reduced, irrespective of the cost, except in extraordinary circumstances. In addition, if the risk falls within the ALARP or tolerable region, then cost is taken into account when determining how far to pursue the goal of minimizing risk or achieving safety. As a consequence, the ALARP or tolerable region (Fig. 4.67) includes a continuous spectrum of conditions ranging from where the cost of risk reduction would exceed the improvement gained adjacent to the lower boundary, to where the risk is only tolerable if risk reduction is impracticable or if the cost of risk reduction is grossly disproportionate to the improvement gained (McQuaid & Le Guen 1998). Risk does not have to be reduced to as low as possible employing best available techniques (BAT), as this will almost certainly involve excessive cost. The benefits to be gained from a reduction in risk are normally expected to exceed the costs of achieving such a reduction.

Although incidents involving multiple fatalities are very rare in the UK (Aberfan in 1966, Lulworth Cove in 1977), this often seems to reflect good fortune rather than the lack of potential. There are no agreed criteria in the UK to gauge whether the calculated levels of landslide risk might be regarded as acceptable or tolerable. However, experience from hazardous industries in the UK and landslide management overseas provides a guide to what could be considered reasonable criteria for both individual risk and societal risk. The HSE (2001) has suggested that, in terms of average individual risk, the boundaries of the ALARP region could be either (1) between the broadly acceptable and tolerable regions; or (2) between the unacceptable and tolerable regions. In the former, there are 10^{-6} fatalities per year (1 in 1 million) for both workers and the public. This figure is regarded as extremely small when compared with the background level of risks that people are exposed to over their lifetimes (typically a risk of death of 10^{-2} per year averaged over a lifetime). In the latter, HSE suggest separate criteria for this boundary depending on who is exposed to the risks. A value of 10^{-4} fatalities per year (1 in 10 000) is suggested for members of the public who have a risk imposed on them 'in the wider interests of society'. For workers in industry, HSE suggested that the boundary should be around 10^{-3} fatalities per year (1 in 1000). The maximum tolerable criterion of 10^{-3} per year was used by the HSE because it approximates to the risk experienced by high-risk groups in mining, quarrying, demolition and deep-sea fishing (HSE

Table 4.16. *Individual risk levels for workers in different industries (from Vinnem 2007)*

Industry	Annual individual risk level for employees
Deep-sea fishermen on UK registered vessels	1 in 750
Extraction of mineral oil and gas	1 in 999
Coal extraction	1 in 7100
Construction	1 in 10 200
Agriculture	1 in 13 500
Metal manufacturing	1 in 17 000

1992; Table 4.16). Table 4.17 shows how many risky activities an individual would need to undertake in 1 year to reach an individual risk per annual of 10^{-3} per year. This illustrates that the 10^{-3} per year upper limit to the tolerable region is actually quite high.

Over the last decade or so there has been increasing interest in the development and application of individual risk criteria for landslide management. In 1997, the Hong Kong Geotechnical Engineering Office (GEO) proposed interim risk guidelines for natural terrain landslides (ERM 1998; Reeves *et al.* 1999; Ho *et al.* 2000; note that the guidelines remain interim in 2014). Average risk levels for individuals most at risk proposed to mark the upper limit of the tolerable region were: a maximum of 10^{-5} per year for new developments, and a maximum of 10^{-4} per year for existing developments.

The Australian Geomechanics Society guidelines for landslide risk management include individual risk criteria identical to those in Hong Kong (AGS 2007; note that these criteria do not represent a regulatory position). In 2009, the District of North Vancouver (DNV), Canada formerly adopted the landslide risk criteria (DNV 2009) of a maximum risk of fatality of 10^{-4} and 10^{-5} per year for redevelopments involving an increase to gross floor area on the property of less than or equal to 25% and greater than 25%, respectively.

This policy gives the District's Chief Building Official the discretion to apply the criteria to building permits, subdivision and development applications for sites exposed to landslide and debris flow hazards.

Table 4.17. *Activities corresponding to a 1×10^{-3} risk of fatalities/year (from HSE 2001; Lewis 2007)*

Activity	Number of activities in 1 year that equals the criteria of 10^{-3} fatalities per year
Hang-gliding	116 flights
Surgical anaesthesia	185 operations
Scuba diving	200 dives
Rock climbing	320 climbs

unacceptable when a large number of people are exposed. So-called societal risk criteria are often presented on F–N curves, where two criteria lines divide the space into three regions: where risk is unacceptable or intolerable; where it is broadly acceptable (negligible); and where it requires further assessment and risk reduction as far as is reasonably practicable. The criteria lines generally slope steeply away from an individual risk value ($N = 1$), reflecting the aversion to multiple fatalities. For the transportation of dangerous goods, the UK has adopted a gradient of -1 for the slope of the criteria line (Fig. 4.68).

In the absence of agreed risk criteria for landslides, Fell & Hartford (1997) considered dam safety to be a good analogy to landsliding. Figure 4.69 presents the F–N curves for dams developed by the Australian National Committee on Large Dams (ANCOLD 2003), who identify separate tolerability lines for new and existing dams. Risks ten times higher are tolerated for existing dams than new dams.

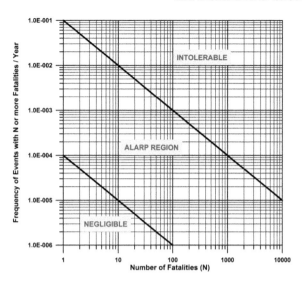

Fig. 4.68. Societal risk criteria: transportation of dangerous goods (adapted from Health and Safety Executive (HSE) 1991).

It is widely believed that society generally tends to be more concerned about multiple fatalities in a single event than a series of smaller events that collectively kill the same number of people (e.g. Horowitz & Carson 1993; Ball & Floyd 1998). While low-frequency high-consequence events might represent a very small risk to an individual, they may be seen as

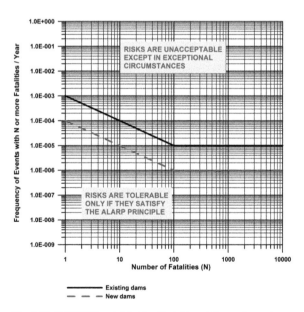

Fig. 4.69. ANCOLD societal risk criteria for existing and new dams. Note that the ALARP region lies below the tolerability lines (adapted from ANCOLD 2003).

4.13 Concluding remarks

The Tees–Exe line has long been used to divide the UK into lowland and upland regions, reflecting differences in the underlying geology, glacial history, climate and landscape (e.g. Monkhouse 1971). Not surprisingly, there are notable variations in landslide activity between these regions.

The landslide environment of lowland Britain is dominated by ancient pre-existing landslides and cambering involving sequences of Cenozoic- and Mesozoic-aged interbedded sedimentary rocks, including landslide-prone clays found in the London Clay Formation, Gault Formation and Lias Group. These features are believed to have developed during the late Pleistocene periods of periglacial activity, during which time widespread landsliding is believed to have occurred on valley sides, flanks of hills and along escarpments. 'New' inland landslides are relatively infrequent events, and are largely the result of localized failure of man-made slopes along the road and railway networks.

Many of the hillsides and mountainsides of upland Britain are mantled with rock slope failures, deep-seated landslides and slope deformation features, often developed in Paleozoic-aged rocks and overlaying glacial deposits. While many features have probably been inherited from the late Pleistocene periods of glacial/periglacial and postglacial activity, large-scale landslide activity has continued through the Holocene, possibly because of the long-term effects of postglacial stress release. Nineteenth and twentieth century mining operations appear to have initiated a new phase of deep-seated landsliding, especially in South Wales. Intermittent debris flow and peat slide activity remains a feature of many upland areas.

Across the UK, the main hazard types can be summarized as follows.

- The cumulative effects of very slow to slow ground movement associated with reactivation of pre-existing

landslides during wet winters. This hazard is widespread in the UK, especially along escarpments and valley-side slopes in England and South Wales, and within major coastal landslide complexes such as the Isle of Wight Undercliff and Lyme Regis. The consequences tend to be limited to property and infrastructure damage, with little threat to people.
- The sudden and often unexpected impacts of high-velocity debris and peat slides, and debris flows, triggered by intense rainstorms. These events can run-out or travel considerable distances beyond the original source area. This hazard tends to be confined to upland UK. Incidents can present a significant threat to people, although the exposed populations are generally not high. The threat to property includes damage to buildings and temporary blockage of infrastructure routes, leading to delays and emergency response costs. Examples of this type of hazard are described in Winter (2019).
- The effects of cliff recession on the unprotected coast and, to a lesser extent, river banks. Rockfalls are a threat to beach and foreshore users, although fatalities and injuries are rare. The impact on property is typically restricted to the loss of agricultural land and, in some cases, isolated houses or small cliff-top communities. However, over the last 100 years or so, many coastal cliffs in urban areas have been protected by seawalls and rock revetments. Failure of these defences would present a significant threat to these coastal communities.

It is important to stress that a significant number of landslide features in the UK are the product of past environmental conditions and probably pose negligible or no hazard under current conditions, unless significantly disturbed by human activity. This is probably true for many of the slope deformation features and major rock slope failures of upland Britain; however, there will always be exceptions to this and it is always wise to evaluate the hazard on a site-by-site basis.

There is now considerable evidence to demonstrate that climate and relative sea level has been changing and the pace of change is predicted to accelerate. However, the effects on the landslide environment of the UK remain very uncertain. It is certainly possible that the changing frequency of intense rainstorms and wet winters will result in a national increase in landslide activity. However, the impact on individual sites is very difficult to quantify (e.g. Dijkstra & Dixon 2010).

Acknowledgments Dr Catherine Pennington from the British Geological Survey is thanked for her significant review and comments on the chapter as well as supplying data and images from the BGS databases. We are also grateful to Dr Peter Brabham, Professor Eddie Bromhead, Paul Maliphant, Emily Farrugia and Professor Mike Winter for the use of images from their landslide collections. The Network Rail Media Centre, Telford District Council and the Isle of Wight County Council are thanked for use of their landslide images. Professor Jim Griffiths once more provided very useful feedback on the text.

Funding This research received no specific grant from any funding agency in the public, commercial, or not-for-profit sectors.

References

AARONS, A., WEEKS, A.G. & PARKES, R.D. 1977. Site investigation for the Channel Tunnel British ferry terminal. *Ground Engineering*, **May 1977**, 43–47.

ACKERMAN, K.J. & CAVE, R. 1967. Superficial deposits and structures including landslip in the Stroud district, Gloucestershire. *Proceedings of the Geologists Association*, **78**, 567–586, https://doi.org/10.1016/S0016-7878(68)80011-2

AGAR, R. 1960. Postglacial erosion of the north Yorkshire coast from Tees estuary to Ravenscar. *Proceedings of the Yorkshire Geological Society*, **32**, 408–425, https://doi.org/10.1144/pygs.32.4.409

ALEXANDER, R., COXON, P. & THORN, R.H. 1985. Bog flows in south-east Sligo and south-west Leitrim. *In*: THORN, R.H. (ed.) *Sligo and West Leitrim*. Field Guide No. 8, Irish Association for Quaternary Studies (IQUA), 58–76.

ALLISON, J.A., MAWDITT, J.M. & WILLIAMS, G.T. 1991. The use of bored piles and counterfort drains to stabilise a major landslip: a comparison of theoretical and field performance. *In*: CHANDLER, R.J. (ed.) *Slope Stability Engineering: Development and Applications*. Thomas Telford, London, 369–376.

ANCOLD 2003. *Guidelines on Risk Assessment*. Australian National Committee on Large Dams Incorporated, Melbourne.

ANON. 1788. The Gazetteer and New Daily Advertiser 5.1 1788 (reproduced in The Whitby Times 28.12.1894).

ANON. 2004. *The Building Regulations 2000 (Structure), Approved Document A, 2004 Edition*. Office of the Deputy Prime Minister. Her Majesty's Stationery Office, London.

AUSTRALIAN GEOMECHANICS SOCIETY (AGS) 2007. Practice Note Guidelines for Landslide Risk Management 2007. *Australian Geomechanics*, **42**.

BACON, J. & BACON, S. 1988. *Dunwich Suffolk*. Segment Publications, Colchester.

BALL, D.J. & FLOYD, P.J. 1998. *Societal risks*. Final Report, Commissioned by the Health and Safety Executive, United Kingdom.

BALLANTYNE, C.K. 1991. Holocene geomorphic activity in the Scottish Highlands. *Scottish Geographical Magazine*, **107**, 84–98, https://doi.org/10.1080/00369229118736815

BALLANTYNE, C.K. 1997. Holocene rock slope failures in the Scottish Highlands. *In*: MATTHEWS, J.A., BRUNSDEN, D., FREZEL, B., GLASER, B. & WEISS, M.M. (eds) *Rapid Mass Movement as a Source of Climatic Evidence for the Holocene*. Gustav Fisher Verlag, 197–206.

BALLANTYNE, C.K. 2003. A Scottish sturzstrom: the Beinn Alligin rock avalanche, Wester Ross. *Scottish Geographical Journal*, **119**, 159–167, https://doi.org/10.1080/0036922031873 7169

BALLANTYNE, C.K. 2012. Chronology of glaciation and deglaciation during the Loch Lomond (Younger Dryas) Stade in the Scottish Highlands: implications of recalibrated 10Be exposure ages. *Boreas*, **41**, 513–526, https://doi.org/10.1111/j.1502-3885.2012.00253.x

BALLANTYNE, C.K. 2013. Lateglacial rock-slope failures in the Scottish Highlands. *Scottish Geographical Magazine*, **129**, 67–84, https://doi.org/10.1080/14702541.2013.781210

BALLANTYNE, C.K. & ECKFORD, J.D. 1984. Characteristics and evolution of two relict talus slopes in Scotland. *Scottish*

Geographical Magazine, **100**, 20–33, https://doi.org/10.1080/00369228418736575

BALLANTYNE, C.K. & HARRIS, C. 1994. *The Periglaciation of Great Britain*. Cambridge University Press, Cambridge.

BALLANTYNE, C.K. & STONE, J.O. 2009. Rock-slope failure at Baosbheinn, Wester Ross, NW Scotland: age and interpretation. *Scottish Journal of Geology*, **45**, 177–181, https://doi.org/10.1144/0036-9276/01-388

BALLANTYNE, C.K. & STONE, J.O. 2013. Timing and periodicity of paraglacial rock-slope failures in the Scottish Highlands. *Geomorphology*, **186**, 150–161, https://doi.org/10.1016/j.geomorph.2012.12.030

BALLANTYNE, C.K., SCHNABEL, C. & XU, S. 2009a. Exposure dating and reinterpretation of coarse debris accumulations ('rock glaciers') in the Cairngorm Mountains, Scotland. *Journal of Quaternary Science*, **24**, 19–31, https://doi.org/10.1002/jqs.1189

BALLANTYNE, C.K., SANDEMAN, G.F., STONE, J.O. & WILSON, P. 2014. Rock-slope failure following Late Pleistocene deglaciation on tectonically stable mountainous terrain. *Quaternary Science Reviews*, **86**, 144–157, https://doi.org/10.1016/j.quascirev.2013.12.021

BARTON, M.E. & COLES, B.J. 1984. The characteristics and rates of the various slope degradation processes in the Barton Clay Cliffs of Hampshire. *Quarterly Journal of Engineering Geology*, **17**, 117–136, https://doi.org/10.1144/GSL.QJEG.1984.017.02.04

BENTLEY, S.P. 2000. Troedrhiwfuwch landslide, Rhymney valley. In: SIDDLE, H.J., BROMHEAD, E.N. & BASSETT, M.G. (eds) *Landslides and Landslide Management in South Wales*, National Museums and Galleries of Wales, Cardiff, Geological Series No. 18, 69–71.

BENTLEY, S.P. & SIDDLE, H.J. 2000a. Bournville landslide, Abertillery. In: SIDDLE, H.J., BROMHEAD, E.N. & BASSETT, M.G. (eds) *Landslides and Landslide Management in South Wales*. National Museums and Galleries of Wales, Cardiff, Geological Series No. 18, 81–84.

BENTLEY, S.P. & SIDDLE, H.J. 2000b. New Tredegar landslide, Rhymney Valley. In: SIDDLE, H.J., BROMHEAD, E.N. & BASSETT, M.G. (eds) *Landslides and Landslide Management in South Wales*. National Museums and Galleries of Wales, Cardiff, Geological Series No. 18, 71–74.

BIRCH, G.P. 2001. Mapping for high pressure gas pipelines in South Wales. In: GRIFFITHS, J.S. (ed.) *Land Surface Evaluation for Engineering Practice*. Geological Society, London, Engineering Group Special Publication, **18**, 73–82.

BISHOP, A.W., HUTCHINSON, J.N., PENMAN, A.D.M. & EVANS, H.E. 1969. *Geotechnical investigation into the causes and circumstances of the disaster of 21st October 1966*. Unpublished Report for the Aberfan Disaster Enquiry.

BISHOP, D.W. & MITCHELL, G.F. 1946. On a recent bog-flow in Meenacharvy Townland, Co. Donegal. *Scientific Proceedings of the Royal Dublin Society*, **24**, 151–156.

BLACKHAM, A., DAVIES, C. & FLENLEY, J. 1981. Evidence for Late Devensian landslipping and Late Flandrian forest degeneration at Gormire Lake, North Yorkshire. In: NEALE, J. & FLENLEY, J. (eds) *The Quaternary in Britain*. Pergamon, 184–199.

BOARDMAN, J. 1982. Glacial geomorphology of the Keswick area, northern Cumbria. *Proceedings of the Cumberland Geological Society*, **4**, 115–134.

BONNARD, C., FORLATI, F. & SCAVIA, C. (eds) 2004. *Identification and Mitigation of Large Landslide Risks in Europe: Advances in Risk Assessment. Imiriland Project*. European Commission, Balkema, Rotterdam.

BOOTH, K.A. & JENKINS, G.O. 2009. *Landslide nature and distribution on the Lincoln 1:50k geological sheet*. British Geological Survey Report, **OR/09/047**.

BOWEN, D.Q., ROSE, J., McCABE, A.M. & SUTHERLAND, D.G. 1986. Correlation of the Quaternary Glaciations in England, Ireland, Scotland and Wales. *Quaternary Science Reviews, Quaternary Glaciations in the Northern Hemisphere*, **5**, 299–340.

BOYLAN, N., LONG, M. & JENNINGS, P. 2008. Peat slope failure in Ireland. *Quarterly Journal of Engineering Geology and Hydrogeology*, **41**, 93–108, https://doi.org/10.1144/1470-9236/06-028

BRADFORD, B. 2005. Why does God allow disasters. *The Good News*, **10**, 4–7, http://www.gnmagazine.org/issues/gn57/gn05ma.pdf-

BRAY, M.J. 1996. *Beach budget analysis and shingle transport dynamics in West Dorset*. Unpublished PhD thesis, London School of Economics, University of London.

BRIGGS, D. & COURTNEY, F. 1972. Ridge and trough topography in the North Cotswolds. *Proceedings of the Cotteswold Naturalists Field Club*, 94–103.

BRITISH GEOLOGICAL SURVEY http://www.bgs.ac.uk/research/earthHazards/naturalHazardsPartnership.html [accessed August 2014]

BROMHEAD, E.N. 1986. *The Stability of Slopes*. Surrey University Press, London.

BROMHEAD, E.N. 1992. *The Stability of Slopes*. 2nd edn. CRC Press, London.

BROMHEAD, E.N. 2005. Geotechnical structures for landslide risk reduction. In: GLADE, T., ANDERSON, M.G. & CROZIER, M.J. (eds) *Landslide Hazard and Risk*. John Wiley, 549–594.

BROMHEAD, E.N. & IBSEN, M.-L. 1997. Land use and climate-change impacts on landslide hazards in SE Britain. In: CRUDEN, D. & FELLS, R. (eds) *Landslide Risk Assessment*. Balkema, Rotterdam, 65–176.

BROMHEAD, E.N., HOPPER, A.C. & IBSEN, M.-L. 1998. Landslides in the Lower Greensand escarpment in south Kent. *Bulletin of Engineering Geology and Environment*, **57**, 131–144, https://doi.org/10.1007/s100640050029

BROMHEAD, E.N., HUGGINS, M. & IBSEN, M.-L. 2000. Shallow landslides in Wadhurst Clay at Robertsbridge, Sussex, UK. In: BROMHEAD, E.N., DIXON, N., IBSEN, M.-L. (eds) *Landslides: In Research, Theory and Practice*. Thomas Telford, London, 183–188.

BROOKS, S. 1997. Modelling the role of climatic change in landslide initiation for different soils during the Holocene. In: MATTHEWS, J.A., BRUNSDEN, D., FRENZEL, B., GLASER, B. & WEISS, M.M. (eds) *Rapid Mass Movement as a Source of Climatic Evidence for the Holocene*. Gustav Fischer Verlag, 207–222.

BROOKS, S., RICHARDS, K.S. & ANDERSON, M.G. 1993. Shallow failure mechanisms during the Holocene: utilisation of a coupled slope hydrology-slope stability model. In: THOMAS, D.S.G. & ALLISON, R.J. (eds) *Landscape Sensitivity*. Wiley and Sons, 149–175

BRUNSDEN, D. & CHANDLER, J.H. 1996. Development of an episodic landform change model based upon the Black Ven mudslide, 1946–1995. In: ANDERSON, M.G. & BROOKS, S.M. (eds) *Advances in Hillslope Processes, 2*. John Wiley.

BRUNSDEN, D. & JONES, D.K.C. 1980. Relative time scales and formative events in coastal landslide systems. *Zeitschrift für Geomorphologie Supplementband*, **34**, 1–19.

BRUNSDEN, D. & LEE, E.M. 2004. Behaviour of coastal landslide systems: an inter-disciplinary view. *Zeitschrift für Geomorphologie*, **134**, 1–112.

BRUNSDEN, D., COOMBE, K., GOUDIE, A.S. & PARKER, A.G. 1996. The structural geomorphology of the Isle of Portland, southern

England. *Proceedings of the Geologists Association*, **107**, 209–230, https://doi.org/10.1016/S0016-7878(96)80030-7

BUTLER, P.B. 1983. *Landsliding and other large scale mass movements on the escarpment of the Cotswolds Hills*. BA thesis (unpublished), Hertford College, University of Oxford.

CABINET OFFICE 2011. Needs of People Health – Incident The Aberfan Disaster 21 October 1966, Improving the UK's ability to absorb, respond to and recover from emergencies, 2 March 2011, https://www.gov.uk/government/uploads/system/uploads/attachment_data/file/61863/NeedsofPeopleHealthIncidentTheAberfanDisaster21October1966.pdf [last accessed July 2014]

CABINET OFFICE 2012. *National Risk Register of Civil Emergencies 2012 edition*.

CAMBERS, G. 1976. Temporal scales in coastal erosion systems. *Transactions of the Institute of British Geographers*, **1**, 246–256, https://doi.org/10.2307/621987

CARNEY, J.N. 1974. *A photo interpretation of mass movement features along the Antrim coast of Northern Ireland*. British Geological Survey Technical Report WN/EG/74/13.

CASCINI, L., BONNARD, C., COROMINAS, J., JIBSON, R. & MONTERO-OLARTE, J. 2005. Landslide hazard and risk zoning for urban planning and development. *In*: HUNGR, O., FELL, R., COUTURE, R. & EBERHARDT, E. (eds) *Landslide Risk Management*. Balkema, London, 199–236.

CHAMBERS, S. 2000. A4061 Rhigos Mountain Road: road management in a landslide area. *In*: SIDDLE, H.J., BROMHEAD, E.N. & BASSETT, M.G. (eds) *Landslides and Landslide Management in South Wales*. National Museums and Galleries of Wales, Cardiff, Geological Series No. 18, 94–97.

CHANDLER, J.H. 1989. *The acquisition of spatial data from archival photographs and their application to geomorphology*. Unpublished PhD thesis, City University, London.

CHANDLER, J.H. & BRUNSDEN, D. 1995. Steady state behaviour of the Black Ven mudslides: the application of archival analytical photogrammetry to studies of landform change. *Earth Surface Processes and Landforms*, **20**, 255–275, https://doi.org/10.1002/esp.3290200307

CHANDLER, R.J. 1970. The degradation of Lias Clay slopes in an area of the East Midlands. *Quarterly Journal of Engineering Geology*, **2**, 161–181, https://doi.org/10.1144/GSL.QJEG.1970.002.03.01

CHANDLER, R.J. 1971. *Landsliding on the Jurassic escarpment near Rockingham, Northamptonshire*. Institute of British Geographers, Special Publication No. 3, 111–128.

CHANDLER, R.J. 1976. The history and stability of two Lias Clay slopes in the Upper Gwash Valley, Rutland. *Philosophical Transactions of the Royal Society, London*, **A283**, 463–491, https://doi.org/10.1098/rsta.1976.0093

CHANDLER, R.J., KELLAWAY, G.A., SKEMPTON, A.W. & WATT, R.J. 1976. Valley slope sections in Jurassic strata near Bath, Somerset. *Philosophical Transactions of the Royal Society, London*, **A283**, 527–556, https://doi.org/10.1098/rsta.1976.0095

CHESTER, D.K. & DUNCAN, A.M. 2009. The Bible, theodicy and Christian responses to historic and contemporary earthquakes and volcanic eruptions. *Environmental Hazards: Human and Policy Dimensions*, **8**, 304–332.

CLAGUE, J.J. & STEAD, D. 2012. *Landslides: Types, Mechanisms and Modelling*. Cambridge University Press, Cambridge.

CLARK, A.R. & GUEST, S. 1991. The Whitby cliff stabilisation and coast protection scheme. *In*: CHANDLER, R.J. (ed.) *Slope Stability Engineering, Developments and Applications*. Thomas Telford, London, 283–290.

CLARK, A.R. & GUEST, S. 1994. The design and construction of the Holbeck Hall landslide coast protection and cliff stabilisation emergency works. *Proceedings of the 29th MAFF Conference of River and Coastal Engineers*, Loughborough, 3.3.1–3.3.6.

CLARK, A.R., PALMER, J.S., FIRTH, T.P. & MCINTYRE, G. 1991. The management and stabilisation of weak sandstone cliffs at Shanklin, Isle of Wight. *In*: CRIPPS, J.C. & MOON, C.F. (eds) *The Engineering Geology of Weak Rock*. Geological Society, London, Special Engineering Publications, **8**, 375–384.

CLARK, A.R., MOORE, R. & MCINNES, R.G. 1995. Landslide response and management: Blackgang, Isle of Wight. *In*: *Proceedings of the 30th MAFF Conference of River and Coastal Engineers*. 6.3.1–6.3.23.

CLARK, A.R., LEE, E.M. & MOORE, R. 1996. *Landslide Investigation and Management in Great Britain: A Guide for Planners and Developers*. HMSO, London.

CLARK, R. & WILSON, P. 2004. A rock avalanche deposit in Burtness Comb, Lake District, northwest England. *Geological Journal*, **39**, 419–430, https://doi.org/10.1002/gj.965

CLAYTON, K.M. 1980. Coastal protection along the East Anglian coast. *Zeitschrift fur Geomorpholgie Supplemet*, **34**, 165–172.

CLAYTON, K.M. 1989. Sediment input from the Norfolk cliffs, eastern England – a century of coast protection and its effect. *Journal of Coastal Research*, **5**, 433–442.

CLEMENTS, M. 1994. The Scarborough experience – Holbeck landslide, 3–4 June 1993. *Proceedings of the Institution of Civil Engineers, Municipal Engineers*, **103**, 63–70, https://doi.org/10.1680/imuen.1994.26381

COBB, A.E., JONES, D.B. & SIDDLE, H.J. 2000. The influence of mining on the Bolsover landslide. *In*: BROMHEAD, E.N., DIXON, N. & IBSEN, M.-L. (eds) *Landslides in Research, Theory and Practice*. Thomas Telford, London, 287–292.

COLE, K. & DAVIS, G. 2002. Landslide early warning and emergency planning systems in West Dorset, England. *In*: MCINNES, R.G. & JAKEWAYS, J. (eds) *Instability – Planning and Management*. Thomas Telford, London, 463–470.

COLHOUN, E.A., COMMON, R. & CRUICKSHANK, M.A. 1965. Recent bog flows and debris slides in the North of Ireland. *Scientific Proceedings, Royal Dublin Society*, **2**, 16–174.

CONWAY, B.W., FORSTER, A., NORTHMORE, K.J. & BARCLAY, W. 1980. *South Wales Coalfield Landslide Survey, British*. Geological Survey Technical Report, London.

CONYBEARE, W.D., BUCKLAND, W. & DAWSON, W. 1840. *Ten Plates Comprising a Plan, Sections and Views Representing the Changes Produced on the Coast of East Devon between Axmouth and Lyme Regis by the Subsidence of the Land … etc*. J Murray, London.

COOPER, R.G. 1980. A sequence of landsliding mechanisms in the Hambleton Hills, Northern England, illustrated by features at Peakscar, Hawnby. *Geografiska Annaler*, **62**, 149–156, https://doi.org/10.1080/04353676.1980.11880006

COOPER, R.G. 1981. Four new windypits. *Caves and Caving*, **14**, 3–4.

COOPER, R.G. 1983a. Fissures in the interior of the Isle of Portland. *William Pengelly Cave Studies Trust Newsletter*, **42**, 1–3.

COOPER, R.G. 1983b. Mass movement in caves in Great Britain. *Studies in Speleology*, **4**, 37–44.

COOPER, R.G. 1997. *John Wesley at Whitestone Cliff, North Yorkshire, 1755*. University of York Borthwick Paper No. 91.

COOPER, R.G. 2007. *Mass Movements in Great Britain*. Joint Nature Conservation Committee, Peterborough, Geological Conservation Review Series No. 33

COOPER, R.G. & JARMAN, D. 2007. Mam Tor, Derbyshire. *In*: COOPER, R.G. (ed.) *Mass Movements in Great Britain*. Joint Nature

Conservation Committee, Peterborough, *Geological Conservation Review Series No. 33*, 167–183.
COOPER, R.G., RYDER, P.F. & SOLMAN, K.R. 1976. *The North Yorkshire Windypits: A Review*. Transactions of the British Cave Association, 3, 77–94.
CORNFORTH, D.H. 2005. *Landslides in Practice: Investigation, Analysis, and Remedial/Preventative Options in Soils*. John Wiley & Sons.
COROMINAS, J. 1996. The angle of reach as a mobility index for small and large landslides. *Canadian Geotechnical Journal*, **33**, 260–271, https://doi.org/10.1139/t96-005
COSGROVE, A.R.P., BENNETT, M.R. & DOYLE, P. 1997. The rate and distribution of coastal cliff erosion in England: a cause for concern? *In*: BENNETT, M.R. & DOYLE, P. (eds) *Issues in Environmental Geology: A British Perspective*. Geological Society Publishing House, Bath, 303–330.
COUNTRYSIDE COUNCIL FOR WALES 2010. Assessing the impact of windfarm developments on peatlands in Wales. CCW Guidance Note, CCW.
CRAIG, D. 1979. *Mudslide activity in East County Antrim, Northern Ireland*. PhD thesis (unpublished), Queen's University, Belfast.
CRAIG, D. 1981. Weather and mudslide activity in east Co. Antrim, Northern Ireland. *Weather*, 58–62.
CREIGHTON, R. 2006. *Landslides in Ireland*. A Report of the Irish Landslides Working Group. Geological Survey of Ireland, http://www.gsi.ie/Programmes/Quaternary+Geotechnical/Landslides/
CRISP, D.T., RAWES, M. & WELCH, D. 1964. A Pennine peat slide. *Geographical Journal*, **130**, 519–524, https://doi.org/10.2307/1792263
CROFTS, J.E. & BERLE, D.A. 1972. An earth slip at Tiverton, Devon. *Géotechnique*, **22**, 345–351, https://doi.org/10.1680/geot.1972.22.2.345
CRUDEN, D. & FELL, R. (eds) 1997. *Landslide Risk Assessment*. Balkema, Rotterdam.
CRUDEN, D.M. & VARNES, D.J. 1996. Landslide types and processes. *In*: A.K.TURNER, and R.L.SCHUSTER, (eds) *Landslides: Investigation and Mitigation*. National Research Council, National Academy Press, Washington DC, Transportation Research Board, Special Report **247**, 36–75.
CUMMINGS, S.J., DORAN, I.G. & SIVAKUMAR, V. 2000. Case study of a landslip in glacial sands and gravels in N. Ireland. *In*: BROMHEAD, E.N., DIXON, N. & IBSEN, M.-L. (eds) *Landslides: In Research, Theory and Practice*. Thomas Telford, London, 375–380.
DAVENPORT, C.A., RINGROSE, P.S., BECJER, A., HANCOCK, P. & FENTON, C. 1989. Geological investigations of Late- and Postglacial earthquake activity in Scotland. *In*: GREGERSEN, S. & BASHAM, P. (eds) *Earthquakes at North Atlantic Passive Margins: Neotectonics and Postglacial Rebound*. Kluwer Academic Publications, 175–194.
DAVIES, T.R.H., WARBURTON, J., DUNNING, S.A. & BUBECK, A.A.P. 2013. A large landslide event in a post-glacial landscape: rethinking glacial legacy. *Earth Surface Processes and Landforms*, **38**, 1261–1268, https://doi.org/10.1002/esp.3377
DEPARTMENT FOR COMMUNITIES AND LOCAL GOVERNMENT 2012. *National Planning Policy Framework*. The Stationery Office, London.
DEPARTMENT OF ENERGY AND CLIMATE CHANGE 2011. *National Policy Statement for Renewable Energy Infrastructure (EN-3)*. The Stationery Office, London.
DEPARTMENT OF THE ENVIRONMENT 1993. Emergency financial assistance to local authorities: guidance notes for claims. HMSO, London.
DEPARTMENT FOR ENVIRONMENT, FOOD AND RURAL AFFAIRS (DEFRA) 1999. FCDPAG3 Economic appraisal (published December 1999, http://defraweb/environ/fcd/pubs/pagn/fcdpag3/default.htm
DEPARTMENT FOR ENVIRONMENT, FOOD AND RURAL AFFAIRS (DEFRA) 2006. Shoreline management plan guidance Volume 1: Aims and requirements March 2006. HMSO, London.
DEPARTMENT OF THE ENVIRONMENT/WELSH OFFICE 1990. *Planning Policy Guidance: Development on Unstable Land*. PPG14. HMSO, London.
DEPARTMENT FOR TRANSPORT, LOCAL GOVERNMENT AND THE REGIONS 2002. *Planning Policy Guidance 14 (Annex 2): Development on Unstable Land: Subsidence and Planning*. The Stationery Office, London.
DERBYSHIRE, E., PAGE, L.W.F. & BURTON, R. 1975. Integrated field mapping of a dynamic land surface: St Mary's Bay, Brixham. *In*: PHILLIPS, A.D.M. & TURTON, B.J. (eds) *Environment, Man and Economic Change*. Longman, 48–77.
DIBB, LUPTON AND ALSOP 1997. *The Holbeck Hall Hotel Case*. Insurance News.
DIJKSTRA, T.A. & DIXON, N. 2010. Climate change and slope stability in the UK: challenges and approaches. *Quarterly Journal of Engineering Geology and Hydrogeology*, **43**, 371–385, https://doi.org/10.1144/1470-9236/09-036
DIKAU, R. & SCHROTT, L. 1999. The temporal stability and activity of landslides in Europe with respect to climate change (TESLEC): main objectives and results. *Geomorphology*, **30**, 1–12, https://doi.org/10.1016/S0169-555X(99)00040-9
DIKAU, R., BRUNSDEN, D., SCHROTT, L. & IBSEN, M.-L. 1996. *Landslide Recognition*. John Wiley and Sons, Chichester.
DISTRICT OF NORTH VANCOUVER (DNV) 2009. *Natural Hazards Risk Tolerance Criteria*. Report to Council, 10 November 2009.
DIXON, N. & BROMHEAD, E.N. 1991. The mechanics of first time slides in the London Clay cliff at the Isle of Sheppey, England. *In*: CHANDLER, R.J. (ed.) *Slope Stability Engineering*. Thomas Telford, London, 277–282.
DIXON, N. & BROMHEAD, E.N. 2002. Landsliding in London Clay coastal cliffs. *Quarterly Journal of Engineering Geology and Hydrogeology*, **35**, 327–343, https://doi.org/10.1144/1470-9236/2000-53
DIXON, N. & BROOK, E. 2007. Impact of predicted climate change on landslide reactivation: case study of Mam Tor, UK in Landslides. *Journal of the International Consortium on Landslides*, **4**, 137–147, https://doi.org/10.1007/s10346-006-0071-y
DONNELLY, L.J. 2008. Subsidence and associated ground movements on the Pennines, northern England. *Quarterly Journal of Engineering Geology and Hydrogeology*, **41**, 315–332, https://doi.org/10.1144/1470-9236/07-216
DONNELLY, L.J., NORTHMORE, K.J. & JERMY, C.A. 2000a. Fault reactivation in the vicinity of landslides in South Wales, UK. *In*: BROMHEAD, E.N., DIXON, N. & IBSEN, M.-L. (eds) *Landslides: In Research, Theory and Practice*. Thomas Telford, London, 481–486.
DONNELLY, L.J., NORTHMORE, K.J. & SIDDEL, H.J. 2000b. Lateral spreading of moorland in South Wales. *In*: SIDDLE, H.J., BROMHEAD, E.N. & BASSETT, M.G. (eds) *Landslides and Landslide Management in South Wales*. National Museums and Galleries of Wales, Cardiff, Geological Series No. 18, 43–48.
DONNELLY, L.J., NORTHMORE, K.J. & SIDDEL, H.J. 2002. Block movements in the Pennines and South Wales and their association with landslides. *Quarterly Journal of Engineering Geology and Hydrogeology*, **35**, 33–39, https://doi.org/10.1144/qjegh.35.1.33

DOUGLAS, I. 1985. Cities and geomorphology. *In*: PITTY, A. (ed.) *Themes in Geomorphology*. Croom Helm, 233–234.

DUGGLEBY, J.C., ARGHERINOS, P.J. & POWDERHAM, A.J. 1991. Channel Tunnel: foundation engineering at the UK Portal. *In*: *Proceedings of the IVth International Conference on Piling and Deep Foundations*, Stresa, Italy.

DYKES, A.P. & KIRK, K.J. 2000. Morphology and interpretation for a recent multiple peat slide event on Cuilcagh Mountain, Northern Ireland. *In*: BROMHEAD, E.N., DIXON, N. & IBSEN, M.-L. (eds) *Landslides: In Research, Theory and Practice*. Thomas Telford, London, 494–500.

DYKES, A.P. & WARBURTON, J. 2007. Mass movements in peat: a formal classification scheme. *Geomorphology*, **86**, 73–93 https://doi.org/10.1016/j.geomorph.2006.08.009

DYKES, A.P. & WARBURTON, J. 2008. Characteristics of the Shetland Islands (UK) peat slides of 19 September 2003. *Landslides*, **5**, 213–226, https://doi.org/10.1007/s10346-008-0114-7

EARLY, K.R. & SKEMPTON, A.W. 1972. Investigations of the landslide at Walton's Wood Staffordshire. *Quarterly Journal of Engineering Geology*, **5**, 19–41, https://doi.org/10.1144/GSL.QJEG.1972.005.01.04

EDWARDS, G.H. 1988. *Subsidence, Landslip and Ground Heave: With Special Reference to Insurance*. Chartered Institute of Loss Adjusters, London.

ERM-HONG KONG 1998. *Landslides and Boulder Falls from Natural Terrain: Interim Risk Guidelines. GEO Report No. 75*. Report prepared for the Geotechnical Engineering Office, Hong Kong.

EVANS, M. & WARBURTON, J. 2007. *Geomorphology of Upland Peat*. Blackwell Publishing.

FABEL, D., BALLANTYNE, C.K. & XU, S. 2012. Trimlines, blockfields, mountain-top erratics and the vertical dimensions of the last British-Irish Ice Sheet in NW Scotland. *Quaternary Science Reviews*, **55**, 91–102, https://doi.org/10.1016/j.quascirev.2012.09.002

FELL, R. 1994. Landslide risk assessment and acceptable risk. *Canadian Geotechnical Journal*, **31**, 261–272, https://doi.org/10.1139/t94-031

FELL, R. & HARTFORD, D. 1997. Landslide risk management. *In*: RUDEN, D. & FELL, R. (eds) *Landslide Risk Assessment*. Balkema, Rotterdam, 51–108.

FELL, R., HO, K.K.S., LACASSE, S. & LEROI, E. 2005. A framework for landslide risk assessment and management. *In*: HUNGR, O., FELL, R., COUTURE, R. & EBERHARDT, E. (eds) *Landslide Risk Management*. Balkema, Rotterdam, 3–26.

FELL, R., COROMINAS, J., BONNARD, C., CASCINI, L., LEROI, E. & SAVAGE, W. ON BEHALF OF THE JTC-1 JOINT TECHNICAL COMMITTEE ON LANDSLIDES AND ENGINEERED SLOPES 2008*a*. Guidelines for landslide susceptibility, hazard and risk zoning. *Engineering Geology*, **102**, 85–98, https://doi.org/10.1016/j.enggeo.2008.03.022

FELL, R., COROMINAS, J., BONNARD, C., CASCINI, L., LEROI, E. & SAVAGE, W. ON BEHALF OF THE JTC-1 JOINT TECHNICAL COMMITTEE ON LANDSLIDES AND ENGINEERED SLOPES 2008*b*. Commentary: guidelines for landslide susceptibility, hazard and risk zoning. *Engineering Geology*, **102**, 99–111, https://doi.org/10.1016/j.enggeo.2008.03.014

FISH, P., MOORE, R. & CAREY, J.M. 2006. Landslide geomorphology of Cayton Bay, North Yorkshire, UK. *Proceedings of the Yorkshire Geological Society*, **56**, 5–14, https://doi.org/10.1144/pygs.56.1.5

FLETCHER, J. 1774. A dreadful phenomenon, described and improved, being a particular account of the sudden stoppage of the River Severn, and of the terrible desolation that happened at the birches between Colebrook Dale and Buildwas Bridge, in Shropshire on the Thursday morning, May 27, 1773; and the substance of a sermon preached the next days on the ruins, to a vast concourse of spectators. *In*: *The Works of the Reverend John Fletcher*, **4**, 60–103. Schmul Publishers, Salem, Ohio.

FORSTER, A. 1998. *The assessment of slope stability for land use planning. A case study on the North East Antrim Coast*. British Geological Survey, Technical Report **WN/98/8**.

FOSTER, C., JENKINS, G.O. & GIBSON, A.D. 2007. *Landslides and mass movement processes and their distribution in the York District (Sheet 63)*. British Geological Survey Open Report, **OR/07/004**.

FOSTER, C., GIBSON, A.D. & WILDMAN, G. 2008. The new national landslide database and landslide hazards assessment of Great Britain. *In*: SASSA, K., FUKUOKA, H. ET AL. (eds) *Proceedings of the First World landslide forum, United Nations University, Tokyo*. The International Promotion Committee of the International Programme on Landslides (IPL), Tokyo, Parallel Session, 203–206.

FOSTER, C., HARRISON, M. & REEVES, H.J. 2011. Standards and methods of hazard assessment for mass-movements in Great Britain. *Journal for Torrent and Avalanche Control*, **166**.

FOSTER, C., PENNINGTON, C.V.L., CULSHAW, M.G. & LAWRIE, K. 2012. The National Landslide Database of Great Britain: development, evolution and applications. *Environmental Earth Sciences*, **66**, 941–953, https://doi.org/10.1007/s12665-011-1304-5

FORSYTH 1883. Report to the Board of Public Works, by Mr. Forsyth, 31st October, 1883, http://www.from-ireland.net/bog-bursts-ireland-county/

FREEBOROUGH, K.A., GIBSON, A.D., HALL, M., POULTON, C.V.L., WILDMAN, G., FORSTER, A. & BURT, E. 2005. Landslide and mass movement processes and their distribution in the Wellington District of Somerset. *Geoscience in South-west England*, **11**, 139–144.

FREW, P. & GUEST, S. 1997. Overstrand coast protection scheme. *Proceedings of the MAFF Conference of River and Coastal Engineers*.

GARDNER, T. 1754. *An historical account of Dunwich, Blithburgh, Southwold with remarks on some places contiguous thereto*. Re-printed by Gale ECCO, Print Editions, 30 May 2010.

GEMSEGE, P. 1756. Letter to the Editor. *Gentleman's Magazine*, **26**, 172–173.

GEOMORPHOLOGICAL SERVICES LIMITED 1986–87. *Review of Research into Landsliding in Great Britain*. Reports to the Department of the Environment, London.

GEOMORPHOLOGICAL SERVICES LIMITED 1988. *Applied Earth Science Mapping for Planning and Development*. Department of the Environment, Torbay.

GIBSON, A.D., CULSHAW, M.G., DASHWOOD, C. & PENNINGTON, C.V.L. 2012. Landslide management in the UK –the problem of managing hazards in a 'low-risk' environment. *Landslides*, July 2012.

GLADE, T., ANDERSON, M. & CROZIER, M.J. (eds) 2005. *Landslide Hazard and Risk*. Wiley.

GOSTLING, R.W. 1756. Land sunk in Kent. *Gentleman's Magazine*, letters, **26**, 160.

GOUDIE, A. & PARKER, A. 1996. *The Geomorphology of the Cotswolds*. Cotteswolds Naturalists' Field Club, Oxford.

GRANTICOLA 1756. Letter to the Editor. *The Gentleman's Magazine*, **26**, 103–104.

GRIFFITHS, J.S. & GILES, D.P. 2017. Conclusions and illustrative case studies. *In*: GRIFFITHS, J.S. & MARTIN, C.J. (eds) *Engineering*

Geology and Geomorphology of Glaciated and Periglaciated Terrains, Engineering Group Working Party Report. Geological Society, London, Engineering Geology Special Publications, **28**, 891–936, https://doi.org/10.1144/EGSP28.9

GRIFFITHS, J.S., BRUNSDEN, D., LEE, E.M. & JONES, D.K.C. 1995. Geomorphological investigations for the Channel Tunnel Terminal and Portal. *The Geographical Journal*, **161**, 275–284, https://doi.org/10.2307/3059832

GROVE, A.T. 1953. The sea flood on the coasts of Norfolk and Suffolk. *Geography*, **38**, 164–170.

GROVE, J.M. 1988. *The Little Ice Age*. Methuen, London.

GUMMER, J. 2000. Country Living, 57–60, March 2000.

HALCROW GROUP LIMITED 2009. *Cayton Bay Cliff Stability Assessment Ground Investigation and Appraisal of Engineering Stabilisation Options*. Report to Scarborough Borough Council.

HALCROW (SIR WILLIAM HALCROW AND PARTNERS) 1988. *Rhondda Landslip Potential Assessment: Inventory*. Department of the Environment and Welsh Office.

HALL, A.P. & GRIFFITHS, J.S. 1999. A possible failure mechanism for the AD 1575 'Wonder Landslide'. *East Midlands Geographer*, **21–22**, 92–105.

HARRAL, T. 1824. Picturesque Views of the Severn with Historical and Topographical Illustrations by Thomas Harral. The Embellishments from Designs by the late Samuel Ireland. Vol.1 and Vol. 2. G and W. B. Whittaker, London.

HARVEY, A.M. 1986. Geomorphic effects of a 100-year storm in the Howgill Fells. *NW England, Zeitschrift für Geomorphologie*, **30**, 71–91.

HAWKINS, A.B. 1977. The Hedgemead landslip, Bath. *In*: GEDDES, J.D. (ed.) *Large Ground Movements and Structures*. Pentech Press, 472–498.

HAWKINS, A.B. 1996. Observation and analysis of the ground conditions in the Jurassic landslip terrain of southern Britain. *In*: SENNESET, K. (ed.) *Landslides. Proceedings of the 7th International Symposium on Landslides*, Balkema, 3–16.

HEALTH AND SAFETY EXECUTIVE (HSE) 1988. *The Tolerability of Risk from Nuclear Power Stations*. HMSO, London.

HEALTH AND SAFETY EXECUTIVE (HSE) 1992. *The Tolerability of Risk from Nuclear Power Stations (Revised)*. HMSO, London.

HEALTH AND SAFETY EXECUTIVE (HSE) 1999. *Reducing Risks, Protecting People*. Discussion Paper, Risk Assessment Policy Unit. HSE, London.

HEALTH AND SAFETY EXECUTIVE, (HSE) 2001. *Reducing Risks, Protecting People*. HSE Books. HMSO, Norwich.

HEALTH AND SAFETY EXECUTIVE, HSE 2006. Guidance on Risk Assessment for Offshore Installations. Offshore Information Sheet No. 3/2006. HMSO, London.

HEMINGWAY, R. & GUNAWAN, O. 2018. The Natural Hazards Partnership: a public-sector collaboration across the UK for natural hazard disaster risk reduction. *International Journal of Disaster Risk Reduction*, **27**, 499–511.

HENKEL, D.J. & SKEMPTON, A.W. 1954. A landslide at Jackfield, Shropshire, in a heavily overconsolidated clay. *Proceedings of the European Conference on Stability of Slopes*, Stockholm, Sweden, 20–25 September 1954, **1**, 99–101.

HIGGETT, D. & LEE, E.M. (eds) 2001. *Geomorphological Processes and Landscape Change: Britain in the Last 1000 Years*. Blackwell.

HIGH-POINT RENDEL 2005*a*. Ironbridge Gorge Instability: The Interpretation of Ground Investigations at Jackfield and the Lloyds, Borough of Telford & Wrekin Council, R/2088/01.

HIGH-POINT RENDEL 2005*b*. Ironbridge Gorge Landslides: Ironbridge and Coalbrookdale Ground Behaviour Study, Borough of Telford & Wrekin Council, R/2320/01.

HO, K., LEROI, E. & ROBERDS, B. 2000. Quantitative risk assessment: application, myths and future direction. *In: Proceedings of the Geo-Eng Conference*, Melbourne, Australia, Publication 1, 269–312.

HOBBS, P.R.N. & JENKINS, G.O. 2008. *Bath's 'foundered strata' –a re-interpretation*. British Geological Survey Physical Hazards Project Research Report **OR/08/052**.

HOBBS, P.R.N., RAINES, M.G., PRATT, W.T. & CULSHAW, M.G. 1994. *St Dogmaels landslip: engineering geomorphological mapping, preliminary geophysical survey and desk study*. British Geological Survey Technical Report **WN/94/9C**.

HOLLINGWORTH, S.E., TAYLOR, J.H. & KELLAWAY, G.A. 1944. Large-scale superficial structures in the Northampton ironstone field. *Quarterly Journal of the Geological Society, London*, **100**, 1–44, https://doi.org/10.1144/GSL.JGS.1944.100.01-04.03

HOLMES, G. & JARVIS, J.J. 1985. Large scale toppling within a sackung type deformation at Ben Attow. *Quarterly Journal of Engineering Geology*, **8**, 287–289, https://doi.org/10.1144/GSL.QJEG.1985.018.03.09

HOROWITZ, J.K. & CARSON, R.T. 1993. Baseline risk and preference for reductions in risk-to-life. *Risk Analysis*, **13**, 457–462, https://doi.org/10.1111/j.1539-6924.1993.tb00746.x

HUNGR, O. 1997. Some methods of landslide hazard intensity mapping. *In*: CRUDEN, D. & FELL, R. (eds) *Landslide Risk Assessment*. Balkema, Rotterdam, 215–226.

HUNGR, O., EVANS, S.G., BOVIS, M.J. & HUTCHINSON, J.N. 2001. A review of the classification of landslides of flow type. *Environmental and Engineering Geoscience*, **3**, 221–238, https://doi.org/10.2113/gseegeosci.7.3.221

HUNGR, O., COROMINAS, J. & EBERHARDT, E. 2005. Estimating landslide motion mechanisms, travel distance and velocity. *In*: HUNGR, O., FELL, R., COUTURE, R. & EBERHARDT, E. (eds) *Landslide Risk Management*. Balkema, London, 99–128.

HUNGR, O., LEROUEIL, S. & PICARELLI, L. 2014. The Varnes classification of landslide types, an update. *Landslides*, **11**, 167–194, https://doi.org/10.1007/s10346-013-0436-y

HUNTER, G. & FELL, R. 2003. Travel distance angle for rapid landslides in constructed and natural slopes. *Canadian Geotechnical Journal*, **40**, 1123–1141, https://doi.org/10.1139/t03-061

HUTCHINS, J. 1803. The History and Antiquities of the County of Dorset. *J Nichols and Son*, **2**, 354–371.

HUTCHINSON, J.N. 1962. *Report on visit to landslide at Lyme Regis, Dorset*. Building Research Station Note c890.

HUTCHINSON, J.N. 1965. *The stability of cliffs composed of soft rocks, with particular reference to the coasts of South East England*. Unpublished Ph.D. thesis. University of Cambridge.

HUTCHINSON, J.N. 1969. A reconsideration of the coastal landslides at Folkestone Warren, Kent. *Géotechnique*, **19**, 6–38, https://doi.org/10.1680/geot.1969.19.1.6

HUTCHINSON, J.N. 1970. A coastal mudflow on the London Clay cliffs at Beltinge, North Kent. *Géotechnique*, **20**, 412–438, https://doi.org/10.1680/geot.1970.20.4.412

HUTCHINSON, J.N. 1973. The response of London Clay cliffs to differing rates of toe erosion. *Geologia Applicata e Idrogeologia*, **8**, 221–239.

HUTCHINSON, J.N. 1976. Coastal landslides in cliffs and Pleistocene deposits between Cromer and Overstrand, Norfolk, England. *In*: JAMBU, N. ET AL. (eds) *Contributions to Soil Mechanics. Bjerrum Memorial Volume*. Norwegian Geotechnical Institute, Oslo, 155–182.

HUTCHINSON, J.N. 1977. Assessment of the effectiveness of corrective measures in relation to geological conditions and types of movement. *Bulletin of the International Association of*

Engineering Geology, **16**, 131–155, https://doi.org/10.1007/BF02591469

HUTCHINSON, J.N. 1980. Various forms of cliff instability arising from coast erosion in the UK. Fjellsprengnings-teknikk-Bergmekanikk-Geoteknikk, Trondheim. *Tapir, for Norsk Jord-og Fjellteknikk Forbund tilknyttet NIF*, **1979**, 19.1–19.32.

HUTCHINSON, J.N. 1982. Slope failures produced by seepage erosion in sands. *In*: SHEKO, A. (ed.) *Landslides and Mudslides*. Centre of International Projects, Moscow.

HUTCHINSON, J.N. 1984. Landslides in Britain and their countermeasures. *Journal of the Japan Landslide Society*, **21**, 1–21, https://doi.org/10.3313/jls1964.21.1

HUTCHINSON, J.N. 1987. Mechanisms producing large displacements in landslides on pre-existing shears. *Memoir of the Geological Society of China*, **9**, 175–200.

HUTCHINSON, J.N. 1988. General report: morphological and geotechnical parameters of landslides in relation to geology and hydrogeology. *In*: BONNARD, C. (ed.) *Landslides*. Balkema, Rotterdam, 3–35.

HUTCHINSON, J.N. 1991a. Periglacial and slope processes. *In*: FORSTER, A., CULSHAW, M.G., CRIPPS, J.C., LITTLE, J.A. & MOON, C.F. (eds) *Quaternary Engineering Geology*. Geological Society, London, Engineering Geology Special Publication No. 7, 283–331.

HUTCHINSON, J.N. 1991b. The landslides forming the South Wight Undercliff. *In*: CHANDLER, R.J. (ed.) *Slope Stability Engineering*. Thomas Telford, London, 157–168.

HUTCHINSON, J.N. 1992. Landslide hazard assessment. *In*: BELL, D.H. (ed.) *Landslides*. Balkema, Rotterdam, **3**, 1805–1841.

HUTCHINSON, J.N. 1995. The assessment of sub-aerial Landslide hazard. *In*: *Landslides Hazard Mitigation*. The Royal Academy of Engineering, London, 57–66.

HUTCHINSON, J.N. 2001. Reading the ground: morphology and geology in site appraisal. *Quarterly Journal of Engineering Geology and Hydrogeology*, **34**, 7–50, https://doi.org/10.1144/qjegh.34.1.7

HUTCHINSON, J.N. & BHANDARI, R.K. 1971. Undrained loading, a fundamental mechanism of mudflows and other mass movements. *Geotechnique*, **18**, 353–358, https://doi.org/10.1680/geot.1971.21.4.353

HUTCHINSON, J.N. & GOSTELOW, T.P. 1976. The development of an abandoned cliff in London Clay at Hadleigh, Essex. *Philosophical Transactions of the Royal Society*, **A283**, 557–604, https://doi.org/10.1098/rsta.1976.0096

HUTCHINSON, J.N. & MILLAR, D.L. 2001. The Graig Goch landslide dam, Meirionnydd, mid Wales. *In*: WALKER, M.J.C. & MCCARROLL, D. (eds) *The Quaternary of West Wales: Field Guide*. Quaternary Research Association, 113–125.

HUTCHINSON, J.N., SOMERVILLE, S.H. & PETLEY, D.J. 1973. A landslide in periglacially disturbed Etruria Marl at Bury Hill, Staffordshire. *Quarterly Journal of Engineering Geology*, **6**, 377–404, https://doi.org/10.1144/GSL.QJEG.1973.006.03.16

HUTCHINSON, J.N., PRIOR, D.B. & STEPHENS, N. 1974. Potentially dangerous surges in an Antrim mudslide. *Quarterly Journal of Engineering Geology*, **7**, 363–376, https://doi.org/10.1144/GSL.QJEG.1974.007.04.08

HUTCHINSON, J.N., BROMHEAD, E.N. & LUPINI, J.F. 1980. Additional observations on the Folkestone Warren landslides. *Quarterly Journal of Engineering Geology*, **13**, 1–31, https://doi.org/10.1144/GSL.QJEG.1980.013.01.01

HUTCHINSON, J.N., CHANDLER, M.P. & BROMHEAD, E.N. 1981. Cliff recession on the Isle of Wight, S W Coast. *Tenth International Conference on Soil Mechanics and Foundation Engineering*, Rotterdam, 429–434.

HUTCHINSON, J.N., COROMINAS, J., PETLEY, D.J. & HENDY, M.S. 2000. Note on some flow slides from industrial tips. *In*: BROMHEAD, E.N., DIXON, N. & IBSEN, M.-L. (eds) *Landslides: In Research, Theory and Practice*. Thomas Telford, London, 755–762.

INNES, J.L. 1983. Lichenometric dating of debris-flow deposits in the Scottish Highlands. *Earth Surface Processes and Landforms*, **8**, 579–588.

INNES, J.L. 1989. Rapid mass movements in upland Britain: a review with particular reference to debris flows. *Studia Geomorphologica Carpatho-Balanica*, **XXIII**, 53–67.

INTERNATIONAL SOCIETY OF SOIL MECHANICS AND GROUND ENGINEERING (ISSMGE) 2004. Risk Assessment – Glossary of Terms. TC32, Technical Committee on Risk Assessment and Management Glossary of Risk Assessment Terms – Version 1, July 2004.

JARMAN, D. 2006. Large rock-slope failures in the Highlands of Scotland: characterisation, causes and spatial distribution. *Engineering Geology*, **83**, 161–182, https://doi.org/10.1016/j.enggeo.2005.06.030

JARMAN, D. 2007. Introduction to mass movements in the older mountain areas of Great Britain. *In*: COOPER, R.G. (ed.) *Mass Movements in Great Britain*. Joint Nature Conservation Committee, Peterborough, Geological Conservation Review, Series No. 33, 33–56.

JARMAN, D. 2009. Paraglacial rock-slope failure as an agent of glacial trough widening. *In*: Knight, J. & Harrison, S. (eds) *Periglacial and Paraglacial Processes and Environments*. Geological Society, London, Special Publications, **320**, 103–131, https://doi.org/10.1144/SP320.8

JENKINS, G.O. & BOOTH, K.A. 2008. *Landslide nature and distribution on the Derby 1:50k geological sheet*. British Geological Survey Internal Report, **OR/08/032**.

JENKINS, G.O., FOSTER, C., GIBSON, A.D. & PRICE, S.J. 2005. *Landslide Survey of North Yorkshire: Reconnaissance Report*. British Geological Survey Internal Report, **IR/06/039**.

JENKINS, G.O., FREEBOROUGH, K.A. & MORGAN, D.J.R. 2010. *Landslide nature and distribution on the Market Rasen 1:50k geological sheet*. British Geological Survey Internal Report, **OR/10/013**.

JOHNSON, R.H. 1987. Dating of ancient, deep-seated landslides in temperate regions. *In*: ANDERSON, M.G. & RICHARDS, K.S. (eds) *Slope Stability*. Wiley, London, 561–600.

JONES, D.B. 2000. Trinant Hall viaduct: bridging a landslide. *In*: SIDDLE, H.J., BROMHEAD, E.N. & BASSETT, M.G. (eds) *Landslides and Landslide Management in South Wales*. National Museums and Galleries of Wales, Cardiff, Geological Series, No. 18, 99–100.

JONES, D.K.C. 1999. Landsliding in the Midlands: a critical evaluation of the contribution of the National Landslide Survey. *East Midlands Geographer*, **21**, 106–125.

JONES, D.K.C. 2001. The evolution of hillslope processes. *In*: HIGGETT, D. & LEE, E.M. (eds) *Geomorphological Processes and Landscape Change: Britain in the Last 1000 Years*. Blackwell, 61–89.

JONES, D.K.C. & LEE, E.M. 1994. *Landsliding in Great Britain*. HMSO, London.

JULIAN, M. & ANTHONY, E. 1996. Aspects of landslide activity in the Mercantour Massif and the French Riviera, SE France. *Geomorphology*, **15**, 275–289, https://doi.org/10.1016/0169-555X(95)00075-G

LACEY, G.N. 1972. Observations on Aberfan. *Journal of Psychometric Research*, **16**, 257–260, https://doi.org/10.1016/0022-3999(72)90007-4

LEE, E.M. 1992. Urban landslides: impact and management. *In*: ALLISON, R. (ed.) *The Coastal Landforms of West Dorset*. Geologists Association Guide No. 47, 80–93.

LEE, E.M. 1999. Coastal planning and management: the impact of the 1993 Holbeck Hall Landslide, Scarborough. *East Midlands Geographer*, **21**, 78–91.

LEE, E.M. 2000. The management of coastal landslide risks in England: the implications of conservation legislation and commitments. *In*: BROMHEAD, E.N., DIXON, N. & IBSEN, M.-L. (eds) *Landslides: In Research, Theory and Practice*. Thomas Telford, London, 893–898.

LEE, E.M. 2001a. Estuaries and coasts: morphological adjustment and process domains. *In*: HIGGETT, D. & LEE, E.M. (eds) *Geomorphological Processes and Landscape Change: Britain in the last 1000 years*. Blackwell, 147–189.

LEE, E.M. 2001b. Living with natural hazards: the costs and management framework. *In*: HIGGETT, D. & LEE, E.M. (eds) *Geomorphological Processes and Landscape Change: Britain in the Last 1000 years*. Blackwell, 237–268.

LEE, E.M. 2002. A dynamic framework for the management of coastal erosion and flooding risks in England. *In*: MCINNES, R.G. & JAKEWAYS, J. (eds) *Instability: Planning and Management*. Thomas Telford, London, 713–720.

LEE, E.M. 2005. Coastal cliff recession risk: a simple judgement based model. *Quarterly Journal of Engineering Geology and Hydrogeology*, **38**, 89–104, https://doi.org/10.1144/1470-9236/04-055

LEE, E.M. 2008. Coastal cliff behaviour: observations on the relationship between beach levels and recession rates. *Geomorphology*, **101**, 558–571, https://doi.org/10.1016/j.geomorph.2008.02.010

LEE, E.M. 2009. Landslide risk assessment: the challenge of estimating the probability of landsliding. *Quarterly Journal of Engineering Geology and Hydrogeology*, **42**, 445–458, https://doi.org/10.1144/1470-9236/08-007

LEE, E.M. 2011. Reflections on the decadal-scale response of coastal cliffs to sea-level rise. *Quarterly Journal of Engineering Geology and Hydrogeology*, **44**, 481–489, https://doi.org/10.1144/1470-9236/10-063

LEE, E.M. & CLARK, A.R. 2002. *Investigation and Management of Soft Rock Cliffs*. Thomas Telford, London.

LEE, E.M. & JONES, D.K.C. 2004. *Landslide Risk Assessment*. Thomas Telford, London.

LEE, E.M. & JONES, D.K.C. 2014. *Landslide Risk Assessment*, 2nd edn Thomas Telford, London.

LEE, E.M. & MOORE, R. 1989. *Landsliding in and around Luccombe Village*. HMSO, London.

LEE, E.M. & MOORE, R. 1991. *Coastal landslip potential assessment: Isle of Wight Undercliff, Ventnor*. Department of the Environment, London.

LEE, E.M. & MOORE, R. 2007. Ventnor Undercliff: development of landslide scenarios and quantitative risk assessment. *In*: MCINNES, R., JAKEWAYS, J., FAIRBANK, H. & MATHIE, E. (eds) *Landslides and Climate Change: Challenges and Solutions*. Balkema, 323–334.

LEE, E.M. & PETHICK, J. 2003. The coast. *In*: BUTLIN, R.A. (ed.) *Historical Atlas of North Yorkshire*. Westbury, 17–19.

LEE, E.M., MOORE, R. & MCINNES, R.G. 1998. Assessment of the probability of landslide reactivation: Isle of Wight Undercliff, UK. *In*: MOORE, D. & HUNGR, O. (eds) *Engineering Geology: The View from the Pacific Rim*. Balkema, 1315–1321.

LEE, E.M., JONES, D.K C. & BRUNSDEN, D. 2000. The landslide environment of Great Britain. *In*: BROMHEAD, E.N., DIXON, N. & IBSEN, M-L. (eds) *Landslides: In Research, Theory and Practice*. Thomas Telford, London, 911–916.

LEE, E.M., HALL, J.W. & MEADOWCROFT, I.C. 2001. Coastal cliff recession: the use of probabilistic prediction methods. *Geomorphology*, **40**, 253–269, https://doi.org/10.1016/S0169-555X(01)00053-8

LEE, E.M., MEADOWCROFT, I.C., HALL, J.W. & WALKDEN, M. 2002. Coastal landslide activity: a probabilistic simulation model. *Bulletin of Engineering Geology and Environment*, **61**, 347–355, https://doi.org/10.1007/s10064-001-0146-x

LEE, K.A., LARK, R.M. ET AL. 2014. *User Guide for the new GeoSure Insurance Product (new GIP)*. British Geological Survey Internal Report, **OR/12/089**.

LÉONE, F., ASTE, J.P. & LEROI, E. 1996. Vulnerability assessment of elements exposed to mass movement: working towards a better risk perception. *In*: SENNESET, K. (ed.) *Landslides*. Balkema, Rotterdam, **1**, 263–268.

Lewis, S. 2007. *Risk Criteria – When is Low Enough Good Enough?* Risktec Solutions Ltd., http://www.risktec.co.uk/media/

LUTAS, J.A. 2002. Land instability and building control. *In*: MCINNES, R.G. & JAKEWAYS, J. (eds) *Instability – Planning and Management*. Thomas Telford, London, 731–737.

MADDISON, J.D. 2000. St Dogmaels landslide: deep drainage by wells. *In*: SIDDLE, H.J., BROMHEAD, E.N. & BASSETT, M.G. (eds) *Landslides and Landslide Management in South Wales*. National Museums and Galleries of Wales, Cardiff, Geological Series No. 18, 106–108.

MADDISON, J.D., SIDDLE, H.J. & FLETCHER, C.J.N. 2000. Investigation and remediation of a major landslide in glacial lake deposits at St Dogmaels, Pembrokeshire. *In*: BROMHEAD, E.N., DIXON, N. & IBSEN, M.-L. (eds) *Landslides: In Research, Theory and Practice*. Thomas Telford, London, 981–986.

MADDOX, R.L. & MADDOX, A.F. 2008. *Hymn on the Lisbon Earthquake 1756*. Duke Center for Studies in the Wesleyan Tradition.

MARSAY, B. 2010. *"T'ills was Fallin' Down"*. *The great Landslip of 1872*. 2nd edn. Bilsdale Study Group.

MATHEWS, E.R. 1934. *Coast Erosion and Protection*. Ch. Griffin.

MAY, V.J. 1971. The retreat of chalk cliffs. *Geographical Journal*, **137**, 203–206, https://doi.org/10.2307/1796740

MAY, V.J. & HEAPS, C. 1985. The nature and rates of change on chalk coastlines. *Zeitschrift fur Geomorphologie Supplementband*, **57**, 81–94.

MCCAHON, C.P., CARLING, P.A. & PASCOE, D. 1987. Chemical and ecological effects of a Pennine peat-slide. *Environmental Pollution*, **45**, 275–289, https://doi.org/10.1016/0269-7491(87)90102-3

MCEWEN, L.J. & WITHERS, C.W. 1989. Historical records and geomorphological events: the 1771 "eruption" of Solway Moss. *Scottish Geographical Magazine*, **106**, 149–157, https://doi.org/10.1080/14702548908554428

MCGREAL, W.S. 1977. *Retreat of the cliff coastline in the Kilkeal area of County Down*. PhD thesis (unpublished), Queen's University, Belfast.

MCGREAL, W.S. 1979. Cliffline recession near Kilkeal, N. Ireland: an example of a dynamic coastal system. *Geografiska Annaler*, **61**, 211–219, https://doi.org/10.1080/04353676.1979.11879992

MCINNES, R.G. 2000. *Managing Ground Instability in Urban Areas: A Guide to Best Practice*. Isle of Wight Centre for the Coastal Environment, Ventnor.

MCINNES, R.G. 2007. *The Undercliff of the Isle of Wight – A Guide to Managing Ground Stability*. Isle of Wight Centre for the Coastal Environment, Ventnor.

MCINNES, R.G. 2009. *Coastal Risk Management: A Non-Technical Guide*. SCOPAC.

MCINNES, R.G. & MOORE, R. 2011. *Cliff Instability and Erosion Management in Great Britain: A Good Practice Guide*. Halcrow Group Ltd.

McQuaid, J. & Le Guen, J-M. 1998. The use of risk assessment in Government. *In*: Hester, R.E. & Harrison, R.M. (eds) *Risk Assessment and Risk Management*. The Royal Society of Chemistry, Issues in Environmental Science and Technology, 9.

Meteorological Office 2013. Natural Hazards Partnership. Met Office web page, http://www.metoffice.gov.uk/nhp/ [last accessed July 2014]

Miller, J. 1974. *Aberfan – A Disaster and its Aftermath*. Constable, London.

Mills, A.J. 2002. *Peat slides: morphology, mechanisms and recovery*. PhD thesis, University of Durham, UK.

Ministry of Agriculture, Fisheries and Food (MAFF) 1993. *Project Appraisal Guidance Notes*. MAFF Publications.

Ministry of Agriculture, Fisheries and Food (MAFF) 1994. *Coast protection survey of England*. Survey Report – Volume 1.

Monkhouse, F.J. 1971. *Principles of Physical Geography*. University of London Press, London.

Moore, R. & McInnes, R.G. 2014. *Living with Ground Instability – Understanding the Risks, Empowering Communities, Building Resilience – A Good Practice Guide*. CH2M Hill, Birmingham.

Moore, R., Lee, E.M. & Longman, F. 1991. The impact, causes and management of landsliding at Luccombe Village, Isle of Wight. *In*: Chandler, R.J. (ed.) *Slope Stability Engineering, Developments and Applications*. 225–230.

Moore, R., Clark, A.R. & Lee, E.M. 1998. Coastal cliff behaviour and management: Blackgang, Isle of Wight. *In*: Maund, J.G. & Eddleston, M. (eds) *Geohazards and Engineering Geology*. Geological Society, London, Special Publications, **15**, 49–59, https://doi.org/10.1144/GSL.ENG.1998.015.01.06

Moore, R., Carey, J. *et al.* 2006. Recent landslide impacts on the UK Scottish road network: investigation into the mechanisms, causes and management of landslide risk. *International Conference on Slopes*, Malaysia.

Moore, R., Rogers, J. & Woodget, A. 2010. *Climate Change Impact on Cliff Instability and Erosion*. FCRM>10 International Centre, Telford.

Morgan, L., Scourfield, J., Wiliams, D., Jasper, A. & Lewis, G. 2003. The Aberfan disaster: 33-year follow-up of survivors. *The British Journal of Psychiatry*, **182**, 532–536, https://doi.org/10.1192/bjp.182.6.532

Morris, S. 2012. Two feared dead in Dorset landslide. *The Guardian*, 17 July 2012.

Newson, M.D. & Bathurst, M.G. 1990. Sediment movement in gravel bed rovers. Department of Geography Seminar Paper 59, University of Newcastle upon Tyne.

Nichol, D., Bassett, M.G. & Deisler, V.K. (eds) 2002. *Landslides and Landslide Management in North Wales*. National Museums & Galleries of Wales.

Nichol, D., Doherty, G.K. & Scott, M.J. 2007. A5 Llyn Ogwen peatslide, Capel Curig, North Wales. *Quarterly Journal of Engineering Geology and Hydrogeology*, **40**, 293–299, https://doi.org/10.1144/1470-9236/06-042

Palm, R.I. 1990. *Natural Hazards: An Integrative Framework for Research and Planning*. John Hopkins University Press, Baltimore.

Parks, C.D. 1991. A review of the possible mechanisms of cambering and valley bulging. *In*: Forster, A., Culshaw, M.G., Little, J.A. & Moon, C. (eds) *Quaternary Engineering Geology 25th Annual Conference of the Engineering Group of the Geological Society*, Edinburgh, 373–380.

Penman, A.D.M. 2000. The Aberfan flow slide, Taff Valley. *In*: Siddle, H.J., Bromhead, E.N. & Bassett, M.G. (eds) *Landslides and Landslide Management in South Wales*. National Museums and Galleries of Wales, Cardiff, Geological Series No. 18, 62–68.

Pennington, C.V.L. & Harrison, A.M. 2013. Landslide year? *Geoscientist*, **23**, 10–15.

Pennington, C.V.L., Evans, H.M. & Foster, C. 2009a. *Landslides of the area around Chesterfield, Geological Sheet 112*. British Geological Survey Open Report, OR/09/022.

Pennington, C.V.L., Foster, C., Chambers, J. & Jenkins, G.O. 2009b. Landslide research at the British Geological Survey: capture, storage and interpretation on a national and site specific scale. *Acta Geologica Sinica (English edition)*, **83**, 801–840, https://doi.org/10.1111/j.1755-6724.2009.00103.x

Pennington, C.V.L., Dijkstra, T.A., Lark, M., Dashwood, C., Harrison, A.M. & Freeborough, K.A. 2014. Antecedent precipitation as a potential proxy for landslide incidence in SW UK. *World Landslides Forum*, **3**, 2–6 June 2014, Beijing.

Pereira, A.S. 2006. The Opportunity of a Disaster: The Economic Impact of the 1755 Lisbon Earthquake. Discussion Paper 06/03, Centre for Historical Economics and Related Research at York, York University.

Pethick, J. 1996. Coastal slope development: temporal and spatial periodicity in the Holderness Cliff Recession. *In*: Anderson, M.G. & Brooks, S.M. (eds) *Advances in Hillslope Processes*. **2**, 897–917.

Phipps, P.J. & McGinnity, B.T. 2001. Classification and stability assessment for chalk cuttings: the Metropolitan Line case study. *Quarterly Journal of Engineering Geology and Hydrogeology*, **34**, 353–370, https://doi.org/10.1144/qjegh.34.4.353

Pitts, J. 1983. The temporal and spatial development of landslides in the Axmouth-Lyme Regis Undercliff, National Nature Reserve, Devon. *Earth Surface Processes and Landforms*, **8**, 589–603, https://doi.org/10.1002/esp.3290080610

Pitts, J. & Brunsden, D. 1987. A reconsideration of the Bindon landslide of 1839. *Proceedings of the Geologists Association*, **98**, 1–18, https://doi.org/10.1016/S0016-7878(87)80014-7

Popescu, M.E. 2002. Landslide causal factors and landslide remediation options. *In*: *3rd International Conference on Landslides, Slope Stability and Safety of Infra-Structures*, 61–81.

Poulsom, A.J. 1995. The application of the concept of brittle-ductile transition to slope behaviour. *In*: Singhal, R.K. *et al.* (eds) *Mine Planning & Equipment Selection*, 1041–1046.

Poulton, C.V.L., Lee, J.R., Hobbs, P.R.N., Jones, L.D. & Hall, M. 2006. Preliminary investigation into monitoring coastal erosion using terrestrial laser scanning: case study at Happisburgh, Norfolk. *The Bulletin of the Geological Society of Norfolk*, **56**, 45–64.

Prior, D.B. & Stevens, N. 1972. Some movement patterns of temperate mudflows: examples from north-east Ireland. *Bulletin of the Geological Society of America*, **83**, 2533–2544, https://doi.org/10.1130/0016-7606(1972)83[2533:SMPOTM]2.0.CO;2

Prior, D.B., Stevens, N. & Archer, D.R. 1968. Composite mudflows on the Antrim coast of north-east Ireland. *Geografiska Annaler*, **50**, 65–78, https://doi.org/10.1080/04353676.1968.11879773

Prior, D.B., Stevens, N. & Douglas, G.R. 1971. Some examples of mudflow and rockfall activity in North-east Ireland. *In*: Brunsden, D. (ed.) *Slopes Form and Process*. Institute of British Geographers, Special Publication no. 3, 129–140.

Reeves, A., Ho, K.K.S. & Lo, D.O.K. 1999. Interim risk criteria for landslides and boulder falls from natural terrain. *Proceedings of the Seminar on Geotechnical Risk Management, Geotechnical Division, Hong Kong Institution of Engineers*, 127–136.

Reid, M.E. 1994. A pore pressure diffusion model for estimating landslide inducing rainfall. *Journal of Geology*, **102**, 709–717, https://doi.org/10.1086/629714

RENDEL GEOTECHNICS 1995. *Applied Earth Science Mapping: Seaham to Teesmouth*. Department of the Environment, London.
RICHMOND, R. 1744/5. A Copy of a Letter from the Reverend Mr. Richmond, to – Leigh, Esq; of Adlington in the County of Chester, concerning a Moving Moss in the Neighbourhood of Church-Town in Lancashire: communicated by Edward Milward. M.D. *Philosophical Transactions of the Royal Society*, **43**, 282–283.
ROBINSON, D.A. & WILLIAMS, R.B.G. 1984. *Classic Landforms of the Weald*. Geographical Association, Sheffield.
ROYAL SOCIETY 1992. *Risk: analysis, perception and management*. Report of a Royal Society Study Group. Royal Society, London.
ROZIER, I.T. & REEVES, M.J. 1979. Ground movements at Runswick Bay, North Yorkshire. *Earth Surface Processes and Landforms*, **4**, 275–280, https://doi.org/10.1002/esp.3290040308
RUTTER, E.H. & GREEN, S. 2011. Quantifying creep behaviour of clay-bearing rocks below the critical stress state for rapid failure: Mam Tor landslide, Derbyshire, England. *Journal of the Geological Society*, **168**, 359–371, https://doi.org/10.1144/0016-76492010-133
SCHOFIELD, J. 1787. An historical and descriptive guide to Scarborough and its environs. W. Blanchard.
SCHUSTER, R.L. (ed.) 1986. *Landslide Dams: Processes, Risk and Mitigation, Geotechnical Special Publication No. 3*, American Society of Civil Engineers.
SCOTTISH EXECUTIVE 2006. *Peat Landslide Hazard and Risk Assessments, Brest Practice Guide for Proposed Electricity Generation Developments*. Scottish Executive.
SCOTTISH GOVERNMENT 2010. Scottish Planning Policy.
SCOTTISH GOVERNMENT 2011. Developments on Peatland: Site Surveys. Guidance Note.
SIDDLE, H.J. 2000a. East Pentwyn landslide, Blaina. *In*: SIDDLE, H.J., BROMHEAD, E.N. & BASSETT, M.G. (eds) *Landslides and Landslide Management in South Wales*. National Museums and Galleries of Wales, Cardiff, Geological Series No. 18, 77–80.
SIDDLE, H.J. 2000b. Blaencwm landslide, Rhondda Fawr. *In*: SIDDLE, H.J., BROMHEAD, E.N. & BASSETT, M.G. (eds) *Landslides and Landslide Management in South Wales*. National Museums and Galleries of Wales, Cardiff, Geological Series No. 18, 55–59.
SIDDLE, H.J., WRIGHT, M.D. & HUTCHINSON, J.N. 1996. Rapid failures of colliery spoil heaps in the South Wales Coalfield. *Quarterly Journal of Engineering Geology*, **29**, 103–132, https://doi.org/10.1144/GSL.QJEGH.1996.029.P2.02
SIDDLE, H.J., BROMHEAD, E.N. & BASSETT, M.G. 2000a. *Landslides and Landslide Management in South Wales*. National Museums and Galleries of Wales, Cardiff, Geological Series No. 18.
SIDDLE, H.J., WRIGHT, M.D. & HUTCHINSON, J.N. , 2000b. Rapid failures of spoil heaps in the South Wales Coalfield. *In*: SIDDLE, H.J., BROMHEAD, E.N. & BASSETT, M.G. (eds) *Landslides and Landslide Management in South Wales*. National Museums and Galleries of Wales, Cardiff, Geological Series No. 18, 32–35.
SIMS, P. & TERNAN, L. 1988. Coastal erosion: protection and planning in relation to public policies – a case study from Downderry, South-east Cornwall. *In*: HOOKE, J.M. (ed.) *Geomorphology in Environmental Planning*. Wiley, Chichester, 231–244.
SKEMPTON, A.W. 1964. Long term stability of clay slopes. *Géotechnique*, **14**, 77–101, https://doi.org/10.1680/geot.1964.14.2.77
SKEMPTON, A.W. 1976. Introductory remarks. *Philosophical Transactions of the Royal Society, London*, **A283**, 423–425. https://doi.org/10.1098/rsta.1976.0091
SKEMPTON, A.W. & HUTCHINSON, J.N. 1969. Stability of natural slopes and embankment foundations. State of the Art Report. *7th International Conference on Soil Mechanics and Foundation Engineering*, Mexico, 291–335.
SKEMPTON, A.W. & WEEKS, A.G. 1976. The Quaternary history of the Lower Greensand escarpment and Weald Clay vale near Sevenoaks, Kent. *Philosophical Transactions of the Royal Society, London*, **A283**, 493–526, https://doi.org/10.1098/rsta.1976.0094
SKEMPTON, A.W., LEADBEATER, A.D. & CHANDLER, R.J. 1989. The Mam Tor landslide, North Derbyshire. *Philosophical Transactions of the Royal Society, London*, **A329**, 503–547, https://doi.org/10.1098/rsta.1989.0088
SMITH, D.I. 1984. *The landslips of the Scottish Highlands in relation to major engineering projects*. Project Report, **09/LS**. British Geological Survey, Edinburgh.
SO, C.L. 1967. Some coastal changes between Whitstable and Reculver, Kent. *Proceedings of the Geologists Association*, **77**, 475–490.
SOLLAS, W.J., PRAEGER, R.L., DIXON, A.F. & DELAP, A. 1897. Report of the Committee appointed by the Royal Dublin Society to investigate the recent Bog-flow in Kerry. *Scientific Proceedings, Royal Dublin Society*, **8**, 475–508.
STANFORD, M. 2014. *The Bellwin Scheme*. House of Commons Library, http://www.parliament.uk/briefing-papers/SN00643.pdf
STATHAM, I. 2002. Failure on the coastal slope above Lon Gâm, Nefyn, Llyn Peninsula: mechanism and remedial measures. *In*: NICHOL, D., BASSETT, M.G. & DEISLER, V.K. (eds) *Landslides and Landslide Management in South Wales*. National Museums and Galleries of Wales, Cardiff, Geological Series No. 22.
STEAD, D. & WOLTER, A. 2015. A critical review of rock slope failure mechanisms: the importance of structural geology. *Journal of Structural Geology*, **74**, 1–23, https://doi.org/10.1016/j.jsg.2015.02.002
STEERS, J.A. 1951. Notes on erosion along the coast of Suffolk. *Geological Magazine*, **88**, 435–439, https://doi.org/10.1017/S001675680007000X
STEVENSON, M. 1994. The Franklands Village landslide. *Quarterly Journal of Engineering Geology*, **27**, 289–292, https://doi.org/10.1144/GSL.QJEGH.1994.027.P4.01
STUART, H. 1984. *A comparative study of Lateglacial and Holocene talus slopes in Snowdonia, North Wales*. Unpublished BSc dissertation, University of St Andrews.
TEBBUTT, T.H.Y. 1998. *Principles of Water Quality Control*. Butterworth Heinemann, Oxford.
THOMPSON, R.P. 1991. Stabilisation of a landslide on Etruria Marl. *In*: CHANDLER, R.J. (ed.) *Slope Stability Engineering: Development and Applications*. Thomas Telford, London, 403–408.
TRIGILA, A. & IADANZA, C. 2008. *Landslides in Italy*. Report, **83/2008**. Italian National Institute for Environmental Protection and Research, Rome.
TURNER, A.K. & SCHUSTER, R.L. (eds) 1996. *Landslides: Investigation and Mitigation, Transportation Research Board*. Special Report, **247**, National Research Council, National Academy Press, Washington DC.
TWIDALE, C.R. 1985. Ancient landscapes: their nature and significance for the question of inheritance. *In*: HAYDEN, R.S. (ed.) *Global Mega-Geomorphology NASA Conference Publication*, **2312**, 29–40.
UNIVERSITY OF STRATHCLYDE 1991. *The assessment and integrated management of coastal cliff systems*. Report to Ministry of Agriculture Fisheries and Food. University of Strathclyde, Glasgow.
VALENTIN, H. 1954. Der landverlust in Holderness, Ostengland von 1852 bis 1952. *Die Erde*, **6**, 296–315.
VAN ASCH, Th.W.J., BUMA, J. & VAN BEEK, L.P.H. 1999. A view on some hydrological triggering systems in landslides.

Geomorphology, **30**, 25–32, https://doi.org/10.1016/S0169-555X(99)00042-2

VARNES, D.J. 1978. Slope movement types and processes. *In*: SCHUSTER, R.L. & KRIZEK, R.J. (eds) *Landslide: Analysis and Control*. Transportation Research Board, National Research Council, Washington DC, 11–33.

VINNEM, J.E. 2007. *Offshore Risk Assessment: Principles, Modelling and Application of QRA Studies*. 2nd edn. Springer.

WALBANCKE, H.J., STEVENSON, M.W. & COE, R.H. 2000. The effect of a surface water runoff attenuation facility on the stability of a slope. *In*: BROMHEAD, E.N., DIXON, N. & IBSEN, M.-L. (eds) *Landslides: In Research, Theory and Practice*. Thomas Telford, London, 1527–1532.

WALKDEN, M.J. & DIXON, M. 2006. The response of soft rock shore profiles to increased sea-level rise. Tyndall Centre for Climate Change Research, Working Paper 105.

WALKDEN, M.J. & HALL, J. 2005. A predictive mesoscale model of the erosion and profile development of soft rock shores. *Coastal Engineering*, **52**, 535–563, https://doi.org/10.1016/j.coastaleng.2005.02.005

WALKER, J. 1772. Account of the irruption of Solway Moss in December 16, 1772. *Philosophical Transactions of the Royal Society*, **62**, 123–127, https://doi.org/10.1098/rstl.1772.0017

WALTHAM, A.C. & DIXON, N. 2000. Movement of the Mam Tor landslide, Derbyshire, UK. *Quarterly Journal of Engineering Geology and Hydrogeology*, **33**, 105–123, https://doi.org/10.1144/qjegh.33.2.105

WARBURTON, J., HOLDEN, J. & MILLS, A.J. 2004. Hydrological controls of surficial mass movement in peat. *Earth Science Reviews*, **67**, 139–156, https://doi.org/10.1016/j.earscirev.2004.03.003

WATERS, C.N., NORTHMORE, K.J. ET AL. 1996. *A Geological Background for Planning and Development in the City of Bradford Metropolitan District*. British Geological Survey Technical Report, **WA/96/1**.

WATSON, P.D.J. & BROMHEAD, E.N. 2000. The effects of waste water disposal on slope stability. *In*: BROMHEAD, E.N., DIXON, N. & IBSEN, M.-L. (eds) *Landslides: In Research, Theory and Practice*. Thomas Telford, London, 1557–1562.

WEEKS, A.G. 1969. The stability of natural slopes in south-east England as affected by periglacial activity. *Quarterly Journal of Engineering Geology*, **2**, 49–61, https://doi.org/10.1144/GSL.QJEG.1969.002.01.04

WESLEY, J. 1755. *Serious Thoughts Occasioned by the Late Earthquake at Lisbon*. London and Bristol.

WHITE, G. 1789. The Natural History and Antiquities of Selborne. Letter XLV. Reprinted by Oxford University Press, Oxford, 12 September 2013.

WILLIAMS, W.W. 1956. The east coast survey: some recent changes in the coast of East Anglia. *Geographical Journal*, **122**, 317–334.

WILLIAMS, A.T., MORGAN, N.R. & DAVIES, P. 1991. Recession of the littoral zone cliffs of the Bristol Channel, UK. *In*: MORGAN, O.T. (ed.) *Coastal Zone '91*. 2394–2408.

WILSON, P. 2003. Landslides in Lakeland. *Conserving Lakeland*, **40**, 24–25.

WILSON, P. 2005. Paraglacial rock-slope failures in Wasdale, western Lake District, England: morphology, styles and significance. *Proceedings of the Geologists Association*, **116**, 349–361, https://doi.org/10.1016/S0016-7878(05)80052-5

WILSON, P. & SMITH, A. 2006. Geomorphological characteristics and significance of Late Quaternary paraglacial rock–slope failures on Skiddaw Group terrain, Lake District, northwest England. *Geografiska Annaler*, **88A**, 237–252, https://doi.org/10.1111/j.1468-0459.2006.00298.x

WILSON, P., GRIFFITHS, D. & CARTER, C. 1996. Characteristics, impacts and causes of the Carntogher bog-flow, Sperrin Mountains, Northern Ireland. *Scottish Geographical Magazine*, **112**, 39–46, https://doi.org/10.1080/00369229618736976

WILSON, P., CLARK, R. & SMITH, A. 2004. Rock-slope failures in the Lake District: a preliminary report. *Proceedings, Cumberland Geological Society*, **7**, 13–36.

WINTER, M.G. 2019. Debris flows. *In*: GILES, D.P. & GRIFFITHS, J.S. (eds) *Geological Hazards in the UK: Their Occurrence, Monitoring and Mitigation*. Geological Society, London, Engineering Geology Special Publication 29.

WINTER, M.G., MACGREGOR, F. & SHACKMAN, L. (eds) 2005. *Scottish Road Network Landslides Study*. Scottish Executive, Edinburgh.

WINTER, M.G., HEALD, A.P., PARSONS, J.A., SHACKMAN, L. & MACGREGOR, F. 2006. Scottish debris flow events of August 2004. *Quarterly Journal of Engineering Geology and Hydrogeology*, **39**, 73–78, https://doi.org/10.1144/1470-9236/05-049

WINTER, M.G., BARKER, K.J., REID, J.M., BOYLAN, N., JENNINGS, P. & LONG, M. 2009. Discussion of 'Peat slope failure in Ireland'. *Quarterly Journal of Engineering Geology & Hydrogeology*, **42**, 129–132, https://doi.org/10.1144/1470-9236/08-065

WINTER, M.G., DENT, J., MACGREGOR, F., DEMPSEY, P., MOTION, A. & SHACKMAN, L. 2010. Debris flow, rainfall and climate change in Scotland. *Quarterly Journal of Engineering Geology and Hydrogeology*, **43**, 429–446, https://doi.org/10.1144/1470-9236/08-108

WONG, H.N. & HO, K.K.S. 1996. Travel distance of landslide debris. *In*: SENNESET, K. (ed.) *Landslides*. Balkema, Rotterdam, **1**, 417–423.

YOUNG, G. & BIRD, J. 1822. *A Geological Survey of the Yorkshire Coast*. Whitby.

Chapter 5   Debris flows

M. G. Winter

Transport Research Laboratory (TRL), 13 Swanston Steading, 109 Swanston Road, Edinburgh, EH10 7DS, UK
Present address: Winter Associates, Kirknewton, Midlothian, EH27 8AF, UK
mwinter@winterassociates.co.uk

Abstract: Fast-moving, rainfall-induced debris-flow events are relatively common in the mountainous areas of the UK. Their impacts are largely, although by no means exclusively, economic and social. They often sever (or delay) access to and from relatively remote communities for services and markets for goods; employment, health and educational opportunities; and social activities. Specific forms of economic impact are described and their extent is defined by the vulnerability shadow. The mechanisms of rainfall-induced, fast-moving debris flows are considered to bridge between slow mass movements and flood phenomena. The occurrence of debris flows is largely restricted to mountainous areas and a series of case studies from Scotland is briefly described. Hazard and risk assessment are briefly considered and a strategic approach to risk reduction is described. The latter allows a clear focus on that overall goal before concentrating on the desired outcomes and the generic approach to achieving those outcomes. The effects of climate change on debris-flow hazard and risk are also considered and it is concluded that, in Scotland, increases in debris-flow frequency and/or magnitude are most likely and that increases in the risks associated with debris flows are also likely.

5.1 Introduction

Debris flows are largely fast-moving and dynamic in nature, and are generally characterized by rapid movement with high proportions of either water or air acting as a lubricant for the solid material that generally comprises the bulk of their mass. Given the right circumstances, they can be highly destructive.

In the UK their presence is largely, although not exclusively, restricted to mountainous areas. Indeed, the UK landslide risk community has historically focused on slow-moving events that, in general, lead to economic losses such as those at Ventnor on the Isle of Wight (e.g. McInnes 2000) and Folkestone Warren (e.g. Bromhead & Ibsen 2007). In contrast, the international landslide risk community has focused on fast-moving, long-run-out events that, while leading to substantial economic losses, also incur a significant risk to life and limb as demonstrated by the loss of 159 lives at Sarno in Italy in May 1998 (Versace 2007).

The fast-moving debris-flow events in Scotland in August 2004, and since, provide a rich source of case study material; it was fortuitous that there were no major injuries to those involved in these events. However, even in the absence of serious injuries and fatalities, the socioeconomic impacts of such events may be serious. These include the severance (or delay) of access to and from relatively remote communities for services and markets for goods; employment, health and educational opportunities; and social activities. The extent of these impacts is described by the vulnerability shadow. The work that has followed has therefore drawn on the more traditional approach to slow-moving landslides, as well as one that typifies the international approach to fast-moving events that pose (as do the events in Scotland) a real risk to life and limb (Winter *et al.* 2008).

In this paper the work undertaken for a hazard and risk assessment for debris flow affecting the Scottish road network is briefly referred to in terms of the following activities: a GIS-based assessment of debris-flow susceptibility; a desk-/computer-based interpretation of the susceptibility and field-based ground-truthing to determine hazard; and a desk-based exposure analysis to determine risk.

A strategic approach to landslide risk reduction allows a clear focus on that overall goal before concentrating on the desired outcomes and the generic approach to achieving those outcomes. Only then are the processes that may be used to achieve those outcomes (i.e. the specific management and mitigation measures and remedial options) addressed. A top-down, rather than a bottom-up, approach is therefore targeted. Risk reduction is considered as: relatively low-cost exposure reduction (management) outcomes that allow specific measures to be extensively applied; or relatively high-cost hazard reduction (mitigation) outcomes that include measures that are targeted at specific sites.

In addition to covering the above themes, this paper also considers the potential effects of future climate change on debris-flow hazard and risk, again using Scotland as an example.

Engineering Group Working Party (main contact for this chapter: M.G. Winter, Winter Associates, Kirknewton, Midlothian, EH27 8AF, UK, mwinter@winterassociates.co.uk)
From: GILES, D. P. & GRIFFITHS, J. S. (eds) 2020. *Geological Hazards in the UK: Their Occurrence, Monitoring and Mitigation – Engineering Group Working Party Report*. Geological Society, London, Engineering Geology Special Publications, **29**, 163–185,
https://doi.org/10.1144/EGSP29.5
© 2020 TRL Limited. Published by The Geological Society of London. All rights reserved.
For permissions: http://www.geolsoc.org.uk/permissions. Publishing disclaimer: www.geolsoc.org.uk/pub_ethics

This paper deals with debris flows; other types of generally slow-moving landslide are dealt with by Lee (2020) and peat slides by Warburton (2020).

5.2 Types of landslide and flow mechanisms

Cruden & Varnes (1996) provide perhaps the most commonly accepted taxonomy of landslide types, dividing these 'movements of a mass of rock, debris or earth down a slope' into falls, topples, slides, spreads and flows. Flows are defined as spatially continuous movements in which shear surfaces are short lived, closely spaced and usually not preserved, and which have velocities resembling a viscous liquid. Importantly, they note that the lower boundary may be either a surface along which appreciable differential movement occurs, or a thick zone of distributed shear indicating a gradation from slides to flows that is dependent upon water content, mobility and the evolution of the movement. The latter issue of evolution is especially important; observations of debris flows often reveal an initial sliding movement that becomes more complex as the flow moves downslope. Cruden & Varnes (1996) describe both open-slope and channelized debris flows.

Hillslope (or open-slope) debris flows form their own path down valley slopes as tracks or sheets (Cruden & Varnes 1996), before depositing material on lower areas with lower slope gradients or where flow rates are reduced (e.g. obstructions, changes in topography; Fig. 5.1). The deposition area may contain channels and levees. The motion of such events is generally considered not to be maintained when the width exceeds five times the average depth (Hungr *et al.* 2001). As the mobilized material in such events rarely persists to either the level of the slope at which transport infrastructure exists, or to the valley floor in Scotland, it is therefore of relatively little practical interest.

Channelized debris flows follow existing channel type features, for example, valleys, gullies, depressions or hollows (Fig. 5.1). The flows are often of high density, comprise 80% solids by weight (Cruden & Varnes 1996), and may have a consistency equivalent to that of wet concrete (Hutchinson 1988). They can therefore transport boulders that are some metres in diameter, for example, a 9-tonne boulder was reported at the debris flow on the A83 at Cairndow (Winter *et al.* 2005*a*). Hungr *et al.* (2014) describe a number of initiating events for debris flows including slides, debris avalanches, rock falls or spontaneous instability of a steep stream bed. Observation in Scotland suggests that slides might be the most frequent form of initiating event (Winter *et al.* 2006). The definition of debris flow adopted in this paper broadly follows that of Hungr *et al.* (2001), who describe such events as extremely rapid movements of saturated material. Hungr *et al.* (2001) allude to the interrelation of debris flow with debris floods (described as extremely rapid movements with free water present). Pierson & Costa (1987) suggest that the divide between debris flow and hyperconcentrated flow (debris flood) (Fig. 5.2) may be at a solids concentration of up to 80%. This is, of course, highly dependent on both sorting and particle size; the non-numerical depiction of this boundary in Figure 5.2 suggests a lower boundary that might be more appropriate for poorly sorted, finer-grained rather than well-sorted, coarse-grained deposits. Hungr *et al.* (2014) specifically refer to the entrainment of material in the stream bed as a result of the rapid undrained loading and associated high pore pressures and even liquefaction that is experienced once material begins to move in a steep

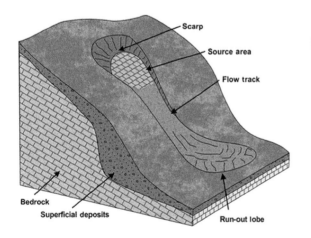

(a) Hillslope debris flow

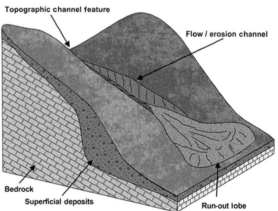

(b) Channelized debris flow

Fig. 5.1. (**a**) Hillslope and (**b**) channelized debris flow (from Nettleton *et al.* 2005*b*).

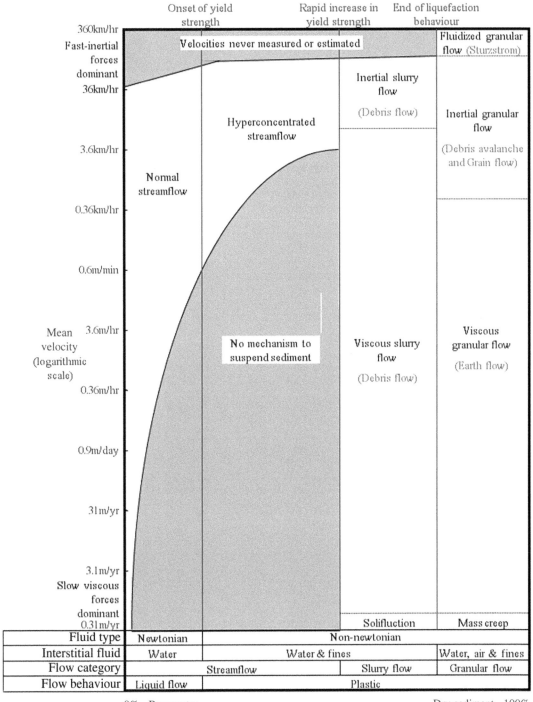

Fig. 5.2. Simplified rheological classification of sediment–water flows (after Pierson & Costa 1987 and as modified by Winter *et al.* 2005*b*). Flow types are provided in green text.

channel. Sassa (1985) notes that such loading can be sufficiently sudden as to merit the description 'impact loading'. This entrainment of stream bed material is typical of debris flows observed in Scotland and the rest of the UK.

Jones & Lee (1994) note that the diagnostic feature of debris flows is the parallel ridges of deposited debris (levées) that flank the track of the flow. However, care is needed with such interpretations on the ground as linear ridges of debris also occur with hyperconcentrated flows or debris floods in the form of boulder bars. The key difference is that debris flow leads to matrix-supported deposits, while debris floods lead to clast-supported deposits (Giroud 2005). It is not uncommon for debris flows to turn into debris floods where additional water is available.

Debris flows mainly occur because of the character of natural slopes (including the presence of a suitable drainage channel or channels), the deposits of which are comprised, and the amount and duration of rainfall (and consequent infiltration) to which they are subject. The presence of alluvial (or debris) cones and/or fans (truncated by subsequent hydrological activity or not) in the lower reaches of valleys can also be a good indicator of past debris-flow and/or debris-flood activity.

Coarser material may form natural levees or accumulate as debris dams at obstacles (e.g. fallen trees or large boulders) or changes in channel gradient, leaving finer material in suspension to continue down the channel. Suspended material in channel flows will typically be deposited in lower-gradient sections of channels, where channels widen and upon emergence from the channel.

In practice, many debris flows may start as the hillslope form but, during the course of flowing downslope, they may enter channel-type features, form their own channel-flow tracks in superficial deposits, or cut through superficial deposits and then be channelled down pre-existing channel features in rockhead (e.g. an infilled stream or gully) (Nettleton et al. 2005a).

5.3 Occurrence

A general overview of debris-flow occurrence in Great Britain is given by Jones & Lee (1994), confirming that they are generally found in mountainous areas. Indeed, they refer to the work of Innes (1983) who undertook a survey of Scotland based upon aerial maps and marked those 10 km by 10 km grid squares that showed some sign of debris-flow activity (Fig. 5.3), clearly indicating that such activity is widespread. As part of the Scottish Road Network Landslides Study, Winter et al. (2009, 2013a) conducted a hazard and risk assessment to determine those parts of the trunk road network that are at highest risk from debris flow (Fig. 5.4). While there are some detailed differences compared with the occurrence map presented by Innes, the images are broadly confirmatory. Perhaps the most striking difference is that the area that includes the Rest and be Thankful site is not included

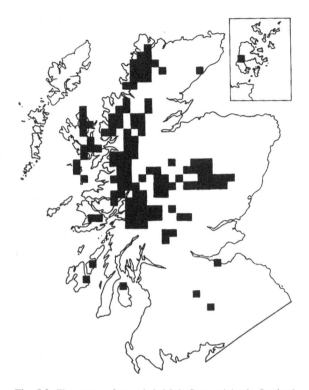

Fig. 5.3. The extent of recorded debris-flow activity in Scotland (from Jones & Lee 1994; after Innes 1983). Note that the figure does not record any activity in the area around the Rest and be Thankful, for example. It seems unlikely that there was no such activity prior to 1983 when the figure was first published, so the dataset should not be seen as exhaustive.

on Innes' map, despite the regular occurrence of such events at this location (Winter & Corby 2012).

The events of August 2004 (Winter et al. 2005a, 2006, 2009) followed rainfall substantially in excess of the norm. Some areas of Scotland received more than 300% of the 30-year average August rainfall, while in the Perth and Kinross area figures of the order of 250–300% were typical. Although the percentage rainfall during August reduced to the west, parts of Stirling and Argyll and Bute still received 200–250% of the monthly average (http://www.metoffice.gov.uk). Subsequent analysis of radar data indicated that at Callander, some 20 km distant from the events at the A85, 85 mm of rain fell during a 4-hour period on 18 August. Some 48 mm fell in just 20 minutes and the storm reached a peak intensity of 147 mm/hour. The 30-year average rainfall for August in Scotland varies between 67 mm on the east coast and 150 mm in the west of Scotland (Anon. 1989).

The rainfall was both intense and long-lasting, and a large number of debris flows were experienced in the hills of Scotland. A small number of these intersected the trunk road network, notably the A83 between Glen Kinglas and to the north of Cairndow (9 August), the A9 to the north of Dunkeld (11

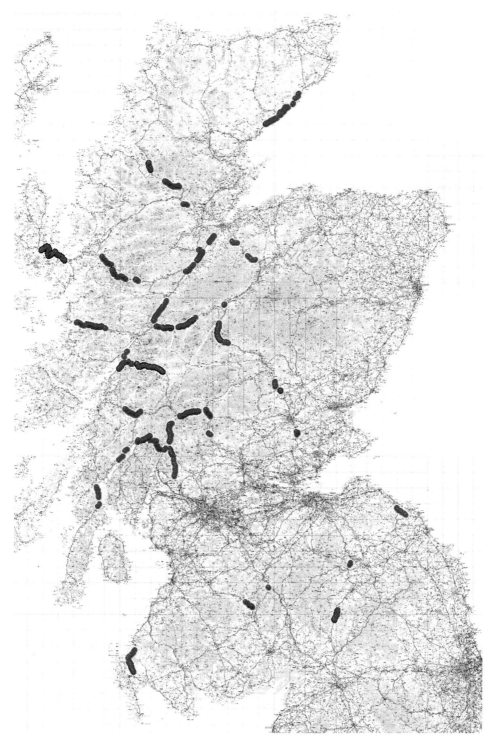

Fig. 5.4. Sites with a hazard ranking score of 100 or greater. (© Crown Copyright. All rights reserved Scottish Government 100020540, 2009; from Winter *et al.* 2009.)

August) and the A85 at Glen Ogle (18 August). These locations are illustrated in Figure 5.5.

While there were no major injuries to those affected, 57 people had to be airlifted to safety when they became trapped between the two main debris flows at Glen Ogle. However, the real impacts of the events were economic and social, in particular the effects of the severance of access to relatively remote communities. The A85, carrying up to 5600 vehicles per day (all vehicles two-way, 24-hour annual average daily flow), was closed for 4 days; the A83, which carries around 5000 vehicles per day, was closed for slightly over a day; and the A9, carrying 13 500 vehicles per day, was closed for 2 days prior to reopening, initially with single-lane working under convoy. The disruption experienced by local and tourist traffic, as well as to goods vehicles, was substantial. The traffic flow figures are for the most highly trafficked month of the year for each of the roads, either July or August. Minimum flows occur in either January or February and are roughly half those of the maxima. The figures reflect the importance of tourism and related seasonal industries to Scotland's economy.

This section provides an overview of the events of August 2004 (Winter *et al.* 2006) and the event at the A83 Rest and be Thankful in October 2007 (Winter *et al.* 2009).

5.3.1 A83 Glen Kinglas/Cairndow: 9 August 2004

The A83 in Argyll and Bute was blocked in Glen Kinglas and to the north of Cairndow. In addition to causing the road to be closed for slightly over a day, the debris flow at Cairndow also had a substantial effect on a residential property immediately upslope from the road. Numerous smaller debris flows were also observed on the hillslopes either side of Glen Kinglas.

In early August 2004 the hillsides were in a saturated condition following a relatively wet spell during the preceding weeks. This was followed by a relatively short period of exceptionally heavy rainfall. The slope is frequently incised with watercourses that are culverted below the road.

Typically the flows commenced in the steep upper reaches of the slopes at around 500 m above ordnance datum (AOD) (Fig. 5.6). A shallow scarp, of less than 1.5 m, was observed at the head of each. The waterlogged material is assumed to have flowed into existing water courses, providing a more erosive sediment charge, resulting in erosion up to 10–15 m either side of the channels. Deposition occurred at the toe of the slope where the gradient slackens. The amounts deposited at the three locations where debris flow occurred were estimated at between 100 tonnes and up to 1000–2000 tonnes. The debris blocking the road comprised very silty sand and gravel with frequent cobbles and boulders, the largest of which was estimated to weigh 9 tonnes.

5.3.2 A9 North of Dunkeld: 11 August 2004

The heavy rain that triggered the A83 events continued for three more days over much of Scotland and precipitated further debris flows, three of which affected the A9 just to the north of Dunkeld in Perth and Kinross.

At this location the A9 passes the foot of a steep slope on its eastern side, with the River Tay a short distance to the west. The old A9, now a minor local road (C502), traverses the hillside above the trunk road. The upper part of the slope between these two roads is wooded, whereas the lower part is vegetated by broom with few trees. The lower part of the slope was steepened at the time of construction of the present trunk road to a gradient close to 1 in 2 (vertical to horizontal), whereas the upper part is slightly less steep, steepening again at the top to form the bench on which the old A9 was constructed. Above the old A9 the wooded hillside continues to rise for approximately 250 m in elevation.

As a result of the exceptional rainfall, a large amount of surface water runoff descended the slope above the old A9, both along the course of existing streams and on the open hillside between. When it reached the old road, the drainage system was unable to contain or disperse such large volumes of water. The surface runoff travelled along the old road, spilling over the edge onto the slope below in a number of places as illustrated in Figure 5.7, and effectively concentrating the flows at these locations.

The upper part of the slope is notable for the presence of a superficial layer of yellow fine sand that is both slightly denser and lighter in colour than the underlying uniformly graded fine sand. This denser material is absent from the lower part of the slope, and it seems likely that material of this nature was removed when the lower part of the slope was steepened to accommodate the A9 trunk road.

As the flood increased, it entered an erosive phase in which gullies, some 3–4 m deep and up to 6 m wide, were scoured. These gullies stretched from the top of the cut slope (or above) to the A9 trunk road verge. Once the vegetation was stripped and the underlying fine sand was mobilized, it flowed freely down the slope and onto the trunk road. The erosion gullies did not extend into the upper part of the slope due to the slightly lesser gradient and the presence of the more erosion-resistant layer of the lighter yellow fine sand (Fig. 5.8). Large amounts of sand were deposited on the trunk road (Fig. 5.9).

The A9 events differ somewhat from the other events described and, indeed, those more typically observed in Scotland. Based on observation, it is considered that the A9 events commenced as straightforward surface flood events that evolved into debris floods, the phase of which is associated with the significant erosion evident. This is perhaps best evidenced by the apparent relative lack of damage of the vehicle entrained in the debris in Figure 5.9, indicating a water-dominated movement. While it is important that such events are not confused with debris flows, it does seem likely that, in their latter phases, these may well have transformed into debris flows. Indeed, King & Williamson (2002) note that in Hong Kong where debris flood and erosion may occur after a debris flow, the volume of erosion associated with such features should not be included with the debris-flow volumes. However, as such events are rarely directly observed in

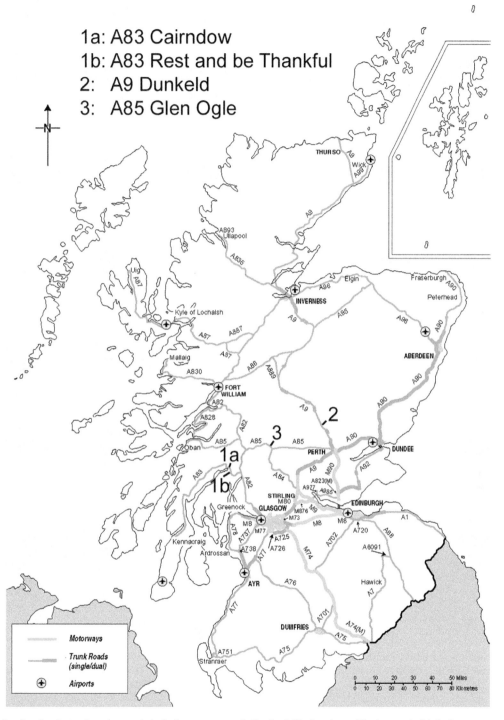

Fig. 5.5. Map showing the trunk road network, including motorways, in Scotland. The locations of the three main debris-flow event groups that affected the trunk road network in Scotland in August 2004 are shown: (1) A83 Glen Kinglas and Cairndow; (2) A9 north of Dunkeld; and (3) A85 Glen Ogle (from Winter *et al.* 2006).

Fig. 5.6. Upland debris slide and flow development at Cairndow on the A83 (from Winter *et al.* 2006). (Courtesy of S Martin, C2HM Hill.)

Scotland, it can be extremely difficult to disassociate the phases of the events and their respective volumes.

It is clear from the A9 events that flat, linear, sidelong slope features such as public roads, forest roads and animal tracks can retard or concentrate the downslope flow of water and thus aid its penetration into the slope below. Such a mechanism has been a factor in a number of previous events such as the washout that blocked the A83 Rest and be Thankful in the vicinity of Roadman's Cottage in 1999. The presence of forest tracks was also identified as a contributory factor in the debris flow which occurred at the A887 at Invermoriston in August 1997 and to debris flow above the A890 at Stromeferry (Nettleton *et al.* 2005*a*, *b*; Winter *et al.* 2005*b*).

5.3.3 A85 Glen Ogle: 18 August 2004

Following the A83 and A9 incidents, the rainfall in the area abated for several days; however, on 18 August a short but exceptionally intense rainstorm occurred in west Stirlingshire, triggering two debris flows that blocked the A85 in Glen Ogle north of Lochearnhead. The southerly slip occurred first and, as advice was being offered to motorists by Trunk Road Operating Company staff, a second debris flow occurred to the north of the first. Some 20 vehicles were trapped between the two debris flows, and 57 occupants were airlifted to safety by RAF and Royal Navy helicopters (Fig. 5.10).

The A85 trunk road climbs NW-wards through Glen Ogle from an elevation of around 100 m AOD to a pass at the head of the glen at around 290 m AOD, predominantly climbing the eastern flank of the valley; it is along this section that the most significant flows occurred. The hillside above the road rises some 400 m at a slope of approximately 1 in 2. Below the road the gradient of the slope decreases rapidly to the Glen Ogle Burn. The two slips followed steep streams that descend this hillside and are culverted beneath the road. The southerly stream descends through an area covered by

Fig. 5.7. Influence of old road on debris flow at A9 Dunkeld. The central flow is shown and the northern flow can also be seen on the left of the picture. Photograph dated 11 August 2004 (from Winter *et al.* 2006). (Courtesy of Alan Mackenzie, BEAR Scotland.)

heather and bracken, while the northerly stream descends a partially rocky area of the hillside.

As a result of the exceptional rainfall, and possibly because of the additive of the high level of antecedent rainfall, the soils in the upper catchments to the streams became saturated, triggering slides into the headwaters of both streams. The culverts rapidly became blocked, and debris spilled across the road (Fig. 5.10) and down the slope beyond (Fig. 5.11). Most of the debris came to rest on the slope below the road, but a small proportion reached the Glen Ogle Burn. This burn was also in spate at the time, and rapidly removed the debris that reached it.

Examination of the northern flow (Fig. 5.11) indicated two independent sources. An arcuate scar indicated the location from which a shallow translational slip broke away, the turf and upper soil travelled over the surface of the vegetation below and entered the upper part of the stream gully. However, scarring also indicated that instability occurred independently at the very top of the gully. It is not known which instability occurred first, although both slides appear to have generated only a relatively small amount of debris. The debris was, however, channelled into and down the steeply inclined bed of the stream and scoured the gully, removing turf and soil. It is likely that the volume of water and debris increased further down the gully and that the consequent damage was increased in areas closer to the road. In the middle and lower parts of the flow, large and small boulders and trees were mobilized in addition to soil and turf. More detailed accounts of the A85 Glen Ogle event is given by Winter *et al.* (2006) and Milne *et al.* (2009).

5.3.4 A83 Rest and be Thankful: 28 October 2007

Events at the A83 Rest and be Thankful, to the east of the events in Glen Kinglas in 2004, occur regularly (Winter & Corby 2012). Figure 5.12 illustrates the event that occurred at approximately 03:30 hours on Sunday 28 October 2007. The photograph is taken from the opposite side of Glen

Fig. 5.8. The top of the central A9 erosion gully at the top of the cut slope. The overhang of the denser yellow fine sand is clearly visible and it appeared that this material was better able to shed the water and debris. Photograph dated 12 August 2004 (from Winter *et al.* 2006).

Croe, and evidence of numerous past events can clearly be seen. Around 400 tonnes of material were deposited at road level. The closely spaced, steeply incised, parallel drainage channels serve to increase the frequency of the events as a high proportion of small slope movements will reach a stream channel; however, to some extent, the size of the events is restricted as the material available to be eroded is limited by both the relatively short distance between channels and recharge time between events.

It does, however, seem likely that many of the most destructive events include a highly mobile phase which may be best described as a debris-flood phase in addition to the debris-flow phase. This appears to follow both the initiating slip and the debris-flow phase (but see also the earlier comments on the A9 events). This may well be the defining difference between those events that cause major damage and those that do not. At the time of writing, the most recent event at the Rest and be Thankful occurred in the early hours of 6 March 2014 (Fig. 5.13). While debris certainly reached the road, the amounts involved were relatively small. This event appears not to have included a debris-flood phase of any great significance and led to a significant drape of material, reportedly up to 2 m thick, being left in a metastable state on the slope.

5.4 Hazard and risk assessment

Hazard and risk assessment is a necessary first step in order to understand the pattern of events and their likely impacts, and to prioritize any risk reduction activities that may be necessary.

Landslide hazard and risk assessment has been revolutionized in recent years by the advent of readily available high-resolution imagery, and by the availability of geographic information systems (GIS) in which disparate datasets can be brought together and interpreted (Corominas *et al.* 2014). This does not negate the need for site-based work, but it can limit the extent and duration of such work that is needed. The hazard and risk assessment undertaken for the Scottish road network was undertaken to help prioritize subsequent actions, and comprised three phases: (1) a pan-Scotland, GIS-based, assessment of debris-flow susceptibility (Fig. 5.14), considering the availability of debris material, hydrogeological conditions, land use, the proximity of stream channels and slope angle; (2) a desk-/computer-based interpretation of the susceptibility and field-based ground-truthing to determine hazard; and (3) a desk-based exposure analysis primarily focusing

Fig. 5.9. The southerly debris flow at the A9 north of Dunkeld. The flow has formed its own channel by erosion. Photograph dated 12 August 2004 (from Winter *et al.* 2006). (Courtesy of Alan Mackenzie, BEAR Scotland.)

upon life and limb risks, but also accounting for socioeconomic impacts (traffic levels and the existence and complexity of the detour were used).

These were then combined to calculate the risk (hazard ranking) and the highest risk locations were illustrated on a map (Fig. 5.4) (Winter *et al.* 2009, 2013*a*). The broad-based approach was, at each stage, to focus on areas highlighted by the preceding stage. The spatial extent of the susceptibility assessment is comparable with the most extensive assessments undertaken internationally (e.g. Castellanos Abella & Van Westen 2007; Dio *et al.* 2010).

The work undertaken to assess hazards and risks of debris flow in relation to the Scottish road network is a regional assessment designed to allow specific higher (and lower) risk sites to be identified. Lee (2020) and Corominas *et al.* (2014) describe more detailed risk analysis and assessment procedures for individual sites and Wong & Winter (2018) and Winter (2018) describe such an approach to more detailed, site-specific quantitative risk assessment on the Scottish road network. Winter *et al.* (2014*a*) describe an approach to determining the vulnerability of road infrastructure to debris flow using fragility curves.

5.5 Risk reduction

The primary purpose of a regional landslide hazard and risk assessment is often to enable the prioritization of sites that will be subject to risk reduction, through management and/or mitigation in the light of defined budgets. However, it is important to note that risk reduction is only applied to cases in which the risk is deemed to be greater than that which is tolerable, or greater than the level at which the risk holder is willing to accept (Winter & Bromhead 2012). There are many forms of mitigation (e.g. VanDine 1996). However, to reduce the risk to acceptable levels, either the magnitude of (1) the hazard or (2) the potential exposure (or vulnerability) or losses that are likely to arise as a result of an event must be addressed. It is therefore possible to consider management strategies that involve exposure reduction outcomes and mitigation strategies that involve hazard reduction outcomes (Fig. 5.15). Further, it is important that those funding such works, including infrastructure owners and local governments, are able to focus clearly on goals of, the outcomes from, and the approaches to such activities rather than the details of individual processes and techniques.

Fig. 5.10. Fifty-seven occupants of the 20 vehicles that were trapped between the two debris flows in Glen Ogle were airlifted to safety (from Winter *et al.* 2005*b*). (By permission, © Perthshire Picture Agency: http://www.ppapix.co.uk.)

To this end, a strategic approach to landslide risk reduction (Fig. 5.15) that incorporates a classification scheme for landslide management and mitigation has been developed in order to provide a common lexicon and to allow a clear focus on goals, outcomes and approaches (Winter 2014).

The scheme is designed to encourage a strategic approach to the selection of landslide management and mitigation processes (specific measures and remedial options). It is intended to aid a focus on the overall goal of landslide risk reduction, what needs to be achieved (the desired outcomes) and the generic approach to achieving that outcome, rather than (initially at least) the specific measure(s) or option(s) (the process or processes) used to achieve that outcome. The focus is therefore first on the desired outcome from risk reduction: whether the exposure, or vulnerability, of those people at risk and their associated socioeconomic activities are to be targeted for reduction or whether the hazard itself is to be reduced (either directly or by affecting the physical elements at risk). In a roads environment the people at risk are road users, whereas in an urban setting those at risk are residents and business people. The secondary focus is then on the approach(es) to be used to achieve the desired outcome before specific measures and remedial options are considered (see also Sections 4 and 5). By these means, a more strategic top-down approach is encouraged in favour of a bottom-up approach.

This approach also provides a common lexicon for the description and discussion of landslide risk reduction strategies, which is especially useful in a multi-agency environment. It also renders a multi-faceted (holistic) approach more viable and easier to articulate, while helping to ensure that the responses to the hazard and risks in play are appropriate. This approach should be especially useful for infrastructure owners and operators who must deal with multiple landslide and other types of risk that are distributed across large networks. Such an approach promotes a considered decision-making process that takes account of both costs and benefits and encourages careful consideration of the correct solution for each location and risk profile, potentially making best use of often limited resources.

The approaches and the specific measures and remedial options (processes) are described in the following in more detail largely, although not exclusively, in the context of landslide hazard and risk management and mitigation on the Scottish trunk road network.

Exposure reduction can take three basic forms: (1) education (and information); (2) geographical (non-temporal)

Fig. 5.11. View of the northern A85 Glen Ogle debris flow two days after the event, showing the sharp bend in the channel just above road level (from Winter *et al.* 2006).

Fig. 5.12. View of the hillside above and below the approach from the east to the Rest and be Thankful (from NGR NN 23160 06559 on the opposite side of Glen Croe). Not only can the event dated 28 October 2007 be clearly seen, but evidence of numerous past events can also be observed on the surrounding hillside (from Winter *et al.* 2009).

warnings; and (3) response (including temporal, or early, warnings).

Typically, education in its broadest sense may form a key part of an information strategy, including leaflets, information signs and other activities. It is important to be clear that education targeted at addressing the understanding of hazards and risks is unlikely to affect the behaviour of, for example, road users. It may influence society's acceptable level of risk, and make a valuable contribution in doing so. However, education and information on the desirable behaviours of, for example, drivers in areas of landslide hazard during periods of higher risk (Winter *et al.* 2013*b*) is intended to influence the desirable behaviours of road users. These include heightening the levels of observation, moderating speeds and excluding certain stopping locations (e.g. bridges) in order to avoid likely areas of hazard and to allow early observation and avoidance of potential hazards.

Geographical warning signs may take the form of the internationally recognized landslide warning signs. The responsive reduction of exposure lends itself to the use of a simple three-part management tool (Winter *et al.* 2005*a*): (1) detection of either the occurrence of an event (e.g. instrumentation, monitoring, observation) or by the forecast of precursor conditions (e.g. rainfall); (2) notification of the likely/actual occurrence of events to the authorities (e.g. in a roads environment the police, road administration and the road operator); and (3) action that reduces the exposure of the elements at risk to the hazard. In a roads environment, the latter could include media announcements, the activation of geographical signs that also have a temporal aspect (e.g. flashing lights), the use of variable message signs, 'landslide patrols' in marked vehicles, road closures and traffic diversions.

This approach of detection, notification and action (or DNA) provides a simple framework for articulating management responses (Winter *et al.* 2005*a*, 2009). Applications to date include the installation of standard landslide warning signs with the addition of flashing lights, otherwise known as wig-wags, on the A83 in the vicinity of the Rest and be Thankful (Winter *et al.* 2013*b*).

The challenge with hazard reduction or mitigation in a relatively low-risk environment is often to identify locations of sufficiently high risk to warrant spending significant sums of money on engineering works. The costs associated with

Fig. 5.13. Debris-flow event at the A83 Rest and be Thankful (6 March 2014, at an estimated time of 01:35 hours). The image was taken from a helicopter at low level above the hillside. Significant boulders can be seen within a significant drape of metastable debris below the initiation area. A work crew wearing fluorescent yellow jackets can be seen for scale. (Image © M.G. Winter.)

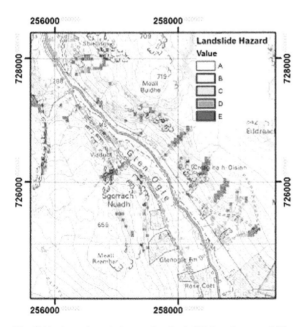

Fig. 5.14. A portion of the pan-Scotland GIS-based susceptibility assessment showing susceptibility for the area around the A85 in Glen Ogle (from Winter *et al.* 2009, 2013*a*).

installing extensive remedial works over very long lengths of road may be both unaffordable and unjustifiable; the costs can be significant even at discrete locations. Moreover, the environmental impact of such engineering work should not be underestimated. Such works often have a lasting visual impact and, potentially, impact upon the surrounding environmental. It is considered that such works should be limited to locations where their worth can be clearly demonstrated.

In addition, simple actions such as ensuring that channels, gullies and other drainage features are kept open and operating effectively are extremely important in terms of hazard reduction. This requires that the maintenance regime is fully effective both in routine terms and also in response to periods of high rainfall, flood and slope movement. It is also important that planned maintenance and construction projects take the opportunity to limit any hazards by incorporating, where suitable, measures such as higher-capacity or improved forms of drainage, or debris traps into the design. In particular, critical review of the alignment of culverts and other conduits close to the road should normally be carried out as part of any planned maintenance or construction activities.

Beyond such relatively low-cost/low-impact options, three categories of hazard reduction measures may be considered: (1) works to engineer, or protect, the elements at risk (e.g. fences, shelters, basins, etc.); (2) remediation of the

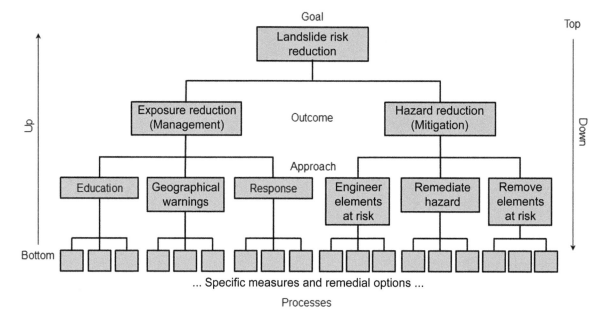

Fig. 5.15. Classification for landslide management and mitigation to enable a strategic approach to risk reduction (from Winter 2014).

hazard to reduce failure probability (e.g. alteration of the slope profile by either cut or fill; improvement of the material strength, most often by decreasing porewater pressures; or providing force systems to counteract the tendency to move; Bromhead 1997); and (3) removal, or evacuation, of the elements at risk (e.g. road realignment or evacuation of a settlement that is deemed to be at risk).

Further details of this approach, based upon Winter *et al.* (2005*a*, 2009), are given by Winter (2014).

5.6 Impacts

The impacts of debris flows can be catastrophic and the loss of life can be substantial. At Sarno, to the east of Naples in Italy, 159 lives were lost in May 1998 due to debris flows that occurred in the volcanic deposits in the foothills of Vesuvius (Versace 2007). In Zhouqu, Gansu Province, China a debris-flow event (Fig. 5.16) that occurred over the night of 7–8 August 2010 claimed the lives of around 1750 people (Dijkstra *et al.* 2014).

In contrast, the UK may be described as a relatively low-risk landslide environment (Gibson *et al.* 2013); major injuries and fatalities are indeed relatively rare, the socio-economic impacts can be substantial. (Notwithstanding this there were four deaths as a result of three separate landslide incidents in SW England during the period July 2012 to March 2013.)

These include the severance (or delay) of access to and from relatively remote communities for services and markets for goods; employment, health and educational opportunities; and social activities. The types of economic impacts were summarized by Winter & Bromhead (2012) as follows.

- *Direct economic impacts*: The direct costs of clean-up and repair and/or replacement of lost and/or damaged infrastructure in the broadest sense and the costs of search and rescue. These are relatively easy to estimate for any given event, providing that the estimate is made contemporaneously.
- *Direct consequential economic impacts*: These generally relate to disruption to infrastructure, and are really about loss of utility. The costs of closing a road (or implementing single-lane working with traffic control) for a given period with a given diversion are relatively simple to estimate using well-established models. The costs of fatal and/or non-fatal injuries may also be included here and taken (on a societal basis) directly from published figures.
- *Indirect consequential economic impacts*: Often landslide events affect access to remote rural areas with economies based upon transport-dependent activities. If a given route is closed for a long period, then how does that affect confidence in, and the ongoing viability of, local businesses? Manufacturing and agriculture (e.g. forestry in western Scotland and coffee production in Jamaica) are a concern as access to markets is constrained, the costs of access are increased and business profits are affected, and short-term to long-term viability may be adversely affected. Further, tourists may be reluctant to travel to areas to which access is restricted.

Fig. 5.16. The channel in which the 7–8 August 2010 Zhouqu debris flow occurred (Gansu Province, PR China). (Image © M.G. Winter.)

The vulnerability shadow (Winter & Bromhead 2012) cast can be extensive, and its geographical extent is determined by the transport network rather than the relatively small footprint of the event itself (Fig. 5.17). In the case of the events at the A83 Rest and be Thankful, the land area affected was estimated to be around 2800 km^2 which is, for example, approximately two-and-a-half times the land area of Hong Kong Special Administrative Region. The magnitude of the economic impacts of such events is the subject of ongoing work (Winter et al. 2014b).

5.7 Climate change

Rainfall is the main cause of debris flow, and rainfall data can be related to known debris flow. Winter et al. (2009, 2010a) used such data to develop a tentative intensity–duration threshold for rainfall-triggered debris flow in Scotland (Fig. 5.18). Taking the central, straight-line portion of the threshold, concluded to be the most important part of the graph (Winter et al. 2010a, 2013b), the threshold is generally reflective of findings from other parts of the world (Winter et al. 2010a), albeit somewhat conservative.

When considering rainfall initiation of debris flow, climate change must be taken into account. Taking Scotland as an example, the climate may be very broadly divided into the relatively dry east and the relatively wet west; on average, almost twice as much rain falls each year in the west compared with the east. In the east, rainfall generally peaks in the summer months of July and August (Anon. 1989). Monthly average rainfall data for Edinburgh (Fig. 5.19), which is broadly representative of the east, indicate that the monthly variations in rainfall are relatively slight. In the wetter west, maximum rainfall levels are reached from September to January (e.g. Tiree in Fig. 5.19). Perhaps most marked is the variation in the monthly averages, with the driest month of May receiving, on average, around half the rainfall experienced in the wettest month of October.

While the evidence of a general trend of global warming and increased climate instability is overwhelming, climate change models are subject to a number of uncertainties (Winter et al. 2010b). It follows that the full range of possible climate change impacts on slope stability, both negative and positive, needs to be considered. It is therefore helpful to draw conclusions from an analysis of a number of complementary datasets (Winter et al. 2010a; Winter & Shearer

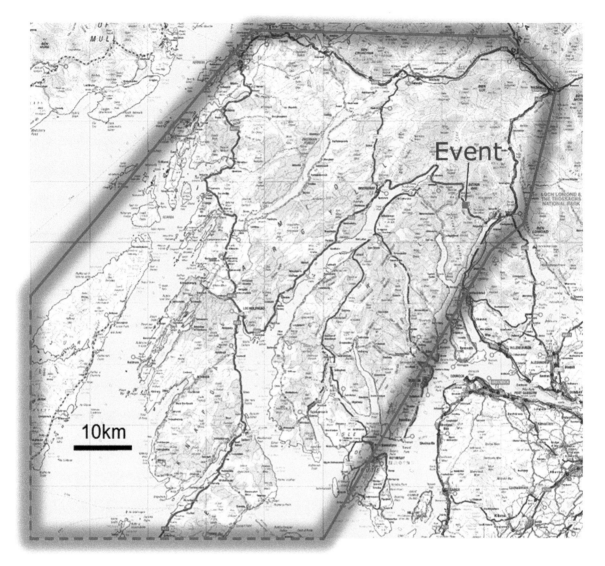

Fig. 5.17. A relatively small debris-flow event (blue square) closed the A83 strategic road in Scotland; the vulnerability shadow that was cast (bordered in red) was extensive (from Winter 2014).

2013), including current climate, recent climate trends and climate change predictions.

The observed recent trends in the climate of the recent past are relevant as well as the predictions of future climate change. Work by Barnett *et al.* (2006) describes trends in climate across Scotland during the last century. The figures for 1914–2004 indicate a definite, but small, increase in the running annual average rainfall for the north and, particularly, the west of Scotland; figures for the east of Scotland are broadly stable.

The deterministic UK Climate Impacts Programme 2002 (UKCIP02) (Hulme *et al.* 2002) climate change predictions for Scotland show little predicted change in annual mean precipitation over the next few decades, with any change being within the range that can be attributed to natural variability. However, Galbraith *et al.* (2005) note that, even by the 2020s, a distinct seasonal pattern can be discerned; while little significant change was predicted for the spring or autumn precipitation amounts, changes were found to be likely for the winter and summer. Winter precipitation was predicted to increase by 10–15% in the east, with changes elsewhere being within the range of that which can be accounted for by natural variability. Decreases in the average summer precipitation were predicted to be widespread, with only the far

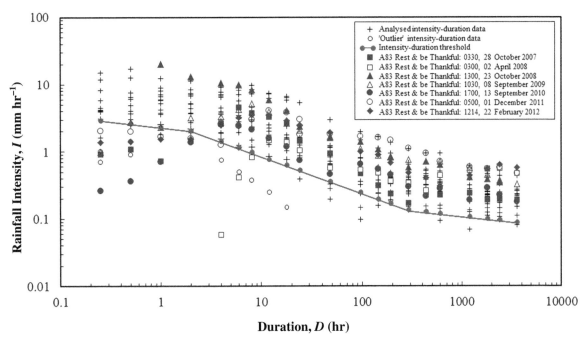

Fig. 5.18. Plot of rainfall intensity and duration of rainfall for debris-flow event. The landslide events (denoted by the black crosses) were used to develop the tentative threshold shown in red; grey circles represent data points which have been determined to be outliers (Winter et al. 2009). The blue symbols indicate recent landslides events used to validate the tentative threshold (from Winter et al. 2013b).

NW seeing little change. By the 2020s, Galbraith et al. (2005) indicate that the decreases are likely to be greatest in the SE of Scotland and could be as much as 20% lower than present-day modelled levels.

The most recent UK Climate Projections 2009 (UKCP09) climate change predictions (Jenkins et al. 2007) allow probabilistic estimates of climate change. These probabilities represent the relative degree to which each climate outcome is supported by the evidence available (Anon. 2011; Walking-the-Talk 2011), and it is possible to build up a picture of the range of possible climate changes and the associated probabilities of occurrence (Winter & Shearer 2013). Rainfall intensity is clearly an important issue for landslides and, particularly, debris flows.

UKCP09 is the most detailed dataset, and it suggests that there is a general propensity for rainfall events of greater intensity (storms) which are intrinsically linked with the triggering of debris flows, and that these rainfall events may become more frequent. Notwithstanding this, it should be noted that the prediction of high-intensity rainfall events is not the purpose of climate models, and such models are not best-suited to the prediction of such phenomena. Changes in rainfall coupled with increased potential evapotranspiration (particularly in the summer) and a longer growing season (leading to increased root water uptake) are expected to substantially affect soil moisture.

The soils that form the slopes subject to debris flow comprise a wide range of materials, albeit that periglacial, glacial and postglacial (including paraglacial) processes dominate their formation and subsequent modification. In terms of the composition of debris flow, the particle sizes range from coarse granular (including boulders) to fine cohesive with most sizes between potentially represented (e.g. McMillan et al. 2005); observations of the characteristics of the materials involved in the August 2004 debris-flow events in Scotland indicate such a diverse composition (Winter et al. 2006). Both Winter et al. (2009) and Milne et al. (2009) acknowledge the importance of water-bearing soils, particularly peat, located on high, relatively flat ground as trigger materials for gulley-constrained debris flows. The materials that are later eroded, entrained and transported in the body of the flow may be of entirely different characteristics; often coarser-grained materials from morainic and related deposits are critical to the development of such flows.

Increased rainfall during the winter months seems likely to increase the prevalence of debris flow in Scotland. This is particularly so when considered in the context of the likelihood of more intense rainfall events.

The reduced soil moisture, as a result of both predicted temperature increases and rainfall decreases, during the summer and autumn may mean that the short-term stability of some slopes, particularly those formed from granular

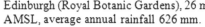
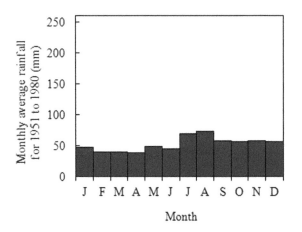
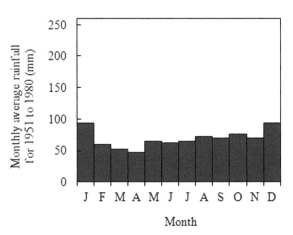
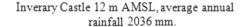
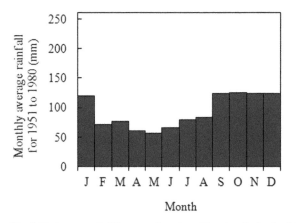
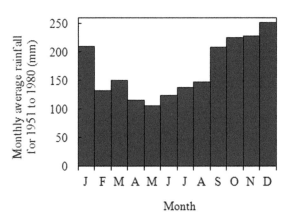

Fig. 5.19. Average rainfall patterns for selected locations in Scotland, based upon 30-year 1951 to 1980 averages (Anon. 1989; from Winter *et al.* 2010*a*).

materials, may be enhanced by the suction pressures that may develop in such conditions. However, although climate models are not best-suited to predicting the frequency of storm events, the available evidence for climate change does point to an increase in short-duration, high-intensity rainfall events, even in summer. Soils under high levels of suction are vulnerable to rapid inundation (Toll 2001) and a consequent reduction in the stabilizing suction pressures, under precisely the conditions that tend to be created by short-duration, localized summer storms.

In addition, non-granular soils may form low-permeability crusts during extended dry periods as a result of desiccation. If these do not experience excessive cracking due to shrinkage, then runoff to areas of vulnerable granular deposits may be increased. However, the formation of drying cracks could lead to the rapid development of instabilities in soil deposits, potentially creating conditions for the formation of debris flows. It is therefore clear that there is a number of potential failure mechanisms that may lead to the initiation of a debris flow.

Vegetation will also be affected by climate change. Lower overall levels and changed patterns of rainfall might be expected to increase the pressure on vegetation and therefore to reduce its beneficial effect upon slope stability. Additionally, extended periods of exceptionally dry weather could potentially lead to wildfires and associated debris flow such as those described by Cannon et al. (2008). This seems less likely to be a factor in the relatively wet climate of Scotland; it seems more likely that an extension to the growing season will predominate in terms of the effects of vegetation, a factor that is most likely to have a positive effect upon instability. However, the possibility of vegetation desiccation cannot be dismissed during prolonged summer dry spells.

The available climate change predictions therefore present a picture that tends to suggest that debris-flow frequency and magnitude may well increase in Scotland, at least in the winter months. A rather more complex picture emerges for the summer months; one possible outcome is that the frequency of events may decrease, but the magnitude of those that do occur may increase.

The debris-flow risks relate primarily, but by no means exclusively, to transport infrastructure. The effects of climate change seem likely to increase the frequency and/or the magnitude of debris-flow hazards; the vulnerability of the elements at risk (i.e. the infrastructure) is also likely to increase in line with road traffic and rail passenger growth. Debris-flow risk is therefore likely to increase as a result of these factors. However, this must be set against planned network upgrades that afford the opportunity to minimize such hazards and risks at the outset. It is, of course, unlikely that this approach of infrastructure upgrade is viable across the network or applicable beyond the strategic road network, for example, and in other cases route-based exposure management schemes and localized hazard mitigation schemes will be necessary. The complex interplay between changing land use and the effects of emerging technology on work patterns will change the way in which society addresses such infrastructure, and may also introduce new infrastructure that will itself be at risk from debris-flow hazards.

Conclusions

Debris-flow events are fast-moving and potentially destructive. Even in situations in which there are no serious injuries or fatalities, the economic and social impacts can be severe. Such impacts include the severance (or delay) of access to remote communities, which can have a negative effect on both businesses in the area and the local population. The extent of these impacts is determined by the vulnerability shadow.

Hazard and risk assessment is necessary in order to prioritize any risk reduction activities that may be necessary. Such activities have been revolutionized in recent years by the wide availability of high-resolution imagery and GIS systems. In order to obtain best value from risk reduction, that goal should be approached strategically with a view to the desired and necessary outcomes and the generic approach to achieving those outcomes. Only then should the processes that may be used to achieve those outcomes (i.e. the specific management and mitigation measures and remedial options) be addressed and the most suitable measures debated and decided upon.

Climate change is an important issue for rainfall-induced debris flow. An analysis of the likely impact of predicted climate change on debris-flow activity in Scotland indicates that climate change is most likely to increase the frequency and/or magnitude of debris-flow hazards, and that the vulnerability of elements at risk (i.e. the infrastructure) is likely to increase in line with road traffic and rail passenger growth. There are many uncertainties associated with such conclusions, including those associated with climate change predictions and those associated with future infrastructure and demographic and social change.

References

ANON. 1989. *The Climate of Scotland – some Facts and Figures*. The Stationery Office, London.

ANON. 2011. *Scottish Road Network Climate Change Study: UKCP09 Update*. Transport Scotland, Edinburgh.

BARNETT, C., PERRY, M., HOSSELL, J., HUGHES, G. & PROCTER, C. 2006. *A Handbook of Climate Trends Across Scotland; Presenting Changes in the Climate Across Scotland over the Last Century*. Sniffer Project CC03. Scotland and Northern Ireland Forum for Environmental Research, Edinburgh.

BROMHEAD, E.N. 1997. The treatment of landslides. *Proceedings, Institution of Civil Engineers (Geotechnical Engineering)*, **125**, 85–96, https://doi.org/10.1680/igeng.1997.29231

BROMHEAD, E.N. & IBSEN, M.L. 2007. Folkestone Warren and the impact of the past rainfall record. *In*: MCINNES, R., JAKEWAYS, J., FAIRBANK, H. & MATHIE, E. (eds) *Landslides and Climate Change: Challenges and Solutions*. Taylor & Francis, London, 17–24.

CANNON, S.H., GARTNER, J.E., WILSON, R.C., BOWERS, J.C. & LABER, J.L. 2008. Storm rainfall conditions for floods and debris flows from recently burned area in southwestern Colorado and southern California. *Geomorphology*, **96**, 250–269, https://doi.org/10.1016/j.geomorph.2007.03.019

CASTELLANOS ABELLA, E.A. & VAN WESTEN, C.J. 2007. Generation of a landslide risk map for Cuba using spatial multi-criteria evaluation. *Landslides*, **4**, 311–325, https://doi.org/10.1007/s10346-007-0087-y

COROMINAS, J., VAN WESTEN, C. ET AL. 2014. Recommendations for the quantitative analysis of landslide risk. *Bulletin of Engineering Geology and the Environment*, **73**, 209–263.

CRUDEN, D.M. & VARNES, D.J. 1996. Landslide types and processes. *In*: TURNER, A.K. & SCHUSTER, R.L. (eds) *Special Report 247: Landslides: Investigation and Mitigation*. Transportation and Road Research Board, National Academy of Science, Washington, DC, 36–75.

DIJKSTRA, T.A., WASOWSKI, J., WINTER, M.G. & MENG, X.M. 2014. Introduction to geohazards of Central China. *Quarterly Journal of Engineering Geology & Hydrogeology*, **47**, 195–199, https://doi.org/10.1144/qjegh2014-054

Dio, S., Forbes, C. & Chiliza, G.S. 2010. Landslide inventorization and susceptibility mapping in South Africa. *Landslides*, **7**, 207–210, https://doi.org/10.1007/s10346-009-0186-z

Galbraith, R.M., Price, D.J. & Shackman, L. (eds) 2005. *Scottish Road Network Climate Change Study*. Scottish Executive, Edinburgh.

Gibson, A.D., Culshaw, M.G., Dashwood, C. & Pennington, C.V.L. 2013. Landslide management in the UK – the problem of managing hazards in a 'low risk' environment. *Landslides*, **10**, 599–610, https://doi.org/10.1007/s10346-012-0346-4

Giroud, R.E. 2005. *Guidelines for the Geologic Evaluation of Debris-Flow Hazards on Alluvial Fans in Utah*. Miscellaneous Publication 05–06. Utah Geological Survey, Salt Lake City, UT.

Hulme, M., Jenkins, G.J. et al. 2002. *Climate Change Scenarios for the United Kingdom: the UKCIP02 Scientific Report*. Tyndall Centre for Climate Change research. University of East Anglia, Norwich.

Hungr, O., Evans, S.G., Bovis, M.J. & Hutchinson, J.N. 2001. A review of the classification of landslides of the flow type. *Environmental & Engineering Geoscience*, **VII**, 221–238, https://doi.org/10.2113/gseegeosci.7.3.221

Hungr, O., Leroueil, S. & Picarelli, L. 2014. The Varnes classification of landslide types, an update. *Landslides*, **11**, 167–194, https://doi.org/10.1007/s10346-013-0436-y

Hutchinson, J.N. 1988. General Report: Morphological and geotechnical parameters of landslides in relation to geology and hydrogeology. *In*: Bonnard, C. (ed.) *Proceedings, Fifth International Symposium on Landslides*. Balkema, Rotterdam, **1**, 3–35.

Innes, J.L. 1983. Lichenometric dating of debris flow deposits in the Scottish Highlands. *Earth Surface Processes and Landforms*, **8**, 579–588, https://doi.org/10.1002/esp.3290080609

Jenkins, G.J., Perry, M.C. & Prior, M.J.O. 2007. *The Climate of the United Kingdom and Recent Trends*. Met Office Hadley Centre, Exeter.

Jones, D.K.C. & Lee, E.M. 1994. *Landsliding in Great Britain*. HMSO, London.

King, J.P. & Williamson, S.J. 2002. Erosion along debris avalanche trails. *Proceedings, Natural Terrain a Constraint to Development*. Institution of Mining & Metallurgy, Hong Kong, 197–205.

Lee, E.M. 2020. Landslides. *In*: Giles, D.P. & Griffiths, J.S. (eds) *Geological Hazards in the UK: Their Occurrence, Monitoring and Mitigation*. Geological Society, London, Engineering Geology Special Publications, **29**, xx–xx.

McInnes, R.G. 2000. *Managing Ground Instability in Urban Areas – a Guide to Best Practice*. Isle of Wight Centre for the Coastal Environment, Isle of Wight.

McMillan, P., Brown, D.J., Forster, A. & Winter, M.G. 2005. Debris flow information sources. *In*: Winter, M.G., Macgregor, F. & Shackman, L. (eds) *Scottish Road Network Landslides Study*. The Scottish Executive, Edinburgh, 25–44.

Milne, F.D., Werritty, A., Davies, M.C.R. & Browne, M.J. 2009. A recent debris flow event and implications for hazard management. *Quarterly Journal of Engineering Geology and Hydrogeology*, **42**, 51–60, https://doi.org/10.1144/1470-9236/07-073

Nettleton, I.M., Martin, S., Hencher, S. & Moore, R. 2005a. Debris flow types and mechanisms. *In*: Winter, M.G., Macgregor, F. & Shackman, L. (eds) *Scottish Road Network Landslides Study*. The Scottish Executive, Edinburgh, 45–67.

Nettleton, I.M., Tonks, D.M., Low, B., MacNaughton, S. & Winter, M.G. 2005b. Debris flows from the perspective of the Scottish Highlands. *In*: Senneset, K., Flaate, K. & Larsen, J.O. (eds) *Landslides and Avalanches: ICFL 2005 Norway*. Taylor & Francis Group, London, 271–277.

Pierson, T.C. & Costa, J.E. 1987. A rheological classification of subaerial sediment-water flows. *In*: Costa, J.E. & Wieczorek, G.F. (eds) *Debris Flow/Avalanches: Process, Recognition and Mitigation*. Geological Society of America, Boulder, CO, *Reviews in Engineering Geology* **VII**, 1–12.

Sassa, K. 1985. Mechanisms of flows in granular soils. *In*: *Proceedings of the 11th International Conference on Soil Mechanics and Foundation Engineering*, **1**, 1173–1176.

Toll, D.G. 2001. Rainfall-induced landslides in Singapore. *Proceedings, Institution of Civil Engineers (Geotechnical Engineering)*, **149**, 211–216, https://doi.org/10.1680/geng.2001.149.4.211

VanDine, D.F. 1996. Debris flow control structures for forest engineering. Ministry of Forests Research Program, Working Paper 22/1996. Victoria, BC: Ministry of Forests.

Versace, P. (ed.) 2007. La mitigazione del rischio da collate di fango: a Sarno e negli altri comuni colpiti dagle eventi del Maggio 1998. Commissariato do Governa per l'Emergenze Idrogeologica in Campania, Napoli (in Italian).

Walking-the-Talk 2011. *Paths and Climate Change – an Investigation into the Potential Impacts of Climate Change on the Planning, Design, Construction and Management of Paths in Scotland*. Scottish Natural Heritage, Commissioned Report No. **436**. Scottish Natural Heritage, Inverness.

Warburton, J. 2020. Compressible soils and peat. *In*: Giles, D.P. & Griffiths, J.S. (eds) *Geological Hazards in the UK: Their Occurrence, Monitoring and Mitigation*. Geological Society, London, Engineering Geology Special Publications, **29**, xx–xx.

Winter, M.G. 2014. A strategic approach to landslide risk reduction. *International Journal of Landslide and Environment*, **2**, 14–23.

Winter, M.G. 2018. *The Quantitative Assessment of Debris Flow Risk to Road Users on the Scottish Trunk Road Network: A85 Glen Ogle*. Published Project Report PPR **799**. Transport Research Laboratory, Wokingham.

Winter, M.G. & Bromhead, E.N. 2012. Landslide risk – some issues that determine societal acceptance. *Natural Hazards*, **62**, 169–187, https://doi.org/10.1007/s11069-011-9987-1

Winter, M.G. & Corby, A. 2012. *A83 Rest and be Thankful: Ecological and Related Landslide Mitigation Options*. Published Project Report PPR **636**. Transport Research Laboratory, Wokingham.

Winter, M.G. & Shearer, B. 2013. *Climate Change and Landslide Hazard and Risk: A Scottish Perspective*. Published Project Report PPR **650**. Transport Research Laboratory, Wokingham.

Winter, M.G., Macgregor, F. & Shackman, L. (eds) 2005a. *Scottish Road Network Landslides Study*. The Scottish Executive, Edinburgh.

Winter, M.G., Shackman, L., Macgregor, F. & Nettleton, I.M. 2005b. Background to Scottish landslides and debris flows. *In*: Winter, M.G., Macgregor, F. & Shackman, L. (eds) *Scottish Road Network Landslides Study*. The Scottish Executive, Edinburgh, 12–24.

Winter, M.G., Heald, A., Parsons, J.P., Macgregor, F. & Shackman, L. 2006. Scottish debris flow events of August 2004. *Quarterly Journal of Engineering Geology & Hydrogeology*, **39**, 73–78, https://doi.org/10.1144/1470-9236/05-049

Winter, M.G., McInnes, R.G. & Bromhead, E.N. 2008. Landslide risk management in the United Kingdom. *In*: Ho, K. & Li, V. (eds) *Proceedings of the 2007 International Forum on Landslide Disaster Management*. The Hong Kong Institution of Engineers, Hong Kong, **I**, 343–374.

Winter, M.G., Macgregor, F. & Shackman, L. (eds). 2009. *Scottish Road Network Landslides Study: Implementation*. Transport Scotland, Edinburgh.

WINTER, M.G., DENT, J., MACGREGOR, F., DEMPSEY, P., MOTION, A. & SHACKMAN, L. 2010a. Debris flow, rainfall and climate change in Scotland. *Quarterly Journal of Engineering Geology & Hydrogeology*, **43**, 429–446, https://doi.org/10.1144/1470-9236/08-108

WINTER, M.G., DIXON, N., WASOWSKI, J. & DIJKSTRA, T.A. 2010b. Introduction to land-use and climate change impacts on landslides. *Quarterly Journal of Engineering Geology & Hydrogeology*, **43**, 3–6, https://doi.org/10.1144/1470-9236/08-200

WINTER, M.G., HARRISON, M., MACGREGOR, F. & SHACKMAN, L. 2013a. Landslide hazard assessment and ranking on the Scottish road network. *Proceedings, Institution of Civil Engineers (Geotechnical Engineering)*, **166**, 522–539, https://doi.org/10.1680/geng.12.00063

WINTER, M.G., KINNEAR, N., SHEARER, B., LLOYD, L. & HELMAN, S. 2013b. *A Technical and Perceptual Evaluation of Wig-Wag Signs at the A83 Rest and be Thankful*. Published Project Report PPR **664**. Transport Research Laboratory, Wokingham.

WINTER, M.G., SMITH, J.T., FOTOPOULOU, S., PITILAKIS, K., MAVROULI, O., COROMINAS, J. & ARGYROUDIS, S. 2014a. An expert judgement approach to determining the physical vulnerability of roads to debris flow. *Bulletin of Engineering Geology and the Environment*, **73**, 291–305, https://doi.org/10.1007/s10064-014-0570-3

WINTER, M.G., PALMER, D., SHARPE, J., SHEARER, B., HARMER, C., PEELING, D. & BRADBURY, T. 2014b. Economic impact assessment of landslide events. *In*: SASSA, K., CANUTI, P. & YIN, Y. (eds) *Landslide Science for a Safer Geoenvironment, Volume 1: The International Programme on Landslides (IPL)*. Springer, New York, 217–222.

WONG, J.F.C. & WINTER, M.G. 2018. *The Quantitative Assessment of Debris Flow Risk to Road Users on the Scottish Trunk Road Network: A83 Rest and be Thankful*. Published Project Report PPR **798**. Transport Research Laboratory, Wokingham.

Chapter 6   Collapsible Soils in the UK

M. G. Culshaw[1,2*], K. J. Northmore[1], I. Jefferson[2],
A. Assadi-Langroudi[2,3] & F. G. Bell[1,4]

[1]British Geological Survey, Keyworth, Nottingham
[2]Department of Civil Engineering, University of Birmingham, UK
[3]School of Architecture, Computing and Engineering, University of East London
[4]University of Kwazulu-Natal, South Africa

*Correspondence: martin.culshaw2@ntlworld.com

Abstract: Metastable soils may collapse because of the nature of their fabric. Generally speaking, these soils have porous textures, high void ratios and low densities. They have high apparent strengths at their natural moisture content, but large reductions of void ratio take place upon wetting and, particularly, when they are loaded because bonds between grains break down upon saturation. Worldwide, there is a range of natural soils that are metastable and can collapse, including loess, residual soils derived from the weathering of acid igneous rocks and from volcanic ashes and lavas, rapidly deposited and then desiccated debris flow materials such as some alluvial fans; for example, in semi-arid basins, colluvium from some semi-arid areas and cemented, high salt content soils such as some sabkhas. In addition, some artificial non-engineered fills can also collapse. In the UK, the main type of collapsible soil is loess, though collapsible non-engineered fills also exist. Loess in the UK can be identified from geological maps, but care is needed because it is usually mapped as 'brickearth'. This is an inappropriate term and it is suggested here that it should be replaced, where the soils consist of loess, by the term 'loessic brickearth'. Loessic brickearth in the UK is found mainly in the south east, south and south west of England, where thicknesses greater than 1 m are found. Elsewhere, thicknesses are usually less than 1 m and, consequently, of limited engineering significance. There are four steps in dealing with the potential risks to engineering posed by collapsible soils: (1) identification of the presence of a potentially collapsible soil using geological and geomorphological information; (2) classification of the degree of collapsibility, including the use of indirect correlations; (3) quantification of the degree of collapsibility using laboratory and/or *in situ* testing; (4) improvement of the collapsible soil using a number of engineering options.

6.1   What are collapsible soils?

Soils that have the potential to collapse generally possess porous textures with high void ratios and relatively low densities. At their natural moisture content these soils possess high apparent strength, but they are susceptible to large reductions in void ratio on wetting, especially under load. In other words, the metastable texture collapses as the bonds between the grains break down as the soil becomes saturated. Jefferson & Rogers (2012) pointed out that as collapse was controlled both microscopically and macroscopically, both these elements need to be understood if the true nature of collapse was to be determined. The potential for soils to collapse is clearly of geotechnical significance, particularly with respect to the potential distress of foundations and services (for example, pipelines) if not recognized and designed for.

The collapse process represents a rearrangement of soil particles into a denser state of packing. Collapse on saturation usually occurs rapidly. As such, the soil passes from an under-consolidated condition to one of normal consolidation.

This definition is similar to the first part of that of Rogers (1995) who gave two basic requirements for a soil to be collapsible:

'A collapsible soil is one in which the constituent parts have an open packing and which forms a metastable state that can collapse to form a closer packed, more stable structure of significantly reduced volume.'

'In most collapsible soils the structural units will be primary, mineral particles rather than clay minerals.'

The latter part of Rogers (1995) second basic requirement of collapsibility is, perhaps, a little confusing (what is meant by 'primary, mineral particles'?) Jefferson & Rogers (2012) made it clearer by defining collapsible soils as: '…soils in which the major structural units are initially arranged in a metastable packing through a suite of different bonding mechanisms.' This can include both individual primary minerals (non-clay) and 'peds' comprising of individual primary minerals with clay mineral coatings and/or clay 'bridges' to other particles.

Engineering Group Working Party (main contact for this chapter: M. G. Culshaw, British Geological Survey, Keyworth, Nottingham and Department of Civil Engineering, martin.culshaw2@ntlworld.com)
From: GILES, D. P. & GRIFFITHS, J. S. (eds) 2020. *Geological Hazards in the UK: Their Occurrence, Monitoring and Mitigation – Engineering Group Working Party Report.* Geological Society, London, Engineering Geology Special Publications, **29**, 187–203,
https://doi.org/10.1144/EGSP29.6
© 2020 [BGS © UKRI All rights reserved]. Published by The Geological Society of London. All rights reserved.
For permissions: http://www.geolsoc.org.uk/permissions. Publishing disclaimer: www.geolsoc.org.uk/pub_ethics

The most widespread naturally collapsible soils are loess or loessic soils of aeolian origin, predominantly of silt size with uniform sorting. The majority of these soils have glacial associations in that it is believed that these silty soils were derived from continental areas where silty source material was produced by glacial action prior to aeolian transportation and deposition.

Vast spreads of loess have accumulated over large areas of North America, Europe, Russia and China over the last two to three million years; for example, the Wucheng loess of China occurs over most of the Lower Pleistocene (c. 2.4–1.15 million years BP) (Liu et al. 1985). However, it is not found exclusively in these regions. For example, it also occurs in Thailand and New Zealand. Jefferson et al. (2001) estimated that loess covered about 10% of the Earth's landmass while Dibben (1998) estimated the figure at 15%.

The term 'loesch' (also 'lŏsz' in Jari & Badura (2013), 'schwemmlöss' or 'löss' in Pye [1995]) was used by countrymen from the Upper Rhine region (SW Germany) to describe the friable silt deposits along the Rhine Valley near Heidelberg. Although the term was literally introduced by Karl Caesar von Leonhard in 1820, it was only brought into the scientific literature in 1834 by Charles Lyell (Smalley et al. 2001). In Britain, loess has been known as 'brickearth' (D'Archiac 1839; Prestwich 1863; Fink 1974). Brickearth was used in Roman buildings (for example, in the Roman London Amphitheatre and timber buildings discovered in the Walbrook Valley [Arkell & Tomkeieff 1953; Lee et al. 1989]) and was described as a homogeneous, structureless loam (Deakin 1986).

Pye (1995) proposed that there were four fundamental requirements necessary for the formation of loess; minor additions have been made to these:

- a dust source;
- adequate wind energy to transport the dust;
- a suitable depositional area (or reduced wind speed [Pye 1987]);
- a sufficient amount of time for its accumulation and epigenetic evolution (Trofimov 1990).

These requirements are not specific to any one climatic or vegetational environment. Whilst much loess was formed in glacial/periglacial environments, derived from the floodplains of glacial braided rivers where glacially ground silts and clays were deposited, windblown deposits can be derived from other environments described by Iriondo & Krohling (2007) as volcanic, tropical, desert and gypsum loesses and climatically controlled ones referred to as trade-wind and anticyclonic.

However, soils other than loess have the potential to collapse. Haskins et al. (1998) referred to collapsible soils from Mpumalanga Province, South Africa, that had been derived from weathered granite. Assadi (2014) observed the formation of a collapsing structure in superficial kaolinite deposits (sub-200 kPa pre-consolidation stress) with a moisture content of between 1–5%, and also in accumulated wet kaolinite (40–45% moisture content) of similar stress history as it dried. Northmore et al. (1992a) summarized the geological and geotechnical characteristics of tropical red clay soils derived from volcanic ash deposits in Kenya, Indonesia and Dominica, including their potential to collapse. In further reports, more detailed information on the geology of the sampled areas (Northmore et al. 1992b), mineralogy (Kemp 1991), geotechnical index properties (Northmore et al. 1992c), strength and consolidation properties (Hobbs et al. 1992), pit sampling (Culshaw et al. 1992) and borehole sampling (Hallam & Northmore 1993) were presented and discussed. Other potentially collapsible soils include rapidly deposited and then desiccated debris flow materials (such as some alluvial fans, for example, in semi-arid basins of inland California [Waltham 2009]), colluvium in Colorado and other semi-arid areas of the western USA [for example, White & Greenman 2008]), cemented, high salt content soils such as some sabkhas (for example, El-Ruwaih & Touma 1986), non-engineered fills (compacted dry of optimum) and waste materials such as fly ash (Jefferson & Rogers 2012). Many of these soils do not occur in the UK and are not considered further here except for non-engineered fills.

6.2 Loess in the UK

In Great Britain, loessic deposits are mapped by the British Geological Survey (BGS) mainly as 'brickearth'. Such deposits occur mainly as a discontinuous spread across southern and eastern England, notably in Essex, Kent, Sussex and Hampshire (Fig. 6.1a, b). The extent of brickearth shown on this map is derived from BGS geological mapping originally at a scale of 1:10 000 or 1:10 560 and includes only those deposits that are at least 1 m thick. This distribution is comparable with the extent of loess greater than 1 m thick shown by Ballantyne & Harris (1994) (their fig. 8.21) after Catt (1977, 1985) and was derived from pedological and geological maps, as well as other sources. This latter map also shows 'Cover sands' and 'Loess 0.3–1.0 m thick.' Figure 6.2a and b shows the distribution of loessic deposits greater than 1 m thick and between 300 mm and 1 m thick, derived from Soil Survey 1:250 000 soil maps. Assadi (2014) has reviewed the geographical distribution of loess soils in the UK. Additionally, material similar to brickearth occurs in widened fissures (gulls) in deposits such as the Hythe Beds in Kent (Bell et al. 2003) and in a sinkhole in the chalk NW of London (Gibbard, P., pers. comm.). Whether such fill is of windblown origin or whether it is derived locally from pre-existing silty clay deposits that have undergone solifluction is still a matter of debate. Patchy loessic deposits have also been described in SW England around the Lizard Peninsula in Cornwall (Ealey & James 2008, 2011) and around Torbay in Devon (Cattell 1997).

The use of the term 'brickearth' for loessic deposits on geological maps may seem strange. However, the reason is quite simple and obvious – the loesses usually had significant clay contents and were very suitable for the manufacture of bricks.

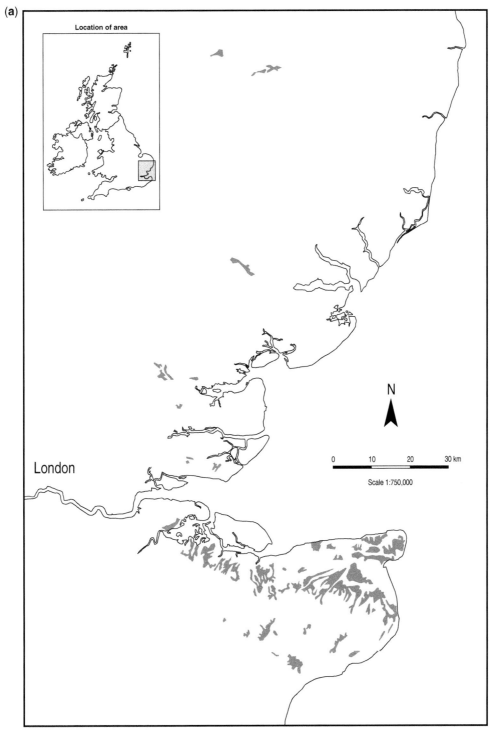

Fig. 6.1. (**a** and **b**) Distribution of brickearth deposits in south eastern and southern England based on British Geological Survey 1:50 000 scale geological maps; (**a**) SE England.

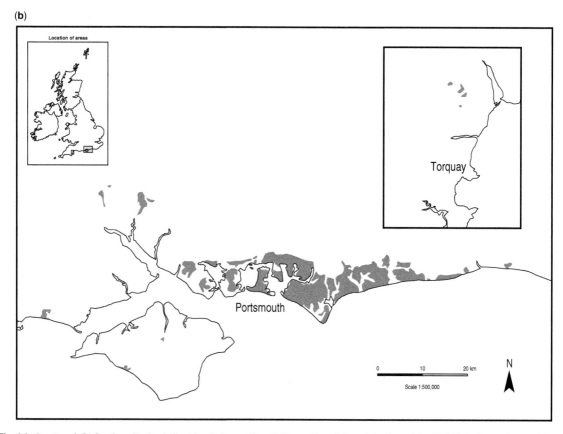

Fig. 6.1. *Continued.* (**b**) Southern England. Contains Ordnance Data © Crown Copyright and database rights [2019]. Ordnance Survey Licence no. 100021290.

However, care is needed when using the term 'brickearth' to identify loesses, as some so-called brickearths are not loess. For example, William Smith used the term on his 1815 geological map of Britain (Smith 1815) – he was referring to what we now know as Lambeth Group deposits and also East Anglian Crag deposits. Similarly, the Norwich Brickearth of East Anglia, UK, is an Anglian sandy glacial till (Rose *et al.* 1999) (the Happisburgh Till Member). The terminology used on geological map legends is confusing as the BGS does not use the term 'loess' and the term 'aeolian' is used only on map sheets 316 (Fareham), 317/332 (Chichester and Bognor) and 331 (Portsmouth). As well as tills, 'brickearth' may also refer to glaciolacustrine deposits and colluvial deposits. Based on the BGS Lexicon of Rock Names, Bell & Culshaw (2001) described the main brickearth units (the first five terms in Table 6.1). All of them should be assumed to be loessic as first deposited, though they may have been reworked. More recently, other terms have been used (see the remainder of Table 6.1). Initially, all of them should be assumed to be loessic as first deposited, though they may have been reworked, and so potentially collapsible. The term 'brickearth' has become so entrenched in the geological literature that it will probably never be superseded entirely, despite new nomenclature being suggested by the BGS(McMillan & Powell 1999). As such, the more appropriate general term of 'loessic brickearth' is recommended for those deposits in the UK that are clearly of aeolian/loessic origin.

According to Kerney (1965), almost all the loessic brickearth soils in the UK are probably of late Devensian age (*c.* 14–30 ka BP). Parks & Rendell (1992) carried out thermoluminescence dating of loessic brickearth deposits from 26 locations in SE England and reported that they seemed to have resulted from at least three depositional phases during the Late Pleistocene (10–25 ka BP, 50–125 ka BP and >170 ka BP). However, most of the samples were of Late Devensian age but with isolated pockets of older material found throughout the area. Using the optically stimulated luminescence (OSL) method, Clarke *et al.* (2007) dated the lower 'calcified' loessic brickearth at Pegwell Bay, Kent, at 17.2 ± 1.3 ka BP and at Ospringe, Kent, at 23.8 ± 1.3 ka BP. The upper 'non-calcareous' loessic brickearth at Pegwell Bay had an age of 15.0 ± 0.9 ka BP. At Ospringe, the age of the upper, non-calcareous loessic brickearths

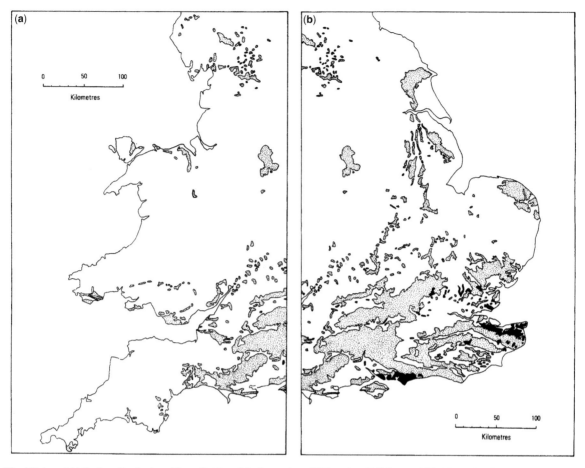

Fig. 6.2. (**a and b**) Surface distribution of loess/brickearth in the southern UK based on Soil Survey 1:250 000 scale soil maps (1983). Loess >1 m thick in black; loess >300 mm thick (and often partly mixed with subjacent deposits) shown stippled (after Catt 1985); (**a**) Wales and western England; (**b**) Eastern England. Whilst every effort has been made to trace the owner of this copyright material it has not proved possible.

was 18.7 ± 2.3 ka BP but Clarke *et al.* (2007) believed this to be an overestimate of the true age because the deposit probably contains grains of mixed ages. The loess of the Lizard Peninsula was dated by Ealey & James (2011) at *c.* 15.9 ka BP. These dates imply that calcareous loessic brickearth was deposited more than once both at around the time of the Late Glacial Maximum and during deglaciation. The deposition of the upper non-calcified material may correspond to a period of rapid climatic warming.

6.3 How to recognize loessic brickearth

6.3.1 Description and mineralogy

Given that identifying loess deposits from geological maps may be problematical, it is essential to obtain and geotechnically test samples suspected of being loessic to determine their index properties and their susceptibility to collapse.

Generally in the UK, loessic brickearth comprises of a discontinuous blanket deposit of yellowish brown, friable, slightly plastic, poorly bedded, clayey and sandy silt with well-developed vertical jointing. It has a very open, low density structure. With regard to loessic brickearths in Kent, UK, Milodowski *et al.* (2015) described their fabric as being characterized by 'an open-packed arrangement of clay-coated, silt-sized quartz particles and pelletized aggregate grains (peds) of compacted silt and clay, supported by an inter-ped matrix of loosely packed, silt/fine-grained sand, in which the grains are held in place by a skeletal framework of illuviated clay.' Similarly, Figure 6.3 shows three silt-size particles with two of them bridged by platy clay particles from South Essex, UK. It now seems that loessic brickearth in southeastern

Table 6.1. *Terminology used on British Geological Survey 1:50 000 scale geological maps that may be indicative of the presence of loess*

Stratigraphic name	Description
Brickearth	Varies from silt to clay, usually yellow-brown and massive
River Brickearth	Varies from silt to clay, usually yellow-brown and massive; of fluvial origin
Head Brickearth	Varies from silt to clay, usually yellow-brown and massive. Poorly sorted and poorly stratified, formed mostly by solifluction and/or hillwash and soil creep.
Head Brickearth, Older	Varies from silt to clay, usually yellow-brown and massive. Poorly sorted and poorly stratified, formed mostly by solifluction and/or hillwash and soil creep. Older than 'Head Brickearth' in the same map area.
Head Brickearth, Younger	Varies from silt to clay, usually yellow-brown and massive. Poorly sorted and poorly stratified, formed mostly by solifluction and/or hillwash and soil creep. Younger than 'Head Brickearth' in the same map area.
Aeolian deposits ('Brickearth')	Mainly fine sandy silt or silt, locally contaminated with gravel.
Langley Silt – 'Brickearth'	Sandy clay and silt.
Enfield Silt – 'Brickearth'	Sandy clay and silt.
Roding Silt – 'Brickearth'	Sandy clay and silt.
Ilford Silt – 'Brickearth'	Sandy clay and silt.
Crayford Silt – 'Brickearth'	Sandy clay and silt.
Dartford Silt – 'Brickearth'	Sandy clay and silt.
Silt	–
Glacial Silt	–
Glacial Silt and Clay	–

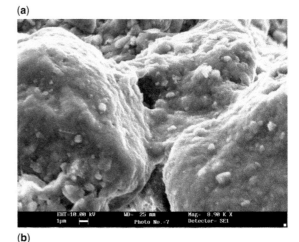

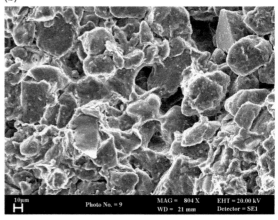

Fig. 6.3. (**a**) Photomicrograph showing silt-size particles with two of them bridged by platy clay particles; sample from Pegwell Bay, Kent. (**b**) Photomicrograph showing large open voids and silt-size particles bridged by platy clay particles; sample from Ospringe, Kent.

England may not consist of an upper part that has been leached of carbonate material and a lower calcareous part. Rather, Clarke *et al.* (2007) and Milodowski *et al.* (2015) argued that in Kent and SE Essex there have been at least two phases of loessic brickearth deposition in the Late Devensian, the older, deeper deposits being calcareous, with the presence of an old soil horizon at the top of it. However, Catt (2008) challenged this interpretation and it seems that systematic dating of various loessic brickearth deposits is needed.

The lower parts of a deposit tend to be more rigid and better consolidated than the upper parts. Quartz is the most abundant mineral in loessic brickearth ranging between 12% and 54%, with an average of 33%, followed by feldspar. Of the clay-type minerals, mica is generally more abundant than montmorillonite that, in turn, is more abundant than illite and kaolinite. For example, Northmore *et al.* (1996) found that montmorillonite averaged some 15% in the loessic brickearth of South Essex with a maximum of 39% and kaolinite constituted generally less than 5%. Calcium carbonate occurs as grains, thin tube infillings, buttress/bridge units and as concretionary nodules. When present it tends to account for less than 10% of the soil. However, in Kent, values quoted by Derbyshire & Mellors (1988) were as high as 20%.

The principal mineral in some gull-fills found in Kent is quartz, it comprising as much as 85% (Bell *et al.* 2003). The quartz grains are sub-angular and sub-rounded to rounded. The degree of roundness tends to increase with increasing size of grains. The remainder of the fill usually consists of sub-angular flint grains, hornblende, some glauconite, and traces of heavy minerals and micaceous material. The carbonate content varies between 0.03 and 0.46%, which is very low compared with loessic brickearths from South Essex and Kent. The small amount of calcium carbonate may be due to the material having been leached.

6.3.2 Geotechnical properties

6.3.2.1 Particle size distribution
Loess deposits, generally, consist of 50–90% silt size particles. The particle size distributions of loessic brickearth from South Essex and Kent are shown in Figure 6.4a, from which it can be seen that they are similar. Clayey, silty and sandy brickearth/loess can be recognized (using Holtz & Gibbs' [1952] chart). The clay content in the loessic brickearth from South Essex ranges from 4–42%, with an average of 21%. This compares with an average silt content of 59% (range 26–84%), silt being the most important size fraction in this deposit (Northmore et al. 1996). Loess with 5–40% clay content may exhibit collapsibility (Dudley 1970; Lawton et al. 1992) but collapsibility potential rises significantly in loess containing 11–24% clay constituents. On Holtz & Gibbs' chart, the South Essex loessic brickearths fall within all three classes (clayey, silty and sandy loess). The particle size distribution of gull-fill material from Allington, Kent, is illustrated in Figure 6.4b. This indicates that it falls within the clayey loess and silty loess zones of Holtz & Gibbs (Bell et al. 2003). This means that it has affinities with loessic brickearth from Pegwell Bay, Kent, as reviewed by Fookes & Best (1969), as well as with those from other locations in Kent and in South Essex. It is well sorted, with uniformity coefficients ranging from 4 to 11.

6.3.2.2 Density
Most bulk densities of loessic brickearth fall within the range 1.7 and 2.1 Mg m^{-3} (Table 6.2), with most dry densities varying from 1.37 to 1.74 Mg m^{-3}. The relatively low densities are similarly reflected in high void ratios (0.56–0.90), indicative of an open microstructure. Grain densities are quite variable, from 2.61 to 2.72. Assallay (1998) noted dry densities and void ratios of 1.15 Mg m^{-3} and 1.61, respectively, for non-British loesses.

6.3.2.3 Plasticity
Loess soils are slightly to moderately plastic, the plasticity increasing as the clay content increases. Loessic brickearth of South Essex has a natural moisture content that usually varies from 13–21%. The gull-fill material from Allington, Kent, is similar but with a slightly higher range; that is, 16–24%. These soils are of variable plasticity, ranging from low to occasionally high but most are of low plasticity. The spread of liquid limits ranged from 27–64% and that of plastic limits was from 17–24% (Table 6.2). These are broadly similar to the values for Kent loessic brickearths quoted by Derbyshire & Mellors (1988). According to their liquid limits (range 31–34%), the gull-fill has a low plasticity. Its plastic limits also have a limited range from 18–23%. Most loessic brickearths have negative liquidity indices indicating that they are in a fairly brittle condition.

Whilst being a function of mineral components of a soil, plasticity changes with the inflow of cationic solutions into the soil. For British loessic brickearth, Boardman et al. (2001) showed an increase in plasticity index when loess is wetted with dilute solutions of $FeCl_3$ or $NaCl$. The increase in plasticity was relatively more marked in non-calcareous sequences when wetted with Na^+-enriched water, and more pronounced in calcareous sequences when wetted with Fe^{+3}-enriched water. The critical water content at which loess collapses is inversely proportional to the plasticity index.

6.3.2.4 Strength, consolidation and permeability of brickearth/loess
The strength of loess is dependent on the initial porosity and moisture content, the degree of deterioration of the bonds and the increase in granular contacts under consolidation, as well as changes in moisture content. When loess with many macro-pores and high water content is loaded, the cementing bonds are first broken resulting in a lowering of the apparent cohesion and, eventually, softening of the soil. With further loading, the grains are brought more and more into contact, thereby increasing friction, so causing a hardening effect. As far as the angle of shearing resistance of loess is concerned, this usually varies between 30° and 34°. As the liquidity index of loess increases, the shear strength decreases, becoming essentially zero at around a liquidity index of one.

The results obtained from undrained triaxial tests on loess from South Essex by Northmore et al. (1996) showed that the shear strength was between 10 and 220 kPa. Such a range indicates the variability in undrained shear strength. However, Northmore et al. (1996) noted a general tendency for shear strength to decrease with increasing depth. They suggested that this might be partly due to the variable composition of the deposit that tends to vary from a stiff sandy silty clay (relatively dry) near the surface, to a clayey silty loam with increasing depth. The higher values of shear strength may reflect the desiccated 'crust-like' nature of the soil near the ground surface. Consolidated drained triaxial tests also indicate variable effective shear strengths. Peak values of internal angle of shearing resistance may be between 19° and 34°, and those of effective cohesion from between 10 and 70 kPa. Residual values drop to between 16° and 25°, and zero, respectively. Recent detailed investigation of the mineralogy and structure of loessic brickearths in Kent (Milodowski et al. 2015) also suggest that the variable formation of secondary calcium carbonate cementation will also influence shear strengths in the lower 'calcareous' loessic brickearths.

Loess can support heavy structures with small settlements if loads do not exceed the apparent preconsolidation stress and natural moisture content is low in comparison with the plasticity index. On the other hand, loess can compress substantially if the apparent preconsolidation stress is exceeded. Primary and secondary compression are similar to that of saturated clay. Primary settlements generally occur rapidly, with

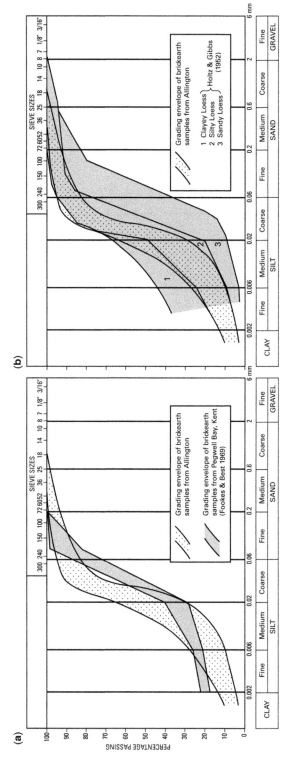

Fig. 6.4. (a) Particle size distribution plots of brickearth/loess from Allington, and Pegwell Bay, Kent. (From Bell *et al.* 2003). (b) Particle size distribution plot for gull-fill material from Allington, Kent. This indicates that it falls mainly within the silty loess zone of Holtz & Gibbs (1952) (From Bell *et al.* 2003).

Table 6.2. Some geotechnical properties of loessic brickearth soils

Property	South Essex[1]	Allington, Kent[2]	Allington, Kent (gull-fill)[3]	Ospringe, E of Faversham Kent[4,5]	Pegwell Bay, Kent[6]	Pegwell Bay, Kent[7]	Ford, NE of Canterbury, Kent[7]	Pine Farm Quarry, E of Maidstone, Kent[7]	Reculver, E of Herne Bay, Kent[7]	Northfleet, Kent[7]	Sturry, NE of Canterbury, Kent[7]
Natural moisture content (%)	13–21	–	16–24	15–20 (9)	2–10 (9)	–	–	–	–	–	–
Particle density (Mg m^{-3})	2.61–2.77	–	2.61–2.62	2.60–2.71 (8)	–	2.69	2.70	2.70	2.68	2.70	2.69
Bulk density (Mg m^{-3})	1.78–2.25	–	1.71–2.04	–	1.55–1.78 (9)	–	–	–	–	–	–
Dry density (Mg m^{-3})	1.43–1.99	–	1.38–1.70	–	1.52–1.65 (9)	1.64–1.73	1.49	1.48	1.62	1.61	1.69
Void ratio	0.57–0.82	–	0.54–0.90	–	0.63–0.77 (9)	0.55–0.64	0.81	0.82	0.65	0.68	0.59
Porosity (%)	36–45	–	35–48	–	–	36–39	45	45	39	41	37
Grain-size distribution (%)											
Sand	4–54	12–31 (3)	5–17	5–20 (9)	12–27 (9)						
Silt	26–84	67–86 (3)	78–86	43–70 (9)	51–70 (9)	>65	>65	>65	>65	>65	<65
Clay	4–42	<3 (3)	5–14	19–39 (9)	14–22 (9)						
Plastic limit (%)	17–24	23–25 (2)	18–23	20–24 (9)	18–21 (11)	17–21	17–20	21–22	19–21	19–20	21–23
Liquid limit (%)	27–64	28–29 (2)	31–34	33–39 (9)	26–32 (11)	28–33	31–45	30–32	32–33	31–33	41–46
Plasticity index (%)	7–40	4.5 (2)	9–16	9–17 (9)	8–12 (11)	11–14	11–28	9–11	12–13	12–13	20–25
Activity	–	–	–	–	0.48–0.59 (2)	–	–	–	–	–	–
Coefficient of collapsibility	-0.009–0.038	–	-0.0003–0.029	0.23–0.68 (9)	–	–	–	–	–	–	–
Angle of friction	11–36	–	–	–	–	–	–	–	–	–	–
Calcium carbonate content (%)	0–16.5 (12)	7.9–8.3	<0.5	–	0–19 (9)	16.2	12.7	14.0	6.0	9.4	–

[1] Northmore *et al.* (1996).
[2] Lill (1976).
[3] Bell *et al.* (2003).
[4] Northmore *et al.* (2008).
[5] Non-calcareous brickearth only.
[6] Fookes & Best (1969).
[7] Derbyshire & Mellors (1988).
The figure in brackets indicates the number of samples tested.

much of the settlement occurring during the actual application of load. Loess also may exhibit creep deformation under loading.

The compression index indicates that the degree of compressibility of loess is high and, generally, the value of coefficient of volume compressibility decreases with increasing load. Tests on loess from South Essex carried out by Northmore *et al.* (1996) exhibited rapid consolidation that is reflected in the high coefficients of consolidation. In many instances, all the primary consolidation may take place within a half minute of loading that is essentially showing collapse of the soil fabric. The primary consolidation, that is, the wetted collapse, for the upper calcareous loess at Pegwell Bay, Kent, Sittingborne, Kent, and Star Lane, South Essex, was as high as 6.3%, 16.9% and 9.4%, respectively (Dibben 1998).

Loess has a much higher vertical than horizontal permeability, which is enhanced by long vertical 'voids' in the loess structure that are formed by fossil root-holes and vertical fissures. Because of this, deposits of loess are better drained (their permeability ranges from 10^{-5} to 10^{-8} m s^{-1}) than are true silts.

6.4 Non-engineered fills

Non-engineered fills are a commonly occurring collapsible deposit found in many parts of the UK as well as across the globe. In the UK, collapse settlements can be particularly problematic for cohesive fills that have been placed dry of optimum or, more generally, on non-engineered fills that have been poorly compacted. The subsequent placement of overburden on these fills, often with an increase in water content, can trigger significant collapse. For example, clayey silt loams, placed dry-of-optimum, were found to exhibit collapse settlement of up to 600 mm (Assadi & Jefferson 2015).

The magnitude of the actual collapse that occurs is a function of the soil properties, the degree of compaction and the thickness of the layer that becomes saturated post-construction settlement (Nowak & Gilbert 2015) and this can be significant for settlement sensitive structures. Case histories recorded by Blanchfield & Anderson (2000); Charles & Skinner (2001) and Charles (2005) indicate that collapse settlements in excess of 1 m can occur in deep, backfilled quarries. In these deposits collapse typically occurs after infilling and is often associated with the re-establishment of true groundwater conditions post-construction or, more localized conditions through leaking drains or poorly located soakaways. Collapse potential can remain for some considerable time post-construction to, for example, the Clifford embankment in York. Here, collapse was trigger by a flooding event hundreds of years after the construction. Further details are provided by Charles & Watts (2001) and Skinner (2012). It would be true to say that almost any soil can exhibit a collapse potential under the right stress environment and, so, collapse potential should be viewed in the light of an open metastable fabric, which has the potential for increased packing of particles (and subsequent volume change) when placed under load and/or through a significant increase in water content through inundation. The next section discusses various aspects that need to be taken account of when identifying potentially collapsible soils.

6.5 Identifying collapsibility

6.5.1 Collapse potential

Soils such as loess have the potential to collapse when wetted or wetted under loading. This process is frequently referred to as 'hydro-consolidation' (for example, Rogers *et al.* 1994). Jefferson & Rogers (2012) summarized the main geotechnical and micro-fabric characteristics required of the most collapsible soils:

- an open, metastable structure;
- a high void ratio and low dry density;
- a high porosity;
- a geologically young or recently altered deposit;
- a soil with inherent low inter-particle bond strength.

In addition to these, favourable conditions for porous granular soils to collapse include:

- a well sorted soil, where grains connect together via two points (Bolton 1999);
- a sub-angular shape and rough texture for silts;
- a low degree of saturation (structure-based) and hence a high apparent cohesion;
- a prolonged application of load smaller than fragmentation load.

In an environmental sense, collapse potential is also related not only to the origin of the material, to its mode of transportation and to the depositional environment, but also to weathering. For example, Gao (1988) pointed out that the weakly weathered loess of the north west of the loess plateau in China has a high potential for collapse, whereas the weathered material of the SE of the plateau is relatively stable and the features associated with collapsible loess are gradually disappearing. In Poland, Grabowska-Olszewska (1988) observed that collapse is most frequent in the youngest loess and that it is almost exclusively restricted to loess that contains slightly more than 10% particles of clay size (less than 0.002 mm). These soils are more or less unweathered and possess a pronounced vertical pattern of fissuring. Assadi & Jefferson (2013) maintained that maximum collapsibility is likely to occur in loess with 10–15% clay and 20% carbonate inclusions. However, this high degree of collapsibility was found to significantly decrease in the presence of amorphous silica precipitates, an indication of quartz weathering. Assadi (2014) reported further observations on the decreased collapsibility after slow precipitation of amorphous silica in clayey silts of <25% clay content (particularly at clay contents below 15% in the presence of metal-based sulphates) and fast precipitation of amorphous silica in silty clays of

>45% clay content. So, it appears that those older deposits, in which the fabric and mineral composition have been altered by weathering, are not nearly as susceptible to collapse as young, unweathered soils.

Popescu (1986) stated that there was a limiting value of pressure, defined as the collapse pressure, beyond which deformation of soil increases appreciably. The collapse pressure varies with the degree of saturation. He defined truly collapsible soils as those in which the collapse pressure is less than the overburden pressure. In other words, such soils collapse when saturated since the soil fabric cannot support the weight of the overburden. When the saturation collapse pressure exceeds the overburden pressure, soils are capable of supporting a certain level of stress on saturation. Popescu defined these soils as conditionally collapsible soils. The maximum load that such soils can support is the difference between the saturation collapse and overburden pressure. Based on observations on loess from Thailand, Phien-wej et al. (1992) concluded that the critical pressure at which collapse of the soil fabric begins was greater in soils with smaller moisture content. So, during the wet season, when the natural moisture content could rise to 12%, there was a reduction in the collapse potential to around 4%.

However, to help understand better which loess soils might be more susceptible to collapse, Jefferson et al. (2003) developed the concept of the provenance (P), transportation (T) and deposition (D) (PTD) model. They argued that it is the PTD sequence that a particular deposit goes through that determines whether the soil will have an open, metastable structure and relatively high void ratio, hence making it more likely to collapse. Jefferson et al. (2003) illustrated the approach with a number of examples. One example that would include the loessic brickearths of South Essex is reproduced in Table 6.3. The different depositional situations explain, in part, the different litho-stratigraphic names used on BGS geological maps (see Table 6.1).

Jefferson & Rogers (2012) summarized the link between the PTD sequence and collapse potential as follows: 'The alternating nature of loess formation significantly influences the engineering behaviour and ultimately the nature of collapse, and the location (depth) where collapse occurs. This will dictate the nature of the infiltration pattern of water into the soil and as a result can yield collapse in unexpected locations. Moreover, this can influence the effectiveness of any ground improvement approach used to remove collapsibility.'

Broadly, loesses have three zones of relative collapsibility (Jefferson & Rogers 2012):

- a surface desiccation crust that doesn't collapse without additional loading;
- a collapsible zone;
- a zone at depth that has collapsed due to overburden pressure.

For collapse to happen, the following are required:

- an open structure with relatively large voids;
- a source of strength to keep the particles in position.

This strength is provided by one or more of the following:

- capillary or matric suction forces;
- clay and silt particles at coarser particle contacts;
- cementing agents, for example carbonates or oxides.

For cemented loesses, Milodowski et al. (2015) said that there were three stages of collapse after inundation (when any suction will be lost):

- first, dispersion and disruption of clay bridges between larger silt particles leading to initial rapid collapse of loose-packed inter-ped silt/sand;
- rearrangement and closer packing of compact aggregate silt/clay peds as the load is taken up via the particle contacts;
- progressive deformation and shearing of the silt/clay peds and collapse into unsupported inter-particle areas.

6.6 Strategies for engineering management: avoidance, prevention and mitigation

Popescu (1986) proposed that there were four steps required when dealing with collapsible soils:

- Identification of the presence of a potentially collapsible soil using geological and geomorphological information. For the UK, the use of geological maps has been discussed in the previous section. Also, the BGS's GeoSure geohazard information system (Culshaw & Harrison 2010) includes a map layer for collapsible soil that shows areas of moderate and significant collapse potential (http://bgs.ac.uk/products/geosure/collapsible.html). Other approaches have been recently presented, for example, a collapse-risk-specific non-conflicting fine soil classification framework such as one recently proposed by Assadi (2014)
- Classification of the degree of collapsibility, perhaps using indirect correlations. A wide range of empirical collapse indices exist. Bell et al. (2003) compared seven of them

Table 6.3. *Progress of a loess particle that falls in the headwater of the River Thames catchment (after Jefferson et al. 2003)*

P1	Particles are formed by cold phase glacial action
T1	Loess material is blown generally southwards
D1	Deposits over midland and southern Britain
D1t	Deposits in River Thames catchment, headwaters region
T2t	Carried into River Thames by slope wash and streams. Short transport may deliver it to the Langley Silt
T3t	Carried by the River Thames into the estuary region
D2t	Deposited on northern bank as a floodplain deposit
T4t	Blown inland to form loess
D3t	Loess deposit formed, perhaps in South Essex

plus liquid limit vs dry density plots of Gibbs & Bara (1962), with the results of oedometer tests to determine the coefficient of collapsibility. They concluded that while many of the indices broadly agreed with the oedometer test results, they should be regarded as only approximate indications of collapsibility. Differences between the indices may be the result of regional differences in the materials used.

- Quantification of the degree of collapsibility using laboratory and/or *in situ* testing. In the laboratory, the double or single oedometer test is used following the methodology of Jennings & Knight (1975) and modified by Houston *et al.* (1988). The amount of collapse strain developed when the test specimen is flooded under a given load indicates the susceptibility to collapse. Although, strictly, these methods should be seen as indicative of actual potential collapse. Table 6.4 shows Jennings & Knight's classification of collapse severity in terms of the percentage collapse derived from the relationship:

$$\text{Coefficient of collapse, } C_{col} = \frac{\Delta e}{1 + e_i} \quad (6.1)$$

In situ tests that have been used include plate loading, pressuremeter and standard penetrometer. These, and other tests have been described by Jefferson & Rogers (2012).

Dynamic cone resistance, q_d, was measured by Northmore *et al.* (2008) for the loessic brickearth at Ospringe, Kent. They showed that q_d takes high values in granular, sandy, gravelly sequences and very low values in desiccated surface layers. Resistance was more or less constant through the loessic brickearth sequences, except a slightly decreasing trend with depth in the upper non-calcareous sequence followed by a slightly increasing trend in the lower calcareous sequence. This may suggest a history of leaching. The 1.5–2.5 MPa resistance through the upper non-calcareous and 3–4 MPa resistance through the lower calcareous sequence could be a means to identify loessic brickearth in a profile with top loam and bottom sand formations of >5 MPa resistance.

In the last ten years, geophysical methods have been increasingly used. Northmore *et al.* (2008) described the use of electrical resistivity, shear wave profiles and electromagnetics (EM) to determine the depth and extent of collapsible and non-collapsible loessic brickearth at Ospringe, Kent. Shear wave velocities through the loessic brickearth sequences were reportedly lower than in the topmost loam and bottom sand formations. Owing to the heterogeneous nature of the lower calcareous loessic brickearth, shear wave velocities followed an erratic pattern. Northmore *et al.* (2008) also argued the capability of EM profiling to distinguish between the lower calcareous and upper non-calcareous loessic brickearth. At Ospringe, Kent, they measured an EM resistivity of 22–24 Ωm for the lower calcareous sequence and greater than 70 Ωm for the underlying chalk formation. They measured an electrical resistivity of 20 Ωm for the upper non-calcareous loessic brickearth, 22–30 Ωm for the lower calcareous loessic brickearth and greater than 75 Ωm for underlying chalk formation.

- Engineering options to improve collapsible soils have been summarized by Jefferson & Rogers (2012) and are summarized in Table 6.5.

For foundations Popescu (1992) suggested four approaches:

Very stiff raft foundations and a rigid superstructure to reduce differential settlement (expensive and not always successful).

Flexible foundation and superstructure to reduce damage.

Use of piles through the collapsible soil.

Soil improvement (see Table 6.5).

For roads and railways, some of the approaches in Table 6.5 may be applicable but care should be taken with roads because the pavement may reduce evaporation and hence alter the water content conditions compared with the surrounding ground.

Slope stability problems are unlikely in the UK because thicknesses of loessic brickearth are mostly not great enough.

More recently, Roohnavaz *et al.* (2011) discussed the use of unsaturated loess fills for earthworks, where standard methods are often insufficient to remove collapse potential (developed through inter-particle bonding). Instead, they suggested that repeated reworking and recompaction is required in combination with greater compactive effort and placement wet of optimum. Further details of the special attention needed when working with collapsible soils, in particular loess, are provided by Jefferson & Rogers (2012).

6.7 Example of damage caused by collapse

Surprisingly little is published about cases of damage caused by collapse of loessic brickearth in the UK. Cattell (1997) described the presence of loess and possibly solifluctued loess in the Torbay area in South Devon. Cattell (2000) said that the susceptibility of the soils to loss of strength on wetting is unusually high. Figure 6.5 shows a derelict house in Torbay, situated on the loess that Cattell described, that has suffered foundation failure probably due to the leaking of a drain or downspout. However, given the extent of loess in SE England, it seems likely that other examples of

Table 6.4. *Collapse percentage as an indication of potential severity (after Jennings & Knight (1975)*

Collapse	Severity of problem
0–1	No problem
1–5	Moderate trouble
5–10	Trouble
10–20	Severe trouble
> 20	Very severe trouble

Table 6.5. *Methods of treating collapsible loess ground (after Jefferson et al. 2005)*

Depth (m)	Treatment method	Comments
0–1.5	Surface compaction with vibratory rollers, light tampers	Economical but requires careful site control, for example, limits on water content.
	Pre-wetting (inundation)	Can effectively treat thicker deposits but needs large volumes of water and time.
	Vibroflotation	Needs careful site control.
1.5–10	Vibrocompaction (stone columns, concrete columns, encased stone columns).	Cheaper than conventional piles but requires careful site control and assessment. If uncased, stone columns may fail with loss of lateral support on collapse.
	Dynamic compaction; rapid impact compaction	Simple and easily understood but requires care with water content and vibrations produced.
	Explosions	Safety issues need to be addressed.
	Compaction pile	Need careful site control.
	Grouting	Flexible but may adversely affect the environment.
	Ponding/inundation/pre-wetting	Difficult to control effectiveness of compression produced.
	Soil mixing lime/cement	Convenient and gains strength with time. Various environmental and safety aspects; the chemical controls on reactions need to be assessed.
	Heat treatment	Expensive.
	Chemical methods	Flexible; relatively expensive.
>10	As for 1.5–10 m, some techniques may have a limited effect.	(see above)
	Pile foundations	High bearing capacity but expensive.

Fig. 6.5. Derelict house in Torbay, situated on loess, which has suffered foundation failure probably due to the leaking of a drain or downspout.

failure exist, though they have not been identified as such, except for a few isolated cases.

However, even in the most loessic parts of SE England the risk of structural collapse effecting foundations, etc. would be low due to the relatively thin deposits of collapsible loess. For example, if foundations are excavated down to 2 m, they would probably cut through many, if not most, areas of collapsible loessic brickearth. Even if loessic brickearth was still found beneath the foundation, the thickness remaining to the underlying chalk, river gravels or London Clay would be insufficient for significant collapse in most cases. Of course, this comment does *not* rule out undertaking adequate site investigation to determine the thickness and measure collapse potential on samples, if the thickness beneath a building or structure warrants it.

Nevertheless, the 'seasonal' collapse of loess-like soils including poorly engineered loam, and, in particular, calcareous loam fills/embankments has been long a well-known challenge for the construction of earthworks. The performance of the heavily used UK transport infrastructure relies, in part, on the performance of underlying embankments, many of which have been in service for over 150 years. Furthermore, new embankments will be built in the coming decades to improve the network, raising the need for a better understanding of the impact of placement conditions on planned maintenance costs. The revised British Code of Practice for Earthworks was published in 2009 and includes compliance with Eurocode 7 (Anon. 2009*a*). This places emphasis on fill classification and compaction specifications, while setting the Specification for Highway Works 600 series (Anon. 2009*b*) as the default approach for earthworks in the

UK. However, the revised earthworks British Standard (and earlier documents such as Charles & Watts [2001]) lack consideration of unexplained ground movements in compacted earthworks, particularly when material from nearby cuttings is used as fill materials. Unexplained settlements include sudden and long-term subsidence particularly in transient loading environments, when fills are built from sand/silts with small clay inclusions, as well as seasonal subsidence in fills with <20% carbonates. Assadi & Jefferson (2015) recently examined dry and wet compressibility of a suite of calcareous and non-calcareous clayey silts and silty clays against the BRE recommendations, as a baseline for UK earthworks practice.

6.8 Conclusions

Globally, a relatively wide range of soil types have the potential for collapse under suitable conditions. However, regardless of provenance, virtually all are characterized by porous textures, high void ratios and low densities with collapse triggered by water inundation and saturation. In the UK, loessic brickearth is the predominant collapsible natural soil along with certain non-engineered (or inadequately engineered) fills. Collapsible soils, including loess, are materials that 'standard' soil mechanics stress-strain principles fail to adequately explain in terms of their engineering behaviour. For the ground engineering industry to avoid and mitigate the risks associated with collapse, a first significant step is to correctly identify the presence of collapsible soils. Once identified, appropriate laboratory testing procedures and, where necessary, follow-up field tests can be applied to assess collapse potential, and the possible need for mitigation measures. This chapter has highlighted current geological-geotechnical-geochemical-mineralogical and geomorphological understanding of UK collapsible soils and may serve as a guide to aid engineering ground investigation in those areas where such natural (loessic) soils and potentially collapsible man-made fills may be present. Current techniques to help mitigate the risks associated with collapse are also described. As planned expansion of the UK's road and rail infrastructure progresses it becomes ever more important that the collapse potential of poorly- or non-engineered fills, including old Victorian railway embankments, is considered by ground engineers, and that the use and appropriate engineered placement of potentially metastable materials is more fully understood and designed for.

Glossary and definitions

Brickearth: a term used in the UK on British Geological Survey maps to describe materials that have been commonly used to make bricks. Deposits mapped as 'brickearth' are mainly, but *not exclusively*, loess deposits.

Loessic brickearth: a new term that applies to the UK. It consists of materials mapped by the British Geological Survey as brickearth, but which comprise deposits of loess.

Collapse: a sudden change in soil structure from an initial open state to a final dense state when grain-to-grain connections fail in the event of wetting or loading (Assadi 2014).

Collapsible soil: a soil in which the constituent parts have an open packing and which forms a metastable state that can collapse to form a closer packed, more stable structure of significantly reduced volume (Rogers 1995).

Loess: a widespread water-sensitive soil with at least one cycle of aeolian deposition and a possible history of post depositional modifications. The metastable aspect of loess is due to its air-fall sedimentation history, while the post-depositional modifications are responsible for the collapsibility of the metastable structure (Rogers 1995; Smalley *et al.* 2006).

Hydrocompaction/hydro-consolidation: a process by which fine grained, low density soils restructure and compact due to the addition of water or the addition of water under load (Rogers *et al.* 1994; Waltham 2009). See also 'Collapse'.

Metastability: a metastable soil has an open structure, that is, the granular frame of the solid particles must be in an open packing that is capable of achieving rapidly a significantly closer packing, producing a stable structure (Assallay 1998).

Non-engineered fill: The term 'fill' (or 'made ground') is used to describe material that has been deposited by human processes. The material could be natural, or have been altered artificially prior to deposition. Fill is either engineered or non-engineered depending on whether any specific treatment during deposition takes place. As a result, a number of engineering challenges may exist and will require different remediation approaches to improve and mitigate problematical behaviour. Non-engineered fill may consist of domestic waste, building waste, slag, mining and quarry waste, industrial waste and soil waste. Non-engineered fills may settle variably, have poor bearing capacity and may suffer significant movements due to causes other than the imposed loading. Other problems associated with some wastes include contamination, spontaneous combustion and the emission of gas. The extent to which non-engineered fill will be suitable as a foundation material depends largely on its age, composition, uniformity, properties and the method by which the material was placed (Bell *et al.* 2012).

Further reading

DERBYSHIRE, E. & MENG, X. 2005. Loess. In: FOOKES, P.G., LEE, E.M. & MILLIGAN, G. (eds) *Geomorphology for Engineers*. Whittles Publishing, Caithness, Scotland, 688–728.

DERBYSHIRE, E., DIJKSTRA, T. & SMALLEY, I.J. (eds) 1995. Genesis and properties of collapsible soils. *Proceedings of the NATO Advanced Workshop*, April 1994, Loughborough, UK, Dordrecht, Springer Science+Business Media.

FALL, D.A. 2003. The geotechnical and geochemical characterisation of the Brickearth of Southern England. Unpublished PhD thesis, University of Portsmouth.

GUNN, D.A., NELDER, L.M. ET AL. 2006. Shear wave velocity monitoring of collapsible loessic brickearth soil. *Quarterly Journal of Engineering Geology and Hydrogeology*, **39**, 173–188.

JACKSON, P.D., NORTHMORE, K.J. ET AL. 2006. Electrical resistivity monitoring of a collapsing metastable soil. *Quarterly Journal of Engineering Geology and Hydrogeology*, **39**, 151–172.

MCKERVEY, J.A. & KEMP, S.J. 2001. *Mineralogical Analysis of Brickearth Samples from Europe*. Internal Report **IR/01/107**. British Geological Survey, Keyworth, Nottingham.

MILLER, H. 2002. Modelling the collapse of metastable loess soils. Unpublished PhD thesis, The Nottingham Trent University.

SMALLEY, I., O'HARA-DHAND, K., WINT, J., MACHALETT, B., JARY, Z. & JEFFERSON, I. 2009. Rivers and loess: the significance of long river transportation in the complex event-sequence approach to loess deposit formation. *Quaternary International*, **198**, 7–18.

ZOURMPAKIS, A., BOARDMAN, D.I. ET AL. 2006. Case study of a loess collapse field trial in Kent, SE England. *Quarterly Journal of Engineering Geology and Hydrogeology*, **39**, 131–150.

Acknowledgements During the early stages of the preparation of this paper the death of one of the authors, Fred Bell, occurred in May 2014. The authors wish to express their sorrow at the loss of such an eminent engineering geologist. This paper is published with the permission of the Executive Director of the British Geological Survey (UKRI).

Funding This research received no specific grant from any funding agency in the public, commercial, or not-for-profit sectors.

References

ANON. 2009a. *Code of Practice for Earthworks*. BS 6031. British Standards Institution, London.

ANON. 2009b. *Manual of contract documents for highway works. Volume 1: Specification for highway works*. Series 600: Earthworks. The Stationery Office, London.

ARKELL, W.J. & TOMKEIEFF, S.I. 1953. *English rock terms: chiefly used by miners and quarrymen*. Oxford University Press, London.

ASSADI, A. 2014. *Micromechanics of collapse in loess*. Unpublished PhD thesis, University of Birmingham.

ASSADI, A. & JEFFERSON, I. 2013. Collapsibility in calcareous clayey loess: a factor of stress-hydraulic history. *International Journal of Geomaterials*, **5**, 619–626.

ASSADI, A. & JEFFERSON, I. 2015. Constraints in using site-won calcareous clayey silt (loam) as fill materials. In: WINTER, M.G., SMITH, D.M., ELDRED, P.J.L. & TOLL, D.G. (eds) *'Geotechnical Engineering for Infrastructure and Development,' Proceedings of the 16th European Conference on Soil Mechanics and Geotechnical Engineering. Edinburgh*. Volume 4: Slopes and Geohazards, ICE Publishing, London, 1947–1952.

ASSALLAY, A.M. 1998. *Structure and hydrocollapse behaviour of loess*. Unpublished PhD thesis, University of Loughborough.

BALLANTYNE, C.K. & HARRIS, C. 1994. *The Periglaciation of Great Britain*. Cambridge University Press.

BELL, F.G. & CULSHAW, M.G. 2001. Problem soils: a review from a British perspective. In: JEFFERSON, I., MURRAY, E.J., FARAGHER, E. & FLEMING, P.R. (eds) *Problematic Soils*. Thomas Telford Publishing, London, 1–35.

BELL, F.G., CULSHAW, M.G. & NORTHMORE, K.J. 2003. The metastability of some gull fill materials from Allington, Kent, UK. *Quarterly Journal of Engineering Geology and Hydrogeology*, **36**, 217–229, https://doi.org/10.1144/1470-9236/02-005

BELL, F.G., CULSHAW, M.G. & SKINNER, H.D. 2012. Non-engineered fills. In: BURLAND, J., CHAPMAN, T., SKINNER, H. & BROWN, M. (eds) *ICE manual of Geotechnical Engineering*. ICE Publishing, London. **1** (Geotechnical engineering principles, problematic soils and site investigation), 443–461.

BLANCHFIELD, R. & ANDERSON, W.F. 2000. Wetting collapse in opencast coalmine backfill. *Proceedings of the Institution of Civil Engineers, Geotechnical Engineering*, **143**, 139–149, https://doi.org/10.1680/geng.2000.143.3.139

BOARDMAN, D.I., ROGERS, D.F., JEFFERSON, I. & ROUAIGUIA, A. 2001. Physico-chemical characteristics of British loess. *Proceedings of the 15th International Conference on Soil Mechanics and Foundation Engineering*. 1, 39–42, Istanbul, Turkey, Lisse, A. A. Balkema Publishers.

BOLTON, M.D. 1999. The role of micro-mechanics in soil mechanics. In: HYODO, M. & NAKATA, Y. (eds) *Proceedings of an International Workshop on Soil Crushability, Ube, Japan*. Yamaguchi University, Ube, Japan, 58–82. Also available at: http://www-civ.eng.cam.ac.uk/geotech_new/people/bolton/mdb_pub/94_TR313.pdf

CATT, J.A. 1977. Loess and coversands. In: SHOTTON, F.W. (ed.) *British Quaternary Studies, Recent Advances*. Clarendon Press, Oxford, 221–229.

CATT, J.A. 1985. Particle size distribution and mineralogy as indicators of pedogenic and geomorphic history: examples from soils of England and Wales. In: RICHARDS, K.S., ARNETT, R.R. & ELLIS, S. (eds) *Geomorphology and Soils*. George Allen and Unwin, London, 202–218.

CATT, J.A. 2008. Comment on Clarke, M., Milodowski, A. E., Bouch, J., Leng, M. J. & Northmore K. J. 2007. New OSL dating of UK loess: indications of two phases of Late Glacial dust accretion in SE England and climate implications. *Journal of Quaternary Science*, **23**, 305–306, https://doi.org/10.1002/jqs.1145

CATTELL, A.C. 1997. The development of loess-bearing soil profiles on Permian Breccias in Torbay. *Proceedings of the Ussher Society*, **9**, 168–172.

CATTELL, A.C. 2000. Shallow foundation problems and ground conditions in Torbay. *Geoscience in South-West England*, **10**, 68–71.

CHARLES, J.A. 2005. *Geotechnics for building professionals*. Building Research Establishment, Watford, UK.

CHARLES, J.A. & SKINNER, H.D. 2001. Compressibility of foundation fills. *Proceedings of the Institution of Civil Engineers, Geotechnical Engineering*, **149**, 145–157, https://doi.org/10.1680/geng.2001.149.3.145

CHARLES, J.A. & WATTS, K.S. 2001. *Building on Fill: Geotechnical Aspects*. 2nd edn. Construction Research Communication Ltd, Watford. By permission of BRS Ltd.

CLARKE, M., MILODOWSKI, A.E., BOUCH, J., LENG, M.J. & NORTHMORE, K.J. 2007. New OSL dating of UK loess: indications of two phases of Late Glacial dust accretion in SE England and climate implications. *Journal of Quaternary Science*, **22**, 361–371, https://doi.org/10.1002/jqs.1061

CULSHAW, M.G. & HARRISON, M. 2010. Geo-information systems for use by the UK insurance industry for 'subsidence' risk. In: WILLIAMS, A.L., PINCHES, G.M., CHIN, C.Y., MCMORRAN, T.J. & MASSEY, C.I. (eds) *'Geologically Active'. Proceedings of the 11th Congress of the International Association for Engineering Geology and the Environment*, September 2010, Auckland, New Zealand, (on CD-ROM, 1043-1051). CRC Press/Balkema, Leiden, The Netherlands.

CULSHAW, M.G., HOBBS, P.R.N. & NORTHMORE, K.J. 1992. *Engineering Geology of Tropical Red Clay Soils: manual Sampling Methods*. Technical Report **WN/93/14**. British Geological Survey, Keyworth, Nottingham.

D'ARCHIAC, A. 1839. *Observations sur le groupe moyen de la formation crétacée*. F. G. Levrault, Paris.

DEAKIN, W.H. 1986. *Kent minerals subject plan, Brickearth, Written Statement*. Springfield, Maidstone.

DERBYSHIRE, E. & MELLORS, T.W. 1988. Geological and geotechnical characteristics of some loess soils from China and Britain – a comparison. *Engineering Geology*, **25**, 135–175, https://doi.org/10.1016/0013-7952(88)90024-5

DIBBEN, S.C. 1998. *A microstructure model for collapsing soils*. Unpublished PhD thesis, Nottingham Trent University.

DUDLEY, J.H. 1970. Review of collapsing soils. *American Society of Civil Engineers, Journal of the Soil Mechanics and Foundations Division*, **96**, 925–947.

EALEY, P.J. & JAMES, H.C.L. 2008. Countybridge Quarry, The Lizard Peninsula, Cornwall – historical and geological significance of an abandoned quarry. *Geoscience in South-West England*, **11**, 22–26.

EALEY, P.J. & JAMES, H.C.L. 2011. Loess of the Lizard Peninsula, Cornwall, SW Britain. *Quaternary International*, **231**, 55–61, https://doi.org/10.1016/j.quaint.2010.06.018

EL-RUWAIH, I.E. & TOUMA, F.T. 1986. Assessment of the engineering properties of some collapsible soils in Saudi Arabia. *Proceedings of the 5th International Congress of the International Association of Engineering Geology, Buenos Aires*. Rotterdam, A. A. Balkema, **2**, 685–693.

FINK, J. 1974. *INQUA Loess Commission Circular Letter*. Beilage 1 Quadrenniel Report.

FOOKES, P.G. & BEST, R. 1969. Consolidation characteristics of some late Pleistocene periglacial metastable soils of east Kent. *Quarterly Journal of Engineering Geology*, **2**, 103–128, https://doi.org/10.1144/GSL.QJEG.1969.002.02.02

GAO, G. 1988. Formation and development of the structure of collapsing loess in China. *Engineering Geology*, **25**, 235–245, https://doi.org/10.1016/0013-7952(88)90029-4

GIBBS, H.J. & BARA, J.P. 1962. *Predicting surface subsidence from basic soil tests*. Special Technical Publication, **322**. American Society for Testing Materials, Philadelphia, 231–246.

GRABOWSKA-OLSZEWSKA, B. 1988. Engineering-geological problems of loess in Poland. *Engineering Geology*, **25**, 177–199, https://doi.org/10.1016/0013-7952(88)90025-7

HALLAM, J.R. & NORTHMORE, K.J. 1993. *Engineering Geology of Tropical Red Clay Soils: geotechnical Borehole Sampling of Tropical Red Clay Soil*. Technical Report **WN/93/27**. British Geological Survey, Keyworth, Nottingham.

HASKINS, D.R., BELL, F.G. & SCHALL, A. 1998. The evaluation of granite saprolite as a founding medium at Injaka Dam, Mpumalanga Province, South Africa. *In*: MOORE, D.P. & HUNGR, O. (eds) *Proceedings of the 8th Congress of the International Association of Engineering Geology*. A. A. Balkema, Rotterdam, **1**, 461–468.

HOBBS, P.R.N., ENTWISLE, D.C., NORTHMORE, K.J. & CULSHAW, M.G. 1992. *Engineering Geology of Tropical Red Clay Soils: geotechnical Characterisation: mechanical Properties and Testing Procedures*. Technical Report **WN/93/13**. British Geological Survey, Keyworth, Nottingham.

HOLTZ, W.G. & GIBBS, H.J. 1952. *Consolidation and Related Properties of Loessial Soils*. Special Technical Publication, **126**. American Society for Testing Materials, Philadelphia, 9–33.

HOUSTON, S.L., HOUSTON, W.N. & SPADOLA, D.J. 1988. Prediction of field collapse of soils due to wetting. *Journal of Geotechnical Engineering*, **114**, 40–58, https://doi.org/10.1061/(ASCE)0733-9410(1988)114:1(40)

IRIONDO, M.H. & KROHLING, D.M. 2007. Non-classical types of loess. *Sedimentary Geology*, **202**, 352–368, https://doi.org/10.1016/j.sedgeo.2007.03.012

JARI, Z. & BADURA, J. 2013. Distribution and characteristics of loess in Lower Silesia with reference to the Great Odra Valley. *In*: O'HARA-DHAND, K. & MCLAREN, S. (eds) *INQUA Loess2013 Workshop: Loess & Dust: Geography-Geology-Archaeology*. Print Services, University of Leicester, Leicester.

JEFFERSON, I. & ROGERS, C.D.F. 2012. Collapsible soils. *In*: BURLAND, J., CHAPMAN, T., SKINNER, H. & BROWN, M. (eds) *ICE Manual of Geotechnical Engineering, Volume 1: Geotechnical Engineering Principles, Problematic Soils and Site Investigation*, **391–411**. ICE Publishing, London.

JEFFERSON, I., TYE, C. & NORTHMORE, K.J. 2001. Behaviour of silt: the engineering characteristics of loess in the UK. *In*: JEFFERSON, I., MURRAY, E.J., FARAGHER, E. & FLEMING, P.R. (eds) *Problematic Soils, Proceedings of the Symposium Held at the Nottingham Trent University*, **37–52**. Thomas Telford Ltd, London.

JEFFERSON, I., SMALLEY, I. & NORTHMORE, K. 2003. Consequences of a modest loess fall over southern and midland England. *Mercian Geologist*, **15**, 199–208.

JEFFERSON, I., ROGERS, C.D.F., EVSTATIEV, D. & KARASTANEV, D. 2005. Treatment of metastable loess soils: lessons from Eastern Europe. *In*: INDRARATNA, B. & CHU, J. (eds) '*Ground Improvement – case Histories*.' Elsevier Geo-Engineering Book Series, Volume **3**. Elsevier, Amsterdam, 723–762.

JENNINGS, J.E. & KNIGHT, K. 1975. A guide to construction on or with materials exhibiting additional settlement due to collapse of grain structure. *In*: PELLS, P.J.N., MAC, A. & ROBERTSON, G. (eds) *Proceedings of the 6th African Conference on Soil Mechanics and Foundation Engineering, Durban*. A. A. Balkema, South Africa **1**, 99–105.

KEMP, S.J. 1991. *Mineralogical analysis of tropical red clay soils*. Technical Report **WN/91/7**. Keyworth, Nottingham, British Geological Survey.

KERNEY, M.P. 1965. Weichselian deposits in the Isle of Thanet, east Kent. *Proceedings of the Geologists' Association*, **76**, 269–274, https://doi.org/10.1016/S0016-7878(65)80029-3

LAWTON, E.C., FRAGASZY, R.J. & HETHERINGTON, M.D. 1992. Review of wetting-induced collapse in compacted soil. *Journal of Geotechnical Engineering*, **118**, 1376–1394, https://doi.org/10.1061/(ASCE)0733-9410(1992)118:9(1376)

LEE, D., WOODGER, A. & ORTON, C. 1989. Excavations in the walbrook valley. *London Archaeologist*, **6**, 115–119.

LILL, G.O. 1976. *The nature and distribution of loess in Britain*. Unpublished PhD thesis. Department of Civil Engineering, University of Leeds.

LIU, T.S., AN, Z.S., YUAN, B. & HAN, J. 1985. The loess-palaeosol sequence in China and climatic history. *Episodes*, **8**, 21–28.

MCMILLAN, A.A. & POWELL, J.H. 1999. *BGS Rock Classification Scheme, Volume 4. Classification of Artificial (Man-Made) Ground and Natural Superficial Deposits – applications to Geological Maps and Datasets in the UK*. British Geological Survey Research Report, RR **99-04**. British Geological Survey, Keyworth, Nottingham.

MILODOWSKI, A.E., NORTHMORE, K.J. ET AL. 2015. The mineralogy and fabric of 'Brickearths' in Kent and their relationship to engineering behaviour. *Bulletin of Engineering Geology and the Environment*, **74**, 1187–1211, https://doi.org/10.1007/s10064-014-0694-5

NORTHMORE, K.J., BELL, F.G. & CULSHAW, M.G. 1996. The engineering properties and behaviour of the brickearth of south Essex. *Quarterly Journal of Engineering Geology*, **29**, 147–161, https://doi.org/10.1144/GSL.QJEGH.1996.029.P2.04

NORTHMORE, K.J., CULSHAW, M.G. & HOBBS, P.R.N. 1992*a*. *Engineering geology of tropical red clay soils: summary findings and their application for engineering purposes.* Technical Report **WN/93/15**. Keyworth, Nottingham: British Geological Survey.

NORTHMORE, K.J., CULSHAW, M.G., HOBBS, P.R.N., HALLAM, J.R. & ENTWISLE, D.C. 1992*b*. *Engineering Geology of Tropical Red Clay Soils: project Background, Study Area and Sampling Sites.* Technical Report **WN/93/11**. British Geological Survey, Keyworth, Nottingham.

NORTHMORE, K.J., ENTWISLE, D.C., HOBBS, P.R.N., CULSHAW, M.G. & JONES, L.D. 1992*c*. *Engineering Geology of Tropical Red Clay Soils: geotechnical Characterisation: index Properties and Testing Procedures.* Technical Report **WN/93/12**. British Geological Survey, Keyworth, Nottingham.

NORTHMORE, K.J., JEFFERSON, I. *ET AL.* 2008. On-site characterisation of loessic brickearth deposits at Ospringe, Kent, UK. *Proceedings of the Institution of Civil Engineers, Geotechnical Engineering*, **161**, 3–17, https://doi.org/10.1680/geng.2008.161.1.3

NOWAK, P. & GILBERT, P. 2015. *Earthworks: a Guide.* 2nd edn. (originally created by N. A. Trenter). ICE Publishing, London.

PARKS, D.A. & RENDELL, H.M. 1992. Thermoluminescence dating and geochemistry of loessic deposits in southeast England. *Journal of Quaternary Science*, **7**, 99–107, https://doi.org/10.1002/jqs.3390070203

PHIEN-WEJ, N., PIENTONG, T. & BALASUBRAMANIAN, A.S. 1992. Collapse and strength characteristics of loess in Thailand. *Engineering Geology*, **32**, 59–72, https://doi.org/10.1016/0013-7952(92)90018-T

POPESCU, M.E. 1986. A comparison between the behaviour of swelling and collapsing soils. *Engineering Geology*, **23**, 145–163, https://doi.org/10.1016/0013-7952(86)90036-0

POPESCU, M.E. 1992. Engineering problems associated with expansive and collapsible soil behaviour. *Proceedings of the 7th International Conference on Expansive Soils*, Dallas, Texas, **2**, 25–56.

PRESTWICH, J. 1863. On the loess of the valleys of the south of England and of the Somme and the Seine. *Proceedings of the Royal Society of London*, **12**, 170–173, https://doi.org/10.1098/rspl.1862.0034

PYE, K. 1987. *Eolian Dust and Dust Deposits.* Academic Press, London.

PYE, K. 1995. The nature, origin and accumulation of loess. *Quaternary Science Reviews*, **14**, 653–667, https://doi.org/10.1016/0277-3791(95)00047-X

ROGERS, C.D.F. 1995. Types and distribution of collapsible soils. *In*: DERBYSHIRE, E., DIJKSTRA, T. & SMALLEY, I.J. (eds) *Genesis and properties of collapsible soils. Proceedings of the NATO Advanced Research Workshop*, Loughborough, 1–17. NATO Science Series C. Kluwer Academic Publishers, Dordrecht, The Netherlands.

ROGERS, C.D.F., DIJKSTRA, T.A. & SMALLEY, I.J. 1994. Hydroconsolidation and subsidence of loess: studies from China, Russia, North America and Europe. *Engineering Geology*, **37**, 83–113, https://doi.org/10.1016/0013-7952(94)90045-0

ROOHNAVAZ, C., RUSSELL, E.J.F. & TAYLOR, H.F. 2011. Unsaturated loessial soils: a sustainable solution for earthworks. *Proceedings of the Institution of Civil Engineers, Geotechnical Engineering*, **164**, 257–276, https://doi.org/10.1680/geng.10.00041

ROSE, J., LEE, J.A., MOORLOCK, B.S.P. & HAMBLIN, R.J.O. 1999. The origin of the Norwich Brickearth: micromorphological evidence for pedological alteration of sandy Anglian Till in northeast Norfolk. *Proceedings of the Geologists' Association*, **110**, 1–8.

SKINNER, H.D. 2012. Building on fills. *In*: BURLAND, J., CHAPMAN, T., SKINNER, H. & BROWN, M. (eds) *ICE Manual of Geotechnical Engineering, Volume 2: Geotechnical Design, Construction and Verification.* ICE Publishing, London, 899–910.

SMALLEY, I.J., JEFFERSON, I.F., DIJKSTRA, T.A. & DERBYSHIRE, E. 2001. Some major events in the development of the scientific study of loess. *Earth-Science Reviews*, **54**, 5–18, https://doi.org/10.1016/S0012-8252(01)00038-1

SMALLEY, I.J., JEFFERSON, I.F., O'HARA-DHAND, K. & EVANS, R.D. 2006. An approach to the problem of loess deposit formation: some comments on the 'in situ' or 'soil-eluvial' hypothesis. *Quaternary International*, **152**, 109–117, https://doi.org/10.1016/j.quaint.2005.12.011

SMITH, W. 1815. *A Delineation of the Strata of England and Wales with Part of Scotland; Exhibiting the Collieries and Mines, the Marshes and Fen Lands Originally Overflowed by the Sea, and the Varieties of Soil According to the Variations in the Substrata, Illustrated by the Most Descriptive Names.* J. Cary, London.

TROFIMOV, V.T. 1990. Some experimental evidence of formation of syngenetic collapsibility of aeolian loess rocks. *Inzhenernaya Geologiya*, **6**, 11–24 [in Russian].

WALTHAM, A.C. 2009. *Foundations of Engineering Geology.* 3rd edn. Taylor & Francis, Abingdon, Oxfordshire, UK.

WHITE, J.L. & GREENMAN, C. 2008. *Collapsible soils in Colorado.* Engineering Geology Publication 14. Colorado Geological Survey, Department of Natural Resources, Denver.

Chapter 7 Quick clay behaviour in sensitive Quaternary marine clays – a UK perspective

David Peter Giles

CGL, 4 Godalming Business Centre, Woolsack Way, Godalming, GU7 1XW, UK

DavidG@cgl-uk.com

Abstract: The term quick clay has been used to denote the behaviour of highly sensitive Quaternary marine clays that, due to post depositional processes, have the tendency to change from a relatively stiff condition to a liquid mass when disturbed. On failure these marine clays can rapidly mobilise into high velocity flow slides and spreads often completely liquefying in the process. For a clay to be defined as potentially behaving as a quick clay in terms of its geotechnical parameters it must have a sensitivity (the ratio of undisturbed to remoulded shear strength) of greater than 30 together with a remoulded shear strength of less than 0.5 kPa. The presence of quick clays in the UK is unclear, but the Quaternary history of the British islands suggests that the precursor conditions for their formation could be present and should be considered when undertaking construction in the coastal zone.

7.1 Introduction

The term quick clay has been used to denote the behaviour of highly sensitive Quaternary marine clays that, due to post depositional processes, have the tendency to change from a relatively stiff condition to a liquid mass when disturbed. On failure these marine clays can rapidly mobilize into high velocity flow slides and spreads often completely liquefying in the process (Torrance 1975, 1979, 1983, 1999, 2012; Geertsema 2013). A clay is defined as potentially behaving as a quick clay in terms of its geotechnical parameters, specifically through its sensitivity (the ratio of undisturbed to remoulded shear strength). The physical structure of these clay deposits completely collapses on remoulding with their shear strength being reduced to virtually zero (Rankka *et al.* 2004). Three key factors are required for the formation of these sensitive clays: a flocculated structure with a high void ratio, low activity mineral content dominant and salt (NaCl) in the pore fluid removed through post-depositional leaching (Hutchinson 1991, 1992).

Potential quick clay behaving soils can be found in areas of former marine boundaries that have been uplifted through isostatic rebound after Quaternary glaciations. Examples have been cited in Norway (Holmsen 1953; Bjerrum 1954; Rosenqvist 1960a, b, 1966; Gregersen & Løken 1979; Aas & Lunne 1981; Gregersen 1981; Stevens *et al.* 1991; Eilertsen *et al.* 2008; L'Heureux *et al.* 2012a, b) Sweden (Rankka *et al.* 2004; Andersson-Sköld *et al.* 2005; Malehmir *et al.* 2013), Canada (Crawford 1968; Gillott 1979; Ells 1908; Quigley *et al.* 1985; Evans & Brooks 1994; Geertsema & Torrance 2005; Locat *et al.* 2003, 2012), Alaska (Kerr & Drew 1965, 1968; Torrance 2012) and Japan (Ohtsubo *et al.* 1982; Torrance & Ohtsubo 1995; Hong *et al.* 2005) although southern hemisphere references are rare with no confirmed examples (Torrance 2012). Formal stratigraphic examples include the Leda Clay from the former Champlain Sea in Canada and the Ariake Bay Clay in Kyushu, Japan. All known quick clay behaving deposits are post-glacial in age. Quick clays have given rise to some significant landslide events, for example Rissa, Norway, in 1978 (Gregersen 1981; L'Heureux *et al.* 2012a, b); Notre Dame de la Salette, Quebec, 1908 (Ells 1908; Lemieux, Ontario, 1993 (Evans & Brooks 1994); and Saint-Jude, Quebec, 2010 (Locat *et al.* 2012). Quick clays present a major hazard due to their ability to generate large, destructive and high velocity landslides (Torrance 2012). Many of these landslide events have caused fatalities due to the very rapid nature of the slope failures with over 100 deaths being reported in Scandinavia in modern times (Malehmir *et al.* 2013) and in excess of 1150 in Norway in historic times (Solberg 2007).

The presence of quick clays in the UK is unclear, but the Quaternary history of the British islands suggests that the precursor conditions for their formation could be present and should be considered when undertaking construction in the coastal zone.

7.2 Mode of formation

Geotechnical research into the formation, properties and occurrence of fine-grained sediments that exhibit quick clay behaviour has been carried out since the 1940s, mainly in Scandinavia and Canada. Work by Rosenqvist (1953; 1960a, b; 1966; 1977; 1984) established a correlation with their potential behaviour with the post-depositional leaching of the salt porewater content from former marine clays.

Quick clay behaviour prone deposits develop from initially marine clays deposited from rock flour rich meltwater streams feeding into a near shore marine environment (Fig. 7.1). On glacial retreat crustal rebound (isostatic recovery) uplifts the marine sediments above current sea level eventually exposing them to a temperate weathering environment and soil leaching by fresh water (Fig. 7.1). In Norway for example the former syn glacial sea level can be found up to 220 m higher than present day sea levels. (Hutchinson 1991; 1992).

For a clay to develop 'quick' properties the sediment must have a flocculated structure and a high void ratio (Rosenqvist 1977; Quigley 1980). This flocculated structure (Fig. 7.2) would be the normal state in which fine grained sediments formed from glacial erosion had been deposited in marine and brackish subaqueous environments. In this setting silt- and clay-sized particles would rapidly flocculate to form these high void ratio sediments (Torrance 1983). Generally, in freshwater sedimentary environments clay-sized particles settle even more slowly than silt grains and tend to accumulate in a dispersed structure with a parallel orientation of particles. In more saline conditions silt and clay particles form aggregates (small flocculates) and settle together in a random pattern (Torrance 1983). This random alignment of particles (in effect a 'house of cards' structure) gives the flocculated material a higher than normal void space and hence potentially higher moisture content. Figure 7.3 demonstrates these different depositional structures developed in freshwater and marine clays. Table 7.1 summarizes a general model for quick clay development as proposed by Torrance (1983).

The mineral composition of quick clays is such that they are dominated by non-swelling clay particles with a low activity, contain a high proportion of fine quartz and hence contain very little actual clay minerals (Locat et al. 1985). The clay sized faction typically consists of quartz, feldspars, amphiboles, micas and chlorite. Only trace amounts of swelling clay minerals are usually present, as the liquid limit of high activity swelling clay minerals increases during leaching which would prevent the development of potential quick clay behaviour within the deposits (Gillott 1979; Torrance 1983). This lack of clay minerals results in a soil which has a very low plasticity index and this low plasticity facilitates the critical solid to liquid state transition which occurs upon failure (Smalley 1971, 1977; Smalley et al. 1984). Most clay sized

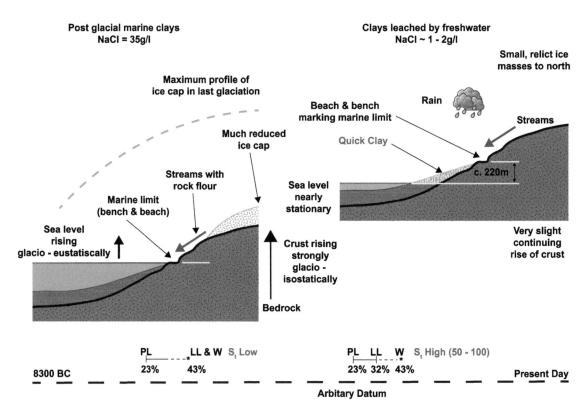

Fig. 7.1. The development of quick clays through the Holocene. (Adapted from Hutchinson 1992.)

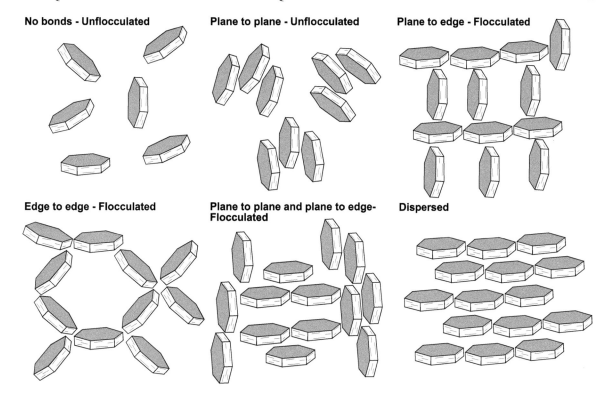

Fig. 7.2. Flocculated and non-flocculated clay structures.

minerals have activities of less than 1, for example: illite and chlorite (1 – <0.5); kaolinite (<0.5); and iron oxides, gibbsite, quartz, feldspars and amphiboles (<0.2). Activity of the smectite clay minerals ranges from 1 to 7 (Torrance 2012). Clays that exhibit quick clay behaviour can also vary greatly in their degree of cementation. This can be significant in that cementation increases the undisturbed strength of the soil and therefore increases the sensitivity of the deposit. Scandinavian quick clays are generally uncemented whereas the Canadian Leda Clay has varying degrees of cementation which can give sensitivities as high as 1500 (Smalley 1971).

Quick clay sediments originally deposited in marine or brackish conditions initially had a porewater geochemistry of up to 35 g/l sodium chloride (Bjerrum 1954). This high cationic strength porewater brought about a flocculation of the clay particles which then formed links between the silt grains (Torrance 1983). Subsequent uplift of the strata to above sea level resulted in them being subject to temperate weathering conditions where soil leaching by fresh water occurred. This weathering created a top crust of leached material with a subsequent reduction in the strength of the former marine clays. The sodium chloride porewaters were progressively leached by rainwater and freshwater streams reducing the salt content to around 1–2 g/l. This had the effect of generating very sensitive clay dominated soils that exist in a metastable state.

The leaching process is a critical factor in the development of the sensitivity of the soil. Leaching occurs through rain and snow meltwater percolating through the fine grained initially marine sediments. Groundwater can also seep upwards through the deposit, due to artesian pressure, mainly through higher permeability soils that may be present as well as diffusion of salts towards ones with lower ion concentrations (Rankka *et al.* 2004). Leaching affects the forces between the soil particles but normally not the soil structure; namely, the flocculated structure of the initially sub aqueously deposited silts and clays.

Whereas soil leaching is critical in the development of potential quick clay behaviour, Torrance (1983) identified other factors which would inhibit quick clay prone soil development; namely, the presence of swelling clay minerals; the weathering process where organic material is produced which can lead to an increase in the remoulded strength and hence a reduction in sensitivity; deep burial with the associated increase in consolidation; and the presence of high valence cations such as Fe^{3+} and Al^{3+}. This is important in the UK context given that for Scandinavian quick clays the predominant sediment source is derived from mainly quartz

Fig. 7.3. Freshwater and marine clay depositional structure. In freshwater, clay minerals settles in a rather compact structure. The structure consists of the clay mineral which binds a small amount of water (bound water) and free water between the minerals. In a marine environment where the salt content is higher than 3%, clay minerals will form a more chaotic structure than freshwater clay due to a polarization of the clay particles. (Adapted from C. Olsen.)

and feldspar rich igneous and metamorphic rocks as opposed to the more sedimentary source rocks found in the UK.

7.3 Geotechnical properties and behaviour

Potential quick clay behaving sediments can be identified by their geotechnical properties in particular by their sensitivity (S_t), the ratio of undrained shear strength to remoulded shear strength at the same moisture content (equation 7.1), and by their activity (A_c) (equation 7.2) (Skempton & Northey 1952; Rosenqvist 1953; Skempton 1953; Skempton & Henkel 1953; Bjerrum & Kenney 1968; Torrance 1983, Reeves et al. 2006).

$$\text{Sensitivity } S_t = \frac{\text{Undisturbed Strength}}{\text{Remoulded Strength}} = \frac{c_u}{c_r} \quad (7.1)$$

$$\text{Activity } A_c = \frac{I_p}{\% \text{ Clay Particles}} \quad (7.2)$$

One of the key characteristics of fine grained sediments prone to quick clay behaviour is that the sediment is dominated by non-swelling minerals which give the material a low activity (Dumbleton & West 1966; Rankka et al. 2004). This concept of activity defined by Skempton (1953) considers the relationship between the plasticity index (I_P) and the clay faction (<2 μm) of the sediment. Quick clay prone deposits are usually inactive with an activity less than 0.5 (Gillott 1979; Geertsema & Torrance 2005). Figure 7.4 from Table 7.2 demonstrates the plasticity index – clay content relationship for a variety of quick clay behaving deposits in Scandinavia and Canada with the majority of these materials having an activity of less than 0.5 Skempton (1953) and Bjerrum (1954) separated activity into three classes: inactive clays, normal clays and active clays (Table 7.3).

Skempton & Northey (1952) proposed a series of sensitivity classes (Table 7.4), work which was further refined with respect to Norwegian quick clays by Rosenqvist (1953) to give greater subdivision of quick clay prone sediments (Table 7.5). The Norwegian Geotechnical Institute define a q*uick clay* as having a sensitivity greater than 30 and having a remoulded shear strength of less than 0.5 kPa (Norsk Geoteknisk Forening 1974; Torrance 1983, 2012). To exhibit quick clay behaviour the soils would normally have low plasticity indices (8–12%), a liquidity index (equation 7.3) that normally exceeds 1 and an average liquid limit of less than 40% (Reeves et al. 2006).

$$\text{Liquidity Index } I_L = \frac{W - W_p}{W_l - W_p} \quad (7.3)$$

Due to their flocculated structure quick clays have a very open fabric and high void ratio and have the potential to have high moisture contents compared with sediments with orientated structures (Rankka et al. 2004). Figure 7.5 illustrates the variation of key defining geotechnical properties with depth for a Canadian quick clay deposit.

7.4 Failure mechanisms

Numerous studies have been undertaken on quick clay landslides with a view to understanding and categorizing the failure mechanisms (e.g. Brown & Paterson 1964; Bjerrum et al. 1969; Cabrera & Smalley 1973; Smalley 1976; Carson 1977; Penner & Burn 1978; Janbu 1979; Kerr 1979; Ter-Stepanian 2000; Hungr et al. 2001; Thakur et al. 2006; Khaldoun et al. 2009). Gregersen (1981) reported on a quick clay failure at Rissa in Norway which occurred in April 1978 and which was caught on film (Norwegian Geotechnical Institute 2008). Gregersen observed that quick clay failures occurred in two ways, either as retrogressive slides, developing relatively slowly, or as 'flake-type', spreading failures which fail instantly. The state of stress that exists in the quick clay prior to instability was observed as a key factor in controlling the failure. When loading of these clays beyond a critical stress level occurs there is a tendency for a volume decrease to take place with a resulting pore pressure increase. To

Table 7.1. *General model for quick clay development (Torrance 1983, 2012; Torrance & Ohtsubo 1995)*

Factors producing a high undisturbed strength	
Depositional	*Post-depositional*
Flocculation[*,†]	**Cementation bonds**
Salinity[*]	Rapidly developed
Divalent cation adsorption[†]	Slowly developed
High suspension concentration	**Slow load increase**
	Time for cementation
	Thixotropic processes
	Other time dependent processes
	Diagenetic changes

Factors producing a high undisturbed strength	
Depositional	*Post-depositional*
Material properties	**Salt removal**
Low activity minerals dominate[*,†]	By leaching and/or diffusion[*]
	Decrease in liquid limit is greater than decrease in water content
	Dispersants[†]
	Minimal consolidation

[*]Essential in marine clays.
[†]Essential in freshwater clays.

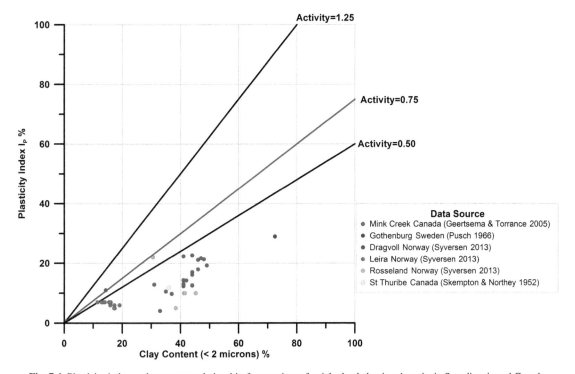

Fig. 7.4. Plasticity index – clay content relationship for a variety of quick clay behaving deposits in Scandinavia and Canada.

Table 7.2. *Geotechnical data for a variety of quick clay behaving soils*

Location	Plasticity index I_P %	Clay content L_c %	Activity A_c	Reference
Mink Creek, Canada	22.4	41.0	0.55	Geertsema & Torrance (2005)
	22.7	44.0	0.52	
	17.1	44.0	0.39	
	21.2	46.0	0.46	
	21.8	47.0	0.46	
	21.4	48.0	0.45	
	19.3	49.0	0.39	
	12.6	44.0	0.29	
	14.3	42.0	0.34	
	14.3	41.0	0.35	
	12.7	44.0	0.29	
	18.0	46.0	0.39	
	16.2	44.0	0.37	
	13.1	41.0	0.32	
	4.1	33.0	0.12	
	14.4	41.0	0.35	
	12.4	41.0	0.30	
	9.8	37.0	0.26	
	12.9	31.0	0.42	
	10.6	35.0	0.30	
	9.8	37.0	0.26	
Gothenberg, Sweden	29.0	72.5	0.40	Pusch (1966)
Dragvoll, Norway	11.0	14.3	0.77	Syversen (2013)
	7.0	13.5	0.52	
	7.0	15.5	0.45	
	5.0	17.4	0.29	
Leira, Norway	6.0	17.3	0.35	Syversen (2013)
	6.0	15.9	0.38	
	7.0	16.1	0.43	
	7.0	14.0	0.50	
	5.0	17.2	0.29	
	7.0	12.9	0.54	
	7.0	15.8	0.44	
	7.0	11.5	0.61	
	6.0	19.1	0.31	
Rosseland, Norway	10.0	41.2	0.24	Syversen (2013)
	10.0	41.5	0.24	
	10.0	45.2	0.22	
	22.0	30.6	0.72	
	5.0	38.4	0.13	
St Thuribe, Canada	11.9	36.0	0.33	Skempton & Northey (1952)
Various, Norway	12.6	36.0	0.36	Bjerrum (1954)
	38.3	60.0	0.64	
	13.4	65.0	0.21	
	10.6	44.0	0.24	
	15.4	46.0	0.33	
	24.5	51.0	0.48	
	11.8	56.0	0.21	
	24.4	52.0	0.47	
Hawkesbury, Canada	35.5	86.0	0.41	Quigley *et al.* (1985)
	48.8	80.0	0.61	
	55.5	89.0	0.60	
	60.6	85.0	0.71	
	41.3	81.0	0.50	
	33.4	88.0	0.37	

(*Continued*)

Table 7.2. *Continued.*

Location	Plasticity index	Clay content	Activity	Reference
	I_P %	L_c %	A_c	
	33.0	82.0	0.40	
	39.5	75.0	0.52	
	46.1	86.0	0.53	
	45.6	84.0	0.54	
	33.7	78.0	0.43	
	48.4	80.0	0.60	
	39.7	91.0	0.43	
	34.1	91.0	0.37	
	36.7	82.0	0.44	
	36.8	95.0	0.38	
	34.5	89.0	0.39	
	32.2	83.0	0.38	
	38.0	93.0	0.40	

Table 7.3. *Designation of clay with regard to activity (Skempton 1953; Bjerrum 1954)*

Activity, A_c	Designation
0–0.75	Inactive clays
0.75–1.4	Normal clays
>1.4	Active clays

Table 7.4. *Sensitivity scale (Skempton & Northey 1952)*

Sensitivity	Designation
1	Insensitive clays
1–2	Clays of low sensitivity
2–4	Clays of medium sensitivity
4–8	Sensitive clays
8–16	Extra sensitive clays
>16	Quick clays

Table 7.5. *Sensitivity scale (Rosenqvist 1953)*

Sensitivity	Designation
1	Insensitive clays
1–2	Slightly sensitive clays
2–4	Medium sensitive clays
4–8	Very sensitive clays
8–16	Slightly quick clays
16–32	Medium quick clays
32–64	Very quick clays
>64	Extra quick clays

obtain an increase in the soil's shear strength the effect of the increased mobilized effective internal angle of friction must be greater than the effective stress reduction due to the increase in porewater pressure. If a quick clay is loaded undrained beyond this critical stress level the porewater pressures will increase dramatically as the metastable 'house of cards' clay particle structure starts to collapse. This will result in a catastrophic decrease in shear strength. As a consequence of this process, failure takes place almost instantaneously, long before the internal angle of friction is fully mobilized. This dramatic failure of quick clay deposits is what accounts for the frequent loss of life due to these particular landslides. For the flake-like spreading failures to occur the initial stress levels in the quick clay deposits must be very close to the critical stress level. Any small increase of stress, due to loading, vibration or erosion, will result in a failure of a large area simultaneously (Gregersen 1981).

Description of progressive failures in both Canadian and Scandinavian sensitive clays (Locat *et al.* 2011) recognized three types of retrogressive landslide involving flows, translational movements and spreading failures (Fig. 7.6). Investigation of the 1989 Saint-Liguori, Quebec, landslide showed a typical morphology of spreads with horsts of clay and grabens covered with grass (Fig. 7.7).

Work by Torrance (2012) describes the factors that influence the development of a quick clay failure and the final state of the debris (Table 7.7). Geomorphologically the failures were described as:

(1) *Stepwise landslides*, where the failure was retrogressive and multidirectional occurring in thick quick clay deposits with a high liquidity index. The extent of the quick clay that liquefies is sufficient to transport overlying sediment as blocks.
(2) *Uninterrupted landslides*, where there is an uninterrupted sequence of failures. These include flake slides

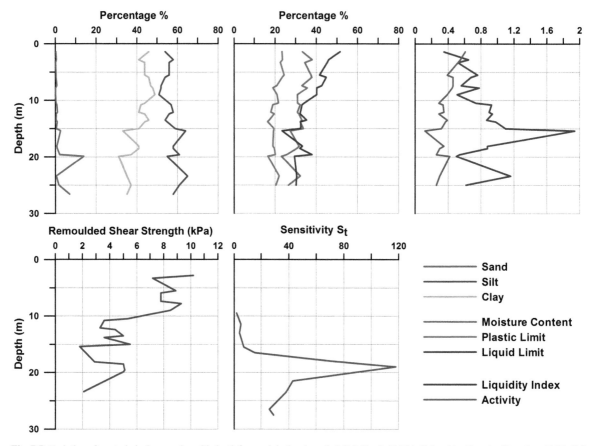

Fig. 7.5. Variation of geotechnical properties with depth for a quick clay deposit, Mink Creek, British Columbia, Canada. (Data from Table 7.6 after Geertsema & Torrance 2005.)

where the failure is initiated when a relatively thin layer of quick clay collapses and liquefies.

(3) *Ribbed landslides* where a spreading failure caused by the sediment liquefying leaves the overlying material as a series of elongate ridges orientated parallel and perpendicular to the spreading direction.

7.5 The UK context

Various studies on post-glacial isostatic recovery and eustatic sea level adjustment indicate that parts of the UK coastal zone have been elevated above former sea levels; for example, over 40 m in northern Scotland (Lambeck 1993a, b; Shennan et al. 2000; Smith et al. 2000, 2010; Shennan & Horton 2002; Milne et al. 2006). Effects of rising sea water levels due to the melting of the ice sheets were matched and finally reversed by the isostatic uplift of the land. Changes in relative sea-level due to glacioisostatic readjustment of the Earth's surface continue to affect the UK to the present day, with NW Scotland rising and SE England sinking. The uplift in Scotland reflects a continuing rebound of the Earth's surface following the melting of the ice sheet that covered NW Britain during the last glacial period. The sinking of SE England is a little more complex with much of the change resulting from a glacioisostatic collapse of a 'fore-bulge' that developed in front of the area of the Earth's surface that was depressed by the adjacent ice sheet. The rates of uplift in Scotland are in the order of 1.0–1.6 mm a^{-1}, while the rate of sinking of south and SE England are in the order of 0.5–1.0 mm a^{-1} (Fig. 7.8). While this mechanism is a significant factor in the changes in relative sea levels in SE England, it cannot explain all the observed variations in sea level reflected in the nearshore sediments deposited across SE England during the Holocene. Neotectonic influences are also identified (Culshaw et al. 2017).

The possibility that former fine-grained marine sediments have subsequently been elevated above sea level and have been subject to weathering processes and potential pore

Table 7.6. *Mink Creek Geotechnical Data (Geertsema & Torrance 2005)*

Depth (m)	W (%)	W_L (%)	W_P (%)	I_P	I_L	Activity	Sand (%)	Silt (%)	Clay (%)	S_R (kPa)	c_u	c_r	S_t
1.4	33.4	51.5	23.4	28.1	0.35	0.61	0.3	54	46				
2.8	38.0	46.0	23.6	22.4	0.65	0.55	0.6	58	41	10.2			
3.3	34.5	45.3	22.6	22.7	0.52	0.52	0.2	56	44	7.2			
5.5	37.5	41.6	24.4	17.1	0.76	0.39	0.4	56	44	8.9			
5.9	37.9	44.8	23.5	21.2	0.68	0.46	0.4	54	46	7.8			
7.4	33.1	42.7	20.9	21.8	0.56	0.46	0.8	52	47	7.8			
7.8	35.7	40.3	19.0	21.4	0.78	0.45	0.1	52	48	9.3			
9.0	30.9	40.3	21.0	19.3	0.51	0.39	0.2	51	49	8.5			
9.5											59.8	31.50	2
10.5	31.0	34.3	21.7	12.6	0.74	0.29	0.4	56	44	5.3			
10.8	32.3	33.2	19.0	14.3	0.93	0.34	0.8	57	42	3.6			
11.5											36.5	7.20	5
12.1	30.9	32.4	18.0	14.3	0.90	0.35	1.1	58	41	3.3			
12.4	31.6	32.3	19.6	12.7	0.94	0.29	0.5	56	44	4.4			
13.0											27.0	6.70	4
13.5	33.1	35.4	17.3	18.0	0.87	0.39	0.4	54	46	5.0			
13.8	32.3	32.6	16.5	16.2	0.98	0.37	1.0	55	44	3.6			
15.0	33.8	32.6	19.5	13.1	1.10	0.32	0.5	59	41	5.5			
15.4	27.4	23.4	19.3	4.1	1.94	0.12	2.6	64	33	1.8			
15.5											34.0	4.70	7
16.5											37.0	2.40	15
18.0											46.0	0.65	72
18.1	31.7	33.4	19.0	14.4	0.88	0.35	0.8	58	41	2.9			
18.5	30.6	32.1	19.7	12.4	0.88	0.30	1.0	58	41	5.0			
19.0											26.0	0.22	118
19.6	25.6	38.0	20.2	9.8	0.54	0.26	2.1	61	37	5.1			
19.9	23.1	29.5	16.6	12.9	0.50	0.42	14.0	55	31	5.0			
21.5											47.0	1.10	43
23.4	32.3	30.6	22.0	10.6	1.16	0.30	0.5	65	35	2.1			
23.5											65.0	1.70	38
25.0	26.5	30.3	20.5	9.8	0.62	0.26	1.7	61	37				
26.5											76.0	2.90	26
26.6							7.0	58	35				
27.5											76.0	2.60	29

water leaching potentially exists in these now onshore coastal areas. The uplifted zones will have experienced the preconditions for quick clay prone behaving sediments to have been developed. In terms of ground investigation in these areas the geotechnical properties of any fine-grained sediments encountered need to be considered with respect to potential quick clay behaviour, specifically with respect to the sensitivity and activity of the deposit as well as the nature of the mineral content of the soil. An awareness that these soils could be prone to rapid failure coupled with a complete remoulding of the soil with the associated liquefaction needs to be taken into account for the design and implementation of the construction works and needs to form part of the hazard assessment and project risk management via the *Risk Register* for the project.

Despite the UK literature being mainly void of any references to quick clay behaviour in investigated sites some possible examples can be found. Quick clay behaviour was reported during the construction of the Portavadie dry dock on Loch Fyne on the west coast of Scotland (Clark *et al.* 1979; Evans 1987). Several minor and major slope failures occurred during the muck shifting operations. The material at the site was described as behaving in a manner similar to the quick clay liquefaction slides of Scandinavia yet no sensitivities had been tested for or recorded in the ground investigation reports prior to the failures at the construction site. Subsequent testing reported sensitivity values in a red laminated silt horizon of between 2.5 and 22 with one test result reported at 49 (Clark *et al.* 1979), well above the threshold value of 30 for quick clay definition. Some of the samples tested reported having natural moisture contents higher than the liquid limit. These laminated silts were deposited under marine conditions (thought to be in about 20 m depth of sea water) around the Late Glacial period 12 500–11 000 BP. Similar deposits had been reported 19 km east of the site at Ardyne Point (Graham & Wilkinson 1978; Peacock *et al.* 1978). Laboratory test on these deposits indicated that their salinity had been reduced by freshwater leaching, one of the key conditions for quick clay development.

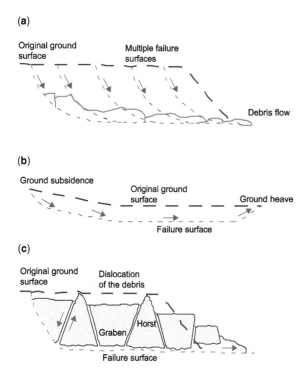

Fig. 7.6. Three types of retrogressive landslide in sensitive clays: (a) flow, (b) translational progressive landslide, and (c) spread. (Adapted from Locat *et al.* 2011.)

Work on the geotechnical properties of the Belfast postglacial estuarine clay known locally as *sleech* (Crooks & Graham 1976; Gregory & Bell 1991) has suggested that these soft sensitive clays may be prone to quick clay behaviour. The post glacial isostatic recovery history of the area suggests a potential quick clay hazard zone. Isostatic upwarping exposed large areas of the estuarine deposits thus forming the site for the city of Belfast (Gregory & Bell 1991). The area is dominated by these fine grained deposits with reported sensitivities of between 5 and 15 with the average around 8, the clay having very high natural moisture contents of up to 93% with associated high void ratios with evidence of the reduction of the Mg cation through leaching (Gregory & Bell 1991).

Work on the Clyde Alluvium (McGown & Miller 1984) highlights a glacial and post-glacial history which is concordant with a potential quick clay forming conditions, with similarities to Champlain Sea in Canada (Parent & Occhietti 1988). The Clyde Valley was glaciated and subsequently isostatically uplifted on ice retreat. A complex pattern of subsequent sediment infill took place with fluvioglacial, lacustrine, marine, brackish and freshwater deposits, all undergoing and subject to post-glacial weathering and potential leaching (Finlayson 2012; Browne & McMillan 1989). Raised glaciomarine deposits (Devensian Clyde Clay Formation of the British Coastal Deposits Group) are found at or close to the ground surface across western parts of the basin that lie below 40 m a.s.l., the relative sea-level at about 15 ka BP (Finlayson 2012). During deglaciation local relative sea-level was high in central Scotland and it is not uncommon to find raised marine sediments at about 40–45 m AOD (Browne 1991; Browne *et al.* 1984). Sediments that are characteristic of the deglaciated sea lochs are rather massive looking silty clays or clayey silts that usually contain many marine fossils (Browne 1991). As a result of the development of the Clyde Sea Loch a uniform layer of marine shell bearing silty clay was deposited directly onto freshwater lacustrine silts within the area of the central Clyde Basin. Figure 7.9 shows the approximate extent of the Clyde Sea Loch through the raised superficial deposits.

Investigations for the Port Glasgow Eastern Bypass (McGown & Miller 1984) encountered soil profiles that were typical of other soft sediment sites around the Clyde Estuary. Marine clays resting directly on glacial tills with fluvial silts and sands above. A large number of vane tests gave sensitivities between 2 and 6 as well as high liquidity indices (in excess of 100%) near the surface. These could be considered as *Sensitive Clays* after Skempton & Northey (1952) or *Very Sensitive Clays* after Rosenqvist (1953) but certainly not classified as quick clays with the 30 sensitivity boundary.

Investigations into the estuarine alluvium deposits of the UK (Hawkins 1984, 1994) found significant sedimentary deposits with a flocculated 'card house' structure with high initial moisture contents but did not report sensitivity or activity values. Work on the sediments from the Avonmouth area in Bristol (Hawkins 1984) found that average plasticity indices were calculated as 20% with clay factions measured as greater than 50% implying possible activities of 0.4 or less. These results would suggest scope for further investigating such deposits for possible quick clay behaviour. These sediments would have experienced at least 20 m of post-glacial uplift.

Geotechnical data on the soft clays of the Somerset Levels Formation (*Somerset Alluvium*) acquired during the investigation for the M5 motorway (Cook & Roy 1984) presented remoulded shear strength values in the range of 1.2–14 kPa, well above the suggested upper limit of 0.5 kPa for Torrance's quick clay definition (Torrance 1983).

In the area around Grangemouth (Sissons 1970) in Scotland extensive late glacial fine-grained marine deposits can be encountered known locally as *carse* (Carse Clay Formation) (0) which can be found up to elevations of 37 m AOD, the maximum level of the isostatically raised shoreline (Fig. 7.10). The silts and clays of the *carse* have typically high moisture contents and have a 1 m plus firm weathered crust. Skempton (1953) reports activity values from the late glacial silts and clays as being in the region of 0.75, just on the typical quick clay defining boundary although there is discussion of these deposits being possibly tsunami related (Smith *et al.* 2007).

During the 1980s a highway embankment construction across the Strathearn Valley to the south of Perth encountered

Fig. 7.7. Photograph of the 1989 landslide at Saint-Liguori, Quebec, Canada, showing the typical morphology of spreads with horsts of clay and grabens covered with grass. (Serge Leroueil, Université Laval, with permission).

Table 7.7. *Factors influencing the development of quick clay failures (after Torrance 2012)*

Factor	Influence
1	Undisturbed shear strength profile, as affected by consolidation, cementation, and weathering, determines the slope height and angle that are stable, the depth where the initial failure of quick clay occurs, the ease of structural breakdown during failure, and the extent to which quick clay above the initial failure zone experiences post-failure remoulding.
2	Remoulded shear strength, related to the liquidity index determines the flow properties of the debris.
3	Thickness of the quick clay layer that initially fails and the proportion of the depth profile above the initial failure zone that liquefies influence how rapidly debris can exit the failure zone and how far it can travel.
4	Thickness of the surface crust and other overburden at the site influences the difficulty of transporting the debris away from the landslide scar.
5	Adversely dipping bedding planes can facilitate debris transport.

a buried valley system which was excavated in the Late Glacial (18 000–10 000 BP) to a depth of 70 m below present day sea level (Cochrane & Carter 1991). The valley had been infilled with thick deposits of soft brown laminated clay overlain by very soft grey clayey silt and fine silty sand. Subsequent isostatic uplift exposed these sediments to weathering and leaching but despite the glacial and post-glacial history of these sediments no quick clay behaviour was reported.

One of the most interesting potential UK sites for the presence of quick clay behaviour prone sediments was at the SERC Soft Clay Test Bed Site at Bothkennar, Scotland (Hawkins *et al.* 1989), which was fully described in the 1992 Geotechnique Symposium in Print (Institution of Civil Engineers 1992). The site presented a remarkably uniform soft silty clay (Carse Clay Formation and *Came Clay*) which was considered to have been deposited under stable estuarine conditions between 8500 and 6000 BP. The site had been isostatically uplifted by at least 30 m with evidence of a 13 000 BP palaeo shoreline present (Nash *et al.* 1992). Clays at the site had a high undrained strength with a reported activity of 1.34 and sensitivity of about 5 but geotechnical testing on one of the Carse Clay Formation subdivisions (*Came Clay*) found interesting results (Paul *et al.* 1992).

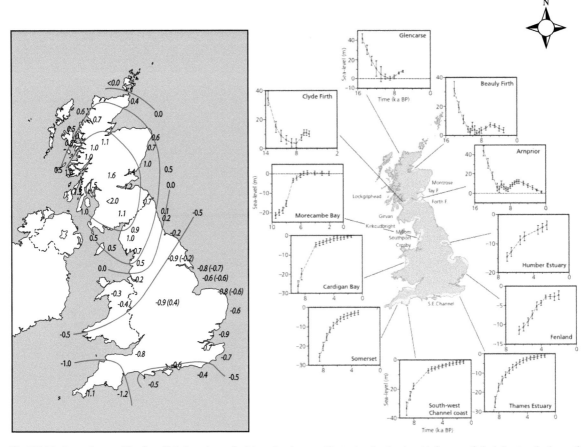

Fig. 7.8. The isostatic record for Great Britain and postglacial sea-level curves illustrating the dominant influence of glacioisostasy in the north and glacioeustasy in the south. (Culshaw *et al.* 2017; Lukas *et al.* 2017.) Rates of late Holocene relative sea- and land-level changes (in mm a^{-1}) due to the glacioisostatic adjustment. Positive values indicate relative land uplift or sea-level fall; negative values are relative land subsidence or sea-level rise. Figures in brackets are the trends that take into account modelled changes in tidal range during the Holocene.

The mineralogical nature of the *Carse Clay* was an illite-chlorite-rock flour mixture suggesting a low activity of around 0.5 placing it in the 'quick clay' potential domain but due to the presence of smectite (with its swelling potential) the activity values were pushed up to 1.25 although the smectite from XRD analysis was not found to be significant. A higher organic content was thought to have influenced these activity values. When this organic content was removed by hydrogen peroxide the activity value dropped to 0.4 (Paul *et al.* 1992) with other results as low as 0.25 (Hight *et al.* 1992). Sensitivities were reported of 5–8 as well as from 7 to 13 yet quick clay behaviour was not discussed in this research. The sediments had a high in-situ void ratio with a flocculated structure being deposited in brackish waters. Salinities of 21 g/l were reported. Some post depositional leaching had taken place. The data from this test site would be interesting to revisit within a quick clay context.

7.6 Geohazard management and mitigation

Due to the high hazard presented by quick clay failures many countries with these deposits have undertaken hazard assessment programmes mainly through mapping and delineating their spatial extent (Gagnon 1975). Landslide inventories have been generated where failures in former marine sediments have been encountered. Additionally to mapping programmes some geophysical techniques have been trialled in order to identify areas of low porewater salinity which may potentially indicate quick clays (Carson 1977; Dahlin *et al.* 2005; Lundström *et al.* 2009 Hunter *et al.* 2010; Donohue *et al.* 2012; Solberg *et al.* 2012). Other more experimental schemes have been explored to prevent failure, investigating the chemical modification of the leached sediments by reintroducing salt, via injection wells, into the subsurface system

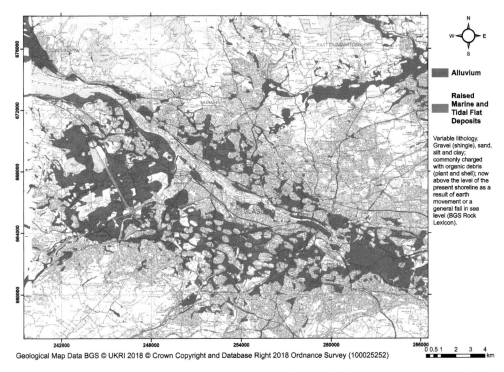

Fig. 7.9. Raised marine superficial deposits around the former Clyde Sea Loch, Glasgow, Scotland (Geological Map Data BGS © UKRI 2018).

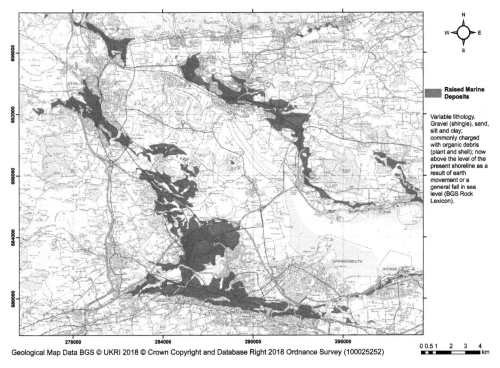

Fig. 7.10. Raised marine superficial deposits around Grangemouth, Scotland (Geological Map Data BGS © UKRI 2018).

(Moum et al. 1968). The use of lime columns has also been explored. The nature of the quick clay failures suggests that the normal physical methods for landslide control and prevention such as introduced drainage, toe support, etc. are impractical (Torrance 2012).

7.7 Conclusions

Where encountered in the coastal zone former marine clays which have been isostatically uplifted above present-day sea levels present a considerable geohazard. Their ability to completely liquefy on disturbance or overloading produces large volume, high velocity landslides which pose a significant risk to life and property. The Quaternary conditions for quick clay development in the UK are certainly present yet there are very few candidate sites reported for such behaviour. For construction sites within the coastal zone which would have experienced marine depositional conditions within the Quaternary and that have been subsequently uplifted the possibility that quick clay behaving sediments may be present should be considered and should be tested for through the soils laboratory testing programme. The sensitivity and activity of any fine-grained sediments should be reported.

Acknowledgments Professor Jim Griffiths, University of Plymouth, is thanked for his helpful review comments. Professor Jim Graham, University of Manitoba, is also thanked for his comments on Bothkennar, the Belfast estuarine deposits and other aspects of quick clays.

Funding This research received no specific grant from any funding agency in the public, commercial, or not-for-profit sectors.

References

AAS, G. & LUNNE, T. 1981. *Stability of Natural Slopes in Quick Clays*. Norges Geotekniske Institutt.

ANDERSSON-SKÖLD, Y., TORRANCE, J.K., LIND, B., ODÉN, K., STEVENS, R.L. & RANKKA, K. 2005. Quick clay – a case study of chemical perspective in Southwest Sweden. *Engineering Geology*, **82**, 107–118, https://doi.org/10.1016/j.enggeo.2005.09.014

BJERRUM, L. 1954. Geotechnical properties of Norwegian marine clays. *Geotechnique*, **4**, 49–69, https://doi.org/10.1680/geot.1954.4.2.49

BJERRUM, L. & KENNEY, T.C. 1968. Effect of structure on the shear behaviour of normally consolidated quick clays. *Proceedings of the Geotechnical Conference, shear strength properties of natural soils and rocks*, 1967, Oslo, Norwegian Geotechnical Institute, **2**, 19–27.

BJERRUM, L., LØKEN, T., HEIBERG, S. & FOSTER, R. 1969. A field study of factors responsible for quick clay slides. In *Proc. Int. Conf. on Soil Mechanics and Foundation Engineering*, Mexico City, Mexico, B31–B40.

BROWN, J.D. & PATERSON, W.G. 1964. Failure of an oil storage tank founded on a sensitive marine clay. *Canadian Geotechnical Journal*, **1**, 205–214, https://doi.org/10.1139/t64-016

BROWNE, M.A.E. 1991. An introduction to the geology of central Scotland. *In*: FORSTER, A., CULSHAW, M.G., CRIPPS, J.C. & MOON, C.F. (eds) *Quaternary Engineering Geology*. Geological Society, London, Engineering Geology Special Publications, **7**, 63–70, https://doi.org/10.1144/GSL.ENG.1991.007.01.03

BROWNE, M.A.E. & MCMILLAN, A.A. 1989. *Quaternary Geology of the Clyde valley*. British Geological Survey Research Report, **SA/89/1**.

BROWNE, M.A.E., GRAHAM, D.K. & GREGORY, D.M. 1984. *Quaternary Estuarine Deposits in the Grangemouth Area, Scotland*. Report of the British Geological Survey, 16, 3. HM Stationery Office.

CABRERA, J.G. & SMALLEY, I.J. 1973. Quick clays as products of glacial action: a new approach to their nature, geology, distribution and geotechnical properties. *Engineering Geology*, **7**, 115–133, https://doi.org/10.1016/0013-7952(73)90041-0

CARSON, M.A. 1977. On the retrogression of landslides in sensitive muddy sediments. *Canadian Geotechnical Journal*, **14**, 582–602, https://doi.org/10.1139/t77-059

CLARK, A.R., HAWKINS, A.B. & GUSH, W.J. 1979. The Portavadie dry dock, west Scotland: a case history of the geotechnical aspects of its construction. *Quarterly Journal of Engineering Geology and Hydrogeology*, **12**, 301–317, https://doi.org/10.1144/GSL.QJEG.1979.012.04.07

COCHRANE, G.A. & CARTER, P.G. 1991. Highway embankment construction across the Strathearn buried Valley. *In*: FORSTER, A., CULSHAW, M.G., CRIPPS, J.C. & MOON, C.F. (eds) *Quaternary Engineering Geology*. Geological Society, London, Engineering Geology Special Publications, **7**, 203–209, https://doi.org/10.1144/GSL.ENG.1991.007.01.18

COOK, D.A. & ROY, M.R. 1984. A review of the geotechnical properties of Somerset alluvium using data from the M5 motorway and other sources. *Quarterly Journal of Engineering Geology and Hydrogeology*, **17**, 235–242, https://doi.org/10.1144/GSL.QJEG.1984.017.03.07

CRAWFORD, C.B. 1968. Quick clays of eastern Canada. *Engineering Geology*, **2**, 239–265, https://doi.org/10.1016/0013-7952(68)90002-1

CROOKS, J.H.A. & GRAHAM, J. 1976. Geotechnical properties of the Belfast estuarine deposits. *Geotechnique*, **26**, 293–315, https://doi.org/10.1680/geot.1976.26.2.293

CULSHAW, M.G., ENTWISLE, D.C., GILES, D.P., BERRY, T., COLLINGS, A., BANKS, V.J. & DONNELLY, L.J. 2017. Material properties and geohazards. *In*: GRIFFITHS, J.S. & MARTIN, C.J. (eds) *Engineering Geology and Geomorphology of Glaciated and Periglaciated Terrains – Engineering Group Working Party Report*. Geological Society, London, Engineering Geology Special Publications, **28**, 599–740, https://doi.org/10.1144/EGSP28.6

DAHLIN, T., LEROUX, V., LARSSON, R. & RANKKA, K. 2005, September. Resistivity imaging for mapping of quick clays for landslide risk assessment. In *11th European Meeting of Environmental and Engineering Geophysics*, Palmero, Italy, 2005, http://www.earthdoc.org/publication/publicationdetails/?publication=799.

DONOHUE, S., LONG, M. & O'CONNOR, P. 2012. Multi-method geophysical mapping of quick clay.

DUMBLETON, M.J. & WEST, G. 1966. Some factors affecting the relation between the clay minerals in soils and their plasticity. *Clay Minerals*, **6**, 179–193, https://doi.org/10.1180/claymin.1966.006.3.05

EILERTSEN, R.S., HANSEN, L., BARGEL, T.H. & SOLBERG, I.L. 2008. Clay slides in the Målselv valley, northern Norway: characteristics, occurrence, and triggering mechanisms. *Geomorphology*, **93**, 548–562, https://doi.org/10.1016/j.geomorph.2007.03.013

ELLS, R.W. 1908. *Report on the landslide at Notre-Dame de la Salette, Lièvre River, Quebec*. Geological Survey Branch, Department of Mines, Ottawa, Ontario, 1–15.

EVANS, N. 1987. Road construction on soft and compressible soils. *Quarterly Journal of Engineering Geology and Hydrogeology*, **20**, 101–102, https://doi.org/10.1144/GSL.QJEG.1987.020.01.12

EVANS, S.G. & BROOKS, G.R. 1994. An earthflow in sensitive Champlain Sea sediment at Lemieux, Ontario, June 20, 1993, and its impact on the South Nation River. *Canadian Geotechnical Journal*, **31**, 384–394, https://doi.org/10.1139/t94-046

FINLAYSON, A.G. 2012. Ice dynamics and sediment movement: last glacial cycle, Clyde basin, Scotland. *Journal of Glaciology*, **58**, 487–500, https://doi.org/10.3189/2012JoG11J207

GAGNON, H. 1975. Remote sensing of landslide hazards on quick clays of eastern Canada. In: *International Symposium on Remote Sensing of Environment*, 10th, Ann Arbor, Michigan, 803–810.

GEERTSEMA, M. 2013. Quick clay. In *Encyclopaedia of Natural Hazards*. Springer, Netherlands, 803–804.

GEERTSEMA, M. & TORRANCE, J.K. 2005. Quick clay from the Mink Creek landslide near Terrace, British Columbia: geotechnical properties, mineralogy, and geochemistry. *Canadian Geotechnical Journal*, **42**, 907–918, https://doi.org/10.1139/t05-028

GILLOTT, J.E. 1979. Fabric, composition and properties of sensitive soils from Canada, Alaska and Norway. *Engineering Geology*, **14**, 149–172, https://doi.org/10.1016/0013-7952(79)90082-6

GRAHAM, D.K. & WILKINSON, I.P. 1978. *A detailed Investigation of a Late-Glacial Faunal Succession at Ardyne, Argyll, Scotland*. Report Institute of Geological Sciences, London. 78/5. HM Stationery Office.

GREGERSEN, O. 1981. The quick clay landslide in Rissa, Norway. The sliding process and discussion of failure modes. In: *Proceedings of the 10th International Conference on Soil Mechanics and Foundation Engineering*. A. A. Balkema, Rotterdam, **3**, 421–426.

GREGERSEN, O. & LØKEN, T. 1979. The quick-clay slide at Baastad, Norway, 1974. *Engineering Geology*, **14**, 183–196, https://doi.org/10.1016/0013-7952(79)90084-X

GREGORY, B.J. & BELL, A.L. 1991. Geotechnical properties of Quaternary deposits in the Belfast area. In: FORSTER, A., CULSHAW, M.G., CRIPPS, J.C. & MOON, C.F. (eds) *Quaternary Engineering Geology*. Geological Society, London, Engineering Geology Special Publications, **7**, 219–228, https://doi.org/10.1144/GSL.ENG.1991.007.01.20

HAWKINS, A.B. 1984. Depositional characteristics of estuarine alluvium: some engineering implications. *Quarterly Journal of Engineering Geology and Hydrogeology*, **17**, 219–234, https://doi.org/10.1144/GSL.QJEG.1984.017.03.06

HAWKINS, A.B. 1994. Construction on recent alluvial: the importance of a correct interpretation of Quaternary geology. *Engineering Geology*, **37**, 67–77, https://doi.org/10.1016/0013-7952(94)90083-3

HAWKINS, A.B., LARNACH, W.J., LLOYD, I.M. & NASH, D.F.T. 1989. Selecting the location, and the initial investigation of the SERC soft clay test bed site. *Quarterly Journal of Engineering Geology and Hydrogeology*, **22**, 281–316, https://doi.org/10.1144/GSL.QJEG.1989.022.04.04

HIGHT, D.W., BOND, A.J. & LEGGE, J.D. 1992. Characterization of the Bothkennar clay: an overview. *Géotechnique*, **42**, 303–347, https://doi.org/10.1680/geot.1992.42.2.303

HOLMSEN, P. 1953. Landslips in Norwegian quick-clays. *Geotechnique*, **3**, 187–194, https://doi.org/10.1680/geot.1953.3.5.187

HONG, Z., LIU, S. & NEGAMI, T. 2005. Strength sensitivity of marine Ariake clays. *Marine Georesources and Geotechnology*, **23**, 221–233, https://doi.org/10.1080/10641190500218329

HUNGR, O., EVANS, S.G., BOVIS, M.J. & HUTCHINSON, J.N. 2001. A review of the classification of landslides of the flow type. *Environmental & Engineering Geoscience*, **7**, 221–238, https://doi.org/10.2113/gseegeosci.7.3.221

HUNTER, J.A., BURNS, R.A., GOOD, R.L., PULLAN, S.E., PUGIN, A. & CROW, H. 2010. Near-surface geophysical techniques for geohazards investigations: some Canadian examples. *The Leading Edge*, **29**, 964–977, https://doi.org/10.1190/1.3480011

HUTCHINSON, J.N. 1991. Periglacial and slope processes. In: FORSTER, A., CULSHAW, M.G., CRIPPS, J.C. & MOON, C.F. (eds) *Quaternary Engineering Geology*. Geological Society, London. Engineering Geology Special Publications, No. 7, 283–331.

HUTCHINSON, J.N. 1992. Engineering in relict periglacial and extraglacial areas in Britain. In: GRAY, J.M. (ed.) *Applications of Quaternary Research, Quaternary Proceedings No. 2*. Quaternary Research Association, Cambridge, 49–65.

INSTITUTION OF CIVIL ENGINEERS 1992. Symposium In Print, Geotechnique, 42, 2.

JANBU, N. 1979. *Failure Mechanism in Quick Clays. NGM-79*. Nordiska Geoteknikermotet, Helsinki.

KERR, P.F. 1979. Quick clays and other slide-forming clays. *Engineering Geology*, **14**, 173–181, https://doi.org/10.1016/0013-7952(79)90083-8

KERR, P.F. & DREW, I.M. 1965. *Quick clay Movements, Anchorage, Alaska (No. Scientific-5)*. Columbia Univ, New York.

KERR, P.F. & DREW, I.M. 1968. Quick-clay slides in the USA. *Engineering Geology*, **2**, 215–238, https://doi.org/10.1016/0013-7952(68)90001-X

KHALDOUN, A., MOLLER, P. *ET AL.* 2009. Quick clay and landslides of clayey soils. *Physical Review Letters*, **103**, 188301, https://doi.org/10.1103/PhysRevLett.103.188301

LAMBECK, K. 1993a. Glacial rebound of the British Isles – I. Preliminary model results. *Geophysical Journal International*, **115**, 941–959, https://doi.org/10.1111/j.1365-246X.1993.tb01503.x

LAMBECK, K. 1993b. Glacial rebound of the British Isles – II. A high-resolution, high-precision model. *Geophysical Journal International*, **115**, 960–990, https://doi.org/10.1111/j.1365-246X.1993.tb01504.x

L'HEUREUX, J.-S., EILERTSEN, R.S., GLIMSDAL, S., ISSLER, D., SOLBERG, I.-L. & HARBITZ, C.B. 2012a. The 1978 quick clay landslide at Rissa, Mid Norway: subaqueous morphology and tsunami Simulations. In: YAMADA, Y., KAWAMURA, K. *ET AL.* (eds) *Submarine Mass Movements and Their Consequences. Advances in Natural and Technological Hazards Research*, Springer, Netherlands, **31**, 507–516.

L'HEUREUX, J.S., LONGVA, O. *ET AL.* 2012b. Identification of weak layers and their role for the stability of slopes at Finneidfjord, northern Norway. In: YAMADA, Y., KAWAMURA, K. *ET AL. Submarine Mass Movements and Their Consequences*. Springer, Netherlands, 321–330.

LOCAT, A., LEROUEIL, S., BERNANDER, S., DEMERS, D., JOSTAD, H.P. & OUEHB, L. 2011. Progressive failures in eastern Canadian and Scandinavian sensitive clays. *Canadian Geotechnical Journal*, **48**, 1696–1712, https://doi.org/10.1139/t11-059

LOCAT, J., BERUBE, M.A., CHAGNON, J.Y. & GELINAS, P. 1985. The mineralogy of sensitive clays in relation to some engineering geology problems – an overview. *Applied clay science*, **1**, 193–205, https://doi.org/10.1016/0169-1317(85)90573-3

LOCAT, J., LEROUEIL, S. & LOCAT, P. 2003. On the mobility of quick clays: the cases of the St. Jean-Vianney flowslides of 1663 and 1971. In: *2nd Symposium On Rapid Mass Movements*, Naples.

LOCAT, P., DEMERS, D. ET AL. 2012. The Saint-Jude landslide of May 10, 2010, Québec, Canada. In: EBERHARDT, E., FROESE, C., TURNER, A.K. & LEROUEIL, S. (eds) *Landslides and Engineered Slopes: Protecting Society Through Improved Understanding. Proceedings of the 11th International and 2nd North American Symposium on Landslides*, CRC Press, Boca Raton, USA, **2**, 635–640.

LUKAS, S., PREUSSER, F., EVANS, D.J.A., BOSTON, C.M. & LOVELL, H. 2017. The quaternary. In: GRIFFITHS, J.S. & MARTIN, C.J. (eds) *Engineering Geology and Geomorphology of Glaciated and Periglaciated Terrains – Engineering Group Working Party Report*. Geological Society, London, Engineering Geology Special Publications, **28**, 31–57, https://doi.org/10.1144/EGSP28.2

LUNDSTRÖM, K., LARSSON, R. & DAHLIN, T. 2009. Mapping of quick clay formations using geotechnical and geophysical methods. *Landslides*, **6**, 1–15, https://doi.org/10.1007/s10346-009-0144-9

MALEHMIR, A., BASTANI, M., KRAWCZYK, C.M., GURK, M., ISMAIL, N., POLOM, U. & PERSSON, L. 2013. Geophysical assessment and geotechnical investigation of quick-clay landslides–a Swedish case study. *Near Surface Geophysics*, **11**, 341–350, https://doi.org/10.3997/1873-0604.2013010

MCGOWN, A. & MILLER, D. 1984. Stratigraphy and properties of the Clyde alluvium. *Quarterly Journal of Engineering Geology and Hydrogeology*, **17**, 243–258, https://doi.org/10.1144/GSL.QJEG.1984.017.03.08

MILNE, G.A., SHENNAN, I. ET AL. 2006. Modelling the glacial isostatic adjustment of the UK region. *Philosophical Transactions of the Royal Society A: Mathematical, Physical and Engineering Sciences*, **364**, 931–948, https://doi.org/10.1098/rsta.2006.1747

MOUM, J., SOPP, O.I. & LOKEN, T. 1968. *Stabilization of Undisturbed Quick Clay by Salt Wells*. Norwegian Geotechnical Institute Publ.

NASH, D.F.T., POWELL, J.J.M. & LLOYD, I.M. 1992. Initial investigations of the soft clay test site at Bothkennar. *Geotechnique*, **42**, 163–181, https://doi.org/10.1680/geot.1992.42.2.163

NORSK GEOTEKNISK FORENING 1974. *Retningslinjer for presentasjon av geotekniske undersökelser*. Oslo, Norway, **16** [in Norwegian].

NORWEGIAN GEOTECHNICAL INSTITUTE 2008. The Rissa landslide: Quick clay in Norway. (DVD recording). (24 minutes).

OHTSUBO, M., TAKAYAMA, M. & EGASHIRA, K. 1982. Marine quick clays from Ariake Bay area, Japan. Soils and foundations.

PARENT, M. & OCCHIETTI, S. 1988. Late Wisconsinan deglaciation and Champlain sea invasion in the St. Lawrence valley, Québec. *Géographie physique et Quaternaire*, **42**, 215–246, https://doi.org/10.7202/032734ar

PAUL, M.A., PEACOCK, J.D. & WOOD, B.F. 1992. The engineering geology of the Carse clay at the National Soft Clay Research Site, Bothkennar. *Geotechnique*, **42**, 183–198, https://doi.org/10.1680/geot.1992.42.2.183

PEACOCK, J.D., GRAHAM, D.K. & WILKINSON, I.P. 1978. *Late-Glacial and post-Glacial Marine Environments at Ardyne, Scotland, and their Significance in the Interpretation of the History of the Clyde Sea Area*. Report Institute of Geological Sciences, London. 78/17. HM Stationery Office.

PENNER, E. & BURN, K.N. 1978. Review of engineering behaviour of marine clays in Eastern Canada. *Canadian Geotechnical Journal*, **15**, 269–282, https://doi.org/10.1139/t78-024

PUSCH, R. 1966. Quick-clay microstructure. *Engineering Geology*, **1**, 433–443, https://doi.org/10.1016/0013-7952(66)90019-6

QUIGLEY, R.M. 1980. Geology, mineralogy, and geochemistry of Canadian soft soils: a geotechnical perspective. *Canadian Geotechnical Journal*, **17**, 261–285, https://doi.org/10.1139/t80-026

QUIGLEY, R.M., HAYNES, J.E., BOHDANOWICZ, A. & GWYN, Q.H.J. 1985. *Geology, Geotechnique, Mineralogy and Geochemistry of Leda Clay from Deep Boreholes, Hawkesbury Area, Prescott County*; Ontario Geological Survey, Study 29.

RANKKA, K., ANDERSSON-SKÖLD, Y., HULTÉN, C., LARSSON, R., LEROUX, V. & DAHLIN, T. 2004. Quick clay in Sweden. *Swedish Geotechnical Institute Report*, **65**, 145.

REEVES, G.M., SIMS, I. & CRIPPS, J.C. (eds). 2006. Clay materials used in construction. In: *Appendix B. Properties Data*. Geological Society, London, Engineering Geology Special Publications, **21**, 461–474, https://doi.org/10.1144/GSL.ENG.2006.021.01.21

ROSENQVIST, I.T. 1953. Considerations on the sensitivity of Norwegian quick-clays. *Geotechnique*, **3**, 195–200, https://doi.org/10.1680/geot.1953.3.5.195

ROSENQVIST, I.T. 1960a. Marine Clays and Quick clay Slides in South and Central Norway: Guide to Excursion; Internat. Geolog. Congress, 21. Session Norden 1960.

ROSENQVIST, I.T. 1960b. Marine clays and quick clays slides. Geology of Norway, Nor. Geol. Unders, 463–471.

ROSENQVIST, I.T. 1966. Norwegian research into the properties of quick clay – a review. *Engineering Geology*, **1**, 445–450, https://doi.org/10.1016/0013-7952(66)90020-2

ROSENQVIST, I.T. 1977, June. A general theory for quick clay properties. In *Proc. Third European Clay Conference*, Oslo, 215–228.

ROSENQVIST, I.T. 1984. The importance of pore water chemistry on mechanical and engineering properties of clay soils. *Philosophical Transactions of the Royal Society of London. Series A, Mathematical and Physical Sciences*, **311**, 369–373, https://doi.org/10.1098/rsta.1984.0034

SHENNAN, I. & HORTON, B. 2002. Holocene land-and sea-level changes in Great Britain. *Journal of Quaternary Science: Published for the Quaternary Research Association*, **17**, 511–526.

SHENNAN, I., LAMBECK, K. ET AL. 2000. Late Devensian and Holocene records of relative sea level changes in northwest Scotland and their implications for glacio-hydro-isostatic modelling. *Quaternary Science Reviews*, **19**, 1103–1136, https://doi.org/10.1016/S0277-3791(99)00089-X

SISSONS, J.B. 1970. Geomorphology and foundation conditions around Grangemouth. *Quarterly Journal of Engineering Geology and Hydrogeology*, **3**, 183–191, https://doi.org/10.1144/GSL.QJEG.1970.003.03.03

SKEMPTON, A.W. 1953. Soil mechanics in relation to geology. *Proceedings of the Yorkshire Geological Society*, **29**, 33–62, https://doi.org/10.1144/pygs.29.1.33

SKEMPTON, A.W. & HENKEL, D.J. 1953, August. The post-glacial clays of the Thames Estuary at Tilbury and Shellhaven. In *Proceedings of the 3rd International Conference on Soil Mechanics and Foundation Engineering*. Zürich, Switzerland, **1**, 302–308.

SKEMPTON, A.W. & NORTHEY, R.D. 1952. The sensitivity of clays. *Geotechnique*, **3**, 30–53, https://doi.org/10.1680/geot.1952.3.1.30

SMALLEY, I.J. 1971. Nature of quickclays. *Nature*, **231**, 310, https://doi.org/10.1038/231310a0

SMALLEY, I. 1976. Factors relating to the landslide process in Canadian quickclays. *Earth Surface Processes*, **1**, 163–172, https://doi.org/10.1002/esp.3290010206

SMALLEY, I.J. 1977. Landslides in sensitive soils. *Nature*, **266**, 408, https://doi.org/10.1038/266408a0

SMALLEY, I.J., FORDHAM, C.J. & CALLANDER, P.F. 1984. Towards a general model of quick clay development. *Sedimentology*, **31**, 595–598, https://doi.org/10.1111/j.1365-3091.1984.tb01822.x

SMITH, D.E., CULLINGFORD, R.A. & FIRTH, C.R. 2000. Patterns of isostatic land uplift during the Holocene: evidence from mainland Scotland. *The Holocene*, **10**, 489–501, https://doi.org/10.1191/095968300676735907

SMITH, D.E., FOSTER, I.D., LONG, D. & SHI, S. 2007. Reconstructing the pattern and depth of flow onshore in a palaeotsunami from associated deposits. *Sedimentary Geology*, **200**, 362–371, https://doi.org/10.1016/j.sedgeo.2007.01.014

SMITH, D.E., DAVIES, M.H. ET AL. 2010. Holocene relative sea levels and related prehistoric activity in the Forth lowland, Scotland, United Kingdom. *Quaternary Science Reviews*, **29**, 2382–2410, https://doi.org/10.1016/j.quascirev.2010.06.003

SOLBERG, I.L. 2007. *Geological, geomorphological and geophysical investigations of areas prone to clay slides: Examples from Buvika, Mid Norway*. PhD thesis.

SOLBERG, I.L., HANSEN, L., RØNNING, J.S., HAUGEN, E.D., DALSEGG, E. & TØNNESEN, J.F. 2012. Combined geophysical and geotechnical approach to ground investigations and hazard zonation of a quick clay area, mid Norway. *Bulletin of Engineering Geology and the Environment*, **71**, 119–133, https://doi.org/10.1007/s10064-011-0363-x

STEVENS, R.L., ROSENBAUM, M.S. & HELLGREN, L.G. 1991. Origins and engineering hazards of Swedish glaciomarine and marine clays. *In*: FORSTER, A., CULSHAW, M.G., CRIPPS, J.C. & MOON, C.F. (eds) *Quaternary Engineering Geology*. Geological Society, London, Engineering Geology Special Publications, **7**, 257–264.

SYVERSEN, F.S. 2013. Et studie av den mineralogiske sammensetningen i den norske sensitive leirer: med et geoteknisk perspektiv.

TER-STEPANIAN, G. 2000. Quick clay landslides: their enigmatic features and mechanism. *Bulletin of Engineering Geology and the Environment*, **59**, 47–57, https://doi.org/10.1007/s100640000052

THAKUR, V., GRIMSTAD, G. & NORDAL, S. 2006. Instability in soft sensitive clays.

TORRANCE, J.K. 1975. On the role of chemistry in the development and behaviour of the sensitive marine clays of Canada and Scandinavia. *Canadian Geotechnical Journal*, **12**, 326–335, https://doi.org/10.1139/t75-037

TORRANCE, J.K. 1979. Post-depositional changes in the pore-water chemistry of the sensitive marine clays of the Ottawa area, eastern Canada. *Engineering Geology*, **14**, 135–147, https://doi.org/10.1016/0013-7952(79)90081-4

TORRANCE, J.K. 1983. Towards a general model of quick clay development. *Sedimentology*, **30**, 547–555, https://doi.org/10.1111/j.1365-3091.1983.tb00692.x

TORRANCE, J.K. 1999. Physical, chemical and mineralogical influences on the rheology of remoulded low-activity sensitive marine clay. *Applied Clay Science*, **14**, 199–223, https://doi.org/10.1016/S0169-1317(98)00057-X

TORRANCE, J.K. 2012. Landslides in quick clay. *In*: CLAGUE, J.J. & STEAD, D. (eds) *Landslides: Types, Mechanisms and Modelling*. Cambridge University Press, Cambridge, 83–94.

TORRANCE, J.K. & OHTSUBO, M. 1995. Ariake Bay quick clays: a comparison with the general model. *Soils and foundations*, **35**, 11–19, https://doi.org/10.3208/sandf1972.35.11

Chapter 8 Swelling and shrinking soils

Lee Jones[1]*, Vanessa Banks[1] & Ian Jefferson[2]

[1]British Geological Survey, Environmental Science Centre, Nicker Hill, Keyworth, Nottingham, NG12 5GG
[2]School of Civil Engineering, University of Birmingham, Edgbaston, Birmingham, B15 2TT

*Correspondence: ldjon@bgs.ac.uk

Abstract: Swelling and shrinking soils are soils that can experience large changes in volume due to changes in water content. This may be due to seasonal changes in moisture content, local site changes such as leakage from water supply pipes or drains, changes to surface drainage and landscaping, or following the planting, removal or severe pruning of trees or hedges. These soils represent a significant hazard to structural engineers across the world due to their shrink–swell behaviour, with the cost of mitigation alone running into several billion pounds annually. These soils usually contain some form of clay mineral, such as smectite or vermiculite, and can be found in humid and arid/semi-arid environments where their expansive nature can cause significant damage to properties and infrastructure. This chapter discusses the properties and costs associated with shrink–swell soils, their formation and distribution throughout the UK and the rest of the world, and their geological and geotechnical characterization. It also considers the mechanisms of shrink-swell soils and their behaviour, reviewing strategies for managing them in an engineering context, before finally outlining the problem of trees and shrink–swell soils.

8.1 Introduction

Shrink–swell soils are one of the most costly and widespread geological hazards globally, with costs estimated to run into several billion pounds annually. These soils present significant geotechnical and structural challenges to anyone wishing to build on, or in, them. Shrink–swell occurs as a result of changes in the moisture content of clay-rich soils. This is reflected in a change in volume of the ground through shrinking or swelling. Swelling pressures can cause heave, or lifting, of structures, while shrinkage can cause differential settlement.

This chapter aims to provide the reader with a basic understanding of shrink–swell soils. To do this, we will review the nature and extent of shrink–swell soils, both in the UK and worldwide, and discuss how they form, how they can be recognized, the mechanisms and behaviour of shrink–swell soils, and strategies for their management (including avoidance, prevention and mitigation). This chapter also includes a glossary of terms (see Appendix), and is illustrated throughout with references and recommendations for further reading.

8.2 Properties of shrink–swell soils

A shrink–swell soil is one that changes in volume in response to changes in its moisture content. The extent of the volumetric change reflects the type and proportion of swelling clay in the soil. More specifically, expansive clay minerals expand by absorbing water and contract, or shrink, as they release water and dry out. Clays range in their potential to absorb water according to their different structures (Table 8.1). For the most expansive clays, expansions of 10% are common. (Chen 1988; Nelson & Miller 1992).

In practice, the amount by which the ground shrinks and/or swells is determined by the water content in the near-surface (active) zone. Soil moisture in this zone responds to changes in the availability of atmospheric recharge and the effects of evapotranspiration. These effects usually extend to about 3 m depth, but this may be increased by the presence of tree roots (Driscoll 1983; Biddle 1998, 2001). Characteristically fine-grained clay-rich soils soften, becoming sticky and heavy, following recharge events such as rainfall; commonly, they can absorb significant volumes of water. Conversely, as they dry, shrinking and cracking of the ground is associated with a hardening of the clay at surface. Structural changes in the soil during shrinkage (e.g. alignment of clay particles) ensure that swelling and shrinkage are not fully reversible processes (Holtz & Kovacs 1981). For example, the cracks that form during soil shrinkage are not perfectly annealed on re-wetting. This volume increase results in a decrease in the soil density, thereby providing enhanced access by water for subsequent episodes of swelling. Over geological timescales, shrinkage cracks may

Engineering Group Working Party (main contact for this chapter: L. Jones, British Geological Survey, Environmental Science Centre, Nicker Hill, Keyworth, Nottingham, NG12 5GG, ldjon@bgs.ac.uk)
From: GILES, D. P. & GRIFFITHS, J. S. (eds) 2020. *Geological Hazards in the UK: Their Occurrence, Monitoring and Mitigation – Engineering Group Working Party Report*. Geological Society, London, Engineering Geology Special Publications, **29**, 223–242, https://doi.org/10.1144/EGSP29.8
© 2020 [BGS © UKRI All rights reserved]. Published by The Geological Society of London. All rights reserved.
For permissions: http://www.geolsoc.org.uk/permissions. Publishing disclaimer: www.geolsoc.org.uk/pub_ethics

Table 8.1. *Clay minerals: their properties and sources*

	Group	Clay minerals	Physical properties	Principal sources
↑ Increasing swelling potential	Smectite 2:1 phyllosilicates	Montmorillonite, beidellite, nontronite, talc, hectorite, saponite and sauconite	Weakly linked by cations (Na and Ca); Na montmorillonite particularly prone to swelling	Alteration of mafic igneous rocks rich in Ca and Mg
	Vermiculite 2:1 phyllosilicates		Also expansive upon heating	Decomposition of micas
	Illite 2:1 phyllosilicates	Phengite, brammalite, celadonite, glauconite and hydrous micas	Predominate in marine clays and shales; characteristic of weathering in temperate climates or high altitudes in the tropics	Decomposition of micas and feldspars
	Kandites 1:1 phyllosilicates	Kaolinite, dickite, nacrite and halloysite		Decomposition of orthoclase feldspar

become infilled with sediment, thus imparting heterogeneity to the soil. Once the cracks have been infilled in this way, the soil is unable to move back, leaving a zone with a network of higher-permeability infills.

When supporting structures, the effects of significant changes in water content on soils with a high shrink–swell potential can be severe. In practical civil engineering applications in the UK, there are three important time-dependent situations, each with different boundary conditions, where shrink–swell processes need to be considered.

(1) Following a reduction in mean total stress, the most notable effects are found adjacent to cut slopes, excavations and tunnels (Vaughan & Walbancke 1973; Grob 1976; Burland *et al.* 1977; Einstein 1979; Madsen & Muller-Vonmoos 1985).

(2) Subsurface groundwater abstraction or artificial/natural recharge under conditions of constant total stress in both unconfined and confined aquifers. Regional subsidence or heave can be induced by this process.

(3) Surface climatic and/or water balance fluctuations related to land-use change under conditions of constant total stress. The most notable effects follow the development of seasonally desiccated soils (shrinkage), which can cause structural damage to existing shallow foundation (Driscoll 1983; Taylor & Smith 1986).

As well as effective stress changes, some deformation may be caused by biogeochemical alteration and dissolution of minerals, as a result of 'steady-state' fluid transport processes. Although surface movements and engineering problems can occur due to a loss or addition of solid material, these are not strictly shrink–swell soils. However, these processes are often combined with effective stress changes and/or fluid movements, and may be difficult to separate from true shrink–swell processes that might be taking place at the same time.

The chief factors controlling shrink–swell susceptibility in geological formations are material composition (clay mineralogy), initial *in situ* effective stress state and stiffness of the material. Variations in the initial condition caused through processes such as original geological environment, climate, topography, land use and weathering affect *in situ* effective stresses and stiffness, and hence shrink–swell susceptibility. Clays belonging to the silicate family comprise the major elements silicone, aluminium and oxygen. There are many other elements that can become incorporated into the clay mineral structure (hydrogen, sodium, calcium, magnesium, sulfur). The presence and abundance of these dissolved ions can have a large impact on the behaviour of the clay minerals. The clay minerals are defined by the ratio of silica tetrahedra to alumina, iron or magnesium octahedra.

Subsidence also occurs in superficial deposits such as alluvium, peat and laminated clays that are susceptible to consolidation settlement (e.g. in the Vale of York east of Leeds, and in the Cheshire Basin), but these are not true shrink–swell soils.

8.3 Costs associated with shrink–swell clay damage

Clay shrink–swell is a global problem, and many of the world's major towns and cities are founded on clay-rich soils and rocks. In the UK the effects of shrinkage and swelling of clay soils with respect to foundation and building damage were first recognized by geotechnical specialists following the dry summer of 1947. Following the drought of 1975–1976, insurance claims in the UK came to over £50 million and the cost has risen dramatically since then. After a preceding drought in 1991, claims peaked at over £500 million. Over the past 10 years the adverse effects of shrink–swell behaviour has cost the economy an estimated £3 billion, making it the most damaging geohazard in Britain today; as many as one in five homes in England and Wales are at risk from ground that swells when it becomes wet and shrinks as it dries out (Jones 2004), although susceptible ground conditions are perhaps less severe under a temperate UK climate than in some other countries. The Association of British Insurers (ABI) has estimated that the average cost of shrink–swell-related subsidence to the insurance industry stands at over £400 million a year (Driscoll & Crilly 2000). The ABI stated that the 350% increase in the value of claims during July to September 2018 was the highest quarterly jump since records started more than 25 years ago. In the USA, the estimated damage to buildings and infrastructure exceeds $15 billion annually. The American Society of Civil Engineers estimates that one in four homes have some damage caused by shrink–swell soils; in a typical year, such soils cause a greater financial loss to property owners than earthquakes, floods, hurricanes and tornadoes combined (Nelson & Miller 1992).

8.4 Formation processes

Clay minerals (Table 8.1) are a product of weathering. They mostly form on land but are often transported to the oceans (Eberl 1984). They can form through three mechanisms: inheritance, or sediment transport; neoformation, from solution or reaction of amorphous material; or transformation, retaining some of the inherent structure while undergoing chemical reaction. As weathering proceeds silica and potassium are gradually leached; Eberl (1984) cites a number of case studies that demonstrate how wet climates are associated with kaolinite-rich soils while dry environments are characterized by smectite clays. The distribution of clay minerals at the surface therefore reflects both the underlying geology and the nature of weathering. In addition to the material composition (clay mineralogy; Table 8.1), the main factors controlling shrink–swell susceptibility in geological formations are: depositional environment, diagenesis, stress history and weathering. These factors influence the *in situ* effective stress state and stiffness of the material, and therefore its propensity to swell.

Ultimately, the clay minerals derived from weathering are transported to the ocean where they undergo little reaction, except for ion exchange and neoformation of smectite, which is associated with volcanic activity in the oceans and demonstrates the influence of geotectonic setting on clay mineral distribution. Clay mineralogy is also affected by diagenesis and stress history (Keller 1963; Eberl 1984; Merriman 2005). Broadly, diagenesis takes the form of smectite to illite transformations. Illitization proceeds through a series of reactions involving intermediate mixed layers of illite and smectite of varying compositions (Lanson *et al.* 2009). Near-surface weathering can reverse this process, leading to the formation of mixed layer clays. While the stress history can be important in influencing clay mineralogy, it has been argued that the influences of burial depth on engineering properties is less significant than that of stratigraphical position (e.g. Jackson & Fookes 1974).

8.5 Distribution

Shrink–swell soils are found throughout many regions of the world, particularly in arid and semi-arid regions, as well as where wet conditions occur after prolonged periods of drought. Their distribution is dependent on geology (parent material), climate, hydrology and geomorphology and vegetation, and is susceptibile to environmental change (e.g. Harrison *et al.* 2012).

Countries where shrink–swell soils occur and give rise to major construction costs include: Ethiopia, Ghana, Kenya, Morocco, South Africa and Zimbabwe in Africa; Burma, China, India, Iran, Israel, Japan and Oman in Asia; Argentina, Canada, Cuba, Mexico, Trinidad, USA and Venezuela in the Americas; Cyprus, Germany, Greece, Norway, Romania, Spain, Sweden, Turkey and the UK in Europe; and Australia (Fig. 8.1). In these countries, or significant areas of them, the evaporation rate is higher than the annual rainfall so there is usually a moisture deficiency in the soil; soil suction then increases the potential for heave. In semi-arid regions a pattern of short periods of rainfall followed by long dry periods (drought) can develop, resulting in seasonal cycles of swelling and shrinkage.

In humid climates problems with shrink–swell soils trend to be limited to those soils containing higher plasticity clays. In arid or semi-arid climates, soils that exhibit moderate shrink–swell potential can cause distress to residential property. This occurs as a direct result of the relatively high suction that exists in these soils, and the larger changes in water content regimes that results when water level changes.

Reflecting the broad younging of the UK stratigraphy to the SE, the clay-rich soils most susceptible to shrink–swell behaviour predominate in the SE of the country, primarily distributed to the east of a line from Dorset to Birmingham, Nottingham and North Yorkshire (Fig. 8.2). In this zone the near-surface Jurassic, Paleocene and Eocene clays are less indurated than their older counterparts, and they retain a greater propensity to absorb and lose moisture. These mudrocks are normally firm to very stiff clays or very weak mudstones that weather to firm to stiff clays near the surface. Indurated clays are less prone to shrink–swell behaviour because of the clay transformations that occur during diagenesis.

The transformations associated with diagenesis are attributable to the changes in the stress, fluid pressure, geochemistry and temperature as a consequence of burial. Associated with the clay transformation are compaction and fluid migration, development of diagenetic bonds, mineralization and cementation, recrystallization and pressure solution. In some areas of the UK (e.g. around the Wash, NW of Peterborough, and under the Lancashire Plain) the mudrocks are deeply buried beneath other (superficial) soils that are not as susceptible to shrink–swell behaviour, although consideration should be given to the potential for swelling and heave in deep excavations in these soils.

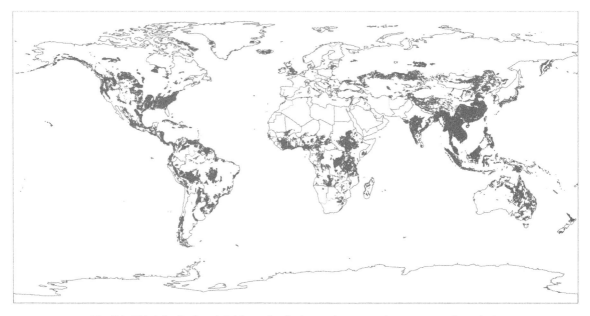

Fig. 8.1. Global distribution of shrink–swell soil where major construction costs occur (by region).

While the two-dimensional (2D) distribution of the UK clay soils is relatively well known (e.g. Loveland 1984; Wilson *et al.* 1984; Jeans 2006a, b), the 3D distribution is less well known. There is significant potential for furthering this understanding through the use of 3D geological models (e.g. Jones & Terrington 2011). Indications are that climate change will have an increasingly adverse effect on the moisture conditions that UK soils experience and therefore on the damage caused to the homes, buildings, roads and services founded on them, with further potential for the application of 3D geological modelling.

A meaningful assessment of the shrink–swell potential of a soil in the UK requires a considerable amount of high-quality and well-distributed spatial data of a consistent standard (Jones & Jefferson 2012), and from this a volume change potential (VCP) map can be constructed. However, although giving a good indication of potential problem areas, looking at soils on a national scale does not tell the whole story. No two clay soils are the same in terms of their behaviour or their shrink–swell potential; it is therefore better to look at them on a more regional scale. Jones & Terrington (2011) discuss a methodology for creating a 3D VCP interpolation of the London Clay, visualizing plasticity values at a variety of depths relative to ground level across the outcrop (Fig. 8.3).

8.6 Characterization of shrink–swell soils

Shrink–swell soils can be recognized from either geological and geotechnical characterization, or from the damage incurred by buildings and infrastructure. Typically, swelling pressures can cause heave or lifting of structures, while shrinkage can cause differential settlement. Damage to a structure is possible when as little as 3% volume expansion takes place (Jones 2002). Failure results when the volume changes are unevenly distributed beneath the foundation. For example, water content changes in the soil around the edge of a building can cause swelling pressure beneath the perimeter of the building (Fig. 8.4), while the water content of the soil beneath the centre remains constant. This results in a failure known as end lift. Conversely, soil shrinkage around the perimeter may result in centre lift. Subsidence problems are easier to recognize than heave problems; this may occur on a dry clay soil and the only obvious sign might be irregular crack patterns, wider at the bottom than at the top, with no obvious cause.

Another major contributing factor to ground shrinkage is tree growth, more specifically tree roots. Roots grow in the direction of least resistance and where they have the best access to water, air and nutrients (Roberts 1976). The actual pattern of root growth depends upon, among other factors, the type of tree, depth to water table and local ground conditions. Trees will tend to maintain a compact root system. However, when trees become very large, or where trees are under stress, they can send root systems far from the trunk. Damage to foundations resulting from tree growth occurs in two principal ways: physical disturbance or shrinkage of the ground by removal of water. Physical disturbance of the ground caused by root growth is often seen as damage to pavements and broken walls (Fig. 8.5). Shrinkage caused by water removal can lead to differential settlement of building foundations.

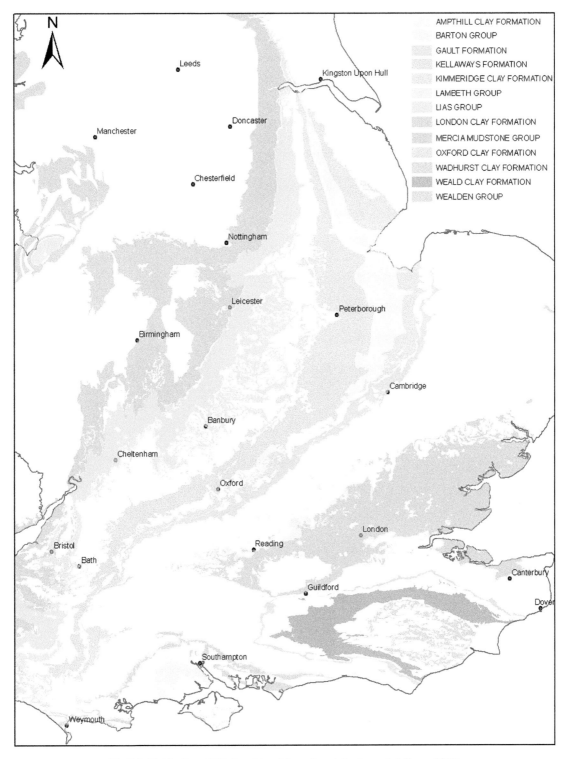

Fig. 8.2. Distribution of UK clay-rich soil formations (after Jones & Jefferson 2012).

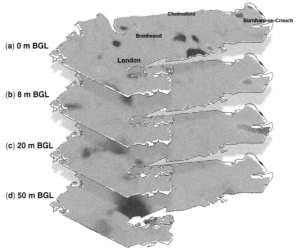

Fig. 8.3. Interpolation of 'Area 3' showing surfaces at 0 m, 8 m, 20 m and 50 m (**a–d**) below ground level (after Jones & Terrington 2011). Blue: medium; green: high; red: very high.

Vegetation-induced changes to water profiles can also have a significant impact on other underground features, including utilities. Tree-induced movement has the potential to be a significant contributor to failure of old pipes located in clay soils near deciduous trees (Clayton *et al.* 2010), and can also cause physical disturbance to services.

Potential shrinkage and/or swelling from these causes can usually be anticipated in most engineering circumstances. However, because of the differences between natural and tree-induced shrink–swell, and varying initial conditions, the relative susceptibility to volume change at any place may not necessarily always be the same for a given geological formation or soil type. Spatial generalization is therefore

Fig. 8.4. Swelling beneath perimeter of building causing heave in porch (© David Noe).

Fig. 8.5. Damage to kerbstone and paving caused by root growth.

difficult, and a prediction of the severity of potential deformation often requires site-specific subsurface investigations and *in situ* monitoring.

To summarize, the main environmental causes of shrink–swell are: (1) normal seasonal movements associated with changes in rainfall and vegetation growth; (2) enhanced seasonal movement associated with trees, severe pruning or removal of trees or hedges; (3) long-term subsidence, as a persistent water deficit develops; (4) long-term heave as a persistent water deficit dissipates; or (5) increased susceptibility of the near-surface soils as its density is reduced.

While much research has been carried out worldwide to infer shrink–swell behaviour from soil index properties, few direct data are publically available in UK geotechnical databases (Hobbs *et al.* 1998). Two schemes that are commonly used to assess shrink–swell properties within the UK are based on the Building Research Establishment (BRE) and National House-Builders Council (NHBC) schemes. High-shrinkage soils may not behave very differently from low-shrinkage soils, because conditions in the UK do not allow full potential to be realized (Reeve *et al.* 1980).

Potential shrink–swell soils are initially identified by soils engineers from particle size analyses to determine the percentage of fine particles in a sample. If more than 35% of the particles are able to pass through a 63 μm sieve, the sample is classified as a fine soil comprising either silt or clay, or a combination of both (BSI 1999). Clay-sized particles are considered to be less than 2 μm (although this value varies slightly throughout the world), but the difference between clays and silts is more to do with origin and particle shape. Silt particles (generally comprising quartz particles) are products of mechanical erosion, whereas clay particles are products of chemical weathering and are characterized by their sheet structure and composition.

Although there are a number of methods available to identify shrink–swell soils, each with their relative merits, there are no universally reliable methods available (Jones & Jefferson 2012), and they are rarely employed in the course of routine site investigations in the UK. This means that few data are available for compiling the directly measured shrink–swell properties of the major clay formations, and reliance has to be placed on estimates based on index parameters such as liquid limit, plasticity index and density (Reeve *et al.* 1980; Holtz & Kovacs 1981). No consideration has been given to the saturation state of the soil and therefore to the effective stress or pore pressures within it.

The most widely used parameter for determining the shrinkage and swelling potential of a soil is the plasticity index (I_P). Such plasticity parameters, being based on remoulded specimens, cannot precisely predict the shrink–swell behaviour of an *in situ* soil. However, they do follow established procedures, being performed under reproducible conditions to internationally recognized standards. A modified plasticity index (I_P') is proposed in the Building Research Establishment Digest 240 (BRE 1993) for use where the particle size data, specifically the fraction passing a 425 µm sieve, is known or can be assumed as 100% passing (BRE 1993). I_P' takes into account the whole sample and not just the fines fraction; it therefore gives a better indication of the real plasticity value of an engineering soil (Jones & Terrington 2011).

Such empirical correlations may be based on a small dataset, using a specific test method, and at only a small number of sites. Variation of the test method would probably lead to errors in the correlation. The reason for the lack of direct shrink–swell test data is that few engineering applications have a perceived requirement for these data for design or construction. However, the stages of investigation needed for shrink–swell soils follow those used for any site (see Leroueil 2001; Simons *et al.* 2002). Indications are that climate change will have an increasingly adverse effect on the moisture conditions that UK clay soils experience, and therefore on the damage caused to the structures founded on, or within, them. The government has recognized that climate change is one of the biggest problems that the UK faces and, if current predictions are correct, we can expect hotter, drier summers and milder, wetter winters (UKCIP 2009), with as many as one in five homes in England and Wales likely to be damaged by the shrinking and swelling behaviour of these clay soils (Jones 2004). If the UK were to experience an increase in extended periods of dry weather prior to rainfall events, costs caused by shrink–swell damage could rise significantly. It is therefore important to recognize the existence of shrink–swell soils at the earliest stage possible, during site and laboratory investigations, in order to ensure that the correct design procedures are put in place, before costly remediation is required.

Soil suction is a measure of the free energy or the relative vapour pressure (relative humidity) of the soil moisture. It can be defined by the suction required to remove the water from above the water table. There are two components: the matric potential, which is the moisture held in soil pores by capillary action; and the much smaller solute potential, which is the osmotic effect of dissolved salts. Soil suction characteristics vary between clay soils in accordance with the composition of the soil, particularly its particle size and clay mineral content. Figure 8.6 shows typical moisture content suction profiles for a variety of UK and North American mudrocks.

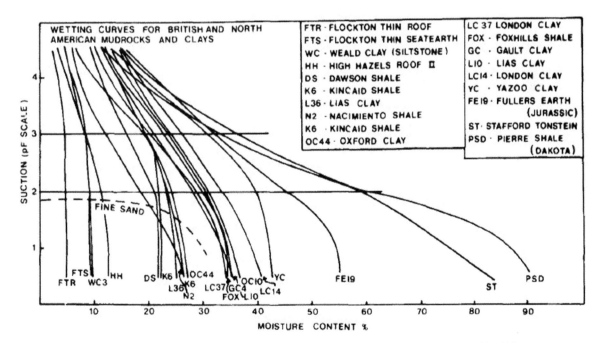

Fig. 8.6. Suction versus moisture content for UK and North American soils (after Taylor & Smith 1986).

The hydraulic conductivity of a soil also varies with the suction, both seasonally and over longer timescales. A secondary permeability can be induced through fabric changes, tension cracking and shallow shear failure during the shrink–swell process which may influence subsequent moisture movements. For example, in a micro fabric study of clay soils, Scott *et al.* (1986) showed that compression (swelling) cracks tended to parallel ground contours and dip into the slope at *c.* 60°, and could usually be distinguished from shrinkage cracks that were randomly distributed. In the London Clay Formation soils studied, they found that the ratio between shrinkage and swelling discontinuities was about 2:1. The actual mechanism of volume increase (or decrease), and the differences in susceptibility to these processes that have been found between different sediments, has been attributed to a combination of both mechanical and physicochemical processes (Bolt 1956; Mitchell 1976; Taylor & Smith 1986).

8.7 Mechanisms of shrink–swell

Ground deformation in response to a reduction in total stress can be considered in terms of: (a) an immediate, but time-dependent, elastic rebound; and (2) swelling due to the change in effective stress.

Changes in effective stress drive fluid movement into or out of the geological formation or soil. The magnitude of strains associated with this process depends on the drained stiffness, the extent of the stress change, the water pressures that are set up in the soil or rock, and the new boundary conditions. The rate of volume change depends on the compressibility, expansibility and hydraulic conductivity of the sediment and surrounding materials. In stiff homogeneous materials with a low hydraulic conductivity, several decades may be necessary to complete the process. The same physical parameters and processes control the swelling that accompanies atmospheric-driven aquifer recharge; however, without a total stress change there is no elastic rebound of the soil and/or rock structure.

Shrinkage by evaporation is similarly accompanied by a reduction in water pressure and development of negative capillary pressures; deformation follows the same principles of effective stress. The processes of shrinkage due to evaporation have been reviewed in detail using effective stress concepts by Sridharan & Venkatappa Rao (1971), Bishop *et al.* (1975) and Alonso *et al.* (1990).

At any time, the equilibrium water content in a geological formation or weathered soil represents a balance between the clay minerals needed to suck in water, and the tendency for applied stresses to squeeze out water. A clay and/or soil mass initially in equilibrium will swell if a transient change in geoenvironmental boundary conditions results in unloading or an increase in water pressure. As corollaries to the three situations of shrink–swell associated with anthropogenic interference (engineering works), there are also three comparable groups of natural causes: (1) a total stress change following erosion, tectonism or mass movement; (2) climatically controlled hydrodynamic processes associated with sea-level change, and groundwater recharge and discharge; and (3) biogenic, chemical and/or physical weathering under conditions of almost constant total stress.

Freeze–thaw involves effective stress changes, and is a natural cause of shrink–swell. It is extremely important in northern latitudes; the top few metres of most weathered materials in the UK were affected by freeze–thaw processes during the late glacial period. However, examples of heave are known to have occurred next to some refrigeration plants and beneath road pavements, and it may also have contributed to some slope instability problems. A review of research progress in this field, as well as a good starting point and bibliography for the subject, is provided by Morgenstern (1981).

Civil engineering and other anthropogenic interference essentially mimic these natural effects through environmental changes connected with construction, artificial groundwater recharge, agriculture and/or waste disposal, but their effects are superimposed onto a range of initial time-dependent conditions.

8.8 Shrink–swell behaviour

The shape of clay particles is determined by the arrangement of the thin crystal lattice layers that they form (Fig. 8.7). The molecular structure and arrangement of the clay crystal sheets in shrink–swell clays guides how water is attracted and held between the crystalline layers (and on their surfaces) in a strongly bonded 'sandwich' (Fig. 8.7), an example of which is shown in Figure 8.8. The electrical dipole structure of water molecules drive an electrochemical attraction to the microscopic clay sheets. The mechanism by which these molecules become attached to each other is called adsorption. Na-smectite clays (such as montmorillonite) have the greatest affinity for water, and can adsorb very large amounts of water molecules between their clay sheets. In theory, they can expand in volume by a factor of 800, causing dispersion of clay platelets by the elimination of repulsive interlayer forces. They therefore have a large shrink–swell potential, as do vermiculite and chlorite, which exhibit crystalline swelling. For further details of mineralogy of clay minerals and their influence of engineering properties of soils, see Mitchell & Soga (2005). Driscoll (1983), Taylor & Cripps (1984) and Taylor & Smith (1986) provide useful reviews of the controls that clay mineralogy has on the drained compressibility and/or expansibility of geological materials, and hence their susceptibility to large deformations from effective stress changes which lead to shrinkage and/or swelling.

Saturated clay shrink–swell soils contain water molecules between the clay sheets, causing the bulk volume of the soil to increase or swell with changes in water content. This process

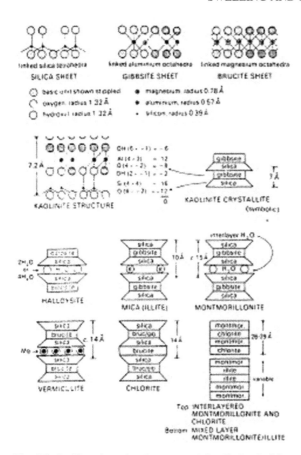

Fig. 8.7. Swelling clay mineral structures (after Taylor & Cripps 1984).

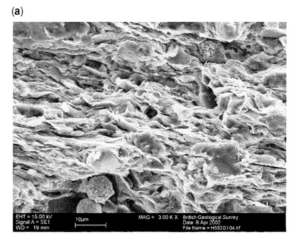

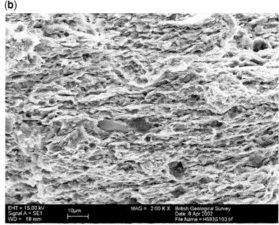

Fig. 8.8. Scanning electron microscope (SEM) images showing typical features of Lias Group clay: (a) tightly packed, flat-lying, curved clay flakes, c. 10 μm in diameter and < 1 μm thick; and (b) well-laminated, tightly packed, flat-lying clay flakes.

of absorption weakens the inter-clay bonds and causes a reduction in the strength of the soil. As moisture is reduced by evaporation or gravitational forces, the water between the clay sheets is released, causing the overall volume of the soil to decrease or shrink. Features such as voids or desiccation cracks are associated with this process.

Shrinkage and swelling usually occurs in the near-surface to depths of about 3 m, but this can vary depending on climatic conditions. The shrink–swell potential of expansive soils is determined by its initial water content, void ratio, internal structure and vertical stresses, as well as the type and amount of clay minerals in the soil (Bell & Culshaw 2001). These minerals determine the natural expansiveness of the soil, and include smectite, montmorillonite, nontronite, vermiculite, illite and chlorite. Generally, the greater the concentration of these minerals in the soil, the greater the shrink–swell potential. However, these effects may become diluted by the presence of other non-swelling minerals such as quartz and carbonite (Kemp et al. 2005).

Soils with high shrink–swell potential will not usually cause problems as long as their water content remains relatively constant. This is controlled by the soil properties (mineralogy), suction and water conditions, water content variations, and geometry and stiffness of a structure founded on it (Houston et al. 2011). In a partially saturated soil, suction or water content changes increase the likelihood of damage occurring. In a fully saturated soil, the shrink–swell behaviour is controlled by the clay mineralogy.

Seasonal volume change generally takes place with the assistance of fractures and secondary macropores within the weathered horizons. If undersaturation is present at very shallow depths, slaking by air breakage can also occur. Taylor & Smith (1986) described these processes, and the physico-chemical techniques for predicting structural breakdown. The depth of active groundwater flow is therefore an important control on the susceptibility of soils to volume change, because this affects the magnitude of that change in both the saturated and unsaturated groundwater zones.

Shrink–swell soil problems typically occur due to water content changes in the upper few metres, with deep-seated heave being rare (Nelson & Miller 1992). The water content in this zone, which is known as the active layer, is significantly influenced by climatic and environmental factors; it is generally termed the zone of seasonal fluctuations or active zone, as shown in Figure 8.9.

In the zone of seasonal fluctuations, moisture content ranges between negative and positive porewater pressures (Fig. 8.9); see Nelson *et al.* (2001) for further details. It is important to determine the depth of the active zone during a site investigation in order to ensure adequate foundation design. This can vary significantly with climate conditions with depths of 5–6 m in some countries, whereas in the UK 1.5–2.0 m is typical (Biddle 2001), extending to 3–4 m in some areas of the London Clay Formation (Biddle 2001). Depending on the relative significance of heave or subsidence, the active zone is defined in one of a number of ways (e.g. Nelson *et al.* 2001).

(1) *Active zone*: the zone of soil that contributes to soil expansion at any particular time.
(2) *Zone of seasonal moisture fluctuation*: the zone in which water content changes due to climatic changes at the ground surface.
(3) *Depth of wetting*: the depth to which water content has increased due to the introduction of water from external sources, used to estimate heave by integrating the strain produced over the zone in which water contents change (Walsh *et al.* 2009).
(4) *Depth of potential heave*: the depth at which the overburden vertical stress equals or exceeds the swelling pressure of the soil. This is the maximum depth of the active zone.

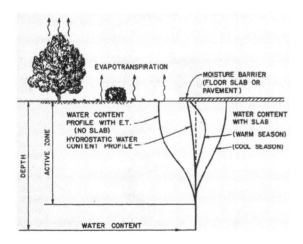

Fig. 8.9. Water content profiles in the active zone (after Nelson & Miller 1992).

The natural, time-dependent physical and chemical changes which occur in a geological formation are inherent weathering processes. In a stiff clay, the result is one of softening (i.e. an increased water content), while in a soft clay a hardened crust is often formed in association with desiccation. Mesri *et al.* (1978) provide a useful summary of the reasons why softening and weathering processes are dependent on time.

(1) *Hydraulic conductivity of the clay*: the increase in water content, including the adsorbed double layer water, requires flow into the clay either from internal adjustments or from an external hydrological input.
(2) *Particle rearrangement*: structural readjustment, including particle deformation and reorientation, is a chain reaction process and hence time-dependent.
(3) *Progressive breakdown of diagenetic bonds*: interlayer bonding and cementation is common in many stiff clays and shales, as evidenced by the difficulty of dispersing soils for index testing.
(4) *Progressive development of structural discontinuities*: inhomogeneous swelling and shear distortion may produce structural discontinuities, fissures and slickensides, which in turn affect the breakdown of diagenetic bonding and mass hydraulic conductivity.
(5) *Chemical changes*: chemical changes of inorganic and organic compounds, including bacterial oxidation.

Superimposed on these widespread climatic influences are local influences such as tree roots and leakage from water supply pipes and drains. The removal or severe pruning of trees may result in swelling problems, as extraction through root systems is the most efficient method of supplying water to desiccated soils (Cheney 1986). The swelling of shrinkable clay soils after trees have been removed can produce either very large uplifts or very large pressures (if confined), and the ground recovery can continue over a period of many years (Cheney 1986).

Building or paving on previously open areas of land, such as the building of patios and driveways, can cause major disruption to the soil–water system. Sealing the ground in this way cuts off the infiltration of rainwater; the trees that are dependent upon this water will have to send their roots deeper or further afield in order to find water. The movement of these root systems will cause a major ground disturbance, and will lead to the removal of water from a larger area around the tree (Jones & Jefferson 2012). Problems occur when structures are situated within the zone of influence of a tree (Fig. 8.10).

Occasionally this situation may worsen as nearby trees continue to extract water during the growing months, when rainfall is low. If a more permeable type of surface, such as block paving, is used, more rainwater can enter the ground and supply nearby tree roots. If an impermeable method of paving is used, it may prevent water from infiltrating into the ground. This can affect the shrink–swell behaviour of the ground and also the growing patterns of nearby trees. A well-designed impermeable paving system, and one that is

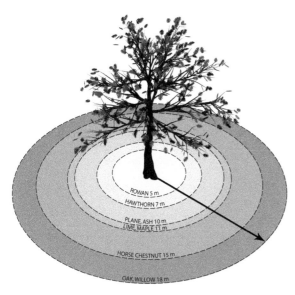

Fig. 8.10. The zone of influence of some common UK trees (after Jones *et al.* 2006).

and other shallow services are especially vulnerable to damage because they are less able to suppress differential movements than heavier multi-storey structures. Due to the global distribution of shrink–swell soils, many different ways to tackle the problem have been developed (Radevsky 2001). The preferred methodology depends on technical developments, country-specific legal frameworks and regulations, insurance policies and the attitude of insurers, the experience of the engineers and other specialists dealing with the problem and, most importantly, the sensitivity of the owner of the property affected. A summary of these issues is provided by Radevsky (2001) in his review of how different countries deal with shrink–swell soil problems, and a detailed informative study from the USA has more recently been presented by Houston *et al.* (2011).

In order to make a proper characterization of a site where shrink–swell soils exist, two factors need to be identified: (1) the properties of the soil (e.g. mineralogy, soil water chemistry, suction, soils fabric); and (2) environmental conditions that can contribute to changes in water contents of the soil, for example, water conditions and their variations (climate, drainage, vegetation, permeability and temperature) and stress conditions (history and *in situ* conditions, loading and soil profile).

Shrink–swell soils require extensive site investigation in order to provide sufficient information, normal investigations, relating to the structures most affected by shrink–swell soils, are often not adequate. These investigations may involve specialist test programmes even for relatively lightweight structures (Nelson & Miller 1992).

While the selection of the type of foundation can be critical in the mitigation of clay shrink–swell, there are a number of factors that influence the selection of foundation types and design methods including climate, local experience, financial and legal aspects, and technical issues. If shrink–swell clays have been identified, the design should take account of the depth of the active layer and the fact that shrink–swell behaviour often does not manifest itself for several months; early communication with all relevant stakeholders is essential. Often higher initial costs are offset many times over by a reduction in post-construction maintenance costs dealing with expansive soils (Nelson & Miller 1992).

Foundation alternatives when dealing with potentially expansive soils are likely to require one of a number of structural alternatives designed to either isolate the structure from soil movement (e.g. using drilled pier and beam foundations) or introduce sufficient stiffness to resist movement (e.g. using stiffened slab on grade or modified continuous perimeter footings) (Jones & Jefferson 2012). The use of ground improvement techniques in isolation or in combination with these approaches may be a good alternative (Chen 1988; Nelson & Miller 1992; Nelson *et al.* 2011; Jones & Jefferson 2012). Of particular importance is the aim to minimize total and differential movements. Globally, the most common types of foundations used in expansive soils are drilled pier and beam systems, reinforced slabs-on-grade and modified continuous perimeter spread footings.

in good condition, may actually reduce the amount of shrink–swell activity in the ground immediately below it. Paving moderates variations in water content of the soil and therefore the range of shrink–swell behaviour that might be expected. However, if the paving seal is broken water can suddenly enter the system, causing swelling of the ground.

Trees are a vital component of the environment and play an important role valued by the public at large for many different reasons. However, trees are also a very real problem when they shed their leaves, petals and fruit or block out sunlight, and their roots are associated with cracked pavements and blocked drains; further, where tree roots are close to buildings in the presence of shrink–swell soils, building subsidence can occur. If vegetation is involved, it produces a characteristic seasonal pattern of foundation movement of subsidence during summer, reaching a maximum usually in September, followed by upwards recovery during winter (Crilly & Driscoll 2000; Driscoll & Chown 2001). This pattern provides clear evidence of clay shrink–swell as as no other cause produces a similar pattern, and it can be concluded that soil drying by vegetation must be involved (unless the foundations are less than 300 mm).

8.9 Strategies for engineering management: avoidance, prevention and mitigation

The structures most susceptible to damage caused by shrink–swell soils are usually lightweight in construction. Houses and other low-rise buildings, pavements, pylons, pipelines

Table 8.2. *Soil stabilization approaches applied to expansive soils (after Jones & Jefferson 2012)*

Improvement method		Outline of approach	Advantage	Disadvantage
Avoidance	Removal and replacement	Expansive soil removed and replaced by non-expansive fill to a depth necessary to prevent excessive heave; depth governed by weight needed to prevent uplift and mitigate differential movement	Non-expansive fill may increase bearing capacities; simple and easy to undertake; often quicker than alternatives	Impervious fill to prevent water ingress; thickness required may be impractical; failure can occur during construction due to water ingress
Stabilization (physical and chemical)	Remoulding and compaction	Less expansion observed for soil compacted at low densities above oil/water contact and below oil/water contact; standard compaction methods and control can be used to achieve target densities	Onsite resources eliminate cost of imported fill; relatively impermeable fill achievable to minimize water ingress; swell potential reduced without introducing excess water	Low-density compaction may be detrimental to bearing capacity; may not be effective for soil of high swell potential; requires close and careful quality control
	Remoulding and mixing	Variations on the above include mixing with non-swelling material to dilute swell potential, e.g. sand (Hudyma & Avar 2006) or granulated tyre rubber (Patil et al. 2011)	Swell potential reduced without introducing excess water	Requires close and careful quality control
	Cation exchange	Exchangeable cations that originate in pore waters are attracted to the external and internal surfaces of the clay; main cations in natural waters in order of abundance are calcium (Ca^{2+}), magnesium (Mg^{2+}), potassium (K^+) and sodium (Na^+)	Cation exchange has been used in some situations to reduce swelling, e.g. using calcium to replace sodium; Rimmer & Greenland (1976) have also shown how calcium carbonate suppresses the swelling behaviour of a clay through its effect on the diffuse double layer	Requirement to understand soil chemistry and undertake laboratory trials; requires close and careful quality control
	Chemical stabilization: lime and/or cement	Lime (3–8 wt%) common with cements (2–6 wt%) sometimes used, and salts, fly ash and organic compounds less commonly used; generally lime mixed into surface (c. 300 mm), sealed, cured and then compacted; lime may also be injected in slurry form; lime generally best when dealing with highly plastic clays	All fine-grained soils can be treated by chemical stabilizers; effective in reducing plasticity and swell potential of an expansive soil; chemical stabilization can be used to provide a cushion immediately below foundation placed on expansive soils, e.g. pavements (Murty & Praveen 2008)	Soil chemistry may be detrimental to chemical treatment; potential health and safety issues and environmental risks; curing inhibited in colder temperatures; requires site specific design and quality control; Beetham et al. (2014) found that if durability and/or long-term performance are to be optimized, treatment of medium–high-plasticity clay soils may require adaptations to standard working practices

	Chemical stabilization: byproducts	Byproduct stabilizers include cement and/or lime kiln dust and pulverized fuel ash (PFA)	As above, with the added potential for integrated sustainability	As above
	Chemical stabilization: non-traditional stabilizers	Non-traditional stabilizers include sulfonated oils, potassium compounds, ammonium compounds and polymers; electrochemical soil treatment approaches are being developed that utilize electrical current to inject stabilizing agents into the soils (Barker et al. 2004)	Some of these methods are less bulky, reducing transport costs	Requires site-specific design and quality control; some emerging technologies can be expensive; health and safety considerations associated with some techniques
Water control methods	Pre-wetting or ponding	Water content increased to promote heave prior to construction; dykes or berms used to impound water in flooded area; alternatively, trenches may be used and vertical drains can be used to also speed infiltration of water into soil	Moves the edge effects away from the foundation and/or pavement to minimize seasonal fluctuation effects; used when soils have sufficiently high permeabilities to allow relatively quick water ingress, e.g. with fissured clays Lengthens the time for water content changes to occur due to longer migration paths under foundations	May require several years to achieve adequate wetting; loss of strength and failure can occur; ingress limited to depth less than the active zone; water redistribution can cause heave after construction
	Vertical barriers: polyethylene, concrete, impervious semi-hardening slurries Horizontal barriers: membranes, bituminous membranes or concrete			

Pavements are particularly susceptible to damage caused by shrink–swell soils. Their vulnerability stems from their relative lightweight nature extended over a relatively large area. For example, Cameron (2006) described problems with railways built on shrink–swell soils where poor drainage exists, and Zheng et al. (2009) describes problems with highway subgrade construction on embankments and slopes in China. Damage to pavements on shrink–swell soils manifests in a number of ways, including severe unevenness along significant lengths of pavement (sometimes evidenced by cracks); longitudinal cracking; lateral cracking developed from significant localized deformations, and localized pavement failure associated with disintegration of the surface. A detailed account of a range of treatment approaches is described by Chen (1988) and Nelson & Miller (1992), and Petry & Little (2002) provide a detailed review of the use of stabilization techniques over the last 60 years.

The philosophy of pavement design is essentially the same as that used for foundations. However, a range of different approaches are required to overcome the difficulty of it not being possible to isolate the pavement from the soils, and it is impractical to make pavements stiff enough to avoid differential movements (Jones & Jefferson 2012). Pavement designs are considered based on either flexible or rigid pavement systems; see the *Manual of Contract Documents for Highway Works* (DETR 1993). Decision-making with respect to pavement design in shrink–swell soils should be based on comprehensive ground understanding, in the context of the project requirements.

Conceptual ground models should include, but not be limited by: the depth of the active zone, the potential for volume change, soil and soil water chemistry, the distribution and seasonality of moisture in the soil, soil permeability and soil structure. Failure to carry out an adequate site investigation can lead to false diagnoses and the use of inappropriate remedial measures (Nelson & Miller 1992).

A number of solutions are available for the management of shrink–swell soils (Table 8.1). As fluctuations in water contents are one of the primary drivers of shrink–swell problems, with non-uniform heave occurring due to non-uniformity of water contents, shrink–swell problems could be mitigated by minimizing moisture content change over time. Moreover, if moisture content changes can be slowed down and moisture distributions made uniform, then differential movement could also be reduced. The lowest-technology approach is avoidance and the selection of alternative routes. Engineered solutions include: replacement of expansive soil with a non-expansive alternative; a modified design based on a low-strength soil with a requirement for regular maintenance; improvement of the expansive soils through disturbance and recompaction; stabilization of the expansive soil through chemical additives (e.g. lime treatment); or the advanced-technology approach of moisture control over the life of the pavement. Details of these are provided by Chen (1988) and Nelson & Miller (1992). There is a legacy of negative perceptions associated with some techniques, for example, Houston et al. (2011) found that many geotechnical and structural engineers considered chemical stabilization approaches such as the use of lime as ineffective for the pre-treatment of shrink–swell soils for foundations.

As with new construction on shrink–swell soils, mitigation projects should be appraised in the context of a number of scoping questions (Jones & Jefferson 2012) feeding into a good conceptual ground model. For example, are remedial measures needed or is damage severe enough to warrant treatment? What are the ground conditions and processes driving the damage? Has the problem stabilized and when should intervention take place? Is the full extent of the problem known? In the context of the ground model, the detail with respect to the options available for remediation and its specification in terms of design and quality assurance can be addressed. Further, the economic questions around funding sources and post-remedial residual risks (e.g. in the context of localized treatment) can also be answered.

Examples of remedial measures employed for mitigation of foundation-related failures in shrink–swell soils include (Jones & Jefferson 2012): (1) repair and replace structural elements or correct improperly designed features; (2) underpinning, particularly of key parts of the foundations (Buzzi et al. 2010); (3) provision of structural adjustments of additional structural support (e.g. post tensioning); (4) stiffening of foundations; (5) provision of drainage control; (6) stabilization of water contents of foundation soils; or (7) installation of moisture barriers to control water content fluctuations.

For pavements, the most common remedial measures are either removal and replacement or construction of overlays. Whichever method is used, care is needed to ensure that the causes of the original distress are dealt with. Remedial measures should aim to minimize volume changes, and options include: removal and replacement of expansive soil; use of additives to reduce volume change capacity; application of surcharge to confine soils if swell pressures are not too great; or minimizing moisture content changes in the subgrade. Many of the pre-construction approaches can also be applied to post-construction treatments, and for pavements these include: moisture barriers; removal, replacement and compaction; and drainage control (Table 8.2).

8.10 Shrink–swell soils and trees

Different problems are faced when considering the distinctly separate areas of designing new build structures or remediating existing damaged buildings. New build guidelines in terms of domestic dwellings recognizes the need for thorough ground investigations to design systems to cope with the hazards presented by the presence of existing trees or building following their recent removal. See NHBC Standards chapter 4.2 on building near trees (NHBC 2011) and NHBC Efficient Design of Foundations for Low Rise Housing, Design Guide (NHBC 2010). In the case of existing dwellings, a range of reports and digests (BRE 1991, 1995, 1996, 1999, 2002) and a summary of good technical practice (Driscoll & Skinner 2007) are available.

In the case of existing structures, the main cause of distress results from the effects of differential settlement where different parts of the building move by varying amounts due to differences in the properties of the underlying soil. Although significant in terms of vertical movement, equal or proportionate movements across the plan area of a building may result in little structural damage (ISE 1994). Biddle (2001) suggests the following remedial options to deal with the adverse actions of trees: (1) fell the offending tree to eliminate all future drying; (2) prune tree to reduce drying and the amplitude of seasonal movement; (3) control root spread to prevent drying under foundations; or (4) provide supplementary watering to prevent soil from drying.

According to Biddle (2001), in most situations underpinning is unnecessary and foundations can be stabilized by appropriate tree management, usually felling the offending tree or carrying out heavy crown reduction. Site investigations should reflect this change, and be aimed at providing information to allow appropriate decisions on tree management, in particular: (1) confirmation that vegetation-related subsidence is involved; (2) identification of which tree(s) or shrub(s) are involved; (3) assessment of the risk of heave if a tree is felled or managed; (4) identification of the need for any other site investigations; (5) if the tree warrants retention, assessment of whether partial underpinning would be sufficient; (6) confirmation that vegetation management has been effective in stabilizing the foundations; and (7) provision of information within an acceptable timescale.

Tree pruning to reduce its water use and therefore its influence on the surrounding soil is often employed. However, if the trees are thereafter not subjected to a frequent and ongoing regime of management, the problem will very quickly return and problems continue. While tree removal will ultimately provide an absolute solution in the majority of cases, there are situations where this is not an option (e.g. protected trees, adverse risk of heave, incomplete evidence in contentious issues, and physical proximity of trees). In addition, pruning is only an appropriate option in some circumstances with details discussed by Biddle (2001). If vegetation is involved, the greatest movement is usually closest to the culprit. Unless trees are very close to each other, the spatial distribution of movement can help to identify the culprit. The influence of trees is likely to be widespread, whereas shrubs are localized. The pattern of movement detected by level monitoring is sometimes unusual, for example, no movement where it might be expected, or vice versa. If so, other investigations can be formulated to try to explain the anomaly. For instance, if part of a building closest to a tree is not moving whereas other parts are, it might be because of previous partial underpinning. Trial pits at relevant locations can provide an answer. However, these must be appropriate as poorly obtained information is not reliable (Biddle 2001).

If a mature tree is felled, consequential heave of a building on a clay soil can occur. Unfortunately, the evidence is rarely available. However, a number of clues can help and include: (1) the house is new (less than 20 years old); (2) there is expansive soil present; (3) the crack pattern might appear a bit odd (e.g. wider at the bottom than the top, with no obvious cause); or (4) cracks continue to open, even during wet months.

Heave problems can be costly and always require thorough investigation, involving soil sampling, precise levels and aerial photographs. Heave is a threat but rarely a reality where established existing properties are involved and the structure predates the planting of the tree. Ultimately, if the offending tree can be accurately targeted and dealt with rapidly before another growing season, the extent of any damage and need for remedial work will be kept to a minimum (Biddle 2001).

8.11 Conclusions

Subsidence and heave caused by shrink–swell soils are probably the most significant geological hazards to affect domestic properties and other low-rise structures worldwide, costing billions of pounds annually. In a typical year, they cause a greater financial loss to property owners than earthquakes, floods, landslides, hurricanes and tornadoes combined. They are found throughout many regions of the world, particularly in arid and semi-arid regions, as well as in more humid regions such as the UK, where their problematic behaviour can be evident in highly plastic clay soils, especially after prolonged periods of drought. This shrink–swell effect is induced by changes in the water content of the ground causing subsidence or heave to occur, which can be exacerbated by external influences such as the presence of tree roots.

The shrink–swell hazard is controlled by a number of factors, primarily the geology and mineralogy (high-plasticity clays), and the climate (prolonged periods of dry weather). The seasonal volumetric behaviour of a desiccated soil is complex, and this complexity increases with severity of the shrinkage phenomena. The actual mechanism of volume increase (or decrease) and the differences in susceptibility to these processes can be attributed to a combination of both mechanical and physicochemical processes. Shrinkage and swelling usually occurs in the near-surface to depths of about 3 m; water content in this upper layer, generally termed the active zone, is significantly influenced by climatic and environmental factors. The shrink–swell potential of expansive soils is determined by its initial water content, void ratio, internal structure and vertical stresses, as well as the type and amount of clay minerals in the soil. Clay particles are very small and their shape is determined by the arrangement of the thin crystal lattice layers that they form. In an expansive clay, the molecular structure and arrangement of these clay crystal sheets has a particular affinity to attract and hold water molecules between the crystalline layers (and on their surfaces) in a strongly bonded 'sandwich', giving them a large shrink–swell potential. Seasonal volume change generally takes place with the assistance of fractures and secondary macropores within the weathered horizons.

Houses and other low-rise buildings, pavements, pylons, pipelines and other shallow services are especially vulnerable to damage from shrink–swell clays because they are less able to suppress differential movements than heavier multi-storey structures. Normal site investigations are often not adequate, and a more extensive examination is required to provide sufficient information. This may involve specialist test programmes even for relatively lightweight structures. A large number of factors influence foundation types and design methods, including climate, financial and legal aspects as well as technical issues. Often higher initial costs are offset many times over by a reduction in post-construction maintenance costs dealing with expansive soils. Pavements are highly susceptible to damage caused by shrink–swell soils; their vulnerability stems from their relative lightweight nature extended over a relatively large area. Building or paving on previously open areas of land can cause major disruption to the soil–water system. Sealing the ground in this way cuts off the infiltration of rainwater, meaning that the trees that are dependent upon this water will have to send their roots deeper or further afield. The movement of these root systems will cause a major ground disturbance and lead to the removal of water from a larger area around the tree. Tree pruning to reduce its water use and therefore its influence on the surrounding soil is often employed. However, if the trees are thereafter not subjected to a frequent and ongoing regime of management, the problem will very quickly return and problems continue. While tree removal will ultimately provide an absolute solution in the majority of cases, there are situations where this is not an option.

Acknowledgements The authors would like to thank colleagues at the British Geological Survey and the University of Birmingham for their contributions to the development of this chapter.

Funding This research received no specific grant from any funding agency in the public, commercial, or not-for-profit sectors.

Appendix: Definitions and glossary

Absorb: Take in or soak up (energy or a liquid or other substance) by chemical or physical action.

Adsorb: To hold molecules as a thin film on the outside surface or on internal surfaces within the material.

Atterberg limits: Consistency criteria for defining key water contents of a clay soil: liquid limit, plastic limit and shrinkage limit.

Bearing capacity: The ability of a material to support an applied load. Ultimate bearing capacity is the pressure at which shear failure of the supporting soil immediately below and adjacent to a foundation occurs. A foundation is usually designed with a working load that is some proportion of the bearing capacity.

Clay: A naturally occurring material that is a plastic material at natural water content and hardens when dried to form a brittle material. It is the only type of soil and/or rock susceptible to significant shrinkage and swelling. It is made up mainly, but not exclusively, of clay minerals. It is defined by its particle-size range (<0.002 mm). Clay does not have to be the dominant component of a soil in order to impart clay-like properties to it.

Clay minerals: A group of minerals with a layer lattice structure that occurs as minute platy or fibrous crystals. These tend to have a very large surface area compared with other minerals, giving clays their plastic nature and the ability to support large suction forces. They have the ability to take up and retain water and to undergo base exchange.

Density: The mass of a unit volume of a material. Often used (incorrectly) as synonym for unit weight. Usually qualified by condition of sample (e.g. saturated, dry).

Diagenesis: The physical and chemical changes occurring during the conversion of sediment to sedimentary rock.

Discontinuity: Any break in the continuum of a rock mass (e.g. faults, joints).

Drained: Condition applied to strength tests where pore fluid is allowed to escape under an applied load. This enables an effective stress condition to develop.

Drought: A prolonged period of abnormally low rainfall, leading to a shortage of water.

Effective stress: The total stress minus pore pressure. The stress transferred across the solid matter within a rock or soil.

Elasticity: Deformation where strain is proportional to stress, and is recoverable.

Expansive: A clay that is prone to large volume changes that are directly related to changes in water content. The mineral make-up of this type of soil is responsible for the moisture retaining capabilities. Expansive soils typically contain one or more of these clay minerals: montmorillonite, smectite, or illite.

Fill: Material used to make engineered earthworks such as embankments, capable of acquiring the necessary engineering properties during placement and compaction.

Formation: The basic unit of subdivision of geological strata, comprising strata with common, distinctive, mappable geological characteristics.

Glacial: Of, or relating to, the presence of ice or glaciers; formed as a result of glaciation.

Groundwater: Water contained in saturated soil or rock below the water table.

Group: A stratigraphical unit usually comprising one or more formations with similar or linking characteristics.

Heave: Upwards movement of the ground and the corresponding movement of the affected foundations. Heave of low-rise structures occurs when a clay is able to absorb more water than it had previously contained and expands. This can occur when trees or hedges, which take moisture from the soil, are severely pruned, or removed, resulting in the soil increasing in moisture content in the affected area.

Hydraulic continuity: Juxtaposition of two or more permeable deposits or rock units such that fluids may pass easily from one to another.

Illite: A 2:1 clay mineral, common in sedimentary rocks, not noted for susceptibility to shrink–swell behaviour.

Index tests: Simple geotechnical laboratory tests which characterize the properties of soil (usually) in a remoulded, homogeneous form; distinct from 'mechanical properties', which are specific to the conditions applied.

Indurate: The process of making a hardened mass of material, possibly in distinct horizons.

Jurassic: The middle period of the Mesozoic (208.0–145.6 Ma).

Linear shrinkage (LS): The percentage length reduction of a prism of remoulded clay subjected to oven drying at 105°C.

Liquid limit (LL): The moisture content at the point between the cliquid and the plastic state of a clay. An Atterberg limit.

Mineral: A naturally occurring chemical compound (or element) with a crystalline structure and a composition which may be defined as a single ratio of elements or a ratio that varies within defined end-members.

Moisture content: See water content.

Montmorillonite: A s:1 clay mineral, member of the smectite family, highly susceptible to shrinkage and/or swelling.

Mudrock: A term used by engineers, synonymous with mudstone.

Mudstone: A fine-grained, non-fissile, sedimentary rock composed of predominately clay and silt-sized particles.

Outcrop: The area over which a particular rock unit occurs at the surface.

Overburden: Material, or stress applied by material, overlying a particular stratum. Unwanted material requiring removal (quarrying).

Overconsolidated (OC): Deposit such as clay that in previous geological times was loaded more heavily than the present day, and consequently has a tendency to expand if it has access to water and is subject to progressive shear failure. The moisture content is less than that for an equivalent material that has been normally consolidated.

Particle-size analysis (PSA): The measurement of the range of sizes of particles in a disaggregated soil sample. The tests follow standard procedures with sieves being used for coarser sizes, and various sedimentation, laser or X-ray methods for the finer sizes usually contained within a suspension.

Particle-size distribution (PSD): The result of a particle-size analysis. It is shown as a 'grading' curve, usually in terms of percentage by weight passing particular sizes. The terms 'clay', 'silt', 'sand' and 'gravel' are defined by their particle sizes.

Permeability: The property or capacity of a rock, sediment or soil for transmitting a fluid; frequently used as a synonym for 'hydraulic conductivity' (engineering). The property may be measured in the field or in the laboratory using various direct or indirect methods.

Plasticity index (PI): The difference between the liquid and plastic limits. It shows the range of water contents for which the clay can be said to behave plastically. It is often used as a guide to shrink–swell behaviour, compressibility, strength and other geotechnical properties.

Plastic limit (PL): The water content at the lower limit of the plastic state of a clay. It is the minimum water content at which a soil can be rolled into a thread 3 mm in diameter without crumbling. The plastic limit is an Atterberg limit.

Pores: The microscopic voids within a soil or rock. The non-solid component of a soil or rock. May be filled with liquid or gas.

Pore pressure: The pressure of the water (or air) in the pore spaces of a soil or rock, which equals total stress minus effective stress. The pore pressure may be negative.

Quartz: The most common silica mineral (SiO_2).

Sand: A soil with a particle-size range 0.06–2.00 mm. Typically consists of quartz particles in a loose state.

Saturation: The extent to which the pores within a soil or rock are filled with water (or other liquid).

Settlement: The lowering of the ground surface due to an applied load (see consolidation).

Shale: A fissile mudstone.

Shear strength: The maximum stress that a soil or rock can withstand before failing catastrophically or being subject to large unrecoverable deformations.

Shrinkage: The volume reduction of a clay (or clay-rich soil or rock) resulting from reduction of water content. Shrinkage may cause subsidence of shallow foundations.

Shrinkage limit (SL): The water content below which little or no further volume decrease occurs during drying of a clay (or clay-rich soil or rock). The laboratory tests which measure shrinkage limit have largely fallen into disuse in the UK. An Atterberg limit.

Silt: A soil with a particle-size range 0.002–0.060 mm (between clay and sand).

Smectite: A group of 2:1 clay minerals noted for their high plasticity and susceptibility to shrink–swell behaviour (e.g. montmorillonite).

Stiffness: The ability of a material to resist deformation.

Strain: A measure of deformation resulting from application of stress.

Stress: The force per unit area to which it is applied. Frequently used as synonym for pressure.

Subsidence: The settling of the ground or a building in response to physical changes in the subsurface such as underground mining, clay shrinkage or drained response to overburden (consolidation).

Suction: The force exerted when fluid within pores in a soil or rock is subjected to reduced atmospheric (or other environmental) pressure.

Superficial deposits: A general term for usually unlithified deposits of Quaternary age overlying bedrock; formerly called 'drift'.

Swelling: The volume increase of a clay (or clay-rich soil or rock) resulting from an increase in water content. Swelling behaviour may cause heave of shallow foundations.

Swelling index (SI): The rebound (unloading) equivalent of the compression index.

Till: An unsorted mixture which may contain any combination of clay, sand, silt, gravel, cobbles and boulders (diamict) deposited by glacial action without subsequent reworking by meltwater.

Water content: In a geotechnical context, this is the mass of water in a soil and/or rock as a percentage of the dry mass (usually dried at 105°C). Synonymous with moisture content.

Water table: The level in the rocks at which the porewater pressure is at atmospheric pressure, and below which all voids are water filled; it generally follows the surface topography but with less relief, and meets the ground surface at lakes and most rivers. Water can occur above a water table.

Weathering: The physical and chemical processes leading to the breakdown of rock materials (e.g. due to water, wind, temperature).

Recommended further reading

AL-RAWAS, A.A. & GOOSEN, M.F.A. (eds). 2006. *Expansive Soils: Recent Advances in Characterization and Treatement*. Taylor & Francis, London.
BRE 1993. Low-rise buildings on shrinkable clay soils. *Building Research Establishment Digests*, **240**, 241–242.
BRAB 1968. Criteria for selection and design of residential slabs-on-ground. Building Regulations Advisory Body, London.
FREDLUND, D.G. & RAHARDJO, H. 1993. *Soil Mechanics for Unsaturated Soils*. Wiley, New York.
TRB 1985. Evaluation and control of expansive soils. Transportation Research Board, London.
VIPULANANDAN, C., ADDISON, M.B. & HASEN, M. [eds] 2001. *Expansive Clay Soils and Vegetative Influence on Shallow Foundations*. Geotechnical Special Publication **115**, Virginia: American Society of Civil Engineers.

Useful web addresses

ASSOCIATION OF BRITISH INSURERS www.abi.org.uk.
BRITISH GEOLOGICAL SURVEY www.bgs.ac.uk.
INSTITUTION OF CIVIL ENGINEERS www.ice.org.uk.
INTERNATIONAL SOCIETY OF ARBORICULTURE www.isa-arboriculture.org.
ROYAL INSTITUTION OF CHARTERED SURVEYORS www.rics.org.
SUBSIDENCE CLAIMS ADVISORY BUREAU www.subsidencebureau.com.
THE CLAY RESEARCH GROUP www.theclayresearchgroup.com.
THE GEOLOGICAL SOCIETY www.geolsoc.org.uk.
THE SUBSIDENCE FORUM www.subsidenceforum.org.
UNITED STATES GEOLOGICAL SURVEY www.usgs.gov.

References

ALONSO, E.E., GENS, A. & JOSA, A.A. 1990. Constitutive model for partially saturated soils. *Geotechnique*, **40**, 405–430, https://doi.org/10.1680/geot.1990.40.3.405
BARKER, J.E., ROGERS, C.D.F., BOARDMAN, D.I. & PETERSON, J. 2004. Electrokinetic stabilisation: an overview and case study. *Ground Improvement*, **8**, 47–58, https://doi.org/10.1680/grim.2004.8.2.47
BEETHAM, P., DIJKSTRA, T.A., DIXON, N., FLEMING, P., HUTCHINSON, R. & BATEMAN, J. 2014. Lime stabilisation for earthworks: a UK perspective. *Proceedings of the Institution of Civil Engineers: Ground Improvement*, **168**, 81–95, https://doi.org/10.1680/grim.13.00030
BELL, F.G. & CULSHAW, M.G. 2001. Problem soils: a review from a British perspective. *In*: JEFFERSON, I., MURRAY, E.J., FARAGHER, E. & FLEMING, P.R. (eds) *Problematic Soils Symposium*, November 2001, Nottingham, 1–35.
BIDDLE, P.G. 1998. Tree roots and foundations. *Arboriculture Research and Information Note 142/98/EXT*.
BIDDLE, P.G. 2001. Tree Root Damage to Buildings. Expansive Clay Soils and Vegetative Influence on Shallow Foundations. *ASCE Geotechnical Special Publications*, **115**, 1–23.
BISHOP, A.W., KUMAPLEY, N.K. & EL-RUWAYIH, A.E. 1975. The influence of pore-water tension on the strength of clay. *Philosophical Transactions of the Royal Society London*, **278**, 511–554, https://doi.org/10.1098/rsta.1975.0034
BOLT, G.H. 1956. Physico-chemical analysis of the compressibility of pure clays. *Geotechnique*, **6**, 86–93, https://doi.org/10.1680/geot.1956.6.2.86
BRE. 1991. *Why Do Buildings Crack?* BRE Digest, **361**. CRC, London.
BRE. 1995. *Assessment of Damage in Low-Rise Buildings*. BRE Digest, **251**. CRC, London.
BRE. 1996. *Desiccation in Clay Soils*. BRE Digest, **412**. CRC, London, 12.
BRE. 1999. *The Influence of Trees on House Foundations in Clay Soils*. BRE Digest, **298**. CRC, London, 8.
BRE. 2002. *Low-Rise Building Foundations on Soft Ground*. BRE Digest, **471**. CRC, London.
BSI. 1999. *BS 5930:1999 + Amendment 2:2010 Code of practice for site investigations*. BSI, London.
BURLAND, J.B., LONGWORTH, T.I. & MOORE, J.F.A. 1977. A study of ground movement and progressive failure caused by a deep excavation in Oxford Clay. *Geotechnique*, **27**, 557–591, https://doi.org/10.1680/geot.1977.27.4.557
BUZZI, O., FITYUS, S. & SLOAN, S.W. 2010. Use of expanding polyurethane resin to remediate expansive soil foundations. *Canadian Geotechnical Journal*, **47**, 623–634, https://doi.org/10.1139/T09-132
CAMERON, D.A. 2006. The role of vegetation in stabilizing highly plastic clay subgrades. *In*: GHATAORA, G.S. & BURROW, M.P.N. (eds) *Proceedings of First International Conference on Railway Foundations, RailFound 06*, September 2006, Birmingham, 165–186.
CHEN, F.H. 1988. *Foundations on Expansive Soils*. Elsevier, Amsterdam.
CHENEY, J.E. 1986. 25 years' heave of a building constructed on clay, after tree removal. 1988 edn. *Ground Engineering*, July, 13–27.
CLAYTON, C.R.I., XU, M., WHITER, J.T., HAM, A. & RUST, M. 2010. Stresses in cast-iron pipes due to seasonal shrink–swell of clay soils. *Proceedings of the Institution of Civil Engineers: Water Management*, **163**, 157–162, https://doi.org/10.1680/wama.2010.163.3.157
CRILLY, M.S. & DRISCOLL, R.M.C 2000. The behaviour of lightly loaded piles in swelling ground and implications for their design. *Proceedings of the Institution of Civil Engineers: Geotechnical Engineering*, **143**, 3–16, https://doi.org/10.1680/geng.2000.143.1.3
DETR. 1993. *Manual of Contract Documents for Highway Works Volume 1: Specification for Highway Works*. Department of Transport, HMSO, London.
DRISCOLL, R. 1983. The influence of vegetation on the swelling and shrinking of clay soils in Britain. *Geotechnique*, **33**, 93–105, https://doi.org/10.1680/geot.1983.33.2.93
DRISCOLL, R. & CRILLY, M. 2000. *Subsidence Damage to Domestic Buildings. Lessons Learned and Questions Asked*. BRE Press, London.
DRISCOLL, R.M.C. & CHOWN, R. 2001. Shrinking and swelling of clays. *In*: JEFFERSON, I., MURRAY, E.J., FARAGHER, E. & FLEMING, P.R. (eds) *Problematic Soils Symposium*, November 2001, Nottingham, 53–66.
DRISCOLL, R.M.C. & SKINNER, H. 2007. *Subsidence Damage to Domestic Building – A Good Technical Practice Guide*. BRE Press, London.
EBERL, D.D. 1984. Clay mineral formation and transformation in rocks and soils. *Philosophical Transactions of the Royal Society of London, A*, **311**, 241–257, https://doi.org/10.1098/rsta.1984.0026
EINSTEIN, H.H. 1979. Tunnelling in swelling rock. *Underground Space*, **4**, 51–61.
GROB, H. 1976. Swelling and heave in Swiss tunnels. *Bulletin of the International Association of Engineering Geology.*, **13**, 55–60, https://doi.org/10.1007/BF02634759

HARRISON, A.M., PLIM, J., HARRISON, M., JONES, L.D. & CULSHAW, M.G. 2012. The relationship between shrink–swell occurrence and climate in southeast England. *Proceedings of the Geologists' Association*, **123**, 556–575, https://doi.org/10.1016/j.pgeola.2012.05.002

HOBBS, P.R.N., HALLAM, J.R. *ET AL*. 1998. *Engineering Geology of British Rocks and Soils: Mercia Mudstone*. British Geological Survey, Technical Report No. **WN/98/4**.

HOLTZ, R.D. & KOVACS, W.D. 1981. *An Introduction to Geotechnical Engineering*. Prentice-Hall, New Jersey.

HOUSTON, S.L., DYE, H.B., ZAPATA, C.E., WALSH, K.D. & HOUSTON, W.N. 2011. Study of expansive soils and residential foundations on expansive soils in Arizona. *Journal of Performance of Constructed Facilities*, **25**, 31–44, https://doi.org/10.1061/(ASCE)CF.1943-5509.0000077

HUDYMA, N.B. & AVAR, B. 2006. Changes in swell behaviour of expansive soils from dilution with sand. *Environmental Engineering Geoscience*, **12**, 137–145, https://doi.org/10.2113/12.2.137

ISE 1994. *Subsidence of Low Rise Buildings*. Thomas Telford, London.

JACKSON, J.O. & FOOKES, P.G. 1974. The relationship of the estimated former burial depth of the Lower Oxford Clay to some soil properties. *Quarterly Journal of Engineering Geology and Hydrogeology*, **7**, 137–179, https://doi.org/10.1144/GSL.QJEG.1974.007.02.03

JEANS, C.V. 2006a. Clay mineralogy of the Cretaceous strata of the British Isles. *Clay Minerals*, **41**, 47–150, https://doi.org/10.1180/0009855064110196

JEANS, C.V. 2006b. Clay mineralogy of the Jurassic strata of the British Isles. *Clay Minerals*, **41**, 187–307, https://doi.org/10.1180/0009855064110198

JONES, L.D. 2002. Shrinking and swelling soils in the UK: Assessing clays for the planning process. British Geological Survey, UK, Earthwise vol. **18**.

JONES, L.D. 2004. Cracking open the property market. *Planet Earth*, Autumn **2004**, 30–31.

JONES, L.D. & JEFFERSON, I. 2012. Expansive soils. *In*: BURLAND, J., CHAPMAN, T., SKINNER, H. & BROWN, M. (eds) *ICE manual of Geotechnical Engineering. Volume 1 Geotechnical Engineering Principles, Problematic Soils and Site Investigation*. ICE Publishing, London, 413–441.

JONES, L.D. & TERRINGTON, R. 2011. Modelling volume change potential in the London Clay. *Quarterly Journal of Engineering Geology and Hydrogeology*, **44**, 1–15, https://doi.org/10.1144/1470-9236/08-112

JONES, L.D., VENUS, J. & GIBSON, A.D. 2006. *Trees and foundation damage*. British Geological Survey Commissioned Report **CR/06/225**.

KELLER, W.D. 1963. Diagenesis in clay minerals: a review. *Clays and Clay Minerals*, **13**, 136–157.

KEMP, S.J., MERRIMAN, R.J. & BOUCH, J.E. 2005. Clay mineral reaction progress – the maturity and burial history of the Lias Group of England and Wales. *Clay Minerals*, **40**, 43–61, https://doi.org/10.1180/0009855054010154

LANSON, B., SAKHAROVE, B.A., CLARET, F. & DRITS, V.A. 2009. Diagenetic smectite-to-illite transition in clay-rich sediments: a reappraisal of X-ray diffraction results using the multi-specimen method. *American Journal of Science*, **309**, 476–516, https://doi.org/10.2475/06.2009.03

LEROUEIL, S. 2001. No problematic soils, only engineering solutions. *In*: JEFFERSON, I., MURRAY, E.J., FARAGHER, E. & FLEMING, P.R. (eds) *Problematic Soils Symposium*, November 2001, Nottingham, 191–211.

LOVELAND, P.J. 1984. The soil clays of Great Britain: I. England and Wales. *Clay Minerals*, **19**, 681–707, https://doi.org/10.1180/claymin.1984.019.5.02

MADSEN, F.T. & MULLER-VONMOOS, M. 1985. Swelling pressure calculated from mineralogical properties of a Jurassic Opalinum shale, Switzerland. *Clays and Clay Minerals*, **33**, 501–509, https://doi.org/10.1346/CCMN.1985.0330604

MERRIMAN, R. 2005. Clay minerals and sedimentary basin history. *European Journal of Mineralogy*, **17**, 7–20, https://doi.org/10.1127/0935-1221/2005/0017-0007

MESRI, G., ULLRICH, C.R. & CHOI, Y.K. 1978. The rate of swelling of overconsolidated clays subjected to unloading. *Geotechnique*, **28**, 281–307, https://doi.org/10.1680/geot.1978.28.3.281

MITCHELL, J.K. 1976. *Fundamentals of Soil Behaviour*. Wiley, New York.

MITCHELL, J.K. & SOGA, K. 2005. *Fundamentals of Soil Behaviour*. 3rd edn. Wiley, New York.

MORGENSTERN, N.R. 1981. Geotechnical engineering and frontier resource development. *Geotechnique*, **31**, 305–365, https://doi.org/10.1680/geot.1981.31.3.305

MURTY, V.R. & PRAVEEN, G.V. 2008. Use of chemically stabilized soil as cushion material below light weight structures founded on expansive soils. *Journal of Materials in Civil Engineering*, **20**, 392–400, https://doi.org/10.1061/(ASCE)0899-1561(2008)20:5(392)

NELSON, J.D. & MILLER, D.J. 1992. *Expansive Soils: Problems and Practice in Foundation and Pavement Engineering*. Wiley, New York.

NELSON, J.D., OVERTON, D.D. & DURKEE, D.B. 2001. Depth of wetting and the active zone. Expansive clay soils and vegetative influence on shallow foundations. *ASCE Geotechnical Special Publications*, **115**, 95–109.

NELSON, J.D., CHAO, K.C. & OVERTON, D.D. 2011. Discussion of 'Method for evaluation of depth of wetting in residential areas' by Walsh *et al*. 2009. *Journal of Geotechnical and Geoenvironmental Engineering*, **173**, 293–296, https://doi.org/10.1061/(ASCE)GT.1943-5606.0000213

NHBC 2010. *Efficient Design of Foundations for Low Rise Housing – Design Guide*. National House-Building Council, London.

NHBC STANDARDS 2011. *Building near trees. NHBC Standards Chapter 4.2*. National House-Building Council, London.

PATIL, U., VALDES, J.R. & EVANS, M.T. 2011. Swell mitigation with granulated tire rubber. *Journal of Materials in Civil Engineering*, **25**, 721–727, https://doi.org/10.1061/(ASCE)MT.1943-5533.0000229

PETRY, T.M. & LITTLE, D.N. 2002. Review of stabilization of clays and expansive soils in pavement and lightly loaded structures – history, practice and future. *Journal of Materials in Civil Engineering*, **14**, 447–460, https://doi.org/10.1061/(ASCE)0899-1561(2002)14:6(447)

RADEVSKY, R. 2001. Expansive clay problems – how are they dealt with outside the US? Expansive clay soils and vegetative influence on shallow foundations. *ASCE Geotechnical Special Publications*, **115**, 172–191.

REEVE, M.J., HALL, D.G.M. & BULLOCK, P. 1980. The effect of soil composition and environmental factors on the shrinkage of some clayey British soils. *Journal of Soil Science*, **31**, 429–442, https://doi.org/10.1111/j.1365-2389.1980.tb02092.x

RIMMER, D.L. & GREENLAND, D.J. 1976. Effects of calcium carbonate on the swelling behaviour of a soil clay. *Journal of Soil Science*, **27**, 129–140, https://doi.org/10.1111/j.1365-2389.1976.tb01983.x

ROBERTS, J. 1976. A study of root distribution and growth in a *Pinus Sylvestris L.* (Scots Pine) plantation in East Anglia. *Plant and Soil*, **44**, 607–621, https://doi.org/10.1007/BF00011380

SCOTT, G.J.T., WEBSTER, R. & NORTCLIFF, S. 1986. An analysis of crack pattern in clay soil: its density and orientation. *Journal of Soil Science*, **37**, 653–668, https://doi.org/10.1111/j.1365-2389.1986.tb00394.x

SIMONS, N.E., MENZIES, B.K. & MATTHEWS, M.C. 2002. *A Short Course in Geotechnical Site Investigation*. Thomas Telford, London.

SRIDHARAN, A. & VENKATAPPA RAO, G. 1971. Mechanisms controlling compressibility of clays. *Journal of Soil Mechanics and Foundations*, **97**, 940–945.

TAYLOR, R.K. & CRIPPS, J.C. 1984. Mineralogical controls on volume change. *In: Ground Movements and Their Effects on Structures*. Surrey University Press, Surrey, 268–302.

TAYLOR, R.K. & SMITH, T.J. 1986. The engineering geology of clay minerals: swelling, shrinking and mudrock breakdown. *Clay Minerals*, **21**, 235–260, https://doi.org/10.1180/claymin.1986.021.3.01

UKCIP 2009. *UK Climate Projections*. UKCP09, London.

VAUGHAN, P.R. & WALBANCKE, H.J. 1973. Pore pressure changes and delayed failure of cutting slopes in overconsolidated clay. *Geotechnique*, **23**, 531–539, https://doi.org/10.1680/geot.1973.23.4.531

WALSH, K.D., COLBY, C.A., HOUSTON, W.N. & HOUSTON, S.L. 2009. Method for evaluation of depth of wetting in residential areas. *Journal of Geotechnical and Geoenvironmental Engineering*, **135**, 169–176, https://doi.org/10.1061/(ASCE)1090-0241(2009)135:2(169)

WILSON, M.J., BAIN, D.C. & DUTHIE, D.M.L. 1984. The soil clays of Great Britain: II. Scotland. *Clay Minerals*, **19**, 709–735, https://doi.org/10.1180/claymin.1984.019.5.03

ZHENG, J.L., ZHENG, R. & YANG, H.P. 2009. Highway subgrade construction in expansive soil areas. *Journal of Materials in Civil Engineering*, **21**, 154–162, https://doi.org/10.1061/(ASCE)0899-1561(2009)21:4(154)

Chapter 9   Peat hazards: compression and failure

Jeff Warburton

Department of Geography, Durham University, Lower Mountjoy, South Road, Durham DH1 3LE, UK
0000-0003-3589-9039

jeff.warburton@durham.ac.uk

Abstract: Peat is a highly compressible geological material whose time-dependent consolidation and rheological behaviour is determined by peat structure, degree of humification and hydraulic properties. This chapter reviews the engineering background to peat compression, describes the distribution of peat soils in the UK, provides examples of the hazards associated with compressible peat deposits and considers ways these hazards might be mitigated. Although some generalizations can be made about gross differences between broad peat types, no simple relationship exists between the magnitude and rate of compression of peat and loading. Based on examples described here, land failures resulting from peat compression are locally generated, but due to the sensitive nature of peat these can result in runaway failures that pose great risk. Understanding the geological hazards associated with compressible peat soils is challenging because peat is geotechnically highly variable and the mapped extent of peat in the UK is subject to considerable error due to inconsistencies in the definition of peat. Mitigating compression hazards in peat soils is therefore subject to considerable uncertainty; however, a combination of improved understanding of the properties of compressible peat, better mapping and land use zoning, and appropriate construction will help to mitigate risk.

9.1 Introduction and scope

Peat is a low-density, highly compressible soil that occurs at the surface or may be buried at depth. Peat is essentially an organic, non-mineral soil resulting from the decay of organic matter. Peat deposits are widespread in the UK, occurring in a wide variety of upland and lowland environments covering all parts of the country (Fig. 9.1). Peat accumulates wherever suitable conditions occur such as in areas of high (excess) rainfall and where ground drainage is poor leading to high water tables. In these waterlogged areas, peat develops where the rate of dry vegetative matter accumulation exceeds the rate of decay. Physiochemical and biochemical processes associated with wetland conditions ensure that the accumulating organic matter decays very slowly, safeguarding plant structures that remain partially intact for long periods of time (Bell 2000). In the UK, temperate peat accumulates slowly, typically at 0.2–1 mm a^{-1} with local rates varying depending on the topography and hydrology of the peat mire (Charman 2002).

In the engineering community, peats and organic soils are well known for their high compressibility and long-term settlement; in terms of engineering properties, peat is notoriously difficult to deal with, which prompted Powrie (1997, p.16) to comment that: '…[organic soils] should not be relied on for anything, except to cause trouble'. To an engineer peat and organic soils are extremely soft, wet, unconsolidated surficial deposits that pose a range of geotechnical problems for sampling, settlement, stability, *in situ* testing, stabilization and construction.

The link between the compressibility of peat, its shear strength properties and the risk of bearing capacity failure has not been explored in detail, although the mechanism has been suggested for some peat failures (Lindsay & Bragg 2005). Peat soils are highly organic, highly compressible and generally possess low undrained strength; their compression and/or settlement may take a considerable amount of time to stabilize (Huat *et al.* 2014). Estimating the geotechnical properties of peat is notoriously difficult because published values are relatively few and the testing of peat using standard geotechnical tests is fraught with problems (Long 2005; Long & Boylan 2012). Nevertheless, published data (e.g. Dykes 2008) suggest that peat in its undisturbed state has little strength with undrained shear strength values typically varying over 5–20 kPa (Long 2005; Huat *et al.* 2014). These values vary with the vegetation composition of the peat (particularly fibre content) and the degree of humification, but are also affected by the method of testing (Boylan *et al.* 2008). Given the high compressibility and low strength of peat, local shear failure may occur when compression and/or compaction gives rise to vertical displacements that exceed the shear strength (bearing capacity) of the soil (Knappett & Craig 2012). Shear failure may result where differential displacements of surface peat occur between the area experiencing compression (loading) and the adjacent unloaded peat. In

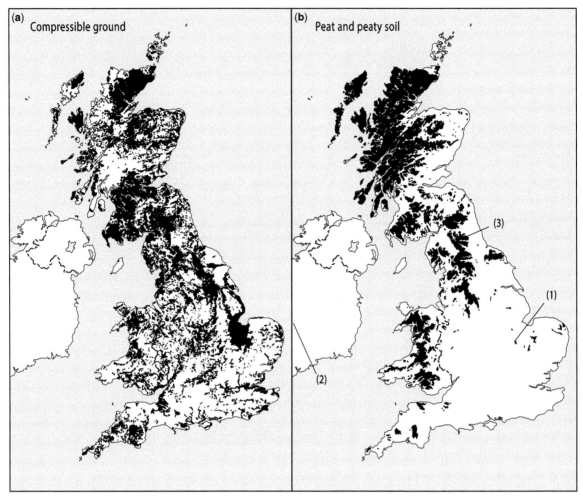

Fig. 9.1. (a) Compressible ground potential map (adapted from British Geological Survey 2014) and (b) peat and peaty soils of the UK (map adapted from JNCC 2011). Numbers indicate key sites discussed in the text: (1) Holme Fen, (2) Derrybrien and (3) North Pennines (Burnhope and Harthope).

peatlands, such sites typically include construction embankments and/or waste heaps, roads and tracks, and foundations such as wind turbine bases. Although such failures are local in origin due to the sensitive nature of peat stability, under the right site conditions these may rapidly propagate to runaway failures.

In engineering practice there is therefore a tendency to either avoid construction on these soils or, if this is not possible, remove or replace the peat material. However, in many countries including the UK, peat extends over a substantial part of the terrestrial biosphere and peatlands are under increasing pressure for their land use (Fig. 9.1). In lowland areas, particularly in the distal parts of populated deltas and estuaries, peat is common and, due to compaction, may cause land subsidence resulting in damage to infrastructure and land inundation by the sea (van Asselen *et al.* 2009).

In the UK, there is growing public awareness of the effect of ground conditions on safety and property values and increasing pressure from the government to provide environmental information (Royse 2011). Information about geological hazards and, in particular, the identification of areas which are susceptible to ground movement, is needed (BGS 2010; Figure 9.1a). As part of its UK hazard assessment programme, the British Geological Survey (BGS) has summarized key information on compressible ground. Their definition is:

> Ground is compressible if an applied load, such as a house, causes the fluid in the pore space between its solid components to be squeezed out causing it to decrease rapidly in thickness

(compress). Peat, alluvium and laminated clays are common types of deposits associated with various degrees of compressibility. The deformation of the ground is usually a one-way process that occurs during or soon after construction.

Peatlands are considered areas of compressible ground and, given the widespread occurrence of peat deposits in the UK (Fig. 9.1b), pose a large potential hazard as a compressible soil.

Peat soils are well known for landslide-related hazards and these have been widely reported and documented in the UK and Ireland (Warburton et al. 2004; Dykes & Warburton 2007; Boylan et al. 2008; Dykes 2009). However, far less is known about the hazards posed by peat compression and the potential problems associated with this. The aim of this chapter is therefore to briefly review the engineering background to peat compression, describe the occurrence of peat soils in the UK, provide examples of the compression hazards associated with these deposits, and consider some of the ways these hazards can be mitigated.

9.2 Engineering background: peat consolidation and compression

A number of characteristics distinguish peat as an engineering material. These include a high but variable natural water content (c. 500–1500%), very high organic content (loss on ignition 25–100%), significant fibre content, low specific gravity (bulk density), high voids ratio (5–15), high initial permeability, high compressibility and low strength (Edil 2001). Although peat deposits are highly variable, the degree of humification (the extent of biochemical decomposition of plant remains) is a key factor determining the overall behaviour of peat. Table 9.1 outlines the von Post scheme of characterizing peat deposits based on humification. The end-members of this scale go from highly fibrous deposits with insignificant decomposition to amorphous peat with no discernible plant remains. This distinction has been used by MacFarlane & Radford (1965) and MacFarlane (1969) to broadly categorize the engineering behaviour of peat into fibrous and amorphous granular deposits. Further division of the scale into three categories which characterize broad divisions of peat are based on fibre content (Fc) and the von Post scale (Edil 2001; Table 9.1): (1) fibric, >67% Fc, von Post H_1–H_3; (2) hemic, 33–67% Fc, von Post H_4–H_6; and (3) sapric, <33% Fc, von Post H_7–H_{10}.

Numerous other classifications of organic soils exist, and there is no overall standardized scheme (Myślińska 2003).

In many peats, water content by volume may typically vary over the range 75–98% (Hobbs 1986), making peat extremely susceptible to rapid compression. Water may be held in peat in three main phases: intercellular water in macropores; interparticle (intracellular) water in micropores; and adsorbed or bound water. The typical proportions of these will vary with the peat type. Commonly, active bog peat exhibits a two-layered structure (Ivanov 1981; Ingram 1982). The lower layer, which is in general continually saturated and is composed of older, more humified peat, is known as the catotelm; the more aerated upper layer, which typically lies above the lowest water table limit, is called the acrotelm.

Figure 9.2a shows the gas, water and solid components, and their volume relationships, for a peat core under sphagnum vegetation (Rydin & Jeglum 2006). In the upper layers (acrotelm), gas volume near the surface is around 85% but this decreases to 0% at just below the water table. Below the water table, extracellular and intracellular water make up 90% of the volume. The proportion of solid peat increases

Table 9.1. *Humification of peat*

Degree of humification	Decomposition	Plant structure	Content of amorphous material	Material extruded when squeezing (passing between fingers)
H_1	None	Easily identified	None	Clear, colourless water
H_2	Insignificant	Easily identified	None	Yellowish water
H_3	Very slight	Still identifiable	Slight	Brown, muddy water; no peat
H_4	Slight	Not easily identified	Some	Dark brown, muddy water; no peat
H_5	Moderate	Recognizable, but vague	Considerable	Muddy water and some peat
H_6	Moderately strong	Indistinct (more distinct after squeezing)	Considerable	About one-third of peat squeezed out; water dark brown
H_7	Strong	Faintly recognizable	High	About half of peat squeezed out; any water very dark brown
H_8	Very strong	Very indistinct	High	About two-thirds of peat squeezed out; also some pasty water
H_9	Nearly complete	Almost unrecognizable	High	Nearly all of peat squeezed out as a fairly uniform paste
H_{10}	Complete	Not discernible	High	All peat passes between the fingers; no free water visible

Modified from von Post (1924) and Hobbs (1986).

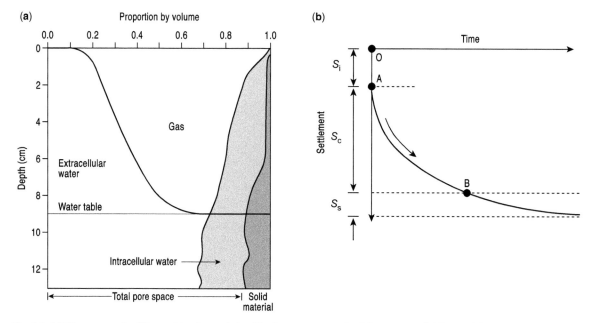

Fig. 9.2. (a) Water, gas and solid material volume relationships in a peat core under *Sphagnum capillifolium* and (b) typical time settlement relationship for a saturated soil under a vertical load (adapted from Aysen 2005; Rydin & Jeglum 2006).

with depth as the partial decayed plant material becomes increasingly humified and compressed. Total pore spaces in peat vary over the range 78–93% (Rydin & Jeglum 2006), with porosity decreasing as humification, bulk density and the degree of compaction increase. The hydrological functioning of peat is strongly controlled by the structure of the peat matrix, which in turn is highly susceptible to compression and deformation (Price & Schlotzhauer 1999). This novel property of peat means that volume changes to a peat mass may occur over timescales which are characteristic of hydrological events due to rapid readjustments of the peat pore structure (Kennedy & Price 2005).

The presence of gas trapped in peat has important implications for the surface hydrology of peat bogs, affecting both microclimate and ecohydrology (Strack *et al.* 2006). Equally, the physical properties of peat itself, including permeability, rate of consolidation and pore pressure, are also affected (Macfarlane 1969). Trapped gas arises due to the slow decomposition of organic matter below the water table which generates CH_4 (methane, a key component of marsh gas) and lesser amounts of N and CO_2. Typically, the free gas content of peat varies over the range 5–10%. Biogenic peat gas was originally thought to originate in deep peat, but more recent studies (e.g. Kellner *et al.* 2005) have demonstrated the presence of shallow pressurized gas pockets in shallow peat (<1 m). Biogenic gases (mainly CH_4) are released from peat by diffusion, vascular plant transport or ebullition. Recent work by Comas *et al.* (2014) has shown that biogenic peat gas can be stored in both deep and shallow peat depending on the physical properties of the peat matrix and, in particular, the presence in the peat stratigraphy of wood layers or abrupt transitions in humification that produce confining layers that temporarily entrap gas. Results comparing the gas content of two bogs, one with woody peat with distinct wood layers and one with a more homogenous peat, showed that the gas contents were 10.8% and 5.7%, respectively, the woody peat having the high biogenic gas content (Comas *et al.* 2014). Although the presence of gas in peat has been recognized for a long time, it is only recently with the advent of more sophisticated subsurface mapping technology (e.g. ground-penetrating radar) that the extent and distribution of gas has been precisely determined.

9.2.1 Compression of peat

When peat is subject to an increase of compressive stress (load), the resulting compression or settlement consists first of immediate elastic compression (immediate settlement), then primary compression (consolidation) and eventually secondary compression. Peat can compress and consolidate both slowly and rapidly. Slow compression and consolidation allow time for the peat body to respond to the applied load, allowing porewater pressures to dissipate and the peat to improve strength and bearing capacity. Alternatively, rapid loading, with or without actual compression, may result in a rapid increase in porewater pressure and potential shear failure (Huat *et al.* 2014). The factors controlling the compressibility of peat include permeability, natural water content, void ratio, fibre content, peat structure and interparticle chemical bonding (Hobbs 1986; Carlsten 1988; Mesri & Ajlouni

2007). Permeability is commonly regarded as the most important engineering property of peat because it controls the rate of consolidation of peat under load and ultimately the strength of the material (Hobbs 1986).

Peat will undergo settlement over time, and Figure 9.2b shows a schematic time settlement relationship in a soil element undergoing vertical loading (Aysen 2005). The consolidation (vertical compression) of a soil can be divided into three main stages (Day 2000):

(1) S_i: Initial compression occurs immediately a load is applied and is often estimated from the observed settlement of structures or predicted from the theory of elasticity. If this occurs without any change in the amount of water in the soil (elastic settlement), this may lead to undrained shear deformation or plastic flow due to loading.
(2) S_c: Primary consolidation, the compression of the soil under load which occurs as excess porewater pressures dissipate over time.
(3) S_s: This third stage represents settlement under a constant effective vertical stress, and is termed secondary consolidation or secondary compression. This is the component of settlement which occurs after all the excess pore pressures have dissipated, and is sometimes referred to as drained creep. The exact mechanisms of secondary consolidation are not well known, but appear related to colloid–chemical interactions and small residual excess pore pressures (Aysen 2005). In peat soils, it is generally accepted that this secondary phase of compression is associated with the rearrangement of vegetative fragments and plant structures into a denser matrix (Huat et al. 2014).

The compression index (C_c) describes variation of the voids ratio (e) as a function of the change in effective vertical stress (σ'_v) plotted on a logarithmic scale (Fratta et al. 2007):

$$C_c = \frac{\Delta e}{\Delta \log \sigma'_v}. \qquad (9.1)$$

C_c represents the deformation character of a particular soil during primary consolidation (S_c); in peat soils, the plotted curves are very steep indicating a high compression index, typically in the range 2–15 (Huat et al. 2014). Clays typically have a compression index of <1.0.

Figure 9.3 shows a plot of percentage natural water content (W_o) and compression index (C_c) for soft clay and/or silt deposits and peats (from Mesri et al. 1997). Peat deposits are clearly distinguishable from clays and silts, displaying significantly higher natural water content and corresponding compression index values. This relates to the high void ratio that is characteristic of peat. As peat accumulates, high void ratios develop because the plant remains that make up the peat consist of low-density particles, fibres and platy structures that create a porous medium with a high water storage capacity (Table 9.2). Because C_c is directly related to the secondary compression index C_α (Huat et al. 2014), peat

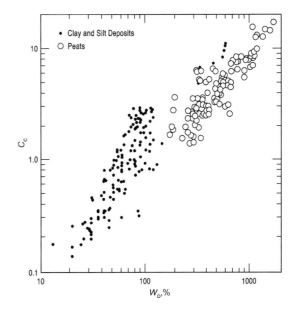

Fig. 9.3. Plot of percentage natural water content (W_o) and compression index (C_c) for soft clay and/or silt deposits and peats (redrawn after Mesri et al. 1997).

Table 9.2. *Peat properties and compressibility ratio data*

Peat type	Water content (% of dry weight)	Vertical coefficient of permeability (m s^{-1})	C_α/C_c
Fibrous peat	850	4×10^{-6}	0.06–0.10
Amorphous to fibrous peat	500–1500	10^{-7}–10^{-6}	0.035–0.078
Canadian muskeg	200–600	10^{-5}	0.090–0.10
Fibrous peat	613–886	10^{-6}–10^{-5}	0.06–0.085
Fibrous peat	660–1590	(5×10^{-7})–(5×10^{-5})	0.06
Fibrous peat	610–850	(6×10^{-8})–10^{-7}	0.052
Long & Boylan (2013) – Irish peat	c. 350–1000		0.072

Modified from Mesri et al. (1997).

deposits also have correspondingly high values of C_α, which occurs during secondary consolidation (S_s), that is:

$$C_\alpha = \frac{\Delta e}{\Delta \log t}. \qquad (9.2)$$

Compressibility of peat can therefore be summarized using the C_α/C_c concept (Huat et al. 2014), with peat yielding the highest values of natural soils (Table 9.2) that appears to depend on the deformability, including the compressibility, of the soil particles and/or matrix (Mesri et al. 1997).

Once loaded, peat settles and consolidates. However, the duration of primary consolidation is very short as a result of the high permeability of the peat deposits; typically, this is completed within a few weeks or months (MacFarlane 1969; Mesri et al. 1997). Peat should therefore be loaded slowly so that the reduction in strength due to raised pore-water pressures is kept to a minimum; if it is loaded too quickly, the peat will shear and fail. Once the initial phase of primary consolidation is complete, consolidation proceeds at a much slower rate in a state of secondary compression that is linear with the logarithm of time (Equation (9.2)). The large magnitude and short duration of primary consolidation and the continuous long-term secondary compression distinguish peat soils from their mineral counterparts (MacFarlane 1969; Fox & Edil 1996; Fox et al. 1999). The high porosity and low bulk density is thought to account for the dramatic phase of initial compression, and the continued deformation of the solid material in the peat results in the long-term secondary compression.

Figure 9.4 shows the variation of the water content (%) of UK mire peats with bulk density (Hobbs 1986) for both bog and fen peats. This serves to emphasize the variability in these properties that commonly occur in UK, which mirrors that from other parts of the world. The presence of mineral soil components (fen peats) generally reduces water content and hence variability; bog peats therefore have higher and more variable water contents. Above about 600% water content the curves in Figure 9.4 flatten, indicating that the specific gravity and water content do not particularly influence bulk density. The primary influence is the degree of saturation or gas content (Bell 2000), which in turn depends on the structure and degree of humification of the peat. For example, amorphous granular peat which has undergone greater humification will have a high bulk density and low void ratio resulting in considerable secondary compression. Conversely, more fibrous, less humified peats will be more susceptible to primary consolidation (Bell 2000; Mesri & Ajlouni 2007).

Variation in the compression index (C_c) with water content (%) is show in Figure 9.5. This diagram compares UK fen and bog peat with other international examples (Hobbs 1986). In Figure 9.5, UK fen peat can be distinguished from UK bog peat, and Hobbs (1986) proposed two criteria based on water content (W):

$$C_c = 0.0065\,W \quad \text{(bog peat)} \qquad (9.3)$$
$$C_c = 0.008\,W \quad \text{(fen peat)}. \qquad (9.4)$$

Although these functions discriminate between the two main categories of peat, it should be noted there is a transition between the two types, there is considerable scatter around these relationships (Fig. 9.5), and the number and range of samples tested is relatively few. Nevertheless, the range of these general relationships are consistent with the behaviour of these types of peat. However, in the transition zone (e.g. 1000% water content), the bog peat has a marginally lower compression index value than fen peat, which is contrary to basic understanding and predictions based on other geotechnical properties (e.g. liquid limit) (Hobbs 1986).

This highlights a significant issue in understanding the consolidation behaviour of peat, identified in early research (e.g. Barden 1968): amorphous well-humified peats have properties that are more closely related to clay soils rather than fibrous peats, whose properties are similar to those described above. However, Berry & Poskitt (1972) and Berry & Vickers (1975) developed a general one-dimensional (1D) consolidation theory for amorphous and fibrous peat that was validated with experimental testing. Understanding of the engineering properties of fibrous peat and peat engineering in general is now well developed as part of engineering science (Carlsten 1988; Mesri & Ajlouni 2007).

More recently, work by Mesri & Ajlouni (2007) has demonstrated an approximate linear relationship between C_c with initial water content (W_i), that is:

$$C_c = W_i/100. \qquad (9.5)$$

A similar relationship was also noted by Long & Boylan (2013), who suggest that for Irish peat samples a relation of $C_c = W_i/125$ was more appropriate, which is equivalent to Equation (9.4) above for fen peat.

Due to these special properties, peat differs greatly from other engineering soils in that: initial settlement may occur

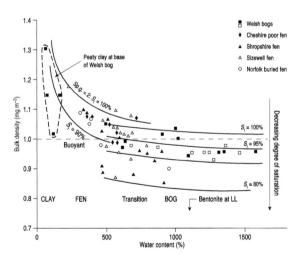

Fig. 9.4. Variation of the water content (%) of UK mire peats with bulk density (adapted from Hobbs 1986).

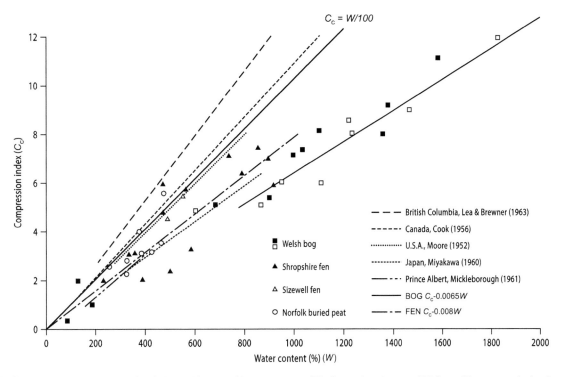

Fig. 9.5. Variation in the compression index (C_c) in peat with water content (%). Comparison between UK fen and bog peat and other international examples (adapted from Hobbs 1986).

very quickly leading to rapid and unpredictable failure; primary consolidation may be large and rapid; and secondary compression, typically negligible in inorganic soils, may be significant over an extended timescale.

9.3 UK peatlands: extent and occurrence

In 2011 the UK Joint Nature Conservation Committee provided an assessment of the state of UK peatlands (JNCC 2011). This important report summarized the extent, location and condition of peat soil and peatlands, vegetation, land cover, land use, management and a range of environmental influences. In their report, the JNCC estimate peat covers around 4 million km^2 or 3% of the world land area; in Europe, recent estimates of the extent of peatlands are approximately 515 000 km^2 (JNCC 2011). In the UK, the extent of peat is not easily defined because there is little consistent UK-wide information on peatlands (maps or statistics); reconciling the various descriptions and classifications to provide a unified picture of the state of UK peatlands represents a significant challenge (JNCC 2011).

Different minimum depth and percentage organic matter content thresholds are used for differentiation between mineral, peaty (organo-mineral) and peat soils in Scotland, England and Wales, and Northern Ireland. In the different soil classification schemes, 'deep peaty soils' in England and Wales and Northern Ireland and 'peat soils' in Scotland are taken to be broadly equivalent, although criteria differ as follows. Minimum depths of 0.4 m, 0.5 m and 0.5 m and minimum organic matter content of 20%, 40% and 60% are the criteria adopted by England and Wales, Northern Ireland and Scotland, respectively.

Figure 9.1b shows the extent of peat and peaty soils in the UK. Although the peat deposits are characterized slightly differently depending on national mapping strategies (e.g. in England there are four main deposits: deep peaty soils, wasted former deep peats, shallow peaty soils and soils with pockets of deep peat), the map shows the general extent of these soils in the UK. Table 9.3 summarizes the extent of such soils.

The striking characteristic of Table 9.3 is that about one-third (79 390 km^2) of the UK is underlain peat- or organic-rich soils. Clearly there is a strong country bias with the majority of these soils occurring in Scotland, but nevertheless significant deposits occur throughout the UK. The distribution clearly reflects areas of high rainfall and poor drainage, with a clear bias to the wetter west of the UK. Peat soils typically occur in the wet uplands (North Pennines) and poorly drained lowland (e.g. Somerset levels) (Fig. 9.1b).

Table 9.3. *Summary of the extent of peat (organic-rich soils) in the UK based on soil map data*

	Shallow peaty or organo-mineral (ha)	Deep peaty or organic soils (ha)	Total peaty soils (ha)	% of UK land area
England	738 618	679 926	1 418 544	5.8
Wales	359 200	70 830	430 030	1.7
Northern Ireland	141 700	160 902	302 602	1.2
Scotland	3 461 200	2 326 900	5 788 100	23.6
TOTAL	4 700 718	3 238 558	7 939 276	32.3

Modified from JNCC (2011).

Figure 9.6 is a schematic cross-sectional view of key hydromorphological mire types found in peatlands (Charman 2002). Although a multitude of different peatland classification systems exist, the scheme shown in Figure 9.6 is useful because it provides a general description of the local water table (hydrological) conditions and underlying topography (morphology) prevalent in key mire types typical of temperate environments such as those in the UK. It also provides a basic illustration of the difference between a bog, which is fed by rain or snow falling directly on its' surface, and a fen, which is influenced by water from outside its own limits (Charman 2002). Typically, but not exclusively, bogs are ombrotrophic and tend to be acid with low nutrient status, while fens are minerotrophic, receiving greater mineral nutrients from groundwater and surface runoff and therefore have a higher trophic status and are more alkaline (Rydin & Jeglum 2006). The importance of this in terms of understanding the engineering behaviour of peats is that the topographic and hydrological variety implicit in Figure 9.6 produces peats which are highly variable in their depth, stratigraphy and microscopic physical properties (Hobbs 1986); these properties affect the compression and settlement characteristics of the peat. As suggested by Gould *et al.* (2002), 'The ability of the engineer to recognize potential problems due to the presence of compressible organic soils requires familiarity with the geology, topography, and development history of an area.'

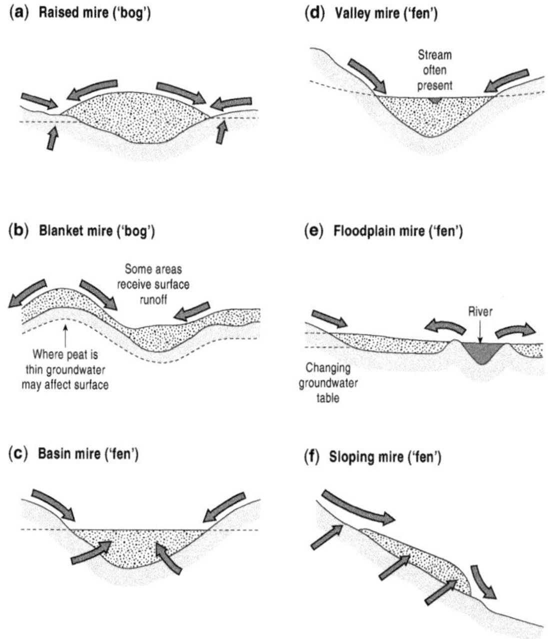

9.4 Geological hazards associated with peat compressibility

There are a range of geological hazards related to compressible soils (BGS 2010, 2014). However, with peat soils, the range of hazards might be broadly classified in terms of upland or lowland settings, and whether the soil is loaded due to the construction of roads, erection of structures (e.g.2 generally buildings, but particularly wind turbines in the uplands), the dumping of waste (e.g. quarrying and/or mining), or implementation of large engineering schemes (e.g. reservoir construction, gas terminal). Although the peat compression process is similar whether on upland or lowland peat (subject to material constraints), the potential consequences can differ in the two environments due to the depth of the peat and local topography (Fig. 9.6). Here we focus on examples from upland environments that illustrate the range of potential hazards associated with compression and bearing capacity failures in upland peats. The following section begins with some general background on the subsidence of organic soils.

9.4.1 Subsidence of peat

Compression of peat is part of the general subsidence of organic soil systems. These systems undergo subsidence

Fig. 9.6. Illustration (cross-section and plan) of key hydromorphological mire types found in temperate peatlands such as the UK (redrawn after Charman 2002). Six typical mire types (**a–f**), with contrasting hydromorphological settings are shown.

from densification (loss of buoyancy, shrinkage and compaction); loss of mass (biological oxidation, burning, hydrolysis and/or leaching, erosion and mining) and direct loading (road construction, utilities, buildings and waste dumping) (Stephens *et al.* 1984; Cooke & Doornkamp 1990; Wösten *et al.* 1997; Kechavarzi *et al.* 2010). The significance of the latter, which forms the focus of this chapter, needs to be viewed in the context of these other factors that have an impact on peat wastage. The best example of this in a UK context is the detailed record of peat wastage recorded at the Holme Post at Holme Fen in the East Anglian Fenlands from 1848 to 1978 (Hutchinson 1980; Waltham 2000) (Fig. 9.1). Using archived evidence, Hutchinson (1980) was able to produce a precise chronology of the lowering of the peat surface in relation to the Holme Post, a cast-iron column footed into clay below the peat, which served as a datum for evaluating surface changes. It was shown that land drainage using artificial pumps resulted in four distinct phases of surface lowering. During initial lowering, shrinkage appeared to be the dominant factor in controlling surface lowering (0.2 m a^{-1}) as the bog was rapidly drained. However, as time progressed, rates of lowering in the partially drained bog slowed (0.01 m a^{-1}) and biochemical oxidation became the predominant process. Such processes continue to affect these low-lying peat soils, resulting in risks of road subsidence (Pritchard *et al.* 2013).

Evaluating the relative importance of the factors resulting in mire surface lowering and compression is a complex problem, and one which is common in many peatland settings. Kool *et al.* (2006) considered the importance of oxidation and compaction of a collapsed peat dome in Central Kalimantan caused by logging and agricultural drainage. Although it was difficult to quantify these effects precisely, Kool *et al.* concluded that compaction appeared to be a more important factor in governing the loss of peatland structure than oxidation. It is therefore important in any study of compressible peat soils to acknowledge that this is only one factor that governs the dynamics and stability of these complex organic soil systems.

9.4.2 Derrybrien landslide, wind farm construction, County Galway 2003

The catastrophic peat mass movement that occurred at the Derrybrien wind farm development on 16 October 2003 captured the attention of all stakeholders with interests in the uplands, and put on hold a wind farm industry that was rapidly expanding across many upland areas (Lindsay & Bragg 2005) (Fig. 9.7). The initial peat slide (estimated volume 450 000 m^3) failed on 16 October and travelled 2.5 km from the south-facing slope of Cashlaundrumlahan Mountain (365 m), originating in the vicinity of two partially constructed turbine bases (Fig. 9.7a). The peat came to rest on 19 October at 195 m altitude, but was reactivated by heavy rain on 28 October. It then moved another 1.5 km to the Owendalulleegh River where it became highly fluidized and continued to flow downstream into Lough Cutra, which was the source of the domestic water supply. The initial impacts along the downstream run-out track (Fig. 9.7b) included loss of land, obstruction of roads, pollution of the domestic water supply and the death of an estimated 100 000 fish (Lindsay & Bragg 2005; Bragg 2007).

Although engineering investigations were undertaken to evaluate the cause of the failure, there was no clear

Fig. 9.7. The Derrybrien wind farm development site: (**a**) disturbed ground in the vicinity of a wind turbine foundation showing the exposed peat and mineral substrate; and (**b**) the main landslide run-out track downstream of the failure site. (Photographs courtesy of Olivia Bragg.)

conclusion reached amongst the consulting engineers. However, investigations by peatland specialists (Lindsay & Bragg 2005) suggest that the failure may have resulted either from loading by excavation machinery or from the release of water into heavily fissured peat or a combination of both. In addition, loading due to the concrete turbine foundations also appears to have triggered local failures (Bragg 2007) (Fig. 9.7a). Following the incident at Derrybrien, there was far greater awareness of the construction problems that are inherent when building on peat. Although these were well known elsewhere (MacFarlane 1969), a greater consciousness was triggered in the UK, resulting in much better planning and implementation of improved guidelines (e.g. MacCulloch 2006; Munro & MacCulloch 2006) and greater research into the mechanisms that can trigger peat failures (Long 2005; Dykes & Warburton 2007; Long & Boylan 2012).

9.4.3 Direct loading by quarry waste, Harthope Quarry, North Pennines, UK

Figure 9.8 shows a historic RAF air photograph of Harthope Quarry, North Pennines, UK (Fig. 9.1) taken in June 1953. The image shows the area of the main quarry (A) and spoil/waste heaps which have been dumped on the surface of the adjacent blanket peat (B). The loading of the peat has triggered a series of six main peat failures (C) emanating from the quarry area. Peat depths in this area are approximately 1.5 m, and the morphology of these failures is still evident on the ground today. Based on the air photograph evidence, the type of peat mass movement appears to be a peat flow as described by Dykes & Warburton (2007), consistent with a head-loading-type failure, implying that the rapid compression of the peat was significant in initiating this pattern of landslides. This is borne out by the general observation that peat slides are a more common form of peat failure in the North Pennines (Warburton et al. 2004) and peat flows are unusual. Three key features identified from Figure 9.8 give clues to the mechanism of peat failure. Firstly, the morphology is clearly a flow rather than a slide, suggesting failure did not extend to the mineral substrate. Secondly, the failure paths are relatively short (c. <100 m) and terminate in lobes, suggesting these represent a local failure at the slope head (site of loading) whose flow path was rapidly attenuated downslope, indicating that the hydrological conditions necessary to promote failure across the whole slope were not present (Warburton et al. 2004). Thirdly, the failures are adjacent to one another emanating from the point of loading by the individual waste heaps at the head of the slope. This contrasts with the majority of peat slides in the region that generally occur along single lines of drainage, and are often loosely clustered in response to local hydrological conditions (Evans & Warburton 2007).

9.4.4 Failure during upland road construction, North Pennines, UK

Figure 9.9 shows a peat slide triggered by moorland road construction over an area of blanket peat, close to Burnhope Seat in the North Pennines, UK (Fig. 9.1). The failure occurred in August 2006 during the construction phase of the road works. The floating road construction was excavated directly into the peat, which in the vicinity of the failure was approximately 1.1–1.3 m deep, consisting of an upper fibrous peat overlying a more humified basal peat unit. The road was approximately 3.8 m wide and back-filled with a coarse aggregate mix to a depth of about 0.5 m over a wire mesh that was laid in the construction trench. Observations at the time of failure suggest that the aggregate truck which was hauling the road fill had just passed over the point of failure when the slope gave way, and the driver was fortunate not to get caught up in the landslide. The failed mass slid downslope and entered the local stream course at the base of the slope, discharging peat debris downstream into a drinking water reservoir. It is clear from the inset shown in Figure 9.9 that a 25 m section of road was transported down the slope during the failure from the top left to lower right of the picture, remaining largely intact while in transit.

In this example the road was of a simple low-cost construction designed for low volumes of traffic (mainly for shooting parties), but illustrates some of the potential challenges facing road building on peat. In this example, high water content, high compressibility and low strength all appear to be significant factors in the failure. Firstly, at the point of failure the road traversed a natural moorland 'flush' where the peat was slightly deeper and was very wet, even during the summer, due to preferred seepage of groundwater along the flush. Secondly, the construction method used conventional coarse aggregate to backfill the road excavation, which was grounded in peat. The construction did not take advantage of lightweight fill materials, and it is estimated that the haulage tricks were running 20 tonne loads over the newly

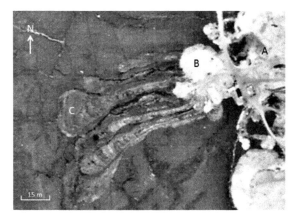

Fig. 9.8. June 1953 RAF air photograph of Harthope quarry, North Pennines, UK. The image shows the main quarry area (A) and spoil heaps which have been dumped on the surface of the adjacent blanket peat (B), resulting in a series of peat failures (C) due to the vertical loading.

Fig. 9.9. (a) A peat slide triggered by moorland road construction over an area of blanket peat in the North Pennines, UK. (b) Inset shows 25 m section of road which was transported downslope during the failure from the top left to lower right of the picture.

constructed road with no period of consolidation (Munro 2004). Thirdly, the peat in this locality consisted of approximately 0.7 m of fibrous peat over about 0.6 m of amorphous peat, which overlaid a coarse stone clayey substrate. There was a pronounced water seepage zone approximately 0.3 m above the base of the amorphous peat, which appeared to be the failure zone. Given these characteristics, it is clear the road was very unstable at this point along its route, and failure was almost inevitable.

These three case studies clearly illustrate the importance of compressible soils in affecting hazards in upland peat areas, and highlight the important interactions between human activities and sensitive peat soils. However, it is also worth emphasizing that more subtle indirect actions on upland peat soils might modify the water table, which can lead to additional problems. For example, in addition to compression caused by direct loading by engineering structures or waste, in natural and drained peatlands changes in the height of the water table can cause changes in the effective stress that are large enough to significantly alter peat volume and hydraulic parameters (Price *et al.* 2005). However, determining how peat compressibility relates to the physical properties of peat and the consequent hydraulic behaviour remains a complex task (Hobbs 1986; Bell 2000; Price *et al.* 2005). This is significant for hazard assessment in two main ways: (1) soil testing of peat to determine the geotechnical and physical properties does not necessarily lead to an enhanced understanding of peat compression, because these properties cannot be readily used to assess the hazard posed (see Discussion); and (2) the sensitive relationship between water table fluctuations and peat consolidation is an important consideration when planning engineering structures over peat as water tables are frequently disturbed, particularly during the construction phase. This may have immediate and possible long-term effects and, as such, represents a secondary hazard related to the primary hazard of direct loading of the peat surface (Munro 2004).

9.5 Mitigation of the hazards posed by compressible peat soils

Peat is highly compressible, has a low bearing capacity and is generally considered one of the worst foundation materials by practising engineers. Typically, the costs of construction over peat are approximately 40% more expensive than over competent ground (Nichol 1998), and ongoing maintenance and/or remediation costs can exceed initial project costs (Gould *et al.* 2002). Furthermore, peat is highly heterogeneous and displays non-linear behaviour which often conflicts with theories and engineering practices used on mineral soils (Carlsten 1988; Long 2005).

In terms of mitigating the hazards posed by compressible peat soils, there are three main approaches: (1) obtaining a greater understanding of the properties of compressible soils and dissemination of good practice to engineers (e.g. review of settlement of peat soils; Long & Boylan 2013); (2) non-structural mitigation such as land-use zoning and development planning (e.g. GeoSure; BGS 2014); or (3) structural mitigation and improved engineering practice (e.g. design of low-volume roads over peat; refer to the ROADEX project; Munro & MacCulloch 2006).

Long & Boylan (2013) provide one of the most recent reviews of predicting the settlement of peat soils. Their aim was to offer guidance to engineers who are required to make predictions of 1D compression of structures founded on peat, and to identify good practice in laboratory testing and the methods of calculation. Using a range of laboratory data (14 sites) and full-scale loading case studies (five sites), they demonstrate that the C_α/C_c law of compressibility generally applies, and the compressibility ratio has an average value of 0.072 (Table 9.2). Long and Boylan concluded that, although peat properties were highly variable, conventional staged construction with surcharge loading could be successfully applied to peat soils.

To some extent, these results confirm earlier laboratory testing by other engineers seeking to understand the *in situ* behaviour of peat as an embankment foundation material (e.g. Lefebvre *et al.* 1984). Geotechnical testing of peat samples in the laboratory is only one of a series of approaches used to determine the site characterization of peat for engineering purposes (Edil 2001; Long & Boylan 2012). Edil (2001) concludes that a combination of methods using extensive sampling to define site variability, *in situ* tests (modified vane shear and penetration tests; Long & Boylan 2012), laboratory testing for mechanical properties and, where possible, the use of test fills (full-scale loading test) provide a reasonable approach to dealing with problematic organic deposits (Magnan 1994).

In the UK, engineers are well aware of the challenges facing construction projects over peat. An early example from Ward *et al.* (1955) describes a slip in a flood embankment in 1948 constructed over a thin peat layer as part of the River Don channel diversion scheme. More recently, Nichol & Farmer (1998) describe the settlement problems associated with the main A5 trunk road as it traverses a peat bog at Pant Dedwydd in central North Wales. Originally constructed in 1819, the road has suffered continuous maintenance problems and safety issues resulting from long-term settlement. The solution has been to add successive layers of asphalt; although this provides a short-term solution, it serves to increases the loading that simply adds to the problems in the longer-term. Using geotechnical tests, Nichol & Farmer (1998) confirmed the sensitive nature of the peat to compression and water loss (shrinkage), but also demonstrated that areas of extreme subsidence were associated with willow scrub patches growing in roadside ditches, which contribute to subsidence by extracting water from the peat beneath the highway. Experience in Ireland is more widespread given the more common occurrence of peat across both upland and lowland environments (Long & Boylan 2013). However, wherever peat occurs in small pockets or over entire regions, engineers must be aware of its hazardous consequences (Nichol 2001).

As part of its UK hazard assessment programme, the BGS provides information about geological hazards and, in particular, identifies areas that are susceptible to ground movement, including compressible soils and peat (Fig. 9.1a). Natural ground stability (GeoSure) national datasets provide geological information on the potential for compressible ground to be a hazard. This assessment is based on 1:50 000 scale digital maps of superficial and bedrock deposits combined with information from the BGS superficial drift thickness data and engineering reports. This is one of six different GIS layers, each representing a different potential natural ground stability hazard in the UK. Layers include shrink–swell (volume change in swelling clays), landslides (slope instability), soluble rocks (dissolution), collapsible deposits (when loaded), running sand (loosely packed sands fluidized by water), and compressible ground (soft materials that compress when loaded). For mapping compressible deposits, a digital geological map is used to select deposits with regard to their compressibility potential; for example, peats and lake deposits are highly compressible, whereas bedrock is unlikely to be compressible (Royse 2011). Each polygon from the digital map is then scored according to its susceptibility to compressibility. This was then combined with thickness of the superficial deposit and the two scores combined to give the overall hazard susceptibility rating (Royse 2011).

Results translate to five categories of hazard rating (A–E; Table 9.4), and each category contains advice for the public and specialist on the appropriate actions and risk control measures that are potentially required. Typically, buildings constructed on compressible soils may experience structural damage to foundations; cracks in the walls, floors or ceilings; tilting of walls; and strains or breaks in connections to water, gas and electricity supplies (BGS 2014).

Engineering methods for peatland areas have evolved considerably over the last few decades, and engineers now have a range of structural techniques and engineering practices at their disposal. These include: preloading and surface reinforcement; excavation and replacement methods; drainage systems; injection and deep mixing stabilization; cement and/or stone columns; and geomaterials and lightweight fill (Huat *et al.* 2014). The choice of method differs depending on the nature of the construction project and the budget available. Typically, preconsolidation or preloading is used in road projects to increase shear strength and reduce the long-term compression of peat soils (Carlsten 1988). The principle involves loading the peat with a load that is in

Table 9.4. *British Geological Survey compressible ground hazard rating key*

	Hazard rating	Advice for public	Advice for specialist
A	No indicators for compressible deposits identified	No actions required to avoid problems due to compressible deposits	No special ground investigation required, and no increased construction costs or increased financial risk due to potential problems with compressible deposits
B	Very slight potential for compressible deposits to be present	No actions required to avoid problems due to compressible deposits	No special ground investigation required; unlikely to be increased construction costs or increased financial risk due to potential problems with compressible deposits
C	Slight possibility of compressibility problems	Take technical advice regarding settlement when planning extensions to existing property	**New build**: Consider possibility of settlement during construction due to compressible deposits; unlikely to be an increase in construction costs due to potential compressibility problems **Existing property**: No significant increase in insurance risk due to compressibility problems
D	Significant potential for compressibility problems	Avoid large differential loadings of ground; do not drain or dewater ground near the property without technical advice	**New build**: Assess the variability and bearing capacity of the ground; may need special foundations to avoid excessive settlement during and after construction; consider effects of groundwater changes; extra construction costs are likely **Existing property**: Possible increase in insurance risk from compressibility if lowered groundwater levels drop due to drought or dewatering
E	Very significant potential for compressibility problems	Avoid large differential loadings of ground; do not drain or dewater ground near the property without technical advice	**New build**: Assess the variability and bearing capacity of the ground; probably needs special foundations to avoid excessive settlement during and after construction; consider effects of groundwater changes; construction may not be possible at economic cost **Existing property**: Probable increase in insurance risk from compressibility due to drought or dewatering unless appropriate foundations are present

After BGS (2010, 2014)

excess of the final load that will be carried by the peat. This is allowed to settle until the design settlement of the planned load is reached. The excess load or surcharge is then removed and construction is completed (MacFarlane 1969). Some of the advantages of this method are that reduced fill material is used and that peat excavation is not necessary (meaning no disposal of excavated peat is required).

A good example where knowledge of loading methods is essential is in the construction of floating roads over peat. As part of the ROADEX project, a technical cooperation between roads organizations across northern Europe, the engineering of low-cost roads over peat was investigated. This is summarized in three reports which deal with bearing capacity problems on low-volume roads constructed on peat (Munro 2004); provide guidelines for the risk management of peat slips on the construction of low-volume, low-cost roads over peat (MacCulloch 2006); and provide a discussion of the main issues to be considered when planning rehabilitation measures for floating roads over peat (Munro & MacCulloch 2006). Collectively, the three reports provide useful practical guidance for the local planners, construction engineers and road maintenance engineers that can be used to address common problems of peatland roads and avoid the failure of such projects (see earlier example).

9.6 Conclusion

It has been demonstrated that peat is a highly compressible geological material whose consolidation and rheological behaviour is dependent on the permeability and distribution of water within the peat. The structure of the peat, degree of humification, water content and type of water (intercellular, interparticular or adsorbed) all influence the time-dependent behaviour of this type of organic deposit. It is also apparent that there is no single, simple relationship between the magnitude and rate of compression of peat and loading. Generalizations can be made about gross differences between broad peat types (fibrous compared with amorphous peat) and some useful progress has been made in developing engineering guidelines, but overall peat is intrinsically a complex and highly variable geotechnical material, adding uncertainty to our understanding of its geological hazards. Coupled to this source of uncertainty is the fact that the extent of mapped peat deposits in the UK is subject to considerable error due to inconsistencies in the definition of peat soils between individual countries and differences in available data (JNCC 2011). Hence, different mapping agencies, steered by different objectives, produce different estimates of the extent and occurrence of significant peat deposit.

Mitigating compression and bearing capacity failure hazards in peat soils is therefore a difficult process; however, a combination of improved understanding of the properties of compressible peat, better mapping and land use zoning, and appropriate construction will mitigate risk. Rather than rely on specific geographical knowledge of peat extent to guide local decisions, engineers and environmental scientists should be aware of the general occurrence of peat, be able to recognize it and have knowledge of its geotechnical and/or environmental behaviour so that appropriate strategies can be selected. Finally, failures resulting from peat compression are locally generated, but due to the sensitive nature of blanket peat these can result in runaway failures that pose a far greater risk.

Acknowledgements Chris Orton (Department of Geography, Durham University) is thanked for his assistance in producing the figures. Olivia Bragg (Geography and Environmental Science, University of Dundee) kindly gave consent to reproduce the photographs used in Figure 9.7.

Funding This research received no specific grant from any funding agency in the public, commercial, or not-for-profit sectors.

References

AYSEN, A. 2005. *Soil Mechanics: Basic Concepts and Engineering Applications*. Balkema, Netherlands.

BARDEN, L. 1968. Primary and secondary consolidation of clay and peat. *Géotechnique*, **18**, 1–24, https://doi.org/10.1680/geot.1968.18.1.1

BELL, F.D. 2000. Organic soils: peat [Chapter 7]. *In*: *Engineering Properties of Soils and Rocks*. 4th edn. Blackwell Science, Oxford, 202–222.

BERRY, P.L. & POSKITT, T.J. 1972. The consolidation of peat. *Géotechnique*, **22**, 27–52, https://doi.org/10.1680/geot.1972.22.1.27

BERRY, P.L. & VICKERS, B. 1975. The consolidation of fibrous peat. Proceedings American Society of Civil Engineers. *Journal Geotechnical Engineering Division*, **101**, 741–753.

BOYLAN, N., JENNINGS, R. & LONG, M. 2008. Peat slope failure in Ireland. *Quarterly Journal of Engineering Geology and Hydrogeology*, **41**, 93–108, https://doi.org/10.1144/1470-9236/06-028

BRAGG, O. 2007. Derrybrien: where the questions began. *International Mires Conservation Group Newsletter*, **4**, 3–8.

BRITISH GEOLOGICAL SURVEY. 2010. *User Guide for the British Geological Survey GeoSure Dataset*. Internal Report **OR/10/066**, British Geological Survey, NERC, Keyworth, Nottingham.

BRITISH GEOLOGICAL SURVEY. 2014. *Compressible Ground*. NERC, Keyworth, Nottingham. http://www.bgs.ac.uk/products/geosure/compressible.html [last accessed January 2014]

CARLSTEN, P. 1988. *Geotechnical Properties of Peat and up-to-Date Methods for Design and Construction*. Varia, No. 215, Swedish Geotechnical Institute (SGI).

CHARMAN, D. 2002. *Peatlands and Environmental Change*. John Wiley & Sons, Chichester.

COMAS, X., KETTRIDGE, N. *ET AL.* 2014. The effect of peat structure on the spatial distribution of biogenic gases within bogs. *Hydrological Processes*, **28**, 5483–5494, https://doi.org/10.1002/hyp.10056

COOKE, R.U. & DOORNKAMP, J.C. 1990. *Geomorphology in Environmental Management: A New Introduction*. 2nd edn. Clarendon Press, Oxford.

DAY, R.W. 2000. *Geotechnical Engineer's Portable Handbook*. McGraw-Hill, New York.

DYKES, A.P. 2008. Properties of peat relating to instability of blanket bogs. *In*: CHEN, Z., ZHANG, J.-M., HO, K., WU, F.-Q. & LI, Z.-K. (eds) *Landslides and Engineered Slopes: From the Past to the Future*. CRC Press, Boca Raton, USA, 339–345.

DYKES, A.P. 2009. Geomorphological maps of Irish peat landslides created using hand-held GPS. 2nd edn. *Journal of Maps*, **5**, 179–185, https://doi.org/10.4113/jom.2009.1091

DYKES, A.P. & WARBURTON, J. 2007. Mass movements in peat: a formal classification. *Geomorphology*, **86**, 73–93, https://doi.org/10.1016/j.geomorph.2006.08.009

EDIL, T.B. 2001. Site characterization in peat and organic soils. *International Conference on Insitu Measurement of Soil properties and Case Histories*. Parhyangan Catholic University, Bandung, 49–60.

EVANS, M.G. & WARBURTON, J. 2007. *Geomorphology of Upland Peat: Erosion, Form and Landscape Change*. Blackwell Publishing, Oxford.

FOX, P.J. & EDIL, T.B. 1996. Effects of stress and temperature on secondary compression of peat. *Canadian Geotechnical Journal*, **33**, 405–415, https://doi.org/10.1139/t96-062

FOX, P.J., ROY-CHOWDHURY, N. & EDIL, T.B. 1999. Secondary compression of peat with or without surcharging – Discussion. *Journal of Geotechnical and Geoenvironmental Engineering*, ASCE 125, **2**, 160–162, https://doi.org/10.1061/(ASCE)1090-0241(1999)125:2(160)

FRATTA, D., AGUETTANT, J. & ROUSSEL-SMITH, L. 2007. *Introduction to Soil Mechanics Laboratory Testing*. CRC Press Taylor Francis Group, Boca Raton.

GOULD, R., BEDELL, P.R. & MUCKLE, J.G. 2002. Construction over organic soils in an urban environment: four case histories. *Canadian Geotechnical Journal*, **39**, 345–356, https://doi.org/10.1139/t01-090

HOBBS, N.B. 1986. Mire morphology and the properties and behaviour of some British and foreign peats. *Quarterly Journal of Engineering Geology*, **19**, 7–80, https://doi.org/10.1144/GSL.QJEG.1986.019.01.02

HUAT, B.K., PRASAD, A., ASADI, A. & KAZEMIAN, S. 2014. *Geotechnics of Organic Soils and Peat*. Taylor & Francis Group, London.

HUTCHINSON, J.N. 1980. The record of peat wastage in the East Anglian Fenlands at Holme Post, 1848–1978 AD. *Journal of Ecology*, **68**, 229–249, https://doi.org/10.2307/2259253

INGRAM, H.A.P. 1982. Size and shape in raised mire systems: a geophysical model. *Nature*, **297**, 300–303, https://doi.org/10.1038/297300a0

IVANOV, K.E. 1981. *Water Movement in Mirelands*. Academic Press.

JNCC. 2011. *Towards an Assessment of the State of UK Peatlands*. Joint Nature Conservation Committee report No **445**. ISSN 0963 8901

KECHAVARZI, C., DAWSON, Q. & LEEDS-HARRISON, P.B. 2010. Physical properties of low-lying agricultural peat soils in England. *Geoderma*, **154**, 196–202, https://doi.org/10.1016/j.geoderma.2009.08.018

KELLNER, E., WADDINGTON, J.M. & PRICE, J.S. 2005. Dynamics of biogenic gas bubbles in peat: potential effects on water storage and peat deformation. *Water Resources Research*, **41**, 8, https://doi.org/10.1029/2004WR003732

KENNEDY, G.W. & PRICE, J.S. 2005. A conceptual model of volume-change controls on the hydrology of cutover peats. *Journal of Hydrology*, **302**, 13–27, https://doi.org/10.1016/j.jhydrol.2004.06.024

KNAPPETT, J.A. & CRAIG, R.F. 2012. *Craig's Soil Mechanics*. 8th edn. Spon Press, London.

KOOL, D.M., BUURMAN, P. & HOEKMAN, D.H. 2006. Oxidation and compaction of a collapsed peat dome in central Kalimantan. *Geoderma*, **137**, 217–225, https://doi.org/10.1016/j.geoderma.2006.08.021

LEFEBVRE, G., LANGLOIS, P. & LUPIEN, C. 1984. Laboratory testing and in situ behaviour of peat as embankment foundation. *Canadian Geotechnical Journal*, **21**, 322–337, https://doi.org/10.1139/t84-033

LINDSAY, R.A. & BRAGG, O.M. 2005. *Wind Farms and Blanket Peat: The Bog Slide of 16th October 2003 at Derrybrien, Co. Galway, Ireland*. The Derrybrien Development Cooperative Ltd., Galway.

LONG, M. 2005. Review of peat strength, peat characterisation and constitutive modelling of peat with reference to landslides. *Studia Geotechnica et Mechanica*, **XXVII**, 67–90.

LONG, M. & BOYLAN, N. 2012. In-Situ Testing of Peat – a Review and Update on Recent Developments. *Geotechnical Engineering Journal of the SEAGS & AGSSEA*, **43**, 41–55.

LONG, M. & BOYLAN, N. 2013. Predictions of settlement in peat soils. *Quarterly Journal of Engineering Geology and Hydrogeology*, **46**, 303–322, https://doi.org/10.1144/qjegh2011-063

MACCULLOCH, F. 2006. *Guidelines for the Risk Management of Peat Slips on the Construction of Low Volume/Low Cost Roads Over Peat*. Roadex II Project Report. https://www.roadex.org/wp-content/uploads/2014/01/Guidelines-for-the-Risk-Management-of-Peat-Slips.pdf

MACFARLANE, I.C. (ed.) 1969. *Muskeg Engineering Handbook*. University of Toronto Press, Toronto.

MACFARLANE, I.C. & RADFORD, N.W. 1965. A study of the physical behaviour of derivatives under pressure. *Proceedings of 10th Muskeg Research Conference, NRC of Canada*, Montreal, Canada.

MAGNAN, J.P. 1994. Construction on peat: state of the art in France. In: DEN HAAN, E., TERMAAT, R. & EDIL, T.B. (eds) *Advances in Understanding and Modelling the Mechanical Behaviour of Peat*. Balkema, Rotterdam, 369–379.

MESRI, G. & AJLOUNI, M.A. 2007. Engineering properties of fibrous peat. *Journal of Geotechnical and Geoenvironmental Engineering*, ASCE, **133**, 850–866, https://doi.org/10.1061/(ASCE)1090-0241(2007)133:7(850)

MESRI, G., STARK, T.D., AJLOUNI, M.A. & CHEN, C.S. 1997. Secondary compression of peat with or without surcharging. *Journal of Geotechnical and Geoenvironmental Engineering*, ASCE, **123**, 411–421, https://doi.org/10.1061/(ASCE)1090-0241(1997)123:5(411)

MUNRO, R. 2004. *Dealing with Bearing Capacity Problems on Low Volume Roads Constructed on Peat: Including Case Histories from Roads Projects within the ROADEX Partner Districts*. Roadex II Project Report. https://www.roadex.org/wp-content/uploads/2014/01/2_5-Roads-on-Peat_l.pdf

MUNRO, R. & MACCULLOCH, F. 2006. *Managing Peat Related Problems on Low Volume Roads – Executive Summary*. Roadex III Project Report. http://www.roadex.org/wp-content/uploads/2014/01/Roads-on-Peat_English.pdf

MYŚLIŃSKA, E. 2003. Classification of organic soils for engineering geology. *Geological Quarterly*, **47**, 39–42.

NICHOL, D. 1998. Construction over peat in Greater Vancouver, British Columbia. *Proceedings Institution of Civil Engineers – Municipal Engineer*, **127**, 109–119, https://doi.org/10.1680/imuen.1998.30986

NICHOL, D. 2001. Geo-engineering along the A55 North Wales coast road. *Quarterly Journal of Engineering Geology and Hydrogeology*, **34**, 51–64, https://doi.org/10.1144/qjegh.34.1.51

NICHOL, D. & FARMER, I.W. 1998. Settlement over peat on the A5 at Pant Dedwydd near Cerrigydrudion, North Wales. *Engineering Geology*, **50**, 299–307, https://doi.org/10.1016/S0013-7952(98)00027-1

POWRIE, W. 1997. *Soil Mechanics: Concepts and Applications*. 2nd edn. Spon Press, London.

PRICE, J.S. & SCHLOTZHAUER, S.M. 1999. Importance of shrinkage and compression in determining water analyse changes in peat: the case of a mined peatland. *Hydrological Processes*, **13**, 2591–2601, https://doi.org/10.1002/(SICI)1099-1085(199911)13:16<2591::AID-HYP933>3.0.CO;2-E

PRICE, J.S., CAGAMPAN, J. & KELLNER, E. 2005. Assessment of peat compressibility: is there an easy way? *Hydrological Processes*, **19**, 3469–3475, https://doi.org/10.1002/hyp.6068

PRITCHARD, O.G., HALLETT, S.H. & FAREWELL, T.S. 2013. *Road subsidence in Lincolnshire: Soils and road condition*. NSRI, Cranfield University, UK.

ROYSE, K.R. 2011. The handling of hazard data on a national scale: a case study from the British Geological Survey. *Surveys in Geophysics*, **32**, 753–776, https://doi.org/10.1007/s10712-011-9141-3

RYDIN, H. & JEGLUM, J. 2006. *The Biology of Peatlands*. Oxford University Press, Oxford.

STEPHENS, J.C., ALLEN, L.H.Jr & CHEN, E. 1984. Organic soil subsidence. *Reviews in Engineering Geology*, **6**, 107–122, https://doi.org/10.1130/REG6-p107

STRACK, M., KELLNER, E. & WADDINGTON, J.M. 2006. Effect of entrapped gas on peatland surface level fluctuations. *Hydrological Processes*, **20**, 3611–3622, https://doi.org/10.1002/hyp.6518

VAN ASSELEN, S., STOUTHAMER, E. & VAN ASCH, T.W.J. 2009. Effects of peat compaction on delta evolution: a review on processes, responses, measuring and modelling. *Earth-Science Reviews*, **92**, 35–51, https://doi.org/10.1016/j.earscirev.2008.11.001

VON POST, L. 1924. Das genetische System der organogenen Bildungen Schwedens. Comité International de Pédologie IV Commission 22.

WALTHAM, T. 2000. Peat subsidence at the Holme Post. *Mercian Geologist*, **15**, 49–51.

WARBURTON, J., HOLDEN, J. & MILLS, A.J. 2004. Hydrological controls on surficial mass movements in peat. *Earth Science Reviews*, **67**, 139–156, https://doi.org/10.1016/j.earscirev.2004.03.003

WARD, W.H., PENMAN, A. & GIBSON, R.E. 1955. Stability of a bank on a thin peat layer. *Géotechnique*, **5**, 154–163, https://doi.org/10.1680/geot.1955.5.2.154

WÖSTEN, J.H.W., ISMAIL, A.B. & VAN WIJK, A.L.M. 1997. Peat subsidence and its practical implications: a case study in Malaysia. *Geoderma*, **78**, 25–36, https://doi.org/10.1016/S0016-7061(97)00013-X

Chapter 10   Periglacial geohazards in the UK

T. W. Berry[1]*, P. R. Fish[1], S. J. Price[2] & N. W. Hadlow[1]

[1]Jacobs UK Limited, 1 City Walk, Leeds, LS11 9DX, UK
[2]Arup, 13 Fitzroy Street, London, W1 T 4BQ, UK

*Correspondence: tom.berry2@jacobs.com

Abstract: Almost all areas of the UK have been affected by periglaciation during the Quaternary and, as such, relict periglacial geohazards can provide a significant technical and commercial risk for many civil engineering projects. The processes and products associated with periglaciation in the relict periglacial landscape of the UK are described in terms of their nature and distribution, the hazards they pose to engineering projects, and how they might be monitored and mitigated. A periglacial landsystems classification is applied here to show its application to the assessment of ground engineering hazards within upland and lowland periglacial geomorphological terrains. Techniques for the early identification of the susceptibility of a site to periglacial geohazards are discussed. These include the increased availability of high-resolution aerial imagery such as Google Earth, which has proved to be a valuable tool in periglacial geohazard identification when considered in conjunction with the more usual sources of desk study information such as geological, geomorphological and topographical publications. Descriptions of periglacial geohazards and how they might impact engineering works are presented, along with suggestions for possible monitoring and remediation strategies.

10.1 Introduction

The aim of this chapter is to describe the specific geological and geotechnical geohazards associated with relict periglacial processes and products in the UK. The nature of the geohazards and their potential distribution in terms of landsystems and terrains are described. Their potential impacts on ground engineering and the consequences of failing to identify them during the investigation phase of the project are discussed. Potential monitoring and mitigation strategies are suggested as ways of reducing the risk of encountering unexpected ground conditions as a result of the past effects of periglaciation.

This chapter summarizes and builds on the landsystem approach developed by Higginbottom & Fookes (1970), Hutchinson (1991), Ballantyne & Harris (1994), Fookes (1997), Brunsden (2002), Fookes et al. (2005), Fookes et al. (2007), Booth et al. (2015), Evans (2017), Murton & Ballantyne (2017) and Privett (2019).

The Quaternary Period (2.6 Ma to present day) represents a geological interval of intensification in the frequency and magnitude of high- to middle-latitude glaciation. Cyclical variations in the amount and aspect of orbitally controlled solar radiation reaching the Earth have been recognized from atmospheric proxy records preserved in marine sediments and ice cores (Lisiecki & Raymo 2005). The Quaternary Period has been characterized by marine isotope stages (MIS) representing over 100 alternating cold (glacial) and temperate (interglacial) events (Fig. 10.1). The most significant period of the Quaternary Period in terms of geomorphology and geohazards in the UK is the last c. 500 ka BP, when climate fluctuations were most marked. These fluctuations led to the most extensive glaciation of the UK (the Anglian, MIS 12, c.460 ka) and intervening interglacial periods that were often warmer than today, with higher relative sea-levels.

During cold stages in middle and high latitudes and altitudes, ice sheets expanded and glacial environments prevailed, with a characteristic suite of landforms, sediments and associated geohazards. Areas beyond the limits of glacier ice were subjected to periglacial activity, which has less easily recognizable landforms (Ballantyne & Harris 1994; Lowe & Walker 2015; Murton & Ballantyne 2017). The significance of these cyclical changes in climate to landscape development and engineering geology are the temporal and spatial superposition of cold and temperate processes, sediments and landforms. This means that for any geographic area, the geotechnical properties and geomorphology of the ground represents the combined effect of a complex interaction of cyclical sedimentation and erosion, and the growth and decay of ground ice. This ultimately induces changes in effective stress, which has the potential to overprint or modify the stress history of an engineering soil or rock.

Engineering Group Working Party (main contact for this chapter: T. W. Berry, Jacobs UK Limited, tom.berry2@jacobs.com)
From: GILES, D. P. & GRIFFITHS, J. S. (eds) 2020. *Geological Hazards in the UK: Their Occurrence, Monitoring and Mitigation – Engineering Group Working Party Report*. Geological Society, London, Engineering Geology Special Publications, **29**, 259–289,
https://doi.org/10.1144/EGSP29.10
© 2020 The Author(s). Published by The Geological Society of London. All rights reserved.
For permissions: http://www.geolsoc.org.uk/permissions. Publishing disclaimer: www.geolsoc.org.uk/pub_ethics

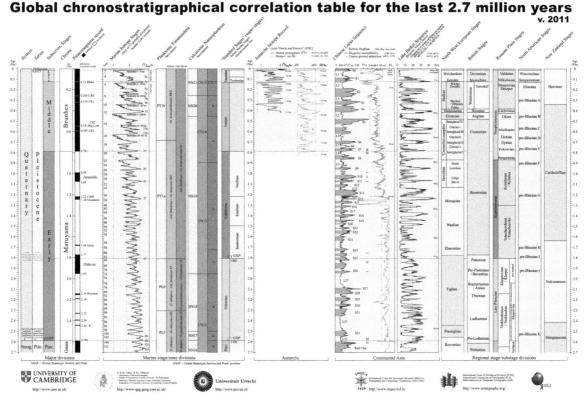

Fig. 10.1. Global chronostratigraphical correlation for the last 2.6 Ma (Cohen & Gibbard 2011). The magnitude and frequency of changes in oxygen isotope ratio provides a proxy for global changes in the volume of ice in the cryosphere and temperature.

The most important climatically driven events affecting the UK over the last 500 ka, with respect to ground engineering, were:

- the Anglian glaciation (MIS12), where ice advanced as far south as north London and Bristol, with the areas further south experiencing an intense periglacial climate;
- several periods of cold climate between the Anglian and Devensian glaciations (including MIS 10, 8 and 6), where much of the UK would have experienced periods of periglacial conditions;
- the Devensian glaciation (Last Glacial Maximum, MIS 5–2), where ice advanced to north Norfolk and the Midlands and an intense periglacial climate was experienced further south;
- the Loch Lomond Readvance, which was a very brief period of intense cold c. 13 ka, where small glaciers re-established in mountain areas and high-latitude/altitude areas of the UK may have experienced periglacial climates; and
- the Little Ice Age of the sixteenth to nineteenth centuries, which was associated with particularly cold climates and the periodic freezing of the River Thames; glaciers were not re-established and it is unlikely that permafrost developed in lowland areas, but freeze–thaw processes would have been enhanced.

The geohazards associated with these past periglacial environments in the UK are the focus of this chapter. The geohazards associated with present-day periglacial climates are not discussed here, but summaries can found in, for example, French (2007) and Ballantyne (2018).

Quaternary engineering soils vary from silt and clay with cobbles and boulders, to dense to loose sand and gravel, to laminated clay and silt, and to organic sediments. Their degree of consolidation reflects their stress history that in turn reflects the combined processes of erosion, transport, deposition and weathering. Geotechnical problems often arise because of the passage from one sediment type to another over short vertical and lateral distances, leading to unforeseen ground conditions if the ground model is not suitably characterized (Eyles 1983). Perched groundwater and saturated 'running sands' also contribute to subsurface variability in Quaternary sediments. The range and heterogeneity of grain sizes that are characteristic of Quaternary sediments mean that they do not conform to simplified models of soil behaviour based on clay and sand end-members.

Periglacial environments describe those elements of the landscape affected by cold conditions that border or have bordered ice sheets, and include processes as well as the resultant sediments, structures and landforms (Ballantyne & Harris 1994; Walker 2005). They are characterized by the widespread occurrence of ground freezing and the development of ground ice which may result in the formation of permafrost. Permafrost is defined as ground that remains at or below 0°C for two or more consecutive years. It is therefore a temperature definition, and the presence of ice is not necessarily implied.

In the 1970s, using the principles of soil mechanics, a body of evidence began to emerge from field studies that past ground freezing and subsequent thaw could be a significant factor in accounting for the presence of landslides on low-angle (<3°) slopes in southern Britain. The reduction of clay-rich soils to their residual shear strengths and the presence of polished shear surfaces was shown to be most likely the result of mechanical failure during periods of elevated porewater pressure following ground ice melting in former areas of permafrost development (Higginbottom & Fookes 1970; Hutchinson 1974, 1991; Skempton & Weeks 1976; Paul et al. 1981). This was highlighted in 1984 with the catastrophic failure of the Carsington Dam, UK. Skempton et al. (1991) concluded that failure occurred along shear planes in Quaternary solifluction sediments that were at their residual strength.

The increasing recognition of sedimentological and geomorphological evidence for former permafrost development in the geological record (Kemp et al. 1993; Murton et al. 1995, 2015; Murton & Kolstrup 2003) was used to support the periglacial origin of features including bedrock brecciation (Chandler 1970; Murton et al. 2006), low- and high-angle shear planes (Hutchinson 1991; Spink 1991), cambering and superficial valley disturbances (Parks 1991) and rockhead anomalies (Berry 1979; Hutchinson 1980). Despite these advances, the development of landsystems approaches for engineering geological mapping focused on formerly glaciated landscapes (e.g. Dearman & Fookes 1974). The recognition of periglacial landsystems was not considered with the same level of attention until the publications of Ballantyne & Harris (1994), Fookes (1997) and, most recently, Murton & Ballantyne (2017).

The processes of ground freezing, precipitation and subsequent decay of ground ice is known to change the geotechnical properties of engineering soils (Boulton & Paul 1976; McGown & Derbyshire 1977; Chamberlain & Gow 1979; Boulton & Dobbie 1993). In relict periglacial landscapes, the former effects of ground-freezing permafrost may be recognized by characteristic sediments, structures and geomorphological features. In many cases, the effects of ground freezing are subtle, and may be a geohazard if not recognized because ground engineering effects, including changes in effective stress brought about by the growth and degradation of ground ice, is significant. The former extent and modelled depth of past permafrost zones in Great Britain is depicted in Murton & Ballantyne's (2017) illustration (Fig. 10.2).

Factors that influence the susceptibility of a geological material to the formation of ice within it are particle size, allowing capillary action to take place, and a supply of ground water (Williams & Smith 1989). Gravels have pore spaces that are too large to promote capillary flow. In contrast, clay- and silt-rich soils have pore spaces that promote capillary flow of water and the development of ice, making them susceptible to frost (Fig. 10.3). The implications of this for relict periglacial landsystems in the UK are that frost-susceptible bedrock lithologies, including Mesozoic mudrocks and chalk, are likely to have been affected by the repeated growth and decay of ground ice in response to Quaternary climate change.

The main types of ground ice of relevance for understanding former ground conditions in lowland Britain are pore, segregated, intrusive and wedge ice (Murton & Ballantyne 2017). All form because groundwater migrates down a thermal gradient towards cold areas, causing ice to accumulate. Ground ice growth is controlled by grain size, mineralogy, hydraulic conductivity and porosity. A grain size of <0.075 mm (fine silt) is critical for the development of segregated ice. Pore ice is crystalline and generally forms in the pore space of granular materials, while segregated ice forms in fine-grained sediments and rocks rich in silt and clay. Segregated ice lenses observed in contemporary periglacial environments (Fig. 10.4) may be <10 mm to >10 m thick. Intrusive ice forms as veins injected under pressure into frozen or partially frozen ground. Wedge ice forms by the aggradation of ice in pre-existing thermal contraction cracks.

The geotechnical effects of freezing and thawing on the consolidation characteristics of undisturbed and reconstituted fine-grained soils are described experimentally by Morgenstern & Nixon (1971), Nixon & Morgenstern (1973) and Konrad & Morgenstern (1981). The potential wider effects on geotechnical properties and behaviour are illustrated qualitatively in Figure 10.5.

In contrast to present-day periglacial environments, the areal extent of former periglacial environments was much greater. Consequently, relict periglacial features are likely to have once covered the whole of Great Britain, including offshore continental shelves of the North Sea, English Channel and Irish Sea that were exposed by low relative sea-levels (Walker 2005).

Here, the hierarchical classification system of Murton & Ballantyne (2017) is adopted. Periglacial processes are initially described in terms of upland and lowland terrain systems based on relative elevation, and then further classified into landsystems. Within both upland and lowland terrains there are the following four landsystems: plateaus, sediment-mantled hillslopes, rock slopes, and slope-foot landsystems. There are two additional landsystems described in lowland terrains only: valley and buried landsystems. We also comment on the significance for marine engineering of the influence of past changes in sea-level and its impact on the

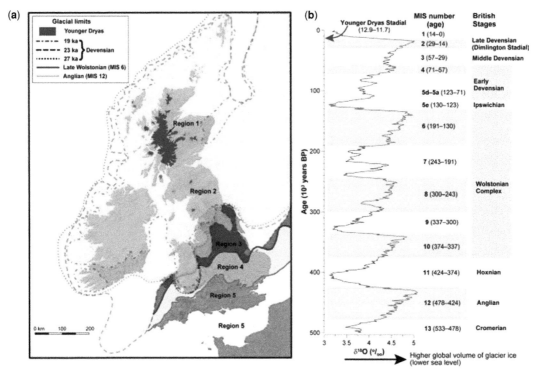

Fig. 10.2. Figure 5.7 from Murton & Ballantyne (2017) (a) Periglacial regions of the UK and Ireland superimposed on a digital elevation model. Region 1 (blue) was largely covered by glacier ice during the Younger Dryas Stadial, and periglacial activity within this zone has been confined to high ground during the Holocene. Region 2 (green) was covered by the last (Late Devensian) British–Irish Ice Sheet and experienced periglacial conditions during ice-sheet recession in the Late Devensian and during the Younger Dryas. Region 3 (brown and red) was subject to periglacial conditions during the 'Wolstonian' deglaciation and throughout the Devensian. Region 4 (yellow and olive green) experienced periglaciation during Anglian deglaciation and during the 'Wolstonian' and Devensian. Region 5 (dark green) experienced recurrent periglacial conditions throughout the Quaternary. Putative glacial limit on Dartmoor is not shown (see text).

submergence of land that was previously subject to periglaciation.

Some periglacial processes and deposits are a significant geohazard because they are a serious risk to engineering projects in areas of present-day development. In areas beyond the limit of Devensian glaciation, frost-susceptible mudrocks and older fine-grained glaciogenic deposits are likely to have been affected by the growth and subsequent decay of ground ice. Other periglacial processes are intrinsically less hazardous to engineering work, or are only found in areas that are rarely the focus of projects, for example in mountainous terrain.

This chapter begins with a discussion of the most significant periglacial features in the UK in terms of their geohazard significance and likelihood of being encountered. Periglacial features that pose a less significant geohazard, or that are located in areas where little or no development activity occurs and are unlikely to be encountered, are summarized with significant periglacial geohazards in Table 10.1. Subsidiary geohazards are differentiated from those sediments, landforms and processes, which are a direct result of preconditioning of the landscape by glaciation and whose subsequent response is collectively referred to as 'paraglacial' (Ballantyne 2002). Subsidiary geohazards include processes that are not true periglacial geohazards but which are significantly influenced by periglacial climates and processes, including deep landslides, carbonate karst development, buried valleys and submerged landsystems, and are described in more detail in other chapters of this volume. Tables 10.2–10.5 summarize tools and techniques which may be used to investigate and mitigate the effects of ground affected by past periglacial processes.

10.2 Relict periglacial geohazards

10.2.1 Deep weathering

Deep weathering and irregular engineering rockhead are geohazards associated with the periglacial frost weathering

PERIGLACIAL GEOHAZARDS

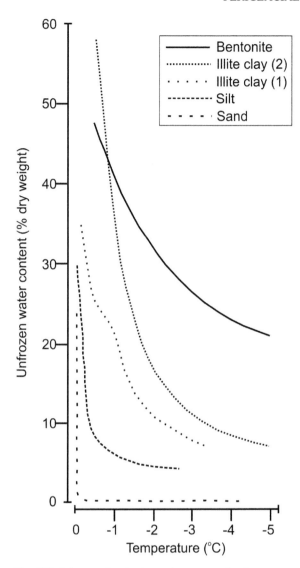

Fig. 10.3. Influence of grain size on the amount of unfrozen water remaining during freezing. Modified after Williams & Smith (1989). Below c. 1.5°C water is adsorbed onto dominantly clay minerals.

Fig. 10.4. Segregated ice lenses grading downwards to massive ice in icy sandy-silt, valley fill sediments, Adventdalen, Svalbard. (a) Transition from thin ice laminations in the seasonally frozen active layer (<42 cm), thick laminations in the transition zone (40–49 cm) to thin beds of segregated ice in the permafrost zone; and (b) massive-ice >165 cm (depths are centimetres below ground level).

processes of ice segregation, heave, and thermal contraction and expansion (Hutchinson 1991; Murton & Ballantyne 2017; Murton 2018). We define engineering rockhead as the boundary between material that behaves as a weathering derived engineering soil and that which has properties consistent with the primary engineering properties of the parent bedrock unit, where that parent bedrock unit may comprise rock or overconsolidated soil. In practice, this boundary may be dependent on the particular engineering situation.

The ultimate effect of these processes, and consequent cause of the geohazard, is spatial variability in material and hydrogeological properties due to mechanical alteration and breakdown of the bedrock. While these geohazards principally result from mechanical weathering processes, chemical weathering processes also produce irregular rockhead in soluble rocks that may be enhanced during periglacial conditions. Chemical weathering and carbonate karst are discussed in Chapter 15 of this publication.

Deep periglacially weathered profiles and irregular engineering rockhead can affect all rock types, and should be anticipated across the UK south of the limits of Quaternary glaciation in the non-glaciated province of Booth *et al.* (2015) or Region 5 of Murton & Ballantyne (2017). Periglacial weathered profiles may also be preserved beneath glacial sediments, and older glacial sediments (e.g. MIS 12) may be affected by periglacial weathering in regions 1–4 of Murton & Ballantyne (2017) and Figure 10.2.

The features associated with deep weathered rock profiles include: alteration of the pore structure (e.g. Bloomfield 1999); increased and enhanced discontinuities within the rock mass (illustrated in Giles *et al.* 2017); brecciated bedrock where the original material has become structureless, commonly with an upper layer deformed by involutions (illustrated in Giles *et al.* 2017); and, ultimately, the original bedrock reduced to a comminuted fine soil with lithorelicts. The latter material is often seen within the shallowest portion of the weathered profile or filling discontinuities, and is common in fine-grained porous rocks such as chalk,

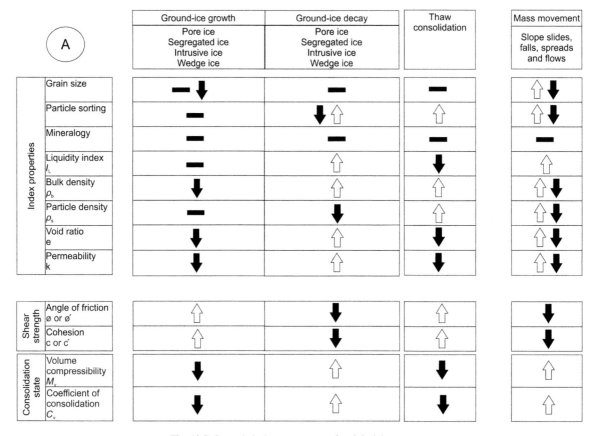

Fig. 10.5. Geotechnical consequences of periglacial processes.

siltstone or mudstone. In the case of chalk, the material has historically been referred to as coombe deposits and 'putty chalk' (e.g. Younger 1989). Brecciated bedrock also typically occurs within the shallow portion of the profile (the type 2 brecciation of Murton & Ballantyne 2017) and transitions with depth into the zone of increased and enhanced discontinuities (the type 1 brecciation of Murton & Ballantyne 2017), which progressively transitions into bedrock with only primary discontinuities.

Near-surface bedrock with a higher level of saturation during periglacial conditions would have been particularly susceptible to this form of weathering. The spatial variability in bedrock saturation, and consequently weathering, relates to the geomorphological, hydrogeological and hydrological setting. For example, in the case of chalk at outcrop under periglacial conditions, the near-surface bedrock in valleys was likely to have a higher level of saturation than interfluves and plateau areas due to the concentration of rainwater and snow meltwater. This would have favoured ice segregation, frost heave and thaw consolidation (Murton 2018), and consequently led to brecciation and enhancement of discontinuities in the valleys compared with interfluves and plateau areas; the latter typically exhibit much shallower depths of weathering.

Published descriptions of periglacially weathered rock profiles include those of Williams (1986, 1987) and Murton (1996) for the White Chalk Subgroup; Bradshaw & Ingle Smith (1963) and Harris (1989) for the Mercia Mudstone Group; Woodward (1912) for the Great Oolite Group; Higginbottom & Fookes (1970) for the Borrowdale Volcanic Group; and Ealey (2012) for the Lower Palaeozoic rocks of the Lizard Peninsula. Periglacial weathered profiles also occur within clay bedrock units. Spink (1991) identified discontinuities within the weathered profile of the Eocene London Clay Formation and Reading Formation that he considered related to a thawing front, and which are in a similar position to discontinuities generated in the chalk freezing experiments of Murton et al. (2016) and Murton (2018) from the formation of segregated ice. Such periglacially weathered profiles may occur to depths in excess of 10 m and it is not uncommon for 20 m of such material to be encountered in valley bottoms (e.g. Williams 1987).

Structureless and brecciated bedrock may also have been subjected to periglacial slope movement processes forming derived soils (illustrated in Giles *et al.* 2017) that may be intermixed with loess and coversand and may include palaeosol layers, which may pose an instability hazard (Hutchinson 1991). These periglacial slope movements processes, combined with the effect of slope aspect and seasonal fluvial erosion, are also considered to have led to the development of asymmetry in valley slope gradients and, to an extent, asymmetry in the associated deep weathering profile. Slopes with NW- and west-facing slopes are commonly seen as steeper with thinner head deposits; SE- and east-facing slopes are commonly seen as shallower with thicker head deposits (Lord *et al.* 2002; Murton & Ballantyne 2017; Murton 2018).

The depth of weathering, and associated lateral variation in engineering rockhead, has consequences for foundations, slopes, excavations and tunnels. Specifically, deep weathering of bedrock can lead to reduced bearing capacity, increased settlement, increased permeability and reduced density (Higginbottom & Fookes 1970; Eyles & Paul 1983; French 2018). Weathered materials also present problems for ground investigations, and it may be difficult to recover samples for accurate description and testing.

At the desk study stage, a geomorphological assessment, with reference to the ground models of Murton & Ballantyne (2017), and walkover survey combined with a review of existing ground investigation records in the general site area, will permit an initial evaluation of the likely depth of weathering and engineering rockhead (e.g. Hadlow *et al.* 2018). This work may also allow identification of superficial deposits which may mantle the bedrock. Ground investigations should be informed by the desk study, and aim to have a phased approach that proportionally employs a combination of non-intrusive (e.g. geophysical) and intrusive techniques together with an assessment of analogous exposures within the area (e.g. cliff or quarry sections).

A particular issue related to this geohazard is the misidentification of engineering rockhead due to poor description or the presence of unweathered 'core stones'; consequently, careful description of the sampled materials is important. There are a number of logging schemes designed to aid geologists in their description of bedrock units that are known to have deep weathering profiles and irregular engineering rockhead, including chalk (Lord *et al.* 2002), weak rocks and overconsolidated clays (Spink & Norbury 1993), and the Mercia Mudstone Group (Chandler 1969). Once these materials have been identified, they should be described to a detail sufficient for the intended design purposes. If a ground investigation has not been adequate and/or the material encountered not described correctly, then the subtle textures present due to periglacial weathering will not be noted. Consequently, if highly weathered material is encountered when the contractor was expecting 'engineering' rock (or vice versa), it could lead to health and safety issues as well as severe delays as redesign is carried out and alternative plant and equipment is mobilized.

Once the ground model has been sufficiently developed and the nature and extent of weathering is understood, various remediation strategies can be adopted to avoid the features or prevent them being an issue. Layouts of buildings can sometimes be moved or altered to avoid features of concern. A shallow foundation solution could be adapted to include extensions to strip footings to bridge or lessen the effect of variations in material type; strip and pad foundations can also be expanded to raft solutions for the same reasons. Alternatively, extending foundations to below the soils using a piled foundation solution may be appropriate. Similarly, suitable slope designs can be developed and the alignments for tunnels and settlement can be assessed.

10.2.2 Shallow-slope movements

This section describes various former cold-climate-related mass movement processes associated with two main processes: (1) the relatively slow downslope movement of soil as a result of cyclic freezing and thawing of the ground; this collection of processes has traditionally been referred to as solifluction; and (2) more rapid mass movement processes including active-layer detachment failures, which often take the form of translational landslides.

In the context of this section, the term solifluction includes four main processes that often act together to result in the slow, downslope movement of soil particles. The four processes are needle-ice creep, frost creep, gelifluction and plug-like deformation. Murton & Ballantyne (2017) describe gelifluction, possibly the most significant of the four processes, as seasonal gravity-induced shear deformation of thawing soil due to the generation of high porewater pressure during thaw consolidation of soil containing excess ice (ice lenses). Murton & Ballantyne (2017) also describe frost creep as the downslope movement of soil resulting from repeated cycles of volumetric expansion (frost heave) and contraction. Net downslope movement occurs as a result of expansion normal to the slope, but contraction has a gravity-imposed vertical component.

Active layer detachment slides are a more rapid translational type of slope instability in the upper layer of permafrost. They are triggered by water pressures generated by melting ice at the base of the active layer (Murton & Ballantyne 2017).

In summary, active-layer detachment slides are a rapid downslope flow or sliding of the active layer above permafrost, whereas solifluction encompasses processes involved in the slow downslope movement of soil due to cyclic freezing and thawing of the ground (see Fig. 10.6). Regardless of process, the significance for engineering geology is that they create low-shear-strength horizons, including specific shear surfaces; the more rapid slab slides or skin flows along basal shears were identified by Spink (1980).

The terminology and classifications of processes and their resulting landforms and deposits discussed below is based on those proposed by Ballantyne (2018). It should be noted that the British Geological Survey (BGS) use the genetic term

Table 10.1. *Summary of periglacial geohazards, their occurrence, engineering implications, investigation, monitoring and mitigation*

Geohazard	Landsystem*									
	Lowland						Upland			
	Plateau	Rock slope	Foot slope	Valley	Buried	Sub-merged	Plateau	Sediment mantled	Rock slope	Foot slope
Irregular rockhead and deep weathering (including tors)	W	R	W	W	W	W	W	W	R	W
Solifluction and active-layer detachment slides	N	R	W	W	W	W	N	W	R	R
Cambering and valley superficial valley disturbances	N	W	W	W	R	R	N	N	R	R
Rockhead anomalies (scour features, segregated ice, discontinuities)	N	N	N	R	N	N	N	N	N	N
Cryogenic wedges (ice wedges, sand wedges and composite wedges), ice wedge pseudomorphs	W	N	R	W	R	R	W	R	R	R
Patterned ground	W	R	R	R	R	R	R	R	R	R
Periglacial influence on deep-seated landslides (see Chapter 4)	N	N	R	R	W	R	N	W	N	R

Engineering implication	Investigation and monitoring (see Tables 10.2–10.5 for more information)†	Planning considerations and engineering mitigation
Variable or unforeseen ground conditions resulting in rockhead being deeper or shallower than expected; impact on anticipated groundwater regime	Preliminary ground model to document processes of weathering and likely depths; geophysical survey calibrated with logged boreholes and/or trial pits to develop a detailed understanding of the weathering profile	Ground modelling to understand relationships between topography, materials and engineering rockhead; ensure engineering manages ground risk (i.e. foundations on consistent formation)
Shallow-slope instability following reactivation of shallow shears (<5 m below ground) that may occur on very-low-angle slopes; instability may be triggered by engineering works (slope loading or cuts to slope, etc.) and changing groundwater levels	Geomorphological mapping to identify subtle lobes; interpretation of LiDAR particularly useful; trial pits to confirm depths of shears; monitoring of slope movements and groundwater	Avoid areas of mapped potential instability if possible; dig and replace sheared soils if possible or consider harder engineering solutions (shear keys, soils nails, geotextile, etc.)
Variable or unforeseen ground conditions (<20 m below ground); competent cap rocks at valley margin and on slopes not *in situ* and blocks may be separated by voids or sediment fills (gulls); dips often represent downslope block movements, not geological structure; stresses caused by disturbance of underlying incompetent strata changes geotechnical properties; deep slope instability, relict shear surfaces, disturbance of strata; atypical hydrogeology	Ground model to define likely affected areas; geomorphological mapping to identify location of cambered blocks and intervening gulls and superficial valley disturbances; geophysics, boreholes and/or trial pits to confirm depth of affected strata and impact on geotechnical properties; cambering unlikely to be active under temperate climates, but localized block movements may occur in response to elevated groundwater levels or engineering; monitor using laser scanning or GPS	Design to appropriate support to cambered rock slopes (stiffened extended foundations, separate structures); consider variable ground conditions associated with gulls and hydrogeology; consider ongoing stresses from disturbance of clays on slopes and valley bottoms; relict shear surfaces need consideration
Variable or unforeseen ground conditions associated with bedrock hollows that are filled with sediment and often bounded by a zone of weaker bedrock; atypical hydrogeology associated with sediment fills; differential settlement of structures built upon hollows	Develop an appropriate ground model in areas of high susceptibility; design ground investigation to detect (including geophysics) and characterize hollows with boreholes or trial pits; settlement monitoring of structures built upon hollows (e.g. using persistent scatter InSAR); goundwater monitoring	Appropriate design to manage ground conditions when excavating (dig and replace, ground improvement, extend foundations) or tunnelling (control face conditions)
Range of landforms that are wedge-shaped in section and polygonal in plan that were formed either by thermal contraction of bedrock and/or superficial sediments or opening of cracks through growth of ice veins; cracks subsequently fill with sediment that falls, or is blown, into the void from the land surface, giving rise to variable ground conditions; sediments adjacent to the wedge cast are often deformed and consequently affected strata extend beyond the cast fill	Patterned ground may be visible in aerial photographs; wedge casts can be recognized in sections and demonstrate that host sediments have been affected by periglacial climates and geotechnical properties may be atypical; features are relict and do not require monitoring	Ensure implications of variable ground conditions and impacts of periglacial climate on materials are considered when designing ground investigations and structures
Landform resulting from the process of frost-heave that pushes gravels towards the land surface; gravel forms net-like patterns separated by frost-heaved domes on flat ground, and stripes aligned downslope at angles of >6°; variable ground conditions in the upper section indicates that sediments have been affected by frost	Superficial feature often visible in aerial photographs where land has not been ploughed; unlikely to be hazardous; thickness of affected strata may be characterized in trial pits	Consider presence of patterned ground when developing ground models; may prohibit development at environmental impact assessment and planning stage of a project
Periglacial slopes are often mantled by thick sequences of soliflucted material that comprise downslope transported frost-shattered material; these sequences may obscure landforms developed in past climate stages, including deep-seated landslides that are particularly prevalent in higher-relief terrain (e.g. the Peak District) and along the coast, where that may extend offshore; costly engineering failures can be avoided by ensuring buried landforms are recognized and their implications for slope stability are understood	Geomorphological mapping in the field and/or using aerial photos and LiDAR; ground investigation to develop a model of the buried landforms/landslides; ground investigations comprise conventional ground investigations including boreholes and trial pits, etc.; monitoring of movements using inclinometers, permanent ground markers or laser scanning; groundwater monitoring can help predict possible new movements	Engineering designs that consider the impact of variable ground conditions associated with buried landslides, the impacts of changes in the pattern of ground movement associated with climate change (increased rainfall), and exhumation of the landslide toe by coastal erosion

(*Continued*)

Table 10.1. *Continued*

Geohazard	Landsystem*									
	Lowland						Upland			
	Plateau	Rock slope	Foot slope	Valley	Buried	Sub-merged	Plateau	Sediment mantled	Rock slope	Foot slope
Aeolian deposits (loess and coversand)	N	N	R	R	R	R	W	W	R	R
Cryogenic involutions	W	N	W	W	W	R	W	W	N	N
Carbonate karst (see Chapter 15)	W	N	W	W	R	R	W	W	N	N
Sediment mantled hillslope systems (including 'head'; see Chapter 4)	N	R	W	W	W	R	N	W	R	R
Debris flows and cones (see Chapter 5)	N	R	R	R	R	R	N	R	W	W
Rock slope failure (see Chapter 4)	N	N	N	N	N	N	R	N	W	R

*W: widespread/likely; R: rare/unlikely; N: not present/feasible
InSAR: synthetic aperture radar interferometry; LiDAR: light detection and ranging

'head' on geological maps to describe the sedimentary products of multiple processes including solifluction and/or hillwash and soil creep. The BGS Lexicon of Named Rock Units (British Geological Survey 2019*b*) describes 'head' as poorly sorted and poorly stratified deposits; a polymict deposit comprises gravel, sand and clay depending on upslope source and distance from source.

In relict periglaciated landscapes, typically only the upper 1–2 m of soils are affected by relict periglacial shallow-slope movements, but the resulting downslope accumulation of sediment may be much thicker. For example, at a site near Stoke Hammond, Buckinghamshire, up to 5 m of stacked debris was noted on the lower flanks of valley sides and valley bottoms. Solifluction lobes and features interpreted as active-layer slides (Giles *et al.* 2017)at Sevenoaks in Kent were recorded as being between 3 and 5 m (Skempton & Weeks 1976). The internal structure of the products of relict periglacial shallow-slope movements also includes discontinuities that can form as a product of post-depositional shearing or extremely to closely spaced accommodation shears

Engineering implication	Investigation and monitoring (see Tables 10.2–10.5 for more information)†	Planning considerations and engineering mitigation
UK periglacial environments are often associated with thick covers of blown sand or silt (loess) deposits derived from adjacent glacial landsystems; blown sand and loess are often compressible, prone to collapse and may be soluble; they may obscure underlying landforms/hazards	Ensure ground investigation is appropriately designed to characterize the nature and thickness of wind-blown sediments; consider impact of sediment collapse.=	Pre-loading and wetting can mitigate some of the settlement followed by appropriate engineering design
Diapiric soft-sediment contortions brought about by density gradients formed by seasonal thaw in the active layer of former permafrost; may be present at multiple levels/depths where sedimentation has been rapid and sediments experienced repeated periglacial climates and active-layer processes; variable ground conditions in the upper sequence of ground (typically <5 m); elevation of rockhead likely to be variable	Suitably spaced boreholes to determine range of depths of affected strata; log available sections in region to characterize involutions; trial trenches to characterize specific site areas	Consider presence of involutions when developing ground models; caution when interpreting level of rockhead from borehole data; may prohibit development at EIA and planning stage of a project
Karst develops due to additional water in geosphere from climate warming and thaw of permafrost; colder water accelerates karst formation; hazards limited to carbonate rocks, and comprise rapid change in ground conditions, poor-quality soils and high groundwater tables	Desk study, review of LiDAR and aerial photos to identify susceptible areas; geomorphological mapping and walkover to confirm the location of features; geophysics to investigate presence and extent; boreholes/trial pits to characterize the ground	Collect sufficient ground investigation data to sufficiently characterize the ground; voids can be spanned or bridged using stiffened oversized foundations or piles; weaker soils can be improved, replaced or incorporated in the design
Periglacial slopes are often mantled by thick sequences of material derived from a variety of sources often described collectively as 'head'; source of material includes: solifluction of frost-shattered materials, blown sand and loess, and slope wash; these sequences may include relict shear surfaces on very shallow-angled slopes, obscure landforms developed in past climates stages and comprise variable and/or collapsible sediment	Ensure ground investigations characterize the variability in depth and geotechnical properties of head materials; geomorphological mapping to identify subtle lobes; interpretation of LiDAR particularly useful; trial pits to confirm depths of shears; slope movement and groundwater monitoring	Ensure relict shear surfaces are considered when interpreting GI data and in engineering design, particularly slopes; ensure geotechnical variability is characterized
Shallow-slope instability following reactivation of past instability; may occur on low-angle slopes; instability may be triggered by engineering works (slope loading or cuts to slope, etc.) and changing groundwater levels	Geomorphological mapping and LiDAR interpretation to identify features of interest; trial pits to confirm depths of shears; slope movement and groundwater monitoring	Avoid areas of mapped potential instability if possible; dig and replace sheared soils if possible or consider harder engineering solutions (shear keys, soils nails, geotextile, etc.)
Stress relief failures associated with deglaciation and loss of support by glacier ice; not strictly periglacial in origin, but strongly influenced by periglacial climates; rock slope instability geohazards	Geomorphological mapping, rock slope and discontinuity mapping and characterization	Appropriate design of rock slopes, such as slackening slopes, bolting, shotcrete, netting and rockfall traps

that can be separated by only a few millimetres or centimetres (Fig. 10.7).

One key feature of this geohazard is the shallow angle at which the processes can occur. Relict periglacial shallow-slope movements have been observed on slopes of around 1.5–2.0° (Skempton & Weeks 1976) and, although stable in their current condition, relict shears are prone to reactivation; this is often because of construction activities, changes in loading regimes (loading or cutting into a slope) or alterations of hydrological or drainage conditions that cause increased porewater pressure. The geoprofessional should be aware that slope stability can be a problem on sites with shallow topographical gradients, that is, as low as a few degrees.

Shearing and disturbance of soils associated with the product of relict periglacial shallow-slope movements has been described as a geohazard since the 1940s (Dines et al. 1940) and has been intensively investigated since the 1960s (Weeks 1969; Early & Skempton 1972). Unfortunately, it still causes issues for modern construction projects and is considered the dominant major relict periglacial

Table 10.2. *Sources of desk study information*

Subject	Examples of sources of information
Topography	Maps, nautical charts and plans; aerial photographs; satellite and aerial imagery
Geology, geomorphology and hydrogeology	Maps, nautical charts and plans; memoirs; unpublished reports; aerial photographs; satellite and aerial imagery; published papers and books; mine and quarry records; thematic databases; previous site investigations
Environment and planning	Maps; county and local authority plans; aerial photographs; satellite and aerial imagery; archaeological site and historic building records; contaminated land records; environmental impact assessments; climate records; river and coastal information
Site condition, land use and history	County and local authority plans; land-use maps; historic maps; historical documents; aerial photographs; satellite and aerial imagery
Local knowledge	Local historical societies; site residents; previous site users; construction records; building control office
Precedent	Case histories; construction records
Field reconnaissance	'Skilled-eye' inspection of the site and its surrounds; ground truth reconnaissance; visits to specific localities
Codes, standards, regulations and guidance	Professional bodies and institutes; government departments; research organizations and universities

After de Freitas *et al.* (2017). Adapted for the UK from Shilston *et al.* (2012).

geohazard in the UK. Occasional cases still occur where failures of embankments, slopes and foundations are attributed to the 'unforeseen' presence of existing slip surfaces (Gabriel 2008).

Figure 10.8 illustrates the distribution of rocks most susceptible to relict periglacial shallow-slope movements, which includes overconsolidated mudrocks of the English South Midlands and southern England including the London Clay, Gault Clay, Oxford Clay, Fuller's Earth and Weald Clay formations and the Lias Group. Relict periglacial shallow-slope movements can affect other rock types outside of the geographical areas described above, including, but not limited to, coal measures strata and, importantly, older glacial deposits including till.

The properties of the products of relict periglacial shallow-slope movements are the result of the cyclic sequence of freezing and thawing in the ground and its subsequent movement. The depth and extent of freezing and thawing is influenced by factors including air temperature, snow depth, thermal properties of the soil and the geothermal gradient. Cycles of freezing and thawing that result in the growth and decay of ice within the ground have the potential to modify the previous stress history of the soil. Firstly, the soil will have been disturbed by cryoturbation and frost heave, resulting in sediment mixing. Secondly, thawing of ice-rich permafrost is likely to have resulted in moisture, in excess of the soil's natural moisture content. Thirdly, the soil will have been reconsolidated. Finally, it will have

Table 10.3. *Walkover survey and assessment likely to require specialist input to confirm presence of features and determine scope of any additional work*

Topic	Objectives
Site use	Hazards, boundaries, topography including cut and fill, access constraints and ownership
Topography	Aspect, elevation, slope angles, slope curvature
Geology	Descriptions of exposures of bedrock and other evidence (burrow arisings), superficial sediments and organic soils
Geomorphology	Terrain characteristics and principal landforms at site and wider setting; location relative to Quaternary glacial limits
Movement indicators	Slope instability (debris, tension cracks, seepage, hummocky ground), subsidence, cracking of bulging of structures, deformation to trees
Water and groundwater	Flood trash lines, springs and/or seepage lines, waterlogged ground
Vegetation	Indicators of movement, hydrophilic vegetation, vegetation dieback, vegetation maintenance (e.g. rootballs that may lead to slope instability), invasive species
Ecology	Evidence of protected species
Contamination	Evidence of spills, fly tipping, old storage tanks, landfills

Table 10.4. *Ground investigation: non-intrusive techniques*[*,†]

	Electromagnetic			Electrical			Seismic				Potential field methods	
	GPR	Frequency domain	Very low frequency	ERT	Resistivity	Induced polarization	Reflection	Refraction	MASW	MAM	Microgravity	Magnetics
Fines-dominated sediments	SO	SO	NA	O	O	SO	O	SO	O	O	NA	NA
Coarse-grained sediments and gravels	O	SO	NA	O	O	SO	O	SO	O	O	NA	NA
Sedimentary structures (e.g. bedding)	O	SO	NA	SO	SO	SO	O	NA	NA	NA	NA	NA
Fractures and faults	O	SO	O	SO	SO	O	O	SO	SO	SO	SO	SO
Depth to bedrock	O	SO	NA	O	SO	SO	O	O	O	O	O	NA
Depth to water table	O	SO	NA	O	SO	SO	SO	SO	SO	SO	NA	NA
Sediment and rock properties (e.g. strength)	SO	SO	NA	SO	SO	SO	SO	O	O	O	SO	NA
Voids	SO	SO	NA	SO	SO	NA	SO	SO	SO	SO	O	NA

[*] O: optimal; SO: suboptimal; NA: not applicable.
[†] ERT: electrical resistivity tomography; GPR: ground-penetrating radar; MAM: microtremor array measurement; MASW: multi-channel analysis of surface waves
After de Freitas *et al.* (2017).

Table 10.5. *Ground investigation: intrusive techniques** *(de Freitas et al. 2017)*

Ground conditions	Trial pits	Cable percussive	Window sampling	Soft ground rotary drilling	Rotary sampling	Rotosonic drilling	Cone penetration test	Probing	Hand auger
General soils									
Coarse and/or heterogeneous fill/made ground	A	B	B	U	U	A	B	B	B
Fine homogeneous fill/made ground	A	A	A	U	U	A	A	A	A
Unconsolidated sands	B	B	B†	U	U	B	A	B	B
Unconsolidated gravel	A	B	B	U	U	B	B	B	B
Boulders and cobbles	A	B	U	U	U	B	U	U	U
Soft sediments (alluvial deposits, marine muds)	A	A	A	A	A	B	A	A	B
Medium sediments (alluvial deposits, Tertiary age deposits)	B	B	B	U	A	A	B	B	U
Specific glacial and/or periglacial deposits									
Till	A	A†	U	U	A	A	B	U	U
Overconsolidated subglacial tills	B	B†	U	U	A	B	U	U	U
Till sequences of variable consolidation	B	A†	u	U	A	A	B	U	U
Waterlain deposits (laminated, varved deposits)	A	A†	A	B	U	B	A	B	B
Thick granular glaciofluvial deposits (medium dense to very dense)	U	B†	B†	U	B	U	U	B/U	U
Interbedded diamicton and/or granular deposits	B	B†	B	U	A	B	B	B	U
Slope and/or mass movement deposits	A	B†	A†	U	A†	A	B	B	B
Solifluction/gelifluction deposits	A	B†	A†	U	A†	A	B	B	B

*A: method should be suitable; B: method may be suitable depending on working method adopted; U: method is unlikely to be suitable.
†Dependent on quality of sample required.

Fig. 10.6. Active-layer detachment slides (previously described as solifluction lobes) in Cretaceous Lower Greensand Formation, Weald Group, adjacent to the Sevenoaks Bypass, Kent. © Google Earth Pro, Getmapping PLC 2012.

Fig. 10.7. Solifluction shears in Middle Jurassic Oxford Clay Formation, Stoke Hammond (photograph Tom Berry/Mott MacDonald).

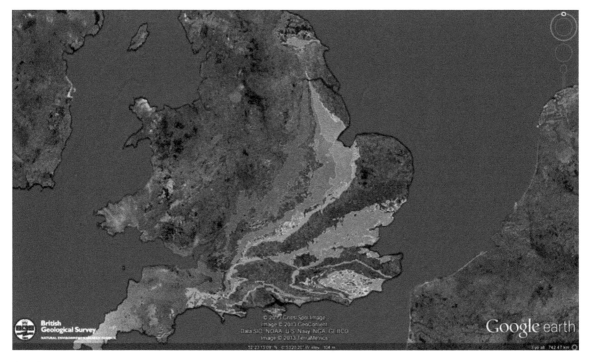

Fig. 10.8. Geographical distribution of rocks that are susceptible to solifluction in the UK. Contains British Geological Survey materials ©NERC 2019. © Google Earth Pro, TerraMetrics, SIO, NOAA, US Navy, GEBCO, 2012. Reproduced with permission.

been transported downslope to be deposited (Hutchinson 1991). Because of the disturbance and movement, the soil disturbed by relict periglacial shallow-slope movements may have lower shear strength parameters (cohesion and friction) and be more compressible than its *in situ* undisturbed counterparts (e.g. Chamberlain & Gow 1979). The magnitude of the reduction in shear strength and increase in compressibility will depend on the physical and thermal properties of the soil and factors including air temperature. In general, coarser soils are usually less affected than fine-grained plastic soils, which are more sensitive to a reduction in shear strength parameters to 'residual' levels.

When clays move downslope under the influence of gravity, platy particles realign in response to changes in shear stress and their physical interaction. Particles on the slip surface undergo reorientation from the sliding, which creates a preferred parallel orientation resulting in the lowest possible shear strength, known as the residual shear strength (Fig. 10.9). Although it may only be the material on the shear surfaces that has a shear strength reduced to the lowest residual friction levels, the remaining soil mass will also have reduced shear strength to some level below the peak.

In summary, relict periglacial shallow-slope movement material will have poorer soil properties than *in situ* unaffected material. The effects of a reduction in shear strength parameters can result in lower bearing capacities, as well as reduced slope stability and increased permeability.

A desk study and walkover survey should be enough to identify whether there is a potential for, or even the presence of, relict periglacial shallow-slope movements at a site. The desk study, walkover survey and ground investigations are of vital importance; the reactivation of relict shears, for example, means extensive redesign, significant cost increases and considerable time delays on engineering projects.

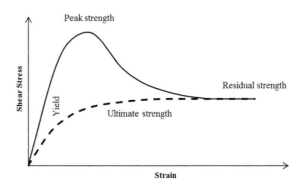

Fig. 10.9. Peak, ultimate and residual shear strength.

A review of the near-surface geology and geographical location will allow the determination of whether a site is susceptible to relict periglacial shallow-slope movements. A review of publicly available aerial photography from archive photographic libraries, or even Google Earth, and natural hazards databases (e.g. British Geological Survey 2019a) will indicate if slope instability is present or has been previously identified and recorded. A site walkover to look for signs of slope movement such as dislocated or tilted fence, posts, telegraph poles, etc. and geomorphological mapping may indicate signs of movement. It should be noted that the best time to see low-relief features such shallow periglacial slope movements is when the sun is low in the sky and vegetation growth is at a minimum; if possible and safe to do so, site visits should be scheduled at these times. In addition, oblique aerial photography or air photos taken in the evening or morning might be most useful in identifying such shallow periglacial slope movements.

Google Earth Pro provides access to good-quality aerial photography that allows the user to see monoscopic (two-dimensional, 2D) vertical (but not necessarily orthorectified) aerial photography draped over a digital terrain model, allowing a 3D perspective (a version of oblique aerial images). When combined with more traditional reference material, including geological maps and memoirs, these aerial images allow researchers to explore remote areas easily and quickly to investigate the potential presence of features associated with periglaciation in the UK. Any aerial photography accessed from internet sources may include some errors, such as inaccurate rectification and poor mosaic joints.

The absence of features at the desk study/walkover stage does not mean the absence of relict periglacial shallow-slope movements at the site, as activities such as ploughing may obliterate any surface expression or vegetation may mask any movement. It may be prudent to examine adjacent areas or areas near the study site with similar geology and topography to see if there are shallow periglacial slope movement features. Field investigations are always recommended to confirm, amplify and supplement the findings of any desk-based investigation.

On completion of the desk study and walkover survey, the ground investigation phase of the works can proceed. Ground investigations are usually approached in a phased manner (as necessary) in order to confirm and supplement any features identified in the desk study or walkover survey, allow the definition and classification of the site stratigraphy and groundwater generally, and collect geotechnical information as well as investigate the ground conditions at the specific locations of the proposed engineering works. More detailed information about the nature and phasing of ground investigations can be found in de Freitas *et al.* (2017).

Determining parametric information for relict periglacial shallow-slope movement deposits can be challenging due to the inherent difficulties in sampling and testing material with pre-existing discontinuity (shear) surfaces in them. The most common sampling method is to carefully prepare a block samples (Fig. 10.10), making sure the discontinuities are clearly marked on the sample and wrappings so that shear box tests can be carried out to determine the soil parameters. In the authors' experience, for planning and budgetary purposes, it can take up to a week to complete one test pit with detailed fabric logging and block sampling. It should be noted that logging soils and collecting samples from trial pits in shallow periglacial slope movement material poses some significant health and safety issues, not least regarding side stability. The installation and maintenance of an adequate shoring system while logging and sampling in relict periglacial shallow-slope movement soils is an essential part of any method statement. In lieu of being able to sample and test blocks, ring shear testing on remoulded samples can be an effective alternate option for establishing values of residual strength for design.

The geographical spread of material affected by relict periglacial shallow-slope movements in certain areas means that avoidance in engineering works may not be possible; the geoprofessional should therefore prepare preventative designs to mitigate the risks posed by soils affected by periglacial mass movement. Remedial strategies and construction works are similar to the preventative measures that can be taken, but remedial works usually cost more and add to the overall construction programme.

The mitigation measures for foundations proposed on shallow periglacial slope movement material include methods for

Fig. 10.10. Careful preparation of a block sample, Pound Green Embankment, Berkshire (photograph Tom Berry/Mott MacDonald).

lowering the stress imposed by the structure by increasing foundation sizes (strip footing widths and pad dimensions) or by placing the structure on a raft. While such measures may account for the weakness of the foundation soils, a global slope stability check is also required to ensure the whole structure does not slide down the slope. Alternatively, depending on the nature of the structure, it may be prudent to take the loads through the relict periglacial shallow-slope movements layer and into *in situ* material using a deep-foundation solution such as piling.

If increasing the foundation areas or founding deeper is not viable, again depending on the nature of the structure, a dig-and-replace solution where the relict periglacial shallow-slope movement material is excavated and replaced with a suitably compacted fill (which could be the sheared material) may be appropriate. However, care may be needed in removing soil because the ground upslope of the excavation may be destabilized. Regardless of the mitigation solution selected, it would be prudent to design the structure and services entering and leaving the structure to be tolerant to movements above and beyond what would normally be considered acceptable to ensure the long-term serviceability of the structure.

The mitigation measures for earthworks for both the original design and redesign can be divided into three broad categories. The simplest, but not always the most cost-effective or practical, solution for both cuttings and embankments is flattening side slopes. The basis of the works is to reduce the face gradient of the cutting to a level where the slope angle is less than the residual angle of shearing resistance. Because this friction angle of the *in situ* material could be very low, this option is unlikely to be viable. The aim of regrading works on an embankment would be to sufficiently spread the load to less than the bearing capacity of the soil. Again, this is rarely a viable option due to higher costs and constraints of available land and material.

The second option, applicable to both embankments and cuttings, is to excavate and replace the sheared material. In the case of embankments, the sheared material beneath part, or all, of the embankment and slightly outside of the embankment footprint is excavated to form a shear key. The excavated material is then replaced with suitable engineered fill (which could include the excavated material). A similar programme of excavation of sheared material and replacement with suitable engineered fill can be applied to cutting side slopes. As mentioned above, care is required on sloping ground to avoid destabilization.

Engineered fill material used for these remediation solutions can take many forms, for example imported granular fill, lightweight fill or the excavated sheared material. If using the excavated sheared material as fill, care should be taken to ensure that the backfill material has been sufficiently reworked such that the *in situ* shear planes have been broken up. As such, when recompacting the previously sheared material back into the excavation, a 'sheep's foot roller' or similar, rather than a smooth drum roller, should be used to ensure new planes of weakness are not reintroduced to the fill material.

The third broad set of mitigation measures that could be considered are 'harder' engineering solutions. Basal reinforcement of embankments can be adopted where other solutions are either impractical or unfeasible, for example, where space is limited such that shear keys or regrading solutions cannot be used. The principal purpose of the basal reinforcement is to provide sufficient tensile strength and pull-out resistance at the base of the embankment to prevent global bearing capacity failure mechanisms from mobilizing. Piles can also be installed beneath embankments to increase the overall strength properties of the formation material as well as transfer some of the load down to unsheared material. 'Harder' engineering solutions for cuttings include piling (sheet piles, contiguous or secant) and soil nails that will provide a rigid barrier or element to prevent future movement from taking place.

Whichever mitigation (or remediation) options are considered, careful consideration of the long-term impact of groundwater should be considered a priority. Control of groundwater into the project and control of drainage within the project should be carefully considered to ensure the most effective solution is adopted for the project. Consideration of the long-term operational effectiveness of the drainage and/or groundwater controls, plus any maintenance requirements, should be in the forefront of the minds of the designers.

10.2.3 Cambering and superficial valley disturbances

Cambering is the lateral extension and downwards displacement of usually interbedded competent and less-competent, subhorizontal sedimentary strata. Cambering is often associated with superficial valley disturbances, sometimes referred to as valley bulging (Hutchinson 1991), which is upwards, plastic deformation, also known as clay extrusion (Brunsden & Jones 1972) of commonly argillaceous rock in the bottom or lower side slopes of the valleys with cambered upper and middle side slopes (Hutchinson 1991; Parks 1991; Ballantyne & Harris 1994). The processes occur on and above hillslopes in response to valley formation (often rapid down cutting) and unloading (stress relief). The development of cambers and superficial valley disturbances involves several interrelated but poorly understood processes that operate most efficiently under periglacial climatic conditions. Although cambering and superficial valley disturbances are not periglacial processes *per se*, periglacial conditions make the processes much more efficient. While cambering and superficial valley disturbances are not thought to be active under temperate climates, the processes have prepared contemporary slopes for a range of landslide processes, and many active landslides include cambered material.

Superficial valley disturbances comprise the extrusion of less-competent strata, typically silts or clays, into valley sides or bottoms. The strata become progressively thinner towards valley bottoms and are elevated above their true stratigraphical position to produce non-tectonic folds that may be gentle folds to tighter chevron folds (Gallois 2010). Cambering occurs where there is a more competent caprock, such as a limestone, which will move downslope to drape the land surface. Large blocks may become detached and develop into mass movement, with material ultimately reducing to a drape of brashy rubble at the most distant downslope edge. Partially rotated (steeper than the regional dip of the strata) blocks tend to dip downslope but can be back-tilted and are known as dip-and-fault structures. The gaps between blocks, usually formed by opening of joints, are known as gulls. They may appear as voids or be filled with fallen blocks or soil from horizons above. Secondary cementation of void fills may form a variable conglomerate material known as gull rock. Subsequent erosion and dissolution processes can form 'gull caves' that may extend several hundred metres into the slope (Self 1986; Self & Boycott 2000).

Cambering and superficial valley disturbances occur in periglacial environments when there is rapid valley erosion combined with the release of lateral stress in valley sides, cryogenic disturbance and weakening of the mudstones or clays in the valley bottoms, concentration of ground ice in the mudstones or clays, and the continued erosion of material from the valley floor (Fig. 10.11). On amelioration of the climate, ground ice melts and the mudstones or clays become saturated and porewater pressures are elevated; combined with the stress relief and cryogenic disturbance, weakening of the mudstones or clays results. These weakened mudstones or clays deform and cause heave in valley bottoms and lower side slopes, resulting in subsequent settlement, rotation and downslope movement (cambering) of the competent caprocks above (Higginbottom & Fookes 1970; Eyles & Paul 1983; Hutchinson 1991; Ballantyne & Harris 1994).

The age of valley disturbances has been inferred from stratigraphical relationships of cambered strata to river terraces, which suggest they predate the last interglacial (Chandler 1970; Hutchinson 1991). Worssam (1981) also suggested a Middle Pleistocene age for cambering near Maidstone in Kent on the basis of the relationship of clay linings of gulls to glacially derived loess that infilled gulls. Notwithstanding these specific observations, it is likely that cambering and superficial valley disturbance processes were active at multiple times during cold climate periods of the Pleistocene, with the origins of features closely associated with drainage development and valley incision.

The geohazards associated with cambering are principally the gulls in the competent caprocks, which may be open or sediment-filled voids up to a few metres wide and tends of metres deep. In addition, weakening, disturbance and shearing of the underlying mudrocks and increased permeabilities pose a hazard. The geohazards associated with superficial valley disturbances are deep disturbance of the mudstones or clays in the sides and bottom of valleys, which Hutchinson (1991) observed up to 62 m below the valley floor at the Ladybower Dam, Derbyshire. Although not active in the UK at present, these relict features can still have a significant influence on current construction projects.

Rocks susceptible to cambering and superficial valley disturbances are widespread in the UK (Fig. 10.12). The features are most commonly associated with the Jurassic strata of Northamptonshire and Rutland, but are also recorded in Carboniferous, Permo-Triassic and Cretaceous rocks (Higginbottom & Fookes 1970; Skempton & Weeks 1976; Hutchinson 1991; Parks 1991; Ballantyne & Harris 1994). On historical BGS maps, the term 'foundered strata' was used to describes areas where extensive landsliding and cambering have occurred, but beneath which the solid geology could not be determined by the mapping geologist. Confusion can occur when interpreting more recent maps where the same term is used to describe areas of collapsed ground, either natural or by human activity, with no inferred relationship to cambering or valley bulging.

Gulls pose a hazard to foundations due to the potential for voids or relatively soft infill when competent rock is expected (Fig. 10.13).

In the absence of pre-existing structures, gulls will generally form parallel and perpendicular to the long axis of the valley. Gulls can be metres wide, tens of metres deep and hundreds of metres long (Eyles & Paul 1983; Ballantyne & Harris 1994). Hawkins & Privett (1981) described three main types of extensional movement observed within the Blue Lias (Fig. 10.14): Type A, opening to create a single joint to create a straight, parallel sided gull; Type B, extension of a broader zone where bed-over-bed movement produces a large number of small voids; and Type C, Movement at depth which forms a large cavity and almost undisturbed roof.

Voided and infilled gulls can lead to a rapid change in ground conditions (unexpected but not unforeseen) leading to reduced bearing capacities (zero in the case of large voids), increased settlement and higher permeabilities than anticipated if intact rock was present. The variation in ground conditions means gulls can be visible in aerial photographs as

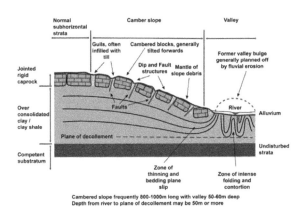

Fig. 10.11. Figure 9.24 from Griffiths & Giles (2017). Schematic section showing cambering and superficial valley disturbances (after Hutchinson 1989).

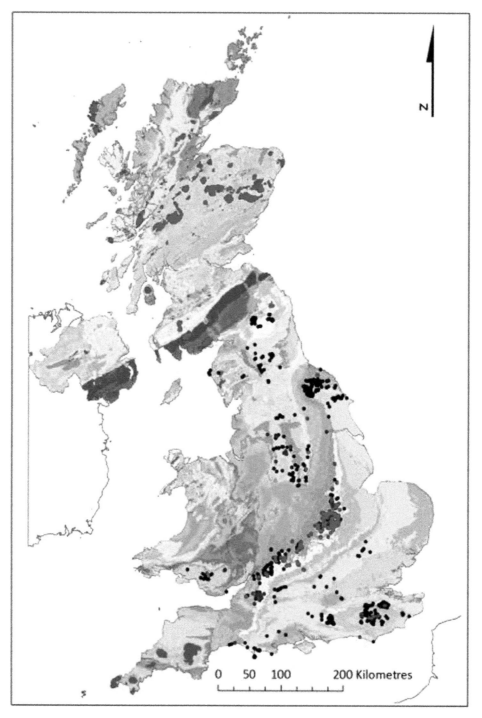

Fig. 10.12. Distribution of rocks potentially susceptible to cambering and valley bulge from Culshaw *et al.* (2017). Recorded occurrences of cambered strata from the literature (30) (purple dots), in the British Geological Survey's National Landslide Database (136) (red dots) and in the PhD thesis of Parks (1991) (639) (black dots) overlain on a geological map of the UK (derived from the British Geological Survey 1:625,000 scale map, reproduced with permission of the British Geological Survey © NERC).

Fig. 10.13. Infilled gull in Upper Permian Cadeby Formation (formally Permo-Triassic Lower Magnesium Limestone) in an excavation for a reservoir near Clifton, Yorkshire (photograph Tom Berry/Mott MacDonald).

CLASS	NAME	TYPE, MOVEMENT	CHARACTERISTICS	SKETCH
I	Infilled	A-Type movement, small to large displacement	Head sags into gull	
IIa	Open	B-Type movement, small displacement	Head not affected. Medium to large voids at depth	
IIb	Open	B-Type movement, large displacement	No Head. Many tilted blocks. Large voids at depth	
III	Mixed	B-Type movement, large displacement	Head sags into gull. Large voids at depth	
IVa	Intact Roofed	C-Type movement, medium to large displacement	Level limestone roof. If present, Head is undisturbed. Large voids and cavity at depth	
IVb	Collapsed Roofed	C-Type movement, (+ B-Type at top), large displacement	Roof of fallen blocks wedged in the top of a large cavity. If present, Head may sag a little	

Fig. 10.14. Classification of gulls (from Hawkins & Privett 1981). Reproduced with permission.

crop marks, particularly in a period of drought when the vegetation is under stress.

Superficial valley disturbances pose a hazard to construction works due to the disturbance of the material in the valley bottom and valley side. The disturbance results in folds, fissures and faults, with a subsequent reduction in the soil and/or rock strength properties of the material, juxtaposition of differing materials and increased permeability. This reduction in strength leads to reduced bearing capacities, increased settlement, and decreased stability of excavations and cut slopes.

Cambering and superficial valley disturbances can affect large areas of a site. Eyles & Paul (1983) suggest that cambering can occur up to 1 km back from the crest of valleys, meaning that avoidance of the hazard may not be possible. Geoprofessionals should therefore investigate this potential thoroughly to develop preventative designs to manage or mitigate the geohazards.

A desk study, including review of aerial photography and light detection and ranging (LiDAR) imagery, and a walkover survey should be carried out to identify if there is field evidence for cambering and superficial valley disturbances. However, as gulls can be obscured by a mantle of superficial deposits, masked by vegetation or bridged by caprock, the desk study should consider the potential for cambering and superficial valley disturbances to be present, based on the geomorphological setting. Further investigations should be planned to determine the location and degree of associated geohazards.

The ground investigation techniques to identify the nature and extent of gulls should be carried out in a phased manner to identify, classify and sample the features of concern. Geophysical techniques could be employed to identify the presence of gulls but may not be sensitive enough to identify the smaller features or features with low density contrasts. Trial trenching perpendicular to slope contours offers the potential for the identification of gulls, although voids bridged by a thin horizontal layer of rock may not be detected (Hawkins & Privett 1981). It may therefore be necessary to hand-clean trenches to look for evidence of possible voids at depth; however, there are significant health and safety issues associated with entering excavations with the potential for voids beneath the base, and each entry should be risk-assessed by a competent person before entering. It should also be noted that such trial pitting investigations can themselves be destructive, in that the action of the pitting can disturb, damage or remove any 'good ground' that is present. A comprehensive program of trial pitting and conventional borings should be sufficient to identify and quantify superficial valley disturbances.

Assuming that avoidance is not an option, the mitigation measures for gulls include infilling voids or adapting foundation solutions to bridge the gulls. Care should be exercised when filling gulls so that drainage paths are not obstructed, which might cause flow paths to change or cause a build-up of water that could reactivate relict shear surfaces. Although cambering and superficial valley disturbances are periglacial processes and cannot be reactivated by engineering works, rainfall and engineering works along with other factors can cause reactivation of cambered strata and strata affected by superficial valley disturbances. Cambered blocks can be destabilized and caused to move downslope as part of landslide processes, and material affected by superficial valley disturbances can also be subject to slope instability. If infilling of gulls is proposed, a permeable medium is recommended such as suitable compacted granular fill or permeable concrete to avoid changing the hydrogeology of the site and surrounding area. The complex and variable pattern of gulls will necessitate bespoke foundation solutions for each site, if not each structure. Foundation solutions proposed on cambered blocks include: moving the location of the structure to avoid any known gulls, positioning so that the gull is under the central third of the foundation, or increasing the size of foundations and adding reinforcement and/or geotextiles to bridge any potential gulls (voids or infilled). Services to structures should be designed and constructed to accommodate movement above and beyond what would normally be considered acceptable to ensure the variations in properties between rock and gull do not affect performance.

Care should be taken to ensure that groundwater flow within and through gulls does not become a problem. Gulls can be pathways for groundwater entering or leaving a site, making the site either a new receptor for offsite potentially contaminated groundwater to migrate to or a new source of potentially contaminated groundwater for other receptors down the hydraulic gradient. Soakaway drainage should be avoided because concentrations of water entering gulls can reactivate relict shear surfaces or wash out material, resulting in sink holes appearing at the surface.

The mitigation measures for foundations proposed on superficial valley disturbance material include methods for lowering the stress imposed by the structure by increasing foundation sizes (strip footing widths and pad dimensions) or by placing the structure on a raft and providing support to slopes cut or loaded as part of the engineering works. Alternatively, if the superficial valley disturbance material cannot be engineered, and depending on the nature of the structure, it may be prudent to take the loads through the disturbed material and into *in situ* material using a deep-foundation solution such as piling.

The temporary stability of excavations and slopes during construction may require additional support beyond what would be normally required (or allowed for in costs) during the construction period, due to the disturbed nature of the ground. Mitigation measures for excavations could include the provision of more robust temporary shoring in trenches or other temporary supports such as sheet piling and bracing.

Mitigation measures for cut slopes in material affected by superficial valley disturbances will be like those employed on slopes where residual strength conditions are anticipated. Common stability measures employed elsewhere can be applicable in the short term. They include control of water on and within a slope, profile modification and slope support, which includes toe loading, berms and retaining structures, bolting, nailing and anchoring, netting and vegetating. Further details on slope stabilization are presented in Chapters 4 and 5.

To mitigate hazards associated with increased permeabilities, both in gulls and superficial valley disturbance material, and depending on the nature of the project, it may be necessary to managed unexpected localized inflows of groundwater. This may be managed through construction of sumps and pumping, with consideration of the associated issues of storage and discharge consents. On larger projects, it may be necessary to install drainage, cut-off trenches, grouting or grout curtains.

Whichever mitigation or remediation option(s) are considered, careful consideration of the long-term impact of groundwater should be considered a priority. Control of groundwater into the project and control of drainage within the project should be carefully considered to ensure the most effective solution is adopted for the project. Consideration of the long-term operational effectiveness of the drainage or ground-water controls, together with any maintenance requirements, should be a priority for designers.

10.2.4 Rockhead anomalies

A rockhead anomaly is the collective term proposed for features exhibiting rapid deepening of bedrock levels over short distances. Conventionally, rockhead anomalies, especially in central London, have been variously referred to as pingos, scour hollows, anomalous buried hollows or drift-filled hollows. These terms imply one or more modes of origin to local over-deepening. Given the complexity of the features, their equivocal mode(s) of formation and the variability of their sedimentary fill (including Quaternary sediments and deformed bedrock), the non-genetic term 'rockhead anomalies' is preferred here, and its use is recommended for future publications.

The features are generally oval or circular in plan up to around 500 m in diameter on their long axis, and they have a downward narrowing funnel-shaped profile typically 15–25 m deep, but can be more than 60 m deep. Rockhead anomalies can be wholly within the superficial deposits or extend deep into bedrock (Banks *et al.* 2015). More than 80% of the rockhead anomalies presented in Banks *et al.* (2015) occur beneath the Devensian-aged Kempton Park Gravel Member, which indicates they may predate all or part of that member. It is probable that rockhead anomalies developed many times throughout the Quaternary (Banks *et al.* 2015) and can therefore be anticipated in a wide range of settings. The frequent observation of these features in London probably reflects the high number of deep excavations in areas susceptible to rockhead anomalies compared with other regions of the country.

The engineering issues associated with rockhead anomalies are that the infill material is often saturated and has a significantly different composition and nature than the surrounding bedrock. The infill does not therefore behave as anticipated when constructing through, in, under or on these features. The infill is generally composed of saturated superficial deposits and disturbed weakened bedrock. The nature and performance of the rockhead anomaly 'walls' are usually also disturbed and weakened. The zone of disturbed bedrock may extend for a few metres from the margin of the Quaternary sedimentary fill.

Hutchinson (1980) noted that rockhead anomaly features were being described in as early as the late nineteenth century. Rockhead anomalies have been described by many authors and have been attributed to various formational processes over the subsequent years, including: scour-hollows (Berry 1979), pingos (Hutchinson 1980), scour and pressure release (Hutchinson 1991), chalk dissolution (Gibbard 1985), discontinuous gulley formation (Rose et al. 1980) and clay diapirism (Berry 1979; Hutchinson 1980).

There are several possible mechanisms proposed for the formation of rockhead anomalies, and it has also been considered that their formation is the result of multiple processes operating through time. The current principal proposed mechanisms of formation are as follows.

- *Fluvial or glaciofluvial scouring*: the physical erosion of material from a river bed, where increased volume of water and flow velocity can accelerate scour. Berry (1979) identified 26 drift-filled hollows in London and suggested that they can be related to shallow buried channels as they appear to coincide with stream junctions in recent drainage patterns (Zeuner 1959). Although scour at tributary junctions might be responsible for some of the rockhead anomalies, scour alone cannot explain all the features associated with them. It is also noted that the rate and magnitude of scour is likely to be influenced by changes in Quaternary sea-levels that have fluctuated by up to 150 m in the last 2.6 Ma.
- *Pingo remnants*: open-system pingos have been proposed as a mechanism for rockhead anomaly formation (possibly associated with scour) by Hutchinson (1980). In periglacial areas ground ice can coalesce and grow, displacing soil upwards to form mounds up to 50 m high and 600 m in diameter called pingos (Ballantyne & Harris 1994). On melting, the previously upwardly displaced soils collapse into the melted ice. The result is a deep crater-like formation infilled with disturbed peaty soil. Open-system pingos form in areas of discontinuous permafrost when the downslope movement of groundwater through permeable soils rises near the surface through artesian pressures, where it freezes. The frozen water, fed from beneath, grows upwards, displacing the soil and forming the pingo. This process gives rise to the characteristic funnel-shaped geometry of a rockhead anomaly, described above. Closed-system pingos form in areas of continuous permafrost when isolated lakes or ponds drain, leaving saturated ground that then freezes from the surface, forming a lens of pingo ice. This lens of pingo ice covers a pocket of unfrozen soil and groundwater. As the surrounding permafrost encroaches on the unfrozen soil and groundwater, hydrostatic pressures draw groundwater into the pingo ice, allowing it to grow, displace the soil upwards and form the pingo (Paul 1983; Ballantyne & Harris 1994).
- *Faulting*: Banks et al. (2015) note that some rockhead anomalies have long axes aligned with faults, which implies that melting of deep permafrost could release pressurized groundwater that preferentially flows along faults or fracture discharge points that may be associated with unloading of overburden.

A rockhead anomaly hazard susceptibility map for London has been compiled (Fig. 10.15) based on various contributory and advisory factors (Banks et al. 2015).

Collins (2013) presented case studies of 'anomalous depressions' in the Thames Valley west of London at Brimpton-Woolhampton and Ashford Hill in the Kennet Valley (Fig. 10.16). Rockhead anomalies are likely to be more extensively distributed than currently mapped; rockhead anomalies should therefore be considered in other areas of the London Basin and beyond.

Having reviewed the location and nature of 40 of these features, Banks et al. (2015) proposed the following contributory factors that can be used to constrain their likely distribution: potential for historical artesian groundwater levels, London Clay thickness <35 m and the presence of Kempton Park Gravels. In addition, they suggested advisory factors including the presence of geological faults that focus groundwater flow, and an association with present-day river networks, which have progressively eroded sediment to unload the ground. These observations can be used to make a preliminary prediction of the distribution of such features in other parts of the UK.

As rockhead anomalies often cluster together over many square kilometres, it may be difficult to avoid such features when planning construction works, particularly if sites or alignments are constrained, as is often the case in London. The geohazard associated with rockhead anomalies is primarily the rapid change in expected ground conditions to soils with poorer engineering properties than anticipated and localized elevated groundwater levels. Such rapid changes in ground conditions and subsequent engineering properties of the soils will have to be mitigated in the design element of the works. The resultant geotechnical problems include high groundwater, the presence of organic soils, low-bearing-capacity soils, high-compressibility soils, high-permeability soils and differential soil properties. In addition, these features can be collection points and pathways for contamination. Cut faces may not be stable in the very short term and may discharge (potentially contaminated) groundwater into working areas that will need to be controlled and managed.

Rockhead anomalies formed by relict pingos outside of London often form ponds or wetlands that provide a rich mosaic of habitats with high ecological value, which may constrain a project.

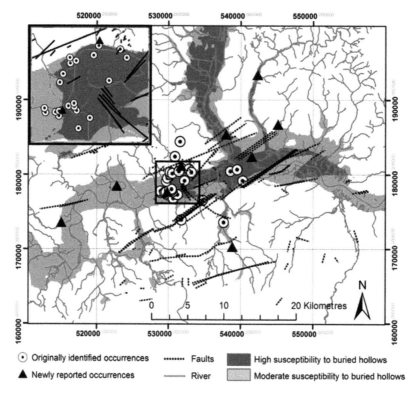

Fig. 10.15. Figure 5 from Banks *et al.* (2015). Zoned hazard susceptibility map. BGS©NERC. Contains OS Open data ©Crown Copyright and database rights 2014. Lost rivers of London reproduced from Barton (1992). Reproduced with permission.

The level of geohazard that rockhead anomalies pose can potentially be high if they are not identified prior to design and construction, and the financial and programme consequences can be significant. Rockhead anomaly geohazards should therefore be identified through desk studies, including susceptibility mapping, walkover surveys of more rural areas, and ground investigations that allow features to be identified and preventative designs prepared to mitigate the hazard. Allowance should be made in the schedule, budget and construction works to accommodate these features.

The mitigation measures for foundations proposed over rockhead anomalies include ground improvement methods (vertical drains, soil mixing, etc.) or lowering the stress imposed by the structure by increasing foundation sizes, for example, by placing structures on rafts. Alternatively, and depending on the nature of the structure, it may be prudent to take loads through the soft compressible soils into *in situ* undisturbed material using a deep-foundation solution such as piling. Similarly, excavations (including tunnelling) into rockhead anomalies will also have to deal with the rapid deterioration of ground conditions and possible ingress of groundwater. Mitigation measures for excavations and tunnels include possible pre-excavation groundwater control (including cut-off structures (e.g. grouting), pumping, ground freezing, etc.) and groundwater management during excavation. Excavation stability can be managed using techniques including ground improvement and side support (shoring) placed simultaneously with excavation. The stability of the tunnel face can be managed by ground improvement or

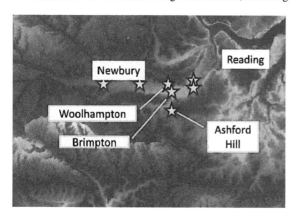

Fig. 10.16. Anomalous depressions in the Thames Valley west of London at Brimpton-Woolhampton and Ashford Hill in the Kennet Valley (from Collins 2013). Reproduced with permission.

providing a positive pressure to the face during mining using air or fluids.

10.2.5 Cryogenic wedges (ice-wedge pseudomorphs)

Ground ice that has impacted the UK can generally be divided into the following four categories: pore ice, segregated ice, intrusive ice and wedge ice. Regardless of the form of ground ice, they have a similar impact on soils; the growth and degradation of ground ice changes the fabric of soils, which can result in a reduction of shear strength and an increase in compressibility (e.g. Morgenstern & Nixon 1971; Nixon & Morgenstern 1973). One of the most significant but often overlooked impacts of ground ice is that it changes the stress history of the soil, particularly in the weak mudrocks (e.g. Oxford Clay and London Clay formations). Stress history influences fundamental geotechnical behaviour that may change from regional- to site-scale, change within similar geological units and overprint the depositional and erosional histories of bedrock.

Cryogenic wedges are the remnant expressions of ice, sand or composite wedges active during periglacial periods, often referred to as ice-wedge pseudomorphs. Cryogenic wedges form when the rate and magnitude of ground cooling is sufficient to overcome the tensile strength of the soils, causing cracking. Any water or groundwater in the crack freezes and expands, exacerbating vein development. Cracks formed during the winter are subsequently filled with water and/or mineral or organic debris sand. Repeated cycles of warming and cooling allow the vein to develop into water-filled ice wedges or detritus-filled sand wedges that can grow to be up to 3 m wide, 10 m deep (Pewe 1974) and tens of metres long (Ballantyne & Harris 1994). Cryogenic wedges can be active for as little as a season, or can be active for many hundreds to thousands of years (Murton & Ballantyne 2017). The resultant relict periglacial features are a network of infilled cracks often manifested as a loose polygonal or angular network (on a scale of metres to tens of metres) of connected wedges visible from above (Fig. 10.17). Cryogenic casts (ice-wedge pseudomorphs) can be found across the UK and, when encountered, are often present over many square kilometres, making it difficult to avoid such features when planning construction works.

The level of geohazard is considered here to be low. The main geotechnical consequence of encountering such features is the rapid change in soil composition and subsequent engineering properties of the soils. Morgan (1971) described geohazards associated with cryogenic casts as 'lines of weakness' that transmitted groundwater into the trench they were excavating, causing flooding and having an adverse effect on trench stability. This instability can extend to trial pit excavation during ground investigation,

Fig. 10.17. Cryogenic casts (ice-wedge pseudomorphs) seen in Google Earth Images near Doncaster (2.3 km west of Carcroft). © Google Earth Pro, Getmapping PLC. 2012.

instability in foundation and other excavations and even, in extreme cases, to slope instability. In addition, the rapid transmission of groundwater from potentially significant distances away has the potential to cause contamination of a site from a nearby source of contamination, as well as transmit any contamination offsite to vulnerable receptors not considered originally at risk from the works.

Although the level of geohazards that cryogenic casts pose is reasonably low, if they are not identified prior to construction, the financial and temporal costs of remediation can be unexpected. As such (and as always), cryogenic casts geohazards should be identified through desk studies and ground investigations so that preventative designs to mitigate the risks can be incorporated into the schedule, budget and construction works. Once the nature and extent of the problem is understood, various remediation strategies can be adopted to prevent the features becoming an issue during construction.

The mitigation measures for excavations include allowing for localized additional side support, even in shallow excavations, and the provision of equipment to control the ingress and outflow of groundwater as needed (Morgan 1971). The control of groundwater could incorporate the use of sumps and pumps or barriers to prevent excavations being inundated or the egress of potentially contaminated groundwater offsite.

10.3 Subsidiary relict periglacial geohazards

Subsidiary relict periglacial geohazards include processes that are not truly periglacial in origin, but which are significantly influenced by periglacial climates and processes, including the activation of deep landslides, the development of carbonate karst, and the formation of buried valleys and submerged landsystems.

10.3.1 The influence of periglacial climates and processes on deep-seated landslide systems

The influence of former periglacial climates on shallow landslides is well-documented (e.g. Murton & Ballantyne 2017). The influence of periglacial climates on deeper-seated landslides has received less attention, but the release of additional water associated with seasonal melt or glacier retreat, rising sea-levels, freeze–thaw and frost shattering are all important processes in preparing slopes for failure. Deep-seated landslides are described in more detail in Chapter 4 so only the influence that periglaciation could have on their occurrence and behaviour is discussed briefly below.

Research has considered the role of permafrost in enhancing mass-movement processes of sagging, cambering and block toppling in interbedded clay–limestone sequences on the Isle of Portland (Brunsden et al. 1996); the role of frost in reducing the strength of sandstone and shale bedrock, leading to multiple rotational failures and blockslides in Longdendale (Johnson 1980); and the reactivation of pre-existing landslides by the erosion of thick drapes of periglacial sediments at Ventnor (Moore et al. 2007, 2010) and along the Dorset coast (Brunsden & Jones 1972; Gallois 2010). Rising sea-levels since the end of the last cold stage has eroded sea cliffs and caused deep-seated slope instability and cliff recession. Empirical data suggest that erosion rates could be doubled by 2080 as a result of the effects of sea-level rise and more frequent storms (Moore et al. 2010).

When considering deep-seated rock slope failures, it is important to understand the significance of paraglacial geomorphology; paraglacial geomorphology describes processes that are conditioned, but not directly caused, by glaciers (Ballantyne 2002), and which are consequently often a characteristic of former periglacial upland environments in the UK. Paraglacial rock slope failures are caused by stress relief following retreat of glacier ice, with subsequent movements associated with freeze–thaw processes (Ballantyne 2002; Wilson 2005). Rock slope failures are marginally stable under contemporary climate conditions, but can be hazardous if disturbed.

The significance of these observations to engineering practice is twofold: firstly, it is important to document the full extent of deep-seated landslide complexes and ensure areas buried beneath periglacial drapes are recognized; and secondly, while many large landslide complexes have their origins in Pleistocene climates and are only periodically active under contemporary climate regimes, they can still pose a significant hazard to construction if not understood when designing groundworks. As rock slope failures are associated with upland areas, they are rarely encountered by engineering projects. However, upland wind farms and cable routing projects should ensure that the hazards are appropriately characterized and mitigated.

10.3.2 Carbonate dissolution

In contemporary periglacial environments, thermokarst is a landform that develops in response to melting of permafrost either in response to changes in land use or to climate change. The processes result in groundwater drainage, leading to differential settlement as shallow lake basins formed, and the release of greenhouse gases from thawed peaty soils. The geohazards occur during the initial phase of permafrost thaw, which in the UK will have occurred during the transition from glacial to interglacial climate and are therefore not expected to be a geohazard under the current climate regime of the UK.

Carbonate karst is a geochemical process that results from the dissolution of limestone by the action of water and dissolved carbon dioxide to form uneven rockhead and bedrock voids. The process is discussed in detail in Chapter 15. While this process is not typically considered as periglacial in origin, it is likely to have been significantly more effective in a periglacial climate. This is because the solubility of calcium carbonate becomes more effective with decreasing water temperature, and because enhanced carbon dioxide production in response to the thaw of peat soils makes the process more aggressive; both are characteristic of the seasonal melting of the active layer of permafrost. Evidence for the enhanced formation of karst during periglacial climates in the UK is shown by loess infilling karst features in chalk in southern England.

10.3.3 Buried terrains

Buried periglacial terrains in the UK are typically associated with sediment-mantled slopes and infilled valleys that obscure rockhead. Relict landforms associated with surplus water production during seasonal melt of permafrost or glacial meltwater streams operating in a periglacial environment can form challenging ground conditions (Woodland 1970; Cox 1985). An example is the significant tunnel valleys of East Anglia and adjacent North Sea areas that form anastomosing networks of channels tens of metres wide and up to 100 m deep (Woodland 1970; Lonegran et al. 2006; Bricker et al. 2012). Many other areas of the UK contain buried terrains, particular between the limits of the Anglian and Devensian glaciations, and lowland areas north of the limit of the Anglian glaciation. For example, Case Study 9.2 in Griffiths & Giles (2017) describes a significant occurrence of subglacial channels and tunnel valleys that extend for many kilometres in length near Doncaster, South Yorkshire that incised the Sherwood Sandstone bedrock by up to 40 m.

Buried terrains can also preserve periglacial formations (e.g. cryogenic wedge casts) from previous cold periods that can help to determine the geological history of a site to understand the likely engineering behaviour of the soils (Murton & Ballantyne 2017). These features pose a geohazard because they are not all mapped; they also represent rapid changes in ground conditions, and therefore have variable material properties and behaviour as well as rapid changes in the anticipated hydrogeology.

10.3.4 Submerged periglacial terrains

During glacial maxima of the Middle and Late Pleistocene, global sea-level fell by c. 120–150 m, exposing the floor of most of the southern North Sea Basin and English Channel to coeval periglacial conditions (Murton & Ballantyne 2017). Areas currently in a shallow-marine environment (water levels <150 m) were above sea–level; as well as being exposed to periglacial conditions, they were also were subject to fluvial incision and dissolution (>100 m below current sea-levels in Galway, Ireland). As such, some (if not all) of the periglacial and paraglacial geohazards described above could have been active during the Quaternary in the shallow-marine environment, and represent a relict submerged periglacial landscape.

Periglacial deposits and processes should be considered as part of any engineering works proposed in the shallow-marine environment such as the laying of marine cables and installation of wind farm foundations, etc. Shallow-marine-specific periglacially related geohazards (some of which are discussed in more detail in on-land scenarios elsewhere in this chapter and Chapter 10 include dense granular deposits in periglacial channels that could be an impediment to driven piles or drive-drill monopile, soft compressible deposits in periglacial channels, loose periglacial materials susceptible to scour, boulders (erratics) or other hard surfaces that could impede jetting or ploughing for cables, stability of gravity bases and penetration of piles, and deep weathering of chalk (and other) bedrock (Murton & Ballantyne 2017; Winter et al. 2017).

10.3.5 Loess and coversand

Loess is an aeolian deposit commonly associated with cold-climate lowland plateau landsystems. In the UK, loess is exclusively associated with periods of cold climate when the absence of vegetation allowed wind to erode, transport and redeposit the abundant sediment released by glacial processes. Loess is predominately composed of silt-sized particles (with minor constituents of clay and fine sand) that have become weakly cemented and have an open metastable structure prone to collapse when disturbed. Loess is widespread in southern and eastern England in two main areas: an eastern province that extends from Yorkshire to Kent and westwards to east Devon, and a western province in Cornwall (Murton & Ballantyne 2017). Loess can be present as substantial well-jointed deposits many metres thick, or preserved in other periglacial features such as cryogenic wedges and gulls. On British geological maps the term 'brickearth' was used to describe areas where loess was recorded.

The main geohazard associated with loess is collapse and consolidation of the low-density, high-porosity, high-void-ratio, open soil structure due to loading and/or wetting of the soil (Northmore et al. 1996; Jefferson & Rogers 2012; Murton & Giles 2016). Mitigation is by pretreatment to densify the soil prior to construction by filling the voids or collapsing the structure.

10.4 Conclusions

Periglacial features develop in environments exposed to persistent cold climate and beyond the limits of glacier ice. Such conditions rarely occur in the UK today, but were widespread during the repeated cold-climate phases of the Quaternary. Relict periglacial features are therefore a potentially significant geohazard across the UK. They occur both north and south of the limits of Quaternary glaciations, including in shallow offshore areas, and the depth of influence of ground freezing may extend to depths in excess of 100 m below ground level. Periglacial processes can therefore be expected to have impacted material across most of the UK, and should therefore be anticipated in most UK ground engineering schemes. However, there is a sufficient body of experience, literature and tools readily available for the professional geoscientist to identify the potential for these geohazards and mitigate against them as part of the normal investigation and design process.

An understanding of the geomorphology, geology and evolutionary history of a site enables identification of the potential for and nature and extent of possible relict periglacial geohazards. Correct identification of the geohazard is the critical first step, and allows recommendations for further detailed investigations, risk assessment and design of mitigation measures to be implemented. Early warning allows realistic project budgeting and scheduling.

The most significant periglacial geohazards are probably reduced-shear-strength layers (including relict shear surfaces) on very shallow slopes associated with cold-climate solifluction and active-layer detachment slide processes. In all aspects of periglacial geohazards, likely stress release and changes in groundwater during construction should be considered a priority. In this context, it is vitally important to consider the pre-engineering groundwater regime, the groundwater regime during construction and the post-engineering groundwater regime, as groundwater has a significant influence on engineering design in general and periglacial geohazards affect slope stability specifically. This process is essential for understanding the effect of the groundwater on the project and vice versa. Very rarely do engineering projects and groundwater interact without consequence.

Acknowledgements The authors would like to express thanks to Dr Dave Giles, Professor Jim Griffiths, an anonymous reviewer, Professor Roger Moore, Professor Julian Murton, Peter Gilbert and Jon Ashton for their constructive comments that enhanced the content of the manuscript. Tom Berry would also like to thank Tim Spink.

Funding This research received no specific grant from any funding agency in the public, commercial, or not-for-profit sectors.

References

BALLANTYNE, C.K. 2002. Paraglacial geomorphology. *Quaternary Science Reviews*, **21**, 1935–2017, https://doi.org/10.1016/S0277-3791(02)00005-7

BALLANTYNE, C.K. 2018. *Periglacial Geomorphology*. Wiley-Blackwell, Oxford.

BALLANTYNE, C.K. & HARRIS, C. 1994. *The Periglaciation of Great Britain*. Cambridge University Press, Cambridge.

BANKS, V.J., BRICKER, S.H., ROYSE, K.R. & COLLINS, P.E.F. 2015. Anomalous buried hollows in London: development of a hazard susceptibility map. *Quarterly Journal of Engineering Geology and Hydrogeology*, **48**, 55–70, https://doi.org/10.1144/qjegh2014-037

BARTON, N.J. 1992. *The Lost Rivers of London: A Study of Their Effects Upon London and Londoners, and the Effects of London and Londoners on Them*. Historical Publications Ltd.

BERRY, F.G. 1979. Late Quaternary scour-hollows and related features in central London. *Quarterly Journal of Engineering Geology and Hydrogeology*, **12**, 9–29, https://doi.org/10.1144/GSL.QJEG.1979.012.01.03

BLOOMFIELD, J.P. 1999. *FRACFLOW – Geological State-of-the-Art Review*. British Geological Survey, Keyworth, Nottingham.

BOOTH, S., MERRITT, J. & ROSE, J. 2015. Quaternary provinces and domains – a quantitative and qualitative description of British landscape types. *Proceedings of the Geologists' Association*, **126**, 163–187, https://doi.org/10.1016/j.pgeola.2014.11.002

BOULTON, G.S. & DOBBIE, K.E. 1993. Consolidation of sediments by glaciers: relations between sediment geotechnics, soft-bed glacier dynamics and subglacial ground-water flow. *Journal of Glaciology*, **39**, 26–44, https://doi.org/10.1017/S0022143000015690

BOULTON, G.S. & PAUL, M.A. 1976. The influence of genetic processes on some geotechnical properties of glacial tills. *Quarterly Journal of Engineering Geology and Hydrogeology*, **9**, 159–194, https://doi.org/10.1144/GSL.QJEG.1976.009.03.03

BRADSHAW, R. & INGLE SMITH, D. 1963. Permafrost structures on Sully Island, Glamorgan. *Geological Magazine*, **100**, 556–564, https://doi.org/10.1017/S0016756800059100

BRICKER, S.H., LEE, J.R., BANKS, V.J., MORIGI, A.N. & GARCIA-BAJO, M. 2012. Woodland revisited: East Anglia's buried channel network brought to life in 3D. *Geoscientist*, **22**, 14–19.

BRITISH GEOLOGICAL SURVEY 2019a. BGS National Landslide Database. http://www.bgs.ac.uk/landslideswww.bgs.ac.uk/landslides

BRITISH GEOLOGICAL SURVEY 2019b. Lexicon of named rock units. http://www.bgs.ac.uk/lexiconwww.bgs.ac.uk/lexicon

BRUNSDEN, D. 2002. Geomorphological roulette for engineers and planners: some insights into an old game. *Quarterly Journal of Engineering Geology and Hydrogeology*, **35**, 101–142, https://doi.org/10.1144/1470-92362001-40

BRUNSDEN, D. & JONES, D.K.G. 1972. The morphology of degraded landslide slopes in South West Dorset. *Journal of Engineering Geology*, **5**, 205–222, https://doi.org/10.1144/GSL.QJEG.1972.005.03.01

BRUNSDEN, D., COOMBE, K., GOUDIE, A.S. & PARKER, A.G. 1996. The structural geomorphology of the Isle of Portland, southern England. *Proceedings of the Geologists' Association*, **107**, 209–230, https://doi.org/10.1016/S0016-7878(96)80030-7

CHAMBERLAIN, E.J. & GOW, A.J. 1979. Effect of freezing and thawing on the permeability and structures of soils. *Engineering Geology*, **13**, 73–92, https://doi.org/10.1016/0013-7952(79)90022-X

CHANDLER, R.J. 1969. The effect of weathering on the shear strength properties of Keuper Marl. *Geotechnique*, **19**, 321–334, https://doi.org/10.1680/geot.1969.19.3.321

CHANDLER, R.J. 1970. The degradation of Lias clay slopes in an area of the east Midlands. *Quarterly Journal of Engineering Geology and Hydrogeology*, **2**, 161–181, https://doi.org/10.1144/GSL.QJEG.1970.002.03.01

COHEN, K.M. & GIBBARD, P.L. 2011. *Global Chronostratigraphical Correlation Table for the Last 2.7 Million Years*. Subcommission on Quaternary Stratigraphy (International Commission on Stratigraphy), Cambridge, UK.

COLLINS, P. 2013. Anomalous depressions – a view from the west. *In: Engineering Geology of Scour Features*, 22nd January 2013. Geological Society, London, Engineering Group Meeting, https://www.geolsoc.org.uk/~/media/shared/documents/specialist%20and%20regional%20groups/EngineeringGroup/Anomalous%20depressions%20%20%20a%20view%20from%20the%20west.pdf?la=en

Cox, F.C. 1985. The tunnel-valleys of Norfolk, East Anglia. *Proceedings of the Geologists' Association*, **96**, 357–369, https://doi.org/10.1016/S0016-7878(85)80024-9

Culshaw, M.G., Entwisle, D.C., Giles, D., Berry, T., Collings, A., Banks, V.J. & Donnelly, L.J. 2017. Material properties and geohazards. In: Griffiths, J.S. & Martin, C.J. (eds) *Engineering Geology and Geomorphology of Glaciated and Periglaciated Terrains. Engineering Group Working Party Report*. Geological Society of London, Engineering Geology Special Publication No. 28, 599–740, https://doi.org/10.1144/EGSP28.6

Dearman, W.R. & Fookes, P.G. 1974. Engineering geological mapping for civil engineering practice in the United Kingdom. *Quarterly Journal of Engineering Geology and Hydrogeology*, **7**, 223–256, https://doi.org/10.1144/GSL.QJEG.1974.007.03.01

de Freitas, M.H., Griffiths, J.S. et al. 2017. Engineering investigation and assessment. In: Griffiths, J.S. & Martin, C.J. (eds) *Engineering Geology and Geomorphology of Glaciated and Periglaciated Terrains. Engineering Group Working Party Report*. Geological Society of London, Engineering Geology Special Publication No. 28, 741–830.

Dines, H.G., Hollingworth, S.E., Edwards, W., Buchan, S. & Welch, F.B. 1940. The mapping of head deposits. *Geology Magazine*, **77**, 198–226, https://doi.org/10.1017/S0016756800071302

Ealey, P.J. 2012. Periglacial bedrock features of the Lizard Peninsula and surrounding area. *Geoscience in South-West England*, **13**, 52–64.

Early, K.R. & Skempton, A.W. 1972. The landslide at Walton's Wood, Staffordshire. *Quarterly Journal of Engineering Geology*, **5**, 19–41, https://doi.org/10.1144/GSL.QJEG.1972.005.01.04

Evans, D.J.A. 2017. Conceptual glacial ground models: British and Irish Case Studies. In: Griffiths, J.S. & Martin, C.J. (eds) *Engineering Geology and Geomorphology of Glaciated and Periglaciated Terrains. Engineering Group Working Party Report*. Geological Society of London, Engineering Geology Special Publication No. 28, 369–500.

Eyles, N. 1983. Glacial geology: a landsystems approach. In: Eyles, N. (ed.) *Glacial Geology: An Introduction for Engineers and Earth Scientists*. Pergamon Press, Trowbridge, 1–18.

Eyles, N. & Paul, M.A. 1983. Landforms and sediments resulting from former periglacial climates. In: Eyles, N. (ed.) *Glacial Geology: An Introduction for Engineers and Earth Scientists*. Pergamon Press Limited, Trowbridge, 111–139, https://doi.org/10.1016/B978-0-08-030263-8.50011-7

Fookes, P.G. 1997. Geology for engineers: the geological model, prediction and performance. *Quarterly Journal of Engineering Geology and Hydrogeology*, **30**, 293–424, https://doi.org/10.1144/GSL.QJEG.1997.030.P4.02

Fookes, P.G., Lee, M. & Milligan, G. 2005. *Geomorphology for Engineers*. Whittles Publishing, Caithness.

Fookes, P.G., Griffiths, J.S. & Lee, M. 2007. *Engineering Geomorphology. Theory and Practice*. Whittles Publishing, Caithness.

French, H.M. 2007. *The Periglacial Environment*, 3rd edn. John Wiley & Sons, Inc., Chichester

French, H.M. 2018. The periglacial environment. 4th edition, John Wiley & Sons Ltd, Chichester.

Gabriel, K. 2008. Soliflucted clay. *Ground Engineering*, **April**, 30–32.

Gallois, R.W. 2010. Large-scale periglacial creep folds in Jurassic mudstones on the Dorset coast, UK. *Geoscience in South-West England*, **12**, 223–232.

Gibbard, P.L. 1985. *The Pleistocene History of the Middle Thames Valley*. Cambridge University Press, Cambridge.

Giles, D.P., Griffiths, J.S., Evans, D.J.A. & Murton, J.B. 2017. Geomorphological framework: glacial and periglacial sediments, structures and landforms. In: Griffiths, J.S. & Martin, C.J. (eds) *Engineering Geology and Geomorphology of Glaciated and Periglaciated Terrains. Engineering Group Working Party Report*. Geological Society of London, Engineering Geology Special Publication No. 28, 59–368.

Griffiths, J.S. & Giles, D.P. 2017. Conclusions and illustrative case studies. In: Griffiths, J.S. & Martin, C. (eds) *Engineering Geology and Geomorphology of Glaciated and Periglaciated Terrains. Engineering Group Working Party Report*. Geological Society of London, Engineering Geology Special Publication No. 28, 891–936.

Hadlow, N.W., Lawrence, J.A. & Mortimore, R.N. 2018. Evaluation and prediction of anticipated depths of weathering (engineering rockhead) as a function of geomorphology in areas of chalk outcrop in southern England and northern France. In: Lawrence, J.A., Preene, M., Lawrence, U.L. & Buckley, R. (eds) *Engineering in Chalk: Proceedings of the Chalk 2018 Conference*. ICE Publishing, London, 711–720, https://icevirtuallibrary.com/doi/full/10.1680/eiccf.64072.711

Harris, C. 1989. Some possible Devensian ice-wedge casts in Mercia Mudstone near Cardiff, South Wales. *Quaternary Newsletter*, **58**, 11–13.

Hawkins, A.B. & Privett, K.D. 1981. A building site on cambered ground at Radstock, Avon. *Quarterly Journal of Engineering Geology and Hydrogeology*, **14**, 151–167, https://doi.org/10.1144/GSL.QJEG.1981.014.03.02

Higginbottom, I.E. & Fookes, P.G. 1970. Engineering aspects of periglacial features in Britain. *Quarterly Journal of Engineering Geology and Hydrogeology*, **3**, 85–117, https://doi.org/10.1144/GSL.QJEG.1970.003.02.02

Hutchinson, J.N. 1974. Periglacial solifluxion: an approximate mechanism for clayey soils. *Géotechnique*, **24**, 438–443, https://doi.org/10.1680/geot.1974.24.3.438

Hutchinson, J.N. 1980. Possible late Quaternary pingo remnants in central London. *Nature*, **284**, 253–255, https://doi.org/10.1038/284253a0

Hutchinson, J.N. 1989. General report: morphological and geotechnical parameters of landslides in relation to geology and hydrogeology. *Proceedings of the 5th International Symposium on Landslides*. Lausanne, 10–15 July 1988, A.A. Balkema, Rotterdam, V1, 3–35.

Hutchinson, J.N. 1991. Theme lecture: periglacial and slope processes. In: Forster, A., Culshaw, M.G., Cripps, J.C., Little, J.A. & Moon, C.F. (eds) *Quaternary Engineering Geology. Proceedings of the 25th Annual Conference of the Engineering Group of the Geological Society*, 10–14 September 1989, Heriot-Watt University, Edinburgh. Geological Society of London, 283–331, https://doi.org/10.1144/GSL.ENG.1991.007.01.27

Jefferson, I. & Rogers, C.D.F. 2012. Collapsible soils. In: Burland, J.B., Chapman, T., Skinner, H. & Brown, M. (eds) *ICE Manual of Geotechnical Engineering. Volume 1, Geotechnical Engineering Principles, Problematic Soils and Site Investigation*. Institution of Civil Engineers, London, 391–412.

Johnson, R.H. 1980. Hillslope stability and landslide hazard – a case study from Longdendale. *Proceedings of the Geologists' Association*, **91**, 315–325, https://doi.org/10.1016/S0016-7878(80)80026-5

Kemp, R.A., Whiteman, C.A. & Rose, J. 1993. Palaeoenvironmental and stratigraphic significance of the Valley Farm and Barham Soils in Eastern England. *Quaternary Science Reviews*, **12**, 833–848, https://doi.org/10.1016/0277-3791(93)90022-E

Konrad, J.-M. & Morgenstern, N.R. 1981. The segregation potential of a freezing soil. *Canadian Geotechnical Journal*, **18**, 482–491, https://doi.org/10.1139/t81-059

Lisiecki, L.E. & Raymo, M.E. 2005. A Pliocene-Pleistocene stack of 57 globally distributed benthic δ18O records. *Paleoceanography*, **20**, https://doi.org/10.1029/2004PA001071.

Lonegran, L., Maidment, S.C.R. & Collier, J.S. 2006. Pleistocene subglacial tunnel valleys in the central North Sea basin: 3-D morphology and evolution. *Journal of Quaternary Science*, **21**, 891–903.

Lord, J.A., Clayton, C.R.I. & Mortimore, R.N. 2002. *Engineering in Chalk*. CIRIA, London, UK.

Lowe, J. & Walker, M. 2015. *Reconstructing Quaternary Environments*. 3rd edn. Routledge, London, New York.

McGown, A. & Derbyshire, E. 1977. Genetic influences on the properties of tills. *Quarterly Journal of Engineering Geology and Hydrogeology*, **10**, 389–410, https://doi.org/10.1144/GSL.QJEG.1977.010.04.02

Moore, R., Turner, M.D., Palmer, M.J. & Carey, J.M. 2007. The Ventnor Undercliff: landslide model, mechanisms and causes, and the implications of climate change induced ground behaviour and risk. *In: Proceedings of the International Conference on Landslides and Climate Change*, Ventnor, Isle of Wight, 365–375.

Moore, R., Carey, J.M. & McInnes, R.G. 2010. Landslide behaviour and climate change: predictable consequences for the Ventnor Undercliff, Isle of Wight. *Quarterly Journal of Engineering Geology and Hydrogeology*, **43**, 365–375, https://doi.org/10.1144/1470-9236/08-086

Morgan, A.V. 1971. Engineering problems caused by fossil permafrost features in the English Midlands. *Quarterly Journal of Engineering Geology and Hydrogeology*, **4**, 111–114, https://doi.org/10.1144/GSL.QJEG.1971.004.02.02

Morgenstern, N.R. & Nixon, J.F. 1971. One-dimensional consolidation of thawing soils. *Canadian Geotechnical Journal*, **8**, 558–565, https://doi.org/10.1139/t71-057

Murton, J.B. 1996. Near-surface brecciation of chalk, Isle of Thanet, south-east England: a comparison with ice-rich brecciated bedrocks in Canada and Spitsbergen. *Permafrost and Periglacial Processes*, **7**, 153–164, https://doi.org/10.1002/(SICI)1099-1530(199604)7:2<153::AID-PPP215>3.0.CO;2-7

Murton, J.B. 2018. Frost weathering of chalk. *In*: Lawrence, J.A., Preene, M.A., Lawrence, U.L. & Buckley, R. (eds) *Engineering in Chalk: Proceedings of the Chalk 2018 Conference*. ICE Publishing, London, 497–502, https://icevirtuallibrary.com/doi/abs/10.1680/eiccf.64072.497

Murton, J.B. & Ballantyne, C.K. 2017. Periglacial and permafrost ground models for Great Britain. In: Griffiths, J.S. & Martin, C.J. (eds) *Engineering Geology and Geomorphology of Glaciated and Periglaciated Terrains. Engineering Group Working Party Report*. Geological Society of London, Engineering Geology Special Publication No. 28, 501–597.

Murton, J.B. & Giles, D.P. 2016. *The Quaternary Periglaciation of Kent Field Guide*. Quaternary Research Association, London.

Murton, J.B. & Kolstrup, E. 2003. Ice-wedge casts as indicators of palaeotemperatures: precise proxy or wishful thinking? *Progress in Physical Geography*, **2**, 155–170, https://doi.org/10.1191/0309133303pp365ra

Murton, J.B., Whiteman, C.A. & Allen, P. 1995. Involutions in the Middle Pleistocene (Anglian) Barham Soil, eastern England: a comparison with thermokarst involutions from arctic Canada. *Boreas*, **24**, 269–280, https://doi.org/10.1111/j.1502-3885.1995.tb00779.x

Murton, J.B., Peterson, R. & Ozouf, J.-C. 2006. Bedrock fracture by ice segregation in cold regions. *Science*, **314**, 1127–1129, https://doi.org/10.1126/science.1132127

Murton, J.B., Bowen, D.Q. et al. 2015. Middle and Late Pleistocene environmental history of the Marsworth area, south-central England. *Proceedings of the Geologists' Association*, **126**, 18–49, https://doi.org/10.1016/j.pgeola.2014.11.003

Murton, J.B., Ozouf, J. & Peterson, R. 2016. Heave, settlement and fracture of chalk during physical modelling experiments with temperature cycling above and below 0°C. *Geomorphology*, **270**, 71–87, https://doi.org/10.1016/j.geomorph.2016.07.016

Nixon, J.F. & Morgenstern, N.R. 1973. The residual stress in thawing soils. *Canadian Geotechnical Journal*, **10**, 571–580, https://doi.org/10.1139/t73-053

Northmore, K.J., Bell, F.G. & Culshaw, M.G. 1996. The engineering properties and behaviour of the brickearth of south Essex. *Quarterly Journal of Engineering Geology and Hydrogeology*, **29**, 147–161, https://10.1144/GSL.QJEGH.1996.029.P2.04

Parks, C.D. 1991. A review of the mechanisms of cambering and valley bulging. *In*: Forster, A., Culshaw, M.G., Cripps, J.C., Little, J.A. & Moon, C.F. (eds) *Quaternary Engineering Geology. Proceedings of the 25th Annual Conference of the Engineering Group of the Geological Society*, 10–14 September, 1989. Heriot-Watt University, Edinburgh. Geological Society of London, 373–380, https://doi.org/10.1144/GSL.ENG.1991.007.01.33.

Paul, M.A. 1983. The supraglacial landsystem. *In*: Eyles, N. (ed.) *Glacial Geology: An Introduction for Engineers and Earth Scientists*. Pergamon Press, Trowbridge, 71–90.

Paul, M.A., Pole, E.L. & Gostelow, T.P. 1981. *Soil Mechanics in Quaternary Science*. Quaternary Research Association, London.

Pewe, T.L. 1974. Geomorphic processes in polar deserts. In: Smiley, T.L. & Zumberge, J.H. (eds) *Polar Deserts and Modern Man*. University of Arizona Press, Tuscon, 33–52.

Privett, K.D. 2019. The lines of evidence approach to challenges faced in engineering geological practice. *Quarterly Journal of Engineering Geology and Hydrogeology*, **52**, 141–172 https://doi.org/10.1144/qjegh2018-131

Rose, J., Turner, C., Cooper, G.R. & Bryan, M.D. 1980. Channel changes in a lowland river catchment over the last 13000 years. *In*: Davidson, D.A. & Lewin, J. (eds) *Timescales in Geomorphology*. Wiley-Blackwell, Chichester, 159–175.

Self, C.A. 1986. Two gull caves from the Wiltshire/Avon border. *University of Bristol Speleological Society*, **17**, 153–174.

Self, C.A. & Boycott, A. 2000. Landslip caves of the southern Cotswolds. *University of Bristol Speleological Society*, **21**, 197–214.

Shilston, D.T., Teeuw, R.M. & West, G. 2012. Desk study, remote sensing, geographical information systems and field evaluation. *In*: Walker, M.J. (ed.) *Hot Deserts: Engineering Geology and Geomorphology*. Geological Society, London, Engineering Group Geology Working Party Report, **25**, 159–200, https://doi.org/10.1144/EGSP25.06

Skempton, A.W. & Weeks, A.G. 1976. The Quaternary history of the Lower Greensand Escarpment and Weald Clay Vale near Sevenoaks, Kent [and Discussion]. *Philosophical Transactions of the Royal Society A: Physical, Mathematical and Engineering Sciences*, **283**, 493–526, https://doi.org/10.1098/rsta.1976.0094

Skempton, A.W., Norbury, D., Petley, D.J. & Spink, T.W. 1991. Solifluction shears at Carsington, Derbyshire. *In*: Forster, A., Culshaw, M.G., Cripps, J.C., Little, J.A. & Moon, C.F. (eds)

Quaternary Engineering Geology. Proceedings of the 25th Annual Conference of the Engineering Group of the Geological Society, 10–14 September, 1989. Heriot-Watt University, Edinburgh. Geological Society of London, 381–387, https://doi.org/10.1144/GSL.ENG.1991.007.01.34

SPINK, T.W. 1980. *Mechanisms of Periglacial Disturbance*. Imperial College, London.

SPINK, T.W. 1991. Periglacial discontinuities in Eocene clays near Denham, Buckinghamshire. *In*: FORSTER, A., CULSHAW, M.G., CRIPPS, J.C., LITTLE, J.A. & MOON, C.F. (eds) *Quaternary Engineering Geology. Proceedings of the 25th Annual Conference of the Engineering Group of the Geological Society*, 10–14 September, 1989. Heriot-Watt University, Edinburgh. Geological Society of London, 389–396, https://doi.org/10.1144/GSL.ENG.1991.007.01.35

SPINK, T.W. & NORBURY, D.R. 1993. The engineering geological description of weak rocks and overconsolidated soils. *In*: CRIPPS, J.C., COULTHARD, J.M., CULSHAW, M.G., FORSTER, A., HENCHER, S.R. & MOON, C.F. (eds) *The Engineering Geology of Weak Rock: Proceedings of the 26th Annual Conference of the Engineering Group of the Geological Society*. Rotterdam, Balkema, Engineering Geology Special Publications, **8**, 289–301.

WALKER, J. 2005. Periglacial form and processes. *In*: FOOKES, P.G., LEE, E.M. & MILLIGAN, G. (eds) *Geomorphology for Engineers*. Whittles Publishing, Caithness.

WEEKS, A.G. 1969. The stability of natural slopes in south-east England as affected by periglacial activity. *Quarterly Journal of Engineering Geology*, **2**, 49–61, https://doi.org/10.1144/gsl.qjeg.1969.002.01.04

WILLIAMS, P.J. & SMITH, M.W. 1989. *The Frozen Earth. Fundamentals of Geocryology*. Cambridge University Press, Cambridge.

WILLIAMS, R.B.G. 1986. Periglacial phenomena in the South Downs. *In*: SIEVEKING, G. & HART, M.B. (eds) *The Scientific Study of Flint and Chert*. Cambridge University Press, Cambridge, 161–167.

WILLIAMS, R.B.G. 1987. Frost weathered mantles on the chalk. *In*: BOARDMAN, J. (ed.) *Processes and Landforms in Britain and Ireland*. Cambridge University Press, Cambridge, 127–133.

WILSON, P. 2005. Paraglacial rock-slope failures in Wasdale, western Lake District, England: morphology, styles and significance. *Proceedings of the Geologists' Association*, **116**, 349–361, https://doi.org/10.1016/S0016-7878(05)80052-5

WINTER, M.G., TROUGHTON, V., BAYLISS, R., GOLIGHTLY, C., SPASIC-GRIL, L., HOBBS, P.R.N. & PRIVETT, K.D. 2017. Design and construction considerations. *In*: GRIFFITHS, J.S. & MARTIN, C.J. (eds) *Engineering Geology and Geomorphology of Glaciated and Periglaciated Terrains. Engineering Group Working Party Report*. Geological Society of London, Engineering Geology Special Publication No. 28, 831–890.

WOODLAND, A.W. 1970. The buried tunnel-valleys of East Anglia. *Proceedings of the Yorkshire Geological Society*, **37**, 521–578, https://doi.org/10.1144/pygs.37.4.521

WOODWARD, H.B. 1912. *Geology of Soils and Substrata*. Arnold, London.

WORSSAM, B.C. 1981. Pleistocene deposits and superficial structures. Allington Quarry, Maidstone, Kent. *In*: NEALE, J. & FLENLEY, J. (eds) *The Quaternary in Britain*. Pergamon Press, Oxford, 20–31.

YOUNGER, P.L. 1989. Devensian periglacial influences on the development of spatially variable permeability in the Chalk of south-east England. *Quarterly Journal of Engineering Geology and Hydrogeology*, **22**, 343–354, https://doi.org/10.1144/GSL.QJEG.1989.022.04.07

ZEUNER, F.E. 1959. *The Pleistocene period: its climate, chronology, and faunal successions*. Hutchinson Scientific & Technical.

Chapter 11 Coal mining subsidence in the UK

Laurance Donnelly

International Union of Geological Sciences, Initiative on Forensic Geology,
398 Rossendale Road, Burnley, Lancashire, BB11 5HN, UK

Correspondence: geologist@hotmail.co.uk

Abstract: One of the geohazards associated with coal mining is subsidence. Coal was originally extracted where it outcropped, then mining became progressively deeper via shallow workings including bell pits, which later developed into room-and-pillar workings. By the middle of the 1900s, coal was mined in larger open pits and underground by longwall mining methods. The mining of coal can often result in the subsidence of the ground surface. Generally, there are two main types of subsidence associated with coal mining. The first is the generation of crown holes caused by the collapse of mine entries and mine roadway intersections and the consolidation of shallow voids. The second is where longwall mining encourages the roof to fail to relieve the strains on the working face and this generates a subsidence trough. The ground movement migrates upwards and outwards from the seam being mined and ultimately causes the subsidence and deformation of the ground surface. Methods are available to predict mining subsidence so that existing or proposed structures and land developments may be safeguarded. Ground investigative methods and geotechnical engineering options are also available for sites that have been or may be adversely affected by coal mining subsidence.

11.1 Introduction

Many of the major cities and conurbations owe their existence and expansion to the presence of coal and associated mineral deposits (Donnelly 2018*a*). Coal mining in the UK peaked in 1912–15, and then experienced a wave of expansion and contraction. The last deep coal mine in the UK closed in December 2015. Coal mining has left behind a legacy of mining hazards (geohazards), which if not properly managed and investigated can represent a risk to new construction and development. One of these hazards is subsidence. This chapter provides an overview of the occurrence, prediction and control of coal mining subsidence and is intended to be of interest to other engineering geologists, geotechnical engineers, civil-engineers, planners and developers, and others interested in building, construction and the development of land in the abandoned (and still active) coal mining fields of the UK (Figs 11.1 & 11.2).

11.2 Subsidence characteristics

In the context of this chapter, subsidence may be considered to be the ground movements that occur following the underground mining of coal, mainly the lowering of the ground surface. It should be noted, however, that the Coal Measures provided other minerals, which were also mined by underground methods, such as fireclay, ganister, ironstones, clays, shales, mudstones and sandstones for building purposes, etc. There may be no or only an incomplete record of the existence of such mine workings, and these can also generate subsidence. The effects of subsidence depend on several factors, such as the geology, thickness and depth of the coal seam, the mining methods and in particular the types of roof supports used, the engineering characteristics and behaviours of the strata and soils (superficial deposit) and any mitigative or engineering methods used to reduce subsidence. Coal mining subsidence can cause serious, often (but not always) dramatic and catastrophic consequences on houses, buildings, engineered structures, underground utilities and services, and agricultural land. The inability to accurately predict the effects of ground subsidence has in the past resulted in the sterilization of coal mining reserves in some urban areas. This was in part associated with the expected subsidence compensation costs for damage to land, houses, roads and structures (Anon. 1959, 1977; Walton & Cobb 1984; Peng 1986; Whittaker & Reddish 1989; Bell & Donnelly 2006).

11.3 Overview of mining methods

11.3.1 Adits, drifts (inclines) and shafts

The date when coal mining first took place in the UK is not known for certain. However, coal has probably been mined since Roman times. It was not until about the thirteenth century, that coal mining began to develop. Initially, drifts and

Fig. 11.1. Subsidence trough caused by the extraction of a shallow coal seam (Donnelly 1994; Donnelly & Melton 1995).

adits were sunk into hill sides along coal outcrops. Adits were extended as far as the natural ventilation and light allowed or until the roof failed.

Mine shafts, sunk to access coal seams that occurred below the ground surface, were originally circular or rectangular and typically less than a few metres in diameter. Many were unlined, or lined with timber. Later, mine shafts were brick or masonry lined. More modern shafts had steel or reinforced concrete linings and often reached more than 5.0 m in diameter. In 1842 it became a requirement for collieries to have at least two means of access into and egress from a mine, for safety reasons and ventilation purposes (Mines Act of 1842).

Upon the abandonment of a coal mine, shaft positions were often not recorded and the shaft was backfilled with debris or waste onto a wooden staging, below which the shaft was usually unfilled. The eventual deterioration and rotting of the timber would result in the collapse of the shaft. In the coal fields of the UK there are probably many adits, drifts and mine shafts whose locations and conditions remain unknown. It was not until 1875 that it became a statutory requirement for mine shafts to be recorded and mien plans lodged with the appropriate authority. These open or partially filled shafts present a possible subsidence hazard (Dean 1967; Bell 1988; Donnelly & McCann 2000; Donnelly 2018b; Donnelly et al. 2019) (Fig. 11.3).

11.3.2 Bell pits

Bell pits became more common from about the fourteenth century. Generally, the shafts do not exceed about 12 m deep (depending on the depth of the water table) and their diameters are around 1.5 m. Bell pits were sunk to extract coal seams at a shallow depth and followed the dip of the seam. As such, they tend to be located along the outcrop of the coal seam and may be observed as a series of collapses of small depressions, or may be recognizable from the displaced debris around the shafts. Numerous bell pits may exist in one location and it can be difficult to estimate the void space remaining (Fig. 11.4).

11.3.3 Room-and-pillar

Fig. 11.2. Fissures in the zone of tension on the flanks of a subsidence trough (Donnelly 1994; Donnelly & Melton 1995).

Room-and-pillar mining (also known as 'pillar-and-stall' or 'bord-and-pillar') become more common from the sixteenth

thickness and type. The coal was cut on a 'longwall face', loaded onto a conveyor belt and transported to the surface of the mine for processing. Hydraulic roof supports propel forward and the roof was encouraged to collapse and form an area known as the 'goaf'. This subsequently led to the collapse of the overlying strata and approximately simultaneous transmission of the ground movements to the ground surface causing (trough) subsidence.

11.3.5 Subsidence associated with partial extraction of coal

11.3.5.1 Mine shafts and bell pits
Abandoned mine shafts can suddenly and unexpectedly collapse. Predicting the timing of the location and magnitude of shaft collapses is not possible. Generally, a shaft collapse may produce a crown hole and may also be preceded by subtle ground movements (such as concentric fissuring). It should be noted that the determination of the safe stand-off zone around a mine shaft will require consideration of the geology, mining type, shaft age, geometry and condition, groundwater, and the type, thickness and strength of the superficial deposits and soils (Donnelly *et al.* 2003). The potential hazards presented by abandoned shafts include the following:

- collapse or movement of the ground, which may occur suddenly, or gradually;
- the discharge of acid or ochrous minewaters, which can result in pollution of water courses and aquifers;
- flooding of basements, building and structural foundations;
- emission of mine gases; and
- accidental entry – injury or death is almost inevitable following a fall down an open shaft, and flooded shafts compound the risk with the added danger of drowning, suffocation or poisoning by gas; collapse of the shaft is also possible, once disturbed.

11.3.5.2 Room-and-pillar workings
Shallow, abandoned pillars can degrade over time, often many years after mining ceased, which may cause subsidence. The failure mechanisms will be controlled by, for example, the depth and thickness of extraction, pillar size and geometry, whether pillar robbing took place, pillar weathering, deterioration of any timber roof supports, the presence of minewater and the rock mass discontinuities in the pillar. Densely jointed (cleat) coal may cause pillars to spall under low overburden stresses. If several pillars yield, for example by roof failure or punching into the floor of a weak mudrock or fireclay, this can lead to the development of a broad subsidence 'sag' (Culshaw *et al.* 2000). Where pillars fail by crushing or squeezing, this may generate seismicity in some abandoned coal fields (Lovell *et al.* 1996). The ingress of minewater or groundwater may also cause the accelerated weathering of pillars, by slaking and swelling or creep and strain softening (Briggs 1929; Piggott & Eynon 1978; Sizer & Gill 2000)

Fig. 11.3. An exposed mine shafts in Leeds city centre exposed during a ground investigation.

to seventeenth centuries (Fig. 11.5). The coal seam was accessed via shafts, drifts and adits before being extracted via a series of roadways, with coal pillars left to support the roof and the use of timber props. Before the completion of mining some of the pillars were commonly 'robbed', which later resulted in the failure of the roof and the migration of the void upwards to generate subsidence (Thorburn & Reid 1978). The timing of these collapses cannot be accurately predicted and may take place tens to hundreds of years after mining took place. In the UK, the width of the rooms varied from about 6 to 14 ft (1.83–4.57 m) in the nineteenth century, with an extraction ratio of 30–70% (Wardell & Wood 1965), although this varied considerably (Carter *et al.* 1981; Fig. 11.6).

11.3.4 Longwall mining

Longwall mining became the dominant type of underground coal mining from the start of the twentieth century (Bruhn *et al.* 1981; Yokel *et al.* 1982). Here, coal was extracted following the development of two parallel roadways, usually abut 200–250 m apart and several hundreds of metres long. These were joined by a third roadway at one end and the intervening coal seam was extracted by mechanized mining methods using a shearer or plough, depending on the coal

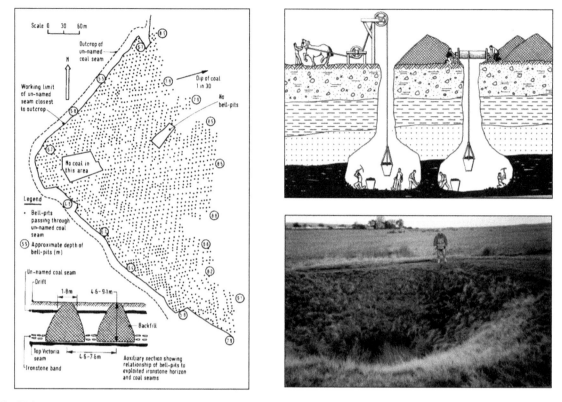

Fig. 11.4. (Left) Bell pits close to Sprouts opencast site, Northumberland (Bell 1975; Gregory 1982). (Right) Exposed bell pit, Baildon Moor, Bradford, West Yorkshire (left and top right reproduced from CIRIA 32, 1984, with kind permission from The Coal Authority).

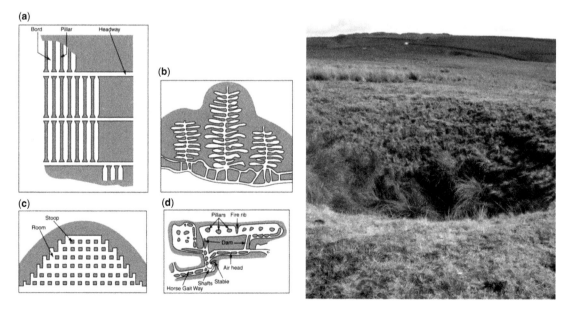

Fig. 11.5. (Left) Different arrangement of room-and-pillar workings (reproduced from CIRIA 32, 1984, with kind permission from The Coal Authority). (Right) Collapsed mine entry, Hameldon Hill, Burnley, Lancashire.

Fig. 11.6. Shallow room-and-pillar workings exposed in Tinley Park opencast site, Sheffield, South Yorkshire (Donnelly 1994; Bell & Donnelly 2006).

When a mining void develops roof rock may fall into the void and the void may migrate upwards towards the ground surface. Whether the void will reach the ground surface to generate a crown hole or to develop into a subsidence trough depends upon the geology, depth of workings, exaction ratio, seam dip, depth and thickness, mining characteristics, the width of the unsupported span, the height of the workings, the bulking characteristics of the displaced rock, the shear strength of discontinuities and groundwater conditions (Orchard 1954, 1957, 1964). The strata and roof materials spalling and falling into a void will bulk and the upward migration of the void may stop as the void becomes arrested by choking. The upward-migrating void may also be stopped by a stronger bed of rock, such as a sandstone horizon (beam) and/or the migrating void may self-stabilize if an arch roof develops (Garrard & Taylor 1988). The size and geometry of a crown hole may vary considerably and will be influenced by the thickness of the overlying strata and the type and strength of the soils or superficial deposits. Generally, crown holes tend to be roughly circular to elliptical in plan.

11.3.6 Subsidence associated with total extraction of coal

During the extraction of a horizontal coal seam by longwall mining, the forward advancement of the hydraulic roof supports causes the strata immediately above the coal seam to collapse into the void created, following the redistribution of the stress field around the mine opening. The stratal movements migrate upwards and outwards to form a subsidence trough, which is virtually contemporaneous with the longwall mining of coal (Brauner 1973*a*, *b*; Shadbolt 1978; Bruhn *et al.* 1981; Farmer & Altounyan 1981; Whittaker & Reddish 1989; Fig. 11.7).

New fractures may be induced above the goaf propagating into the overlying strata, whilst existing discontinuities may dilate and/or or shear and bedding planes may become separated. In general, this results in a zone of voids being created as the fracture zone extends approximately half the width of the face, above the level of the coal seam. Some of the factors that influence subsidence caused by the total extraction of coal (e.g. by the longwall mining method) and the characteristics of that subsidence are summarized below.

11.3.6.1 Tilt
Ground tilts develop above the limit of the area of extraction and points located along the ground surface subside downwards, with horizontal displacements propagating inwards towards the central axis of the excavation. The amount of tilt can be determined by dividing the distance between two fixed points. Tilt decreases towards the central part of the

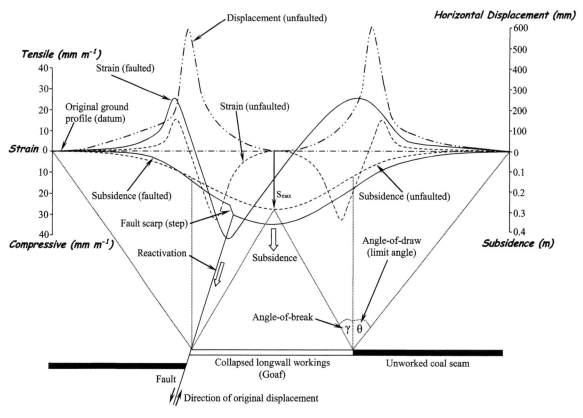

Fig. 11.7. Conceptual illustration to show the development of subsidence (modified after Donnelly 2009).

subsidence trough and the maximum tilt occurs at the point of inflection, usually at $2.75 S_{max}/d$, where S_{max} represents the maximum subsidence and d represents the seam depth.

11.3.6.2 Slope
The slope profile of a subsidence trough deepens in relation to the width–depth ratios of an area of excavation. The maximum slope occurs when the width–depth ratio of an excavation is 0.45 (Whittaker & Reddish 1989).

11.3.6.3 Curvature
Curvature can be expressed by the difference in slope between two survey stations defining a radius of curvature. For a given maximum subsidence, the curvature of the ground surface is more marked over shallow workings than over deep workings owing to the smaller distance over which the subsidence curve is spread. The horizontal strains are proportional to subsidence and inversely proportional to the depth of workings. The maximum slope of the ground in a subsidence trough is also proportional to the subsidence at the bottom of the trough and inversely proportional to the depth of the workings. It follows that maximum slope is proportional to the maximum horizontal strain (Orchard & Allen 1965).

11.3.6.4 Strain
Strain is proportional to the ratio between the differential slope and the distance between observation points.

11.3.6.5 Horizontal displacements
The convex part of a subsidence trough may experience extension (+ve), whereas the concave section develops compression (−ve).

11.3.6.6 Strain
As a result of differential horizontal displacements, tensile strain ($\varepsilon+$) and compressive strain ($\varepsilon-$) occur on both sides of the subsidence profile. The point of inflection occurs where the compression and tensile strains meet, at the point of half-maximum subsidence. Generally, the maximum compressive strain occurs above the goaf and the maximum compressive strain is above the rib sides (edges) of the longwall panel.

11.3.6.7 Width–depth ratio

The width–depth ratio is one of the most critical factors influencing subsidence. In the UK coalfields, maximum subsidence occurs at a width–depth ratio of 1.4:1. This represents the 'critical condition' (Drent 1957).

11.3.6.8 Angle-of-draw (limit angle)

The area affected by subsidence on the ground surface exceeds the area of coal seam extraction. The outer limit is known as the 'limit angle' or 'angle-of-draw'. This can vary from coalfield to coalfield by 8–45°, depending on seam depth, seam thickness and geology, in particular the existence and location of self-supporting strata, such as a thick, strong, sandstone layer above the worked coal seam. If the coal seams being mined are dipping, the resultant subsidence trough is skewed towards the downdip side of the workings. Where this occurs, the 'angle-of-draw' is greatest on the dip side and least on the rise side.

11.3.6.9 Area-of-influence

From a fixed point on the ground surface (P) and a horizontal coal seam, the area-of-influence represents a circle at the base of an imaginary cone that has its axis pointing upwards. The diameter of the area of influence is 1.4 × seam depth. Any point on the ground surface that falls outside this area does not influence this fixed point. All of the coal within this critical area has to be extracted before the fixed point can complete maximum subsidence, otherwise only partial subsidence will take place. 'Sub-critical' subsidence occurs when the width is less than 1.4 × depth, 'critical' subsidence when the width is equal to 1.4 × depth and 'supercritical' subsidence when the width is greater than 1.4 × depth'. The large width–depth ratios required to cause 90% subsidence are normally achieved for shallow workings. In deeper workings, narrow coal pillars left between neighbouring longwall panels to protect roadways reduced the subsidence effects. As such, some UK coalfields have not experienced subsidence in excess of 75–80%. However, where shallow workings are supported or packed, deeper workings can cause greater subsidence than shallow workings with the same width–depth ratio. For a horizontally mined coal seam, the largest amount of subsidence will occur above the centre of the longwall panel and reduces to zero at $c.$ 0.7 × the depth beyond the boundary of the longwall panel. However, observable ground movements may only be detected at a distance of $c.$ 0.5 times the depth, but this is variable (Orchard & Allen 1970).

11.3.6.10 Maximum subsidence

Generally, the thicker the coal seam is, the larger the subsidence. Where more than one seam is extracted or if seams are simultaneously mined, the resultant subsidence will be cumulative. The maximum possible subsidence (S_{max}) is:

$$S_{max} = Ha$$

where H is the seam thickness (extraction thickness) and a is the subsidence factor, which ranges from 0.1 to 0.9.

11.3.6.11 The subsidence factor

The subsidence factor will be influenced by the particular mining methods employed, for example, whether the area of excavation has been backfilled (packed or stowed) and the type and characteristics of the backfilling materials. Therefore, a qualitative assessment of the subsidence factor is required and this is independent of the extraction depth. The subsidence factor decreases with an increasing proportion of strong rocks in the overburden above a coal seam. Stronger sandstones and carbonates may behave as cantilever beams and take longer to respond to ground movements; non-uniform subsidence may result (Orchard 1954; Wardell & Webster 1957; Orchard & Allen 1965, 1970; Peng & Geng 1982).

11.3.6.12 Dip of seam

The dip of a coal seam influences the direction in which longwall mining takes place in that, when the dip exceeds 30°, coal working commonly takes place along the strike direction. The principal method of working such seams is by horizon mining. This involves driving several horizontal roadways through the strata in which the coal seams occur. Subsidence attributable to horizon mining layouts tends to be concentrated, with most subsidence troughs having their major axes parallel to the strike of the seam. Caving of the roof behind the longwall in dipping seams tends to displace the maximum subsidence from over the centre line of the face towards the dip side, giving rise to an asymmetrical development of ground movement. Maximum subsidence occurs at the point normal to the centre of the goaf and the angle-of-draw depends on the dip of the seam, it being least at the rise side and increasing towards the dip side. Hence, the area-of-influence at the dip side is broader, which means that the area of tensile strain is wider. The dip side experiences appreciably more lateral displacement for equal vertical subsidence on each side of the longwall panel and hence correspondingly increased tensile strain. If the seam is at a shallow depth, then the subsidence at the rise side of the face has a more marked effect at the surface than at the dip side, especially with the ground strain being concentrated. Steeply dipping coal seams (e.g. over 75°) can generate large strains on the sides of a longwall working that, in turn, can lead to discontinuities being opened at the surface, along with the development of stepping (Fig. 11.8).

11.3.6.13 Bulking

There may be a significant difference between the volume of coal mined and the subsidence of the ground caused by the bulking of the displaced and subsided strata (Garrard & Taylor 1988).

11.3.6.14 Time-dependent subsidence and residual subsidence

The time taken for subsidence at a surface point (P) to be completed is more or less inversely proportional to the rate

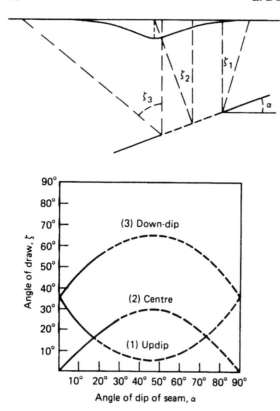

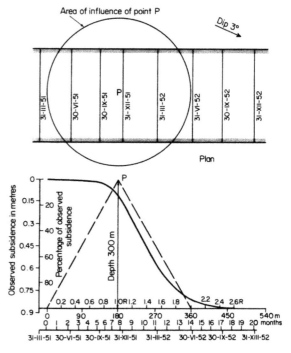

Fig. 11.8. (Left) The influence of seam dip on the angle-of-draw (after Anon. 1975, reproduced from CIRIA 32, 1984, with kind permission from The Coal Authority).

Fig. 11.9. The relationship between subsidence and time (after Wardell 1954, reproduced from CIRIA 32, 1984, with kind permission from The Coal Authority).

of forward advance of the workings. The propagation of movement to the ground surface is almost instantaneous, commencing when the strata at seam horizon begin to relax. Although the precise beginning and end of subsidence are difficult to determine, measurable subsidence occurs when the face is within a distance of $0.75d$ (d = depth to coal seam) and reaches c. 15% of the maximum when the face is directly below a given point (P) on the surface. For all practical purposes, it is complete when the face has advanced $0.8d$ beyond this given point (P). Residual subsidence then occurs, those points that are subsiding fastest experiencing the most residual subsidence. The time factor is minimal under normal conditions. The magnitude of residual subsidence is proportional to the rate of subsidence of the surface and is related to the mechanical properties of the rocks above the coal seam concerned. For instance, strong rocks produce more residual subsidence than weaker ones and some faults appear to increase residual ground movements, but this is rare and there are only few observations of this (Donnelly et al. 2008). Residual subsidence rarely exceeds 10% of total subsidence if the face is stopped within the critical width, but falls to 2–3% if the face has passed the critical width. Very occasionally values greater than 10% have been recorded. Residual subsidence normally takes about 2 years to complete. In some cases, residual subsidence has ranged from 8 to 45% of total subsidence and continued for up to 11 years after mining operations ceased. Residual subsidence also may be influenced by, for example, minewater and groundwater, the collapse of old pillars, multiple interacting workings and settlement of goaf (Donnelly 2018a,b,c; Wardell 1954; Drent 1957; Pottgens 1979; Karfakis 1993; Ferrari 1997; Mason et al. 2019; Fig. 11.9).

11.3.6.15 Multiple seams
The interaction effects of adjacent areas of extraction can give rise to erratic subsidence. If a seam has been worked previously by total extraction methods, this frequently causes the effective movements to increase by about 10% when another seam is worked subsequently above or below. This is because the first seam extracted will have disturbed the ground and induced a network of dilated rock mass discontinuities.

11.3.7 Subsidence and the engineering properties of soils and rocks

11.3.7.1 Soils/superficial deposits
Clay, silt and sandy soils tend to deform uniformly during a subsidence event. However, on slopes soils may shear along bedrock interfaces, which can result in downslope

soil creep or compressional heave/folding. During subsidence soils might consolidate or become loosened or may become drained if fractures are induced. Soils and superficial deposits tend to obscure small to moderate ground movements, but may allow movement over larger areas.

11.3.7.2 Rock

As noted above, the presence of strong rock layers in the roof above a worked coal seam, such as a sandstone or limestone, can influence the deformation of the strata and subsidence, especially when the width–depth ratios are small. Some sandstone beds can resist and reduce the amount of subsidence, often resulting in bed separation. However, if the extractions become sufficiently large to collapse then 'normal' subsidence may occur. Large collapses into the goaf may also induce seismicity caused by the sudden collapse of a sandstone horizon.

In the UK, geological strata that commonly overlie the Coal Measures, such as the Sherwood Sandstone (Triassic) and the Magnesian Limestone (Permian), have a tendency to produce less subsidence than predicted for width–depth ratios exceeding 1.0. This is due to the two formations behaving as block-jointed rock, whereby the individual blocks did not return exactly to their former positions after compressional ground movement had ceased. The rock type did not appear to have a marked influence when the width–depth ratio was less than 1.0. These well-jointed rock masses also had a tendency to develop fissures owing to the concentration of tensile strains on vertical and subvertical joints, causing their dilation. Often, fissures may be bridged by semiconsolidation surface soils or compacted fill and only revealed when the soil bridge (arch) collapses many years after they were generated. Subsidence associated with longwall mining can also significantly change the groundwater regime and hydraulic properties of aquifers. Induced fractures in particular influence aquifer recharge, well yield and the potential for opening up new contaminant pathways.

11.3.8 Subsidence prediction

Subsidence effects associated with shallow, abandoned and often complex patterns of mine workings with numerous mine entries cannot be accurately predicted. However, subsidence associated with the more geometrically simple longwall panel mining technique can be better predicted with a relatively higher degree of accuracy and precision. The three-dimensional predictions of ground movements are more difficult to predict than two-dimensional ground movements. There are three categories of subsidence prediction, namely empirical methods, analytical or theoretical methods and semi-empirical methods. First, empirical methods attempt to fit subsidence functions to field measurements (from surveying) and observations. Secondly, analytical or theoretical methods of prediction derive subsidence functions from elastic theory and rock mechanics, and are based entirely on theory. Thirdly, semi-empirical methods develop subsidence functions based on theory but that are related to field data

by the use of constants and correlation coefficients (Salamon 1964; Stacey 1972; Anon. 1975; Marr 1975; Berry 1978; Burton 1978; Burton 1981, 1985; Hood *et al.* 1983; Sutherland & Munson 1984; Ren *et al.* 1987; Coulthard & Dutton 1988; Yao & Reddish 1994; Donnelly *et al.* 2001).

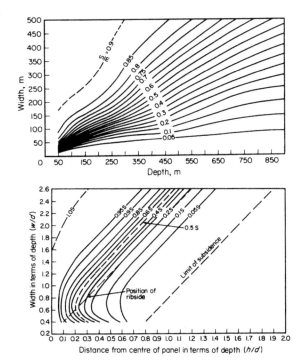

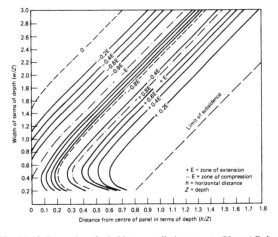

Fig. 11.10. Examples of subsidence prediction curves. (Upper) Relationship between subsidence and depth to width, where 'S' is the maximum subsidence and 'm' is the seam thickness. (Middle) Prediction of subsidence profiles. (Lower) Prediction of strain profiles (after Anon. 1975, reproduced from CIRIA 32, 1984, with kind permission from The Coal Authority).

11.3.8.1 Empirical methods

Empirical methods of subsidence prediction fit subsidence functions to observed ground subsidence measurements. This technique was developed by the former National Coal Board (then the British Coal Corporation, now The Coal Authority) for selected UK coalfields (Anon. 1975). A limitation of this approach is that it does not consider all UK coalfields, where the geology and topography significantly differs. In some cases, the use of empirical methods can predict subsidence in the UK within a 10% accuracy, although this is not always the case (Figs 11.10 & 11.11).

11.3.8.2 Analytical or theoretical

Analytical or theoretical methods of subsidence prediction are based on the premise that the behaviour and deformation of uniform and homogenous strata and ground are consistent with the constitutive equations of continuum mechanics and elastic ground conditions. These approaches allow quantitative analysis of subsidence. As such, finite element modelling has been used and this can also include non-homogeneous media, non-linear material behaviour and complicated mine geometries.

11.3.8.3 Semi-empirical methods

Generally, the majority of mining subsidence prediction methods employed are semi-empirical as they fit observational field data from recorded cases of mining subsidence and provide a good correlation between the 'predicted' and 'observed'. The main methods of semi-empirical prediction are 'profile functions' and the 'influence functions' methods. The profile function method basically consists of deriving a function that describes a subsidence trough. The equation produced is normally for one-half of the subsidence profile and is expressed in terms of maximum subsidence and the location of the points of the profile. In supercritical extraction the central position of the curve is S_{max} and changes to zero subsidence at the edges of the critical area. For sub-critical extraction the profile is determined from the critical profile produced from an empirical/mathematical relationship (Fig. 11.12).

11.3.8.4 Void migration

There are a number of methods used to predict the collapse and migration of voids, including the following:

- clamped beam theory, considers the tensile strength of the immediate roof rocks;

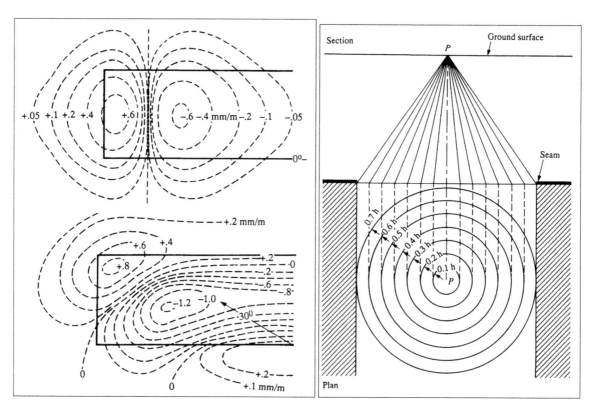

Fig. 11.11. Examples of subsidence prediction. (Left) Strain contours for the same panel but a different layout (after Burton 1978). (Right) Zone/circle method of subsidence prediction (after Marr 1975, reproduced from CIRIA 32, 1984, with kind permission from The Coal Authority).

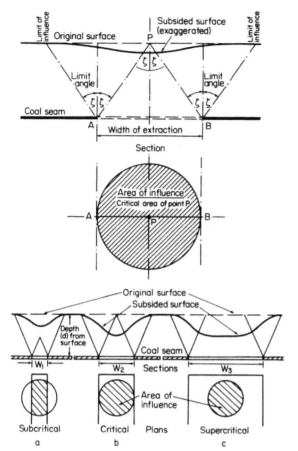

Fig. 11.12. Subcritical, critical and supracritical subsidence with varying width–depth ratio (after Anon. 1975, reproduced from CIRIA 32, 1984, with kind permission from The Coal Authority).

- bulking equations, considers the maximum height of collapse before a void is choked;
- arching theory, considers the estimated height at which a collapse will occur before a stable arch is generated; and
- coefficients based on experience and field observations, which act as multipliers of seam thickness or span width.

Migrating voids may adopt a number of geometrical forms including conical, wedge and rectangular collapses and many combinations thereof. Different expressions have been devised to calculate void migration and often require modelling from mine abandonment plans or an estimation of the pillar and void geometries. (Tincelin 1956; Wardell & Eynon 1968; Price *et al.* 1969; Piggott & Eynon 1978; Goodman *et al.* 1980; Carter 1985; Garrard & Taylor 1988; Palchik 2000) (Figs 11.13 & 11.14). With regards to void migration:

- For a given width of mine opening (void) the height of collapse or migration is determined by the height of the mine opening and the bulking factor of the overlying strata.

- Void migration is directly proportional to the thickness of seam mined, assuming that the total thickness is worked (but note this may not always be the case) and inversely proportional to the change in volume of the collapsed material.
- The height of collapse in pillared workings in coal is frequently proportional to the width of the excavation and the larger the span is, the more likely it is that collapse will occur.
- The maximum height of migration in exceptional cases might extend to 10 times the height of the original void; however, it is generally 3–5 times the void height (but will depend on the geology).
- The depth of cover should not include superficial deposits or made-ground since low bulking factors are characteristic of these materials.
- Weak, unlithified superficial deposits and soils may flow into voids that have reached the rockhead, thereby forming features that may vary from a gentle dishing of the surface to inverted cone-like depressions of large diameter.
- In a sequence of alternating rock types, if a strong rock layer is to span an opening, then its thickness should be equal to twice the span width in order to allow for natural arching to develop.
- A thick bed of strong sandstone usually will arrest a void, especially if it is located some distance above the immediate roof of the working.
- Small voids may be bridged until the span increases through spalling/slabbing or if a more competent bed is encountered.
- Chimney-type collapses can propagate larger distances towards the ground surface in massive strata in which the natural joints diverge downwards.
- Exceptionally, void migrations in excess of 20 times the worked height of a coal seam have been recorded (but are thought to be rare).
- In open mine workings with hydraulic minewater flow paths, self-choking may not be fulfilled, especially in dipping seams, if they are affected by moving groundwater that can redistribute the fallen material.
- The redistribution of collapsed material can lead to the formation of supervoids.
- The migration of a void from a coal seam into a worked seam directly above can lead to pillar collapse in that seam with the formation of voids that are larger than the original stalls. Migration of super voids up to rockhead may generate larger-scale subsidence at ground level. Under such circumstances simple analysis according to bulking factors proves inadequate.

11.4 Managing subsidence risks

The development of land and new construction projects taking place in the UK coalfields must first properly assess the likely subsidence risks (Donnelly 2018c). In the UK, over 2

Fig. 11.13. Void migration above a coal seam and shallow coal seam (Bell & Donnelly 2006, reproduced with kind permission from The Coal Authority).

million buildings (about 8%) are affected by shallow coal mine workings and 130 000 are recorded to be located within 20 metres of one or more mine shafts (The Coal Authority). However, quite surprisingly, only about 15 new mine shaft collapses are reported each year (Sharpe 2010). The responsibility for assessing the suitability of a site for any purpose rests with the developer and/or landowner. The developer will need to consult The Coal Authority (the successor of the British Coal Corporation, formerly the National Coal Board), which was set up in 1994 to manage the legacies and liabilities arising from past coal mining (Healy & Head 1984; Department of the Environment 1990; South African Institution of Civil Engineers 1992; HA 1997; Forster & Freeborough 2006; Pennell & Banton 2008, Parry & Chiverrell 2019). When required, specialist engineering geological advice should be sought to properly assess subsidence and associated mining hazards, liabilities and risks.

11.4.1 Desk study

The commissioning of a Phase 1 desk study allows the Coal Measures geology, past mining history and any current or predicted subsidence to be properly assessed. This information will be required before a meaningful appraisal of subsidence can be made. Numerous sources of data and information will be required to be identified, collated and analysed. Typical data sources include, but should not necessarily be limited to, the following:

- the Coal Authority's mine abandonment plans and referral maps;
- CIRIA abandoned mineworkings manual (Parry & Chiverrell 2019);
- Law Society & Coal Authority Directory & Guidance (2006);
- geological maps, memoirs, technical and scientific reports published by the British Geological Survey;
- British Geological Survey; ground stability report and/or 'Geosure' database and/or 'Georeports' service;
- published and peer-reviewed scientific literature, including journals, conference proceeding, industry magazines, best practise guidance notes, aid memoirs and working party reports;
- past and current topographic maps produced by the Ordnance Survey;
- publications of the Engineering Group of the Geological Society of London;

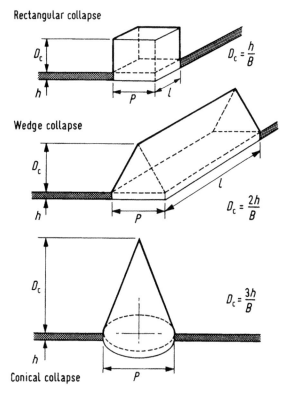
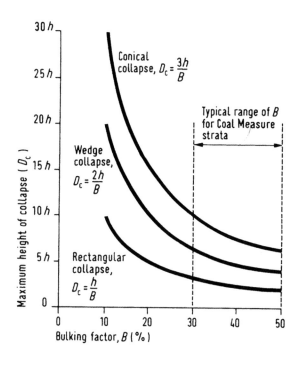

(a) Diagram showing notation relating to maximum height of collapse and geometry

(b) Postulated variation in maximum height of collapse for different modes of failure and bulking factors

Fig. 11.14. Void migration types and estimates (after Piggott & Eynon 1978, reproduced from CIRIA 32, 1984, with kind permission from The Coal Authority).

- Building Research Establishment Environmental Code of Practice (BRE 2006);
- local interest groups, such as mining clubs and mine explorers;
- public records such as libraries and museums;
- Department of the Environment land instability reports;
- natural mining cavities database (maintained under licence from DEFRA, by Peter Brett Associates);
- consultants and contractor archives;
- university departments;
- air photo archives;
- LiDAR imagery; and
- remote sensing imagery.

11.4.2 Reconnaissance (walk-over) survey

A reconnaissance survey involves a walk-over visit of the site to find any evidence of former mining and subsidence and to confirm any anecdotal and documentary evidence for past mining. This usually takes place as part of the initial 'desk study' and should include an engineering geologist with a 'trained eye' and other specialist as might be appropriate with experience in observing hazards associated with subsidence. Subtle variations in the topography may be observed together with evidence of past land use. 'Walk overs' are best carried out in low-light winter conditions where the foliage is minimal, although his is not always practical.

11.4.3 Ground investigations

Abandoned mine workings occur beneath many urban and rural areas in the UK. On a site due to be developed it is essential that, if abandoned mine workings are suspected, they are accurately located and suitable remediation measures taken to make them safe, as determined form the desk study (above). A properly procured, supervised and interpreted site investigation should be undertaken to fully characterize the ground conditions, abandoned mine entries and the location of shallow (less than 30 mbgl (metres below ground level))

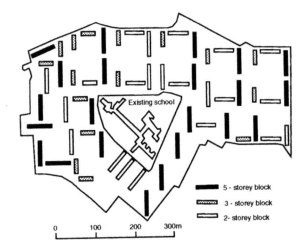

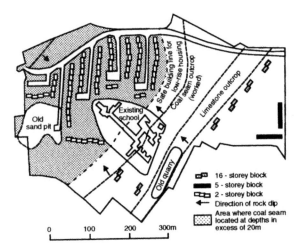

Fig. 11.15. Avoidance of coal mining hazards and the redesign of a new development (Price *et al.* 1969, reproduced from CIRIA 32, 1984, with kind permission from The Coal Authority).

voids associated with past subsidence (Taylor 1968; Bell 1975, 1986).

If coal of potentially workable thickness occurs at shallow depth (usually taken to be within about 30–50 m of the rockhead, and say *c.* 0.3 m or thicker), it should be assumed that it has been worked at some time, although this may not have been in a systematic way. Even if there are no mine plans or records of past workings, the ground investigation should be planned on the assumption that mining has taken place until it is possible to prove otherwise. In particular, the location of subsurface voids owing to mineral extraction is of prime importance in this context. In other words, an attempt should be made to determine the number and depth of mined horizons, the extraction ratio, the geometry of the layout and the condition of any old pillar-and-stall workings. The stratigraphic sequence and type of roof rocks may provide some indication as to whether void migration has taken place, and if so, its possible extent. Of particular importance is the condition of the old workings; careful note should be taken of whether they are open and air-filled, waste-filled, flooded, partially collapsed or collapsed, and the degree of fracturing, joint dilation and bed separation in the roof rocks should be recorded, if possible. This helps to provide an assessment of past and future collapse and subsidence. For active subsidence a ground surveying (such as precise levelling or GPS total station surveying) and monitoring programme might be recommended (Dean 1967; Price 1971; Littlejohn 1979; Gregory 1982; McCann *et al.* 1982, 1987; Statham *et al.* 1987; Cripps *et al.* 1988; Anon. 1988, 1999*a*, b; Culshaw *et al.* 2000, Merad *et al.* 2004; BRE 2006; The Law Society & Coal Authority 2006; AGS 2006, AGGS 2009). Ground investigative techniques may include one or a combination of the following, and should generally proceed from the non-invasive to the invasive and form the macro-scale to the micro-scale:

- remote sensing (Earth Observation), thermal imagery and LiDAR (Anon. 1976; Gunn *et al.* 2004; Donnelly *et al.* 2008);
- airborne imagery and high-resolution aerial photos (Mooijman *et al.* 1998) (possibly carried out using inexpensive drones);
- geomorphological observation;
- engineering geological mapping, hazard zonation and risk assessments;
- geophysical surveys;
- trial pitting;
- drilling; and
- monitoring.

11.5 Mitigation and remediation

Sites that contain shallow mine workings and voids can be managed in a variety of ways to manage the subsidence risks. The most suitable method employed should be a pragmatic balance between costs, and what may be achievable in the time frames, budgets and resources available. The amount of effort and money employed is usually in context with the intended land use or development. For example, the construction of a tarmac car park area may be treated differently from that of a new housing estate or school, where the risks for the latter are significantly greater. One of the most difficult assessments to make is related to the possible effects of progressive deterioration of the workings and associated potential subsidence risk. The placement of any new building structure must be carefully considered to ensure that the ground is not adversely affected by the additional load and/or that the ground is sufficiently stabilized that it will not suffer distress during the anticipated life of the structure.

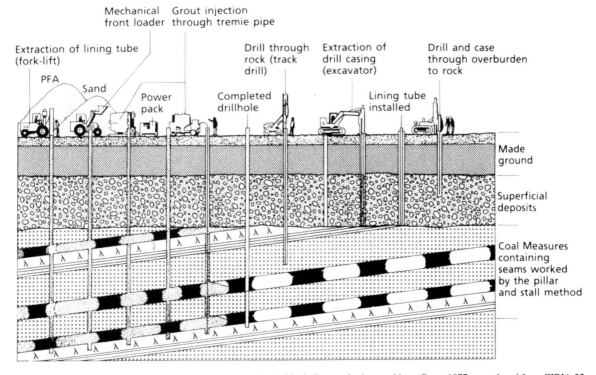

Fig. 11.16. Illustration to show the grouting of voids associated with shallow coal mine workings (Patey 1977, reproduced from CIRIA 32, 1984, with kind permission from The Coal Authority).

Furthermore, induced ground vibrations or loads from heavy plant and changes to the groundwater could also contribute to the failure of mine entries or shallow workings (Geddes 1984; Bell 1987; Bell & Fox 1988). Some typical mitigative measures include the following:

- Avoidance – a cost-effective strategy is the avoidance of mining hazards. For instance, the establishment of a suitable standoff zone around a mine shaft or the delineation of a zone of influence downdip from an outcropping coal seam, where room-and-pillar workings could exist. An

Fig. 11.17. (Left) Substantial cap across a mine shaft (Bell & Donnelly 2006). (Right). Construction of a reinforced concrete cap over a mine shaft in Lancashire, UK.

avoidance approach will almost certainly require a good understanding of the positions and conditions of the shallow workings and mine entries, their conditions and the geology so that areas of possible subsidence may be delineated (Fig. 11.15).
- Precautionary and preventative methods were used in room-and-pillar longwall mining whereby panels were designed with a layout avoiding densely built-up areas and sensitive or historical structures. This involved the establishment of a support pillar beneath the sites.
- If high ground strains are expected a building may be designed to prevent subsidence, for example by the use of one or a combination of the following:
    o rigid foundations and rafts;
    o reinforced earth;
    o geogrids and geowebbing;
    o bearing and expansion joints on bridges;
    o flexible joints on pipelines;
    o the design of trenches to accommodate the predicted ground movements associated with subsidence;
    o flexible structures (such as the CLASP, 'Consortium of Local Authorities Special Programme', system);
    o piled foundations (although this may not always be advisable);
    o introduction of spaces into buildings to accommodate ground movements;
    o excavations and trenches, backfilled with a compressible material;
    o the use of strapping and bolts on some structures.
- Narrow shortwall panels, approximately 40 m wide, with 50 m pillars between each panel, were used to reduce subsidence in urban areas in the UK Midlands.
- A reduction in the width–depth extraction ratio may help to reduce subsidence.
- The goaf can be packed (stowing) with mining waste or by pneumatic stowing.
- If mine workings are shallow they could be backfilled with bulk materials, although this is not common for workings deeper than about 7 mbgl.
- Shallow pillars can also be excavated and sold to offset the development costs (note that special permissions will be required from the Coal Authority).
- Grouting involves the backfilling of old mine workings, usually hydraulically filled with sand, crushed rock, pulverized fuel ash, fluidized-bed combustion ash or coarse colliery discard, or pneumatically with some suitable material. This is a common option to consolidate voids and shallow mine workings to prevent subsidence (Gray & Meyers 1970; Lloyd et al. 1995) (Fig. 11.16).
- Mine entries can be treated and capped (Anon 1982; Fig. 11.17).

11.6 Summary

Both recorded and unrecorded coal mine workings exist in the UK in urban and rural settings. Even where past coal mines have been mapped, their exact locations, extent and conditions cannot always be known with confidence and relied upon.

The methods of underground coal mining has varied since mining began. In general, older coal mines tended to extract coal via the driving of adits, bell-pits and room-and-pillar workings, whereas more 'modern' mines employed the longwall mining method.

Coal mining can lead to the subsidence of the ground surface, causing damage or disruption to land, homes, engineered structures, transport networks and utilities. Although underground coal mining has now ceased in the UK, subsidence can still potentially occur.

Subsidence associated with longwall mining generates a subsidence trough, which is generally completed within a few years of mining taking place, although residual ground movements can take longer to be completed in some cases. However, subsidence associated with the collapse of mine entries (shafts and adits), bell pits or shallow, abandoned, room-and-pillar workings may occur many years after mining ceased.

Subsidence induced by the total extraction (longwall) mining may be predicted using empirical, analytical or semi-empirical methods. In contrast, it is more difficult and in many cases not possible to predict the likelihood, timing, duration and magnitude of subsidence associated with the past partial extraction of coal.

Those developing land and undertaking new construction in the UK must consider if the sites are located on coal fields. If so, the hazards and associated subsidence risks must be properly assessed before the project goes ahead. This may be conveniently and cost-effectively undertaken as part of a desk study, followed if necessary by a ground investigation as part of a coal mining and subsidence risk assessment.

References

AGGS (ASSOCIATION OF GEOTECHNICAL AND GEOENVIRONMENTAL SPECIALISTS) 2009. Loss Prevention Alert No. 40. Drilling into Coal Authority Assets: a Review of the Coal Authority's Terms and Conditions for Entering or Disturbing Coal Authority Mining Interests. Retrieved from http://www.ags.org.uk/site/aboutlosprevention/lp40.cfm

AGS 2006. Guidelines for Good Practice in Site Investigation Issue 2, March 2006.

ANON. 1959. *Report on Mining Subsidence*. Institution of Civil Engineers, London.

ANON. 1975. *Subsidence Engineers Handbook*. National Coal Board, London.

ANON. 1976. *Reclamation of Derelict Land: Procedure for Locating Abandoned Mine Shafts*. Department of the Environment, London.

ANON. 1977. *Ground Subsidence*. Institution of Civil Engineers, London.

ANON. 1982. *Treatment of Disused Mine Shafts and Adits*. National Coal Board, London.

ANON. 1988. Engineering geophysics: report by the Geological Society Engineering Group Working Party. *Quarterly Journal*

Engineering Geology, **21**, 207–271, https://doi.org/10.1144/GSL.QJEG.1988.021.03.01

ANON. 1999a. *Code of Practice for Site Investigations*. BS5930. British Standards Institution, London.

ANON. 1999b. *Standard Guide for Selecting Surface Geophysical Methods. Designation D-6429*. American Society for Testing and Materials, Philadelphia.

BELL, F.G. 1975. *Site Investigation in Areas of Mining Subsidence*. Newnes Butterworth.

BELL, F.G. 1986. Location of abandoned workings in coal seams. *Bulletin International Association of Engineering Geology*, (**30**) 123–132, https://doi.org/10.1007/BF02594714

BELL, F.G. 1987. The influence of subsidence due to present day coal mining on surface development. *In*: CULSHAW, M.G., BELL, F.G., CRIPPS, J.C. & O'HARA, M. (eds) *Planning and Engineering Geology*. Engineering Geology Special Publication **4**, The Geological Society, London, 359–368.

BELL, F.G. 1988. The history and techniques of coal mining and the associated effects and influence on construction. *Bulletin Association Engineering Geologists*, **24**, 471–504.

BELL, F.G. & DONNELLY, L.J. 2006. *Mining and its Impact on the Environment*. Taylor/Francis (Spon).

BELL, F.G. & FOX, R.M. 1988. Ground treatment and foundations above discontinuous rock masses affected by mining subsidence. *Mining Engineering*, **148**, 278–283.

BERRY, D.S. 1978. Progress in the analysis of ground movements due to mining. *In*: GEDDES, J.D. (ed.) *Proceedings of the First International Conference on Large Ground Movements and Structures*, Cardiff. Pentech Press, London, 781–811.

BRAUNER, G. 1973a. *Subsidence due to Underground Mining. Part I: Theory and Practice in Predicting Surface Deformation*. Bureau of Mines, Department of the Interior, United States Government Printing Office, Washington, DC.

BRAUNER, G. 1973b. *Subsidence due to Underground Mining. Part II: Ground Movements and Mining Damage*. Bureau of Mines, Department of the Interior, United States Government Printing Office, Washington, DC.

BRE (BUILDING RESEARCH ESTABLISHMENT) 2006. *Stabilising mine workings with pfa grouts. Environmental Code of Practice*. Report **488**.

BRIGGS, H. 1929. *Mining Subsidence*. Arnold, London.

BRUHN, R.W., MAGNUSON, M.O. & GRAY, R.E. 1981. Subsidence over abandoned mines in the Pittsburg Coalbed. *In*: GEDDES, J.D. (ed.) *Proceedings of the Second International Conference on Ground Movements and Structures*, Cardiff. Pentech Press, London, 142–156.

BURTON, D.A. 1978. A three dimensional system for the prediction of surface movements due to mining. *In*: GEDDES, J.D. (ed.) *Proceedings of the First International Conference on Large Ground Movements and Structures*, Cardiff. Pentech Press, London, 209–228.

BURTON, D.A. 1981. The introduction of mathematical models for the purpose of predicting surface movements due to mining. *In*: GEDDES, J.D. (ed.) *Proceedings of the Second International Conference on Ground Movements and Structures*, Cardiff. Pentech Press, London, 50–64.

BURTON, D.A. 1985. Program in BASIC for the analysis and prediction of ground movement above longwall panels. *In*: GEDDES, J.D. (ed.) *Proceedings of the Third International Conference on Ground Movements and Structures*, Cardiff. Pentech Press, London, 338–353.

CARTER, P.G. 1985. Case histories which break the rules *In*: FORDE, M.C., TOPPING, B.H.V. & WHITTINGTON, H.W. (eds) *Mineworkings '84. Proceedings of the International Conference on Construction in Areas of Abandoned Mineworkings*, Edinburgh. Engineering Technics Press, Edinburgh, 20–29.

CARTER, P., JARMAN, D. & SNEDDON, M. 1981. Mining subsidence in Bathgate, a town study. *In*: GEDDES, J.D. (ed.) *Proceedings of the Second International Conference on Ground Movements and Structures*. Pentech Press, London, 101–124.

COULTHARD, M.A. & DUTTON, A.J. 1988. Numerical modelling of subsidence induced by underground coal mining. *of the Proceedings Twenty Ninth United States Rock Mechanics Symposium, Key Questions in Rock Mechanics*, Rolla, MO, 529–536.

CRIPPS, J.C., McCANN, D.M., CULSHAW, M.G. & BELL, F.G. 1988. The use of geophysical methods as an aid to the detection of abandoned shallow mine workings. In: Minescape '88. *Proceedings of Symposium on Mineral Extraction, Utilisation and Surface Environment*, Harrogate. Institution Mining Engineers, Doncaster, 281–289.

CULSHAW, M.G., McCANN, D.M. & DONNELLY, L.J. 2000. Impacts of abandoned mineworkings on aspects of urban development. *Transactions of the Institution of Mining and Metallurgy*, **109**, A131–A130.

DEAN, J.W. 1967. Old mine shafts and their hazard. *Mining Engineer*, **127**, 368–377.

DEPARTMENT OF THE ENVIRONMENT 1990. Planning Policy Guidance 14: Development on Unstable Land. Including Annex 2: Subsidence and Planning 2002. Retrieved from http://www.communities.gov.uk/

DONNELLY, L.J. 1994. *Predicting the reactivation of geological faults and rock mass discontinuities during mineral exploitation, mining subsidence and geotechnical engineering*. PhD Thesis, University of Nottingham, Department of Mining Engineering.

DONNELLY, L.J. 2009. A review of international cases of fault reactivation during mining subsidence and fluid abstraction. *Quarterly Journal of Engineering Geology & Hydrogeology*, **42**, 73–94, https://doi.org/10.1144/1470-9236/07-017

DONNELLY, L.J. 2018a. Coal. *In*: BOBROWSKY, P.T. & MARKER, B. (eds). *Encyclopaedia of Engineering Geology*. Springer, Berlin, 148–149.

DONNELLY, L.J. 2018b. Mining. *In*: BOBROWSKY, P.T. & MARKER, B. (eds). *Encyclopaedia of Engineering Geology*. Springer, Berlin, 629–649.

DONNELLY, L.J. 2018c. Mining hazards. *In*: BOBROWSKY, P.T. & MARKER, B. (eds). *Encyclopaedia of Engineering Geology*. Springer, Berlin, 649–655.

DONNELLY, L.J. & MELTON, N.D. 1995. Compression ridges in the subsidence trough. *Geotechnique, Journal of the Institute of Civil Engineers*, **45**, 555–560.

DONNELLY, L.J. & McCANN, D.M. 2000. The location of abandoned mine workings using thermal techniques. *Engineering Geology*, **57**, 39–52, https://doi.org/10.1016/S0013-7952(99)00146-5

DONNELLY, L.J., DE LA CRUZ, H., ASMAR, I., ZAPATA, O. & PEREZ, J.D. 2001. The monitoring and prediction of mining subsidence in the Amaga, Angelopolis, Venecia and Bolombolo Regions, Antioquia, Columbia. *Engineering Geology*, **59**, 103–114, https://doi.org/10.1016/S0013-7952(00)00068-5

DONNELLY, L.J., McCANN, D.M. & CULSHAW, M.G. 2003. Abandoned mineshafts and the problem of defining a reasonable 'search' distance for conveyance purposes. *In*: FORDE, M.C. (ed.) *Proceedings of the 10th International Conference on Structural Faults and Repair. Commonwealth Institute*, 1st to 3rd July 2003, London. Engineering Technics Press, Edinburgh. CD-ROM publication only.

DONNELLY, L.J., CULSHAW, M.G. & BELL, F.G. 2008. Longwall mining-induced fault reactivation and delayed subsidence ground movement in British Coalfields. Subsidence-Collapse, Symposium-in-Print, *Quarterly Journal of Engineering Geology and Hydrogeology*, **41**, 301–314. https://doi.org/10.1144/1470-9236/07-215

DONNELLY, L.J., PARRY, D.N. & DENNEHY, J.P. 2019. Review of mining methods. *In*: PARRY, D. & CHIVERRELL, C. (eds) *Abandoned Mine Workings Manual, C758D*. CIRIA, London, UK, 99–116.

DRENT, S. 1957. Time curves and thickness of overlying strata. *Colliery Engineer*, **34**, 271–278.

FARMER, I.W. & ALTOUNYAN, P.F.R. 1981. The mechanics of ground deformation above a caving longwall face. *In*: GEDDES, J.D. (ed.) *Proceedings of the Second International Conference on Ground Movements and Structures*, Cardiff. Pentech Press, London, 75–91.

FERRARI, C.R. 1997. Residual coal mining subsidence – some facts. *Mining Technology*, **79**, 177–183.

FORSTER, A. & FREEBOROUGH, K. 2006. *A guide to the communication of geohazards information to the public. Urban Geoscience and Geohazards Programme British Geological Survey, Internal Report*, **IR/06/009**.

GARRARD, G.E.G. & TAYLOR, R.K. 1988. Collapse mechanisms of shallow coal mine workings from field measurements. *In*: BELL, F.G., CULSHAW, M.G., CRIPPS, J.C. & LOVELL, M.A. (eds) *Engineering Geology of Underground Movements, Engineering Geology Special Publication No. 5*. The Geological Society, London, 181–192.

GEDDES, J.D. 1984. Structural design and ground movements. *In*: ATTEWELL, P.B. & TAYLOR, R.K. (eds) *Ground Movements and their Effects on Structures*. Surrey University Press, 243–267.

GOODMAN, R.E., KORBAY, S. & BUCHIGNANI, A. 1980. Evaluation of collapse potential over abandoned room and pillar mines. *Bulletin of the Association Engineering Geologists*, **17**, 27–37.

GRAY, R.E. & MEYERS, J.F. 1970. Mine subsidence support methods in the Pittsburg area. Proceedings American Society Civil Engineers. *Journal of the Soil Mechanics and Foundations Division*, **96**, 1267–1287.

GREGORY, O. 1982. Defining the problem of disused coal mine shafts. *Chartered Land Surveyor/Chartered Mine Surveyor*, **4**, 4–15.

GUNN, D.A., AGER, G.J. ET AL. 2004. *The development of thermal imaging techniques to detect mineshafts. British Geological Survey, Internal Report*, **IR/01/24**. Keyworth, Nottinghamshire.

HA (HIGHWAYS AGENCY) 1997. *Manual of Contract Documents for Highway Works, (MCHW) Volume 5 Section 3 Part 2 SA9/97 Ground Investigation Procedure, Site Investigation procedure on land in which the existence of abandoned mines and shafts is known or suspected*. Amendment 1 May 2005.

HEALY, P.R. & HEAD, J.M. 1984. *Construction over Abandoned Mine Workings*. Construction Industry Research and Information Association (CIRIA), London, Special Publication **32** (revision pending, CIRIA RP940, Abandoned Mine Workings Manual).

HOOD, M., EWY, R.T. & RIDDLE, R.L. 1983. Empirical methods of subsidence prediction. A case study from Illinois. *International Journal of Rock Mechanics and Mining Science and Geomechanical Abstracts*, **20**, 153–170, https://doi.org/10.1016/0148-9062(83)90940-3

KARFAKIS, M.G. 1993. Residual subsidence over abandoned coal mines. *In*: HOEK, E. (ed.) *Comprehensive Rock Engineering, Volume 5, Surface and Underground Case Histories*. Pergamon Press, Oxford, 451–476.

LITTLEJOHN, G.S. 1979, Consolidation of old coal workings. *Ground Engineering*, **12**, 1–21.

LLOYD, B.N., CRIPPS, J.C. & BELL, F.G. 1995. The estimation of grout take for small scale developments in areas of shallow abandoned coal mining: some examples from the East Pennine Coalfield. *In*: CULSHAW, M.G., CRIPPS, J.C. & WALTHALL, S. (eds) *Engineering Geology and Construction, Engineering Geology Special Publications*, **10**. Geological Society, London, 135–141.

LOVELL, J.H., FORD, G.D., HENNI, P.H.O., BAKER, C., SIMPSON, I. & PETTITT, W. 1996. *Recent Seismicity in the Stoke-on-Trent Area, Staffordshire. British Geological Survey (NERC) Technical Report*, **WL/96/20**, Keyworth, Nottingham.

MARR, J.E. 1975. The application of the zone area system to the prediction of mining subsidence. *Mining Engineer*, **135**, 53–62.

MASON, D.A., DENNEHY, J.P., DONNELLY, L.J., PARRY, D.N. & CHIVERRELL, C. 2019. Residual impacts of past mining. *In*: PARRY, D. & CHIVERRELL, C. (eds) *Abandoned Mine Workings Manual, C758D*. CIRIA, London, UK, 117–156.

MCCANN, D.M., SUDDABY, D.L. & HALLAM, J.R. 1982. *The use of geophysical methods in the detection of natural cavities, mineshafts and anomalous ground conditions. Open File Report*, **EG82/5**, Institute of Geological Sciences, London.

MCCANN, D.M., JACKSON, P.D. & CULSHAW, M.G. 1987. The use of geophysical surveying methods in the detection of natural cavities and mineshafts. *Quarterly Journal Engineering Geology*, **20**, 59–73, https://doi.org/10.1144/GSL.QJEG.1987.020.01.06

MERAD, M.M., VERDAL, T., ROY, B. & KOUNIALI, S. 2004. Use of multi-criteria decision-aids for risk zoning and management of large area subjected to mining induced hazards. *Tunnelling and Underground Space Technology*, **19**, 125–138, https://doi.org/10.1016/S0886-7798(03)00106-8

MOOIJMAN, O.P.M., VAN DER KRUK, J. & ROEST, J.P.A. 1998. The detection of abandoned mine shafts in the Netherlands. *Environmental and Engineering Geoscience*, **4**, 307–316, https://doi.org/10.2113/gseegeosci.IV.3.307

ORCHARD, R.J. 1954. Recent developments in predicting the amplitude of subsidence. *Journal of the Royal Institution Chartered Surveyors*, **86**, 864–876.

ORCHARD, R.J. 1957. Prediction of the magnitude of surface movements. *Colliery Engineer*, **34**, 455–462.

ORCHARD, R.J. 1964. Partial extraction and subsidence. *Mining Engineer*, **123**, 417–427.

ORCHARD, R.J. & ALLEN, W.S. 1965. Ground curvature due to coal mining. *The Chartered Surveyor*, **34**, 86–93.

ORCHARD, R.J. & ALLEN, W.S. 1970. Longwall partial extraction systems. *Mining Engineer*, **129**, 523–535.

PALCHIK, V. 2000. Prediction of hollows in abandoned underground workings at shallow depth. *Geotechnical and Geological Engineering*, **18**, 39–51, https://doi.org/10.1023/A:1008943310457

PARRY, D. & CHIVERRELL, C. (eds) 2019. Abandoned mine workings manual, C758D. CIRIA, London, UK (ISBN 978-0-86017-765-4).

PENG, S.S. 1986. *Coal Mine Ground Control*. 2nd edn. John Wiley & Sons, New York.

PENG, S.S. & GENG, D.Y. 1982. Methods of predicting the subsidence factor, angle of draw and angle of critical deformation. *In*: *State-of-the-Art of Ground Control in Longwall Mining and Mining Subsidence*, American Institute Mining Engineers, New York, 211–215.

PENNELL, S. & BANTON, C. 2008. Ground Rules. *The Land Journal*, Royal Institution of Chartered Surveyors.

PIGGOTT, R.J. & EYNON, P. 1978. Ground movements arising from the presence of shallow abandoned mine workings. *In*: GEDDES, J.D. (ed.) *Proceedings of the First International Conference on Large Ground Movements and Structures*, Cardiff. Pentech Press, London, 749–780.

PRICE, D.G. 1971. Engineering geology in the urban environment. *Quarterly Journal of Engineering Geology*, **4**, 191–208, https://doi.org/10.1144/GSL.QJEG.1971.004.03.03

PRICE, D.G., MALKIN, A.B. & KNILL, J.L. 1969. Foundations of multi-storey blocks on Coal Measures with special reference to old mine workings. *Quarterly Journal of Engineering Geology*, **1**, 271–322, https://doi.org/10.1144/GSL.QJEG.1969.001.04.03

POTTGENS, J.J.E. 1979. Ground movements by coal mining in the Netherlands. *In*: SAXENA, S.K. (ed.) *Proceedings Speciality Conference of the American Society Civil Engineers, Evaluation and Prediction of Subsidence*, Gainesville, FL, 267–282.

REN, G., REDDISH, D.J. & WHITTAKER, B.N. 1987. Mining subsidence and displacement prediction using influence function methods. *Mining Science and Technology*, **5**, 89–104, https://doi.org/10.1016/S0167-9031(87)90966-2

SALAMON, M.D.G. 1964. Elastic analysis of displacements and stresses induced by mining of seam and reef deposits. Part I: fundamental principles and basic solutions as derived from idealized models. *Journal of the South African Institute of Mining and Metallurgy*, **64**, 128–149.

SHADBOLT, C.H. 1978. Mining subsidence. *In*: GEDDES, J.D. (ed.) *Proceedings of the First International Conference on Large Ground Movements and Structures*, Cardiff. Pentech Press, London, 705–748.

SHARPE, L. 2010. *Civil Engineering and Mining Related Geohazards. A Clients Guide to the Regulatory Process*. The Coal Authority. Retrieved from http://www.emcouncils.gov.uk/write/TCA%20Permissions%20Process.pdf

SIZER, K.E. & GILL, M. 2000. Pillar failure in shallow coal mines – a recent case history. *Transactions Institution of Mining and Metallurgy, Section A, Mining Technology*, **109**, A146–A152.

STACEY, T.R. 1972. Three dimensional finite element stress analysis applied to two problems in rock mechanics. *Journal South African Institute Mining and Metallurgy*, **72**, 251–256.

STATHAM, I., GOLIGHTLY, C. & TREHARNE, G. 1987. The thematic mapping of the abandoned mining hazard – a pilot study for the South Wales Coalfield. *In*: CULSHAW, M.G., BELL, F.G., CRIPPS, J.C. & O'HARA, M. (eds) *Planning and Engineering Geology, Engineering Geology Special Publication No. 4*. The Geological Society, London, 255–268, https://doi.org/10.1144/GSL.ENG.1987.004.01.31

SUTHERLAND, H.J. & MUNSON, D.E. 1984. Prediction of subsidence using complementary influence functions. *International Journal Rock Mechanics and Mining Science and Geomechanical Abstracts*, **21**, 195–202, https://doi.org/10.1016/0148-9062(84)90796-4

South African Institution of Civil Engineers 1992. *Symposium on Construction Over Mined Areas*. Pretoria, South African Institution of Civil Engineers, Yeoville, 1–20.

TAYLOR, R.K. 1968. Site investigation in coalfields. *Quarterly Journal of Engineering Geology*, **1**, 115–133, https://doi.org/10.1144/GSL.QJEG.1968.001.02.03

THE LAW SOCIETY AND COAL AUTHORITY 2006. *Coal Mining and Brine Subsidence Claim Searches: Directory and Guidance*. 6th Edition England & Wales (2006 Edition Scotland).

THORBURN, S. & REID, W.H. 1978. Incipient failure and demolition of two-storey dwellings due to large ground movements. *In*: GEDDES, J.D. (ed.) *Proceedings of the First International Conference on Large Ground Movements and Structures*, Cardiff. Pentech Press, London, 87–99.

TINCELIN, E. 1956. *Pression et Deformations de Terrain dans les Mines de Fer de Lorraine*. Jouve Editeurs, Paris.

WALTON, G. & COBB, A.E. 1984. Mining subsidence. *In*: ATTEWELL, P.B. & TAYLOR, R.K. (eds) *Ground Movements and Their Effects on Structures*. Surrey University Press, London, 216–242.

WARDELL, K. 1954. Some observations on the relationship between time and mining subsidence. *Transactions Institution Mining Engineers*, **113**, 471–483 and 799–814.

WARDELL, K. & EYNON, P. 1968. Structural concept of strata control and mine design. *Transactions Institution of Mining and Metallurgy*, **77**, A125–A150.

WARDELL, K. & WEBSTER, N.E. 1957. Surface observations and strata movement underground. *Colliery Engineer*, **34**, 329–336.

WARDELL, K. & WOOD, J.C. 1965. Ground instability problems arising from the presence of shallow old mine workings. *Proceedings Midland Society Soil Mechanics and Foundation Engineering*, **7**, 7–30.

WHITTAKER, B.N. & REDDISH, D.J. 1989. *Subsidence: Occurrence, Prediction and Control*. Elsevier, Amsterdam.

YAO, X.L. & REDDISH, D.J. 1994. Analysis of residual subsidence movements in U.K. coalfields. *Quarterly Journal Engineering Geology*, **27**, 15–23, https://doi.org/10.1144/GSL.QJEGH.1994.027.P1.04

YOKEL, F.Y., SALOMONE, L.A. & GRAY, R.E. 1982. Housing construction in areas of mine subsidence. Proceedings American Society Civil Engineers, *Journal Geotechnical Engineering Division*, **108**, 1133–1149.

Chapter 12 Subsidence – chalk mining

Clive N. Edmonds

Stantec UK Ltd (Stantec), Caversham Bridge House, Waterman Place, Reading, Berkshire RG1 8DN, UK

clive.edmonds@stantec.com

Abstract: Old chalk and flint mine workings occur widely across southern and eastern England. Over 3500 mines are recorded in the national Stantec Mining Cavities Database and more are being discovered each year. The oldest flint mines date from the Neolithic period and oldest chalk mines from at least medieval times, possibly Roman times. The most intensive period for mining was during the 1800s, although some mining activities continued into the 1900s. The size, shape and extent of the mines vary considerably with some types only being found in particular areas. They range from crudely excavated bellpits to more extensive pillar-and-stall styles of mining. The mines were created for a series of industrial, building and agricultural purposes. Mining locations were not formally recorded so most are discovered following the collapse of the ground over poorly backfilled shafts and adits. The subsidence activity, often triggered by heavy rainfall or leaking water services, poses a hazard to the built environment and people. Purpose-designed ground investigations are needed to map out the mine workings and carry out follow-on ground stabilization after subsidence events. Where mine workings can be safely entered they can sometimes be stabilized by reinforcement rather than infilling.

12.1 Introduction

Underground mining within the chalk outcrop of southern and eastern England has taken place over a long time span. The earliest flint mine workings created date from the Neolithic period while some of the latest chalk mines to be dug date from the 1930s. The peak times for widespread mining activity and volume of output appear to have been from the late 1700s through to the late 1800s. There are a wide variety of mining styles that were created for a range of purposes. The scale of mining may differ according to purpose, comprising for example a single small-scale mine working covering an area of <100 m^2, expanding up to larger-scale workings covering an area of say 10 000 m^2 or more.

Many of the old chalk mine workings were left open on abandonment, with just their shaft or adit entrances filled and sealed. The locations of abandoned mines are not well recorded and so they pose a serious ground subsidence hazard. As urban development extends outwards around the historical centres of towns and cities, construction activities are revealing more mines each year as collapse of the ground occurs. Some examples of chalk mines are shown in Figures 12.1 and 12.2.

12.2 Geographical occurrence

Data collected on the geographical occurrence of chalk and flint mines over the last 30 years has been compiled and

Fig. 12.1. Hanover chalk mine, Reading. Courtesy of Stantec UK Ltd (Stantec).

Engineering Group Working Party (main contact for this chapter: C. N. Edmonds, Stantec UK Ltd (Stantec), Caversham Bridge House, Waterman Place, Reading, Berkshire RG1 8DN, UK, clive.edmonds@stantec.com)
From: GILES, D. P. & GRIFFITHS, J. S. (eds) 2020. *Geological Hazards in the UK: Their Occurrence, Monitoring and Mitigation – Engineering Group Working Party Report.* Geological Society, London, Engineering Geology Special Publications, **29**, 311–319, https://doi.org/10.1144/EGSP29.12
© 2020 The Author(s). Published by The Geological Society of London. All rights reserved.
For permissions: http://www.geolsoc.org.uk/permissions. Publishing disclaimer: www.geolsoc.org.uk/pub_ethics

Fig. 12.2. Chalk mine, Kintbury. Courtesy of Stantec.

input to the national Mining Cavities (non-coal) Database held by Stantec. The published statistics by Edmonds (2008) have been updated to the end of 2012, subdivided into regional outcrop areas of the chalk, and are reproduced in Table 12.1. A plan showing the spatial distribution of recorded chalk and flint mines is presented in Figure 12.3.

There are no mines recorded in the chalk of Yorkshire and Lincolnshire. Southwards in East Anglia the chalk mines tend to be located in and around major urban centres such as Norwich, Bury St Edmunds, Sudbury and Thetford. The rural mines are associated with flint extraction, such as Grimes Graves, near Thetford, and Lingheath, near Brandon. To the SW, across the Chiltern Hills, the chalk has been mined to provide building stone (e.g. Totternhoe), to make whiting (e.g. Kintbury), for lime burning (e.g. Hatfield, Hemel Hempstead, Stevenage), for paper coatings (e.g. Hemel Hempstead) and for use in the manufacture of bricks, tiles and pottery (e.g. Reading, Hermitage, Nettlebed, Walter's Ash). In the west in Wiltshire and on Salisbury Plain almost no mines are known. Hampshire contains small numbers of mines (e.g. Andover), as does Devon (e.g. Beer). Eastwards along the South Downs are concentrations of flint mines associated with ancient hill forts (e.g. Cissbury Ring). South of London, along the Hog's Back, a few mines are associated with urban centres (e.g. Guildford). Others occur within certain parts of London, especially Bexley, Blackheath, Bromley, Chislehurst, Pinner and Woolwich. Large numbers of chalkwells and deneholes are found along the North Downs (e.g. Dartford, Gravesend, Chatham, Faversham, Gillingham,

Table 12.1. *Recorded geographical distribution of chalk and flint mines in England*

Area	County	Number of chalk mine cavities (C)	Number of flint mine cavities (F)	Total number of mine cavities
Northern outcrops	Yorkshire	0	0	0 F, 0 C
	Lincolnshire	0	0	
East Anglia	Norfolk	75	602	1115 F, 115 C
	Suffolk	38	513	
	Cambridgeshire	2	0	
Chilterns, Berkshire and Marlborough Downs	Bedfordshire	4	0	36 F, 276 C
	Hertfordshire	109	0	
	Buckinghamshire	40	1	
	Oxfordshire	19	3	
	Berkshire	100	30	
	Wiltshire	4	2	
Hampshire Basin	Hampshire	20	4	4 F, 22 C
	Isle of Wight	2	0	
Dorset and Devon outliers	Dorset	0	0	0 F, 8 C
	Devon	8	0	
Greater London, Essex, Surrey and North Downs	Greater London	77	0	1 F, 1415 C
	Essex	131	0	
	Surrey	23	1	
	Kent	1184	0	
South Downs	West Sussex	0	511	522 F, 3 C
	East Sussex	3	11	

Notes: (a) There are a total of at least 1678 confirmed flint mines and at least 1839 chalk mines – more are suspected and new ones discovered each year; (b) based on data contained in the archive of the national Stantec Mining Cavities Database to the end of 2012.

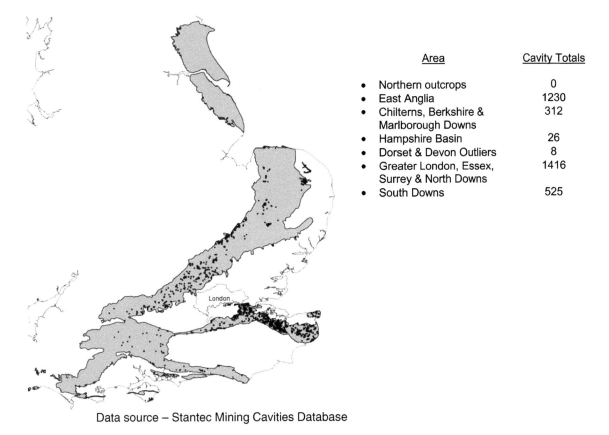

Area	Cavity Totals
• Northern outcrops	0
• East Anglia	1230
• Chilterns, Berkshire & Marlborough Downs	312
• Hampshire Basin	26
• Dorset & Devon Outliers	8
• Greater London, Essex, Surrey & North Downs	1416
• South Downs	525

Data source – Stantec Mining Cavities Database

Fig. 12.3. The spatial distribution of chalk and flint mines in England. Based upon BGS Geology 1:625 K Data, with the permission of the BGS © UKRI. http://www.nationalarchives.gov.uk/doc/open-government-licence/version/3; https://www.bgs.ac.uk/products/digital-maps/DiGMapGB_625.html

Rochester, Sittingbourne) to the Isle of Thanet (e.g. Margate, Broadstairs) and in Essex (e.g. Grays). The Stantec database also includes many records of underground bunkers and tunnels used for access and storage and by smugglers, plus military fortifications. These latter records have not been included in the above assessment and map.

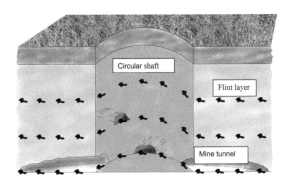

Fig. 12.4. Neolithic flint mine – schematic cross-section.

12.3 Characteristics of the mine workings

12.3.1 Flint mine workings

Flint mine workings date from two distinct periods of time – the Neolithic (*c.* 4000 to 2500 BC) and the early ninetteenth century up to the late 1930s.

12.3.1.1 Neolithic flint mines

The Neolithic mines have been dated by archaeological investigations (Russell 2000). The miners used deer antlers (as picks), hammer stones and scapulas (as shovels) to create the excavations. The mine workings took the form of large-diameter shafts (up to 20 m across) and irregular tunnels were dug outwards from the shaft (*c.* 1 m high, up to 2 m wide and usually <10 m in length) to extract fine-quality flints from the flint seams and bands exposed in the sides of the shaft. The flints were then knapped and shaped to produce flint tools and weapons. The number of shafts dug was often large (e.g. 254 shafts at Grimes Graves, Norfolk) and could cover large areas. A typical schematic cross-section through a Neolithic flint mine is shown in Fig. 12.4.

12.3.1.2 Modern flint mines

The modern flint mines were dug using a spade, pick and heavy hammer (Mason 1978). Initially a rectangular shaft (about 3 m × 1 m) was dug through the superficial deposits to reach the chalk. The shaft was continued down into the chalk in a reduced size (about 1 × 1 m) and stepped with levels to form a crude stairwell, to permit the miner to climb up and down the shaft, while also removing the loads of extracted flint. The final shaft was inclined steeply at about 80° to the vertical. A typical cross-section is shown in Figure 12.5. Low-level tunnels (about 1 m high) were dug outwards from the shaft to extract the flints contained in the chalk. The best quality flints extracted were knapped or flaked to produce gun-flints for flintlock muskets and pistols, especially during the Napoleonic war period. Lesser-quality flints were used for building and road making.

12.3.1.3 Chalk mine workings

Chalk mine workings also have quite a long history, possibly from Roman times onwards. Mining styles show regional variation and both simple and more complex mine forms appear to co-exist through time. Chalk occurs widely and has been extracted by surface quarrying as well as by mining. In understanding the reasons for carrying out chalk mining it needs to be realized that historically labour was available and cheap on farms and estates, plus carrying and moving heavy loads of chalk by horse-drawn wagons along poor-quality tracks for any distance was difficult, particularly in wet weather. Therefore, it was more convenient to obtain chalk from depth by means of mining at the point where it was needed rather than to obtain it from a distant source.

The main uses of mined chalk included:

- chalking (raw chalk), liming (burnt chalk) and marling (with a mix of chalk or lime with clay and manure) of farmland (usually spread across clay soils) to improve soil texture for ploughing and crop yield;
- production of lime for producing lime mortar for building purposes;
- production of bricks, tiles, drain pipes and pottery ware (ground chalk mixed with clay and fired in a kiln) for building purposes;
- manufacture of whiting (powdered and washed chalk), whitewash and bleaching agents (ground chalk mixed with urine) used to dye cloth and sail cloth;
- manufacture of products to whiten, stiffen and coat paper;
- production of building stone for private and public buildings.

12.3.1.4 Bellpits

Sinking a shaft down through superficial deposits into unweathered chalk at depth and forming a bell-shaped chamber at the shaft base is the simplest form of mining (see Fig. 12.6). It is probably the oldest form of mining but the style has also continued through to more recent times (e.g. Elliott 1887). These pits occur widely, generally as isolated mines or in small numbers, often associated with obtaining chalk for agricultural purposes as described above. Investigations of chalk mines within Reading, Berkshire, have identified a number of bellpits whose positions are associated with old field boundaries (Field Road area) and a particularly large grouping of medieval bellpits at Palmer Park.

12.3.1.5 Deneholes

Deneholes are another ancient form of chalk mine, recorded from Roman times onwards, the most accurately dated ones being contemporary with earthworks in NW Kent, considered to be from the first half of the thirteenth century (Spurrell 1881; Caiger 1953, 1964; Tester & Caiger 1958). The most commonly occurring deneholes comprise a 1 m diameter shaft dug down into fresh chalk, opened out at the shaft base, initially as two tunnels on opposite sides. A pair of short-length tunnels was then formed on either side of the initial tunnel. This style of mine, having double sets of three chambers (trefoils) on each side of the shaft, is referred to as a double trefoil type of denehole. Later, more complex forms of pillared deneholes (see Fig. 12.6) were developed as mining skills improved (Spurrell 1881, 1882; Le Gear 1976). A common feature of deneholes is that they are

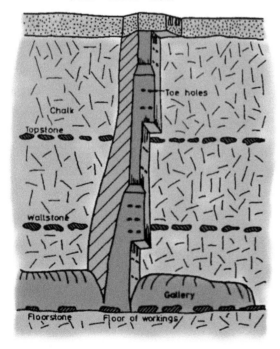

Fig. 12.5. Modern flint mine – schematic cross-section.

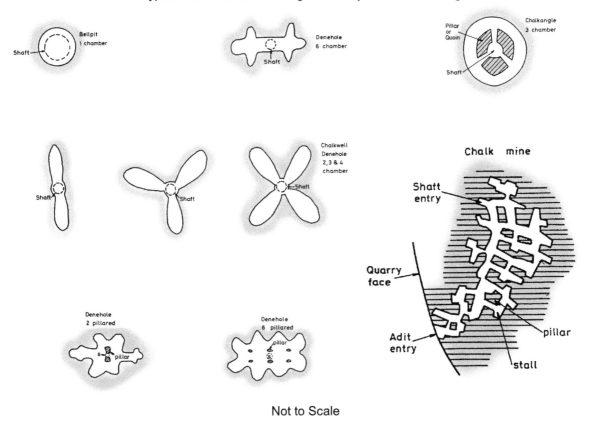

Fig. 12.6. Typical schematic plan sections through a variety of chalk mine workings (after Edmonds *et al.* 1990).

relatively compact (not normally extending beyond 15 m radius from the shaft) and that the tunnels are carefully formed, having a smooth finish to the walls and roof.

Deneholes tend to be found across the North Downs and in the Grays area of Essex. They were most likely dug for obtaining chalk for agricultural purposes as described above and generally occur singly or in small clusters associated with ancient land boundaries. However, some larger groupings are recorded at particular locations in Kent and Essex, including: Hangman's Wood area, Grays (50+ mines) and Joyden's Wood to Cavey's Spring areas, near Dartford (80+ mines).

12.3.1.6 Chalkwells
Chalkwells mainly date from the 1700s and 1800s, although some were dug as recently as the early 1900s. Shaft diameters typically vary from 1.5 to 3 m and were dug down through superficial deposits into the underlying unweathered chalk. The shafts were either dug vertically down through the superficial deposits as shown in Figure 12.7, or where the

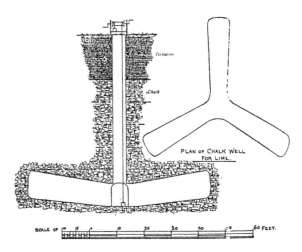

Fig. 12.7. Plan and section through a chalkwell (after Bennett 1887).

Fig. 12.8. Mine entry shaft, South Oxhey, Hertfordshire. Courtesy of Stantec.

superficial deposits were thinner the chalk shaft walls were sometimes stepped to permit access without the use of a windlass (see Fig. 12.8).

At the base of the shaft anything from two to four tunnels were excavated radially outwards from the shaft base (see Fig. 12.6). The tunnel floors were often inclined towards the shaft base level and increased in both height and width away from the shaft. The design was to assist with movement of mined chalk (often by wheelbarrow) towards the shaft from where it was hauled up to the surface by windlass. In some instances the base of the shaft was tapered or belled out within the chalk. The mining process and agricultural use of the chalk are described in detail by Young (1804). The tunnels created could be 3 m or more wide, 3 m or more high and extend radially for up to 20 m or more. A detailed plan and section through a typical chalkwell prepared by Bennett (1887) is reproduced in Figure 12.7.

Chalkwells were principally excavated to obtain chalk for lime burning in connection with agricultural uses, but sometimes were associated with brick making. The mines mostly occur singly and are aligned with old land boundaries. They are located widely across many parts of the chalk outcrop.

12.3.1.7 Chalkangles

Chalkangles date from the 1800s and represent a localized variant of the chalkwell. In this case (Bennett 1887; Money 1906) a shaft about 1.5 m in diameter is dug through the cover of superficial deposits down into the unweathered chalk at depth. Three or four horizontal tunnels are then driven radially outwards from the shaft base for a short distance before 'angling' begins. Angling consists of digging upwards at a steep angle for a distance before establishing a level platform above the shaft floor level. At the new platform level a second set of angles is then dug upwards in opposing directions until they meet with the angles being dug from adjacent platforms. These second sets of angles therefore become interlinked with the adjacent angles to create a ring-shaped pattern (see Fig. 12.6). The steeply constructed angles allowed the mined chalk to fall to the floor level at the base of the angle where it was collected into boxes for hauling to the surface. Finally, on completion of chalk extraction, the pillars (quoins) located between the initially dug horizontal tunnels at the shaft floor level were dug away purposefully to destabilize the mine working. The aim was for the mine to gradually collapse to form a shallow dell at the surface. A detailed plan and section through a typical chalkangle taken from Bennett (1887) is reproduced in Figure 12.9.

Chalkangles were usually excavated to obtain chalk for lime burning and spreading over farmland with clay soils. The mines were often positioned within the centre of a field where a shallow dell was created on completion as described or sometimes on the field margins or in adjacent woodland. The mine style is only recorded within west Berkshire.

12.3.1.8 Pillar-and-stall mines

These larger-scale mines appear to date mainly from the early 1800s through to the early 1900s. They were usually accessed via a main working shaft, or sometimes an adit entrance, and tended to comprise an irregular network of interlinked tunnels (see Fig. 12.6). The efficiency of chalk extraction varied, hence mines have larger or smaller pillars to support the mine roof. Tunnels were typically 2–5 m wide and could be between about 3 and 10 m high or more. Mined chalk was often moved by wheelbarrows from the worked face to the

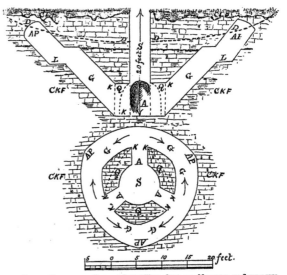

Fig. 12.9. Plan and section through a chalkangle (after Bennett 1887).

adit entrance or working shaft for hauling to the surface. These larger mines often had a secondary shaft created for ventilation. For practical purposes it seems that mines did not generally extend more than 100 m from the working shaft position, otherwise the underground movement of mined chalk became too onerous.

Some of the largest known chalk mines are associated with areas where the brick-, tile- and pottery-making industries were long established. In the past clays found within the Lambeth Group deposits were favoured and this resulted in many brick, tile and pottery works being set up in places where accessible thicknesses of these deposits were available. Notable centres where significant numbers of brickworks operated over a long period included Reading in Berkshire (see Fig. 12.1), Nettlebed in Oxfordshire and Plumstead in London. Elsewhere, smaller numbers of works are scattered across the Chiltern Hills, concentrated in and around towns and villages underlain by the Lambeth Group deposits (e.g. Hermitage and Curridge, north of Newbury).

The shaft or adit entry mines were often located below surface clay pits and associated building complexes that were purpose built for the mass production of bricks, tiles and pottery ware. The clay dug at the surface was put into a pug mill to grind the clay and destroy its texture. The mined chalk was also ground and powdered in a similar way with stones and flints removed during the process. The pug mills were operated using either horse gins or steam engines to drive the drum rollers. Ground chalk was mixed with the ground clay (25% chalk: 75% clay), shaped, cut and fired in a kiln to produce the building wares. The chalk reduced shrinkage of the raw product prior to firing and also acted as a binding agent and chemical flux during the firing and drying process, making the product stronger and more durable. Permanent kiln structures were built to fire the bricks, tiles and pottery and also lime kilns operated to produce lime mortar. These sites were continuously worked for tens of years and hence the volumes of chalk required and removed underground by mining were much greater.

Fig. 12.10. Chalk mine workings, Merton Road, Norwich. Courtesy of Stantec.

Fig. 12.11. Chalk mine chamber, Guildford. Courtesy of Stantec.

Other pillar-and-stall chalk mines were created to service larger-scale industrial enterprises set up to meet market demand where a suitable quality of chalk was available. A selection of these include: lime burning at Hatfield, Hemel Hempstead and Stevenage in Hertfordshire, Norwich in Norfolk (see Fig. 12.10), also Bromley and Chislehurst in London; whiting manufacture in Kintbury, Berkshire (see Fig. 12.2); paper treatment at Hemel Hempstead in Hertfordshire; and building stone at Totternhoe, Bedfordshire and Guildford, Surrey (see Fig. 12.11).

12.4 Engineering management strategy

Depending upon the geological setting and the type of chalk mine present, it is possible for significant large ground collapses to occur (see Fig. 12.12). Not only vertical downward movement of the collapse debris can occur, but also lateral flow movement of the debris along open connected mine tunnels, particularly in the presence of water (e.g. leaking water mains, sewers and soakaways). The principal engineering management strategies are: avoidance, prevention and mitigation.

If the preferred strategy is *avoidance* then the first step is to check the likelihood of the occurrence of chalk or flint mines at the site by carrying out a desk study and site walkover survey. Information about mines in an area can sometimes be obtained through historical map and document research (e.g. visiting the local studies library or county records office) or from other sources (e.g. local council archives, press cuttings). There are also other sources such as websites and publications produced by underground exploration enthusiasts, e.g. Subterranea Britannica and Chelsea Spelaeological Society. However, such research is time consuming and therefore it is usually quickest to obtain available information recorded within comprehensive

Fig. 12.12. Crown hole causing property damage at Field Road, Reading. Courtesy of Stantec.

national databases (e.g. Stantec Mining Cavities Database) that have been collated by experienced specialists over many years. The Stantec database was originally collected for government-funded research published in the 1990s (the primary relevant documents are listed in the Appendix: Further Reading). The current Stantec database has been considerably updated and expanded since the early 1990s. Alternatively, if a mine is known to exist on or adjacent to the site then more detailed information about the mine type, depth and size may be obtained from the database archive to assess a suitable stand-off distance.

In many cases the extent of mined ground may be unknown and therefore the only way to prove the presence or absence of mined ground may be to carry out a close-centred line (a centre distance of 1.5 m is typical) of exploratory holes.

Depending upon the geology sequence and the depth and condition of the mine workings (open voids or collapsed, backfilled tunnels), geophysical surveying can also provide useful methods for detecting the presence of mine workings. Other methods of inspecting the condition and extent of mines present as open voids can include borehole cctv surveys and downhole laser surveys.

Most mine workings reveal their presence after the entry shaft or tunnel roof collapses. On occasions it may be possible to enter the mine and survey its condition, but more frequently the health and safety considerations and difficulty of access exclude this.

Prevention of instability over old chalk mine workings can be challenging. Experience shows that with time there is a tendency for old chalk mine workings to become increasingly unstable as the ground around and above the mine 'relaxes' and loses strength. This situation can be exacerbated by changes to ground loading and drainage, leaking water-carrying services and long-term downward percolation of water from rainfall. Often mine entrances are backfilled or collapsed and so they cannot be easily entered and inspected. Where entry is possible after venting and taking suitable health and safety precautions, then the walls and roof of the mine and entrance can be inspected directly for signs of instability. The signs may include spalled blocks of chalk upon the mine floor and upwards ravelling of the roof above, open joints and bedding planes and collapsed sections of tunnel. In some instances it may be possible to support the tunnel roof by constructing a wall or pier or lining the tunnel with shotcrete and mesh. Other possibilities include rock bolting and soil nailing depending on conditions. Unstable sections of mine workings can be walled off and sealed by grouting while maintaining access to other stable parts. However, in most situations the only feasible way to stabilize the mine workings is by grouting from the surface to infill open voids and densify the overlying degrading, collapsing ground.

Presently there is a readily accessible wealth of information about chalk and flint mines across southern and eastern England owing to the existence of databases and published papers, books and research. However, such information has only become more widely available since the 1990s, hence many past developments have been constructed over mined ground in ignorance. Consequently each year the positions of former mine workings suddenly re-appear by collapse of the ground, forming crown holes. Mine entrances that have been poorly infilled in the past tend to cause most of the subsidence events, but collapse of mine tunnel roofs can also occur. For such events it is important that certain mitigation measures are taken.

Following the initial appearance of a crown hole there is a tendency for the sides of the hole to ravel back and degrade, increasing the plan size of the hole with time. Therefore it is important to take early remedial action to infill the hole to prevent continuing enlargement, further damage and associated

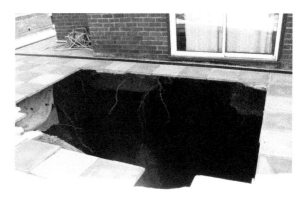

Fig. 12.13. Crown hole collapse at Hemel Hempstead. Courtesy of Stantec.

Fig. 12.14. Grout infilling of crown hole at Hemel Hempstead. Courtesy of Stantec.

health and safety risks. This is especially important if the crown hole occurs beside and/or below a building.

In Fig. 12.13 the appearance of crown hole is shown beside a house. The hole extends below the property, revealing the foundations. Within days of the collapse the house was suffering from the development of increasing structural crack damage as the gable end wall was cantilevered over the void. Left untreated the wall would almost certainly have failed and collapsed into the hole in a short time. In this case, as Fig. 12.14 shows, remedial measures were taken to infill the hole with foamed concrete. The lightweight fill choked the collapse and gave support to the hole sides and the underside of the house foundations, preventing further degradation of the property and the ground around. Once secured it was then possible to safely carry out a detailed investigation to map and treat the old chalk mine workings that had caused the subsidence problem. Open mine voids can be bulk infilled with grout while the disturbed weakened/collapsed ground over and above the mine can be stabilized using other grouting techniques such as pressure and compaction grouting depending upon the particular circumstances.

Appendix: Further reading

ARUP GEOTECHNICS 1990. Review of mining instability in Great Britain. Contract No. PECD 7/1/271 for the Department of the Environment. Arup Geotechnics, Newcastle upon Tyne.

DEPARTMENT OF THE ENVIRONMENT 1990. Planning Policy Guidance PPG14. *Development on unstable land*.

DEPARTMENT OF THE ENVIRONMENT 1996. Planning Policy Guidance PPG14 (Annex 1). *Development on unstable land: Landslides and planning*.

DEPARTMENT OF THE ENVIRONMENT, TRANSPORT AND THE REGIONS 2000. Planning Policy Guidance PPG14 (Annex 2). *Development on unstable land: Subsidence and planning*.

HOWARD HUMPHREYS AND PARTNERS LIMITED 1993. *Subsidence in Norwich*, Report of the Study on the Causes and Mechanisms of Land Subsidence, Norwich, Department of the Environment Research Contract PECD7/1/362. HMSO, London

Websites

www.chelseaspelaeo.org.uk
www.subbrit.org.uk

References

BENNETT, F.J. 1887. On chalk wells. *Essex Naturalist*, **1**, 260–265.

CAIGER, J.E.L. 1953. Researches and discoveries in Kent. *Archaeologia Cantiana*, **66**, 147–149.

CAIGER, J.E.L. 1964. Darenth Wood; its earthworks and antiquities. *Archaeologia Cantiana*, **79**, 77–94.

EDMONDS, C.N. 1988. Induced subsurface movements associated with the presence of natural and artificial underground openings in areas underlain by Cretaceous Chalk. *In*: BELL, F.G., CULSHAW, M.G., CRIPPS, J.C. & LOVELL, M.A. (eds) *Engineering Geology of Underground Movements*. Geological Society, London, Engineering Geology Special Publications, 5, 205–214.

EDMONDS, C.N. 2008. Karst and mining geohazards with particular reference to the Chalk outcrop, England. *Quarterly Journal of Engineering Geology and Hydrogeology*, **41**, 261–278, https://doi.org/10.1144/1470-9236/07-206

EDMONDS, C.N., GREEN, C.P. & HIGGINBOTTOM, I.E. 1990. Review of underground mines in the English chalk: form, origin, distribution and engineering significance. *In*: *Chalk*. Thomas Telford, London.

ELLIOTT, F.W. 1887. Notes on some chalk-wells in Buckinghamshire, Addenda (A) in Report on the denehole exploration at Hangman's Wood, Grays, by T. V. Holmes and W. Cole. *Essex Naturalist*, **1**, 254–255.

LE GEAR, R.F. 1976. Pillared deneholes at Stankey Wood, Bexley. *Archaeologia Cantiana*, **92**, 137–144.

LORD, J.A., CLAYTON, C.R.I. & MORTIMORE, R.N. 2002. *Engineering in Chalk*. CIRIA, London, **C574**.

MASON, H.J. 1978. *Flint – The Versatile Stone*. Providence Press.

MONEY, W. 1906. A Berkshire denehole. *Journal of the British Archaeological Association*, **12–13**, 212.

RUSSELL, M. 2000. *Flint mines in Neolithic Britain*. Tempus, Stroud.

SPURRELL, F.C.J. 1881. Deneholes and artificial caves with vertical entrances. *Archaeological Journal*, **38**, 391–409, https://doi.org/10.1080/00665983.1881.10851996

SPURRELL, F.C.J. 1882. Deneholes and artificial caves with vertical entrances (continued). *Archaeological Journal*, **39**, 1–22, https://doi.org/10.1080/00665983.1882.10852019

TESTER, P.J. & CAIGER, J.E.L. 1958. Medieval buildings in the Joyden's Wood Square Earthwork. *Archaeologia Cantiana*, **72**, 18–39.

YOUNG, A. 1804. *Agriculture of Hertfordshire*. Macmillan, London.

Chapter 13 Hazards associated with mining and mineral exploitation in Cornwall and Devon, SW England

B. Gamble[1], M. Anderson[2] & J. S. Griffiths[2]*

[1]Independent consultant to UNESCO, Plymouth, UK
[2]SoGEES, University of Plymouth, Drake Circus, Plymouth, UK

Correspondence: j1griffiths@yahoo.co.uk

Abstract: The largest UNESCO World Heritage Site in the UK is found in Cornwall and west Devon, and its designation is based specifically on its heritage for metalliferous mining, especially tin, copper and arsenic. With a history of over 2000 years of mining, SW England is exceptional in the nature and extent of its mining landscape. The mining for metallic ores, and more recently for kaolin, is a function of the distinctive geology of the region. The mining hazards that are encountered in areas of metallic mines are a function of: the Paleozoic rocks; the predominant steeply dipping nature of mineral veins and consequent shaft mining; the great depth and complexity of some of the mines; the waste derived from processing metallic ores; the long history of exploitation; and the contamination associated with various by-products of primary ore-processing, refining and smelting, notably arsenic. The hazards associated with kaolin mining are mainly related to the volume of the inert waste products and the need to maintain stable spoil tips, and the depth of the various tailings' ponds and pits. The extent of mining in Cornwall and Devon has resulted in the counties being leaders in mining heritage preservation and the treatment and remediation of mining-related hazards.

13.1 Introduction

The importance of mining in the history of Cornwall is demonstrated by the county hosting the second oldest geological society in the world (established 1814; http://geologycornwall.com), and with Cornwall and west Devon being selected in 2006 as a World Heritage Site by UNESCO for its mining landscape (http://cornwall-mining.org.uk) (Fig. 13.1). The World Heritage designation was specifically related to Cornwall and West Devon's long history of metallic mining (mainly copper, tin and arsenic). However, whilst the last Cornish tin mine closed in 1998 (South Crofty), the 10th edition of the *Directory of Mines and Quarries* (Cameron *et al.* 2014) listed nearly 70 active mines in Cornwall and Devon that were still extracting and processing china clay, china clay waste, clay and shale (including ball clay), igneous and metamorphic rocks, sandstone, sea salt, silica sand, slate and tungsten. There was even a small tin-streaming operation. In Cornwall, china clay alone has yielded 165 million tons of marketable clay since mining began in the mid-eighteenth century (Cornwall County Council 2015). The mineral resources for Cornwall were summarized in a 1:100 000 scale map produced by the British Geological Survey in 1997 to support Cornwall development plans (Scrivener *et al.* 1997) and were shown still to be extensive. Today, mining remains an integral part of the West Country economy, not least now because the heritage of mining is a source of revenue from tourism. Nevertheless, in 1966 Government figures showed Cornwall to be the county with the highest degree of dereliction in the UK as a result of the demise of most of the mining industry (Barr 1970). Therefore, Cornwall and Devon can be considered as exceptional in the UK for their long history of mining (suggested to have started in Phoenician times *c.* 1550–300 BC; Hobbs 2005), the temporal and spatial coverage of its mining infrastructure, its changing history of mineral exploitation, the range of mining related hazards and the nature and extent of remedial works that have been undertaken.

In this paper, the geological basis for the mining industry in Cornwall and Devon is briefly summarized followed by a description of the history of mining and the environmental consequences. Then the approaches to assessing the hazards associated with the mining are examined along with the varied methods of remediation. Throughout, reference will be made to case studies that demonstrate the various facets of the mining heritage of Cornwall and Devon. The emphasis will be on abandoned mines associated with the polymetallic Cornubian orefield, as much of the china clay, ball clay and aggregate workings are on-going or the ownership of abandoned workings is well established, so any hazards associated with these remain the responsibility of the quarry operators.

Engineering Group Working Party (main contact for this chapter: J. S. Griffiths, SoGEES, University of Plymouth, Drake Circus, Plymouth, UK, j1griffiths@yahoo.co.uk)
From: GILES, D. P. & GRIFFITHS, J. S. (eds) 2020. *Geological Hazards in the UK: Their Occurrence, Monitoring and Mitigation – Engineering Group Working Party Report.* Geological Society, London, Engineering Geology Special Publications, **29**, 321–367,
https://doi.org/10.1144/EGSP29.13
© 2019 The Author(s). Published by The Geological Society of London. All rights reserved.
For permissions: http://www.geolsoc.org.uk/permissions. Publishing disclaimer: www.geolsoc.org.uk/pub_ethics

Fig. 13.1. Aerial view of part of Consolidated Mines, near St Day, formed in 1790 and destined to achieve Cornwall's largest output of copper ore. The mines were to be reworked in the 1980s by Carnon Consolidated Tin Mines Ltd.

13.2 The geological model and the setting for mining-related hazards

The extent and nature of the mining-related hazards can only be evaluated based on a thorough and detailed understanding of the geology, thus creating a 'geological model' for SW England.

13.2.1 Geological overview

Early work on the geology of SW England was conducted by Sedgwick & Murchison (1837), De la Beche (1839), Hendriks (1937, 1959) and Dearman (1963). Systematic remapping and revaluation of generally poor inland exposure and well-exposed coastal sections has been largely ongoing since then (e.g. Sanderson & Dearman 1973; Badham 1982; Shackleton *et al.* 1982; Leveridge *et al.* 1984; Isaac 1985; Holder & Leveridge 1986*a*, *b*; Whalley & Lloyd 1986; Andrews *et al.* 1988; Shail 1989, Hartley & Warr 1990; Rogers 1997; Selwood *et al.* 1998). Based on these combined studies, the bedrock geology of SW England (Fig. 13.2) can be subdivided into a structurally complex Paleozoic (Variscan) basement and a less-deformed cover sequence extending from the Permian into the Paleogene. Quaternary deposits mantle both bedrock units.

The Variscan basement of SW England (Fig. 13.2) is dominated by Upper Paleozoic (Devonian and Carboniferous) successions that are correlated with the Rhenohercynian Zone of the Variscan orogen elsewhere in NW Europe (e.g. Shail & Leveridge 2009; Nance *et al.* 2010, 2012). These successions were deposited during the development of a southwards-facing passive margin to either a short-lived marginal or a successor basin to the Rheic Ocean (Leveridge & Hartley 2006; Shail & Leveridge 2009).

SW England occupied a lower plate position during Late Devonian convergence; continental collision initiated in the earliest Carboniferous and Variscan shortening migrated northwards through the passive margin (Franke 2000; Shail & Leveridge 2009). The youngest deformed successions are Late Carboniferous (Moscovian) in age (Edwards *et al.* 1997). Regional metamorphism of all Devonian–Carboniferous successions seldom exceeds very low grade (epizone–anchizone) (Warr *et al.* 1991). The post-collisional Cornubian Batholith was emplaced into this Variscan basement in the Lower Permian and is broadly coeval with the oldest post-Variscan sedimentary cover sequences (Simons *et al.* 2016).

13.2.2 Paleozoic rocks of the Variscan (Rhenohercynian) basement

13.2.2.1 Upper Paleozoic rift basins of the Rhenohercynian passive margin

Upper Paleozoic metasedimentary successions throughout SW England were deposited in a series of east–west-trending

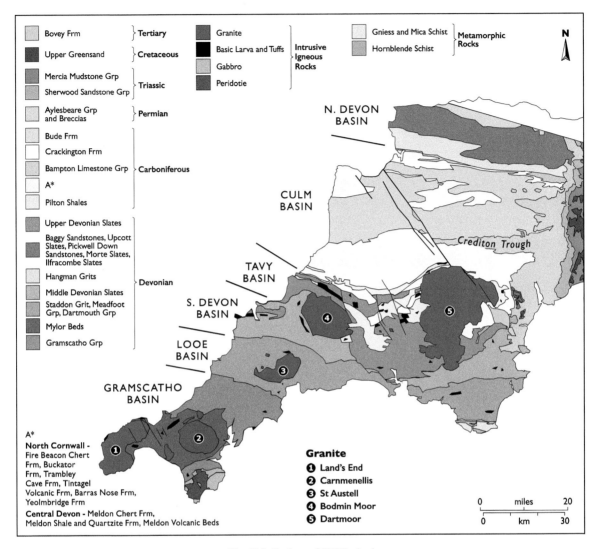

Fig. 13.2. Geology of SW England.

extensional basins (Leveridge & Shail 2011a). Coeval igneous rocks vary from early, bimodal rift-related (Merriman et al. 2000) to those with a relatively evolved transitional or normal mid-ocean ridge basalt geochemistry (T- and N-MORB; Floyd et al. 1993a, b; Leveridge & Hartley 2006). Each basin has been involved in subsequent Variscan contractional tectonics with regional-scale folds and thrusts locally repeating stratigraphy and juxtaposing basin fill of differing ages (Fig. 13.2). Structural restorations, however, facilitate evaluation of original structural and stratigraphic relationships. These reveal differences in the timing and nature of the basin fills. The oldest sedimentary fill in each basin varies from Lower Devonian (Looe and South Devon Basins) through Middle Devonian (Tavy Basin) and into the Upper Devonian (Culm Basin). Shail & Leveridge (2009) suggest that rift basins initiated progressively northwards during the development of the Upper Paleozoic Rhenohercynian passive margin. Shail & Leveridge (2009) present a number of alternative relationships between the proposed passive margin and the Rheic Ocean which lay to the south of SW England throughout the Paleozoic in many palaeogeographic reconstructions (Fig. 13.3; Nance et al. 2010; Domeier 2016). Shail & Leveridge (2009) concluded that the passive margin probably developed in a marginal basin north of the main Rheic Ocean.

The most southerly Upper Paleozoic basin in SW England, the Gramscatho Basin, comprises a rarely preserved Middle Devonian syn-rift (Leveridge & Shail 2011b). The majority

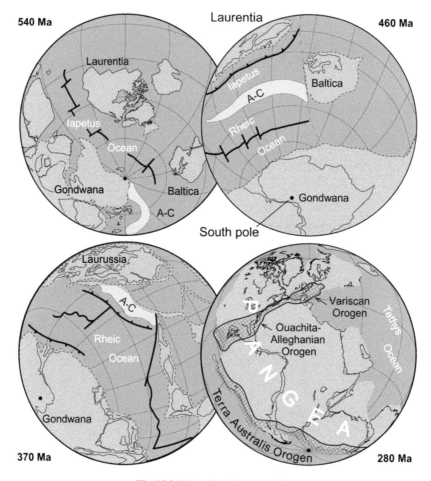

Fig. 13.3. Paleozoic palaeogeography.

of the remaining Devonian depositional fill in the Gramscatho Basin is synchronous with Variscan contraction. Contractional deformation propagated into the basin from the south, a precursor to SW England becoming an active convergent margin in the early Carboniferous Period.

The oldest basin fill in the North Devon Basin is Lower Devonian. A 6000 m-thick stratigraphic succession extends into the early Carboniferous and is considered to represent deposition in a more proximal part of the Rhenohercynian passive margin (Whittaker & Leveridge 2011). Sediments are generally considered to have a northerly source. Whether the provenance is from the Lower Paleozoic and older successions of southern Britain, which presently lie immediately to the north, is disputed. Hundreds of kilometres of translation of the Rhenohercynian passive margin sequences along major dextral strike-slip faults (such as the Bristol Channel Bray Fault) into their current position have been suggested by several authors (e.g. Holder & Leveridge 1986b; Woodcock et al. 2007). The timing of such translations is poorly constrained but would post-date pervasive Acadian (Middle Devonian) deformation recorded further north in Wales and England and be largely completed by the late Carboniferous (Shail & Leveridge 2009).

13.2.2.2 Upper Paleozoic mafic and ultramafic rocks of the Lizard Complex

The Lizard Complex comprises mantle peridotites and a structurally dismembered crustal sequence comprising various types of gabbro, a gabbro–dyke transition zone, a sheeted dyke complex and rare basaltic pillow lavas. Although the contacts between many units are faulted (Jones 1997), variable MORB geochemistry suggests that the complex represents an ophiolite (see Floyd et al. 1993a updated in Leveridge & Shail 2011b). The ophiolite preserves a textural, structural and geochemical record of mantle melting and crustal accretion, most likely during continental rifting throughout the Lower and Middle Devonian (Clark et al. 1998a, b, 2003; Nutman et al. 2001). Sheared and metamorphosed equivalents of primary mantle and crustal units are structurally juxtaposed at several levels within the complex. Pre-rift, continental basement rocks are locally exposed towards the base of the complex (see below). Available age data from these units (e.g. Sandeman et al. 2000) suggest a 30 Ma history between initial rifting, accretion of new oceanic crust and subsequent detachment and emplacement of the ophiolite (Leveridge & Shail 2011b). Whether the Lizard Complex was generated in an orthogonal rift setting, similar to that suggested for the development of other, broadly coeval, basins along the Rhenohercynian passive margin (Leveridge et al. 2002) or as a result of localized extension within a zone of dextral strike-slip shear along the margin (e.g. Barnes & Andrews 1986; Holdsworth 1989; Cook et al. 2002) remains unresolved.

A second extensive outcrop of basic igneous rocks is exposed around Start Point approximately 100 km east of the Lizard Complex. Rocks of the Start Complex are pervasively deformed throughout and comprise interlayered, metabasic, greenschists and metasedimentary grey schists. The geochemistry of the greenschists was interpreted by Floyd et al. (1993b) to have N-MORB affinity and was, therefore considered to represent a second oceanic component to the Variscan (Rhenohercynian) basement of SW England. Although the Start Complex has not been dated radiometrically, Shail & Leveridge (2009) suggest that a comparison with rocks of the Gramscatho Basin is probably more appropriate than a direct correlation with ophiolitic rocks of the Lizard Complex.

13.2.2.3 Upper Paleozoic allochthons
The majority of rocks described above are considered to have been developed along, or in close proximity to, the Rhenohercynian passive margin and may therefore be regarded as largely autochthonous despite significant subsequent Variscan deformation. The Gramscatho Basin should most probably be regarded as parautochthon, possibly originating further south along a broadly coeval north-facing passive margin. The Gramscatho Basin received a significant thickness of syn-tectonic sediment fill from an emergent southerly source area throughout the latter part of the Middle and Upper Devonian. Leveridge et al. (1990) suggest that this basin fill reflects the earliest record of Paleozoic convergent tectonics in SW England, with thrusting eventually migrating northwards into the southern margin of the Gramscatho Basin. A unit of magmatic and metamorphic rocks has been interpreted from seismic records offshore (south) of SW England and are collectively referred to as the Normannian Nappe (Holder & Leveridge, 1986a; Leveridge et al. 1984). An exposed correlative at Eddystone Rocks, the Eddystone gneiss, has been dated as Upper Devonian (Frasnian) using K–Ar on biotite (Miller & Green 1961). The Normannian Nappe is consiþred to be allochthonous with respect to the Rhenohercynian passive margin and represents the upper plate during convergence in the evolutionary models of Leveridge et al. (1990), Shail & Leveridge (2009) and Leveridge & Shail (2011b).

13.2.2.4 Lower Paleozoic (pre-rift) basement
Lower Paleozoic assemblages are rarely observed, and most are associated spatially with the Lizard Complex, namely (1) the Man of War Gneiss and Old Lizard Head Series, exposed structurally beneath mantle peridotites, (2) zircons recovered from metasedimentary rocks beneath, and deformed mafic rocks within, the Lizard Complex and (3) xenoliths within late Devonian syn-rift basalts of the Gramscatho Basin.

13.2.3 Regional structure

Rogers & Anderson (1996) and Rogers (1997) identified the main lineament trends in SW England from Landsat Thematic Mapper imagery (dated 11 September 1985; path/row 205/024). Major east–west lineament trends are thought to reflect the trend of regional Variscan contractional structures including outcrop patterns around major folds and thrust faults and associated fabrics such as pervasive cleavages developed in the Paleozoic basin fill. Subordinate north–south, NW–SE and NE–SW lineament trends correspond to regional fracture patterns associated with late and post-Variscan (cross-cutting) brittle structures such as late stage faults and joints.

In detail, the pattern of Variscan contractional deformation is more complex. Sanderson & Dearman (1973) and Sanderson (1984) suggest that SW England can be divided into 12 structural zones based on the assemblages of fold axis and facing directions, cleavage fabrics and polyphase deformation histories. Rogers (1997) simplified these into ENE/WSW and NE/SW fold axial trends south of a line running from south Devon to north Cornwall, the Start–Perranporth Line that was a zone of dextral transpression during Variscan convergence according to Holdsworth (1989). Northwards of this line, Rogers (1997) argued that ENE–WSW and east–west fold axial trends dominate.

Contractional deformation is considered to have propagated progressively from south to north across the region from the Middle Devonian until the late Carboniferous. Folding and thrusting causing inversion of the Culm Basin continued until the Moscovian (Westphalian) and may be linked to 'docking' of the Rhenohercynian Zone with southern Britain along the Bristol Channel Bray Fault (Shail & Leveridge 2009). For the most part, folds throughout SW England face and verge to the north and are often associated with north-directed thrusts. However, a zone of dominantly southwards-facing folds occurs from the Padstow area of Cornwall through east Cornwall and to the north of Plymouth in Devon (e.g. Andrews et al. 1988, Seago & Chapman 1988). This so-called regional 'facing confrontation' is considered by some authors to be caused by under-thrusting of the northward propagating thrust sheets (Seago & Chapman 1988). Shail & Leveridge (2009) refine this interpretation in the context of inversion of the Upper Paleozoic rift basins that had developed along the Rhenohercynian passive margin immediately prior to the onset of Variscan convergence. South-dipping half-graben bounding faults are considered to have been reactivated as thrusts, with kilometre-scale hanging-wall antiforms often developed; successions from highs were locally thrust northwards, via footwall shortcuts, over adjacent basinal sequences (Leveridge et al. 2002). Backthrusting, controlled by north-dipping faults defining the southern margin of full graben, e.g. Tavy Basin, Culm Basin, is interpreted to have resulted in facing confrontations with regional persistence (Leveridge et al. 2002; Leveridge & Hartley 2006).

Variscan convergence ceased during the Late Carboniferous and was replaced across NW Europe by a dextral transtensional regime (Ziegler & Dèzes 2006). East–west-trending, moderately to steeply dipping normal faults have been identified cross-cutting Variscan contractional structures affecting Upper Carboniferous rocks within the Culm Basin (Freshney et al. 1972). Reactivation of Variscan thrusts and development of new fault systems during NNW–SSE lithospheric extension have also been documented in South Cornwall (Shail & Alexander 1997). Late Carboniferous to early Permian extension was coeval with emplacement of post-collisional (early Permian) granites of the Cornubian Batholith (Shail et al. 2003; Simons et al. 2016).

Other major structures that cross-cut contractional structures in the Variscan basement of SW England include NW–SE-trending dextral strike-slip faults such as the Sticklepath Lustleigh Fault Zone (SLFZ). These demonstrably influence the development of post-Variscan sequences and have most likely been reactivated on more than one occasion (Holloway & Chadwick 1986). NW–SE dextral strike-slip faults such as the SLFZ are, however, shown by Dearman (1963) to have a regional distribution throughout the Variscan basement. A subordinate set of conjugate NE/SW sinistral strike-slip faults (Freshney et al. 1979, Edmonds et al. 1985) is also considered to be regional in extent.

13.2.4 Post-Variscan cover, magmatism, mineralization and alteration

Relatively undeformed post-Permian strata rest over the deformed Paleozoic rocks of the Variscan basement with a distinct angular unconformity which runs approximately north–south from Torbay, through Exeter, to north Somerset (Fig. 13.2). Upper Cretaceous rocks progressively overstep westwards from the older Mesozoic succession of the Wessex Basin, across the Permian and Variscan basement of SW England. The general dip of these strata is to the east, forming north–south to NNE–SSW outcrop trends. To the west of the main post-Variscan unconformity, in Cornwall and Devon, additional exposures of Permian and Triassic sedimentary successions occur in isolated, fault-controlled basins, often too small to illustrate on regional maps (Shail & Wilkinson 1994). An angular unconformable relationship with the Variscan basement rocks is again observed in many cases. The largest basin, the Crediton trough, has a pronounced east–west structural trend, extending westwards slightly beyond the SLFZ and eastwards to link with the main outcrop of post-Variscan cover in east Devon. These 'red bed' basins record thinning and exhumation of lower-plate SW England, during extensional reactivation of the Rhenohercynian suture, accompanied by Early Permian post-collisional magmatism and magmatic–hydrothermal mineralization (Shail & Wilkinson 1994; Scrivener 2006).

The Early Permian Cornubian Batholith is a composite peraluminous batholith that was generated and emplaced over 20 Ma, between 293.9 ± 0.6 Ma (Carnmenellis Granite) and 274.4 ± 0.4 Ma (Land's End Granite) (Chesley et al. 1993; Clark et al. 1994). Observed intrusive relationships post-date all Variscan contractional structures but may overlap, in part, with late-Variscan extensional structures. Contemporaneous mafic lavas occur towards the base of post-Variscan Permian 'red-beds' and as intrusive sheets within Variscan basement rocks (Dupuis et al. 2015). Lamprophyres and basalts are coeval with granite magmatism and together suggest progressive partial melting of Variscan continental crust during post-Variscan extension, influenced by the coeval emplacement of mafic igneous rocks into the lower and mid-crust, possibly with a minor direct contribution from mantle-derived melts (Simons et al. 2016).

Chesley et al. (1993) recognize four major stages of mineralization in Cornwall and Devon, mostly associated with post-collisional (early Permian) magmatism (Table 13.1). Mineralized faults and veins (mode 1 extension fractures) within post-collisional granites and the country rocks of the Variscan basement (Fig. 13.4) trend mainly east–west in Devon and East Cornwall, and ENE–WSW in South Cornwall (Dearman 1963). Darbyshire & Shepherd (1985) proposed the main episode of mineralization (Stage 3) to be 269 ± 4 Ma. The veins carry the main economic metalliferous mineralization in SW England (Chesley et al. 1993). A later set of north–south-trending veins, known as crosscourses, are present throughout the peninsular (Dines 1956). They are considered to be the result of Upper Triassic (236 ± 3 Ma) regional extension and coeval with the formation of several Permo-Triassic 'red bed' basins developed across the Cornubian massif (Scrivener et al. 1994). The fluid source is relatively low temperature and more typical of basinal brines than the granite-derived fluids more typical of mineralization at earlier stages (Table 13.1).

Further alteration of the composite granite batholith probably took place during the Jurassic to Paleogene with the widespread development of internationally important deposits of kaolin (china clay, Fig. 13.4). Kaolinization is a process of hydrolysis by which the feldspars within the granite (and under intense conditions, mica) are altered to kaolinite. In SW England the process was most likely related to the development of convection cells of meteoric water driven by abnormally high radiogenic heat from the decay of thorium and uranium in the granites (Bristow & Exley 1994; Bristow 1998; Bristow et al. 2002). Kaolinization is thought to have been deepest on the downward limbs of the convection cells. Sheeted veins, associated with earlier extensive mineralization from magmatic-derived fluids, and post-orogenic faults probably also provided alteration pathways so that kaolinization could develop more widely across the granite

Table 13.1. *Four stages of mineralization in Cornwall and Devon (after Chesley* et al. *1993)*

	Stage 1	Stage 2	Stage 3	Stage 3	Stage 4
Style	Skarns	Pegmatites	Sheeted veins with greisen borders and Tourmaline–quartz veins	Sn bearing polymetallic veins	Polymetallic cross-course veins
Structural trend				East–west	North–south
Economic metals	Fe, Cu, Sn	Sn, W ± Mo	Sn, W	Sn, Cu, Pb, Zn, As, Fe	Pb, Zn, Ag, Fe, Sb, U
Gangue minerals	Grt, Px, Tur, Am	Qtz, Fsp, Ms, Tur	Qtz, Ms, Tur	Qtz, Fsp, Sn, Chl, Hem, Fl	Qtz, Brt, Dol, Cal, Fl
Fluid source and Th (°C)	Granite derived 375–450°C	Granite derived 300–500°C	Variable mixing of granite and country rock derived 300–500°C	Variable mixing of granite and country rock derived 200–400°C	Fluids from rift basins 100–170°C
Tectonic setting	Lower to Middle Permian (290–255 Ma) post-collisional magmatism				Middle to Upper Triassic (240–220 Ma) rifting

bodies (Manning *et al.* 1996). Individual deposits generally narrow at depth, unaltered granite being present adjacent to the kaolinized areas (Ellis & Scott 2004).

Several isolated outcrops of Paleogene rocks have been identified west of the main post-Variscan unconformity, separated from the main outcrops of East Devon, Somerset and the Wessex Basin. These typically occur along NW–SE strike-slip faults, most notably the SLFZ. The Bovey and Petrockstowe Basins (along the SLFZ) are bound by complex arrays of faults, including dextral strike-slip (in keeping

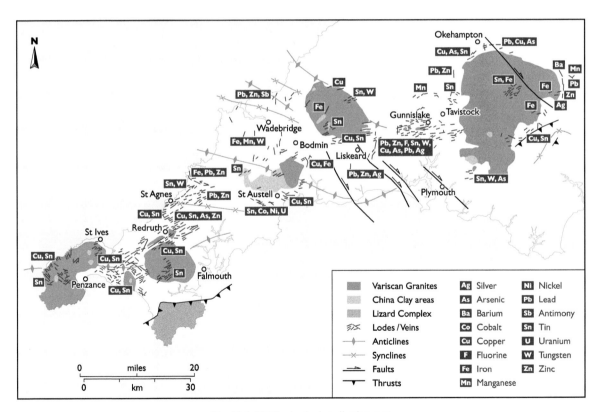

Fig. 13.4. SW England mineralization map.

with the main SLFZ), thrusts and normal faults. Together these structures are interpreted as representing pull-apart basins developed at steps along the main SLFZ in the Middle Eocene (Holloway & Chadwick 1986; Cattell 1996). The youngest basin fill (Middle Oligocene) shows evidence of contractional deformation, suggesting transpressional reactivation after this time (Bristow & Robson 1994). However, the geometrical arrangement of the basin-bounding faults remains unclear, and the sense of strike-slip movement required along the SLFZ to develop pull-apart structures is, therefore, disputed (Holloway & Chadwick 1986, Bristow et al. 1992). Other outcrops of Paleogene and Neogene rocks occur as isolated outliers throughout Cornwall and Devon. These suggest that more extensive deposition, beyond the confines of pull-apart basins, occurred across SW England at several periods during the Cenozoic (Selwood et al. 1998). Both the Bovey and the Petrockstowe Basins contain economic deposits of ball clay, a comparatively rare highly plastic sedimentary clay used in the manufacture of ceramics. The ball clay is thought, at least in part, to be derived from erosion and reworking of kaolinite from the altered granites to the west.

Offshore, Variscan basement rocks are buried beneath an extensive cover sequence of Mesozoic and Cenozoic strata that reaches up to 9 km thick. The thickest sequences have depocentres in small fault-controlled basins, with structural trends that are strongly influenced by those in the underlying Variscan basement. In many, such as the Western Approaches Trough, regional extension, driving fault-controlled subsidence, commenced in the Middle Triassic and continued through the Upper Triassic (Hillis 1988; Chapman 1989; Evans 1990). Thermal subsidence, punctuated by phases of regional thermal uplift, dominated throughout the Mesozoic and early Cenozoic, before far-field Alpine compression in the Middle Eocene and Oligocene generated localized fault-controlled subsidence and uplift along major strike-slip faults such as the SLFZ and more generalized uplift through inversion of offshore rift basins.

13.2.5 Superficial deposits

With the probable exceptions of the high ground on Exmoor (Harrison et al. 2001) and Dartmoor (Evans et al. 2012; Harrison et al. 2015) SW England was unglaciated during the Quaternary period, although it was strongly affected by multiple phases of cold-climate weathering (Harrison & Keen 2005; Murton & Ballantyne 2017). Exhumation of granite tors (Linton 1955; Scourse 1987) was accompanied by enhanced downslope movement of loose surface sediments under periglacial mass movement processes (Murton & Ballantyne 2017), creating stratified valley-fills located around coastal lowlands (e.g. Knight 2005; Scourse 1996). Radiometric dating of organic materials interbedded with these valley-fills shows that the pergiglacial mass movements were episodic between around 35 000 and 12 000 years BP (Scourse 1996). Earlier phases of periglacial erosion and deposition will have occurred, but these can no longer be identified as much of the sedimentary material will have been reworked. Extensive loess deposits (wind-blown sediments) have been dated to around 15 000 years BP using thermoluminescence techniques and indicate that the environment during the later stages of the Devensian glaciation was both cold and dry (Ealey & James 2011). Climatic and environmental changes during the Holocene have had limited regional impact on the landscape, but locally have led to the coastal sand dunes and sand bodies becoming established within estuaries (Knight & Harrison 2013). In contrast, disposal and river reworking of anthropogenic mining spoil has demonstrably impacted river and estuary sedimentation patterns and water quality (Pirrie et al. 2002, 2003; Yim 1981).

13.3 History of mining

Much of the landscape of Cornwall and Devon has been transformed by mining (Fig. 13.5) and this is a key factor in determining the mining geohazards. Up to 15 different metals have been exploited commercially, but particularly tin, from ancient times to the end of the twentieth century, and copper, especially from 1750 to 1870. In addition, china clay was extracted from around 1750 and mining for this material continues to the present day.

The early industrial development made a key contribution to the evolution of an industrialized economy and society in Britain, and throughout the world. The outstanding survival of the wider cultural landscape of metal mining, in a coherent series of 10 highly distinctive cultural landscape areas, is testimony to this achievement and was inscribed by UNESCO as a World Heritage Site in 2006. At 19 710 hectares, it is the largest World Heritage Site in the UK (http://cornish-mining.org.uk)

During the period of the industrial revolution, the metalliferous mining region of SW England dominated the world's output, successively, in tin metal, copper ore and refined arsenic. Demand for these resources and the ever more difficult exploitation of ore bodies at increasing depth, with corresponding challenges of water and hard rock, spurred an overall technological contribution that was crucial to the development of our modern industrial society. The industry also played a leading role in the diffusion of both metal mining and steam technology around the globe, and thus played a key role in the growth of a global capitalist economy (Gamble et al. 2005).

Mining in the region has a long pedigree. Until around 1700, except for some historically important silver-lead mining in the Bere Peninsula in West Devon during the Middle Ages (Rippon et al. 2009), mineral output was essentially represented by one metal, namely tin. Cornwall and Devon contained Europe's principal tin deposits, and placer mining from the series of eroded granite outcrops, followed by shaft mining of veins (known locally as lodes) in hard rock, satisfied substantial demand over four millennia (Penhallurick

Fig. 13.5. St Just Mining Coast (photo credit: B. Gamble & Cornwall County Council).

1986) (Fig. 13.6). In fact, until the late 1870s Cornwall and West Devon produced more tin than any country in the world (Umhau 1932).

Larger-scale tin smelting began during the early eighteenth century following the introduction of reverberatory furnace technology (Barton 1967). Ultimately this replaced the blowing houses that had been in operation since the Middle Ages, particularly in the 'tin streaming' districts on the granite moorlands of Dartmoor, Bodmin Moor, Hensbarrow Moor, Carnmenellis and West Penwith (Fig. 13.2). Virtually all the tin ore from Cornwall and Devon was smelted within the region until the twentieth century owing to an ancient

Fig. 13.6. 'Old mens' workings' in the 100 m high cliffs at Cligga Mine, Perranzabuloe, Cornwall (photo credit: B. Gamble & Cornwall County Council).

tradition of taxation, called 'coinage'. Coinage originated in the 1100s and persisted until 1838, initially under the crown and subsequently under the Duchy of Cornwall (Lewis 1907). Tin was used in pewterware and solder, and in the industrial period to make the tin plate on which the canning industry was built. It was alloyed with copper to make bronze for industrial applications, including machine bearings and ship's propellers.

By 1750, copper was becoming increasingly important and soon took centre stage, and Cornwall broadly led world production for nearly a century (Barton 1961). Eighteenth-century copper mining was principally confined to a small area between Hayle and Gwennap, where several copper ores occurred in varying grades, some exceedingly rich. This became Cornwall's core industrial district, bounded by the towns of Truro, Penzance and Falmouth. Copper production, mostly from West Cornwall, during the first three decades of the nineteenth century amounted to two-thirds of the world's supply. The success resulted in the eastward extension of copper mining to include the St Austell–Par district in the 1810s and 1820s and Caradon Hill in the 1830s, with renewed activity across the Tamar to Tavistock in the 1840s when Devon Great Consols in West Devon became the largest single producer in Western Europe (Gamble *et al.* 2005).

Copper smelting, always comparatively small in Cornwall, ceased in 1819, the only notable commercially successful smelter being in Copperhouse, next to Hayle, in West Cornwall (Pascoe 1982). Swansea in South Wales then became the global centre for the trade, although much of it was under the control of Cornish entrepreneurs (Hughes 2005). Copper was used to sheath the hulls of British ships, provide coinage and make hollowware boilers, vats and piping for the sugar and dyeing industries, and it is also the principal constituent of brass and so provided crucial fittings for steam engines, gun cartridges and brass trading goods. Cornish copper formed the basis of the Bristol, and subsequently Birmingham, brass industries, which were the largest producers in the world at that time. The copper industry was exceptional in that its growth rate exceeded that of every other major national industrial sector. During the second half of the nineteenth century copper became the essential metal of the electrical and communications industries. Output of copper ore from the region declined during the late 1860s owing to the exhaustion of ore bodies and a global fall in the price of the metal caused by oversupply that coincided with an international banking crisis. It was arsenic production and the continued production of tin that provided a stay of execution for the next quarter-century for some mines (Burt *et al.* 1987).

Arsenic production in Britain began as a by-product of tin and copper mining in West Cornwall during the early nineteenth century (Fig. 13.7). This toxic semi-metal was used extensively in the dyes and pigments of the Lancashire cotton industry, and in many other applications, including medicine. Demand grew during the last quarter of the century when it became a popular insecticide to help control the Colorado

Fig. 13.7. Tin and arsenic processing remains, Levant Mine, St Just (photo credit: B. Gamble & Cornwall County Council).

Beetle that had devastated potato, tobacco and other crops across North America. The key factor was that the cost of mining was abnormally low and this made the region, and specific mines, highly competitive. This was because some exhausted copper mines had their workings already opened for copper exploitation, and the arsenic ore occurred as an outer layer on the same lode structure and was left standing on the walls of the old stopes. In the 1870s Devon Great Consols, and a few other mines in West Devon and East Cornwall, produced half of the world's supply of arsenic ore. All arsenic ore was refined near the mines, with mines such as East Pool and South Crofty producing well into the twentieth century.

Rapid industrialization of the mining industry required unprecedented levels of technological innovation in the use of power, transport and processing techniques. The Cornish mining industry was characterized by prolific innovation, sustained by an influx of capital, attracted to what had become a crucible of industrial development. The near-vertical lode deposits could be exploited only by deep-shaft mining. Local pioneers invented the Newcomen atmospheric steam engine and first applied it to a metal mine, probably between 1710 and 1716, at Wheal Vor in West Cornwall (Kitsikopoulos 2016). The expense of shipping coal to the Cornish mining region from Bristol and South Wales stimulated the need for energy efficiency. During the second half of the eighteenth century Newcomen's engines were vastly improved by Cornish engineers, and during the last quarter, large-scale copper mining attracted Boulton & Watt to Cornwall, a region that became their principal market. Their patent expired in 1800 and Cornish engineers went on to develop high-pressure steam pumping technology that resulted in the Cornish beam engine and boiler, the most efficient equipment of its kind at that time anywhere in the world (Barton 1969). It also led to the development of steam as a method of motive power by pioneers elsewhere whose experiments eventually resulted in the mass movement of goods and people.

In the nineteenth century foundries and engineering works produced the steam engines, compressors, rock drills and other mining equipment which pushed the technological barriers, enabling mines to be dug deeper and made larger, and to process their ore ever more efficiently. The impact of these developments was felt throughout the mining world. Transport was crucial as supplies had to be brought in and minerals, particularly copper ore, had to be moved out from the mines to the new purpose-built mineral ports. A high-capacity transport network developed during the early nineteenth century to meet this demand. Technical aspects of ore processing ('dressing') were pioneered, and imported techniques were improved and this enabled ore to be mined which had previously been considered unworkable. The diffusion of such technology to mines overseas proved to be of international significance.

The impact of mining on the Cornwall and West Devon landscape during the period 1700–1914 was large-scale and the speed at which the industry was abandoned combined with a general lack of redevelopment pressure resulted in an unparalleled relict primary mining landscape that features thousands of mine shafts, numerous waste tips and over 200 engine houses, together with the widespread remains of tin and arsenic processing (Gamble et al. 2005) (Fig. 13.8).

Cornish mining expertise and products began to be exported throughout the world from the second decade of the nineteenth century, wherever mining operators sought the latest technology and working practices; often these mines were developed with the help of British capital. The core of the export trade consisted of steam engines, the engineers needed to install and operate them, mining equipment and the miners needed to superintend mining operations. 'Cornish' engine houses, which are among the most distinctive industrial buildings in the world, survive in Spain, Mexico, Australia and South Africa (Fig. 13.9). They are striking evidence of this worldwide impact, and in Cornwall and West Devon they have not only become cultural icons but also represent the largest concentration of such technological monuments anywhere in the world.

A number of scientific societies were established, the most significant being the Royal Geological Society of Cornwall in Penzance (1814), the Royal Institution of Cornwall in Truro (1818) and the Royal Cornwall Polytechnic Society in Falmouth (1833); all three continue to contribute to Cornwall's rich cultural life. Mineralogy and geology, and their practical application to the mining and mineral processing industries, were studied extensively. The combination of engineering and scientific endeavour associated with the development, by deep mining, of one of the world's most mineralogically diverse orefields stimulated the ground-breaking efforts of Cornish scientists whose contributions helped to lay the foundations of geological, chemical and physical science. Camborne School of Mines was founded in 1888, obtained its own teaching mine in 1895 (South Condurrow, renamed King Edward Mine) and by 1900 became internationally renowned. Today, this status is maintained with the school joining the University of Exeter in 1993, but it continues to be based in Cornwall near Falmouth.

The china clay industry is less well known but, with a production of some 170 million tons of china clay exploited from large opencast mines for over 250 years, its landscape impact assumes an even greater scale, particularly in the St Austell district in mid-Cornwall, which produced 90% of British china clay (http://wheal-martyn.com), and SW Devon at Lee Moor on the edge of Dartmoor. Like metal mining, the primary industry was based on the export of one of Cornwall and Devon's natural resources linked to the granite 'spine' of SW England. China clay was, and still is, washed out of blasted kaolinized granite and pumped away for processing. Waste mica is settled in large lagoons whilst prolific quartz sand is tipped, formerly in pyramidal heaps (given the name 'Cornish Alps') and subsequently, for safety reasons following the Aberfan tailings tip disaster in Wales in 1966, in terraced and profiled hills that ironically comprise Cornwall's highest point (Fig. 13.10).

Fig. 13.8. Remains of tin and arsenic processing on the site of Poldice Mine, St Day, Cornwall (Gamble/Cornwall Council 2006).

Fig. 13.9. O'Okiep Mine, Namaqualand, South Africa, with contaminated land, flooded open pit and a Cornish engine house containing a nineteenth-century Cornish beam engine (Gamble/Cornwall Council 2007).

Fig. 13.10. Flooded china clay pit with 'traditional' conical waste tip, St Dennis, Cornwall (photo credit: B. Gamble and Cornwall County Council).

By 1910 Cornwall was producing 50% of the world's china clay, and today, whilst production is much reduced, the Cornish deposits remain a world-class resource. Only 30% of china clay production is now used in ceramics, whilst some 50% is used in paper.

13.4 Environmental legacy of mining

13.4.1 Underground voids and shafts

Entry to underground mines in Cornwall and Devon is normally by means of vertical shafts or subhorizontal tunnels called adits that were commonly used for drainage to a valley or cliffs (Fig. 13.11). Whilst the latter have internal hazards such as collapses, sudden drops (winzes or stopes), 'bad air' and the potential to become lost, they are a less obvious danger and have escaped the degree, and severity, of treatment compared with shafts.

Examine a 1:25 000 scale Ordnance Survey map in any one of the mining districts scattered across the region and the legend 'shaft (dis)' is soon apparent, an indication of a disused shaft head. A legacy of 10 000 or more mineshafts, left abandoned after centuries of mining steeply dipping metalliferous veins, are located on cliffs and in sand dunes, on remote moorland, in farmland and woodland, and in former mining villages, towns and industrial estates. Some visible, some long invisible, they map the extent and morphology of mining, presenting multiple point hazards, as opposed to the laterally extensive subsidence hazard common to coal mining areas. Considering the longevity and high intensity of mining in Cornwall and West Devon, the number of shaft collapses, the most common type of subsidence event, is relatively limited. The broad stability and integrity of the shaft usually remains good. The zone of failure is usually restricted to either the collar (the shaft lining around the throat), which can fail and result in a cone or crater with steeply angled sides that drop into the shaft (Fig. 13.12), or through the failure of the shaft capping and/or its 'fill'. This is a situation that will continue to result in failures for many years as shaft caps that are over a century old begin to fail. Some cavernous stopes where rich bunches of ore were extracted were often wrought perilously close to the surface, and it is not uncommon to find linear gunnises, open to the surface, often in woodland. The lack of more general subsidence is, however, probably due to a combination of 'hard rock' and the narrowness of lodes and corresponding voids.

For centuries, upon abandonment, shafts were commonly left open with little, if any, safety precautions. Following the Metalliferous Mines Regulation Act, 1872, open shafts were usually surrounded with a permanent Cornish shaft hedge (dry-stone wall) to prevent accidental entry, and many of these original structures survive today (Fig. 13.13) and have usually been supplemented by a barbed wire or chain linked/wire mesh fence. Several shafts were 'filled' with mine waste or used by generations of fly-tippers (including domestic, farm and trade waste). This has commonly resulted in sudden slippages of unconsolidated fill that had hung up on timber obstructions, which have subsequently failed, or the fill suddenly compacting or being washed out or displaced by water. Shafts were also capped with lintels of granite, and many examples of these survive as good as the day they were laid, buried under a layer of mine waste. However, more commonly timber capping was used, usually

Fig. 13.11. South Caradon Mine Adit, St Cleer, Cornwall, where the copper riches of Caradon Hill were discovered in the 1830s (photo credit: B. Gamble and Cornwall County Council).

Fig. 13.12. Collapse above stoping at Stowe's section, Phoenix Mine, Linkinhorne, Cornwall (Gamble, 2011a).

Fig. 13.13. Example of a dry-stone wall around an old mine shaft (reproduced with permission from Dave Evans at http://www.southwest coastphotos.com).

followed by burial with mine waste, and over time the timbers weaken as the woods dries out, leading to collapse. This proved to be the problem in the case of the building estate constructed over metalliferous mine workings in East Cornwall at Gunnislake (Fig. 13.14).

In the past most local authorities and private owners dealt with shaft capping individually and on an ad hoc basis when hazards became apparent, especially when the rapid collapse of shaft capping occurred without warning, or when development or new access was planned nearby. However, in 1988, the government announced that derelict land grant funding for the investigation and treatment of abandoned mining land in Cornwall would be made available, especially the capping or back-filling of mine shafts. In 1992, the broad national overview was published in a 'Review of mining instability in Great Britain' that included a case study on metalliferous mines in Kerrier, Cornwall (Arup 1992). The research for this work had already informed the Government's Planning Policy Guidance 14 'Development on unstable land' (Department of the Environment 1990).

During the 1990s, with local authorities committed to bringing derelict land back into use, Derelict Land Grants were made available through English Partnerships (launched in 1993), matched by the European Regional Development Fund and local authority contributions. This enabled widespread and unprecedented large-scale remediation of shaft hazards in Cornwall, especially where a significant threat to the public and/or many properties could be demonstrated, either on land needed for development or redevelopment, or on land with a high degree of public access, such as the SW Coastal Path. Both Kerrier and Carrick district councils were particularly active in the scheme, and around 800 shafts were treated, c. 90% of which were capped using a structural solution that involved mass concrete plugging or reinforced concrete capping (Cornwall Underground Access Advisory Group 2002). Treatment of underground voids and shafts was the preserve of engineers and geologists, which was wholly understandable at that time when the safety of the public, and that of livestock, was essentially the sole concern. In terms of liability, under Common Law, the cause of damage or harm rests with the person who caused it. Problems arose concerning the identity of ownership, as most mine owners were either long-dissolved companies or unknown, or ownership was confused because manorial mineral rights (underground) are commonly separate from the surface freeholder. Responsibility was difficult to determine in many cases, there being no single state-owned operator of mines, unlike in coal mining, for example, where the British Coal Authority took on the liabilities of previous owners and specific legislation was introduced (Coal Mining Subsidence Act 1991). Section 151 of the Mines and Quarries Act 1954 imposes a duty on the owner of an abandoned mine to secure the entrance to a shaft or mine entrance and failure to secure such hazards is a risk as a Statutory Nuisance that is actionable under the Environmental Protection Act 1990. There is also a liability, too, for the current owner of the land on which the shaft is situated, under the Occupiers' Liability

Fig. 13.14. Collapse of the timbers capping a vertical mine shaft creating a surface crown hole in the garden of a new house in Gunnislake, East Cornwall (June 1992).

Act 1984 for injury suffered by persons other than their visitors. The occupier must, of course, be aware of the danger or have reasonable grounds to believe that it exists and have reasonable grounds to believe that persons may come into the vicinity of the risk. Risks to the public further increased with the introduction of new 'right to roam' legislation under the Countryside and Rights of Way Act 2000. This was a new right to walk freely over certain land which was mapped as 'Access Land', that is registered common land, land dedicated by landowners as allowing public access and areas of open countryside such as moorland, heath and down. All public access land is a place of public resort and so all unsecured mine entrances, irrespective of whether worked before or after 9 August 1872, will be deemed a Statutory Nuisance. Under the Land Registration Act 2002, manorial mineral rights, often retained interests when land became freehold, had to be registered by 2013 with HM Land Registry to preserve them for the future, a requirement that will at least go some way towards bringing some clarity in determining ownership and liability in the future.

Unfortunately, at least in terms of conservation, the standard technique involved in shaft treatment was excavation of the shaft collar to bedrock and then plugging either with structural concrete or through the insertion of a prefabricated steel cage known as a Clwyd Cap filled with rock (Fig. 13.15). Each method resulted in what was intended to be a permanent barrier to underground access; although in the case of the Clwyd Cap, long-term performance has become questionable and previously treated shafts may re-open. During treatment works, substantial historic mining landscape, scientific resources and ecological habitat were also negatively and irreversibly impacted, either through access operations for heavy plant and machinery or by design in that the focus of public opinion was usually on the visual impacts of mining. Ecological, scientific, archaeological and historical interests were less of a priority and access was permanently lost to wildlife, particularly bats but also some species of owl and swallows, and to ecologists, geologists and mineralogists, archaeologists and mine explorers. Concerning bats, several species breed in mines (particularly two of the rarest species, the greater horseshoe and lesser horseshoe) and rely on them for at least part of their hibernation period. All bats are protected by the Wildlife and Countryside Act 1981, so it is illegal not only to intentionally kill or to disturb roosting bats but also to damage, destroy or obstruct access to any place used by bats for roosting. More than one shaft had to be expensively uncapped when challenged on the grounds of entombing a bat population. Such treatment in sealing the shaft opening also affected air circulation and consequent ventilation of underground workings (Fig. 13.16). Where shafts were adjacent to a listed structure (such as an engine house), then greater consideration was required not to damage the structure or its setting.

In addition to shafts, usually either vertical or inclined to follow the 'underlie' of the vein or 'lode', there are abandoned stopes underground and narrow excavations that are carried to the surface on the vein or lode and known to miners as a gunnis, coffin or goffen. Whilst most of these are known from old mine plans, many are unrecorded having been exploited well before mines were surveyed and plans kept, or the legal requirement came into force to deposit plans of abandoned mines with the Secretary of State under the Metalliferous Mines Regulation Act, 1872. To help deal with this issue specialist mine search companies operate commercially throughout Cornwall and Devon.

13.4.2 Opencast mines

In terms of metalliferous mines, there are several former opencast tin mines that have left an enduring legacy of gaping linear gashes in the landscape, famously at Hensroost and Vitifer (Dartmoor), Drakewalls (Gunnislake), Great Wheal Whisper (or Treveddoe, St Neot), Mulberry Pit and Great Wheal Prosper (Lanivet) and Great Wheal Fortune (Breage). Several others, including some for copper (e.g. Wheal Music, Porthtowan), have been filled in, usually as rubbish tips that are capped when full. The medieval and early modern

Fig. 13.15. Mineshafts treated with Clwyd Caps and dumps re-profiled, Perran St George Mine, Perranporth, Cornwall (Gamble/Cornwall Council 2004).

opencast mines on Dartmoor (Fig. 13.17), for example, are not considered particularly hazardous as they are not especially deep, but some of the opencasts in central and west Cornwall (having exploited stockwork tin deposits) require fencing as they are sheer-sided chasms up to 35 m deep. A twentieth-century tungsten and tin opencast mine was located at Hemerdon in West Devon that was worked during the 1900s and 1940s. This site was reopened as Drakelands Mine in 2015 (Fig. 13.18).

A far greater legacy of opencast mining is represented by the china clay, china stone, and ball clay industries, primarily on the Hensbarrow (St Austell) Moors, but also on Lee Moor (Dartmoor), Bodmin Moor, Tregonning Hill and in West Penwith. Some of these abandoned mines still contain china clay deposits at depth, but most are water-filled with characteristic turquoise-coloured lakes. Fencing guards the steep faces but some mines are viewed via organized access (the Clay Trails) and one even serves as the host for a tourist attraction – the Eden Project constructed in the former Bodelva China Clay Mine at St Blazey (Fig. 13.19).

13.4.3 Waste tips and contaminated land

Waste tips from metalliferous mining (Fig. 13.20) are likely to be contaminated by heavy metals (Fig. 13.21), although inert waste tips associated with china clay mines are usually environmentally 'benign'. All tips can be a source of secondary aggregates. However, in the World Heritage Site areas of Cornwall and west Devon that relates primarily to metalliferous mining there is a specific protection policy (Policy P7, Cornwall and West Devon Mining Landscape World Heritage Site Management Plan 2013–18) that states: 'There is a presumption against the removal, disturbance or burial of historic mine waste within the Site'. Consequently, secondary aggregate extraction from such tips has been halted, and the tips are actively protected in any site safety and conservation works. In the 1990s, areas of contaminated ground, particularly mine dumps, were evaluated as habitats for rare metallophytes, that is plants that can tolerate high levels of metals such as copper and lead. Several mine sites, therefore, are designated as Sites of Special Scientific Interest for their biological characteristics. Thus, the various categories of waste tips on mine sites that were once heavily remediated with the help of derelict land grants, and/or selectively protected as Sites of Special Scientific Interest for geological, mineralogical or biological significances, are now protected with policies actively promulgated by Natural England, the National Trust and other organizations.

It should be noted that potentially contaminated land is not confined to metalliferous mine waste tips but extends to the ground occupied by mining structures, particularly those associated with ore-processing (dressing), including crushing, concentration, roasting and refining, and sites associated with smelting. Such sites will also be subject to protection within the World Heritage Site Management Plan.

Fig. 13.16. Grilled entrance to open stopes, Bedford United Mine, Tamar Valley, Devon, allowing access to bats and bona fide mine explorers (photo credit: B. Gamble).

13.4.4 Infilled or silted-up estuaries

A lesser-known legacy of mining, particularly of tin streaming or intensive and sustained ore processing of 'mine tin' (stamping and gravity-water concentration), is the infilling or silting up of estuaries and rivers. Examples include: the estuary of the River Par below St Blazey, Cornwall, infilled by the sixteenth century; The River Fowey below Lostwithiel, Cornwall (known in the fourteenth century as 'The Port of Fawi', but silted from tin streaming activity shortly afterwards); the Carnon River below Bissoe, Cornwall (Fig. 13.22); and the Tory Brook that feeds the River Plym below Plympton, Devon, which silted up with tin streaming refuse, causing trade to be relocated to Plymouth.

13.4.5 Slurry lakes or tailings ponds

Large-scale tailings ponds are associated with metalliferous mines (e.g. the inactive Wheal Jane/South Crofty mine, Fig. 13.23) that may be contaminated (Fig. 13.24). There are also numerous china clay operations, where the tailings lagoons form behind 'mica dams' built out of the waste from the extraction process. In these, the micaceous residue from china clay production settles in lagoons to eventually compact. Historically the mica was trapped in wooden 'drags' and disposed of into rivers and out to sea, but from the 1970s this was deemed to be no longer acceptable. The beach at Carlyon Bay in St Austell Bay, Cornwall, that has been present since the mid-nineteenth century, is the result of sand discharged mostly from china clay operations, with contributions of finer material from tinworks (Bristow 2014).

13.4.6 Pollution by contaminated mine water

Adits, used for discharging acidic water from underground, are common to many mines in Cornwall and west Devon (Fig. 13.25). The greatest venture of its kind, the Great County Adit, commenced in 1748 and drained the largest concentration of copper mines in the world at the time (predominantly massive copper–iron sulphide lodes). Water was pumped up from workings that averaged 80–100 m deep and discharged into the adit by more steam-pumping engines than were used at that time by the whole of continental Europe and America combined. The network extended some 65 km and in 1839 its discharge was measured at around 66 million l/day (Buckley 2000). Historically, the high acidity and high level of dissolved metals in the water gave rise to a profitable industry of copper precipitation works and iron ochre works in the valley. The adit still discharges into the Carnon River, Restronguet Creek, and eventually into the Fal Estuary.

Very close to the Great County Adit, the Nangiles Adit (connected to Wheal Jane) is well known for its sudden release in January 1992 of around 50 million litres of very visual (bright rusty orange, from iron hydroxides) polluted mine water into the Fal estuary. In 1991 pumping of the mine ceased and water built up in the workings, something that eventually led to the sudden forced clearance of a blockage in the Nangiles Adit. Discharge was highly acidic (pH 2.8) with a high metal load (cadmium, zinc, copper and iron), with consequent concerns for shellfish and other ecology in the estuary. Long-term management initially included a passive reed bed system, but an active treatment plant based on a high-density sludge alkali dosing plant has since superseded this method (see the section 'Mine water contamination and remediation').

Many adits that drain mines that formerly worked similar copper–iron sulphide deposits discharge into the River Tamar that forms the Devon–Cornwall border. Historically, conditions were imposed in terms of water quality discharging into the river, one of the major mineral lords being particularly mindful of the salmon fishery. The Dukes of Bedford

Fig. 13.17. Hensroost, medieval/early modern openwork for tin on Dartmoor. It is common for shafts to be in the base (Gamble 2015).

Fig. 13.18. Drakelands Tunsten Mine, Plympton, West Devon (opened again in 2015 after previously being worked as Hemerdon Ball Mine in the 1900s and 1940s).

Fig. 13.19. The Eden Project, adaptive re-use of the former Bodelva China Clay Pit, St Blazey, Cornwall. Slope stability and flooding were early issues encountered (Gamble/Cornwall Council 2005).

Fig. 13.20. Contaminated ground surrounding 'burrows' at Cargoll Mine, St Newlyn East, Cornwall (Gamble 2011*a*, *b*).

Fig. 13.21. Prolific 'burrows' of rust-coloured mined rock cascade down the valley sides at United Hills Mine, Porthtowan, Cornwall (Gamble/Cornwall Council 2006).

Fig. 13.22. The heavily silted Restronguet Creek and the former port of Devoran, Cornwall, looking up the once navigable Carnon Valley (Gamble/Cornwall Council 2006).

Fig. 13.23. Tailings dam and pond, Wheal Jane, Baldhu, Cornwall, looking SE down the Carnon Valley and Restronguet Creek into Falmouth Bay (Gamble/Cornwall Council 2009).

Fig. 13.24. Ochrous acidic mine water, contaminated with arsenic, in the Wheal Maid tailings dam: two lagoons separated by three dams that operated during the 1970s and 1980s to serve Mount Wellington mines.

Fig. 13.25. Ochre deposits at Blanchdown Adit, Devon Great Consols, Tamar Valley, Devon. The mature pine tree is used for scale (Gamble/ Tamar Valley AONB 2006).

owned around 19% of SW England copper production around the middle of the nineteenth century.

13.4.7 Flooding

The Environment Agency (2009) recognizes five common forms of flooding: river, coastal, surface water, sewer and groundwater. Sewer flooding occurs when sewers are overwhelmed by heavy rainfall or become blocked, and coastal flooding is primarily a function of sea-levels and coastal topography. However, river, surface water and groundwater flooding are potential environmental legacies of the mining industry that must be considered in Cornwall and Devon. As both opencast and underground mines were kept dry by pumping when they were worked, once operations ceased the groundwater recovered and this floods both the workings and possibly the surface as well, as the water can spill out into nearby rivers.

The problem of flooding from abandoned mines is underlined in the case of the Wheal Jane pollution incident (see the sections 'Pollution by contaminated mine water' and 'Mine water contamination and remediation'). Even though this was caused primarily by the cessation of pumping, prolonged heavy rain quickly infiltrates workings and has subsequently caused periodic cause for concern. In any old mine, the designed water management system via adits often becomes choked owing to obstructions caused by rock falls, even when the mine is in operation. South Crofty Mine, for example, has always regularly maintained a number of adits that serve an extensive interconnected network. Flooding can also cause instability internally in mine workings, and sudden surges through the clearance of a blockage can cause catastrophic effects downstream, internally and externally. Historically, flooded abandoned mines adjacent to operational mines were feared, miners sometimes excavating into what they called a 'house of water', with tragic results.

Elsewhere in the world, dumping metalliferous mine waste into water courses has led to increased sedimentation in rivers and widespread surface water flooding (e.g. the Ok Tedi Mine in Papua New Guinea; Mining Minerals and Sustainable Development 2002). In the UK local councils are the lead local flood authorities under the Flood & Water Management Act (2010). Under this legislation, for example, Cornwall Council was required to produce a 'Preliminary Flood Risk Assessment' to locate areas in which the risk of local flooding was significant and warrants further examination. The Assessment was completed in 2011 (Cornwall County Council 2011*a*) and in Annex 5 of this report (Cornwall County Council 2011*b*) the floods from 1800 to November 2010 are chronicled. None of these floods were identified as being a direct consequence of mining activities; however, locally the increased sedimentation found in estuaries (Fig. 13.21) must be a factor in increased likelihood of surface water flooding.

13.5 Investigating and assessing the hazards

The assessment of the mining hazard for any area follows a standard pattern for site investigation that will be familiar to all engineering geologists and geotechnical engineers involved in the civil engineering industry. Designing and carrying out a site investigation is covered by the requirements of BS5930 (BSI 2015) and the equivalent Eurocodes (Norbury 2016) and is described in standard engineering geological texts (e.g. Simons *et al.* 2002; Griffiths & Bell 2019). The overall aim of a site investigation is to develop a ground model that will facilitate development planning, enable the hazards to be identified and allow construction to proceed in a safe and economic manner (BSI 2015; Parry *et al.* 2014; Fookes 1997). The critical document for both the investigation and treatment of old mine workings in the UK is RP940 (CIRIA 2017; Parry & Chiverrell 2019), the updated version of the Construction Industry Research and Information Association (CIRIA) report first published in 1984 and reprinted in 2002 (Healy & Head 2002). In addition, as former metalliferous mining sites could potentially be classified as contaminated land, the investigation and any remediation of these must be dealt with on a risk management basis as established by the Environment Agency (2004). Drawing on these studies, the main elements of a site investigation, adapted for conditions encountered in Cornwall and Devon, are presented below. However, it must be noted that the legal position in the UK is that no-one can be held liable for the contamination from most mines and it is only since 1999 that the operator of a mine has had any obligation to deal with the consequences of abandonment (Environment Agency 2008; Health and Safety Executive 2013). Therefore, whilst the methods of investigation and remediation in areas of mining-related hazards can be identified, the responsibility for carrying out such works can only be established on a site-specific basis.

13.5.1 Desk studies

Desks studies are carried out prior to any work on site and the contents of a desk study report for assessing the mining hazard in Cornwall and Devon are summarized in Table 13.2 (based on Clayton & Smith 2013). Records of mining had to be recorded and lodged with the Secretary of State following the statute of The Coal Mines Regulation Act and Metalliferous Mines Regulation Act in 1872, although many of the abandoned mines in Cornwall and Devon predate this. However, mine records, and other relevant information, held at the Cornwall Record Office in Truro comprise approximately 3500 plans for 500 mines. The Devon in Focus website (http://www.devoninfocus.co.uk/mining/minedbase.php) lists information on 174 mines in the county. The website of the Cornwall and West Devon Mining World Heritage Site are also an excellent source of information (http://www.cornwall-mining.org.uk) and reference should always be made to the UK National Archive (http://www.nationalarchives.gov.uk) and the British Geological Survey (http://www.bgs.ac.uk).

With so much material available for Cornwall and Devon both as hardcopy and through the internet there is no excuse for failing to carry out a comprehensive desk study of any site of interest and this is likely to be the most cost-effective part of a site investigation.

13.5.2 Remote sensing

Encompassing both airborne and satellite-based scanners, temporal sequences of remote sensing imagery can enable numerous surface features from mining to be identified. Just using images from the visible part of the electromagnetic spectrum, it will be possible to locate both active and disused mining-related features, namely: tin streaming and shode workings (working of tin-rich stones lying on the ground surface); lode-back workings (a series of shallow shaft workings along the mineral lodes); quarries; tailings

Table 13.2. *Main elements of a desk study report for the assessment of mining hazard in Cornwall and Devon (based on Clayton & Smith 2013)*

- Plan and description of topography
- Site history – many mineral rich sites will have a long and varied history of which mining may be only one element. To assess the overall hazard the whole history of the site needs to be evaluated. In Devon and Cornwall, the sources of information may be extensive and many of these are now available online, notably that related to the World Heritage (UNESCO World Heritage Centre 2017). However, there are many older publications that may provide additional details, such as the two volumes by Dines (1956) and Rowe (1953)
- Compilation of existing ground investigation and hydrogeological data – access should be made to the British Geological Survey National Geoscience Date Centre (http://bgs.ac.uk/geoindex)
- Summary of the geotechnical, hydrogeological and geoenvironmental data based on: published articles; any available unpublished reports, notably those held by local or regional authorities; geological maps and memoirs; well records; and material gleaned from the site history on all hazards (subsidence, contamination, flooding, ground instability, seismicity)
- Information on the archaeology of site, which, given the long history of mining in Devon and Cornwall, is likely to be extensive (e.g. see http://www.cornisharchaeology.org.uk; http://www.devoninfocus.co.uk/mining/minedbase.php)
- Creation of an initial ground model that bring together geology, geotechnics, ground contamination, and groundwater data
- An initial risk register that includes all perceived risk related to ground conditions, groundwater, contamination, archaeology and ecology (Clayton 2001)
- Detailed recommendations for subsequent ground investigations

tips and lagoons; mining facilities (pump houses; processing plant, etc.); plus, reinstated former workings, some bell pits or crown hollows from collapsed old workings, and some open and capped shafts. This is demonstrated by the images of the Anna Maria Arsenic Works at the site of the Devon Great Consols Mine presented in Figure 13.26 (Google Earth Image) and Figure 13.27 (Oblique aerial image). However, modern remote sensing techniques operate well beyond the visible spectrum and using multi- and hyper-spectral remote sensing the presence of certain minerals in soils or rocks can be detected, heavy metal pollution in stream sediments can be mapped (Choe *et al.* 2008) or the health of vegetation in proximity to mineral working can be established (Van der Meer & Jong 2001). For example, Ferrier *et al.* (2007) used hyperspectral remote sensing data to monitor the environmental impact of hazardous waste derived from an abandoned gold mining area in SE Spain. A very helpful review of the use of visible and infrared remote imaging for the identification of hazardous waste is provided by Slonecker *et al.* (2010)

An important remote sensing development over the past two decades has been airborne LiDAR (light detection and ranging) and InSAR (interferometric synthetic aperture radar) to measure surface elevation (Komac 2018) and topographic modifications as land subsides (De Mulder *et al.* 2012). Examples of these techniques are limited, but for a natural geohazard investigation Hearn & Fialko (2009) suggested that InSAR deformation measurements may provide information on stress development and anisotropy of stress fields in the subsurface based on an investigation of the 1999 Hector Mine earthquake in California.

There does not appear to be much evidence of the use of remote sensing techniques to assess mining-related hazards in Cornwall and Devon, mainly because the region is so well documented and mapped. Nevertheless, there may be opportunities to use these remote sensing techniques for long-term monitoring of sites, notably for any contamination effects or subsidence.

13.5.3 Geophysics

The investigation for cavities associated with mining using non-invasive geophysical techniques both on the surface and when employed downhole in boreholes represents a very cost-effective method of investigation. McDowell *et al.* (2002) provided comprehensive guidance on the use of geophysics for all aspects of site investigation and Chapter 12 in RP940 (CIRIA 2017) identified the

Fig. 13.26. Google Earth Image of the Anna Maria Arsenic Works located at the site of the Devon Great Consols Mine in West Devon (image copyright Google Earth/CNES/Airbus, Getmapping plc, Infoterra Ltd & Bluesky, Maxar Technologies).

Fig. 13.27. Oblique aerial photograph of the Anna Maria Arsenic Works, Devon Great Consols Mine site (image courtesy of B Gamble).

recommended techniques to locate subsurface mine workings both using surface techniques (Table 13.3) and down boreholes (Table 13.4). No one single technique is applicable to all situations and, as noted by McDowell *et al.* (2002), considerable care is needed in the selection of the techniques to be employed and the design of the survey, to take account of the variable nature of the cavities to be identified (e.g. irregular shape, infilled with debris or water, shaft linings) and the properties of the host rock. All of these factors will affect the amplitude and width of geophysical anomalies associated with subsurface voids that require a specialist ground investigation geophysicist for accurate interpretation. Nevertheless, Culshaw *et al.* (2004) reported that micro-gravity and magnetic surveys had been used to successfully identify underground voids and open spaces such as abandoned shafts or tunnels in mines. Also, Chambers *et al.* (2007) reported on the use of surface and cross-hole electrical resistivity tomography for locating lost mineshafts.

Micro-gravity surveys measuring density difference in the ground, magnetic surveys, which register minute changes in the Earth's magnetic field, and electromagnetic methods that transmit an electromagnetic field to induce a secondary field underground, do not need to be ground-based and can be undertaken by aerial survey (Steuer *et al.* 2009). This is a more expensive option and will inevitably be at a lower resolution than ground-based surveys, but can be a useful approach if large areas are to be evaluated and target areas need to be identified.

13.5.4 Field mapping

No matter how comprehensive the data that are available from desk studies and remote sensing, field mapping of an area is still likely to reveal previously undetected features. Dearman (1991) provides the best summary of the range of techniques that can be utilized for engineering geological mapping, and Griffiths (2017) regarded mapping as an integral part of terrain evaluation to be employed in the creation of engineering geological ground models. Table 13.5 lists the type of features that should be included on an engineering geological map for any area in Cornwall and Devon (based on Griffiths 2015). Chapter 11 of RP940 (CIRIA 2017) also provides guidance on the natural and man-made features that should be identified during field reconnaissance of sites in mining areas.

13.5.5 Ground investigations

Invasive or intrusive ground investigation involves excavating trial pits and trial trenches, drilling boreholes, taking soil and rock samples for description and testing, carrying out *in situ* tests and installing various types of ground monitoring devices such as piezometers and inclinometers. The methods of carrying out all these works safely are laid down in BS5930 (BSI 2015) and the equivalent Eurocodes (Norbury 2016). A summary of the range of techniques that can be employed in intrusive investigations can be found in

Table 13.3. *Surface-based geophysical methods to identify old mine-workings (from CIRIA 2017)*

Method	Applications	Non, or reduced, application	Potential depth of investigation	Potential noise interference sources	Advantages	Limitations
Microgravity	Tunnel-like features, roadways, adits, etc.	Infilled features, vertical shafts. Mapping surface features	5–15 m (for typical target size and density contrast)	Vibration	Readily modelled	Slow coverage rate and cost
Electromagnetic EM31 type; one-person operation	General reconnaissance, shafts, debris, foundation remains	Urban areas, low-conductivity contrast features	5 m (instrumental limitation beyond this depth)	Non-target metal, power lines	Rapid coverage	Noise susceptibility
Electromagnetic EM34 type; two-person operation	Reconnaissance, voids	Urban areas, low-conductivity features	10–20 m	Non-target metal, power lines	High rate of coverage	Low resolution
Magnetic	Capped shafts, brick-lined shafts, debris from former buildings and foundations	Urban areas, low probability of presence of ferrous targets	10 m (depends on intensity of magnetization and mass of target object)	Non-target metal, power lines	Rapid rate of coverage	Magnetic noise
Electrical resistivity tomography	Tunnel-lie features, voids, cavities	Urban areas, use in low-contrast situations. Reduced application on some brownfield sites	20 m (dependent on array configuration and resolution required)	Stray electrical currents	High resolution	Low productivity rate. Does not work well where ground surface is electrically inhomogenous
Ground penetrating radar	Near-surface voids	Clayey or otherwise electrically conductive surface	1 m if mapping foundations; typically <5 m but can be up to 10 m if investigating voids in hard rock areas	Electrical noise. Above-ground metallic non-target objects can be problematic. Shallow voids overlying deeper targets	Rapid and high resolution	Limited depth of investigation

Table 13.4. *Downhole geophysical methods for abandoned mine-workings (from CIRIA 2017)*

Method	Applications	Non, or reduced, application	Potential radius of investigation	Advantages	Limitations
Natural gamma radiation	All types of workings including partially or fully collapsed closed workings	Workings in evaporites and limestone workings with low natural radiation	Typically about 300 mm	Operates in all conditions through plastic or steel casing in wet or dry holes	May exaggerate voiding owing to borehole wall spalling or collapse. Less sensitive in cased holes
Gamma–gamma density using gamma radiation	All workings including backfilled areas		500 mm, varying with density of strata or workings	Operates through plastic or steel casing in wet or dry holes	Less sensitive in cased holes
Resistivity	Water-filled workings only	Dry workings	<1 m		Only effective below water table that is affected by cavitation
Seismic – cross-hole and tomography	Cross-hole for ground stiffness; tomography to image voids and loose ground	Only indicates if workings are present or absent. Collapse structures are difficult to distinguish	Cross-hole measures stiffness of ground between holes c. 5 m apart; tomography typically 10–30 m apart	Strength/elasticity deduction is possible. Can form acoustic image of ground between boreholes. Grouted plastic casing is normally required	Tomography is ineffective in dry broken ground where high frequency seismic signals cannot be transmitted. Resolution of targets limited by wavelength considerations
Acoustic borehole imaging	Limited to water-filled disturbed ground around open workings	Signal lost when borehole wall collapses or open workings are present	Borehole wall	Orientation and spacing of fractures above workings can be determined	Borehole wall collapse and workings are indistinguishable. Strata record can be impeded by borehole wall smearing
Optical borehole imaging	Limited principally to above water table for inspection of strata above workings	Nature of penetrated voids	Borehole wall	Real image available immediately without processing. Gives 360° observation of borehole. Can distinguish borehole collapse form workings	Can be impeded by borehole wall smearing. Not always easy to distinguish between borehole wall collapse and workings

Table 13.5. *Data to be recorded on an engineering geology map*

Geological data
- Map units (chronostratigraphy and lithostratigraphy)
- Geological boundaries (with accuracy indicated)
- Description of soils and rocks (using BS5930 – see Norbury 2016)
- Description of exposures (cross-referenced to field notebooks)
- Description of state of weathering and alteration (note depth and degree of weathering)
- Description of discontinuities (as much detail as possible on the nature, frequency, inclination and orientation of all joints, bedding, cleavage, etc.)
- Structural geological data (folding, faulting, etc.)
- Tectonic activity (notably neotectonics, including rates of uplift)

Engineering geology data
- Engineering soil and rock units (based on their engineering geological properties)
- Subsurface conditions (provision of subsurface information if possible, e.g. rockhead isopachytes)
- Geotechnical data of the engineering soil and rock units
- Location of previous site investigations (i.e. the sites of boreholes, trial pits, and geophysical surveys)
- Location of mines and quarries, including whether active or abandoned, dates of working, materials extracted, and whether or not mine plans are available
- Contaminated ground (waste tips, landfill sites, old industrial sites)
- Man-made features, such as earthworks (with measurements of design slope angles, drainage provision, etc.), bridges and culverts (including data on waterway areas), tunnels and dams

Hydrological and hydrogeological data
- Availability of information (reference to existing maps, well logs, abstraction data)
- General hydrogeological conditions (notes on groundwater flow lines, piezometric conditions, water quality, artesian conditions, potability)
- Hydrogeological properties of rocks and soils (aquifers, aquicludes, and aquitards, permeabilities, perched watertables)
- Springs and seepages (flows to be quantified wherever possible)
- Streams, rivers, lakes, and estuaries (with data on flows, stage heights, and tidal limits)
- Man-made features (canals, leats, drainage ditches, reservoirs)

Geomorphological data (see Smith et al. 2011)
- General geomorphological features (ground morphology, landforms, processes, Quaternary deposits)
- Ground Movement Features (e.g. landslides, subsidence, solifluction lobes, cambering)

Geohazards
- Mass movement (extent and nature of landslides, type and frequency of landsliding, possible estimates of runout hazard, snow avalanche tracks)
- Swelling and shrinking, or collapsible, soils (soil properties)
- Areas of natural and man-made subsidence (karst, areas of mining, over-extraction of groundwater)
- Flooding (areas at risk, flood magnitude and frequency, coastal or river flooding)
- Coastal erosion (cliff form, rate of coastal retreat, coastal processes, types of coastal protection)
- Seismicity (seismic hazard assessment)

Clayton & Smith (2013) and Chapter 12 in RP940 (CIRIA 2017). As noted above, geophysical techniques can also be employed in boreholes to identify cavities associated with mine workings.

In investigating for subsurface voids care must be taken to avoid collapse of the ground during any pitting, trenching or borehole drilling operations. Inclined boreholes are usually required when locating features such as disused shafts and adits. Drilling operations will be much more cost-effective if preceded by mapping and geophysical surveys so that the invasive investigations can be targeted at suspect areas.

A key component of ground investigations in areas of former metalliferous mine workings is the identification and testing of potentially contaminated ground and water bodies.

Given the typical extractable mineral content of metalliferous ore bodies the waste generated is considerable and most of this has been stored in waste tips or behind tailings dams and might potentially be classified as 'contaminated' (see Figs 13.20, 13.23 & 13.24). There is a code of practice for undertaking ground investigations in potentially contaminated sites, BS 10175 (BSI 2011), and the Environment Agency has produced detailed guidance on risk management of contaminated ground. The Institution of Civil Engineers has published guidance specifically for the safe investigation of potentially contaminated land (Site Investigation Steering Group 2013), the Association of Geotechnical and Geoenvironmental Specialists has published guidance on the investigation of derelict land (AGS 2018) and the chemical analyses that need to be carried out in ground and water samples are discussed in Thompson & Nathanail (2003).

With respect to water bodies, Younger *et al.* (2002) identified the sources and pathways of contaminated mine water from quarries, tips and underground mines that can pollute both surface and groundwater bodies. The Environment Agency (2008) acknowledged that in the SW some 153 water bodies were deemed 'at risk' from mining pollution and six groundwater bodies were noted as having 'poor status' owing to contamination. The risk assessment for a ground investigation in potentially contaminated land and groundwater needs to ensure that safe sampling and testing procedures are in place to deal with a wide range of possible toxic substances. In Cornwall and Devon arsenic is widespread in the metalliferous mining waste tips (e.g. Camm *et al.* 2004) and in the mine waste materials washed down into coastal estuaries (Pirrie *et al.* 2003; Rollinson *et al.* 2007). It is not just the obvious metalliferous compounds that might be found; for example, Tatsi & Turner (2014) report on the occurrence of the highly toxic thallium in surface waters in Cornwall that can also be traced back to historical metal mining.

13.5.6 Developing the ground model

In the first Glossop Lecture, Fookes (1997) articulated the concept of the ground model as the means of compiling and representing engineering geological ground conditions. This concept has been developed over the intervening years and was summarized by Parry *et al.* (2014). Ground models are now integral to site investigation practice in the UK as laid down by BS5930 (BSI 2015). A very helpful summary chart of the various forms of model with respect to the stages in a project was provided by Eggers (2016) and is reproduced in Figure 13.28. This shows how the ground model progresses from an initial conceptual form through to actual construction models. This approach is valid for all forms of engineering works including the investigation and remediation of former mine workings and other brownfield sites, that is, previously developed areas that have become disused (definition by Marker 2018).

13.5.7 Hazard and risk assessment

Adopting a risk management approach to geotechnical engineering is now recognized as fundamental to safe and economic design (Van Staveren 2006; Fenton & Griffiths 2008) and compiling the geotechnical risk register is a key component in planning for any new development whether on a greenfield (i.e. where there are no previous engineering structures) or a brownfield site. Given that mining areas are brownfield sites and have the potential to be identified as contaminated land, a risk management

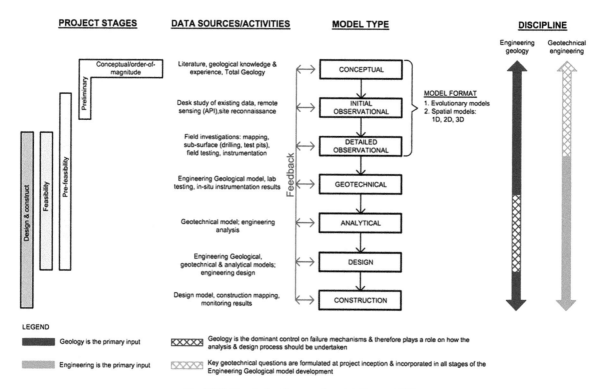

Fig. 13.28. Developing the ground model (Eggers 2016).

approach is fundamental (Environment Agency 2004). For Cornwall and Devon, the elements to be included in the register pertaining to mining hazards are presented in Table 13.6, based on the work of Cornwall County Council (2006), which included this material in its transportation strategy document for Camborne–Pool–Redruth. This list identifies the mining-related hazard, its potential consequences and the recommended remedial or preventative works. Assessing the magnitude of the risk should follow the standard approach used for all geohazards (e.g. Lee & Jones 2014) whereby the likelihood and severity of the risk is assessed using some form of ordinal scaling system both before and after any remedial or preventative works are undertaken.

13.5.8 Monitoring

The installation of various forms of ground, surface water and groundwater, and atmospheric monitoring systems may form part of a ground investigation. Monitoring systems need to be in place and measurements taken well before any construction or remedial works are undertaken to establish the pre-existing conditions. For example, in Cornwall the natural geological conditions mean that radon gas and various heavy minerals will exist in the air, ground and water even without human interference. Monitoring arrangements may also need to continue in the long term as the effects of any remediation or construction work on the environment must be established and checked for their efficacy.

Table 13.6. *Mining elements of a geotechnical risk register appropriate for use in Cornwall and Devon (based on Cornwall County Council 2006)*

Hazard	Consequences	Potential risk control measures or action
Uncapped mine shafts	Collapse at surface with potential fatalities	Design the ground investigation so that geophysics is used first prior to intrusive methods to obtain an indication of whether shafts along route or on development site are suitably capped
Collapse of underground adits or other workings	Collapse at surface with potential fatalities	Design the ground investigation so that geophysics is used first prior to intrusive methods to obtain an indication of whether adits along route or on development site are suitably grouted, or deep enough that bulking following any collapse will be sufficient to ensure void does not progress to surface
Encounter a mine shaft along a route or on a site that was not discovered during the desk studies	Localized settlement or instability requiring localized re-design of the construction works	Design the ground investigation so that geophysics is used first prior to intrusive methods to obtain an indication of where shafts are likely to be found
Encounter unexpected contaminated ground	Potential remediation required incurring additional cost and delaying construction programme	Conduct a detailed chemical testing programme as part of the ground investigation
Local variability in the properties of waste tips and brownfield sites	Potentially conservative or unsafe design	Adequate ground investigation to fully characterize material properties and determine variability
Drill boreholes and trial in contaminated land	Open pathways for contamination that could affect the local environment	Backfill boreholes and trial pits with suitable material to ensure no contaminant pathways exist in the long-term
Inadequate laboratory data to provide design parameters for construction in or over waste tips or brownfield sites	Settlement of structure, inadequate factors of safety, uneconomic design using over-conservative parameters	Sufficient ground investigation and sampling programme to characterize material parameters
Excavate trial pits or cuttings near or in waste tips	Potential slope failure	Design adequate excavation methods and slope angles both for temporary works and in the long-term.
Unexpected mineral or aggregate reserves found under site or along route	Construction prevents future extraction	Advance investigation of reserves and carry out review of old mineral permissions (ROMP). Mitigation may be needed to allow future access
Existing shafts and adits required for access	Late changes in site location or route alignment; prevention of future mining access	Advance planning and liaison with holders of any existing mining and mineral rights

For areas with the potential for subsidence the standard method of monitoring has been to install a topographic ground survey network with a base station on stable ground. This approach has now been effectively superseded by ground and airborne LiDAR and InSAR systems (De Mulder et al. 2012).

The Environment Agency of the UK carries out regular monitoring of surface and groundwater, and the responsibility for maintaining potable water supplies and correct sewage disposal in Cornwall and Devon lies with the company South West Water. Notwithstanding this, all existing active mines have a responsibility to monitor air and water quality in and around their operations. Monitoring of groundwater is undertaken using various types of piezometer installed in boreholes as described in the relevant Eurocode (EN ISO/TS 22475-1 2006a; EN ISO/TS 22475-2 2006b; EN ISO/TS 22475-1 2007). As noted above, groundwater monitoring in and around a potential development needs to be installed before any work starts on site to establish the hydrogeological model and the quality of the groundwater.

13.6 Planning, preservation, treatment and remediation

13.6.1 International and local planning

UNESCO World Heritage Sites are defined as 'designated heritage assets' in the government's National Planning Policy Framework (NPPF) that sets out planning policies for England, and how these are expected to be applied. Information on the status of World Heritage Sites and associated planning considerations can be found in NPPF18 'Conserving and enhancing the historic environment' (Ministry of Housing, Communities & Local Government 2018). Historic England (formerly English Heritage) provides the only official, up-to-date list of all nationally protected historic buildings and sites in England. World Heritage Sites and their settings (including buffer zones if present) are protected in England through the statutory designations and the planning system as outlined in NPPF18.

For planning applications that fall within the boundaries of the ten landscape areas of the Cornwall and West Devon Mining Landscape World Heritage Site there are three responsible local planning authorities that have the duty in plan-making and the determination of planning and related consents to protect the outstanding universal value of the site (cultural significance which is so exceptional as to transcend national boundaries and to be of common importance for present and future generations of all humanity): Cornwall Council (that in 2009 replaced Cornwall County Council and six district councils), West Devon Borough Council and Devon County Council. They are responsible for planning policies that apply to the World Heritage Site and determine planning permissions within it. They are assisted by the World Heritage Site Office, a non-statutory consultee in the planning process, with specialist advice and by the preparation of the World Heritage Site Management Plan and a Supplementary Planning Document, and by helping to prepare Local Development Frameworks. For major development projects, where potential impacts upon the World Heritage Site are likely to be greater, a pre-application advice service is provided. This process ensures, at the earliest possible stage, that there is a consistent approach to the application of Management Plan policies and the further guidance embraced in the Supplementary Planning Document.

The surface features of mine sites, including potentially contaminated land, shaft heads, adits and associated structures, are protected under the World Heritage Site designation. The Management Plan includes the presumption that principal points of underground access expressed at the surface, and any associated mine spoil, are Attributes of Outstanding Universal Value. Concerning the underground environment, Section 5.8 of the World Heritage Site Supplementary Planning Document states: 'The underground workings of the region's mining industry are significant and important context for the attributes which survive at surface. As such they should be afforded special consideration like aspects of setting'. Section 5.11 states that 'Shallow below-ground archaeological remains can play a major role in the significance of a large number of sites and must, therefore, be given careful consideration to fulfil the requirements of the NPPF'. Section 5.12 states that 'Where development could adversely affect underground workings, applicants should be alerted to their significance and encouraged to consider their preservation'.

13.6.2 Preservation

An interesting aspect of the recent treatment of mine hazards is the integration of conservation to the adaptation of techniques in the context of protected landscapes, specifically the Cornwall and West Devon Mining Landscape World Heritage Site.

The shaft-plugging campaigns of the early 1990s, and the pioneering conservation treatment of sites in the later 1990s and 2000s, offer stark contrasts. The former rendered mineshafts permanently (at least in the main) sealed at the interface between the underground environment and surface area of the mine; the latter preserved both the historic fabric and setting and importantly kept open the potential for subsequent access for values connected to geological and mineralogical sciences, ecology, archaeology and historical interests.

Integrated safety and conservation treatment have been systematically applied to numerous sites across the region by owners and stakeholders of the 'Cornish Mining' World Heritage Site, including the former Kerrier District Council, Cornwall Council, the National Trust and others. Characteristic features such as shafts, adits, ore-processing sites including arsenic refineries, various categories of waste tips and general mining relics including waterwheel pits and unstable buildings such as engine houses and chimney stacks, have been made safe whilst at the same time preserved and conserved for future generations. The right to experience and enjoy these sites has been considered, too, with the works being accompanied by enhanced public access, together with interpretation and education. A number of projects have been implemented that have such safety and conservation works as part of their overall delivery, including the St Just Coast Project (£7.5 m, completed 2005, including over 70 mining/industrial sites, 13 engine houses and an estimated 2000 mine shafts), the Geevor Tin Mine (£3.8 m, completed in 2008, including building works with extensive cement-asbestos roof cladding being replaced with modern safer equivalents, and safety and enhanced public access works underground), the East Cornwall Regeneration Project (£2.1 m, completed in 2008, including several significant mine sites on the Cornish side of the Tamar Valley Mining District), the Mineral Tramways Heritage Project (£6 m, completed in 2009, including numerous mine sites accessed by 60 km of trails mostly along former mineral railways), the Heartlands Project (£35 m, completed in 2012, focussed on the Robinson's Shaft section of South Crofty Mine), the Tamar Valley Mining Heritage Project (£7 m, completed 2013, including mine sites and an arsenic works), the Caradon Hill Area Heritage Project (£2.9 m, completed in 2014, including several significant mine sites around Caradon Hill) and King Edward Mine (£1.2 m, completed in 2015).

13.6.3 Treatment and remediation through engineering works

The operation of existing quarries and maintenance of safe mining waste tips throughout the UK fall under the Quarry Regulations 1999 (Health and Safety Executive 2013). Regulation 30 deals with the safety of excavation and tips and states that these must be 'designed, constructed, operated and maintained so as to ensure instability or movement which is likely to give rise to health and safety of any person is avoided'. Regulation 33 and Schedule 1 makes it clear what is expected in a geotechnical assessment to guarantee this is achieved. China clay extraction from the St Austell Granite, the only Cornish area still being worked, falls under this legislation and a 'China Clay and Tipping and Restoration Strategy' was prepared jointly by the mineral planning authority and the industry for the period up to 2050 and beyond (Cornwall County Council 2014). Restoration involves either using the waste to backfill old pits, the preferred option, or re-profiling and vegetating old tips. This process was taken further in Cornwall with the Cornish Heathland Project that created 750 hectares of heathland, which was the largest heathland re-creation project in Western Europe (Kaolin & Ball Clay Association UK 2007). When not backfilled former quarries in the UK have been used for many purposes, including housing estates, nature reserves and marine sports (http://bgs.ac.uk/mineralsUK/sustainability/restoration.html). As noted above, a unique use for a former china clay opencast mine in Cornwall has been the creation of the Eden Project (http://edenproject.com), where a series of massive biomes have been set up that house plant species from around the world to create a successful educational facility and tourist attraction (Fig. 13.19).

With respect to underground mining, all construction work in Cornwall and Devon in areas where old mine workings are identified or suspected should follow the guidelines of RP940 (CIRIA 2017), notably Chapter 13, where treatment and remediation techniques are reviewed, and Chapter 14, on the treatment of abandoned mine entries. Initially the scale of the hazard must be evaluated before any treatment is considered. If the voids from workings are sufficiently deep to make it unlikely that any collapse will to migrate to the surface, then no treatment is needed. For guidance on how to assess the likelihood of migration to the surface, reference should be made to Chapter 5 of RP940 (CIRIA 2017). Typically, there are four broad engineering techniques that can be used to treat or isolate mine workings:

(1) drilling and injection treatments – usually involving grouting;
(2) bulk filling, both underground and surface industrial features;
(3) bridging techniques, either using hard engineering approaches (e.g. concrete rafts) or soft engineering (e.g. geosynthetic mattresses); and
(4) earthworks solutions.

Use of these techniques has a long history that is widely reported in the literature, mainly in relation to construction issues in the extensive UK former coalfields (e.g. Littlejohn 1979*a*, *b*; Waltham 1989). Coal in the UK has been extracted using longwall techniques since the 1900s and this has resulted in widespread ground subsidence. However, subsidence normally stops within two years of mining ceasing and future subsidence in such areas is unlikely. Historically coal and ironstone mining took place using pillar and stall techniques where the roof was supported by natural pillars and 80% of the deposit was mined. Former pillar and stall workings remain a hazard in many of the former UK coalfields. The nature of underground metalliferous mining in Cornwall and Devon is that extraction followed the mineral veins, typically resulting in adit entrances (Fig. 13.10), steep stopes for extraction (Fig. 13.15) and deep vertical ventilation and access shafts. These shafts (and stopes) and adits represent the main hazard and require identification and, where necessary, treatment. As mentioned in the section on underground voids and shafts, some older shafts were covered by timbers (typically railway sleepers) and soil but over time the wood rots and the shafts re-open. This was the case in east Cornwall at Gunnislake, where construction of a building estate had taken place over such shafts and they had not been identified during site investigations (Site Investigation Steering Group 1993) (Fig. 13.14).

The methods of dealing with shafts and adits are dependent on the risk they pose. Table 13.7, taken from Chapter 14 of RP940 (CIRIA 2017) gives a full list of options that are available. There are two aspects of Table 13.7 to emphasize: as adits have become important habits for roosting bats, keeping them open is an important environmental consideration

Table 13.7. *Minimum treatment for shafts and adits related to risk category (RP940, CIRIA 2017)*

Situational risk*	Feature	Characteristics	Treatment options (increasing in standard)
Low	Shaft	Filled with fill slumped <1 m deep	(1) Fence (2) Fence and consider mounding hardcore (3) Lay anchored geosynthetic membrane
	Shaft	Filled with fill slumped >1 m deep	(1) Fence and install grill over shaft top (2) Fence and mound hardcore (3) Lay anchored geosynthetic membrane
	Shaft	Open	(1) Fence and install grill over shaft top (Figure 13.SI.4) (2) Fill shaft and fence (3) Cap at rockhead
	Adit	Damage has occurred to a secure entrance	(1) Metal fence (2) Repair existing entrance
	Adit	Open entrance	(1) Metal fence (2) Repair portal with bricked-up entrance and bat letterbox opening (3) Install grill and bat letterbox
Medium	Shaft	Filled but fill has slumped	(1) Lay anchored geosynthetic membrane and monitor (2) Drill and grout shaft fill (3) Install reinforced concrete cap at rockhead
	Shaft	Open	(1) Install reinforced concrete cap at rockhead or piled reinforced concrete cap (2) Bulk fill shaft with engineering controlled material and install either a reinforced concrete cap at rockhead, a piled reinforced concrete cap, or a grouted cap
	Shaft	Located near to building or structure but beyond their load spread	(1) Install reinforced concrete cap at rockhead (2) Drill and grout, and consider need to install a reinforced concrete cap (3) Install piled reinforced concrete cap or grouted cap
	Adit	Damage to existing secure entrance or water egress	(1) Repair entrance (2) Make drainage provision
	Adit	Open	(1) Repair with bricked-up entrance or grill, both with bat letterbox and drainage outlet/outfall
High	Shaft	Filled and no evidence of movement	(1) Investigate to assess condition (2) Drill and grout, and consider installation of reinforced concrete cap (3) Install reinforced concrete cap at rockhead or piled reinforced concrete cap
	Shaft	Filled but fill has slumped	(1) Investigate to assess condition (2) Drill and grout, and consider installation of reinforced concrete cap (3) Install reinforced concrete cap at rockhead or piled reinforced concrete cap
	Shaft	Located near to building or structure that may transmit stress to shaft	(1) Install reinforced concrete cap at rockhead (2) Drill and grout, and install reinforced concrete cap (3) Install piled reinforced cap with or without grouting (4) In conjunction with 1 to 3 consider zone of external grouting
	Adit	Damage to existing secure entrance	(1) Repair if assessed risk to public allows (2) Excavate out entrance and fill with materials matched to environmental circumstances, and drill and grout under loaded area

(Continued)

Table 13.7. *Continued*

Situational risk*	Feature	Characteristics	Treatment options (increasing in standard)
	Adit	Located near to the new-build and no surface expression	(1) Investigate to locate and assess condition (2) Excavate out portal and length within area potentially stressed by proposed development. Fill with materials matched to environmental circumstances, and drill and grout (3) Drill and grout remotely, including the roof strata (4) In all cases provide for drainage outfall away from new-build.

*Situational risk: low, remote locations, farmland or other non-development; medium, grounds of private dwellings and land with low-use public access such as minor footpaths; high, urban spaces, major roads, railway infrastructure, multiple dwellings and areas with high concentrations of people directly affected by the mine entry.
- The risk category will elevate when unstable ground exists above an adit or settlement has occurred outside a shaft.
- The condition of any lining or internal supports may either mitigate or elevate the risk.
- The need to elevate the risk category should be considered for large features.
- A continued drainage or venting function requirement will impact upon treatment options.

(Figs 13.15 & 13.16); drainage issues from adits must always be considered in any ground treatment.

13.6.4 Derelict land reclamation

The reclamation of derelict land, effectively brownfield sites, is not unique to Cornwall and Devon and the Government describes the UK as a leader in this field with globally recognized expertise (Department for International Trade 2015). Cornwall Council provides specific guidance to all developers intending to redevelop brownfield sites in the county (Cornwall County Council 2010) and these specifically identify mining and mining sites as requiring a contaminated land site investigation. The Cornwall Council guidance also contains checklists for investigations and remediation. Further useful guidance can be found in Strange *et al.* (2016) and a summary of remediation methods for contaminated sites was provided by Wood (1997). Essentially there are two broad approaches to dealing with land that is formally deemed to be contaminated before a new development can go ahead:

(1) Engineering – this primarily involves excavation and removal of contaminated ground and then disposal in landfill.
(2) Process-based techniques – these include physical, biological, chemical stabilization/solidification and thermal processes, the aim being to modify any contaminants to a less toxic, mobile or reactive form. This can sometimes be carried out *in situ* or material is taken away and treated *ex situ*.

Case studies in the use of these techniques are well documented (Department for International Trade 2015) with many firms having specialist expertise to carry out this work. Given that this is not a problem unique to Cornwall, the details of the methods are not discussed further in this paper.

13.6.5 Mine water contamination and remediation

As discussed previously, the drainage of mines and the discharge of the extracted water into the environment in areas around abandoned mines can cause severe environmental pollution if the water is found to be contaminated. In much of the literature this is incorrectly referred to by the collective term 'acid mine drainage', but (McGinness 1999) and Younger (2002) recommended that a better expression would be 'mine water pollution', as many of the most problematic mine drainage waters in the world are alkaline or at least of neutral pH (Younger *et al.* 2002). However it is suggested that 'mine water contamination' is a more appropriate expression. The chemical nature of mine waters is summarized in McGinness (1999). Mine water contamination has become an issue in the UK owing to the widespread closure of mines as previously these were drained by pumps to enable the extraction of the various minerals in zones beneath the groundwater table. Most of the problems are associated with abandoned coal mines (Hughes 1994) and the responsibility of dealing with pollution by contaminated mine water from these mines falls under the responsibility of the UK Coal Authority working in conjunction with various environmental regulators (Johnston *et al.* 2007). However, as noted by Jarvis *et al.* (2007), there is no equivalent body to oversee the management of abandoned non-coal mines, notably the metal mines in Cornwall and Devon, and to date this has been left mainly in the hands of the Environment Agency.

Younger (2002) identified three principal options for dealing with contaminated mine waters:

(1) monitored natural attenuation where natural processes are deemed sufficient to deal with contamination and monitoring is put in place to check this is the case;
(2) prevention or minimization of the contaminant release processes;

(3) mine water treatment by either passive (e.g. use of wetlands) or active (e.g. chemical treatment) methods.

Option one is the most sustainable in the long term and is dependent on the dilution occurring in the water course receiving the mine water discharge and the degree of 'self-cleaning' that takes place in the mine workings over time as water flushes through the mine network. Younger (2002) indicates that option two is not feasible where there are vast labyrinths of underground workings or large surface mine waste repositories. The process involves covering small and moderate size tips with low-permeability covers to inhibit the activities of iron- and sulphur-oxidizing bacteria, although the effects rarely last longer than six months.

Option three is comprehensively documented in Younger *et al.* (2002). Passive treatment involves the use of natural energy sources such as the topographic gradient, microbial metabolic energy, photosynthesis and chemical energy and requires infrequent maintenance to operate. Younger (2002) describes typical passive systems as: aerobic, surface-flow wetlands; compost wetlands with significant surface flow; mixed compost/limestone systems with predominantly subsurface flow; subsurface reactive barriers treating acidic metalliferous groundwaters; closed-system limestone dissolution systems for zinc removal from alkaline waters; and roughing filters for the aerobic treatment of ferruginous mine waters. Active treatment refers to the use of conventional wastewater treatment processes and for acidic and/or ferruginous mine drainage involves three steps: (1) oxidation to help convert soluble ferrous iron to the far less soluble ferric iron, which also allows the pH to rise by venting excess carbon dioxide; (2) dosing with alkali, usually hydrated lime, both to raise the pH and to supply hydroxyl ions for the rapid precipitation of metal hydroxide solids; and (3) accelerated sedimentation using a clarifier or lamellar plate thickener, often aided by the addition of flocculants and/or coagulants.

The Environment Agency (2008) describes the largest mine-water treatment plant in the UK as being found in Cornwall at the Wheal Jane tin mine, and the background and remediation of this site are described by Younger (2002). McGinness (1999) states that between £9.5 million and £14 million had been spent on this project up to 1997 and a BBC News Report (Morris 2014) put the annual cost of the project at £1.5 million. The need for these works followed an incident in January 1992 when 45 million litres of heavily contaminated water gushed out of the recently closed Wheal Jane mine when pumping ceased. The water was contaminated with cadmium, arsenic, zinc and iron and it flowed into the Carnon River, creating a plume of orange water in Falmouth Bay. Temporary remedial works involved adding lime to the discharge to cause the metals to settle behind the existing tailings dam at the mine site. The long-term solution, which became operational in 2000, was to build an active chemical treatment plant (Coulton *et al.* 2003) whereby mine water was pumped from the mine shaft and lime added to raise the pH and cause the metals to create insoluble compounds. These compounds settled with the help of chemical flocculants and the sludge was pumped into a tailings reservoir whilst the treated water was discharged into the river. A passive system of treatment was tried initially using aerobic and anaerobic wetlands, and although this worked, it was found to require too much land to treat the whole discharge, so this active process continues to the present day.

Other solutions to the treatment of contaminated mine water have been trialled in Cornwall, notably the one to treat the Red River and South Crofty mine discharge at Redruth (Potter & Jarvis 2006; Younger 2002). Here the treatment plant uses a combination of anoxic limestone drains and aerobic wetlands to encourage the adsorption properties of the ochre in the mine water discharge to co-precipitate copper and arsenic that is present in the Red River from other abandoned mines in the area. South Crofty was acquired in July 2016 by a Canadian company and the mine has recently been subject to discussion over its reopening. A new mine permit was granted in 2013 and this is valid until 2071 (Strongbow 2016). As part of this discussion the company carried out a trial in the use of a propriety high-density sludge process to treat mine water (Environment Times 2017) which suggests that the mine could be dewatered and the extracted water treated successfully over a 19–24 month period ahead of reopening.

Monitoring and treating surface and groundwater for pollution from mine waters remains an issue throughout Cornwall and mined areas of Devon. The cost of the treatment of contaminated mine water at Wheal Jane suggests that this approach is not an option for most abandoned mines but could be appropriate for any active mines, such as a re-opened South Crofty in the future. However, in terms of long-term environmental sustainability, natural attenuation with monitoring and passive treatment must be regarded as the most suitable course of action.

13.6.6 Case studies of mine site treatment and remediation

Cornwall and West Devon provide a wealth of case studies where mine sites have been treated in various ways. These projects have involved appropriate specialists from the World Heritage Site, Historic Environment advisers from Cornwall Council, Devon County Council and West Devon Borough Council, and Historic England, together with a range of consultants. Funding has been provided by various organizations, including local authorities, the Heritage Lottery Fund, the government's former South West Rural Development Agency and Government Office South West implementing European funding including Objective One (Cornwall qualified for this in 1999) and Objective 2 (Devon) programmes, the European Regional Development Fund, the Rural Development Programme for England, Historic England and the National Trust. Some examples are presented below.

13.6.6.1 Wheal Peevor, Redruth, Cornwall. Kerrier District Council (2003–07)

Shaft safety and buildings conservation works were undertaken, together with public access and interpretation at this important site (Scheduled Monument) (Sharpe 2007). This site contains several shafts, an arsenic works and three listed buildings that form a significant group of engine houses representing the principal functions of pumping, winding and stamping, which present an iconic skyline feature that is clearly visible from the A30 Trunk Road. The site also hosts two areas of rare bryophytes.

Shafts evaluated during assessment were graded according to the archaeological record of important associated features, or the likely presence of their buried remains, and where a shaft was known to connect to a subsurface drainage system (adit), loose material was not allowed to enter the shaft and potentially disrupt water flow.

The most prominent shaft, Sir Frederick's, dates from before 1860 and was sunk to a depth of around 200 m. The shaft collar was in poor condition and had coned owing to subsidence, undermining concrete former head-frame plinths, the masonry balance-bob pit and the condenser housing adjacent to the bob-wall (posing a threat to the listed structure of the engine house itself). Treatment took the form of reshaping the cone to reduce its slope angle, cutting the base of the shaft cone to provide a foundation for a ring beam to stabilize the ground. Stainless steel reinforcement was then applied, shuttered externally and internally in timber to reflect the original shaft timberwork, and a concrete ring beam cast *in situ*. A galvanized steel grille was then fixed over the shaft to permit access for greater horseshoe bats that are known to use the site. Geotextile netting was fixed to stabilize remaining historic masonry and dump material around the shaft. Other shafts were treated depending on their condition. Those that were open were variously treated by the installation of hollow concrete plugs formed at stable bedrock level, followed by stacking pre-cast concrete reinforcing rings (Fig. 13.29) in the collar with the surrounding void being backfilled with concrete, being finally secured with granite-faced 'bat castles' at the surface. Those that were filled or hopelessly choked (including un-named/previously unknown shafts) were secured by the placement of concrete plugs and the re-formation of the ground contour, and the shaft location and dimensions marked with railway sleepers and a shaft marker.

The arsenic works, constructed in the nineteenth century to clean the tin ore and recover arsenic as a by-product, was also treated as part of the scheme. All above-ground masonry was re-pointed, some reconstruction and re-facing was undertaken for structural and safety reasons and limited rubble was cleared to expose arched power vault openings. A specialist decontamination team cleared the arsenic works chimney, and a short section of flue adjoining, and the material was bagged and removed in a closed skip to secure landfill. The interior of the stack was grit-blasted to remove contaminated crusted salts and soot, and again waste was bagged and closed-skipped to landfill. Flue openings were grilled.

Fig. 13.29. Retained shaft access showing stacked pre-cast concrete rings, Wheal Peevor, Redruth, Cornwall (Gamble/Cornwall Council 2007).

The deteriorated top of the stack was demolished and reconstructed, guided by pre-works photographs and using hydraulic lime pointing mixed to match the original (as was used elsewhere on the site).

The rest of the site works comprised consolidation and conservation of the engine houses, boiler houses and other structures, together with access works, improvement of drainage and a programme of interpretation.

13.6.6.2 The National Trust

Conservation and safety works undertaken by the National Trust are exemplified in several Cornish sites, particularly mines and tin and arsenic works along the St Just coast in West Penwith, and Trewavas Mine in the Godolphin area. All are now included in the World Heritage Site.

St Just Coast, West Penwith, Cornwall (1995–2005). The sensitive treatment of shafts and surrounding dumps, and the exemplary safety and conservation treatment of buildings and other structures, are a hallmark of the project carried out by the National Trust in the St Just mining coast in far west Cornwall (Fig. 13.30). Extensive archaeological assessment that underpinned project design mindful of the bid for World Heritage Site status was a key prelude to the works where over 70 industrial sites were treated.

In the case of mineshafts, the general policy was to keep the shaft open (Fig. 13.31) and to implement protective measures to prevent accidental entry. The traditional design of a Cornish 'abandonment safety hedge', once constructed in the thousands to meet the requirements of the Metalliferous

Fig. 13.30. Open gunnies at Wheal Hermon, St Just, Cornwall (Gamble/Cornwall Council 2014).

Mines Regulation Act, 1872, was adopted (Fig. 13.32). These were built using granite and killas walling stone built in drystone fashion. An inconspicuous post and barbed wire fence were installed inside the circular stone wall as a further

Fig. 13.31. Shaft protection at the open Cargodna Shaft, Wheal Owles, St Just, Cornwall, a section in which 19 men and one boy drowned in 1893 through holing into flooded abandoned mine workings (Gamble/Cornwall Council 2015).

Fig. 13.32. Newly-built traditional shaft hedge below the Counthouse, Botallack Mine, St Just, Cornwall (Gamble/Cornwall Council 2005).

precaution, whilst not impinging on access for bats or future bona fide investigations. Some shafts, for example where open immediately adjacent to a listed pumping engine house, were grilled.

The arsenic works at the world-famous Botallack Mine were also treated as part of the project. The well-preserved Brunton calciner, flues, condensing chambers and stack were all decontaminated, consolidated and conserved (Fig. 13.33). At completion of works, a self-guided trail was provided for visitors to follow the route of the arsenic-laden gases *through* the complex.

Trewavas Mine, Breage, Cornwall (2008–09). The National Trust between October 2008 and May 2009 undertook an immensely challenging safety and conservation project at a mine located perilously close to the edge of dramatic sea cliffs (Fig. 13.34). It involved an estimated 39 tonnes of scaffolding and 400 m of steel tethering cable to secure the structures, and 40 tonnes of sand and 8 tonnes of building lime for re-pointing significant structures. This followed the acquisition in April 2008 of a stretch of the coastline at Trewavas, Breage, in SW Cornwall (Sharpe 2009).

Substantial remains of Trewavas Mine lie seaward of the South West Coast Path at Trewavas Head on the southern margin of the Godolphin and Tregonning granite, 3 miles west of Porthleven. Two engine houses sited on the precipitous cliff edge are Grade II listed buildings and, formerly housing pumping-engines, have immediately adjacent (open) shafts. Engine (later Old Engine) Shaft has two shaft collars side-by-side, one for pumping and one for hoisting, sunk on Old Lode that ran out under the seabed in a southeasterly direction. Adits can be seen emerging from the cliff below (Gamble 2011*a, b*). The whole mining cliffscape is designated a Scheduled Monument and the burrows (waste tips) from ore dressing are the Type Locality of the uranium mineral Tristramite (Golley & Williams 1995).

A creative approach to the project ensured an exemplary outcome based on the minimum necessary safety and

Fig. 13.33. Arsenic condensing chambers, Botallack Mine, St Just, following decontamination, safety and conservation works (Gamble/Cornwall Council 2006).

Fig. 13.34. Trewavas Old Engine Shaft, with its granite masonry twin collar below the engine house on the edge of the cliff (Gamble/Cornwall Council 2006).

conservation interventions as outlined in a Conservation Management Statement drawn up in advance of the project. At Trewavas, the engine houses sited so dramatically on the cliff edges have been consolidated and their shafts remain open to nesting 'Cornish' choughs, the red-legged, red-billed member of the crow family that has centuries of association with the Duchy. The previous straggly barbed wire around New Engine Shaft was removed and replaced with a tramway post and recycled scaffolding guy rope cable fence to form an unobtrusive, long-lasting and effective barrier to prevent accidental public access.

Devon Great Consols, Tavistock, Devon (implemented as part of the Tamar Valley Mining Heritage Project, 2006–13, by Tamar Valley AONB, West Devon Borough Council, Devon County Council and others). The Tamar Valley Mining Heritage Project (Fig. 13.35) treated a number of mine sites in the West Devon section of the Tamar Valley; including the 1920s, arsenic works at Wheal Anna Maria, Devon Great Consols (Fig. 13.36). Here the arsenic–iron sulphide ore was crushed, washed and roasted, and the hot gases drawn along flues into condensing chambers where they cooled and deposited crude arsenic on the walls. The highly toxic deposit was then scraped off by men and boys and re-roasted and refined to 99% pure arsenic trioxide. The works were abandoned in 1925 and the important and remarkably complete complex (Scheduled Monument and Grade II listed building) has now been consolidated, contamination selectively cleaned or covered, the site secured and access pathways provided with overlooks and site interpretation (Fig. 13.37). In the above process, a remarkable survival was uncovered and conserved – flatbed reverberatory refining furnace.

13.7 Conclusions

Whilst the abandoned flints mines in Norfolk are older (Topping 2004), the metalliferous mines and pits of Cornwall and West Devon have been continually worked for over 2000 years. This has left a landscape that is unique in the UK and selected areas noted for their exceptional contribution to Britain's Industrial Revolution led to the region being designated a UNESCO World Heritage Site for its metalliferous mining heritage in 2006. Whilst metalliferous mining in Cornwall has now effectively ceased, the Drakelands tungsten opencast mine opened near Plymouth in 2015 with a planned operating life of 10 years (Devon County Council 2015) although changing financial circumstances and the bankruptcy of the main mineral extractor led to the closure of the mine in October 2018 (Waddington 2018). In addition, the extraction of china clay continues in both Cornwall and West Devon, ball clay extraction continues in mid-Devon and there are active aggregate quarries found in both counties.

Fig. 13.35. Devon Great Consols, general aerial view of the principal former dressing areas for copper and arsenic ores (Gamble/Tamar Valley AONB 2011).

Fig. 13.36. Devon Great Consols, Wheal Anna Maria Arsenic Works before safety and conservation works (Gamble/Tamar Valley AONB 2006).

Fig. 13.37. Devon Great Consols, Wheal Anna Maria Arsenic Works after safety and conservation measures (photo credit: B. Gamble).

Thus, the region provides both a historic and an active mining environment and this adds to the unique features of the mining-related hazards. In areas of former metalliferous mines, the associated hazards are a function of the Paleozoic rocks, the predominant steeply dipping nature of mineral veins and consequent shaft mining, the great depth and complexity of some of the mines, the waste derived from processing metallic ores, the long history of exploitation and the contamination associated with various by-products of primary ore-processing, refining and smelting, notably arsenic. The hazards associated with china clay, ball clay and aggregate extraction are mainly related to the volume of the inert waste products and the need to maintain stable spoil heaps, and the depth of the various tailings ponds and pits.

When Cornwall and West Devon were global leaders in tin and copper extraction, the need to develop deep and complex mines led to innovations in mining technology, particularly in the use of steam power, that had an enormous influence on the industrial revolution and in global mining. However, innovation did not end in the nineteenth century and, as shown in this paper, dealing with the long heritage of mining has resulted in the county being a leader in mining heritage preservation and the treatment and remediation of related hazards.

Acknowledgements This paper has drawn heavily on the material prepared for the UNSECO World Heritage bid and various websites and data about the designated sites in Cornwall and West Devon that continue to be maintained. Permission to use this information is gratefully acknowledged. Also, the online open access CIRIA report RP940 (2017) has been an important source of information on the treatment of abandoned mine workings.

Funding This research received no specific grant from any funding agency in the public, commercial, or not-for-profit sectors.

References

AGS 2018. *Guidance on the Investigation of Derelict Sites*. Association of Geotechnical and Geoenvironmental Specialists, Kent, Bromley.

ANDREWS, J.R., BARKER, A.J. & PAMPLIN, C.F. 1988. A reappraisal of the facing confrontation in north Cornwall: fold- or thrust-dominated tectonics? *Journal of the Geological Society of London*, **147**, 777–788, https://doi.org/10.1144/gsjgs.145.5.0777

ARUP 1992. *Mining Instability in Great Britain: Summary Report*. Department of the Environment.

BADHAM, J.P.N. 1982. Strike-slip origins – an explanation for the Hercynides. *Journal of the Geological Society, London*, **139**, 493–504, https://doi.org/10.1144/gsjgs.139.4.0493

BARNES, R.P. & ANDREWS, J.R. 1986. Upper Palaeozoic ophiolite generation in south Cornwall. *Journal of the Geological Society, London*, **143**, 117–124, https://doi.org/10.1144/gsjgs.143.1.0117

BARR, J. 1970. *Derelict Britain*. Penguin, Harmondsworth, England.

BARTON, D.B. 1961. *A History of Copper Mining in Cornwall and Devon*. D. Bradford Barton Ltd, Truro.

BARTON, D.B. 1967. *A History of Tin Mining and Smelting in Cornwall*. D. Bradford Barton Ltd, Truro.

BARTON, D.B. 1969. *The Cornish Beam Engine: Its History and Development*. D. Bradford Barton Ltd, Truro.

BRISTOW, C.M. 1998. China clay. *In*: SELWOOD, E.B., DURRANCE, E.M. & BRISTOW, C.M. (eds) *The Geology of Cornwall*. Exeter University Press, 167–178.

BRISTOW, C.M. 2014. Carlyon Bay watch. http://www.carlyonbaywatch.com (accessed 21 November 2018).

BRISTOW, C.M. & EXLEY, C.S. 1994. Historical and geological aspects of the China-clay industry of Southwest England. *Transactions of the Royal Geological Society of Cornwall*, **21**, 247–314.

BRISTOW, C.M. & ROBSON, J.L. 1994. Palaeogene basin development in Devon. *Transactions of the Institution of Mining and Metallurgy*, **103**, 163–224.

BRISTOW, C.M., PALMER, Q.G. & PIRRIE, D. 1992. Palaeogene basin development: new evidence from the southern Petrockstowe Basin, Devon. *Proceedings of the Ussher Society*, **8**, 19–22.

BRISTOW, C.M., PALMER, Q.G., WITTE, G.J., BOWDITCH, I. & HOWE, J.H. 2002. The ball clay and china clay industries of Southwest England in 2000. *In*: SCOTT, P.W. & BRISTOW, C.M. (eds) *Industrial Minerals and Extractive Industry Geology*. Geological Society, London, 17–41.

BSI 2011. *BS 10175: 2011 Investigation of Potentially Contaminated Sites*. Code of Practice. British Standards Institute, London.

BSI 2015. *BS5930: Code of Practice for Site Investigation*. 4th edn. British Standards Institute, London.

BUCKLEY, J.A. 2000. *The Great County Adit*. Penhellick, Pool, Cornwall.

BURT, R., BURNLEY, R. & WAITE, P. 1987. *Cornish Mines*. Mineral Statistics for the United Kingdom, 1845–1913, **7**, University of Exeter.

CAMERON, D.G., BIDE, T., PARRY, S.F., PARKER, A.S. & MANKELOW, J.M. 2014. *Directory of Mines and Quarries*, 10th edn. British Geological Survey, Keyworth.

CAMM, G.S., GLASS, H.J., BRYCE, D.W. & BUTCHER, A.R. 2004. Characterisation of a mining-related arsenic contaminated site, Cornwall, U.K. *Journal of Geochemical Exploration*, **82**, 1–15, https://doi.org/10.1016/j.gexplo.2004.01.004

CATTELL, A.C. 1996. The connection between the Decoy and Bovey Basins. *Proceedings of the Ussher Society*, **9**, 130–132.

CHAMBERS, J., OGILVY, R., WILKINSON, P., WELLER, A., MELDRUM, P. & CAUNT, S. 2007. Locating lost mineshafts. *Earthwise*, **25**, 26–27.

CHAPMAN, T.J. 1989. The Permian to Cretaceous structural evolution of the Western Approaches Basin (Melville sub-basin), UK. *In*: COOPER, M.A. & WILLIAMS, G.D. (eds) *Inversion Tectonics*. Geological Society, London, Special Publications, **44**, 177–200, https://doi.org/10.1144/GSL.SP.1989.044.01.11

CHESLEY, J.T., HALLADAY, A.N., SNEE, L.W., MEZGER, K., SHEPHERD, T.J. & SCRIVENER, R.C. 1993. Thermochronology of the Cornubian batholith in southwest England: implications for pluton emplacement and protracted hydrothermal mineralization. *Geochimica et Cosmochimica Acta*, **57**, 1817–1835, https://doi.org/10.1016/0016-7037(93)90115-D

CHOE, E., VAN DE MEER, F.D., VAN RUITENBEEK, F.J.A., VAN DER WERFF, H.M.A., DE SMETH, J.B. & KYOUNG-WOONG, K. 2008. Mapping of heavy metal pollution in stream sediments using combined geochemistry, field spectroscopy, and hyperspectral remote sensing: a case study of the Rodalquilar mining area, SE Spain. *Remote Sensing of Environment*, **122**, 3222–3233, https://doi.org/10.1016/j.rse.2008.03.017

CIRIA 2017. *Abandoned Mineworkings Manual (RP940)*. In draft. http://www.ciria.org/Research/.../Abandoned_mineworkings_manual.aspx (accessed 3 February 2017).

CLARK, A.H., CHEN, Y., FARRAR, E., NORTHCOTE, B., WASTENAYS, H.A.H.P., HODGSON, M.J. & BROMLEY, A. 1994. Refinement of the time/space relationships of intrusion and hydrothermal activity in the Cornubian Batholith (abstract). *Proceedings of the Ussher Society*, **8**, 345.

CLARK, A.H., SCOTT, D.J., SANDEMAN, H.A., BROMLEY, A.V. & FARRAR, E. 1998a. Siegenian generation of the Lizard ophiolite: U–Pb zircon age data for plagiogranite, Porthkerris, Cornwall. *Journal of the Geological Society, London*, **155**, 595–598, https://doi.org/10.1144/gsjgs.155.4.0595

CLARK, A.H., SANDEMAN, H.A. ET AL. 1998b. An emerging geochronological record of the construction and emplacement of the Lizard ophiolite, SW Cornwall. (Abstract.) *Geoscience in Southwest England*, **9**, 276–277.

CLARK, A.H., SANDEMAN, H.A., NUTMAN, A.P., GREEN, D.H. & COOK, A. 2003. Discussion on SHRIMP U–Pb dating of the exhumation of the Lizard Peridotite and its emplacement overcrustal rocks: constraints for tectonic models. *Journal of the Geological Society, London*, **160**, 331–335, https://doi.org/10.1144/0016-764901-159

CLAYTON, C.R.I. 2001. *Managing Geotechnical Risk: Improving Productivity in U.K. Building and Construction*. Thomas Telford, London.

CLAYTON, C.R.I. & SMITH, D.M. 2013. *Effective Site Investigation*. Site Investigation Steering Group, ICE, London.

COOK, C.A., HOLDSWORTH, R.E. & STYLES, M.T. 2002. The emplacement of peridotites and associated oceanic rocks from the Lizard Complex, SW England. *Geological Magazine*, **139**, 27–45, https://doi.org/10.1017/S0016756801005933

CORNWALL COUNTY COUNCIL 2010. *Guidance on the Redevelopment of Potentially Contaminated Sites*. Cornwall County Council.

CORNWALL COUNTY COUNCIL 2011a. *Preliminary Flood Risk Assessment (PFRA)*. Cornwall Council.

CORNWALL COUNTY COUNCIL 2011b. *Chronology of Major Flood Events in Cornwall*. Annex 5 of the Preliminary Flood Risk Assessment report. Cornwall Council.

CORNWALL COUNTY COUNCIL 2014. *China Clay*. Technical Paper M1. Cornwall Council.

CORNWALL COUNTY COUNCIL 2015. *Cornwall Local Plan: China Clay*. Cornwall Council.

CORNWALL COUNTY COUNCIL 2006. *Camborne–Pool–Redruth Transportation Strategy*. Cornwall Council.

CORNWALL UNDERGROUND ACCESS ADVISORY GROUP 2002. Strategy Document: 'Preserving Access to Cornwall's Underground Mining, Geological and Ecological Heritage' (unpublished).

COULTON, R., BULLEN, C., DOLAN, J. & MARSDEN, C. 2003. Wheal Jane mine water active treatment plant – design, construction and operation. *Land Contamination and Reclamation*, **11**, 245–252, https://doi.org/10.2462/09670513.821

CULSHAW, M.G., DONNELLY, L. & MCCANN, D. 2004. Location of buried mineshafts and adits using reconnaissance geophysical methods. In HACK, R., AZZAM, R. & CHARLIER, R. (eds) *Engineering Geology for Infrastructure Planning in Europe: A European Perspective*. Lecture Notes in Earth Sciences, **104**. Springer, Berlin, 563–573.

DARBYSHIRE, D.P.F. & SHEPHERD, T.J. 1985. Chronology of granite magmatism and associated mineralisation, SW England. *Journal of the Geological Society, London*, **142**, 1159–1178, https://doi.org/10.1144/gsjgs.142.6.1159

DEARMAN, W.R. 1963. Wrench faulting in Cornwall and South Devon. *Proceedings of the Geologists' Association*, **74**, 265–287, https://doi.org/10.1016/S0016-7878(63)80023-1

DEARMAN, W.R. 1991. *Engineering Geological Mapping*. Butterworth-Heinemann, Oxford.

DE LA BECHE, H.T. 1839. *Report on the geology of Cornwall, Devon and West Somerset*. Memoirs of the Geological Survey of Great Britain. Her Majesty's Stationery Office, London.

DE MULDER, E.F.J., HACK, H.R.G.K. & VAN REE, C.C.D.F. 2012. *Sustainable Development and Management of the Shallow Subsurface*. Geological Society, London, https://doi.org/10.1144/MPSDM

DEPARTMENT FOR INTERNATIONAL TRADE 2015. Land remediation: bringing brownfield sites back to use. http://www.gov.uk/government/publications/land-remediation-bringing-brownfield-sites-back-to-use (accessed 23 February 2017).

DEPARTMENT OF THE ENVIRONMENT 1990. *Planning Policy Guidance: Development on Unstable Land*. PPG14, HMSO, London.

DEVON COUNTY COUNCIL 2015. Future Extension of Drakelands Mine – Background Paper. Published May 2015.

DINES, H.G. 1956. *The Metalliferous Mining Region of South-West England*, **2** volumes. HMSO, London.

DOMEIER, M. 2016. A plate tectonic scenario for the Iapetus and Rheic oceans. *Gondwana Research*, **36**, 275–295, https://doi.org/10.1016/j.gr.2015.08.003

DUPUIS, N.E., BRAID, J.A., MURPHY, J.B., SHAIL, R.K., ARCHIBALD, D.A. & NANCE, R.D. 2015. $^{40}Ar/^{39}Ar$ phlogopite geochronology of lamprophyre dykes in Cornwall, UK: new age constraints on Early Permian post-collisional magmatism in the Rhenohercynian Zone, SW England. *Journal of the Geological Society*, **172**, 566–575, https://doi.org/10.1144/jgs2014-151

EALEY, P.J. & JAMES, H.C.L. 2011. Loess of the Lizard Peninsula, Cornwall, SW Britain. *Quaternary International*, **231**, 55–61, https://doi.org/10.1016/j.quaint.2010.06.018

EDMONDS, E.A., WHITTAKER, A. & WILLIAMS, B.J. 1985. *Geology of the Country Around Ilfracombe and Barnstaple*. British Geological Survey. Memoir for sheets 277 and 293 NS.

EDWARDS, R.A., WARRINGTON, G., SCRIVENER, R.C., JONES, N.S., HASLAM, H.W. & AULT, L. 1997. The Exeter Group, south Devon, England. A contribution to the early post-Variscan stratigraphy of northwest Europe. *Geological Magazine*, **134**, 177–197.

EGGERS, M.J. 2016. Diversity in the science and practice of engineering geology. In: EGGERS, M.J., GRIFFITHS, J.S., PARRY, S. & CULSHAW, M.G. (eds) *Developments in Engineering Geology*. Geological Society, London, Engineering Geology Special Publications, **27**, 1–18, https://doi.org/10.1144/EGSP27.1

ELLIS, R.J. & SCOTT, P.W. 2004. Evaluation of hyperspectral remote sensing as a means of environmental monitoring in the St. Austell China clay (kaolin) region, Cornwall, UK. *Remote Sensing of the Environment*, **93**, 118–130, https://doi.org/10.1016/j.rse.2004.07.004

EN ISO/TS 22475-1 2006a. *Geotechnical Investigation and Testing – Sampling Methods and Groundwater Measurement. Part 1: Technical Principles for Execution*. British Standards Institution, London.

EN ISO/TS 22475-2 2006b. *Geotechnical Investigation and Testing – Sampling Methods and Groundwater Measurement. Part 2: Qualification Criteria for Enterprises and Personnel*. British Standards Institution, London.

EN ISO/TS 22475-1 2007. *Geotechnical Investigation and Testing – Sampling Methods and Groundwater Measurement. Part 3:*

Conformity Assessment of Enterprises and Personnel by Third Parties. British Standards Institution, London.

ENVIRONMENT AGENCY 2004. *Model Procedures for the Management of Contaminated Land*. EA Report **CLR 11**, The Stationery Office, London.

ENVIRONMENT AGENCY 2008. *Abandoned Mines and the Water Environment*. Science Project SC030136-41, Environment Agency, Bristol.

ENVIRONMENT AGENCY 2009. *Flooding in England: A National Assessment of Flood Risk*. Environment Agency, Bristol.

ENVIRONMENT TIMES 2017. Trials on cleaning old Cornish tin mine water prior to reopening, 9 January 2017, http://www.environmenttimes.co.uk

EVANS, C.D.R. 1990. *United Kingdom Offshore Regional Report: the Geology of the Western English Channel and its Western Approaches*. British Geological Survey, NERC. HMSO, London.

EVANS, D.J.A., HARRISON, S., VIELI, A. & ANDERSON, E. 2012. The glaciation of Dartmoor: the southernmost independent Pleistocene ice cap in the British Isles. *Quaternary Science Reviews*, **45**, 31–53, https://doi.org/10.1016/j.quascirev.2012.04.019

FENTON, G.A. & GRIFFITHS, D.V. 2008. *Risk Assessment in Geotechnical Engineering*. John Wiley, Hoboken, NJ.

FERRIER, G., RUMSBY, B. & POPE, R. 2007. Application of hyperspectral remote sensing data in the monitoring of the environmental impact of hazardous waste derived from abandoned mine sites. *In*: TUEEW, R.M. (ed.) *Mapping Hazardous Terrain using Remote Sensing*. Geological Society, London, Special Publications, **283**, 107–116, https://doi.org/10.1144/SP283.9

FLOOD AND WATER MANAGEMENT ACT 2010. HMSO, London.

FLOYD, P.A., EXLEY, C.S. & STYLES, M.T. 1993a. *Igneous Rocks of South-West England*. Geological Conservation Review Series. Chapman and Hall, London, UK, 268.

FLOYD, P.A., HOLDSWORTH, R.E. & STEELE, S.A. 1993b. Geochemistry of the Start Complex greenschists: Rhenohercynian MORB? *Geology Magazine*, **130**, 345–352, https://doi.org/10.1017/S0016756800020021

FOOKES, P.G. 1997. Geology for engineers: the geological model, prediction and performance. The First Glossop Lecture. *Quarterly Journal of Engineering Geology*, **30**, 293–424, https://doi.org/10.1144/GSL.QJEG.1997.030.P4.02

FRANKE, W. 2000. The mid-European segment of the Variscides: tectonostratigraphic units, terrane boundaries and plate tectonics evolution. *In*: FRANKE, W., HAAK, V., ONCKEN, O. & TANNER, D. (eds) *Orogenic Processes: Quantification and Modelling of the Variscan Belt*. Geological Society, London, Special Publication, **179**, 35–61, https://doi.org/10.1144/GSL.SP.2000.179.01.01

FRESHNEY, E.C., MCKEOWN, M.C. & WILLIAMS, M. 1972. *Geology of the Coast Between Tintagel and Bude*. Geological Survey of Great Britain, HMSO, London, Memoirs.

FRESHNEY, E.C., EDMONDS, E.A., TAYLOR, R.E. & WILLIAMS, B.J. 1979. *Geology of the Country around Bude and Bradworthy*. Geological Survey of Great Britain, England and Wales, Memoirs.

GAMBLE, B. 2011a. *Cornish Mines: Gwennap to the Tamar*. Alison Hodge, Penzance.

GAMBLE, B. 2011b. *Cornish Mines: St Just to Redruth*. Alison Hodge, Penzance.

GAMBLE, B. & THE WORLD HERITAGE SITE BID TEAM 2005. Cornwall and West Devon Mining Landscape World Heritage Site Nomination Document (unpublished).

GOLLEY, P. & WILLIAMS, R. 1995. *Cornish Mineral Reference Manual*. Endsleigh Publications, Truro.

GRIFFITHS, J.S. 2015. *Geological Maps: Engineering Geology*. Online Reference Module in Earth Systems and Environmental Science. Elsevier, Amsterdam. https://doi.org/10.1016/B978-0-12-409548-9.09568-3

GRIFFITHS, J.S. 2017. Technical note: terrain evaluation in engineering geology. *Quarterly Journal of Engineering Geology and Hydrogeology*, **50**, 3–11, https://doi.org/10.1144/qjegh2016-090

GRIFFITHS, J.S. & BELL, F.G. 2019. *Environmental and Engineering Geology: Beyond the Basics*. Whittles, Duneath.

HARRISON, S. & KEEN, D.H. 2005. South-west England. *In*: LEWIS, C.A. & RICHARDS, A.E. (eds) *The Glaciations of Wales and Adjacent Areas*. Logaston Press, Almeley, Herefordshire, 165–176.

HARRISON, S., ANDERSON, E. & PASSMORE, D.G. 2001. Further glacial tills on Exmoor, south-west England. Implications for a small ice cap and valley glaciations. *Proceedings of the Geologists Association*, **112**, 1–5.

HARRISON, S., KNIGHT, J. & ROWAN, A.V. 2015. The southernmost Quaternary niche glacier system in Great Britain. *Journal of Quaternary Science*, **30**, 325–334, https://doi.org/10.1002/jqs.2772

HARTLEY, A. & WARR, L.N. 1990. Upper Carboniferous foreland basin evolution in SW Britain. *Proceedings of the Ussher Society*, **7**, 212–216.

HEALTH AND SAFETY EXECUTIVE 2013. *Health and Safety at Quarries, Quarries Regulations 1999*. L118, 1st edn 1999, 2nd edn 2013 amended 2014, HSE Books.

HEALY, P.R. & HEAD, J.M. 2002. Construction over abandoned mine workings. *CIRIA Report SP32*, first printed 1984, reprinted 2002.

HEARN, E.H. & FIALKO, Y. 2009. Can compliant fault zones be used to measure absolute stresses in the upper crust? *Journal of Geophysical Research*, **114**, B04403, https://doi.org/10.1029/2008JB005901

HENDRIKS, E.M. 1937. Rock successions and structure in south Cornwall. *Quarterly Journal of the Geological Society*, **93**, 322–367, https://doi.org/10.1144/GSL.JGS.1937.093.01-04.13

HENDRIKS, E.M. 1959. A summary of present views on the structure of Cornwall and Devon. *Geological Magazine*, **96**, 253–257, https://doi.org/10.1017/S0016756800060246

HILLIS, R.R. 1988. The geology and tectonic evolution of the Western Approaches Trough. PhD Thesis (unpublished), University of Edinburgh.

HOBBS, J.C.A. 2005. Enhancing the contribution of mining to sustainable development. *In*: MARKER, B.R., PETTERSON, M.G., MCEVOY, F. & STEPHENSON, M.H. (eds) *Sustainable Minerals Operations in the Developing World*. Geological Society, London, Special Publications, **250**, 9–24, https://doi.org/10.1144/GSL.SP.2005.250.01.03

HOLDER, M.T. & LEVERIDGE, B.E. 1986a. A model for the tectonic evolution of south Cornwall. *Journal of the Geological Society, London*, **143**, 125–134, https://doi.org/10.1144/gsjgs.143.1.0125

HOLDER, M.T. & LEVERIDGE, B.E. 1986b. Correlation of the Rhenohercynian Variscides. *Journal of the Geological Society, London*, **143**, 141–147, https://doi.org/10.1144/gsjgs.143.1.0141

HOLDSWORTH, R.E. 1989. The Start–Perranporth line: a Devonian terrane boundary in the Variscan orogen of SW England. *Journal of the Geological Society, London*, **146**, 419–422, https://doi.org/10.1144/gsjgs.146.3.0419

HOLLOWAY, S. & CHADWICK, R.A. 1986. The Sticklepath–Lustleigh fault zone: tertiary sinistral reactivation of a Variscan dextral strike-slip fault. *Journal of the Geological Society, London*, **143**, 447–452, https://doi.org/10.1144/gsjgs.143.3.0447

HUGHES, P. 1994. Water Pollution from Abandoned Coal Mines. *Research Paper* 94/43, House of Commons Library.

HUGHES, P. 2005. *Copperopolis: Landscapes of the Early Industrial Period in Swansea*. Royal Commission on the Ancient and Historical Monuments of Wales.

ISAAC, K.P. 1985. Thrust and nappe tectonics of west Devon. *Proceedings of the Geologists' Association*, **96**, 109–127, https://doi.org/10.1016/S0016-7878(85)80062-6

JARVIS, A., FOX, A., GOZZARD, E., HILL, S., MAYES, W. & POTTER, H. 2007. Prospects for effective national management of abandoned metal mine water pollution in the UK. *In*: CIDU, R. & FRAU, F. (eds) *Water in Mining Environments. Proceedings of the IMWA Symposium*, Cagliari, Mako Edizioni, 77–81.

JOHNSTON, D., PARKER, K. & PRITCHARD, J. 2007. Management of abandoned minewater pollution in the United Kingdom. *In*: CIDU, R. & FRAU, F. (eds) *Water in Mining Environments. Proceedings of the IMWA Symposium*, Cagliari, Mako Edizioni, 209–213.

JONES, K.A. 1997. Deformation and emplacement of the Lizard Ophiolite Complex, SW England, based on evidence from the Basal Unit. *Journal of the Geological Society, London*, **154**, 871–885, https://doi.org/10.1144/gsjgs.154.5.0871

KAOLIN AND BALL CLAY ASSOCIATION UK 2007. Kaolin. http://www.kabca.org (accessed 2 March 2017).

KITSIKOPOULOS, H. 2016. *Innovation and Technological Diffusion. An Economic History of Early Steam Engines*. Routledge, Oxford.

KNIGHT, J. 2005. Regional climatic v. local controls on periglacial slope deposition: a case study from west Cornwall. *Geoscience in South-West England*, **11**, 151–157.

KNIGHT, J. & HARRISON, S. 2013. 'A land history of men': the intersection of geomorphology, culture and heritage in Cornwall, south-west England. *Applied Geography*, **42**, 186–194, https://doi.org/10.1016/j.apgeog.2013.03.020

KOMAC, M. 2018. InSAR. *In*: BOBROWSKY, P.T. & MARKER, B. (eds) *Encyclopedia of Engineering Geology*. Springer, Cham, Switzerland, 517–518.

LEE, E.M. & JONES, D.K.C. 2014. *Landslide Risk Assessment*, 2nd edn. ICE, London.

LEVERIDGE, B.E. & HARTLEY, A.J. 2006. The Variscan Orogeny: the development and deformation of Devonian/Carboniferous basins in SW England and South Wales. *In*: BRENCHLEY, P.J. & RAWSON, P.F. (eds) *The Geology of England and Wales*. 2nd edn. The Geological Society, London, 225–255.

LEVERIDGE, B.E. & SHAIL, R.K. 2011*a*. The marine Devonian stratigraphy of Great Britain. *Proceedings of the Geologists' Association*, **122**, 540–567, https://doi.org/10.1016/j.pgeola.2011.03.003

LEVERIDGE, B.E. & SHAIL, R.K. 2011*b*. The Gramscatho Basin, south Cornwall, UK: Devonian active margin successions. *Proceedings of the Geologists' Association*, **122**, 568–615, https://doi.org/10.1016/j.pgeola.2011.03.004

LEVERIDGE, B.E., HOLDER, M.T. & DAY, G.A. 1984. Thrust nappe tectonics in the Devonian of south Cornwall and the western English Channel. *In*: HUTTON, D.H.W. & SANDERSON, D.J. (eds) *Variscan Tectonics of the North Atlantic Region*. Geological Society, London, Special Publications, **14**, 103–112, https://doi.org/10.1144/GSL.SP.1984.014.01.09

LEVERIDGE, B.E., HOLDER, M.T. & GOODE, A.J.J. 1990. *Geology of the Country around Falmouth*. British Geological Survey, Sheet 352, England and Wales, Memoirs.

LEVERIDGE, B.E., HOLDER, M.T., GOODE, A.J.J., SCRIVENER, R.C., JONES, N.S. & MERRIMAN, R.A. 2002. *Geology of the Plymouth and South-East Cornwall Area*. Memoir of the British Geological Survey, Sheet 348, England and Wales.

LEWIS, G.R. 1907. *The Stannaries: A Study of the Medieval Tin Miners of Cornwall and Devon*. Reprinted 1965, D. Bradford Barton Ltd, Truro.

LINTON, D.L. 1955. The problem of tors. *Geographical Journal*, **121**, 470–487, https://doi.org/10.2307/1791756

LITTLEJOHN, G.S. 1979*a*. Surface stability in areas underlain by old coal workings. *Ground Engineering*, **12**, 22–30

LITTLEJOHN, G.S. 1979*b*. Consolidation of old coal workings. *Ground Engineering*, **12**, 15–21.

MANNING, D.A.C., HILL, P.I. & HOWE, J.H. 1996. Primary lithological variation in the kaolinized St. Austell Granite, Cornwall, England. *Journal of the Geological Society, London*, **153**, 827–838, https://doi.org/10.1144/gsjgs.153.6.0827

MARKER, B.R. 2018. Brownfield sites. *In*: BOBROWSKY, P.T. & MARKER, B. (eds) *Encyclopedia of Engineering Geology*, Springer International, Cham, 92–94.

MCDOWELL, P.W., BARKER, R.D. ET AL. 2002. *Geophysics in Engineering Investigations*. CIRIA C562, Engineering Geology Special Publication, **19**, Published by CIRIA, London.

MCGINNESS, S. 1999. Treatment of Acid Mine Drainage. *Research Paper* 99/10, House of Commons Library, London.

MERRIMAN, R.J., EVANS, J.A. & LEVERIDGE, B.E. 2000. Devonian and Carboniferous volcanic rocks associated with the passive margin sequences of SW England; some geochemical perspectives. *Geoscience in South-West England*, **10**, 77–85.

MILLER, J.A. & GREEN, D.H. 1961. Age determination of rocks in the Lizard (Cornwall) area. *Nature*, **192**, 1175–1176.

MINING MINERALS AND SUSTAINABLE DEVELOPMENT 2002. *Mining for the Future: Annex H Ok Tedi Riverine Case Study Case Study*. Report for the project commissioned by the International Institute for Environmental and Development, and the World Business Council for Sustainable Development.

MINISTRY OF HOUSING, COMMUNITIES, AND LOCAL GOVERNMENT 2018. *National Planning Policy Frame Work*. NPPF18, July 2018. HMSO, London.

MORRIS, J. 2014. Pumping the polluted water from mines. BBC News Plymouth, 3 June 2014.

MURTON, J.B. & BALLANTYNE, C.K. 2017. Periglacial and permafrost models for Great Britain. *In*: GRIFFITHS, J.S. & MARTIN, C.J. (eds) *Engineering Geology and Geomorphology of Glaciated and Periglaciated Terrains*. Geological Society, London, Engineering Geology Special Publication, **28**, 501–597, https://doi.org/10.1144/EGSP28.5

NANCE, R.D., GUTIÉRREZ-ALONSO, G. ET AL. 2010. Evolution of the Rheic Ocean. *Gondwana Research*, **17**, 194–222, https://doi.org/10.1016/j.gr.2009.08.001

NANCE, R.D., GUTIÉRREZ-ALONSO, G. ET AL. 2012. A brief history of the Rheic Ocean. *Geoscience Frontiers*, **3**, 125–135, https://doi.org/10.1016/j.gsf.2011.11.008

NORBURY, D. 2016. *Soil and Rock Description in Engineering Practice*. 2nd edn. Whittles, Dunbeath.

NUTMAN, A.P., GREEN, D.H., COOK, A., STYLES, M.T. & HOLDSWORTH, R.E. 2001. SHRIMP U–Pb zircon dating of the exhumation of the Lizard Peridotite and its emplacement overcrustal rocks: constraints for tectonic models. *Journal of the Geological Society, London*, **158**, 809–820, https://doi.org/10.1144/jgs.158.5.809

PARRY, D. & CHIVERRELL, C. (eds) 2019. *Abandoned Mine Workings Manual*. CIRIA Report **C758D**, pp. 546.

PARRY, S., BAYNES, F. ET AL. 2014. Engineering geological models – an introduction: IAEG Commission 25. *Bulletin of the International Association for Engineering Geology and the*

Environment, **73**, 689–706, https://doi.org/10.1007/s10064-014-0576-x

PASCOE, W. 1982. *C.C.C.: The History of the Cornish Copper Company*. Truran, Cornwall.

PENHALLURICK, R. 1986. *Tin in Antiquity*. The Institute of Metals, London.

PIRRIE, D., POWER, M.R., WHEELER, P.D., CUNDY, A., BRIDGES, C. & DAVEY, G. 2002. Geochemical signature of historical mining: fowey Estuary, Cornwall, UK. *Journal of Geochemical Exploration*, **76**, 31–43, https://doi.org/10.1016/S0375-6742(02)00203-0

PIRRIE, D., POWER, M.R., ROLLINSON, G., CAMM, G.S., HUGHES, S.H., BUTCHER, A.R. & HUGHES, P. 2003. The spatial distribution and source of arsenic, copper, tin and zinc within the surface sediments of the Fal Estuary, Cornwall, U.K. *Sedimentology*, **50**, 579–595, https://doi.org/10.1046/j.1365-3091.2003.00566.x

POTTER, H.A.B. & JARVIS, A.P. 2006. Managing mining pollution to deliver the WFD. *Proceedings of the Fourth CIWEM Annual Conference: Emerging Environmental Issues Future Challenges*. 12–14 September 2006, Newcastle upon Tyne.

RIPPON, S., CLAUGHTON, P. & SMART, C. 2009. *Mining in a Medieval Landscape: the Royal Silver Mines of the Tamar Valley*. University of Exeter Press, Exeter.

ROGERS, J.D. 1997. The Interpretation and characterisation of lineaments from Landsat TM Imagery of SW England. Unpublished PhD thesis, University of Plymouth.

ROGERS, J.D. & ANDERSON, M.W. 1996. The Interpretation of lineaments from Landsat TM Imagery of SW England and their Relationship to Regional Structural Trends. (Abstract.) *Proceedings of the Ussher Society*, **9**, 139.

ROLLINSON, G.K., PIRRIE, D., POWER, M.R., CUNDY, A. & CAMM, G.S. 2007. Geochemical and mineralogical record of historical mining, Hayle Estuary, Cornwall, UK. *Geoscience in South-West England*, **11**, 326–337.

ROWE, J. 1953. *Cornwall in the Age of the Industrial Revolution*. Liverpool University Press, Liverpool.

SANDEMAN, H.A.I., CLARK, A.H., SCOTT, D.J. & MALPAS, J.G. 2000. The Kennack Gneiss of the Lizard peninsula, Cornwall, SW England: comingling and mixing of mafic and felsic magmas accompanying Givetian continental incorporation of the Lizard ophiolite. *Journal of the Geological Society, London*, **157**, 1227–1242, https://doi.org/10.1144/jgs.157.6.1227

SANDERSON, D.J. 1984. Structural variation across the northern margin of the Variscides in NW Europe. *In*: HUTTON, D.H.W. & SANDERSON, D.J. (eds) *Variscan Tectonics of the North Atlantic Region*. Geological Society, London, Special Publications, **14**, 149–165, https://doi.org/10.1144/GSL.SP.1984.014.01.15

SANDERSON, D.J. & DEARMAN, W.R. 1973. Structural zones of the Variscan fold belt in SW England, their location and development. *Journal of the Geological Society, London*, **129**, 527–536, https://doi.org/10.1144/gsjgs.129.5.0527

SCOURSE, J.D. 1987. Periglacial sediments and landforms in the Isles of Scilly and Cornwall. *In*: BOARDMAN, J. (ed.) *Periglacial Processes and Landforms in Britain and Ireland*. Cambridge University Press, Cambridge, 225–236.

SCOURSE, J.D. 1996. Late Pleistocene stratigraphy and palaeobotany of north and west Cornwall. *Transactions of the Royal Geological Society of Cornwall*, **22**, 2–56.

SCRIVENER, R.C. 2006. Cornubian granites and mineralisation of SW England. *In*: BRENCHLEY, P.J. & RAWSON, P.F. (eds) *The Geology of England and Wales*. 2nd edn. The Geological Society, London, 257–268.

SCRIVENER, R.C., DARBYSHIRE, D.P.F. & SHEPHERD, T.J. 1994. Timing and significance of crosscourse mineralisation in SW England. *Journal of the Geological Society, London*, **151**, 587–590, https://doi.org/10.1144/gsjgs.151.4.0587

SCRIVENER, R.C., HIGHLEY, D.E., CAMERON, D.G. & LINLEY, K.A. 1997. *Cornwall: a Summary of Mineral Resource Information for Development Plans, Phase One*. 1: 100 000 scale Mineral Resources Map, British Geological Survey, Keyworth.

SEAGO, R.D. & CHAPMAN, T.J. 1988. The confrontation of structural styles and the evolution of a foreland basin in central SW England. *Journal of the Geological Society, London*, **145**, 789–800, https://doi.org/10.1144/gsjgs.145.5.0789

SEDGWICK, A. & MURCHISON, R.I. 1837. Classification of the older stratified rocks of Devonshire and Cornwall. *London and Edinburgh Philosophical Magazine and Journal of Science*, **14**, 241–260 & 354.

SELWOOD, E.B., DURRANCE, E.M. & BRISTOW, C.M. (eds) 1998. *The Geology of Cornwall: and the Isles of Scilly*. University of Exeter Press, Exeter. Reprinted 1999, 2007, 2013.

SHACKLETON, R.M., RIES, A.C. & COWARD, M.P. 1982. An interpretation of the Variscan structures in SW England. *Journal of the Geological Society, London*, **139**, 533–541, https://doi.org/10.1144/gsjgs.139.4.0533

SHAIL, R.K. 1989. Gramscatho–Mylor facies relationships; Hayle, south Cornwall. *Proceedings of the Ussher Society*, **7**, 125–130.

SHAIL, R.K. & ALEXANDER, A.C. 1997. Late Carboniferous to Triassic reactivation of Variscan basement in the western English Channel: evidence from onshore exposures in south Cornwall. *Journal of the Geological Society, London*, **154**, 163–168, https://doi.org/10.1144/gsjgs.154.1.0163

SHAIL, R.K. & LEVERIDGE, B.E. 2009. The Rhenohercynian passive margin of SW England: development, inversion and extensional reactivation. *Comptes Rendus Geoscience*, **341**, 140–155, https://doi.org/10.1016/j.crte.2008.11.002

SHAIL, R.K. & WILKINSON, J.J. 1994. Late- to post-Variscan extensional tectonics in south Cornwall. *Proceedings of the Ussher Society*, **8**, 228–236.

SHAIL, R.K., STUART, F.M., WILKINSON, J.J. & BOYCE, A.J. 2003. The role of post-Variscan extensional tectonics and mantle melting in the generation of the Lower Permian granites and the giant W–As–Sn–Cu–Zn–Pb orefield of Sw England. *Applied Earth Science; Transactions of the Institutions of Mining and Metallurgy: Section B*, **112**, 127–129.

SHARPE, A. 2007. *Wheal Peevor, Cornwall*. Archaeological consultancy and watching briefs during safety and conservation works. Historic Environment Service (Projects), Cornwall County Council.

SHARPE, A. 2009. *Trewavas, Breage, Cornwall*. Historic Environment Consultancy and watching brief during conservation works. Historic Environment Service (Projects), Cornwall County Council.

SIMONS, N., MENZIES, B. & MATTHEWS, M. 2002. *Geotechnical Site Investigation*. Thomas Telford, London.

SIMONS, B., SHAIL, R.K. & ANDERSEN, J.C.Ø. 2016. The Petrogenesis of the Early Permian Variscan granites of the Cornubian Batholith – lower plate post-collisional peraluminous magmatism in the Rhenohercynian Zone of SW England. *Lithos*, **260**, 76–94, https://doi.org/10.1016/j.lithos.2016.05.010

SITE INVESTIGATION STEERING GROUP 1993. *Without Site Investigation Ground is a Hazard*. Thomas Telford, London.

SITE INVESTIGATION STEERING GROUP 2013. *Guidance for Safe Investigation of Potentially Contaminated Ground*. 2nd edn. ICE, London.

SLONECKER, T, FISHER, G.B., AIELLO, D.P. & HAACK, B. 2010. Visible and infrared remote imaging of hazardous waste. *Remote Sensing*, **2**, 2474–2508, https://doi.org/10.3390/rs2112474

SMITH, M.J., PARON, P. & GRIFFITHS, J.S. 2011. *Geomorphological Mapping: Methods and Applications*. Elsevier, Amsterdam.

STEUER, A., SIEMON, B. & AUKEN, E. 2009. A comparison of helicopter-borne electromagnetics in frequency – and time-domain at the Cuxhaven valley in Northern Germany. *Journal of Applied Geophysics*, **67**, 194–205, https://doi.org/10.1016/j.jappgeo.2007.07.001

STRANGE, J., LANGDON, N. & LARGE, A. 2016. *Contaminated Land Guidance*, 3rd edn. ICE, London.

STRONGBOW 2016. *Strongbow South Crofty Water Treatment Trials Commence: surface Planning Conditions Satisfied*. Strongbow Exploration Inc. News Release 17 November 2016.

TATSI, K. & TURNER, A. 2014. Distribution and concentrations of thallium in surface waters of a region impacted by historical metal mining (Cornwall, U.K.). *Science of the Total Environment*, **473–474**, 139–146, https://doi.org/10.1016/j.scitotenv.2013.12.003

THOMPSON, K.C. & NATHANAIL, C.P. (eds) 2003. *Chemical Analysis of Contaminated Land*. Blackwell, Oxford.

TOPPING, P. 2004. *Grime's Graves*. English Heritage Red Guides.

UMHAU, J. 1932. *Summarized data of tin production*. United States Department of Commerce, Bureau of Mines, Washington, DC.

UNESCO WORLD HERITAGE CENTRE 2017. *Cornwall and West Devon Mining Landscape*. https://whc.unesco.org/en/list/1215 (accessed 26 November 2018).

VAN DER MEER, F.D. & JONG, S. 2001. *Imaging Spectrometry: Basic Principles and Prospective Applications*. Kluwer Academic, Dordrecht.

VAN STAVEREN, M.TH. 2006. *Uncertainty and Ground Conditions: A Risk Management Approach*. Butterworth-Heinemann, Elsevier, Oxford.

WADDINGTON, S. 2018. Drakelands mine shuts down and workers sent home. Cornwall Live 10 October 2018, www.cornwalllive.com

WALTHAM, A.C. 1989. *Ground Subsidence*. Blackie, London.

WARR, L.N., PRIMMER, T.J. & ROBINSON, D. 1991. Variscan very low-grade metamorphism in SW England – a diastathermal and thrust-related origin. *Journal of Metamorphic Geology*, **9**, 751–764, https://doi.org/10.1111/j.1525-1314.1991.tb00563.x

WHALLEY, J.S. & LLOYD, G.E. 1986. Tectonics of the Bude Formation, north Cornwall – the recognition of northerly directed décollement. *Journal of the Geological Society, London*, **143**, 83–88, https://doi.org/10.1144/gsjgs.143.1.0083

WHITTAKER, A. & LEVERIDGE, B.E. 2011. The North Devon Basin: a Devonian passive margin shelf succession. *Proceedings of the Geologists Association*, **122**, 718–744, https://doi.org/10.1016/j.pgeola.2011.03.006

WOOD, P.A. 1997. Remediation Methods for Contaminated Sites. *In*: HESTER, R.E. & HARRISON, R.M. (eds) *Contaminated Land and its Reclamation*. Thomas Telford, London, 47–71.

WOODCOCK, N.H., SOPER, N.J. & STRACHAN, R.A. 2007. A Rheic cause for the Acadian deformation in Europe. *Journal of the Geological Society, London*, **164**, 1023–1036, https://doi.org/10.1144/0016-76492006-129

YIM, W.W.S. 1981. Geochemical investigations on fluvial sediments contaminated by tin-mine tailings, Cornwall, England. *Environmental Geology*, **3**, 245–256, https://doi.org/10.1007/BF02473516

YOUNGER, P.L. 2002. Mine water pollution from Kernow to Kwazulu–Natal: geochemical remedial options and their selection in practice. *Geoscience in South-West England*, **10**, 255–266.

YOUNGER, P.L., BANWART, S.A. & HEDIN, R.S. 2002. *Mine Water: Hydrology, Pollution and Remediation*. Kluwer Academic, Dordrecht.

ZIEGLER, P.A. & DÈZES, P. 2006. Crustal evolution of Western and Central Europe. *In*: GEE, D.G. & STEPHENSON, R.A. (eds) *European Lithosphere Dynamics*. Geological Society, London, Memoirs, **32**, 43–56, https://doi.org/10.1144/GSL.MEM.2006.032.01.03

Chapter 14 Geological hazards from salt mining, brine extraction and natural salt dissolution in the UK

Anthony H. Cooper

British Geological Survey, Keyworth, Nottingham NG12 5GG, UK
 0000-0001-8763-8530
ahc@bgs.ac.uk

Abstract: Salt mining along with natural and human-induced salt dissolution affects the ground over Permian and Triassic strata in the UK. In England, subsidence caused by salt mining, brine extraction and natural dissolution is known to have occurred in parts of Cheshire (including Northwich, Nantwich, Middlewich), Stafford, Blackpool, Preesall, Droitwich and Teeside/Middlesbrough; it also occurs around Carrickfergus in Northern Ireland. Subsidence ranges from rapid and catastrophic failure to gentle sagging of the ground, both forms being problematical for development, drainage and the installation of assets and infrastructure such as ground source heat pumps. This paper reviews the areas affected by salt subsidence and details the mitigation measures that have been used; the implications for planning in such areas are also considered.

14.1 Introduction

In the UK, rock salt is present in Triassic and Permian rocks, from which it has been exploited for several millennia (Sherlock 1921; Northolt & Highley 1973). Also called halite or sodium chloride (NaCl), rock salt is not only a valuable industrial commodity, but also a highly soluble material responsible for natural and man-made subsidence geohazards. The Triassic salt-bearing strata are widespread in the Cheshire basin area, but also common in parts of Lancashire, Worcestershire, Staffordshire and Northern Ireland (Fig. 14.1). Permian saliferous rocks are mainly present in the NE of England (Northolt & Highley 1973). This paper looks at the occurrence of salt deposits, the way they either dissolve naturally, or have been extracted by mining, and man-made dissolution. It considers the subsidence problems that have arisen and continue to occur and looks at ways of mitigating the problems by planning and construction/remediation techniques.

Like table salt, rock salt is highly soluble and dissolves very quickly in water to make brine. This process occurs naturally in the UK and as a consequence, salt is not seen anywhere at outcrop. Instead it is present in the subsurface, where the upper part of the sequence is dissolved, producing a buried dissolution surface (salt karst) overlain by collapsed and foundered strata. The natural dissolution processes and groundwater flow are evidenced by the presence of brine springs, many of which have been known and exploited since Roman times. Through the Middle Ages these springs were moderately exploited and gave rise to place names ending in 'wych' or 'wich' (Cooper 2002), but it was in Victorian times that large-scale extraction both by mining and brine extraction accelerated, leading to some large and devastating instances of catastrophic subsidence (Calvert 1915). We are still dealing with this legacy and the effects of subsequent brine and rock salt extraction in many places, especially in parts of Cheshire, Droitwich, Stafford and Preesall. Where shallow brine extraction has occurred it mimics the natural salt karstification processes and the results of natural and man-made events can be difficult to differentiate. In Northern Ireland the salt has been mined traditionally by pillar-and-stall mining. In certain cases severe subsidence has occurred owing to water ingress into the mines causing dissolution of the pillars and catastrophic collapse.

Permian salt occurs at depth beneath coastal Yorkshire and Teeside. Here the salt deposits and the karstification processes are much deeper than in the Triassic salt and the salt deposits are bounded updip by a dissolution front and collapse monocline (Fig. 14.2; Cooper 2002). Salt has been won from these Permian rocks by dissolution mining and some historical to recent subsidence owing to brine extraction has occurred along the banks of the River Tees and to the NE of Middlesbrough (Tomlin 1982).

Modern pillar-and-stall salt mining is deeper than old Victorian mining and located in mudstone and salt sequences that are completely dry. Modern brine extraction is controlled and restricted to deep-engineered cavities that are kept full of brine on completion, or used for other storage such as gas or waste; both methods of extraction have low or zero expectations of subsidence.

Engineering Group Working Party (main contact for this chapter: A. H. Cooper, British Geological Survey, Keyworth, Nottingham NG12 5GG, UK, ahc@bgs.ac.uk)
From: GILES, D. P. & GRIFFITHS, J. S. (eds) 2020. *Geological Hazards in the UK: Their Occurrence, Monitoring and Mitigation – Engineering Group Working Party Report.* Geological Society, London, Engineering Geology Special Publications, **29**, 369–387, https://doi.org/10.1144/EGSP29.14
© 2020 BGS © UKRI All rights reserved. Published by The Geological Society of London. All rights reserved.
For permissions: http://www.geolsoc.org.uk/permissions. Publishing disclaimer: www.geolsoc.org.uk/pub_ethics

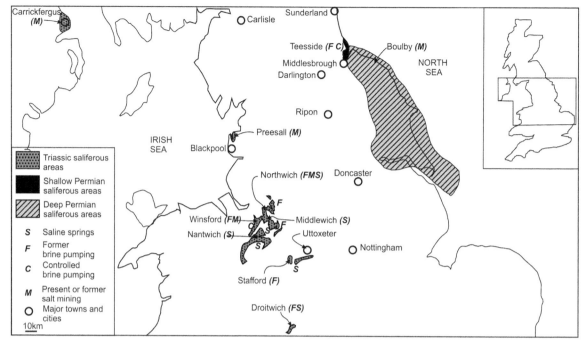

Fig. 14.1. Distribution of salt deposits in the UK showing mined and brine pumping areas.

14.2 Distribution of salt deposits in the Triassic and Permian rocks of the UK

In England and Northern Ireland, Triassic salt deposits are present onshore (Fig. 14.1) in Lancashire, Cheshire, Worcestershire, Staffordshire, Somerset, Dorset and Carrickfergus (Sherlock 1921; Whittaker 1972; Northolt & Highley 1973; Griffith & Wilson 1982; Wilson 1990, 1993; Jackson *et al.*

1995). The deposits formed in a semi-arid environment mainly within mainly fault-controlled land-locked basins linked to the major depositional centres of the North and Irish Seas. The sequence generally comprises thick sandstones and conglomerates overlain by red-brown siltstone and mudstone interbedded with halite and gypsum units. The Permian salt sequence was deposited as part of the Zechstein Group that formed in the extensive Zechstein Sea which extended eastwards from the UK to Holland and Germany

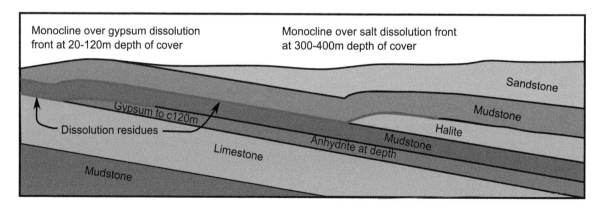

Fig. 14.2. Characteristic structure and subsurface morphology of Permian salt karst in NE England.

(Taylor & Coulter 1975; Smith 1989; Doornenbal & Stevenson 2010). Only the margin of the Zechstein Group occurs onshore in coastal parts of eastern England where the salt is interbedded with thick dolomite, mudstone and anhydrite formations.

14.3 Salt karst and natural dissolution

Salt is a highly soluble mineral; up to 360 g of salt will dissolve in a litre of water at 25°C (Ford & Williams 1989), making a saturated solution with a specific gravity of 1.998 (Thurmond et al. 1984). The dissolution rate of salt is extremely rapid, dissolving about 1000 times faster than gypsum at similar temperatures and pressures (Gutiérrez et al. 2008). The resulting brine is denser than normal groundwater and can, therefore, stay impounded below the ground surface (Howell & Jenkins 1976). However, where this dissolution interface coincides with the surface, or where there is sufficient hydraulic head, brine can flow to surface naturally as brine springs, and it is around these that the historical salt industry has developed. Brine springs are well documented in the Cheshire, Worcestershire and Staffordshire areas and their presence also led to the establishment of the Northern Ireland salt industry (Griffith & Wilson 1982).

In Cheshire the local place name ending in wich or wych derives from the medieval name for a salt spring. In Cheshire, brine springs occurred in many places, including Northwich, Middlewich, Nantwich and Winsford. They were also present near Whitchurch (Shropshire) at Higher Wych and a place formerly called Dirtwich. Natural subsidence has occurred sporadically throughout the salt field, one of the earliest records being in 1533 at Combermere Abbey near Whitchurch (Sherlock 1921). Another was recorded at Bilkely in about 1659 near Cholmondeley Castle (Jackson 1669; Ormerod 1848). Numerous buildings at Wybunbury near Crewe were damaged by salt dissolution subsidence in the late nineteenth century and the cause may have been natural since the nearest brine extraction was over 10 miles away (De Rance 1891). Many natural lakes such as Rostherne Mere near Knutsford were also formed by salt dissolution since the end of the Devensian ice-age and the Cheshire salt-field is dotted with meres, many of which formed in this way (Waltham et al. 1997), although others are more probably of glacial origin.

Where the Triassic saliferous rocks come near to outcrop in the UK, usually at depths of between 30 and 130 m, salt karst may develop (Earp & Taylor 1986; Wilson & Evans 1990) (Fig. 14.3). This karst has been strongly influenced by groundwater flow variability and head induced during the last, and possibly earlier, glaciations. The last (Devensian) glaciation buried the northern and central parts of the UK beneath a thick warm-based ice-sheet producing elevated groundwater levels (Howell & Jenkins 1976; Howell 1984). This elevated groundwater head is thought to have depressed and/or flushed out the saline waters, forcing the dissolution surface of the salt to cut deeper into the sequence. In the Cheshire basin the dissolution of the salt generated a rock-head depression into which the thick glacial sequence was deposited. When the ice-sheet melted, the groundwater regime changed again and further dissolution of the salt occurred as new regional groundwater patterns formed. Because these processes enhanced salt dissolution, the depth of salt dissolution is commonly very deep. This mechanism may also help to explain why the dissolution extends down to about 180 m in the Triassic salt of Cheshire, but to

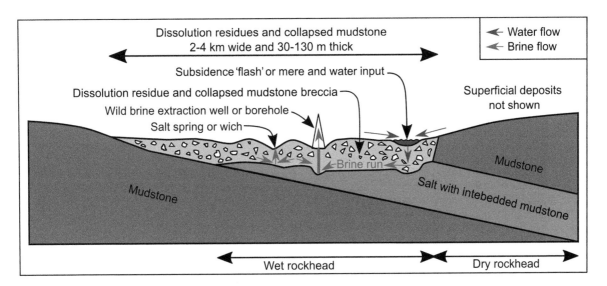

Fig. 14.3. Cross-section through Triassic salt deposits showing wet and dry rockhead along with methods of salt extraction.

a much deeper 3–400 m in the Permian salt of northern England, where the cover of ice was probably much thicker. Here the edge of the preserved salt is marked by a dissolution front and collapse monocline (Fig. 14.2). However, it must also be noted that the salt beds in the Permian probably would not have extended to outcrop because their facies is restricted to the deeper parts of the depositional basin (Smith 1989; Cooper 2002).

The way in which the salt karst has developed is also dependent on the nature of the lithological sequence in which the salt beds were deposited. In the Triassic deposits, most of the salt is interbedded with units of siltstone and mudstone, commonly with gypsum. Where the groundwater has circulated and removed the salt, only the interbedded insoluble rocks remain as a breccia, concealed by the commonly less brecciated, but foundered cover sequence. The traditional term for the salt surface where the dissolution has occurred is 'wet rockhead'. It is called this because in early exploitation by wells and boreholes it was from here that the original brine extraction took place with the borehole sinkers referring to the brine strike as a brine run. The area of wet rockhead encompasses the brine-laden salt karst, but the depth of the karst surface generally lies some 30–130 m below the ground surface and may even reach a depth of 180 m in Cheshire (Howell & Jenkins 1976; Howell 1984). Foundered ground above this zone commonly has numerous collapse features. These range from enclosed hollows 20–200 m across to linear depressions up to several kilometres long that have developed over 'brine runs', leading to the extraction wells.

In contrast, the traditional name for the area where the salt has not dissolved because it is sealed and dry beneath a cover of strata is 'dry rockhead' (Earp & Taylor 1986; Wilson & Evans 1990; Evans et al. 1993; Fig. 14.3). The terms wet and dry rockhead have been used by the drilling industry in the context of salt exploration and appear in borehole logs in the British Geological Survey archives dating from 1948 (David Walker, Sheffield University pers. comm. 18 January 2013); the terms were not used by Sherlock (1921) and thus appear to have originated between these two dates. Wet and dry rockhead must not be confused with the more common term of rockhead, which is the contact between superficial deposits and the underlying rock.

The Cheshire salt field has a long history of exploitation growing from its first pre-Roman development around the salt springs into a major industry (see below). The uncontrolled extraction of brine resulted in an artificial lowering of the brine–freshwater interface. This allied with groundwater abstraction from adjacent aquifers has disturbed the brine above the karst surface and introduced fresh water into some areas (Howell & Jenkins 1976). However, recent geochemical sampling (British Geological Survey 1999) has shown that brine from springs is entering the rivers in the Cheshire saltfield again. From river flow volumes and the content of salt in the water it should be possible to calculate the amount of salt being removed annually from the Cheshire saltfield. However, not all of the salt in the rivers is from the salt beds, since much of the British river chloride solute load is derived from British maritime precipitation (Walling & Webb 1981) and anthropogenic sources such as road de-icing and sewage works (British Geological Survey 1999).

14.4 Mining and dissolution mining of salt

14.4.1 Natural 'wild' brine extraction

Salt mining and brine extraction have played a major role in the modern development of the salt karst in the Cheshire, Worcestershire and Staffordshire areas. The original Roman and Medieval method of getting the salt was to tap into the natural brine springs. However, in the sixteenth to eighteenth centuries increasing demand for salt required that wells were sunk near the brine springs, tapping into the deeper, more concentrated and more reliably available brine. From the late sixteenth century reciprocating pumps were developed to draw up the brine (Woodiwiss 1992).

From the eighteenth century onwards, boreholes with pumps were developed allowing the extraction of brine from greater depths. The technique was to tap into the natural underground brine runs that mainly existed at the interface of the rocksalt and the overlying deposits and collapsed materials. This method of brine extraction was uncontrolled and is referred to as 'wild' brining. The method induced brine to flow towards the extraction boreholes, aggravating any existing subsidence features or causing linear subsidence belts spreading along-strike from the boreholes. This uncontrolled brine extraction enlarged the natural brine runs, causing widespread subsidence for considerable distances from the extraction points. The large-scale abstraction of brine in this way enhanced the development of the natural salt karst into an unnatural form.

The technique of tapping into the natural brine at rockhead was continued well into the twentieth century with extraction in the Droitwich, north Cheshire and Stafford areas. Stafford ceased extraction in 1970 (Coxill 1995) owing to legal pressure over subsidence and Droitwich ceased extraction soon after in 1972 (see Stafford and Droitwich sections). The last wild brine extraction in Cheshire by New Cheshire Salt Ltd stopped as late as 2005 amid concerns of subsidence affecting Northwich (Branston & Styles 2003) and as a result of an industry take over (Competition Commission 2005). The development of the linear subsidence 'flashes' in Cheshire (Fig. 14.4) was documented using air photographic interpretation (Howell 1986) and GIS/map/air photograph interpretation by Cooper et al. (2011). They showed the extent of relatively modern brine extraction subsidence, discussed in detail below in the section on Cheshire.

14.4.2 Shallow salt mining and 'bastard' brining

Two main methods were used to extract salt in the late nineteenth and early twentieth centuries: conventional mining and wild brine solution mining. Conventional mining was carried

Fig. 14.4. View southwards (in 1996) over Moston Flash, a large linear lake lying between Crewe and Middlwich attributed to brine pumping about 2 km to the north (Waltham *et al.* 1997) (photo A.H. Cooper © BGS/UKRI).

out in shallow 'pillar-and-stall' mines with networks of tunnels commonly separated by narrow salt pillars. The salt was mined using very large extraction rates leaving very small pillars for support (Fig. 14.5). Compared with modern mining, the amount of salt left as pillars was minimal. Many of these mines were near to the wet rockhead areas and flooding was a common hazard. To extract additional salt, many of the flooded salt mines were also pumped for brine, a technique referred to as 'bastard' brining. This dissolved the pillars and produced catastrophic mine collapses with surface subsidence on such an enormous and unprecedented scale that it destroyed whole areas of towns, factories and infrastructure such as canals and roads. Around Northwich and Middlewich the resulting subsidence was particularly catastrophic and widespread. New lakes, locally called 'flashes' (Fig. 14.5), appeared almost daily with many extending for hundreds of metres, for example Moston Flash (Fig. 14.4) near Elworth (Earp & Taylor 1986; Waltham 1989; Waltham *et al.* 1997). On 21 July 1907 a canal was disrupted and all the water lost (Calvert 1915). Even where collapse has not been so severe as to produce water-filled flashes, there are linear depressions caused by subsidence that cut through the local topography (Lee & Sakalas 2001). Some of the early mines that have not collapsed have continued to be problematical until recent remediation (see section on mitigation). The damage led to a major report on the subsidence (Dickinson 1873) and the establishment of the Brine Subsidence Compensation Board (discussed below).

No recent major salt mine collapses have been recorded in England. However, in Northern Ireland the collapse of the Carrickfergus mine in 1990 (Griffith 1991) and Maidenmount mine in 2001 (Fig. 14.11) highlight the susceptibility of such workings to water ingress. The collapse of these mines can cause overpressuring of the brine within the remaining caverns and the ejection of material from shafts and other weaknesses (Rigby 1905). Further afield, the loss of the complete Jefferson Island, USA salt mine in 1980 was due to one misplaced oil exploration borehole that drained Lake Peigneur, caused a massive subsidence depression to form and ejected material from the mineshaft (Autin 2002; Johnson 2005). These events are all similar to some of the Victorian mine disasters in Cheshire described by Calvert (1915) that created the Witton Flashes, Marston Big Hole and the Marston Flashes, to name but a few.

14.4.3 Modern salt mining

Now that wild brine extraction has ceased in the UK, modern salt extraction is carried out in two ways – dry mining and controlled brine extraction. Most modern salt extraction

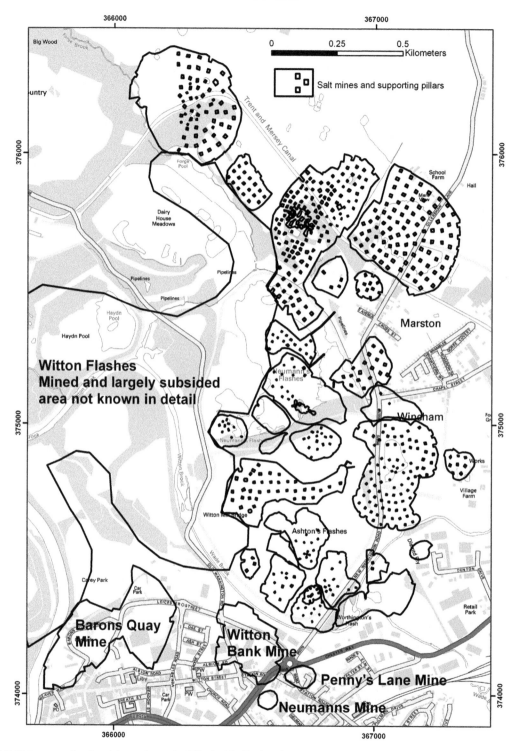

Fig. 14.5. Major areas of undermining to the north of Northwich. The locations of subsidence 'flashes' (both water-filled, and reclaimed by infilling) are also shown. Contains Ordnance Survey data © Crown copyright and database right 2013.

takes place in deep, dry, pillar-and-stall mines at depths of 125–200 m under Cheshire or as an adjunct to potash mining from beneath North Yorkshire (Woods 1979). In Cheshire the salt abstraction rate is between 68 and 75% and the typical void space left behind after extraction is 20 m wide × 8 m high, with 24 m square pillars left to support the over-burden (Swift & Reddish 2005). The Salt Union Ltd mine at Winsford, for example, is completely dry and part of the disused workings is used as an underground document storage facility (Deepstore 2013).

Modern controlled brine extraction, carried out at depths of more than 150–300 m, results in underground chambers that are left flooded and filled with saturated brine in order to minimize subsidence. Some of these chambers are used for waste disposal (Northolt & Highley 1973), but many more are now being utilized for onshore gas storage (Evans 2008). Cavities that have produced salt for industrial use are present in Cheshire, Lancashire (Preesall) and Teesside; salt cavities constructed specifically for gas storage have been made on the Yorkshire coast at Aldbrough (Evans & Holloway 2009).

14.5 Mining of Permian salt deposits

14.5.1 Teesside

In the north of England, salt is present in the Permian strata of the Zechstein Group. This evaporite and carbonate sequence extends from northern England eastwards beneath the North Sea to Germany and beyond. Only the marginal part of the Zechstein Basin encroaches onshore in England, where it includes thick units of anhydrite, halite and potash. The salt and potash deposits are currently mined by pillar-and-stall workings near the coast at Whitby. Here they are up to about 80 m thick and occur at depths of 1100–1250 m beneath the land and near shore (Woods 1979) and up to 1500 m depth 7 km out beneath the adjacent North Sea. The depth and location of the salt and potash workings mean that any subsidence at the surface has very little effect. This mine has now changed ownership and since 2018 ICL extracts polyhalite from deeper in the Permian sequence across a similar area to the former potash mine.

Salt was first found accidentally in a boring for water at Middlesbrough sunk between 1859 and 1863 (Wilson 1888). Subsequent to this, on the north bank of the Tees in Middlesbrough itself and further north around Greatham, nineteenth- and twentieth-century brine wells were sunk and the Teesside salt and chemical industry developed (Sherlock 1921; Morris 1978; Tomlin 1982; Marley 1890). There are some records of subsidence, mainly related to the early wells and those drilled before and up to the 1950s (Morris 1978), but not the catastrophic subsidence seen around the same time in Cheshire (Cooper 2002). In the Teesside area the main brine field extends from Port Clarence northwards through Salthome to Greatham. In addition there is an area that has been exploited for salt to the west of Haverton Hill (British Geological Survey 1987).

At the coast the salt is present at a depth of about 500 m. Westwards and updip there is a dissolution front in the salt (British Geological Survey 1987) and the overlying strata have collapsed and formed a west-facing subsidence monocline (Fig. 14.2 and Cooper (1998).

In the Middlesbrough area the early methods of brine extraction used largely unlined or perforated boreholes and the water from the Sherwood Sandstone aquifer to drive the salt dissolution (Sherlock 1921; Wilson 1888). Sherlock (1921) describes the process:

> In the Middlesbrough district the salt is worked by means of artesian wells sunk by the system adopted in America for petroleum-wells.* Their boring is about 10 inches in diameter until the sandstone is reached, and is lined with steel tubes down to the sandstone to keep the Drift and marl from caving in. When the driving-tube is well embedded into the sandstone no more lining is necessary, and the sinking is continued until the bottom of the salt-bed is reached, after which a suction-pipe is inserted to within two or three feet of the bottom, and the brine extracted by means of a working-barrel and pump-bucket placed at an average depth of 200 feet from the surface. Water contained in the sandstone finds its way outside the suction-tubes to the bottom of the well, dissolves the salt and is lifted by means of the pump. The underground water-level in the district is about 15 to 20 feet below the surface, on the average, and this balances the column of brine to within 150 feet of the surface in the case of the shallow wells, but of course to lower depths in the case of deeper wells.

(*This is what we now call open hole drilling.)

This process preferentially dissolved the top of the salt where the water exited the bottom of the borehole while the denser saturated brine was drawn off from the base of the suction tube. This method produced largely uncontrolled, wide, inverted cone-shaped cavities with a flattish upper surface and pointed lower end situated at the suction pipe (Wilson 1888; Morris 1978). As a consequence large unstable cavities were produced and parts of Middlesbrough north of the Tees and around the Saltholme area suffered subsidence (Fig. 14.6). Further north at Greatham, controlled brine extraction took place between 1894 and 1971 from salt at depths of between 300 and 400 m. The salt here is fairly shallow and the holes lie close to each another. Their presence has led to hydrogeological salt pollution issues affecting the overlying Sherwood Sandstone Group aquifer; consequently, the boreholes are now being remediated for the Environment Agency (Drilcorp 2013). East of the A178 road towards Seal Sands the post-mid 1950s brine extraction has been from deeper and better engineered dissolution cavities that are expected to have good long-term stability, such that some are used for gas storage (Evans 2008). As early as 1888 Wilson expressed concerns about the uncontrolled method of brine extraction and the related subsidence. Morris (1978) suggests that small areas of instability may exist or have already subsided in the vicinity of some of the early brine boreholes located south of the River Tees; these are discussed under planning later in this paper.

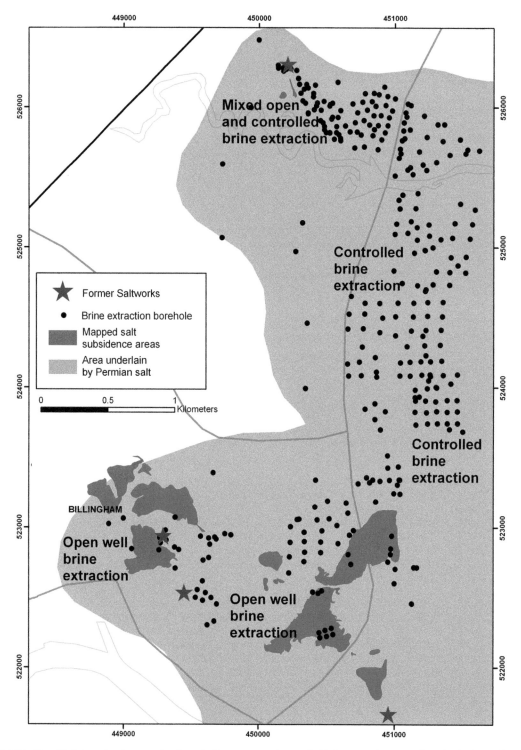

Fig. 14.6. The Teesside brine extraction area showing subsidence areas and controlled brine extraction areas. Contains Ordnance Survey data © Crown copyright and database right 2013; National Grid with 1 km spacing.

14.6 Mining of the Triassic salt deposits

14.6.1 Cheshire

Two salt formations are present in the Triassic Mercia Mudstone Group in Cheshire. These are the Wilkesley Halite Formation (up to 300 m thick) towards the top of the group and the Northwich Halite Formation (up to 200 m thick, with mudstone partings) in the middle of the group (Earp & Taylor 1986; Evans *et al.* 1993; Wilson 1993). The sequence between, below and above the salt formations is dominantly red-brown gypsiferous and dolomitic mudstone and siltstone. The natural salt dissolution has developed a wet rockhead, causing the overlying strata to collapse and founder, resulting in a breccia over the salt (Fig. 14.3).

From Roman times the Cheshire saltfield has been exploited by shallow wild brine extraction then later by mining, bastard brine extraction and modern controlled brine extraction. The area to the north of Northwich (Fig. 14.5) was extensively worked by pillar-and-stall mining and much of this has collapsed, producing lakes or 'flashes' (Calvert 1915), many of them hundreds of metres across. However, areas of undermining remained that threatened some buildings and retail developments. Details of both the shallow mining and brine extraction methods with examples from Cheshire are given in the sections above on natural dissolution and mining methods, illustrated by Figures 14.3–14.5. Figure 14.7 shows the distribution of salt near the surface and the subsidence flashes related to former saltworks in the central part of the Cheshire saltfield. The subsidence flashes form linear belts up to 5 km long approximately radiating out from the former saltworks. Similar, but less extensive, natural subsidence features can be seen in the east of the area. Despite the salt extraction having finished, natural dissolution and continued subsidence over man-made dissolution features affect the area, including the railway (Fig. 14.8). Details of the mitigation in the Cheshire area and the stabilization of salt mines around Northwich are given later in this paper.

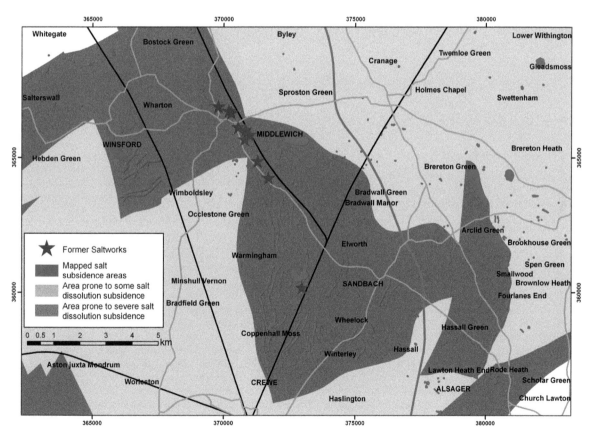

Fig. 14.7. Map of the cental Cheshire brine extraction area showing natural and man-made salt dissolution subsidence features with the locations of salt works. Contains Ordnance Survey data © Crown copyright and database right 2013. See also Figure 14.5 for Northwich mining and Figure 14.12 for the whole Cheshire area.

Fig. 14.8. Subsidence affecting the main railway line north of Crewe (photographed in 1996) near Elton Flashes Nature Reserve. Note the gantries on jackable foundations (photo A.H. Cooper © BGS/UKRI).

14.6.2 Blackpool and Preesall

The Lancashire coast beneath Blackpool and Preesall is underlain by the Triassic Mercia Mudstone Group, which includes several salt units. Two impersistent units, the Rossal Halite and the Mythop Halite, occur low in the sequence and a third more persistent unit, the Preesall Halite, occurs high in the sequence (Wilson 1990; Wilson & Evans 1990). The Preesall Halite was formerly worked in salt mines at Preesall on the east side of the River Wyre. Here the east of the saltfield is marked by the Preesall Fault Zone. Adjacent to this the westerly dipping salt comes near to outcrop, but groundwater circulation has dissolved the salt to a depth of between 50 and 100 m, leaving behind a collapse breccia down to a wet rockhead (Wilson & Evans 1990). Although there are no confirmed saline springs in the area, if they do exist, their most likely position would be immediately offshore, emerging beneath the sea. Sherlock (1921) notes the name of the village of Salwick near Kirkham just south of the district (although, because the coast is nearby, this name could equally relate to the production of sea salt). Probable subsidence areas, in which postglacial peat deposits have formed, are also recorded in the district, especially around Mythop (Wilson & Evans 1990). These subsidence areas range from 30 to 150 m across and contain up to 10 m of peat formed over the last 12 000 years. They give an indication of the amount and rate of salt karstification that has taken place beneath the area. Commercial site investigation in the area of Blackpool south of the football ground suggests salt at rockhead beneath the superficial deposits with shallow brine and some local subsidence owing to salt dissolution. However, few borehole records have been made public and the precise details of the geology of the area remain unknown.

14.6.3 Stafford

The salt deposits under the town of Stafford lie in a synclinal structure faulted along its eastern side (Fig. 14.9 and Arup Geotechnics 1991). The salt occurs interbedded with mudstone in a sequence 50–65 m thick within the Mercia Mudstone Group (Coxill 1995). The near-surface deposits have dissolved to a depth of around 50 m (wet rockhead) and the salt has largely dissolved adjacent to the eastern fault zone. The natural salt springs around Stafford are not well documented, but the town is surrounded by the villages with names ending in wich; these include Baswich, Milwich, Gratwich, Colwich and Shirleywich (along with the village of Salt), all suggesting the presence of brine springs. The springs at Shirleywich have a long history of exploitation from the seventeenth century to the late eighteenth century. The Shirleywich springs, and nearby ones at Weston, appear to be fed from salt deposits lying to the NE in a belt running through Chartley Moss and other small lakes. These two springs produced some 3750 tons in 1873 but production ceased at Shirleywich at the end of the century and at Weston-on-Trent in 1901 (Coxill 1995). Brine still flows to the surface in the Trent valley about 2.5 km SE of Weston feeding the Pasturefields SAC/SSSI saltmarsh meadow (Natural England 1986; Stafford Borough 2012). Lying 5 km to the NE of Shirleywich, Chartley had the strongest saline springs in Staffordshire in the mid sixteenth century (Sherlock 1921). Chartley Moss is a freshwater pond and floating bog formed in a linear salt dissolution subsidence area that is drained in the subsurface towards the salt springs along the banks of the River Trent. Localized collapse here causes minor seismic events that are felt on the surface of the bog, suggesting active salt dissolution and subsidence beneath the site (A. Brandon, pers. comm. 2000).

Brine was discovered in a water borehole north of Stafford town on Stafford Common in 1881 and brine production started in 1893 with production of up to 95 000 tonnes a year in the early 1960s. Subsidence had been reported since 1948, but by 1964 it was becoming a serious problem (Coxill 1995). Lotus Ltd brought a High Court case against British Soda Ltd and in August 1970 they were ordered to cease production, bringing an end to wild brine production here (and coincidentally soon after that in the Droitwich area). Coxill (1995) estimated that 5 million tonnes (2.25 million cubic metres) of rock salt had been extracted as brine from beneath Stafford. Only something like 10% of the volume of salt removed by this brine extraction has been accounted for by recorded subsidence and further subsidence may still occur in the area. The main brine run trends NNE towards the extraction boreholes and about 2 square kilometres of land have been affected by subsidence. Since the 1940s about 20 properties have been demolished and 500 severely damaged (Arup Geotechnics 1991).

14.6.4 Droitwich

In Worcestershire the Triassic saliferous rocks are present in the Mercia Mudstone Group and include the upper and middle sequences of salt beds approximately equivalent to the

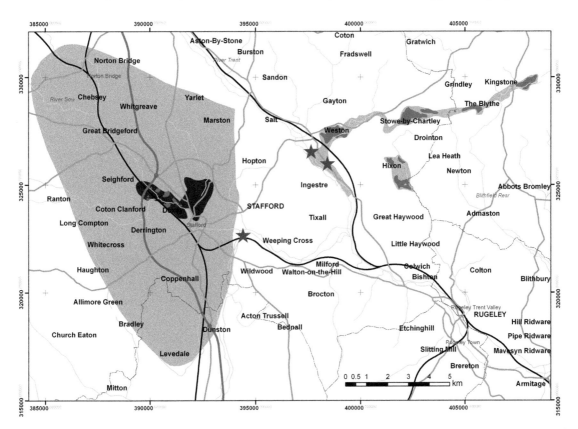

Fig. 14.9. The Stafford salt subsidence areas: those prone to some salt dissolution subsidence (light grey); prone to past severe salt dissolution subsidence (dark grey); mapped subsidence features (red); former saltworks, brine extraction boreholes and brine springs (purple stars). Contains Ordnance Survey data © Crown copyright and database right 2013.

Cheshire salt beds further north. There are about 90 m of saliferous strata including about 40% siltstone and mudstone. The Romans used the name Salinae for what we now call Droitwich; both names are indicative of the brine springs that rose there in the valley of the River Salwarpe (Fig. 14.10). The three main salt-producing springs of Netherwich, Middlewich and Upwich were noted by Rastell 1678 and some of their excavations were described by Woodiwiss (1992) and Hurst (1997). In the seventeenth century shafts were sunk that increased the brine flow from the Droitwich springs. In the eighteenth century deeper pits encountered artesian brine that could not be controlled and largely ran to waste (Poole & Williams 1981). Some salt mining was undertaken in the district, but it encountered natural brine runs and the subsequent exploitation was by brine extraction from several pumps driven by steam engines in the town. Brine extraction in the town ceased in 1922, at which time it was causing serious subsidence, such as that seen in Droitwich High Street today (Wychavon District Council 2013).

In the early nineteenth century salt was discovered some 6 km to the NE of the town and mined at Stoke Prior, where the shaft sequence was recorded by Murchison in 1839 (Wills 1976). In the twentieth century, large-scale brine extraction took place here from Stoke Prior saltworks. Brine extraction here ceased in 1972 as a consequence of the law case that saw the cessation of wild brine extraction in Stafford a few years earlier (Coxill 1995) and because of subsidence problems affecting housing (Pitts et al. 2012). Output in the latter years of operation were some 150 000 tonnes a year (Bloodworth et al. 1999). The brine abstraction caused the natural brine springs at the surface to dry up. It also caused a belt of subsidence along, and spreading out from, the course of the original sinuous brine run.

The natural brine run follows the wet rockhead of the strata which dips to the SE, and it has a NE-trending course through Droitwich, Wychbold and the Upton Warren saline pools to the former salt works at Stoke Prior (Fig. 14.10). The Upton Warren pools date back at least 42 000 years BP (Coope et al. 1961) with a salt marsh biota recorded from then until the present (Worcestershire Biological Records Centre 1999). Where the salt has dissolved, boreholes show that the zone of collapse over the wet rockhead extends down to a depth of about 90 m, forming a belt 1–2 km wide and 12 km long. The route of the original brine run was the area in which the most subsidence and building damage occurred in the nineteenth century. In the 1980s, after the cessation of brine extraction, the brine levels began to rise. This caused Poole & Williams (1981) to speculate that in the future brine would flow again from the sites of the original springs that were used by the Romans. The fact that brine now flows naturally to the surface here is proved by a geochemical survey (British Geological Survey 1999), which showed stream water chloride concentrations of 3100 and 3400 mg l^{-1} on adjacent sites to the east of Droitwich.

There is some evidence that the ongoing flow of brine in the area is responsible for some subsidence and Wychavon District Council has consultants who conduct a regular surveying monitoring programme which measures changes in ground surface levels. The Council points out that the areas affected constantly change, but that with suitable mitigation development can proceed (Wychavon District Council 2006). Figure 14.10 shows the approximate zone affected by ongoing salt subsidence based on work by Poole & Williams (1981) modified in the light of further work.

14.6.5 Northern Ireland

Triassic saliferous rocks are present in the area around Carrickfergus and Larne, bordering on Belfast Loch in Northern Ireland. The Triassic sequence includes up to a nearly 500 m thickness of salt, proved in the Larne Borehole, but in the Carrickfergus salt field only about 40–50 m has been proved. The salt deposits are largely protected from groundwater circulation by the overlying largely impermeable sequence and there is no record of a wet rockhead and extensive brecciation such as that found in Cheshire (Griffith & Wilson 1982). A salt spring was known at Eden, where salt mining subsequently developed, and slight brecciation of the sequence was recorded in a few boreholes, but was not extensive. The most spectacular dissolution feature recorded in the area was the collapse of the Tennant Salt Mine in 1990 (Griffith 1991), which also generated a seismic event of magnitude 2.5 on the Richter scale (Bell et al. 2005). This mine was a conventional pillar-and-stall mine, but subsequent owners removed substantial amounts of brine from the abandoned workings. This dissolved the pillars and produced a large crown hole with concentric failure planes and a depression about 130 m across (Nicholson 2014). Maiden Mount Salt Mine in the same area was mined from 1877 to 1895 when it moved to brine extraction up until 1958 (Griffith & Wilson 1982). It collapsed on 19 August 2001, producing a large crater (Fig. 14.11). This collapse prompted expensive monitoring of the remaining Northern Irish salt mines (Northern Ireland Assembly 2002). These are two cases of modern bastard brining that demonstrate the benefit of outlawing this practice in Cheshire as long ago as 1930. In Northern Ireland other mines, including French Park and Carrickfergus International Mine, are still considered at risk of collapse and to avoid potential subsidence the Carrickfergus Development Plan excludes building over all of the former salt mines (Belfast Metropolitan Area 2013).

14.7 Mitigating salt subsidence problems

14.7.1 Brine Subsidence Compensation Board

In Cheshire, the salt mining allied with 'bastard' and 'wild' brine extraction caused such severe subsidence that a Parliamentary investigation of the problems was commissioned (Dickinson 1873). Subsequently, an Act of Parliament was passed that placed a levy on all locally extracted salt. This levy, which funded building reconstruction and

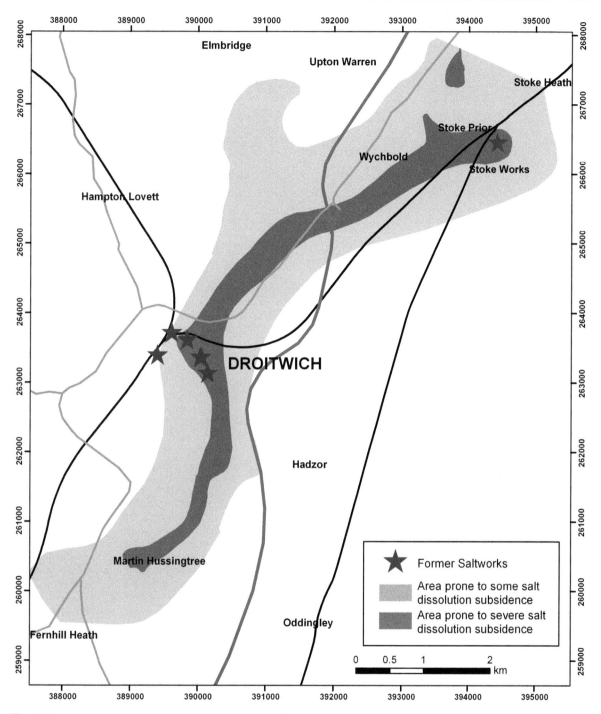

Fig. 14.10. The approximate brine subsidence affected area around Droitwich and locations of saltworks/brine extraction sites. Contains Ordnance Survey data © Crown copyright and database right 2013.

Fig. 14.11. Collapse crater that formed by the failure of Maidenmount Salt Mine near Carrickfergus on 19 August 2001; it was initially 50 m wide and 15 m deep, but continued expanding (BGS photograph P543378 © BGS/UKRI).

compensation payments, is still made by the 'Cheshire Brine Subsidence Compensation Board', but at a lower rate to reflect the reduced risk from modern extraction (Collins 1971). This Act was modified by The Cheshire Brine Pumping (Compensation for Subsidence) Act of 1952, which set up a single compensation district covering all areas of Cheshire where subsidence resulting from the pumping of brine was considered to be a possibility (Fig. 14.12). The 'Cheshire Brine Subsidence Compensation Board' was formed in order to discharge the responsibilities resulting from the passing of this Act.

The main duties of the Cheshire Brine Subsidence Compensation Board are to make compensation for specified categories of damage to land and buildings caused by subsidence owing to brine pumping, and to give advice to the Planning Authorities and Building Control in connection with development on land which may be affected by historic brine pumping (Cheshire Brine Subsidence Compensation Board 2013). The remit of the Board includes the issuance of statutory notices, liaison with the public, maintenance of file records, response to planning enquiries, progressing of subsidence damage claims, maintenance of a claims register, monitoring of ground movements and preparation of search reports for property transactions. With regard to Planning and Building Control consultations, the Board is responsible for prescribing the Consultation Area under Section 38(1) of the 1952 Act and consultations are submitted by the Local Authorities to the Board as part of the planning process. The Board now has no liability for damage resulting from the collapse of current or abandoned salt mine workings or associated shafts.

From 1 May 2006 to 1st June 2018 the duties of the Board were discharged via consultants http://cheshirebrine.com in partnership with the Coal Authority mining search facility. From 1 June 2018 onwards the partnership changed with reporting provided by Groundsure https://groundsure.com/products/cheshire-salt-search/.

14.7.2 Salt mine stabilization

The historical salt mines in the Northwich area have continued to cause problems for development of the town and have threatened some modern constructions. Instability of the old mine workings was aggravated by wild brine pumping, but even after this ceased in 2005 there was still concern about the stability of the mines. In 2002, 32 million pounds was pledged by the government agency of English Partnerships to stabilize the salt mines of Barons Quay, Witton Bank, Neumanns and Penny's Lane (Fig. 14.5). Where possible these mines were ultrasonically scanned to establish their condition and dimensions. They were modelled using this information and historical mine plans. The mines were stabilized using a grout of saturated brine, pulverized fuel ash and cement (Brooks et al. 2007). A void volume of 780 500 m^3 at depths of 90 m was successfully filled and 32 ha of ground stabilized. The exercise was a logistical challenge, bringing in cement and about 1 million tonnes of pulverized fuel ash by rail to Winnington and pumping it down a pipeline that included a 470 m horizontally bored section beneath a river and Site of Biological Interest to the grouting (W S Atkins Consultants Ltd 2001; Brennan 2007; Arup 2013). The stability of other unfilled salt mines in the area, including

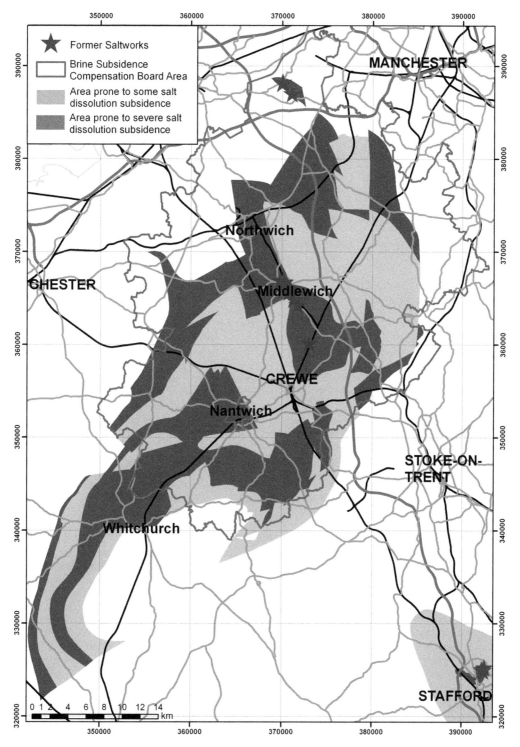

Fig. 14.12. The Cheshire saltfield and the area included in the Brine Subsidence Compensation Board remit. Contains Ordnance Survey data © Crown copyright and database right 2013.

those at Wincham and north of Northwich (Pringle et al. 2012), may be problematical and will require either their avoidance or investigation and remediation by grout filling to make the land suitable for development.

14.7.3 Monitoring and investigation

In certain areas underlain by salt where dissolution has caused subsidence, monitoring by geodetic levelling is undertaken to record the amount and distribution of the movement. Development in Droitwich (Wychavon District Council) is constrained by such levelling. Here it helps to define an active central brine run (Zone A) and a passive marginal brine run area (Zone B); there is also a Zone C where ground movements are suspected, but little evidence is available. Within these zones development may require additional monitoring or levelling. Development is constrained to certain construction types and reinforced structures as designated by the council and their consultants.

In Northern Ireland the monitoring of potentially unstable salt mines has included surface levelling, brine/water level measurements in boreholes, instrumentation to detect underground movements, CCTV and ultrasonic scanning (Northern Ireland Assembly 2002).

In Cheshire, microgravity geophysics has proved to be useful tool for monitoring cavity growth and subsidence owing to salt dissolution. Branston & Styles (2003) imaged a subsiding area to the east of Penny's Lane Mine (Fig. 14.5) and showed that it was not related to undermining. They linked the subsidence to salt dissolution at shallow depths owing to wild brine extraction that was being undertaken from a factory located to the east. This was drawing in brine from a considerable distance away. They showed that the microgravity signature indicated less density over time and the development of brecciated rock associated with brine dissolution and subsidence. Long-term monitoring of the Trent–Mersey Canal where it crosses former salt workings showed microgravity and topographical changes related to upward-migrating cavities and subsiding areas (Pringle et al. 2012). Other geophysics such as ground-probing radar and electrical resistivity tomography also have the potential to image cavities and collapsing ground in similar circumstances. However, electrical methods may be limited in depth penetration if there is shallow conducting brine in the area.

Construction on ground susceptible to salt dissolution subsidence requires the avoidance of actively subsiding areas and similar mitigation measures to those adopted for coal mining regions. Large structures may be particularly susceptible to unwanted movement and it is important to understand the local geology, faulting and collapse mechanisms such as those described by Wilson (2003) for Manchester Airport.

14.7.4 Planning for soluble rock geohazards

Current planning procedures ensure that the modern exploitation of salt deposits lies largely outside of urban areas so that risks are considerably reduced. The salt is now mined either in deep dry mines, or by controlled brine extraction from depth. However, there is still a legacy of problems related to the salt deposits. These include old salt mines that have not collapsed, and compressible or unstable collapsed ground over former salt mines. In addition, natural salt dissolution at the rockhead interface, between the salt deposits and the overlying superficial deposits, can cause ground engineering problems and corrosive saline groundwater. With the ending of most near-surface mining and brine extraction, the hydrological system has re-balanced, or is in the process of re-balancing itself. It may be expected that natural groundwater flows will be re-established through the disrupted saltfields and further subsidence problems may occur. The accurate mapping of the rock salt and associated deposits and salt mines, plus an understanding of salt dissolution and collapse characteristics, can help development and planning in these subsidence-sensitive areas. These problems can then be reduced by careful planning and construction or remediation of former mines.

Examples of such planning include Droitwich, where zonation with specific monitoring and building controls is undertaken. In Teeside, the local planning guidelines for development in former salt dissolution extraction areas (Morris 1978) place a zone that is considered to be susceptible to subsidence of between 150 and 300 m radius around every former brine extraction borehole in the area. The size of this buffer zone appears to relate to the depth and amount of salt extracted from each hole. In these areas, it is recommended that avoidance or special precautions should be taken for certain types of construction. In Northern Ireland development is not permitted over or adjacent to former salt mines (Belfast Metropolitan Area 2013). The potential collapse of Carrickfergus International Salt Mine required remedial measures that in 2003 included moving roads and car parks clear of the threat and monitoring affected areas; these measures cost over £1.1 million (Northern Ireland Audit Office 2004).

In Cheshire the collapse of some former mines is causing subsidence and their remediation is being undertaken. In most of the salt karst areas, there is a legacy of difficult ground conditions produced by natural and man-made causes and restrictions on the places where future mining and brine extraction can be undertaken (Collins 1971). Information to help with large-scale planning is available from the British Geological Survey GeoSure dataset and the karst database (Cooper 2008; Farrant & Cooper 2008; Cooper et al. 2011).

Once dissolution has occurred, there is the possibility of metastable cavities being present in the subsurface. Infiltration and disposal of surface water to the ground can trigger subsidence. Consequently, areas of salt karst are not recommended for the installation of sustainable drainage systems (Dearden et al. 2013) or for the installation of open loop ground source heat pumps (Cooper et al. 2011). In common with building over subsidence-prone ground, the use of strengthened foundations and novel materials such as geogrid and lightweight embankment materials should be considered. In Cheshire, following the formation of the Brine Subsidence

Compensation Board in 1891, construction utilized timber frame foundations with jacking points and brick panel walls within timber frame supports (Northwich Townscape Heritage Project 2019). Subsidence affected many of these buildings and the processes of 'house raising' and 'shoplifting' were common (Calvert 1915). In Northwich steel frames and foundation beams were also used from the 1920s to the 1970s, including the Plaza Cinema (1928; English Heritage 2000), the former Regal Cinema (1939; Northwich History 2019) and modern residential housing (Northwich Guardian 2013). Practices used for development over gypsum karst prone to subsidence, such as reinforced slab foundations, are also suitable for salt karst areas (Cooper & Calow 1998; Cooper & Saunders 2002; Cooper et al. 2011). With careful investigation and planning, development can commonly proceed in salt karst areas; however, where cavities or mines that have not been stabilized are present, avoidance is the best plan.

Acknowledgements The author thanks the following for helpful discussion and for critically reviewing the manuscript: Katy Lee, Hannah Jordan, Dr Vanessa Banks, Dr Helen Reeves and Dr Dave Giles. This article is published with permission of the Executive Director of the British Geological Survey (UKRI).

Funding This research received no specific grant from any funding agency in the public, commercial, or not-for-profit sectors.

References

ARUP GEOTECHNICS 1991. *Review of mining instability in Great Britain, Stafford brine pumping*, Vol. 3, vi. Arup Geotechnics for the Department of the Environment.

ARUP 2013. *Northwich salt mines stabilisation*. http://www.arup.com/Projects/Northwich_Salt_Mines_Stabilisation

AUTIN, W.J. 2002. Landscape evolution of the Five Islands of south Louisiana: scientific policy and salt dome utilization and management. *Geomorphology*, **47**, 227–244, https://doi.org/10.1016/S0169-555X(02)00086-7

BELFAST METROPOLITAN AREA 2013. *Belfast Metropolitan Area Plan 2015 district proposals: Carrickfergus Countryside & Coast (Countryside and Coast) – mineral development – Policy CE 08 Area of Potential Subsidence Carrickfergus*. Northern Ireland Planning Portal, Belfast.

BELL, F.G., DONNELLY, L.J., GENSKE, D.D. & OJEDA, J. 2005. Unusual cases of mining subsidence from Great Britain, Germany and Colombia. *Environmental Geology*, **47**, 620–631, https://doi.org/10.1007/s00254-004-1187-9

BLOODWORTH, A.J., CAMERON, D.G., HARRISON, D.J., HIGHLEY, D.E., HOLLOWAY, S. & WARRINGTON, G. 1999. *Mineral resources information for development plans: phase one. Herefordshire & Worcestershire; resources and constraints*. British Geological Survey Technical Report **WF/99/4**.

BRANSTON, M.W. & STYLES, P. 2003. The application of time-lapse microgravity for the investigation and monitoring of Subsidence at Northwich, Cheshire. *Quarterly Journal of Engineering Geology and Hydrogeology*, **36**, 231–244, https://doi.org/10.1144/1470-9236/03-243

BRENNAN, P. 2007. Northwich – grouting using 1,000,000 tonnes of conditioned PFA from Drax Power Station. *2007 World of Coal Ash (WOCA)*, 7–10 May 2007, Northern Kentucky, USA, 1–7.

BRITISH GEOLOGICAL SURVEY 1987. *Stockton, England and Wales Sheet 33*. Solid and Drift Geology. 1:50,000.

BRITISH GEOLOGICAL SURVEY 1999. *Regional Geochemistry of Wales and Part of West-Central England: Stream Water*. British Geological Survey, Keyworth.

BROOKS, T.G., O'RIORDAN, N.J., BIRD, J.F., STIRLING, R. & BILLINGTON, D. 2007. Stabilisation of Abandoned Salt Mines in North West England – Paper Number 781. Engineering Geology for Tomorrow's Cities IAEG 2006. International Association of Engineering Geology, Nottingham.

CALVERT, A.F. 1915. *Salt in Cheshire*. Spon, London.

CHESHIRE BRINE SUBSIDENCE COMPENSATION BOARD 2013. *Cheshire Brine Subsidence Compensation Board – about us*. http://www.cheshirebrine.com/

COLLINS, J.F.N. 1971. *Salt: a Policy for the Control of Salt Extraction in Cheshire*, Cheshire County Council.

COMPETITION COMMISSION 2005. *British Salt Limited/New Cheshire Salt Works Limited merger inquiry*. Final report and Appendices & Glossary.

COOPE, G., SHOTTON, F. & STRACHAN, I. 1961. A late Pleistocene fauna and flora from Upton Warren, Worcestershire. *Philosophical Transactions of the Royal Society of London (Series B)*, **244**, 379–417, https://doi.org/10.1098/rstb.1961.0012

COOPER, A.H. 1998. Subsidence hazards caused by the dissolution of Permian gypsum in England: geology, investigation and remediation. *In*: MAUND, J.G. & EDDLESTON, M. (eds) *Geohazards in Engineering Geology*. Geological Society, London, Engineering Geology Special Publications, **15**, 265–275, https://doi.org/10.1144/GSL.ENG.1998.015.01.27

COOPER, A.H. 2002. Halite karst geohazards (natural and man-made) in the United Kingdom. *Environmental Geology*, **42**, 505–512, https://doi.org/10.1007/s00254-001-0512-9

COOPER, A.H. 2008. The GIS approach to evaporite karst geohazards in Great Britain. *Environmental Geology*, **53**, 981–992, https://doi.org/10.1007/s00254-007-0724-8

COOPER, A.H. & CALOW, R. 1998. *Avoiding gypsum geohazards: guidance for planning and construction*, British Geological Survey Technical Report, **WC/98/5**, http://nora.nerc.ac.uk/id/eprint/14146/

COOPER, A.H. & SAUNDERS, J.M. 2002. Road and bridge construction across gypsum karst in England. *Engineering Geology*, **65**, 217–223, https://doi.org/10.1016/S0013-7952(01)00131-4

COOPER, A.H., FARRANT, A.R. & PRICE, S.J. 2011. The use of karst geomorphology for planning, hazard avoidance and development in Great Britain. *Geomorphology*, **134**, 118–131, https://doi.org/10.1016/j.geomorph.2011.06.004

COXILL, D. 1995. Geology & extraction history of the non-aggregate mineral resources of Staffordshire. *Shropshire Caving and Mining Club (SCMC) Journal*, **3**, 24–37.

DE RANCE, C. 1891. Subsidence at Wybunbury, near Crewe. *Transactions of the Manchester Geological Society*, **21**, 197–199.

DEARDEN, R.A., MARCHANT, A. & ROYSE, K.E. 2013. Development of a suitability map for infiltration sustainable drainage systems (SuDS). *Environmental Earth Sciences*, **70**, 2587–2602, https://doi.org/10.1007/s12665-013-2301-7

DEEPSTORE 2013. *Deepstore Records Management – a Compass Minerals Company*. World http://www.deepstore.com/about-us

DICKINSON, J. 1873. *Report on the Rock-salt Mines and Brine Pits relative to the Landslips, for Her Majesty's Secretary of State for the Home Department*, **53**.

DOORNENBAL, H. & STEVENSON, A. 2010. *Petroleum Geological Atlas of the Southern Permian Basin Area*. EAGE, Houten.

DRILCORP 2013. *Managed Realignment Scheme Greatham Creek. Brine Well Abandonment*. https://www.drilcorp.com/decommissioning-brine-wells/

EARP, J.R. & TAYLOR, B.J. 1986. *Geology of the country around Chester and Winsford, Sheet 109 (England and Wales)*. British Geological Survey, **119**.

ENGLISH HERITAGE 2000. *Listed building, List entry number 1385195, Plaza Bingo Club, 151 Witton Street, Northwich*. https://historicengland.org.uk/listing/the-list/list-entry/1385195

EVANS, D.J. 2008. *An Appraisal of Underground Gas Storage Technologies and Incidents, for the Development of Risk Assessment Methodology*. Health and Safety Executive.

EVANS, D.J. & HOLLOWAY, S. 2009. A review of onshore UK salt deposits and their potential for underground gas storage. *In*: EVANS, D.J. & CHADWICK, R.A. (eds) *Underground Gas Storage: Worldwide Experiences and Future Development in the UK and Europe*. The Geological Society, London, Special Publications, **313**, 39–80, https://doi.org/10.1144/SP313.5

EVANS, D.J., REES, J.G. & HOLLOWAY, S. 1993. The Permian to Jurassic stratigraphy and structural evolution of the central Cheshire Basin. *Journal of the Geological Society, London*, **150**, 857–870, https://doi.org/10.1144/gsjgs.150.5.0857

FARRANT, A.R. & COOPER, A.H. 2008. Karst geohazards in the UK: the use of digital data for hazard management. *Quarterly Journal of Engineering Geology and Hydrogeology*, **41**, 339–356, https://doi.org/10.1144/1470-9236/07-201

FORD, D.C. & WILLIAMS, P. 1989. *Karst Geomorphology and Hydrology*. Unwin Hyman, London.

GRIFFITH, A.E. 1991. Tennants ills. *Ground Engineering*, **21**, 18–21.

GRIFFITH, A.E. & WILSON, H.E. 1982. *Geology of the Country Around Carrickfergus and Bangor (Counties Antrim and Down): Memoir for one-inch Geological Sheet 29*. Geological Survey of Northern Ireland. HMSO, Belfast.

GUTIÉRREZ, F., JOHNSON, K.S. & COOPER, A.H. 2008. Evaporite karst processes, landforms, and environmental problems. *Environmental Geology*, **53**, 935–936, https://doi.org/10.1007/s00254-007-0715-9

HOWELL, F.T. 1984. Salt karst of the Cheshire Basin, England. *In*: CASTANY, G. & GROBA, E. & ROMIJN, E. (eds) *Hydrogeology of Karstic Terranes. International Contributions to Hydrogeology*. International Association of Hydrogeologists, 252–254.

HOWELL, F.T. 1986. Photogeology of some salt karst subsidence, Cheshire, England. *In*: JOHNSON, A.I., CARBOGNIN, L. & UBERTINI, L. (eds) *Proceedings of the Third International Symposium on Land Subsidence Held at Venice*, March 1984. IAHS, 617–627.

HOWELL, F.T. & JENKINS, P.L. 1976. Some aspects of the subsidences in the rocksalt districts of Cheshire, England. *Proceedings of the Anaheim Symposium. International Association of Hydrogeological Sciences*, 507–520.

HURST, J.D. 1997. *A Multi-period Salt Production Site at Droitwich, Excavations at Upwich*. Council for British Archaeology.

JACKSON, W. 1669. Some inquiries concerning the salt-springs and the way of salt-making at Nantwich in Cheshire. *Philosophical Transactions*, **4**, 1060–1067, https://doi.org/10.1098/rstl.1669.0047

JACKSON, D.I., JACKSON, A.A., EVANS, D., WINGFIELD, R.P., BARNES, R.P. & ARTHUR, M.J. 1995. *United Kingdom Offshore Regional Report: the Geology of the Irish Sea*. HMSO for the British Geological Survey, London.

JOHNSON, K.S. 2005. Salt dissolution and subsidence or collapse caused by human activities. *In*: EHLEN, J., HANEBERG, W.C. & LARSON, R.A. (eds) *Humans as Geologic Agents: Geological Society of America Reviews in Engineering Geology*. Geological Society of America, Boulder, CO, 101–110.

LEE, E.M. & SAKALAS, C.F. 2001. Subsidence map development in an area of abandoned salt mines. *In*: GRIFFITHS, J.S. (ed.) *Land Surface Evaluation for Engineering Practice*. Geological Society, London, Engineering Geology Special Publications, **18**, 193–195.

MARLEY, J. 1890. On the Cleveland and Durham Salt Industry. *Transactions of the Federated Institution of Mining Engineers*, **1**, 229–373.

MORRIS, C.H. 1978. *Report on Abandoned Mineral Workings and Possible Surface Instability Problems*. County of Cleveland, Department of the County Surveyor and Engineer.

NATURAL ENGLAND 1986. *Pasturefields Salt Marsh, Stafford, Site Ref: 15WH7 SSSI designation*. https://designatedsites.naturalengland.org.uk/PDFsForWeb/Citation/1003939.pdf

NICHOLSON, C.A. 2014. Salt mines in the Carrickfergus area of County Antrim. *Journal of the Mining Heritage Trust of Ireland*, **14**, 1–22, https://www.mhti.org/uploads/2/3/6/6/23664026/saltmines_in_the_carrickfergus_area_of_county_antrim.pdf

NORTHERN IRELAND ASSEMBLY 2002. *Written Answers to Questions, Friday 26th July 2002*. International Mine, Carrickfergus. http://archive.niassembly.gov.uk/writtenanswers/020726D.htm

NORTHERN IRELAND AUDIT OFFICE 2004. *Financial Auditing and Reporting: 2002–3003*. General Report by the Comptroller and Auditor General for Northern Ireland. https://www.niauditoffice.gov.uk/sites/niao/files/media-files/financial_audit___reporting_2002-03_-_gen_report_by_c_ag.pdf

NORTHOLT, A.J.G. & HIGHLEY, D.E. 1973. *Salt. Mineral Dossier No. 7, Mineral Resources Consultative Committee, London*. HMSO, London.

NORTHWICH GUARDIAN 2013. *All is not lost for steel framed home owners*. http://www.northwichguardian.co.uk/news/10607362.All_is_not_lost_for_steel_framed_home_owners/

NORTHWICH HISTORY. 2019. The Regal Cinema, Northwich. http://northwichhistory.co.uk/the-regal-cinema-northwich/

NORTHWICH TOWNSCAPE HERITAGE PROJECT 2019. Northwich's unique timber framed buildings, https://www.northwich-th.co.uk/history/northwichs-unique-timber-framed-buildings/

ORMEROD, G.W. 1848. Outline of he principal Geological features of the Salt-Field of Cheshire and the adjoining districts. *Proceedings of the Geological Society of London*, **4**, 262–288, https://doi.org/10.1144/GSL.JGS.1848.004.01-02.37

PITTS, V., JOYNES, M. & DEAN, N. 2012. *Salt and Brine in Worcestershire: Mineral Local Plan Background Document*. Worcestershire County Council.

POOLE, E.G. & WILLIAMS, B.J. 1981. *The Keuper Saliferous Beds of the Droitwich area*. Institute of Geological Sciences Report no. **81/2**.

PRINGLE, J.K., STYLES, P., HOWELL, C.P., BRANSTON, M.W., FURNER, R. & TOON, S.M. 2012. Long-term time-lapse microgravity and geotechnical monitoring of relict salt mines, Marston, Cheshire, U.K. *Geophysics*, **77**, B287–B294, https://doi.org/10.1190/GEO2011-0491.1

RIGBY, J. 1905. Outburst from Duncrue old rock salt mine after being tapped for brine. *Transactions of the Manchester Geological and Mining Society*, **28**, 565–570.

SHERLOCK, R.L. 1921. *Rock-salt and Brine*. Special reports on the Mineral Resources of Great Britain No 18. Memoirs of the Geological Survey. His Majesty's Stationery Office, London, **122**.

SMITH, D.B. 1989. The late Permian palaeogeography of north-east England. *Proceedings of the Yorkshire Geological Society*, **47**, 285–312, https://doi.org/10.1144/pygs.47.4.285

STAFFORD BOROUGH 2012. *Habitat regulations assessment for the Plan for Stafford Borough – publication in Respect of Natura 2000 Sites*, https://www.staffordbc.gov.uk/live/Documents/Forward%20Planning/Examination%20Library%202013/A24-HABITAT-REGULATIONS-ASSESSMENT-FOR-PLAN-FOR-STAFFORD.pdf.

SWIFT, M. & REDDISH, D.J. 2005. Underground excavations in rock salt. *Geotechnical and Geological Engineering*, **23**, 17–42, https://doi.org/10.1007/s10706-003-3159-3

TAYLOR, J.C.M. & COULTER, V.S. 1975. Zechstein of the English Sector of the southern North Sea basin. *In*: WOODLAND, A.W. (ed.) *Petroleum and the Continental Shelf of North-west Europe, Vol. 1 Geology*. Applied Science, Barking, 249–263.

THURMOND, V.L., POTTER, R.W. & CLYNNE, M.A. 1984. The Densities of Saturated Solutions of NaCl and KCl from 10° to 105°. Open File Report 84-253. United States Department of the Interior Geological Survey, https://pubs.usgs.gov/of/1984/0253/report.pdf

TOMLIN, D.M. 1982. *The Salt Industry of the River Tees*. De Archaeologische Pers Nederland.

WALLING, D.E. & WEBB, B.W. 1981. Water quality. *In*: LEWIN, J. (ed.) *British Rivers*. Allen and Unwin, 126–169.

WALTHAM, A.C. 1989. *Ground Subsidence*. Blackie, Glasgow.

WALTHAM, A.C., SIMMS, M.J., FARRANT, A.R. & GOLDIE, H. 1997. *Karst and Caves of Great Britain*. Chapman and Hall, London.

WHITTAKER, A. 1972. The Somerset Saltfield. *Nature*, **238**, 265–266, https://doi.org/10.1038/238265a0

WILLS, L.J. 1976. *The Triassic of Worcestershire and Warwickshire*. Report of the Institute of Geological Sciences Report **76/2**.

WILSON, A.A. 1990. The Mercia Mudstone Group (Trias) of the East Irish Sea Basin. *Proceedings of the Yorkshire Geological Society*, **48**, 1–22, https://doi.org/10.1144/pygs.48.1.1

WILSON, A.A. 1993. The Mercia Mudstone Group (Trias) of the Cheshire Basin. *Proceedings of the Yorkshire Geological Society*, **49**, 171–188, https://doi.org/10.1144/pygs.49.3.171

WILSON, A.A. 2003. The Mercia Mudstone Group (Triassic) of Manchester Airport, Second Runway. *Proceedings of the Yorkshire Geological Society*, **54**, 129–145, https://doi.org/10.1144/pygs.54.3.129

WILSON, A.A. & EVANS, W.B. 1990. *Geology of the country around Blackpool, Sheet 66 (England and Wales)*. British Geological Survey, **82**.

WILSON, E. 1888. Durham salt-district. *Quarterly Journal of the Geological Society, London*, **44**, 761–782, https://doi.org/10.1144/GSL.JGS.1888.044.01-04.48

WOODIWISS, S.G. 1992. *Iron Age and Roman salt production and the medieval town of Droitwich. Excavations at the Old Bowling Green and Friar Street*, Research Report, **81**. Council for British Archaeology.

WOODS, P.J.E. 1979. The geology of Boulby Mine. *Economic Geology*, **74**, 409–418, https://doi.org/10.2113/gsecongeo.74.2.409

WORCESTERSHIRE BIOLOGICAL RECORDS CENTRE 1999. Upton Warren Worcs BRC Survey Day. Application no. APP/2001/0699, http://www.wbrc.org.uk/WORCRECD/Issue7/uptnw.htm

W S ATKINS CONSULTANTS LTD 2001. *Northwich Salt Mines Stabilisation: Environmental Statement Volume 1 Non-technical Summary*. Vale Royal Borough Council, https://pa.cheshirewestandchester.gov.uk/online-applications/

WYCHAVON DISTRICT COUNCIL 2006. *Wychavon District Council Local Plan*: Adopted June 2006, https://www.wychavon.gov.uk/documents/10586/157693/wdc-planning-lp-plan-local_plan2-_for-web_.pdf

WYCHAVON DISTRICT COUNCIL 2013. *Droitwich Spa: the Droitwich Conservation Area Appraisal and Management Plan*. September 2013. https://www.wychavon.gov.uk/documents/10586/157693/Planning-HER-DroitwichCArevisedSept13.pdf

Chapter 15 Dissolution – carbonates

Clive N. Edmonds

Stantec UK Ltd (Stantec), Caversham Bridge House, Waterman Place, Reading, Berkshire, RG1 8DN, UK

clive.edmonds@stantec.com

Abstract: The dissolution of limestone and chalk (soluble carbonates) through geological time can lead to the creation of naturally formed cavities in the rock. The cavities can be air, water, rock or soil infilled and can occur at shallow levels within the carbonate rock surface or at deeper levels below. Depending upon the geological sequence, as the cavities break down and become unstable they can cause overlying rock strata to settle and tilt and also collapse of non-cemented strata and superficial deposits as voids migrate upwards to the surface. Natural cavities can be present in a stable or potentially unstable condition. The latter may be disturbed and triggered to cause ground instability by the action of percolating water, loading or vibration. The outcrops of various limestones and chalk occur widely across the UK, posing a significant subsidence hazard to existing and new land development and people. In addition to subsidence they can also create a variety of other problems such as slope instability, generate pathways for pollutants and soil gas to travel along and impact all manner of engineering works. Knowledge of natural cavities is essential for planning, development control and the construction of safe development.

15.1 Introduction

Limestone and chalk are composed of calcium carbonate that is soluble in the presence of acidic water. Rain water combines with atmospheric and biogenic carbon dioxide to form weak carbonic acid which then dissolves the calcium carbonate. Where the water table level lies at depth within the carbonate sequence, the water can enter the rock via joints and fissures to percolate downwards and cause dissolution. The effects of dissolution tend to be concentrated within the upper surface zone of the rock, especially where permeable overlying deposits are present. The cover deposits influence the acidity and concentration of water flows penetrating the carbonate rock surface. As solution features are formed over time, the cover deposits will tend to settle and collapse down into the enlarging features, often leading to subsidence occurring at the ground surface which can cause damage to buildings and infrastructure in urban areas.

For new construction the challenge is to check whether solution features are present below a site and to understand the karst geohazard setting to ensure that the correct engineering solutions are put in place to permit safe development. This includes addressing not only the safe support of buildings, roads and services but also the effects of surface water drainage disposal. Unfortunately there are many cases where the design of development has not taken karst into account, which can result in subsidence damage taking place in due course. Following a subsidence event it is essential to identify the nature and cause of movement before a suitable remedial solution to stabilize the ground can be executed. Property evacuation may be necessary for safety reasons before the remedial works can be completed and in some instances an economic solution might not be feasible. Blighting and dereliction of property can be an issue arising in certain circumstances.

15.2 Geographical occurrence

Data collected on the geographical occurrence of natural cavities formed by dissolution of limestone and chalk over the last 30 years has been compiled and input to the national Natural Cavities Database held by Stantec UK Ltd (Stantec). The origins of the database are academic research undertaken by Edmonds (1987) and the collection of a large representative number of natural cavity records for a national study funded by the Department of the Environment and undertaken by Applied Geology Limited (1993). The spatial occurrence of different natural cavity types recorded upon soluble carbonate rock units within different regional areas is summarized in Table 15.1. A plan showing the spatial distribution of soluble rocks such as limestone and chalk is presented in Figure 15.1.

As the figure shows, the soluble carbonate rocks are widely distributed across the country. They range in age from the Cambro-Ordovician age Durness Group limestone to the Cretaceous age Chalk Group. Their outcrop areas also vary considerably as does the thickness of the rock units. The most

Table 15.1. *Recorded geographical distribution of natural cavities formed by dissolution of limestone and chalk in Great Britain*

Area	County	Natural cavity location (n)	Natural cavity features (n)	Number of individual natural cavity types			
				Sw Sh	Sp Swj	Pc Vc	Other
SE England	Bedfordshire	9	153	0	153	0	0
	Berkshire	517	1026	525	493	1	7
	Buckinghamshire	267	595	223	369	0	3
	East Sussex	64	170	36	122	9	3
	Essex	31	794	10	784	0	0
	Greater London	103	261	102	149	3	7
	Hampshire	497	1010	313	697	0	0
	Hertfordshire	396	1584	485	1092	0	7
	Isle of Wight	4	6	2	4	0	0
	Kent	620	2391	224	2144	7	16
	Oxfordshire	108	1134	132	1001	1	0
	Surrey	75	267	91	173	0	3
	West Sussex	40	177	117	60	0	0
SW England	Cornwall	1	1	0	0	1	0
	Devon	53	188	4	132	52	0
	Dorset	326	2404	684	1695	25	0
	Gloucestershire	107	189	36	29	124	0
	Somerset	406	1204	324	338	534	8
	Wiltshire	46	135	48	87	0	0
West Midlands	Hereford	8	23	0	0	23	0
	Shropshire	2	2	0	0	1	1
	Staffordshire	57	116	35	25	54	2
East Midlands	Derbyshire	241	620	178	127	314	1
	Leicestershire	3	6	4	2	0	0
	Lincolnshire	22	55	28	23	4	0
	Northamptonshire	4	4	1	3	0	0
	Nottinghamshire	7	13	0	11	2	0
East Anglia	Cambridgeshire	13	31	4	27	0	0
	Norfolk	175	569	155	412	0	2
	Suffolk	35	129	25	102	0	2
NW England	Cheshire	1	1	0	0	1	0
	Lancashire	52	55	10	8	37	0
Yorkshire and Humberside	Humberside	20	49	36	13	0	0
	North Yorkshire	1671	2271	902	390	968	11
	South Yorkshire	21	22	1	3	18	0
Northern England	Cleveland	4	5	0	5	0	0
	Cumbria	351	415	100	1	314	0
	Durham	90	104	32	3	69	0
	Northumberland	16	24	1	3	20	0
Scotland	Central	1	1	0	0	0	1
	Grampian	1	1	0	0	0	1
	Highland	159	171	44	1	108	18
	Lothian	4	8	0	6	2	0
	Strathclyde	21	33	7	2	5	19
	Tayside	13	13	1	0	1	11
Wales	Clwyd	289	429	197	2	229	1
	Dyfed	164	294	153	60	81	0
	Gwent	119	161	68	10	83	0
	Gwynedd	18	570	1	560	9	0
	Mid Glamorgan	304	1009	268	675	59	7
	Powys	107	157	60	21	72	4
	South Glamorgan	137	247	176	11	55	5
	West Glamorgan	137	162	44	27	89	2

Note: Figures based on data contained in the archive of the national PBA Natural Cavities Database to end of 2012 and shown only for counties containing natural cavities developed by dissolution of limestone and chalk. 'Other' category typically refers to decalcification.
Sh, sinkhole; Sw, swallow hole; Sp, solution pipe; Swj, solution widened joint; Pc, phreatic cave; Vc, vadose cave.

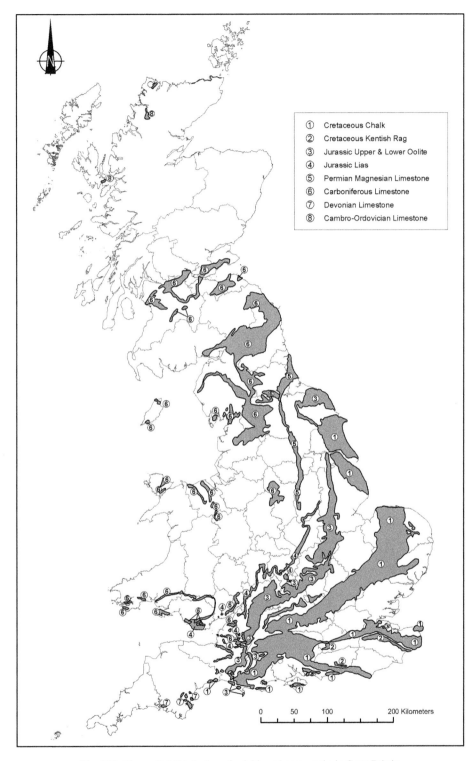

Fig. 15.1. The spatial distribution of soluble carbonate rocks in Great Britain.

significant in terms of both thickness and outcrop area are the Carboniferous-age limestones and Cretaceous-age chalk. The other rock outcrops tend to be more limited in geographical extent, although the thin Jurassic-age limestones form an extensive narrow ribbon outcrop that stretches from SW England to the NE. The Permian-age limestone also forms a prominent outcrop extending in a north–south direction from the East Midlands towards the NE.

As might be expected, in terms of numbers of solution features formed and recorded (Table 15.1) upon the soluble carbonate outcrops, the larger numbers are associated with the more extensive outcrops of limestone and chalk as illustrated by Figure 15.2. For clarity of presentation only those natural cavity locations associated with major soluble rock outcrops are shown and nearby areas of subcrop where other rock strata or superficial deposits overlie the soluble rock stratum. Minor soluble rock outcrops and subcrops with their associated natural cavities are not shown.

15.3 Characteristics of natural cavities formed by dissolution

Natural cavities formed by dissolution of limestone and chalk can be a mix of air-, water-, rock- or soil-infilled. The cavities can form within the surface of the rock and extend downwards, or be present at shallow and deep levels within the rock. They often also impact overlying rock strata and superficial deposits, causing them to subside and collapse into the features as they become progressively enlarged by the dissolution process. Together with characteristic landforms (e.g. spring resurgences, underground drainage and dry valleys) they form karstic landscapes. Many of the National Parks and Areas of Outstanding Natural Beauty in Britain are composed of limestone and chalk.

The typical range of natural cavity forms found upon limestone and chalk is shown in Figure 15.3. Where low-permeability cover deposits are present at the surface then the surface water drainage will tend to collect to form streams that flow across the land surface until they meet the exposed outcrop of the limestone or chalk. At this location the water dissolves the rock surface, leading to the creation of solution-widened joints and bedding planes. As these develop the water readily enters and flows down into the rock mass to form an underground drainage network. Over time a depression is formed at the surface where the overland flow disappears, which is referred to as a swallow hole. A small-scale example is shown in Figure 15.4. In places where permeable cover deposits occur over the limestone and chalk, the water will tend to be absorbed into the surface in a diffuse manner and swallow holes are less prevalent.

Given sufficient time, dissolution of joints at the surface of a limestone will tend to form linear features that extend to depth, widening upwards. Where the bare limestone surface expression of the intersecting dissolution along joints is revealed at the surface they are known as limestone pavements. When dissolution is concentrated at the intersection of joints a point feature may be formed centred on the intersection that becomes pipe shaped with time, extending to depth (see Fig. 15.5). Pipe-shaped features are commonly associated with chalk and referred to as solution pipes (see Fig. 15.6). The subsurface shape of solution pipes formed in chalk can sometimes be irregular and voided (Francescon & Twine 1993). Further discussion of the various forms of features present in chalk is presented by Lord et al. (2002). It is common for overlying cover deposits or rocks to settle down into the enlarging cavities being formed. This leads to downward ravelling of deposits and the development of soil arches which can suddenly collapse, causing subsidence at the surface above. The surface hollows formed are referred to as subsidence sinkholes (see Fig. 15.7). Where cover deposits are absent, sometimes surface hollows can be created by dissolution alone of the limestone surface, focussed on the pattern of joints present or along a fault plane – these surface hollows are referred to as solution sinkholes.

Provided the percolating water remains undersaturated with respect to calcium carbonate, it can continue to dissolve the limestone and chalk rock with depth, producing solution widening of joints (see Fig. 15.8), bedding planes and fault planes. The intensity of the dissolution activity can be enhanced locally where flows of water (of different degrees of acidity) meet and mix, also where temperature/pressure changes occur. Such conditions can result in the formation of caves at depth. Caves formed above the water table are referred to as vadose caves while those formed below the water table are referred to as phreatic caves. Vadose caves tend to enlarge as a result of both dissolution and fluvial erosion as the caves often have streams flowing along the cave floor, cutting down into the limestone and depositing fluvial detritus. It is common for vadose caves to end in water-filled sumps where the cave system continues down into the water table. Phreatic caves tend to form as rounded tubes. Sometimes when the water table level drops, phreatic tube caves can be modified within the vadose zone by streams cutting down through the tube floor (see Fig. 15.9). Cave floor levels generally fall in the direction of the hydraulic gradient. Downward percolation of water entering as drips into vadose caves often precipitates cave deposits – stalagmites and stalactites.

Long-term dissolution and erosion, particularly where streams enter cave systems, can sometimes develop larger vertical conduits referred to as tubular shafts or pots. In some instances caves can enlarge such that large spans can result in collapse of the roof. If the void daylights at the surface and produces a surface depression, it is referred to as a collapse sinkhole. Murphy et al. (2008) discuss the mechanism of breakdown and collapse of cave chamber roofs in the Yorkshire Dales based on studies at Gaping Gill and Hull Pot. They suggest that the mechanism is related to lithology, structure and the glacial climatic cycle. Eventually the underground drainage network, at some location where the water table intersects the surface or maybe at a contact zone with a non-carbonate rock, will cause the water to emerge at the surface as a spring. Sometimes the water gently seeps

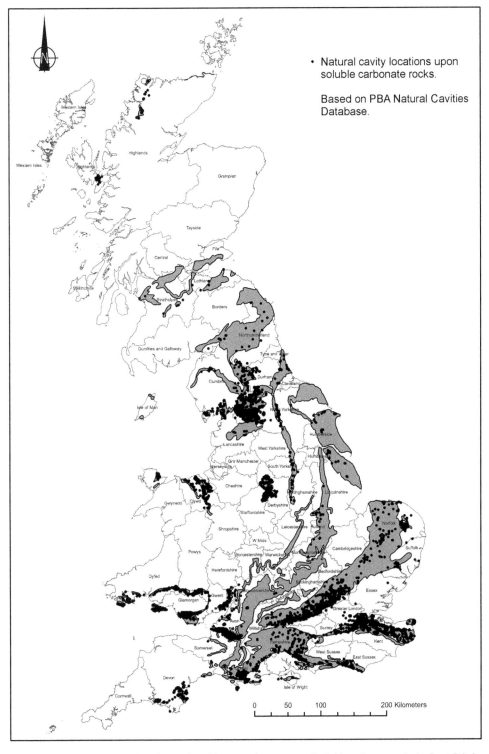

Fig. 15.2. The spatial distribution of natural cavities upon the outcrops of soluble carbonate rocks in Great Britain.

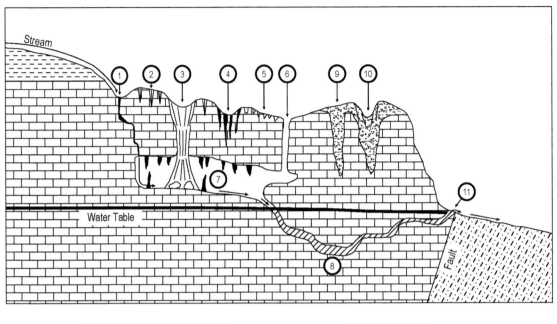

Fig. 15.3. Range of natural cavity types formed on limestone and chalk.

to the surface; other times it may emerge as a major flow from a spring outlet cave.

Useful summaries of the formation of karst features and landforms associated with limestone are presented in a range of textbooks such as Sweeting (1972) and Bögli (1980).

Some forms of natural cavity can be relatively stable over long periods of time while others are prone to being destabilized to cause ground subsidence. Natural cavities created in limestone and chalk take a long geological time period to form (typically tens of thousands

Fig. 15.4. Example of a swallow hole near Hurley, Berkshire.

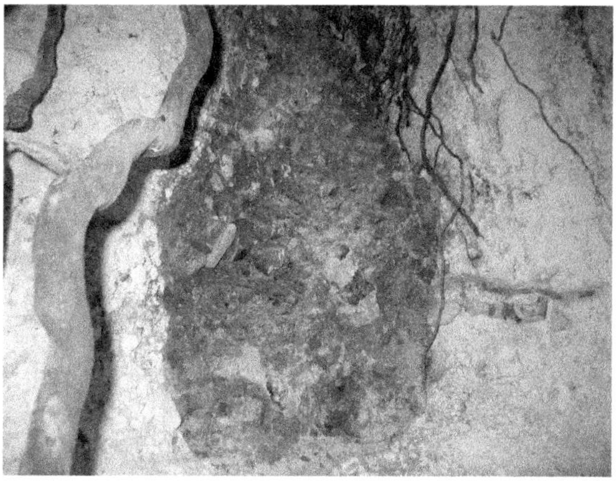

Fig. 15.6. Solution pipe in chalk at Cliveden, Buckinghamshire.

of years and more); therefore the subsidence activity over natural cavities is not usually as a result of 'new' cavities being formed by dissolution but is normally due to the destabilization of the metastable infills to existing cavities.

15.4 Engineering management strategy

In Britain ground instability movement over naturally formed cavities within limestone and chalk usually occurs as a consequence of the presence of water (e.g. heavy rainfall, leaking water mains, sewers and soakaways), and sometimes loading and vibration (e.g. construction, excavation and piling). A particularly notable example of water-induced subsidence occurred at Fontwell in West Sussex where some 70 sinkholes occurred within a 0.2 ha area in November 1985 following a water main burst (McDowell 1989, Waltham 1989, Waltham *et al.* 2005).

Within the UK, study of newly formed subsidence 'holes' over limestone and chalk tends to show that most (probably 95% of cases) sinkholes formed are less than 3–5 m across. Larger sizes are always possible but rely on collapse, flow and loss of ground into larger inter-connected subsurface voids – such cases are less common (probably <5% of cases). More detailed analysis of failure mechanisms, sizes and types of sinkholes generated in areas underlain by chalk is available (Edmonds 1988, 2008*a*). Where detailed investigations are conducted it is often found that the sinkhole has formed within cover deposits overlying the limestone or chalk. These types of sinkhole are commonly referred to as subsidence sinkholes. Waltham *et al.* (2005) provide a useful overview of sinkhole types and

Fig. 15.5. Example of an infilled pipe-like feature in limestone.

Fig. 15.7. Sinkhole at Hermitage, Berkshire.

Fig. 15.8. Example of dissolution activity concentrated along a joint plane in chalk.

Fig. 15.10. Mature sinkhole over chalk at Hermitage, Berkshire.

terminology – subsidence sinkholes can also be referred to as dropout and suffosion sinkholes. At depth the surface sinkhole usually passes down into a loosely infilled and voided natural cavity that extends to depth within the underlying limestone or chalk.

Larger size sinkholes can be observed in limestone and chalk landscapes but these are geologically old or historical 'mature' forms that have usually developed over a long time period to reach their current size. Studies by Sperling *et al.* (1977) and Thomas (1973) illustrate how the nature and size of mature sinkholes can vary. Figure 15.7 above shows an example of a recently formed sinkhole that contrasts in size with the much older, mature sinkhole examples shown in Figures 15.10 & 11.

In addition to subsidence it should also be noted that natural cavities can cause a variety of other problems. They can impact construction such as producing instability of cutting faces in engineering works, whereby the infilling deposits weather and spall outwards and down the face, requiring remedial controlling measures (e.g. Lord *et al.* 2002). These might include the construction of walls to cover and contain the spalling infill (see Fig. 15.12) or the installation of a geotextile layer to protect and stabilize the cutting face (e.g. M25 widening works at Maple Cross, Rickmansworth area).

Another consideration is that limestone and chalk form important aquifers used for public water supply where natural cavities can form conduits along which contaminants may readily travel to cause groundwater pollution. Edwards & Smart (1989) raised concerns in connection with landfill siting and waste disposal upon the Carboniferous-age limestones and showed how both leachate and landfill gas

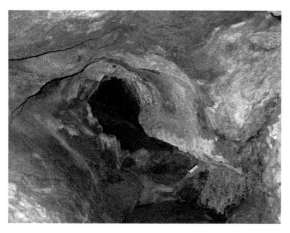

Fig. 15.9. Example of a phreatic tube modified by vadose dissolution/erosion.

Fig. 15.11. Mature sinkhole over limestone at Mynydd Llangynidr, South Wales.

Fig. 15.12. Example of wall constructed to contain solution features infilling deposits in limestone at M4 Motorway, South Wales.

migration might occur along karstic conduits in different circumstances. Major concerns regarding landfill siting over natural cavities in chalk and the potential to cause groundwater pollution led to refusal of a proposed landfill site at Horndean, Hampshire (ENDS 1998). In another location farm effluent caused pollution of a village water supply borehole at Tangley, Hampshire, illustrating the role of natural cavities being able to readily conduct pollutants through chalk (Edmonds 2008b).

The principal engineering management strategies are: avoidance, prevention and mitigation. If the preferred strategy is *avoidance* then the first step is to check the likelihood of the occurrence of natural cavities at the site by carrying out a desk study and site walkover survey. Information on natural cavities is not reliably or consistently shown on topographical maps, although in certain areas where particular concentrations occur swallow holes and springs may be shown. Sometimes local councils have knowledge or press cuttings can reveal information on natural cavity subsidence problems. Most of the detailed information tends to be widely disseminated within technical journals, caving guides and academic studies, for example. There are also other sources such as websites and publications produced by underground exploration enthusiasts, e.g. Chelsea Speleological Society and many caving clubs affiliated to the British Cave Research Association.

Alternatively, natural cavity data can be rapidly provided from information recorded within comprehensive national databases (e.g. Stantec Natural Cavities Database) that have been collated by experienced specialists over many years. The database archive includes maps, photographs and ground investigation data to assist with assessing the risks. While databases are very good at providing an overview of the natural cavity types recorded in a particular geological setting, they can only refer to known features. The question then arises as to what the 'gaps' between recorded features mean – are they areas where no natural cavities are present or simply areas where further unrecorded, but similar natural cavities are present? Interpretation by experienced practitioners is required to understand the karst model that is appropriate to a particular area of interest.

To address this uncertainty a number of researchers have undertaken spatial analysis of natural cavities with a view to predicting their occurrence. The basis of the predictive models is usually finding a series of factors (e.g. geological structure, water table depth or presence of cover deposits, etc.) that can be shown to influence the probability of the occurrence of natural cavities. Some of the earliest work was done in the USA (e.g. McConnell & Horn 1972), starting with point pattern analysis, but the drawback was that the action of a formational process cannot be inferred conclusively from the evidence of pattern alone. With the realization that multiple factors (processes) acted together to generate a pattern, but with differing degrees of influence, interest turned to multiple regression and correlation techniques. These were better for relating the areal association of multiple factors, particularly using numerically expressed factors. An example of this approach used in Florida is published by Thorp & Brook (1984). For large-scale studies a variety of analyses have used remote sensing and/or geophysics (e.g. Littlefield et al. 1984), while others have looked at the control exerted by multiple factors (e.g. Ogden 1984). Ogden in particular outlined the concept of numerically weighting spatial controlling factors and combining them mathematically to produce a predictive model. This approach was used during the 1980s to create a model for predicting the spatial occurrence of natural cavities in chalk that was published some years later (Edmonds 2001). Waltham et al. (2005) confirm the success of using the model for chalk and review it alongside other similar models created for karst areas in South Africa and the USA. Although similarities are apparent, each model is bespoke to particular soluble rock types and geographical area.

Within the UK it is interesting to note that, although the maturity of karst development is greatest for the older, stronger limestones (e.g. Carboniferous age limestone), they tend to occur in upland areas within National Parks and Areas of Outstanding Natural Beauty where fewer urban centres are located. Consequently while the Cretaceous age chalk is younger and possesses less mature karst development, it poses an overall greater level of risk to development because it underlies significant urban areas across SE England.

Consequently background research, searching databases and using hazard mapping models can be useful to avoid the land areas that are most likely to contain natural cavities. Ultimately though, the only way to check for the presence of natural cavities with more certainty is to carry out a ground investigation.

Depending upon the geological sequence and the depth, dimensions and nature of the natural cavity types, geophysical surveying methods may provide useful information regarding the presence of natural cavities. Survey anomalies can then offer targets for other intrusive techniques to investigate. Such techniques may include probing and drilling at centres appropriate to the cavity target size and depth. Inspection of any open voids encountered can be carried out using borehole CCTV surveys and downhole laser surveys. Sometimes it may be possible to enter the natural cavity to inspect its condition subject to prior assessment of the health and safety risks and the use of appropriate equipment and experienced personnel. Weather conditions also need to be considered when entering underground cavities in case of sudden changes in water levels.

As mentioned above, natural cavity instability is mostly due to the destabilization of the metastable infills to existing cavities or roofing soils/rocks above the cavities. Frequently it is the case that the trigger for instability to occur is the passage of water movement down through the ground, whether of natural causation (e.g. rainfall) or man-made via leaking pipes. Consequently one option for the *prevention* of instability over natural cavities is to exclude water. However, in addition to the action of water (reducing friction between soil particles and rock fragments within cavity infilling deposits), there may also be other impacts such as loading and vibration. In practice, therefore, normally the exclusion of water alone will be insufficient and more effort will be required.

Where entry is possible to the natural cavity after taking suitable health and safety precautions, the walls and roof of the cavity can be inspected directly for signs of instability. Depending upon the evidence (e.g. spalled blocks of limestone or chalk and the opening of joints and bedding planes) and considering the depth, height and span of the cavity roof v. the roof thickness and the nature of the construction above, some *in situ* remedial stabilization measures may be appropriate. This could include providing support to the roof by constructing a wall or pier for example. Alternatively the open cavity void could be stabilized by infilling with a cement grout.

In most cases the unstable ground associated with the natural cavity is likely to comprise a deep, loose zone of soil and rock infilling the cavity and extending upwards to the surface as a result of progressive ravelling over time. The volume of open void space will be limited and not safely accessible. These types of ground conditions tend not to be stable for too long before they move and settle or collapse. In such circumstances the only realistic way in which the ground can be stabilized is to carry out ground treatment.

In addition to ground instability concerns, it is also important to protect the quality of the groundwater within limestone and chalk aquifers by preventing the entry of contaminants into the water to cause pollution. However, current groundwater vulnerability maps produced for the Environment Agency (EA) do not take karst into account. Solution features provide readily accessible, enhanced permeability pathways for contaminants to rapidly enter the aquifer without attenuation (see case studies for Horndean and Tangley above) to impact water quality. Drew & Dunne (2004) review a variety of methods for assessing groundwater vulnerability in British karst aquifers for the EA and they make a number of useful observations. However, the EA is yet to revise the set of groundwater vulnerability maps incorporating the impacts of karst. A tool for determining the vulnerability of the chalk aquifer which takes account karst has been developed by Edmonds (2008a). However, the EA does recognize the risk and at a site location near Beaconsfield required ground treatment by grouting to stabilize the ground below a new landfill to mitigate the risks of subsidence damage to the landfill liner, causing the release of pollutants into the groundwater (Parnell 2003).

The land surface across areas underlain by soluble limestone and chalk is prone to ground subsidence. If the risk of structural damage to buildings and infrastructure is to be minimized then it is important that certain *mitigation* measures are taken.

Once a sinkhole has formed, the steep sides of the hole are prone to ravelling back, widening out the hole in plan size with time. Prompt action is required to infill the hole to 'choke' it and prevent continuing enlargement. This remedial action is especially important if the sinkhole occurs beside and/or below a building to limit further damage and mitigate health & safety risks. Suitable fills might include free-draining granular materials, a dense foam grout or foamed concrete, depending on the circumstances.

The next step after securing the sinkhole is to carry out intrusive investigations using a variety of techniques such as dynamic probing and/or rotary drilling to map out the depth and shape of the underlying natural cavity feature. Once the extents of loose and voided zones are known then appropriate stabilization measures can be taken, such as grouting.

Pressure grouting can be used to inject more fluid grouts to penetrate the ground and seek out voids to fill. Compaction grouting using less fluid grouts can be used to inject and locally densify loose infills. In this way further degradation of the ground can be prevented and the strength of the ground improved such that ground support to a structure can be re-instated to prevent further subsidence damage.

Ground improvement techniques such as the installation of vibro stone columns or vibro concrete columns are usually unsuitable because they introduce vibration into the ground which can result in exacerbated ground movement and cannot penetrate deeply enough to fully treat the affected volume of ground. In principle treating the shallow zone, while perhaps gaining some initial benefit, will ultimately fail as the treated zone settles because the untreated ground continues to move downwards at depth and any upward-migrating voids would cause breakdown of the surface as well. To have any beneficial effect a bridging structure would need to be created in the ground spanning between stable support points.

The design and construction of piles is also challenging. Resulting ground vibration can induce more ground movement as mentioned above. Piles passing down through loose, voided zones would most likely need to be bored, cased, cast

Fig. 15.14. Collapse damage to a swimming pool at Beaconsfield, Buckinghamshire.

in situ piles designed to stand as columns end bearing in stable ground at depth below the solution feature and able to withstand vertical and oblique loading as the ground progressively moves down and around the pile shaft with time. Piles are best

Fig. 15.13. Collapse damage to the corner of house at Hertford, Hertfordshire.

Fig. 15. Structural cracking and settlement of a house at Norwich, Norfolk.

installed at positions surrounding solution features at stable ground positions (transferring their loads to depths below the base of the feature), allowing ground beams cast between pile caps to span across the feature.

Another important aspect of maintaining ground stability in areas prone to solution features is not disposing of surface water drainage in close proximity to buildings and infrastructure. The advent of SUDS promotes on-site disposal of drainage rather than disposal via mains drainage systems. For chalk sites where solution features might be present, shallow soakaways are not preferred, but if unavoidable they are recommended to be located no closer than 20 m from a building or road (Lord et al. 2002).

Sometimes sites may contain old, large-diameter, relict sinkholes but while they may not indicate recent active movement it is not recommended that they are infilled and built over. They may be prone to intermittent periods of ground movement and should be avoided. It is often the case that the immediate surrounding zone of ground is also 'relaxed' and prone to future lateral movement towards the feature, therefore a surrounding 'no development buffer zone' of at least 10 m width should be established to exclude buildings, infrastructure and services. Under no circumstances should they be used for surface water drainage disposal. Some examples of structural and other forms of damage caused by the appearance of sinkholes are shown in Figures 15.13–15.

For new-build development in areas underlain by limestone and chalk prone to solution features a strategy needs to be developed for the foundation design taking account of the results of ground investigation data specific to the building footprint. A typical approach for the geohazard setting would be to construct buildings upon rafts or grillages of ground beams designed to span 3–5 m depending upon the conditions present. Following this strategy it is crucial to test the ground beneath the corners of the building where a cantilever of just 1.5–2.5 m is possible. The investigation needs to ensure that the size of any solution features present does not exceed the spanning and cantilevering design criteria or else the features will need to be stabilized by grouting.

Further reading

Department of the Environment 1990. Planning Policy Guidance PPG14 *Development on unstable land*

Department of the Environment 1996. Planning Policy Guidance PPG14 (Annex 1) *Development on unstable land: landslides and planning*

Department of the Environment, Transport and the Regions 2000. Planning Policy Guidance PPG14 (Annex 2) *Development on unstable land: subsidence and planning*

Jefferson, I., Rosenbaum, M., Edmonds, C. & Walton, N. 2008. Subsidence-collapse: occurrence, impact and mitigation. *Quarterly Journal of Engineering Geology and Hydrogeology*, **41**, 259.

Websites

www.chelseaspelaeo.org.uk
www.subbrit.org.uk
www.bcra.org.uk
www.british-caving.org.uk

References

Applied Geology Limited 1993. *Review of Instability due to Natural Underground Cavities in Great Britain*, prepared for the Department of the Environment, Contract No PECD 7/1/280 Regional Reports Volumes 1.1–1.10 with overlays, Technical Reports Volumes 2.1–2.3 & Summary Report.

Bögli, A. 1980. *Karst Hydrology and Physical Speleology*. Springer

Drew, D. & Dunne, S. 2004. *Vulnerability Mapping for the Protection of Karst Aquifers*. R&D Technical Report **W6–032/TR**, Environment Agency

Edmonds, C.N. 1987. Engineering geomorphology of karst development and the prediction of subsidence risk upon the Chalk outcrop in England, PhD thesis, University of London.

Edmonds, C.N. 1988. Induced subsurface movements associated with the presence of natural and artificial underground openings in areas underlain by Cretaceous Chalk. *In*: Bell, F.G., Culshaw, M.G., Cripps, J.C. & Lovell, M.A. (eds) *Engineering Geology of Underground Movements*. Geological Society, London, Engineering Geology Special Publications, **5**, 205–214, https://doi.org/10.1144/GSL.ENG.1988.005.01.20

Edmonds, C.N. 2001. Predicting natural cavities in chalk. *In*: Griffiths, J.S. (ed.) *Land Surface Evaluation for Engineering Practice*. Geological Society, London, Engineering Geology Special Publications, **18**, 29–38, https://doi.org/10.1144/GSL.ENG.2001.018.01.05

Edmonds, C.N. 2008*a*. Karst and mining geohazards with particular reference to the Chalk outcrop, England. *Quarterly Journal of Engineering Geology and Hydrogeology*, **41**, 261–278, https://doi.org/10.1144/1470-9236/07-206

Edmonds, C.N. 2008*b*. Improved groundwater vulnerability mapping for the karstic chalk aquifer of south east England. *In*: Parise, M., De Waele, J. & Gutierrez, F. (eds) *Engineering and Environmental Problems in Karst. Engineering Geology*, **99**, 95–108.

Edwards, A.J. & Smart, P.L. 1989. Waste disposal on karstified carboniferous limestone aquifers of England and Wales. *In*: Beck, B.F. (ed.) *Engineering and Environmental Impacts of Sinkholes and Karst*. Balkema, Rotterdam, 165–182.

Environmental Data Service 1998. *Groundwater risks defeat landfill plan despite thick layer of clay*, ENDS Report **280** (May issue)

Francescon, M. & Twine, D. 1993. Treatment of solution features in Upper Chalk by compaction grouting. *In: Proceedings of Grouting in the Ground Conference*, London, 1992. Thomas Telford, London, 327–347.

Littlefield, J.R., Culbreth, M.A., Upchurch, S.B. & Stewart, M.T. 1984. Relationship of modern sinkhole development to large scale photolinear features. *In*: Beck, B.F. (eds) *Sinkholes: Their Geology, Engineering and Environmental Impact*. Balkema, Rotterdam, 189–195.

Lord, J.A., Clayton, C.R.I. & Mortimore, R.N. 2002. *Engineering in Chalk*. CIRIA, London, C574.

McConnell, H. & Horn, J.M. 1972. Probabilities of surface karst. *In*: Chorley, R.J. (eds) *Spatial Analysis in Geomorphology, Part II Point Systems*. Methuen, 111–133.

McDowell, P.W. 1989. Ground subsidence associated with doline formation in chalk areas of southern England. *In*: Beck, B.F. (eds) *Engineering and Environmental Impacts of Sinkholes and Karst*. Balkema, Rotterdam, 129–134.

Murphy, P., Westerman, A.R., Clark, R., Booth, A. & Parr, A. 2008. Enhancing understanding of breakdown and collapse in the Yorkshire Dales using ground penetrating radar on cave sediments. *In*: Parise, M., De Waele, J. & Gutierrez, F. (eds) *Engineering and Environmental Problems in Karst*. Engineering Geology, Special Issue, **99**, 160–168.

Ogden, A.E. 1984. Methods for describing and predicting the occurrence of sinkholes. *In*: Beck, B.F. (ed.) *Sinkholes: Their Geology, Engineering and Environmental Impact*. Balkema, Rotterdam, 177–182.

Parnell, P. 2003. Ground improvement – three easy pieces. *Ground Engineering*, May issue, 16–17.

Sperling, C.H.B., Goudie, A.S., Stoddart, D.R. & Poole, G.G. 1977. Dolines of the Dorset Chalklands and other areas in Southern Britain. *Transactions of the Institute of British Geographers*, New Series, **2**, 205–223, https://doi.org/10.2307/621858

Sweeting, M.M. 1972. *Karst Landforms*. Macmillan, London.

Thomas, T.M. 1973. Solution subsidence mechanisms and end-products in south-east Breconshire. *Transactions of the Institute of British Geographers*, **60**, 69–85, https://doi.org/10.2307/621506

Thorp, M.J.W. & Brook, G.A. 1984. Application of double Fourier series analysis to ground subsidence susceptibility mapping in covered karst terrain. *In*: Beck, B.F. (ed.) *Engineering and Environmental Impacts of Sinkholes and Karst*. Balkema, Rotterdam, 87–91.

Waltham, A.C. 1989. *Ground Subsidence*. Blackie & Son, London.

Waltham, T., Bell, F. & Culshaw, M. 2005. *Sinkholes and Subsidence. Karst and Cavernous Rocks in Engineering and Construction*, Springer–Praxis, Chichester.

Chapter 16 Geohazards caused by gypsum and anhydrite in the UK: including dissolution, subsidence, sinkholes and heave

Anthony H. Cooper

British Geological Survey, Keyworth, Nottingham NG12 5GG, UK
0000-0001-8763-8530

Correspondence: ahc@bgs.ac.uk

Abstract: Gypsum and anhydrite are both soluble minerals that form rocks that can dissolve at the surface and underground, producing sulphate karst and causing geological hazards, especially subsidence and sinkholes. The dissolution rates of these minerals are rapid and cavities/caves can enlarge and collapse on a human time scale. In addition, the hydration and recrystallization of anhydrite to gypsum can cause considerable expansion and pressures capable of causing uplift and heave. Sulphate-rich water associated with the deposits can react with concrete and be problematic for construction. This paper reviews the occurrence of gypsum and anhydrite in the near surface of the UK and looks at methods for mitigating, avoiding and planning for the problems associated with these rocks.

16.1 Introduction

Gypsum, hydrated calcium sulphate ($CaSO_4 \cdot 2H_2O$), is attractive as satin spar, beautiful as carved alabaster and practical as plasterboard (wallboard) and plaster. However, gypsum is highly soluble and a cause of geological hazards capable of causing severe subsidence to houses, roads, bridges and other infrastructure. Gypsum dissolves rapidly and where this occurs underground it results in caves that evolve and quickly enlarge, commonly leading to subsidence and sometimes to catastrophic collapse. Gypsum is mostly a secondary mineral present in the UK, mainly as fibrous gypsum (satin spar) and alabastrine gypsum (alabaster), which may include large crystals and aggregates of crystals. It occurs near the surface, passing into anhydrite, the dehydrated form ($CaSO_4$) at depths below about 40–120 m or so depending on the local geology and water circulation. The hydration of anhydrite to gypsum in the subsurface causes expansion and heave that are problematic to engineering and hydrogeological installations such as ground source heat pumps. Furthermore, gypsum especially in engineering fills can react with cement, causing heave. Gypsum and anhydrite are present in the Triassic strata of the Midlands and SW of the UK and in the Permian strata of the NE and NW of England. In all these areas various geological hazards are associated with these rocks, the most visible being subsidence and sinkholes. Gypsum and anhydrite also occur to a small extent in the Jurassic of southern England, but no specific problems have been reported related to these deposits.

16.2 The gypsum–anhydrite transition, expansion and heave

Gypsum is hydrated calcium sulphate $CaSO_4 \cdot 2H_2O$, the dehydrated form being anhydrite $CaSO_4$. Gypsum is the most common form found in modern evaporitic sedimentary environments (such as sabkhas in the Gulf; Kendall & Alsharhan 2011), but upon burial it dehydrates to anhydrite. This dehydration generally occurs at depths of between 450 m (Klimchouk & Andrejchuk 1996) and 1200 m (Mossop & Shearman 1973), reaching a theoretical maximum of around 4000 m (dependent on geothermal gradient and the nature of groundwater fluids and overlying strata; Jowett *et al.* 1995). Anhydrite hydrates to gypsum with a significant increase in volume of about 61–63% (Boidin *et al.* 2009; Mossop & Shearman 1973). Upon exhumation and uplift the anhydrite becomes metastable and if sufficient water is available it can hydrate back to gypsum; this process can be enhanced by dissolved salts in the groundwater. This re-hydration tends to happen in the UK at depths of between about 120 and 40 m. However, it can be much deeper in proximity to faults and restricted to shallower depths if the adjacent strata are aquitards. The hydration not only changes *in situ* anhydrite to gypsum (mainly forming alabastrine gypsum), but it also produces fluids containing calcium sulphate that is deposited out mainly as fibrous gypsum. The expansion processes produce high pressures of between 1 and 5 MPa (Einstein 1996), which can fracture the rock, cause folding (James *et al.* 1981) and uplift of the sequence. The pressures generated by the hydration, recrystallization and

---

Engineering Group Working Party (main contact for this chapter: A. H. Cooper, British Geological Survey, Keyworth, Nottingham NG12 5GG, UK, ahc@bgs.ac.uk)
From: Giles, D. P. & Griffiths, J. S. (eds) 2020. *Geological Hazards in the UK: Their Occurrence, Monitoring and Mitigation – Engineering Group Working Party Report*. Geological Society, London, Engineering Geology Special Publications, **29**, 403–423,
https://doi.org/10.1144/EGSP29.16
© 2020 [BGS © UKRI All rights reserved]. Published by The Geological Society of London. All rights reserved.
For permissions: http://www.geolsoc.org.uk/permissions. Publishing disclaimer: www.geolsoc.org.uk/pub_ethics

gypsum growth can be very high, sufficient to cause uplift of more than 80 m of overlying strata. In Staufen (Germany), the misjudging of drilling boreholes through anhydrite and the installation of ground source heat pumps have caused heave that has damaged more than 131 buildings (Goldscheider & Bechtel 2011). Fleuchaus & Blum (2017) reported that ground source heat pumps in anhydrite caused more than €50 million of damage to buildings in Staufen with uplift of 58 cm and lateral movement of 43 cm over an 8-year period. They also noted similar problems with more than €50 million of damage and up to 45 cm of uplift over 3 years affecting the nearby town of Böblingen; heave with similar causes was also noted in Rudersberg, while subrosion and subsidence of sulphate rocks was recorded by them in Wurmlingen. Considerable heave in tunnels through anhydrite has also been reported (Einstein 1996; Boidin et al. 2009), as have the problems of dissolution (Gysel 2002). Despite these observations of expansion, a recent investigation of the abandoned, flooded Warren Anhydrite Mine at Hartlepool (Borehole NZ53SW/147: cores donated to the British Geological Survey in 2001) showed no significant expansion of the country rock and only a thin alteration to gypsum of the surface. This lack of alteration of anhydrite was also noted in anhydrite from French mines, where the lack of porosity was given as the cause for the slow reaction of the anhydrite (Boidin et al. 2009). It is known that small amounts of dissolved elements can have a catalytic effect on the hydration processes (Klimchouk 2000), but the precise details are largely unknown. It is also worth noting that gypsum is also associated with heave where pyritic materials weather and react with carbonate in ground and surface water, producing gypsum (Hawkins & Pinches 1987; Czerewko et al. 2011). This occurs in both natural and anthropogenic deposits.

A further related consideration is the reaction of concrete or cement stabilization binder with sulphate-rich groundwater associated with gypsum, anhydrite and the weathering of pyritic materials. The reactions with the cement can cause the expansive formation of ettringite and thaumasite and detrimental damage to cement and concrete structures, or heave of fill and stabilized soils (Forster et al. 1995; Crammond 2003; Longworth 2004; Czerewko et al. 2011). These problems should be borne in mind when undertaking construction in gypsum karst areas or where sulphate minerals occur.

16.3 The gypsum dissolution problem

Gypsum dissolves readily in flowing water, which next to rivers can be at a rate about 100 times faster than that seen for limestone dissolution. James et al. (1981) observed a 3 m cube of gypsum being dissolved completely by the River Ure near Ripon in about 18 months; the associated gypsum face was then undercut by 6 m in the subsequent 10 years (Fig. 16.1). This dissolution rate is for turbulent unsaturated water at the surface. Underground dissolved sulphate in the water slows the dissolution, but it is still very rapid and caves in gypsum can readily form and expand (Klimchouk 1996; Klimchouk & Andrejchuk 1996, 2005; Klimchouk et al. 1996a). Such caves occur in the Vale of Eden, Cumbria, and are inferred beneath Ripon, North Yorkshire (Ryder & Cooper 1993; Waltham & Cooper 1998), and throughout the gypsum belt. Some of the longest and most complex cave (maze cave) systems in the world developed in the gypsum karst of the Ukraine (Klimchouk 1992; Klimchouk et al. 1996b) and it is thought that similar water-filled (phreatic) caves exist beneath Ripon and other parts of eastern England underlain by similar strata. Under suitable groundwater flow conditions caves in gypsum can enlarge at a rapid rate, resulting in large chambers. Collapse of these chambers produces breccia pipes that propagate through the overlying strata to break through at the surface and form subsidence hollows. Subsidence problems at Ripon are due to this phenomenon (Cooper 1986, 1989, 1998, 2002, 2007; Cooper & Calow 1998).

16.4 Geology of the gypsiferous rocks

16.4.1 Triassic

Triassic Mercia Mudstone Group strata in the Midlands of the UK (Table 16.1 and Fig. 16.2) have several units of gypsum within them, notably the Tutbury Gypsum and the Newark Gypsum. These occur in what is now called the Branscombe Mudstone Formation (formerly the Cropwell Bishop Formation). The Tutbury Gypsum is a massive unit generally 2–3 m thick with diapiric monoliths up to 9 m thick (Noel Worley pers. comm., Yorkshire Geological Society field excursion 24 July 2011). This gypsum is mined in Staffordshire at Fauld Mine and was formerly mined and quarried in the Chellaston area near Derby (Wynne 1906; Smith 1918; Young 1990). It was also mined in the Gotham/East Leake area and is currently mined at Barrow on Soar. In these areas and towards Nottingham the rock is partly dissolved near outcrop and some cavities are present (Cooper 1996; Cooper & Saunders 2002). The Newark gypsum comprises a sequence of thin to thick beds spaced over a thickness of about 25 m of sequence containing about 25% gypsum beds. The Newark gypsum was formerly mined at Orston, 23 km east of Nottingham, and is now opencast mined near Newark, about 25 km NE of Nottingham (Firman 1964, 1984; Worley & Reeves 2007).

16.4.2 Permian

Most of the Permian sequence of eastern England belongs to the Zechstein Group, which includes dolomites, mudstones and gypsum at outcrop with diverse evaporites deeper in the basin to the east. The general stratigraphy at outcrop is shown in Table 16.2 and Figure 16.2. Both the Edlington Formation and Roxby Formation are notable for containing thick gypsum that reaches its maximum around Ripon, thins both

Fig. 16.1. Ripon Parks gypsum in 1979 (top) when it had just collapsed and 1990 (bottom) showing the undercutting of the cliff by about 6 m owing to dissolution from the River Ure. Soon after this lower picture was taken the cliff collapsed again (photos A.H. Cooper © UKRI/BGS).

to the north and south, but thickens again around Darlington. Eastwards towards Billingham and Hartlepool the sulphate occurs mainly as anhydrite having not been converted back to gypsum. To the south the gypsum gets thinner so that it has largely disappeared at Doncaster. The distribution of the gypsum extends from outcrop, where it is partially dissolved, to a depth of around 40–120 m, where it passes into anhydrite. This gypsum zone defines a belt about 2–3 km

Table 16.1. *The gypsiferous Triassic sequence forming part of the Mercia Mudstone Group in the English Midlands*

Chrono-stratigraphy	Lithostratigraphy		Lithology	Approximate thickness
	Formation	Member		
Triassic	Blue Anchor Fm (Mercia Mudstone Group)		Dolomitic mudstone and siltstone	8 m
	Branscombe Mudstone Fm (formerly Cropwell Bishop Fm)	Newark Gypsum	Interbedded red-brown mudstone and mainly white thin to medium bedded alabastrine gypsum	Up to 25 m
			Red-brown mudstone	10 m
		Tutbury Gypsum	Massive to large nodular white and mottled mainly alabastrine gypsum	1–9 m
			Red-brown mudstone	16 m
	Arden Sandstone Fm		Dominantly sandstone	Up to 17 m

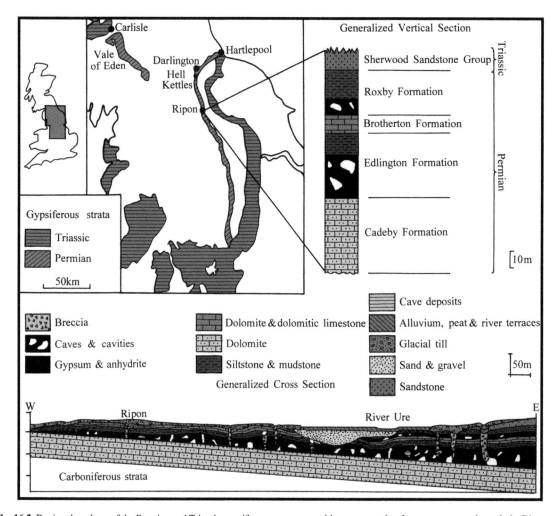

Fig. 16.2. Regional geology of the Permian and Triassic gypsiferous sequences with a cross-section from west to east through the Ripon area showing the easterly dipping dolomite and gypsum sequence cut into by the glacial valley of the River Ure.

Table 16.2. *Generalized Permian sequence of eastern England*

Chrono-stratigraphy	Lithostratigraphy	Lithology	Approximate thickness at outcrop
Triassic	Sherwood Sandstone Group formerly Bunter Sandstone	Red-brown sandstone, pebbly in the south	50–350 m
Permian	Roxby Formation (Zechstein Group) formerly Upper Marl (including gypsum equivalent to the Billingham Anhydrite Formation)	Up to 10 m of gypsum/anhydrite overlain by 0–20 m of red-brown mudstone	0–30 m
	Brotherton Formation (Zechstein Group) formerly Upper Magnesian Limestone	Mainly thin-bedded dolomite and dolomitic limestone	0–30 m
	Edlington Formation (Zechstein Group) formerly Middle Marl (including gypsum equivalent to the Hartlepool Anhydrite Formation)	Up to 35–40 m of gypsum/anhydrite overlain by 0–20 m of red-brown mudstone	0–55 m
	Cadeby Formation (Zechstein Group) formerly Lower Magnesian Limestone	Massive and bedded dolomite with reefs and algal stromatolites overlain by cross-bedded oolitic dolomite	0–50 m
	Yellow Sands Formation and basal breccia (Rotliegendes Group)	Lenticular areas of locally derived breccia overlain by aeolian yellow fine to medium grained sandstone	0–10 m
Carboniferous		Mixed sequence of mudstone, siltstone, coal and sandstone	Up to 3000 m

wide where gypsum can dissolve in the subsurface, causing subsidence.

In the NW of England gypsum/anhydrite is extensively present in the Vale of Eden within the Eden Shales Formation. Here there is a localized sequence of four gypsum beds (A, B, C and D in ascending order – Table 16.3) interbedded with mudstone and overlying the aquifer of the Permian Penrith Sandstone Formation (Sherlock & Hollingworth 1938; Meyer 1965; Burgess & Holliday 1979; Arthurton & Wadge 1981; Holliday 1993). The upper three beds are persistent, but the thick A bed is localized to the middle part of the Vale of Eden. Throughout the area the gypsum and anhydrite have been extensively mined, but very little information is in the public domain (Rogers 1994; Tyler 2000). On the coast and at depth in the Solway basin a different sequence is present with anhydrite of the St Bees Evaporite Formation (Sandwith & Fleshwick cycles) present in the St Bees area; here the Sandwith Anhydrite was formerly mined, but no karstic features have been noted (Akhurst *et al.* 1997; Tyler 2000).

16.5 Subsidence caused by gypsum dissolution

16.5.1 Subsidence geohazards around Ripon

The Permian sequence in Ripon, and the areas to the north and south, contains *c.* 35–40 m of gypsum in the Edlington

Table 16.3. *The Permian evaporite units in the Vale of Eden, Cumbria*

Gypsum units	Characteristics	Thickness in Vale of Eden
D bed	Massive and nodular gypsum/anhydrite with fibrous gypsum; underlain by up to 8 m of the Belah Dolomite	Up to 3.7 m widespread
C bed	Interbedded nodular and bedded gypsum/anhydrite with fibrous gypsum and mudstone	Up to 3.8 m widespread
B bed	Massive fine-grained gypsum and anhydrite	Up to 6.5 m widespread
A bed	Interbedded gypsum/anhydrite and fibrous gypsum with abundant mudstone; closely underlain by the Penrith Sandstone aquifer	Up to 42.4 m limited to central area failing to north and south

Formation and 10 m of gypsum higher up in the Roxby Formation (Powell et al. 1992; Cooper & Burgess 1993; Cooper 1998). These two gypsum sequences rest on two limestone aquifers, the Cadeby Formation and the Brotherton Formation, respectively. The limestone dip slopes act as catchment areas and the water is fed downdip into the gypsiferous sequences, before escaping into a major buried valley along the line of the River Ure (Fig. 16.2 and Cooper & Burgess 1993). Complex cave systems are developed in the gypsum and artesian sulphate-rich springs are locally present (Cooper 1986; Thompson et al. 1996). Because of the thickness of gypsum present, the caves are large and surface collapses up to 30 m across and 20 m deep have been recorded. The subsidence is not random, but occurs in a reticulate pattern related to the jointing in the underlying strata (Cooper 1986). Around Ripon a significant subsidence occurs approximately every few years (Cooper 1986, 1998); the times of the subsidence events show that some zones of subsidence are more active than others. Because of the intersection of the buried valley and the gypsum sequence, Ripon is the most affected area for subsidence. However, the same geological sequence extends north for more than 20 km through the villages of Nunwick, Wath and Sutton Howgrave to Bedale and south for about 15 km through the villages of Littlethorpe and Bishop Monkton (figure 34 of Cooper & Burgess 1993). Subsidence has affected all of these villages, and sinkholes similar to those recorded in Ripon have occurred, especially at Littlethorpe, Bishop Monkton and Nunwick that share a similar hydrogeological setting (Cooper 1986; Thompson et al. 1996).

At Ripon, Littlethorpe, Bishop Monkton and Nunwick the bedrock is cut through by a deep, largely buried valley approximately following the course of the River Ure, and partially filled with up to 22 m of Devensian glacial and postglacial deposits (Powell et al. 1992; Cooper & Burgess 1993). The buried valley intersects the carbonate and gypsum units, creating a hydrological pathway from the bedrock to the river. Considerable groundwater flow occurs along this pathway and artesian water emanates from the Permian strata as springs which issue along the valley sides and up through the Ure valley gravels (Cooper 1986, 1988; Thompson et al. 1996; Cooper et al. 2013). Artesian water, with a head above that of the river, has been encountered in some boreholes. Much of this water is nearly saturated with or rich in dissolved calcium sulphate resulting from the dissolution of gypsum.

The greatest concentration of active subsidence hollows coincides with the areas marginal to the buried valley. Much of the sand and gravel partially filling the valley is cemented with calcareous tufa deposited from the groundwater, which is also rich in dissolved carbonate (Cooper 1988).

Fig. 16.3. Sinkhole that formed in 1997 on Ure Bank, Ripon, destroying four garages (category 7 damage) and damaging the adjacent house (photo P. Tod © UKRI/BGS).

Fig. 16.4. Sinkhole that formed in July 1834 near Ripon Railway Station. The hole is in red Sherwood Sandstone and was 20 m deep and 14 m across when it formed (photo A.H. Cooper © UKRI/BGS).

Subsidence is more extensive than it appears on the floodplain of the River Ure, because many of the subsidence features are infilled by overbank deposits (Cooper 1989, 1998).

The subsidence has been mapped from its surface expression as sinkholes or dolines and by looking at building damage (Figs 16.3–16.5). Several building damage surveys have been carried out; the technique used is based on an extended version of the subsidence damage recording scheme introduced by the National Coal Board (Cooper 2008a) and scores the damage from 1 which is very minor to 7 which is complete collapse. Information from a survey is shown in Figure 16.6. Information about building damage, sinkholes, springs, stream sinks and caves for Ripon has been gathered by the British Geological Survey and stored in a Geographic Information System and associated databases (Cooper et al. 2001, 2011; Cooper 2008b; Farrant & Cooper 2008). This database now has information for gypsum and salt karst features covering most of the country; it also contains information about limestone and chalk for about half of the country.

16.5.2 Subsidence geohazards around Darlington

The subsidence-prone belt extends from Ripon northwards, attenuates east of Catterick and comes back in from Darlington towards Hartlepool (Fig. 16.2). In these areas the sequence is similar to that at Ripon, but the carbonate formations have different names. Around Darlington up to 40 m of gypsum (equivalent to the Hartlepool Anhydrite) is present in the Edlington Formation and up to 7 m (equivalent to the Billingham Anhydrite) in the Roxby Formation. Two types of subsidence occur in this northern area – sinkhole collapse similar to that at Ripon and more widespread settlement. The distinction between the two types is controlled by the hydrogeology and the thickness and lithology of the overlying glacial deposits (Lamont-Black et al. 1999, 2002). The distribution of the major subsidence features is shown on the 1:50 000 scale geological map for the area (British Geological Survey 1987) and by Thompson et al. (1996, their fig. 8.2b).

South of Darlington at Hell Kettles, catastrophic collapse occurred in 1179. Four subsidence hollows up to 35 m in diameter and 6 m deep were formerly present (Longstaffe 1854), but one of these is now filled in. These hollows are very similar to those at Ripon (Cooper 1986, 1998). Artesian water emanates from Hell Kettles and from sulphate-rich springs nearby at Croft (Lamont-Black et al. 2005; Cooper et al. 2013). Like Ripon, the sequence here dips gently eastwards and the outcrop of the carbonate formations is a groundwater recharge area. The ground water moves down-dip to the low ground of the wide, partly buried, valley of the River Tees. The subsidence appears to be associated with the margins of the buried valley, as at Ripon.

The southern half of Darlington has suffered subsidence related to gypsum dissolution. Here most of the subsidence has been prolonged, less severe (generally less than 0.3 m), and spread over subsidence depressions up to several hundred metres in diameter. Local boreholes have proved thick gypsiferous strata similar to those at Ripon, and cavities were encountered in one borehole. The bedrock surface forms a very broad valley filled with around 50 m of glacial and postglacial deposits which include water-saturated sand and plastic laminated clays (Lamont-Black et al. 2002). As the gypsum dissolution proceeds, it appears that for much of the area the overlying water-saturated sand flows into the gypsum cavities. Support is removed over wide areas, causing broad subsidence depressions at the surface. In addition to numerous broad subsidence features, one small sinkhole has been recorded. It was 1.5 m diameter by 1 m deep and appeared near the river in the southern part of Skerne Park on 21 February 2011 (NGR 428294, 513271); it was remediated with free-draining fill and fenced. The subsidence-prone belt continues to the NE of Darlington, extending towards the coast at Hartlepool, where thick deposits of anhydrite underlie part of the town. Here they have been mined at several levels in Warren Mine. Boreholes here, to examine the mine and country rock, found anhydrite with very little gypsum, except at the contact with the overlying superficial deposits and no subsidence has been recorded in the town.

Fig. 16.5. Severe building damage (category 7) that occurred on the night of 17 February 2014 at Magdalens Close, Ripon, caused by a sinkhole, possibly triggered by heavy antecedent rainfall (photo A.H. Cooper © UKRI/BGS).

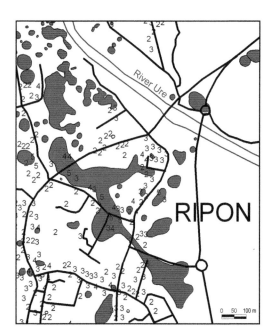

Fig. 16.6. Building damage recording in Ripon by McNerney (2000) and included in the British Geological Survey karst GIS database; subsidence hollows shown in red with horizontal ornament in part of Ripon; from Cooper (2008*a*). See also Figure 16.15.

16.5.3 Subsidence geohazards between Ripon and Doncaster

South of the Ripon area, the gypsum sequences extend to near Doncaster where they become attenuated. Dissolution of the gypsum has caused subsidence, sinkholes and some ancient breccia pipes in the areas around Church Fenton, Sherburn in Elmet, Brotherton and Askern (Smith 1972; Thompson *et al.* 1996; Murphy 2000; Cooper *et al.* 2013). Most of these areas have been assessed by Thompson *et al.* (1996) and are considered to be of fairly low susceptibility to future subsidence. However, while subsidence in these areas is minor when compared with Ripon, future subsidence cannot be ruled out and prudence in development remains important.

16.5.4 Subsidence geohazards in the Vale of Eden

Permian gypsum is present in the Vale of Eden where four gypsum units are present in the Eden Shales (Table 16.3). The thickest unit is the A bed, comprising interlayered gypsum and mudstone and the most uniform high-quality and prominent unit the B bed gypsum. Both the A and B beds have been extensively quarried and mined; the C bed has also been worked in places (Arthurton & Wadge 1981; Mottahed & Szeki 1982; Hughes 2003).

Evidence of gypsum dissolution, in the B bed, was recorded in the Newbiggin gypsum opencast site where caves and a karstified gypsum surface were recorded by (Ryder & Cooper 1993). They also noted an open cavity migrating towards the surface through the overlying Eden Shales from a cavity in the gypsum below. Similar dissolution features were recorded in the A bed where it was formerly opencast at Kirkby Thore (Sherlock & Hollingworth 1938). Dissolution features have also been recorded in the mine workings in the lower part of the A bed with beehive-shaped caves up to 9 m high and horizontal caves about 2 m high and up to five or six mine pillars long (c. 35 m) proved (Dunham & Hollingworth 1947; Rogers 1994). These natural cavities are commonly water-filled and become more common in proximity to faulting, their presence limiting the areas that can be mined (Rogers 1994). They are also more common in the lower part of the A bed sequence where the underlying Penrith Sandstone aquifer provides water that has dissolved the gypsum. The individual outcrops of gypsum in the Vale of Eden are narrow, but within them and the intervening mudstones subsidence has been recorded, both associated with the extensive mining and with areas that have not been mined; the presence of the rock is a concern for development in the areas where it is present at shallow to moderate depths.

16.5.5 Subsidence over Triassic gypsum

Compared with the Permian sequence, dissolution of the Triassic gypsum generally has less catastrophic effects. This is largely due to the sequence being mainly of mudstone/siltstone and gypsum without the aquifers that are present in the Permian rocks. In Staffordshire, Derbyshire, Nottinghamshire and northern Leicestershire the gypsum is thick enough to be extensively mined and massive gypsum may reach 9 m thick (for example in diapiric monoliths in Fauld Mine). Evidence of dissolution of the gypsum is locally shown by the way the old mines exploited the rock and terminated their mining operations when they encountered the partly dissolved and collapsed areas (Cooper 1996; Cooper & Saunders 2002). Evidence of active dissolution is also shown by the water emanating from the gypsiferous sequence, which contains a high sulphate content, as evidenced at Burton on Trent where the water is said to be 'Burtonized' and is used for brewing beer (Cooper et al. 2011). Evidence of gypsum dissolution is also given by sporadic records of subsidence and problems, such as those encountered during the construction of the Derby Southern Bypass (Cooper & Saunders 2002) and A453 improvements west of Nottingham. The dissolution of the Triassic gypsum can lead to upstanding outliers of partially dissolved rock, zones of complete dissolution and collapse or deeper zones of partial dissolution (see section on construction). Similar features are described in the former Chellaston quarries and gypsum of the Dove valley; here cylindrical cavities (wash holes or water washes) up to nearly 5 m across and commonly connecting to the surface have also been recorded (Wynne 1906; Smith 1918; Sherlock & Hollingworth 1938; Young 1990; Cooper 1996). In a few places, such as East Leake and Keyworth to the south of Nottingham, subsidence has been attributed to gypsum dissolution and ingress of surface water washing material into the cavities caused by the dissolution (T. Colman pers. comm. 2012).

16.6 Ground investigation: surveying, geophysics and boreholes in gypsum areas

Following on from a literature review, the starting point for ground investigation is geomorphological and geological surveying that is required to identify the subsidence features that constitute the basis for making sinkhole susceptibility and hazard maps. Field surveying is indispensable and verbal information from farmers, residents and local government officials can add considerably to the data on subsidence features. Multiple tools can be used to produce the best possible cartographic sinkhole inventory. Historical maps and multiple temporally spaced sets of monoscopic or stereoscopic aerial photographs are essential starting points for a survey. These can be complemented by multispectral scanning (Cooper 1989) and geodetic techniques like LiDAR (light detection and ranging) surveys or radar interferometry from aircraft or satellite (DinSar) (Castañeda et al. 2009). A complete karst inventory including subsidence features gives an indication of the spatial distribution and severity of the problem (Cooper 1986, 1998; Galve et al. 2009a). This type of information can then be analysed with respect to other parameters using GIS techniques to produce susceptibility and hazard models (Cooper 2008b; Farrant & Cooper 2008; Galve et al. 2008, 2009b).

The depth of the gypsum sequences over many of the areas in question, and the evolving nature of the subsidence phenomenon, make site investigation and remediation difficult. Generally, only shallow site investigations have been undertaken over the gypsum subsidence belt. Detailed investigations for modern developments are now demanded by the planning authorities (Thompson et al. 1996). If sites are investigated by boreholes alone, the size and spacing of the subsidence features will demand closely spaced boreholes (at around 10 m intervals or less) drilled to the base of the gypsum; commonly this is 40–60 m deep under the city of Ripon. However, because of the rapid dissolution of gypsum and the likelihood of artesian water, any borehole drilled through the sequence has the potential to become a hydraulic pathway. This could encourage enhanced dissolution and possibly become a focus for future subsidence. Consequently, the number of boreholes should be kept to a minimum. The current consensus is to avoid drilling through the gypsum into the aquifer, but to drill shallow holes and undertake geophysical investigations. Great care should be taken to ensure that they are properly grouted with sulphate-proof grout and that the integrity of the ground around them is not compromised.

Another potential problem with site investigation by drilling is the likelihood of triggering a subsidence event in already unstable ground. This could be caused by vibration or circulation of pressurized drilling fluids. These problems should be considered when planning site investigations in the context of the safety of the drill crew, the safety of people in nearby buildings and the associated insurance cover. Where such conditions are suspected it is prudent both to avoid disturbing the ground and development.

Site investigation using cored boreholes through the gypsiferous strata should be logged by a competent geologist. The geologist should be able to identify gypsum in all of its forms, and recognize dissolution and collapse features. In much of the archival site investigation data for this sequence, gypsum (except for satin spar) is commonly mis-identified as limestone. The most abundant form of gypsum encountered at and near outcrop here is grey alabastrine gypsum or alabaster. This gypsum is commonly mis-identified as grey limestone in boreholes. This mistake can potentially lead to disastrous engineering problems with sites underlain by gypsum being designated as having competent limestone present beneath them (Cooper & Calow 1998). In general the gypsum in Yorkshire and Durham (Hartlepool Anhydrite equivalent) is a pale grey compact alabaster with fibrous gypsum veins, whereas the underlying Cadeby Formation (of Yorkshire) or Ford Formation (of Teesside/Durham) is dolomite or dolomitic limestone, which is pale yellowish brown and porous. This is not always the case and particular care is required in areas where the limestone beneath the gypsum has not been dolomitized. At the contact between the dolomite and the overlying gypsum, it is also fairly common for the underlying dolomite to be de-dolomitized. When this happens it is transformed into a poorly cemented mesh of calcitic sand that breaks up easily and has a low bearing strength. Other misleading situations include areas where the upper gypsum in the Roxby Formation (Billingham Anhydrite equivalent) rests on the mainly grey limestones of the Brotherton Formation (of Yorkshire) or Seaham Formation (of Teesside/Durham).

Drilling open holes and collecting chippings instead of coring can be cost effective. This method can be more reliable if automated or manual recording of the drill penetration rate is made (Cailleux & Toulemont 1983; Patterson et al. 1995). However, experience and skill are needed to interpret the drilling rate figures with the identification of the chippings material. Downhole geophysics, cross-hole geophysics and downhole optical and acoustic cameras can also help to understand the local karst geology (Yuhr et al. 2008).

A practical investigation technique is to use geophysics as part of a phased drilling and dynamic probing investigation. At Ripon, microgravity has been used successfully to delineate anomalies that have subsequently been drilled (Patterson et al. 1995). Computer-based modelling and field investigations show that microgravity can delineate breccia pipes and large cavities that breach, or come near to, the surface. However, even large caves, at depth, are difficult to image and edge effects of superficial deposits can partially conceal anomalies. Subsidence features have also been investigated using 2D and 3D resistivity tomography (Cooper 1998;

Gutiérrez et al. 2009). This method has proved a faster survey method than microgravity and has shown many of the anomalies. The downside is that electrical methods are difficult to use in built-up areas where pipes, cables and metal objects may be present.

In summary, areas underlain by gypsum can pose difficult problems for developers and engineers. Drilling on either a grid or on a random sample pattern has little chance of finding all of the anomalies on a site (Fig. 16.7). Breccia pipes and near-surface cavities can be present (Fig. 16.8), and physical investigations of their locations by all but the closest spaced borehole survey is difficult. Geophysics can help to delineate anomalies, which can then be avoided or investigated as part of the site development (Thompson et al. 1996). Successful techniques include microgravity (Patterson et al. 1995) and various forms of resistivity and conductivity survey with 3D resistivity tomography being particularly effective (Cooper 2009; Kaufmann & Romanov 2009). Ground probing radar has also been used in areas where the surface material is not clayey (Benito et al. 1995), but has not been used much in the UK. Geophysics combined with trenching has also proved effective in Spain (Gutiérrez et al. 2009, 2018). Detailed seismic reflection techniques have been used near Darlington and Ripon, but the method is very labour intensive and more expensive than the other methods including microgravity (Sargent & Goulty 2009, 2010). Geophysics is best used in conjunction with limited amounts of drilling; a phased

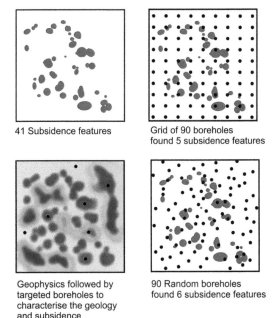

41 Subsidence features

Grid of 90 boreholes found 5 subsidence features

Geophysics followed by targeted boreholes to characterise the geology and subsidence

90 Random boreholes found 6 subsidence features

Fig. 16.7. Actual subsidence features present in a square 650 by 650 m and the potential success of boreholes and geophysics for locating them.

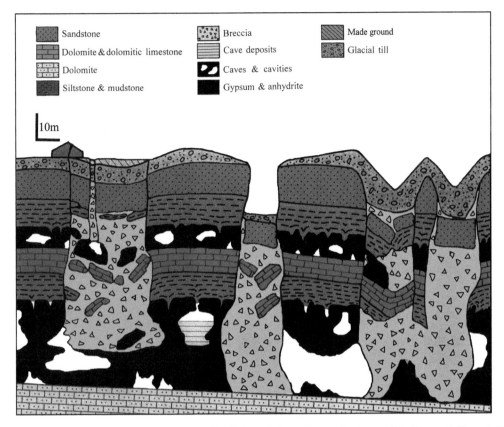

Fig. 16.8. Stylized cross-section through gypsum dissolution subsidence features in the east of the Ripon subsidence belt.

approach of using geophysics to target 'anomalies' and 'normal' areas, followed by drilling has proved effective (Fig. 16.7).

16.7 Gypsum dissolution as a hazard to civil engineering

Subsidence caused by gypsum dissolution produces difficult conditions for building and construction; in many cases the collapses are so severe that little can be done to mitigate the problems (Fig. 16.8). Consequently, good site investigation and hazard avoidance are the best approaches, followed by construction that can cope with any expected dissolution or subsidence. The interaction of gypsum and water in engineering projects can cause severe problems and catastrophic ground and structural failures. In the foundations of hydraulic structures, such as dams and canals, seepage through gypsum can lead to rapidly accelerating leakage, dissolution and failure. In the USA, the presence of gypsiferous beds beneath dam sites has resulted in at least 14 examples of dams losing water or failing (James 1992; Johnson 2008), and at least two dams in China have also been affected (Lu & Cooper 1997). Leakage of canals and irrigation ditches along with irrigation and rainfall events are recorded as triggers for subsidence in the Zaragoza region of Spain (Gutiérrez et al. 2005; Galve et al. 2008). In the UK, at Ratcliffe, SW of Nottingham, power station foundations have been affected by water leakage and dissolution of thin Triassic gypsum beds (Seedhouse & Sanders 1993).

Development of the gypsum areas prone to subsidence is best undertaken with a phased approach of detailed site investigation and careful design. For housing construction in Ripon there is special planning control and buildings are now constructed on reinforced raft foundations (Thompson et al. 1996 and Fig. 16.9); additional protection could be afforded by extending foundations with supporting beams outside the main footprint of the property (Cooper & Calow 1998). In subsidence-prone karstic areas it is important to use flexible service pipe materials and to guard against water loss and infiltration that themselves could trigger subsidence. In some places, service trenches have been lined with waterproof membranes to stop this happening.

Fig. 16.9. Massive rafted foundations at Ripon designed to cope with subsidence, note that the identified subsidence feature (just left of this picture) was avoided in this construction (Patterson *et al.* 1995; photo A.H. Cooper © UKRI/BGS).

Fig. 16.10. Ripon bypass bridge over the River Ure. The bridge span is one continuous structure designed not to collapse if subsidence removes the support of a pillar. Electronic monitoring is incorporated in the bridge bearings (photo A.H. Cooper © UKRI/BGS).

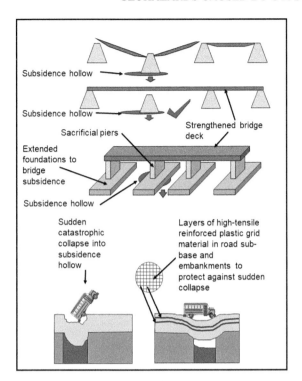

Fig. 16.11. Reinforced bridge and road at Ripon designed to cope with subsidence problems (Thompson *et al.* 1996; Cooper & Saunders 2002; Jones & Cooper 2005).

Linear structures such as roads and bridges are very prone to subsidence damage. At Ripon, the new Ure Bridge is built with redundant strength and the capability for the structure to lose any one of its pillar supports without collapse (Figures 16.10 & 16.11; Cooper & Saunders 2002). On the same stretch of road, the embankments are protected by two layers of strong geogrid material sandwiched in the embankment fill (Thompson *et al.* 1996; Cooper & Saunders 2002; Jones & Cooper 2005). This sandwich of material is designed to span cavities up to about 15 m across and sag rather than fail so that an indication of any problem becomes visible at the surface (Fig. 16.11). Sensitive structures such as bridges and viaducts can be equipped with monitoring and warning systems such as those installed in the Paris road viaducts (Arnould 1970) and the bridge at Ripon (Cooper & Saunders 2002).

Over gypsum dissolution in the Triassic strata, the Derby Southern Bypass used geophysics to locate cavities that were then filled. During construction large blocks of gypsum were excavated and removed while dissolved areas and old mine workings were grouted with sulphate-proof grout. Subsequently, the road was constructed from reinforced concrete to avoid any subsidence problems – Figure 16.12 (Cooper & Saunders 2002). Similar geophysical techniques and the removal of the gypsum have recently been undertaken for the improvement constructions on the A453 near Ratcliffe Power Station in Nottinghamshire (authors observations 2013).

Construction of infrastructure such as high-speed railway lines across gypsum karst areas needs to consider the likelihood and sizes of potential collapses. Such studies have been carried out in southern Germany (Molek 2003) and Spain (Guerrero *et al.* 2008) and may be required for parts of the proposed UK high-speed rail network that crosses Permian and Triassic strata. The construction of high-pressure gas pipelines also needs to consider these problems with appropriate investigation and design (Gibson *et al.* 2005).

Engineers have suggested that grouting can be used to stabilize gypsum caves; this technique has been used in the Paleogene gypsum of the Paris area (Toulemont 1984), but the long-term outcome of the work has not been reported. In general, grouting of a gypsum cave system is not advisable and for caves of the size found under places such as Ripon it would be completely impractical (Fig. 16.8). Unless the caves are small, proved to be abandoned and completely dry, filling them with grout could alter the groundwater regime. This could cause dissolution in the adjacent ground in the same way that natural collapse may block a cave system, and through diversion of water channels, cause dissolution nearby. There could also be problems caused by locally raising the water table, which could trigger subsidence. If a dry abandoned cave system is to be grouted, sulphate-resistant cement would have to be used. Similarly, for any site investigation boreholes it is essential to grout them completely with sulphate proof grout.

Conventional piled foundations, as already practiced in Ripon and elsewhere, are also problematical. Piles through disturbed and unconsolidated deposits may achieve the required bearing strength on the base of the pile in either the superficial deposits or the bedrock below. However, since the bedrock contains gypsum beds, caverns might be present and these may propagate upwards, thereby destabilizing the piled structures. In some areas, it might be feasible to pile through the gypsum sequences, using bored piles. This has been suggested for Ripon, with piles to the carbonate formation below (if the latter is not dedolomitized), but this could involve piling to depths of about 80 m to the east of the city. The use of sulphate-resistant cement would add to the cost and there is the risk of artesian water and danger that dissolution and collapse of the strata could place additional loads on the piles. This would necessitate the use of piles with a negative skin friction. Because of the prohibitive costs and likely difficulties associated with piling, it is largely impractical except in the west of the subsidence belt, where only a small amount of gypsum is present (Thompson *et al.* 1996).

An alternative approach used has been to delineate and avoid any subsidence hollows and breccia pipes (Patterson *et al.* 1995). The constructions have then been placed within the site over the best ground conditions and designed to have minimal impact on the subsurface. They have also been designed to span any subsidence features that

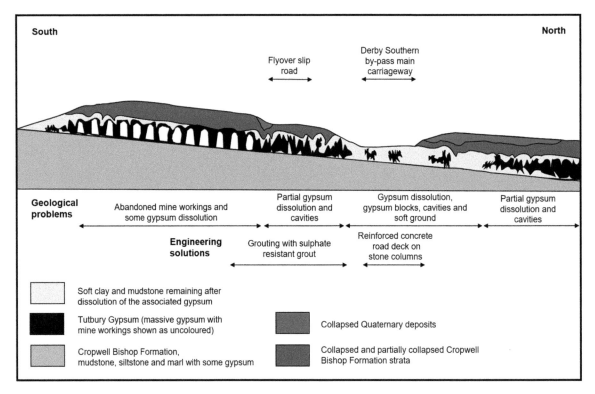

Fig. 16.12. Section through the Tutbury Gypsum and associated strata on the Derby Southern Bypass (Cooper & Saunders 2002).

may potentially develop. This sort of approach requires extensive site investigation by engineering geologists working in close liaison with foundation engineers (Thompson *et al.* 1996).

16.8 Problems related to water abstraction and injection in gypsum areas

Gypsum aquifers, despite their hard sulphate-rich water, are commonly used for water supplies. In some places the availability of the hard sulphate-rich water is considered a benefit, as it is already 'Burtonized' and suitable for use in beer brewing; along with the hops it gives the beer some of its bitter taste – hence the name of English beer – 'bitter'. Several of the important brewing areas in the UK, such as Burton on Trent and Tadcaster, draw water from the gypsiferous sequences. Like all karst water systems gypsum karst can rapidly transmit pollutants (Lamont-Black *et al.* 2005). Gypsum karst aquifers are thus sensitive to both industrial and agricultural pollution and require careful exploitation and protection (Paukstys *et al.* 1999).

Water abstraction in gypsiferous terrains can aggravate the natural dissolution process by removing large volumes of sulphate-rich groundwater and drawing in aggressive recharge (Cooper 1988). Calculations for a major water abstractor pumping from the Permian gypsum and limestone beds of northern England showed some alarming results. The water contained $c.$ 1200 ppm of SO_4 mainly as dissolved $CaSO_4$ and the abstractor pumped 212 Ml of water per annum. This was equivalent to removing $c.$ 200 m^3 of gypsum a year from the area. It is likely that much of the dissolution represented the enlargement of joints over a wide area. However, adjacent to the boreholes where rapid water flow occurs, severe dissolution could occur and result in subsidence around the wellsite; active subsidence was recorded in the vicinity. In addition water pumping can also cause changes in the groundwater level, triggering subsidence in the cover rocks and superficial deposits (Benito *et al.* 1995; Lamont-Black *et al.* 2002).

The development of sustainable drainage systems (SUDS) is being promoted in England and Wales to help mitigate the effects of flooding caused by development, which is increasing the rapidity of surface runoff (Woods-Ballard *et al.* 2007). The Flood and Water Management Act 2010, for England and Wales, includes provision for the implementation of National Standards for SUDS (Department for Environment Food & Rural Affairs 2010). Identifying areas suitable for the safe installation of SUDS, in the light of legislation, will ensure that long-term performance is maintained while minimizing potential environmental impacts. In many areas SUDS can be effective and safe, with various solutions available including soak-aways, retention basins, porous pavements and surface materials. In most gypsum karst areas, the installation of infiltration-based SUDS may be inappropriate. The disposal of surface water into the ground may cause pollution of groundwater or increase the susceptibility to geohazards. Water that infiltrates through the ground has the potential to wash fine materials out of the covering deposits and induce sinkhole development (Fig. 16.13). This is a well-documented phenomenon alongside US highways where drainage ditches commonly cause the development of sinkholes adjacent to the road (Waltham *et al.* 2005; Ford & Williams 2007; Fleury 2009). Sinkholes have been reported alongside some modern British roads, especially where old land drains have been cut or new drainage channels installed. Sinkholes caused by leaking pipes are also well documented (McDowell 2005; Waltham *et al.* 2005).

Soakaways from surface runoff or septic tanks are a well-known anthropogenic trigger that has caused subsidence in many places (Waltham *et al.* 2005). Within England and Wales there are policies for groundwater protection related to foul water soakaways (Environment Alliance 2006; Environment Agency 2010) and information about infiltration testing and system design (British Standards Institution 2005, 2007). Septic tanks are not allowed in Zone 1 groundwater protection areas (which are the areas most susceptible to groundwater pollution), but the only apparent constraints are pollution prevention and suitable ground permeability assessed by percolation testing (Office of the Deputy Prime Minister 2002). No mention is made of potential karstic ground instability problems that can result from changes in the local input of water into the ground; these need to be considered in gypsum karst areas.

Ground source heat pumps and ground cooling systems (Busby *et al.* 2009; Arthur *et al.* 2010) are becoming popular as a source of green energy or an energy-efficient cooling system. However, they could be problematical. Open loop installations must be avoided in gypsum and anhydrite not only because of the dissolution problems, but also because of the potential to hydrate anhydrite to gypsum (see section 'The gypsum–anhydrite transition, expansion and heave'). This has happened in Staufen, Germany, where a well-meaning 'green' heat pump installation has caused severe heave to the recently restored town hall and 131 buildings, causing in excess of €50 million of damage (Goldscheider & Bechtel 2011; Fleuchaus & Blum 2017). In addition to the dissolution and heave problems, the changes in groundwater levels associated with extraction and injection of water in open loop systems also has a potential to cause reactivation of breccia pipe features and subsidence/sinkholes owing to the suffusion of overlying materials (Fig. 16.14).

16.9 Planning for subsidence

The timing and precise location of the sudden, and sometimes catastrophic, subsidence caused by gypsum dissolution cannot yet be predicted. However, within England the main gypsum subsidence belts have been defined and many of their controlling mechanisms described (Cooper 1986, 1998; Thompson *et al.* 1996). Some areas are more at risk than others and deep buried valleys cutting through the gypsiferous beds are major controlling factors at Ripon, south of

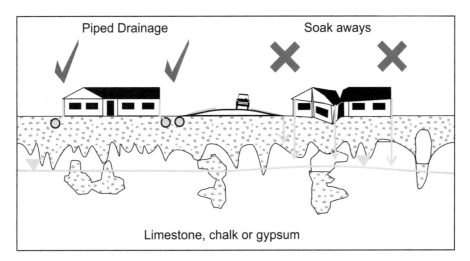

Fig. 16.13. Subsidence problems caused by soakaways and their avoidance by correct drainage.

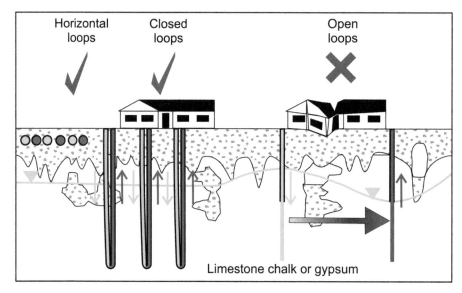

Fig. 16.14. Subsidence caused by open loop ground source heat pump installations and their avoidance by using closed loop systems.

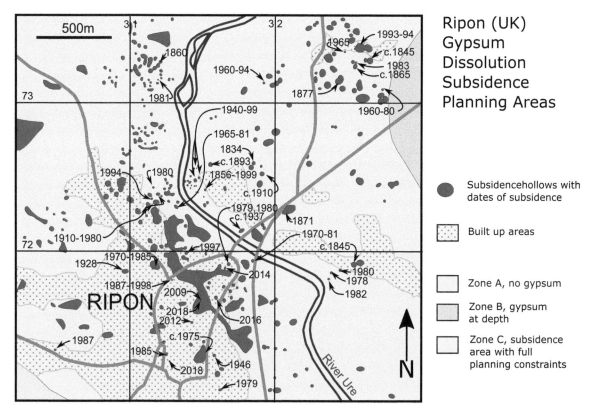

Fig. 16.15. The planning zones and distribution of subsidence hollows in the Ripon area with the dates of collapse where they are known (after Thompson *et al.* 1996; and Cooper 1998).

Darlington and Brotherton. Collapsed areas and existing breccia pipes remain potentially unstable and are best avoided for development (Thompson et al. 1996; Cooper 2008b). Areas adjacent to collapses are also suspect because of dissolution around the bases of the collapse pipes (Cooper 1986).

In the Ripon area, the dates and locations recorded for the historically recent subsidence events suggest concentration of water flow in the cave systems along certain specific paths (Fig. 16.5 and Cooper 1986, 1989, 1998). The close grouping of subsidence hollows suggests that once a collapse has occurred the cave partially chokes and the dissolution continues in the adjacent strata. This commonly produces linear belts of subsidence related to the joint pattern. It also means that localities adjacent to, or in line with, existing subsidence hollows are probably more at risk from future subsidence. From the distribution of the subsidence features and their sizes, the worst areas can be avoided and development in the less-susceptible areas tailored to cope with the magnitude of the likely subsidence events.

In Ripon there is a formal planning policy with checklists and signed documents to help control and protect development in the area (Thompson et al. 1996, 1998). To support this process the Ripon area has been divided into three development control zones: (A) no know gypsum present; (B) some gypsum present at depth; and (C) gypsum present and susceptible to dissolution.

Within zone A no special planning constraints are imposed. In zone B, where the risk of subsidence is small, a ground stability report prepared by a competent person is usually required and the problem is considered in local planning.

The zone C area (which is most of Figure 16.15 except the very SW and NE areas) is subject to significant formal constraints and controls on development, which local planning has to take into account. In this zone, a ground stability report prepared by a competent professional person is required before planning applications for new buildings, or changes of use of buildings, are determined. In most cases this report has to be based on a geotechnical desk study and a site appraisal, followed by a programme of ground investigation designed to provide information needed for detailed foundation design (unless this information, such as boreholes, exists from a previous study). Where planning consent is given it may be conditional on the implementation of approved foundation or other mitigation measures, designed to minimize the impact of any future subsidence activity. One key to the implementation of this approach is the use of a proforma checklist to be completed and signed by a competent professional person. For the UK a competent person is defined in the report as Geotechnical Specialist who is 'A Chartered Engineer or Chartered Geologist, with a postgraduate qualification in geotechnical engineering or engineering geology, equivalent at least to an MSc, and with three years of post-Charter practice in geotechnics; or a Chartered Engineer or Chartered Geologist with five years of post-Charter practice in geotechnics'. In addition to these qualifications it is also desirable that the practitioner has experience of the problems, although this is not formally stated. This procedure has been adopted by Harrogate Borough Council, but is likely to be subject to changes based on experience of its use.

The procedure removes the responsibility from the planners to the developers. However some sites, where stability should have been more thoroughly investigated and assessed, have been developed and some modern houses have suffered subsidence. It is evident that qualifications alone do not guarantee a successful build; some practitioners have developed areas that are demonstrably unsuitable and others have appeared to specify inadequate or unsuitable foundation solutions. There appears to be a reluctance to avoid some of the very difficult areas. If the procedures were working correctly problems should not have happened. Questions have been raised about the standards of investigations and the over-willingness of practitioners to sign off sites; a review of the procedures is required. A second tier of signing off sites by an independent more experienced practitioner appointed by the local council may be a solution to the problem.

Conclusions

Although gypsum is of fairly limited extent within the UK, it causes some of the most severe and localized subsidence that affects a small city, parts of a large town and numerous villages. Because of its solubility and rapid dissolution rate it presents a geological hazard that can evolve quickly in areas where there is natural or induced water flow through the sequence. It poses a challenge to the construction of buildings, roads, railways, pipelines, bridges and dams. In many cases, especially for hydraulic structures, areas underlain by gypsum and are best avoided. Similarly, where sinkholes have developed these areas are also best avoided. However, most of the gypsiferous areas can be utilized with careful site investigation and appropriate mitigation measures; the difficulty is recognizing the areas where instability makes construction untenable.

Acknowledgements The author thanks the following for helpful discussion and for critically reviewing the manuscript at various stages of its production: Dr Vanessa Banks, Dr Helen Reeves, Dr Andy Farrant, Dr Dave Giles and Dr Alan Thompson. This article is published with permission of the Executive Director of the British Geological Survey (UKRI).

Funding This research received no specific grant from any funding agency in the public, commercial, or not-for-profit sectors.

References

AKHURST, M.C., CHADWICK, R.A. ET AL. 1997. Geology of the west Cumbria district. In: Memoir of the British Geological Survey,

Sheets 28, 37 and 47 (England and Wales)*. British Geological Survey, Keyworth, Nottingham, 138.

ARNOULD, M. 1970. Problems associated with underground cavities in the Paris region. Geological and geographical problems of areas of high population density. *Proceedings of the Symposium, Association of Engineering Geologist*, Sacramento, CA, 1–25.

ARTHUR, S., STREETLEY, H.R., VALLEY, S., STREETLEY, M.J. & HERBERT, A.W. 2010. Modelling large ground source cooling systems in the Chalk aquifer of central London. *Quarterly Journal of Engineering Geology and Hydrogeology*, **43**, 289–306, https://doi.org/10.1144/1470-9236/09-039

ARTHURTON, R.S. & WADGE, A.J. 1981. Geology of the country around Penrith. *In: Memoir of the Geological Survey of Great Britain (Sheet 24)*. HMSO, London.

BENITO, G., PÉRES DEL CAMPO, P., GUTIÉRREZ-ELORZA, M. & SANCHO, C. 1995. Natural and human-induced sinkholes in gypsum terrain and associated environmental problems in NE Spain. *Environmental Geology*, **25**, 156–164, https://doi.org/10.1007/BF00768545

BOIDIN, E., HOMAND, F., THOMAS, F. & YVON, J. 2009. Anhydrite–gypsum transition in the argillites of flooded salt workings in eastern France. *Environmental Geology*, **58**, 531–542, https://doi.org/10.1007/s00254-008-1528-1

BRITISH GEOLOGICAL SURVEY 1987. *Stockton, England and Wales Sheet 33. Solid and Drift Geology*. 1:50,000.

BRITISH STANDARDS INSTITUTION 2005. *BS EN 12566-3:2005 + A1:2009, Small wastewater treatment systems for up to 50PT. Packaged and/or site assemble domestic wastewater treatment plants*. The British Standards Institution, 48.

BRITISH STANDARDS INSTITUTION 2007. *BS 6297:2007 + A1:2008, Code of practice for the design and installation of drainage fields for use in wastewater treatment*. 44.

BURGESS, I.C. & HOLLIDAY, D.W. 1979. Geology of the country around Brough-under-Stainmore. *In: Memoirs of the Geological Survey of Great Britain (Sheet 31 Including Parts of Sheets 25 and 30 (England and Wales))*.

BUSBY, J., LEWIS, M., REEVES, H. & LAWLEY, R.S. 2009. Initial geological considerations before installing ground source heat pump systems. *Quarterly Journal of Engineering Geology and Hydrogeology*, **42**, 295–306, https://doi.org/10.1144/1470-9236/08-092

CAILLEUX, J.B. & TOULEMONT, M. 1983. La reconnaissance des cavites souterraines par methods diagraphiques. *Bulletin of the International Association of Engineering Geology*, **26–27**, 33–42.

CASTAÑEDA, C., GUTIÉRREZ, F., MANUNTA, M. & GALVE, J.P. 2009. DInSar measurements of ground deformation by sinkholes, mining subsidence, and landslides Ebro River, Spain. *Earth Surface Processes and Landforms*, **34**, 1562–1574, https://doi.org/10.1002/esp.1848

COOPER, A.H. 1986. Foundered strata and subsidence resulting from the dissolution of Permian gypsum in the Ripon and Bedale areas, North Yorkshire. *In*: HARWOOD, G.M. & SMITH, D.B. (eds) *The English Zechstein and Related Topics*. Geological Society, London, Special Publications, **22**, 127–139, https://doi.org/10.1144/GSL.SP.1986.022.01.11

COOPER, A.H. 1988. Subsidence resulting from the dissolution of Permian gypsum in the Ripon area; its relevance to mining and water abstraction. *In*: BELL, F.G., CULSHAW, M.G., CRIPPS, J.C. & LOVELL, M.A. (eds) *Engineering Geology of Underground Movements*. Geological Society, London, Engineering Geology Special Publications, **5**, 387–390, https://doi.org/10.1144/GSL.ENG.1988.005.01.42

COOPER, A.H. 1989. Airborne multispectral scanning of subsidence caused by Permian gypsum dissolution at Ripon, North Yorkshire. *Quarterly Journal of Engineering Geology*, **22**, 219–229, https://doi.org/10.1144/GSL.QJEG.1989.022.03.06

COOPER, A.H. 1996. *Gypsum: geology, quarrying, mining and geological hazards in the Chellaston area of South Derbyshire*. British Geological Survey Technical Report, **WA/96/30**, http://nora.nerc.ac.uk/id/eprint/19816/

COOPER, A.H. 1998. Subsidence hazards caused by the dissolution of Permian gypsum in England: geology, investigation and remediation. *In*: MAUND, J.G. & EDDLESTON, M. (eds) *Geohazards in Engineering Geology*. The Geological Society, London, Engineering Geology Special Publications, **15**, 265–275, https://doi.org/10.1144/GSL.ENG.1998.015.01.27

COOPER, A.H. 2002. Environmental problems caused by gypsum karst and salt karst in Great Britain. *Carbonates and Evaporites*, **17**, 116–120, https://doi.org/10.1007/BF03176477

COOPER, A.H. 2007. *Gypsum Dissolution Geohazards at Ripon, North Yorkshire, UK. Engineering Geology for Tomorrow's Cities*. International Association for Engineering Geology, Nottingham.

COOPER, A.H. 2008a. The classification, recording, databasing and use of information about building damage caused by subsidence and landslides. *Quarterly Journal of Engineering Geology and Hydrogeology*, **41**, 409–424, https://doi.org/10.1144/1470-9236/07-223

COOPER, A.H. 2008b. The GIS approach to evaporite karst geohazards in Great Britain. *Environmental Geology*, **53**, 981–992, https://doi.org/10.1007/s00254-007-0724-8

COOPER, A.H. 2009. Gypsum dissolution geohazards at Ripon, North Yorkshire, UK: CD IAEG2006 UK Field Trip 2. *In*: CULSHAW, M.G., REEVES, H.J., JEFFERSON, I. & SPINK, T.W. (eds) *Engineering Geology for Tomorrow's Cities*. Geological Society, London, Special Engineering Publications, **22**, CD-ROM supplement, https://doi.org/10.1144/EGSP22.I

COOPER, A.H. & BURGESS, I.C. 1993. *Geology of the Country around Harrogate; Sheet 62 (England and Wales)*. British Geological Survey.

COOPER, A.H. & CALOW, R. 1998. *Avoiding gypsum geohazards: guidance for planning and construction*. British Geological Survey Technical Report, **WC/98/5**.

COOPER, A.H. & SAUNDERS, J.M. 2002. Road and bridge construction across gypsum karst in England. *Engineering Geology*, **65**, 217–223, https://doi.org/10.1016/S0013-7952(01)00131-4

COOPER, A.H., FARRANT, A.R., ADLAM, K.A.M. & WALSBY, J.C. 2001. The development of a national Geographic Information System (GIS) for British karst geohazards and risk assessment. *In*: BECK, B.F. & HERRING, J.G. (eds) *Geotechnical and Environmental Applications of Karst Geology and Hydrology. Proceedings of the Eighth Multidisciplinary Conference on Sinkholes and the Engineering and Environmental Impacts of Karst*, 1–4 April, Louisville, KY. Balkema, Rotterdam, 125–130.

COOPER, A.H., FARRANT, A.R. & PRICE, S.J. 2011. The use of karst geomorphology for planning, hazard avoidance and development in Great Britain. *Geomorphology*, **134**, 118–131, https://doi.org/10.1016/j.geomorph.2011.06.004

COOPER, A.H., ODLING, N.E., MURPHY, P.J., MILLER, C., GREENWOOD, C.J. & BROWN, D.S. 2013. The role of sulfate-rich springs and groundwater in the formation of sinkholes over gypsum in Eastern England. *In*: LAND, L., DOCTOR, D.H. & STEPHENSON, J.B. (eds) *Sinkholes and the Engineering and Environmental Impacts of Karst: Proceedings of the Thirteenth Multidisciplinary Conference*, 6–10 May. National Cave and Karst Research

Institute, Carlsbad, NM, 141–150, https://doi.org/10.5038/9780979542275.1122

CRAMMOND, N.J. 2003. The thaumasite form of sulfate attack in the UK. *Cement and Concrete Composites*, **25**, 809–818, https://doi.org/10.1016/S0958-9465(03)00106-9

CZEREWKO, M.A., CROSS, S.A., DUMELOW, P.G. & SAADVANDI, A. 2011. Assessment of pyritic Lower Lias mudrocks for earthworks. *Geotechnical Engineering*, **162**, 59–77, https://doi.org/10.1680/geng.2011.164.2.59

DEPARTMENT FOR ENVIRONMENT FOOD AND RURAL AFFAIRS 2010. *Flood and Water Management Act*. https://services.parliament.uk/bills/2009-10/floodandwatermanagement.html

DUNHAM, K.C. & HOLLINGWORTH, S.E. 1947. Excursion to Penrith and north-west. *Mineralogical Magazine*, **28**, 248–254, https://doi.org/10.1180/minmag.1947.028.199.04

EINSTEIN, H.H. 1996. Tunnelling in difficult ground-swelling behaviour and identification of swelling rocks. *Rock Mechanics and Rock Engineering*, **29**, 113–124, https://doi.org/10.1007/BF01032649

ENVIRONMENT AGENCY 2010. *Groundwater protection: Policy and practice (GP3 – Part 4)*, http://publications.environment-agency.gov.uk/pdf/GEHO0708BOGU-e-e.pdf [last accessed 22 March 2011]

ENVIRONMENT ALLIANCE 2006. Pollution prevention guidelines: treatment and disposal of dewage where no foul sewer is available: PPG4. *Environment Agency*, **12**.

FARRANT, A.R. & COOPER, A.H. 2008. Karst geohazards in the UK: the use of digital data for hazard management. *Quarterly Journal of Engineering Geology and Hydrogeology*, **41**, 339–356, https://doi.org/10.1144/1470-9236/07-201

FIRMAN, R.J. 1964. Gypsum in Nottinghamshire. *Bulletin of the Peak District Mines Historical Society*, **4**, 189–203.

FIRMAN, R.J. 1984. A geological approach to the history of English alabaster. *Mercian Geologist*, **9**, 161–178.

FLEUCHAUS, P. & BLUM, P. 2017. Damage event analysis of vertical ground source heat pump systems in Germany. *Geothermal Energy*, **5**, 1–15, http://doi.org/10.1186/s40517-017-0067-y

FLEURY, S. 2009. *Land Use Policy and Practice on Karst Terrains – Living on Limestone*. Springer, Berlin.

FORD, D. & WILLIAMS, P. 2007. *Karst Hydrogeology and Geomorphology*. John Wiley and Sons, Chichester.

FORSTER, A., CULSHAW, M.G. & BELL, F.G. 1995. Regional distribution of sulphate in rocks and soils of Britain. *In*: EDDLESTON, M., WALTHALL, S., CRIPPS, J.C. & CULSHAW, M.G. (eds) *Engineering Geology of Construction*. The Geological Society, London, Engineering Geology Special Publications, **10**, 95–104, https://doi.org/10.1144/GSL.ENG.1995.010.01.07

GALVE, J.P., GUTIÉRREZ, F., LUCHA, P., GUERRERO, J., BONACHEA, J., REMONDO, J. & CENDRERO, A. 2008. Probabilistic sinkhole modeling for hazard assessment. *Earth Surface Processes and Landforms*, **34**, 437–452, https://doi.org/10.1002/esp.1753

GALVE, J.P., GUTIÉRREZ, F. *ET AL.* 2009a. Sinkholes in the salt-bearing evaporite karst of the Ebro River valley upstream of Zaragoza city (NE Spain). Geomorphological mapping and analysis as a basis for risk management. *Geomorphology*, **108**, 145–158, https://doi.org/10.1016/j.geomorph.2008.12.018

GALVE, J.P., GUTIÉRREZ, F., REMONDO, J., BONACHEA, J., LUCHA, P. & CENDRERO, A. 2009b. Evaluating and comparing methods of sinkhole susceptibility mapping in the Ebro Valley evaporite karst (NE Spain). *Geomorphology*, **111**, 160–172, https://doi.org/10.1016/j.geomorph.2009.04.017

GIBSON, A.D., FORSTER, A., CULSHAW, M.G., COOPER, A.H., FARRANT, A.R., JACKSON, N. & WILLET, D. 2005. Rapid geohazard assessment system for the UK Natural Gas Pipeline Network. *Geoline 2005: International Symposium on Geology and Linear Developments*, 23–25 May 2005, Lyon, France.

GOLDSCHEIDER, N. & BECHTEL, T.D. 2011. Editors' message: The housing crisis from underground – damage to the historic town by geothermal drillings through anhydrite, Staufen, Germany. *Hydrogeology Journal*, **17**, 491–493, https://doi.org/10.1007/s10040-009-0458-7

GUERRERO, J., GUTIÉRREZ, F., BONACHEA, J. & PEDRO, L. 2008. A sinkhole susceptibility zonation based on paleokarst analysis along a stretch of the Madrid–Barcelona high-speed railway built over gypsum- and salt-bearing evaporites (NE Spain). *Engineering Geology*, **102**, https://doi.org/10.1016/j.enggeo.2008.07.010

GUTIÉRREZ, F., GUTIÉRREZ, M., MARIN, C., DESIR, G. & MALDONADO, C. 2005. Spatial distribution, morphometry and activitiy of La Puebla de Alfindén sinkhole field in the Ebro river valley (NE Spain): applied aspects for hazard zonation. *Environmental Geology*, **48**, 360–369, https://doi.org/10.1007/s00254-005-1280-8

GUTIÉRREZ, F., GALVE, J.P., LUCHA, P., BONACHEA, J., JORDÁ, L. & JORDÁ, R. 2009. Investigation of a large collapse sinkhole affecting a multi-storey building by means of geophysics and the trenching technique (Zaragoza city, NE Spain). *Environmental Geology*, **58**, 1107–1122, https://doi.org/10.1007/s00254-008-1590-8

GUTIÉRREZ, F., ZARROCA, M. *ET AL.* 2018. Identifying the boundaries of sinkholes and subsidence areas via trenching and establishing setback distances. *Engineering Geology*, **231**, 255–268, https://doi.org/10.1016/j.enggeo.2017.12.015

GYSEL, M. 2002. Anhydrite dissolution phenomena: three case histories of anhydrite karst caused by water tunnel operation. *Rock Mechanics and Rock Engineering*, **35**, 1–21, https://doi.org/10.1007/s006030200006

HAWKINS, A.B. & PINCHES, G.M. 1987. Cause and significance of heave at Llandough Hospital, Cardiff – a case history of ground floor heave due to gypsum growth. *Quarterly Journal of Engineering Geology*, **20**, 41–57, https://doi.org/10.1144/GSL.QJEG.1987.020.01.05

HOLLIDAY, D.W. 1993. Geophysical log signatures in the Eden Shales (Permo-Triassic) of Cumbria and their regional significance. *Proceedings of the Yorkshire Geological Society*, **49**, 345–354, https://doi.org/10.1144/pygs.49.4.345

HUGHES, R.A. 2003. *Permian and Triassic rocks of the Appleby district (part of Sheet 30, England and Wales)*. British Geological Survey, Research Report, **RR/02/01**.

JAMES, A.N. 1992. *Soluble Materials in Civil Engineering*. Ellis Horwood, Chichester.

JAMES, A.N., COOPER, A.H. & HOLLIDAY, D.W. 1981. Solution of the gypsum cliff (Permian Middle Marl) by the River Ure at Ripon Parks, North Yorkshire. *Proceedings of the Yorkshire Geological Society*, **43**, 433–450, https://doi.org/10.1144/pygs.43.4.433

JOHNSON, K.S. 2008. Gypsum–karst problems in constructing dams in the United States. *Environmental Geology*, **53**, 945–950, https://doi.org/10.1007/s00254-007-0720-z

JONES, C.J.F.P. & COOPER, A.H. 2005. Road construction over voids caused by active gypsum dissolution, with an example from Ripon, North Yorkshire, England. *Environmental Geology*, **48**, 384–394, https://doi.org/10.1007/s00254-005-1282-6

JOWETT, E.C., CATHLES, L.M. & DAVIS, B.W. 1995. Predicting depths of gypsum dehydration in evaporitic sedimentary basins. *The American Association of Petroleum Geologists Bulletin*, **77**, 402–413.

KAUFMANN, G. & ROMANOV, D. 2009. Geophysical investigations of a sinkhole in the northern Harz foreland (North Germany).

Environmental Geology, **58**, 401–405, https://doi.org/10.1007/s00254-008-1598-0

KENDALL, C.G. & ALSHARHAN, A. 2011. *Quaternary Carbonate and Evaporite Sedimentary Facies and Their Ancient Analogues: A Tribute to Douglas James Shearman: Vol 43*. International Association of Sedimentologists, Tulsa, OK/Blackwell, Oxford, Special Publications, **43**, 496.

KLIMCHOUK, A. 1992. Large gypsum caves in the Western Ukraine and their genesis. *Cave Science*, **19**, 3–11.

KLIMCHOUK, A. 1996. The dissolution and conversion of gypsum and anhydrite. In: KLIMCHOUK, A., LOWE, D., COOPER, A. & SAURO, U. (eds) *Gypsum Karst of the World*. International Journal of Speleology, **25**, 21–36, https://doi.org/10.5038/1827-806X.25.3.2

KLIMCHOUK, A. 2000. Dissolution and conversions of gypsum and anhydrite. In: KLIMCHOUK, A., FORD, D.C., PALMER, A.N. & DREYBRODT, W. (eds) *Speleogenesis: Evolution of Karst Aquifers*. National Speleological Society, 160–168.

KLIMCHOUK, A. & ANDREJCHUK, V. 1996. Sulphate rocks as an arena for karst development. In: KLIMCHOUK, A., LOWE, D., COOPER, A. & SAURO, U. (eds) *Gypsum Karst of the World*. International Journal of Speleology, **25**, 9–20, https://doi.org/10.5038/1827-806X.25.3.1

KLIMCHOUK, A. & ANDREJCHUK, V. 2005. Karst breakdown mechanisms from observations in the gypsum caves of the Western Ukraine: implications for subsidence hazard assessment. *Environmental Geology*, **48**, 336–359, https://doi.org/10.1007/s00254-005-1279-1

KLIMCHOUK, A., CUCCHI, F., CALAFORRA, J.M., AKSEM, S., FINOCCHIARO, F. & FORTI, P. 1996a. The dissolution of gypsum from field observations. In: KLIMCHOUK, A., LOWE, D., COOPER, A. & SAURO, U. (eds) *Gypsum Karst of the World*. International Journal of Speleology, **25**, 37–48, https://doi.org/10.5038/1827-806X.25.3.3

KLIMCHOUK, A., LOWE, D., COOPER, A. & SAURO, U. 1996b. Gypsum karst of the world. *International Journal of Speleology*.

LAMONT-BLACK, J., YOUNGER, P.L., FORTH, R.A., COOPER, A.H. & BONNIFACE, J.P. 1999. Hydrogeological monitoring strategies for investigating subsidence problems potentially attributable to gypsum karstification. *Hydrogeology and Engineering Geology of Sinkholes and Karst*, **1999**, 141–148.

LAMONT-BLACK, J., YOUNGER, P.L., FORTH, R.A., COOPER, A.H. & BONNIFACE, J.P. 2002. A decision-logic framework for investigating subsidence problems potentially attributable to gypsum karstification. *Engineering Geology*, **65**, 205–215, https://doi.org/10.1016/S0013-7952(01)00130-2

LAMONT-BLACK, J., BAKER, A., YOUNGER, P.L. & COOPER, A.H. 2005. Utilising seasonal variations in hydrogeochemistry and excitation-emission fluorescence to develop a conceptual groundwater flow model with implications for subsidence hazards: an example from Co. Durham, UK. *Environmental Geology*, **48**, 320–335, https://doi.org/10.1007/s00254-005-1278-2, https://doi.org/10.1007/s00254-005-1278-2

LONGSTAFFE, W.H.D. 1854. *The History and Antiquities of the Parish of Darlington*. The proprietors of the Darlington and Stockton Times, London: republished by Patric and Shotton.

LONGWORTH, I. 2004. *Assessment of sulfate-bearing ground for soil stabilisation for built development*, http://www.nce.co.uk/assessment-of-sulfate-bearing-ground-for-soil-stabilisation-for-built-development/744870.article

LU, Y.R. & COOPER, A.H. 1997. Gypsum karst geohazards in China. *Engineering Geology and Hydrogeology of Karst Terranes*, **17**, 117–126.

MCDOWELL, P.W. 2005. Geophysical investigations of sinkholes in chalk, U.K. Case Study No 9. In: WALTHAM, A.C., BELL, F.G. & CULSHAW, M.G. (eds) *Sinkholes and Subsidence. Karst and Cavernous Rocks in Engineering and Construction*. Praxis, Chichester, 313–316.

MCNERNEY, P. 2000. *The extent of subsidence caused by the dissolution of gypsum in Ripon, North Yorkshire*. BSc thesis, Sunderland University.

MEYER, H.O.A. 1965. Revision of the stratigraphy of the Permian evaporites and associated strata in north-western England. *Proceedings of the Yorkshire Geological Society*, **35**, 71–89, https://doi.org/10.1144/pygs.35.1.71

MOLEK, H. 2003. Engineering-geological and geomechanical analysis for the fracture origin of sinkholes in the realm of a high velocity railway line. In: BECK, B.F. (ed.) *Sinkholes and the Engineering and Environmental Impacts of Karst. Proceedings of the Ninth Multidisciplinary Conference*, 6–10 September 2003. American Society of Civil Engineers, Huntsville, AL, 551–558.

MOSSOP, G.D. & SHEARMAN, D.J. 1973. Origins of secondary gypsum rocks. *Transactions of the Institution of Mining and Metallugy (Section B Applied Earth Sciences)*, **82**, B147–B154.

MOTTAHED, P. & SZEKI, A. 1982. The collapse of room and pillar working in a shaley gypsum mine due to dynamic loading. In: *Strata Mechanics: Developments in Geotechnical Engineering*. Elsevier, Oxford, 260–264.

MURPHY, P.J. 2000. The karstification of the Permian strata east of Leeds. *Proceedings of the Yorkshire Geological Society*, **53**, 25–30, https://doi.org/10.1144/pygs.53.1.25

OFFICE OF THE DEPUTY PRIME MINISTER 2002. *The Building Regulations 2000; section H: Drainage and waste disposal*, https://webarchive.nationalarchives.gov.uk/20150601175100/http://www.planningportal.gov.uk/uploads/br/BR_PDF_ADH_2002.pdf [last accessed 4th January 2020].

PATTERSON, D.A., DAVEY, J.C., COOPER, A.H. & FERRIS, J.K. 1995. The investigation of dissolution subsidence incorporating microgravity geophysics at Ripon, Yorkshire. *Quarterly Journal of Engineering Geology*, **28**, 83–94, https://doi.org/10.1144/GSL.QJEGH.1995.028.P1.08

PAUKSTYS, B., COOPER, A.H. & ARUSTIENE, J. 1999. Planning for gypsum geohazards in Lithuania and England. *Engineering Geology*, **52**, 93–103, https://doi.org/10.1016/S0013-7952(98)00061-1

POWELL, J.H., COOPER, A.H. & BENFIELD, A.C. 1992. *Geology of the country around Thirsk; Sheet 52 (England and Wales)*. British Geological Survey.

ROGERS, C. 1994. *To be a Gypsum Miner*. Pentland Press, Bishop Auckland, Durham.

RYDER, P.F. & COOPER, A.H. 1993. A cave system in Permian gypsum at Houtsay Quarry, Newbiggin, Cumbria, England. *Cave Science*, **20**, 23–28.

SARGENT, C. & GOULTY, N.R. 2009. Seismic reflection survey for investigation of gypsum dissolution and subsidence at Hell Kettles, Darlington, UK. *Quarterly Journal of Engineering Geology and Hydrogeology*, **42**, 31–38, https://doi.org/10.1144/1470-9236/07-071

SARGENT, C. & GOULTY, N.R. 2010. Shallow seismic reflection profiles over Permian strata affected by gypsum dissolution in NE England. *Quarterly Journal of Engineering Geology and Hydrogeology*, **43**, 221–232, https://doi.org/10.1144/1470-9236/08-115

SEEDHOUSE, R.L. & SANDERS, R.L. 1993. Investigations for cooling tower foundations in Mercia Mudstone at Ratcliffe-on-Soar, Nottinghamshire. In: CRIPPS, J.C., COULTHARD, J.C., CULSHAW, M.G.,

FORSTER, A., HENCHER, S.R. & MOON, C. (eds) *The Engineering Geology of Weak Rock. Proceedings of the 26th Annual Conference of the Engineering Group of the Geological Society, September 1990, Leeds*. Balkema, Rotterdam.

SHERLOCK, R.L. & HOLLINGWORTH, S.E. 1938. *Gypsum and Anhydrite & Celestine and Strontianite*. His Majesty's Stationery Office, London.

SMITH, B. 1918. The Chellaston gypsum breccia and its relation to the gypsum–anhydrite deposits of Britain. *Quarterly Journal of the Geological Society, London*, **77**, 174–273, https://doi.org/10.1144/GSL.JGS.1918.074.01-04.08

SMITH, D.B. 1972. Foundered strata, collapse breccias and subsidence features of the English Zechstein. *In*: RICHTER-BERNBURG, G. (ed.) *Geology of Saline Deposits. Proceedings of the Hannover Symposium (Earth Sciences, No. 7)*. UNESCO, Paris, 255–269.

THOMPSON, A., HINE, P.D., GREIG, J.R. & PEACH, D.W. 1996. *Assessment of Subsidence Arising from Gypsum Dissolution with Particular Reference to Ripon*. Symonds Travers Morgan, East Grinstead.

THOMPSON, A., HINE, P.D., PEACH, D.W., FROST, L. & BROOK, D. 1998. Subsidence hazard assessment as a basis for planning guidance in Ripon. *In*: MAUND, J.G. & EDDLESTON, M. (eds) *Geohazards in Engineering Geology*. The Geological Society, London, Engineering Geology Special Publications **15**, 415–426, https://doi.org/10.1144/GSL.ENG.1998.015.01.42

TOULEMONT, M. 1984. Le karst gypseux du Lutétien supérieur de la région parisienne. Charactéristiques et impacts sur le milieu urbain. *Revue de Géologie Dynamique et de Géographie Physique*, **25**, 213–228.

TYLER, I. 2000. *Gypsum in Cumbria*. Blue Rock Publications.

WALTHAM, A.C. & COOPER, A.H. 1998. Features of gypsum caves and karst at Pinega (Russia) and Ripon (England). *Cave and Karst Science*, **25**, 131–140.

WALTHAM, A.C., BELL, F.G. & CULSHAW, M.G. 2005. *Sinkholes and Subsidence; Karst and Cavernous Rocks in Engineering and Construction*. Praxis, Springer, Chichester.

WOODS-BALLARD, B., KELLAGHER, R., MARTIN, P., JEFFRIES, C., BRAY, R. & SHAFFER, P. 2007. *The SUDS Manual*. CIRIA, London, https://www.ciria.org/ItemDetail?iProductCode=C753F&Category=FREEPUBS

WORLEY, N. & REEVES, H. 2007. *Field Guide: Application of Engineering Geology to Surface Mine Design, British Gypsum, Newark, Nottinghamshire, Sunday 1st April 2007*, Yorkshire Geological Society. http://nora.nerc.ac.uk/3225/

WYNNE, T.T. 1906. Gypsum and its occurrence in the Dove Valley. *Transactions of the Institute of Mining Engineers*, **32**, 171–192.

YOUNG, J. 1990. *Alabaster*. Derbyshire Museums Service.

YUHR, L.B., KAUFMANN, R., CASTO, D., SINGER, M., MCELROY, B. & GLASGOW, J. 2008. Karst characterization of the Marshall Space Flight Center: two years after. *In*: YUHR, L.B., ALEXANDER, E.C. & BECK, B.F. (eds) *Sinkholes and the Engineering and Environmental Impacts of Karst*. ASCE, Geotechnical Special Publications, 98–109.

Chapter 17 Mining-induced fault reactivation in the UK

Laurance Donnelly

International Union of Geological Sciences, Initiative on Forensic Geology, 398 Rossendale Road, Burnley, Lancashire, BB11 5HN, UK

geologist@hotmail.co.uk

Abstract: Faults are susceptible to reactivation during coal mining subsidence. The effects may be the generation of a scarp along the ground surface that may or may not be accompanied by associated ground deformation including fissuring or compression. Reactivated faults vary considerably in their occurrence, height, length and geometry. Some reactivated faults may not be recognizable along the ground surface, known only to those who have measured the ground movements or who are familiar with the associated subtle ground deformations. In comparison, other reactivated faults generate scarps up to several metres high and many kilometres long, often accompanied by widespread fissuring of the ground surface. Mining subsidence-induced reactivated faults have caused damage to roads, structures and land. The objective of this chapter is to provide a general overview of the occurrence and characteristics of fault reactivation in the UK.

17.1 Background

Various documents and publications are available to assist with the prediction of coal mining subsidence, such as the *Subsidence Engineers Handbook* (Anon 1975) and Whittaker & Reddish (1989). However, faults located in areas prone to coal mining subsidence are susceptible to reactivation, and this cannot be forecast. Reactivated faults may result in the generation of a scarp, graben, fissure or compression hump along the ground surface (Fig. 17.1).

Fault reactivation has been documented in the UK since the middle part of the 1800s. However, many of the earlier theories on fault reactivation were somewhat speculative and lacked a fundamental geological appreciation of fault mechanisms. During the 1950s, increased claims for mining subsidence compensation provided the incentive for the British coal mining industry to investigate fault reactivation. However, by the 1980s the exact mechanisms of fault reactivation still remained unclear, although numerous cases were known and documented. As a result, some coal resources, particularly those located in densely populated parts of the UK, were effectively sterilized, since it was not possible to predict the ground movements and to estimate the potential compensation claims. In the 1990s, following continued cases of fault reactivation, recommendations from Government (The Commission on Energy & The Environment 1981) resulted in further research to investigate fault reactivation. As with all faults, it is still not possible to predict exactly if, when and where a fault may reactivate when subjected to mining subsidence. However, this research has now enabled the factors that control fault reactivation and the different styles of ground deformation to be better understood. Further work on fault reactivation may be found in Donnelly (1994, 2000*a*, *b*, 2005, 2006, 2009, 2011, 2013, 2018), Donnelly *et al.* (1993, 1998*a,b*, 2001, 2002, 2006, 2008, 2000*a,b*, 2010*a*, *b*, 2019), Donnelly & Reddish (1993), Donnelly & Melton (1995), Donnelly & Rees (2001), Hellewell (1988), Phillips & Hellewell (1994), Phillips (1991), Wilde & Crook (1984), Whittaker & Reddish (1989), Culshaw *et al.* (2006), Lee (1965, 1966), Marr (1961), Smith & Colls (1996), Wilde & Crook (1984), Wigham (2000), Young & Culshaw (2001), Yu *et al.* (2006, 2007) and Institute of Civil Engineers (1977).

17.2 Occurrence

Fault reactivation has been recorded throughout the exposed and concealed coalfields of the UK. Over the past 150 years there have been over at least 226 documented case examples and several other examples which have not been published. The majority of the case examples of fault reactivation occur in the more densely populated urban parts of the British coalfields. Reactivated faults in these areas were more likely to have been observed and reported to have caused damage to land or structures and therefore subsequently investigated. Reactivated faults have also been reported where they have caused dramatic changes to the landscape and influenced landsliding.

Engineering Group Working Party (main contact for this chapter: L. Donnelly, International Union of Geological Sciences, Initiative on Forensic Geology, 398 Rossendale Road, Burnley, Lancashire, BB11 5HN, UK, geologist@hotmail.co.uk)
From: GILES, D. P. & GRIFFITHS, J. S. (eds) 2020. *Geological Hazards in the UK: Their Occurrence, Monitoring and Mitigation – Engineering Group Working Party Report*. Geological Society, London, Engineering Geology Special Publications, **29**, 425–432, https://doi.org/10.1144/EGSP29.17
© 2020 The Author(s). Published by The Geological Society of London. All rights reserved.
For permissions: http://www.geolsoc.org.uk/permissions. Publishing disclaimer: www.geolsoc.org.uk/pub_ethics

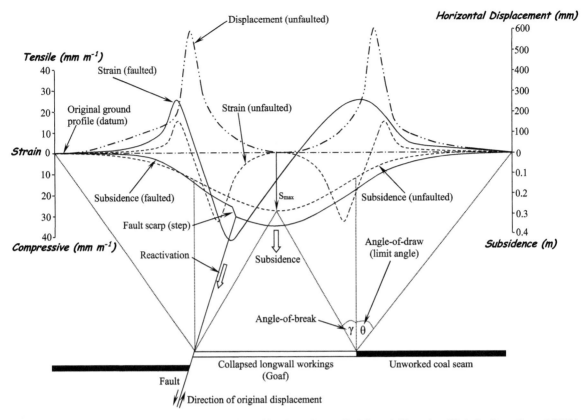

Fig. 17.1. Conceptual illustration to show the possible effects of faulting on longwall mining subsidence (modified after Donnelly *et al.* 2019 in Parry & Chiverrell 2019).

17.3 Diagnostic characteristics

Fault reactivation is significant because it has caused damage and financial losses to land, property, infrastructure, utilities, civil engineered structures, and underground and opencast mining operations. Faults are capable of several phases of reactivation during multiseam mining operations, separated by periods of relative stability (Fig. 17.2).

Reactivated faults may not be easy to recognize. These were historically suspected when linear lines of structural damage (known as 'break lines' by some subsidence engineers and mine surveyors) were observed along relatively brittle ground surface, through densely populated or built up areas, often without any trace on more granular or cohesive grass verges, gardens or agricultural land (Fig. 17.3), but this was not always the case. These have resulted in severe damage to surface structures (buildings, houses, industrial premises, bridges, dams, pylons and towers), services and utilities (sewers, water conveyances, gas mains, pipelines and communications cables) and transport networks (tracks, roads, motorways, railways, rivers and canals) (Fig. 17.4).

Frequently, reactivated faults have also disrupted agricultural land (through alteration of drainage and gradient). Fault reactivation may cause the first-time failure of natural slopes, high-walls in opencast mines, engineered cuttings and embankments, and can influence stream flows, aquifers and the reactivation of ancient (postglacial) landslides.

The surface expressions of reactivated faults, their geometry, morphology and size vary considerably. They range from subtle topographic deflections and flexures merely recognizable across agricultural land or road side verges to distinct, high-angled fault scarp walls 3–4 m high and at least 4 km long across moorland plateau areas of South Wales (Fig. 17.5). More commonly, reactivated faults are observed to be tens of millimetres to less than about a metre high, less than a metre wide and a few hundred metres long.

Where Coal Measures are concealed by younger Permo-Triassic rocks, such as the Sherwood Sandstone, Magnesian Limestone and Pennant Measures Sandstone fissures may become revealed during construction, the development of the ground or following prolonged heavy rainfall. These are likely to have been generated at the time of mining but have been obscured from view owing to the bridging of

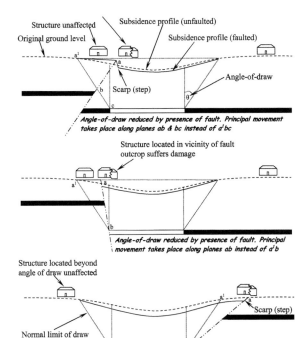

Fig. 17.2. The influence of faults on mining subsidence and the angle-of-draw (**a, b**) Any structures locating in the vicinity of fault outcrops during their reactivation will almost certainly suffer moderate to severe damage. When workings are located in the footwall of the fault any structure located in the hanging wall may be safeguarded, although this is not always the case. In examples (a) and (b) the presence of the fault has reduced the angle-of-draw (and therefore area-of-influence) in the hanging wall. (**c**) Faults may also extend the angle-of-draw, beyond that which would otherwise prevail in the absence of any faults (after Donnelly 2009).

weathered bedrock and soil cover. The width and depth of these may be exacerbated by the subsequent erosion of their walls following their exposure.

Reactivated fault scarps, fissures and compression humps do not always appear at their postulated outcrop position as inferred on geological maps. This may be attributed to the acceptable mapping tolerances (since geological maps provide an estimate of their likely outcrop position on the ground surface). This is complicated by the variable nature of the strata (or made ground) which a fault displaces, resulting in deflections, grabens, fissures, splays and runners.

Fault scarps are normally temporary features of the ground surface and may be destroyed soon after their generation by, for example, repairs to roads and structures, or by the ploughing of agricultural land. Their absence on the ground surface, however, does not eliminate the risks of ground movements if continued or renewed reactivation occurs. Greater thicknesses of superficial deposits, or made ground, will reduce the height and angle of a fault scarp, but influence a much broader area. Where the cover is thin or absent a distinct, high-angled fault scarp may develop but where these are thick (10 m+), a less distinct, broad, open flexure will be generated. In some instances, reactivated faults have reduced the amount of subsidence on the unworked side of a fault by absorbing ground strains and safeguarding houses, structures and land which may have been otherwise damaged, but this is not always the case.

Although fault reactivation, in certain circumstances, may continue for periods of time (weeks to several years) after 'normal' subsidence has been completed, movements along most faults does eventually cease, as shown by ground monitoring (precise levelling and GPS surveying) and field observations. Those faults more prone to reactivation tend to be the master and main faults which define structural blocks in the coalfields.

17.4 Mitigation

Not all faults reactivate during mining subsidence and, given the decline in deep coal mining in Britain over the past few decades, fewer cases are expected. Faults that have reactivated in the past during mining subsidence should be investigated to make sure the ground movements have ceased and associated geotechnical constraints are managed. It should be noted that the absence of information on fault reactivation, in former and currently active coalfields, does not necessarily imply that reactivation has not occurred. It would be prudent on all sites containing geological faults in the coalfield areas to investigate their potential influence on ground stability, before development and construction are carried out. It is recommended that this be undertaken at the desk study and site investigation stage of a project. Data and information on past cases of fault reactivation are available in the published literature cited in this chapter and/or from the British Geological Survey and The Coal Authority.

It is recommended that recently active faults be avoided for the siting of engineered structures, houses, landfill, waste sites, reservoirs, dams, roads, motorways, railways, tunnels and utilities. If alternative sites cannot be found then a ground investigation is recommended to identify the associated potential hazards and to assess their risks, liabilities and consequences, or to determine any necessary ground treatment. This information may be required by geologists, engineers and planners so that the ground can be suitably treated, or appropriate foundations designed, prior to any construction. The results from a desk study and ground investigation may then enable suitable mitigation measures to be designed. The options available will depend on the geology, past mining and structures being built, and may include grouting, the emplacement of reinforced geotextile mattresses, the infilling of fissures or the monitoring of ground movements.

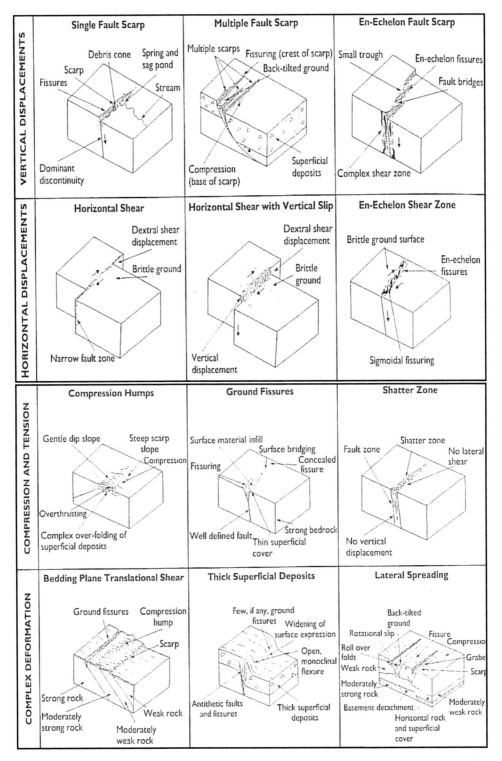

Fig. 17.3. Geological models to illustrate the surface expression reactivated fault scarps (after Donnelly & Rees 2001).

Fig. 17.4. Case examples to demonstrate the surface expression reactivated fault scarps (after Donnelly 1994, 2011).

Fig. 17.5. (Left) Fault scarps, graben and fissures, on moorland plateau beyond the back scarp of the Darren Ddu landslide, Ebbw Fach Valley, South Wales (source, Halcrow in Donnelly 2011). (Right) Tableland fault scarp, exposed beyond the back scarp of the Darren Goch landslide, Ogmore Valley, South Wales (source, Donnelly 2011).

Acknowledgements The author would like to acknowledge the support of former: University of Nottingham, Department of Mining Engineering, Engineering Geology Group of the British Geological Survey and British Coal (now The Coal Authority).

Funding This research was originally funded by British Coal, and implemented as a PhD at the University of Nottingham (1991–94). Further funding was provided by the British Geological Survey (1994–2003) then the investigations continued with private funding (2003–20).

References

ANON. 1975. *Subsidence Engineers Handbook*. National Coal Board, London.

CULSHAW, M.G., TRAGHEIM, D., BATESON, A. & DONNELLY, L.J. 2006. Measurement of ground movements in Stoke-on-Trent (UK) using radar interferometry. *In*: CULSHAW, M.G., REEVES, H., SPINK, T. & JEFFERSON, I. (eds) *Engineering Geology for Tomorrow's Cities*. 10th IAEG, International Congress, 6–10 September 2006, Nottingham, Theme 2, Paper 125.

DONNELLY, L.J. 1994. *Predicting the reactivation of geological faults and rock mass discontinuities during mineral exploitation, mining subsidence and geotechnical engineering*. PhD thesis, University of Nottingham.

DONNELLY, L.J. 2000a. The reactivation of geological faults during mining subsidence from 1859 to 2000 and beyond. *IMM Conference 2000. The Legacy of Mineral Extraction. Transactions of the Institution of Mining and Metallurgy*, September–December 2000, A179–A190.

DONNELLY, L.J. 2000b. Fault reactivation induced by mining in the East Midlands. *Mercian Geologist*, **15**, 29–36.

DONNELLY, L.J. 2005. Fault reactivation in South Wales and the effects of on ground stability. *In*: NICHOL, D., BASSSETT, M.G. & DEISLER, V.K. (eds) *The Urban Geology of Wales 2*. National Museum of Wales Geological Series no 24, Cardiff, 99–117.

DONNELLY, L.J. 2006. A review of coal mining-induced fault reactivation in Great Britain. *Quarterly Journal of Engineering Geology and Hydrogeology*, **39**, 5–50, https://doi.org/10.1144/1470-9236/05-015

DONNELLY, L.J. 2009. A review of international cases of fault reactivation during mining subsidence and fluid abstraction. *Quarterly Journal of Engineering Geology and Hydrogeology*, **42**, 73–94, https://doi.org/10.1144/1470-9236/07-017

DONNELLY, L.J. 2011. *Ground Deformation in the Vicinity of Deep Seated Landslides in the South Wales Coalfield: Mining Induced or Geological?* Geological Society, London, Engineering Group, Field Trip to South Wales, 2 July 2011.

DONNELLY, L.J. 2013. Mining induced fault reactivation. *In*: BOBROWSKY, P. (ed.) *Encyclopaedia of Natural Hazards*. Springer, Berlin, 673–687.

DONNELLY, L.J. 2018. Faults. *In*: BOBROWSKY, P.T. & MARKER, B. (eds) *Encyclopaedia of Engineering Geology*. Springer, Berlin, 329–336.

DONNELLY, L.J. & MELTON, N.D. 1995. Compression ridges in the subsidence trough. *Geotechnique, Journal of the Institute of Civil Engineers*, **45**, 555–560.

DONNELLY, L.J. & REDDISH, D.J. 1993. The development of surface steps during mining subsidence: not due to fault reactivation. *Engineering Geology*, **36**, 243–255, https://doi.org/10.1016/0013-7952(94)90006-X

DONNELLY, L.J. & REES, J. 2001. Tectonic and mining-induced fault reactivation around Barlaston on the Midlands Microcraton. *Quarterly Journal of Engineering Geology and Hydrogeology*, **34**(Part 2), 195–214. May 2001, https://doi.org/10.1144/qjegh.34.2.195

DONNELLY, L.J., WHITTAKER, B.N. & REDDISH, D.J. 1993. Ground deformation mechanisms at fault outcrops during mining subsidence: a geological perspective. *Conference on Mine Subsidence in Urban and Developed Areas*, 9–10 September 1993, Rock Springs, WY

DONNELLY, L.J., DUMPLETON, S., CULSHAW, M.G., SHEDLCOK, S.L. & McCANN, D.M. 1998a. The Legacy of Abandoned Mining in the Urban Environment in the UK. *Ground Engineering*, July 1998, 29.

DONNELLY, L.J., DUMPLETON, S., CULSHAW, M.G., SHEDLCOK, S.L. & McCANN, D.M. 1998b. The legacy of abandoned mining in the urban environment in the UK. *Polluted and Marginal Land – 98, 5th International Conference and Exhibition*. Re-use of Contaminated Land and Landfills, Brunel University, 7–9 July 1998.

DONNELLY, L.J., NORTHMORE, K.J. & JERMY, C.A. 2000a. Fault reactivation in the vicinity of landslides in the South Wales Coalfield. *In*: BROMHEAD, E., DIXON, N. & IBSEN, M.L. (eds) *Landslides in Research, Theory and Practise. ISSMGE and BGS 8th International Symposium on Landslides*, 26–30 June 2000, 481–486.

DONNELLY, L.J., NORTHMORE, K.J. & SIDDLE, H.J. 2000b. Lateral spreading of Moorland in South Wales. *In*: SIDDLE, H.J., BROMHEAD, E.N. & BASSETT, M.G. (eds) *Landslides and Landslide Management in South Wales*. National Museum & Galleries of Wales, Cardiff, Geological Series **18**, 43–48.

DONNELLY, L.J., PEREZ, J. & DE LA CRUZ, N. 2001. Operation Colombia: The Underground Mining Industry in Colombia Prior to Rationalisation in Colombia's Sinifana Coal Basin. *World Coal*, 21–27, April 2001.

DONNELLY, L.J., NORTHMORE, K.J. & SIDDLE, H.J. 2002. Block movements in the Pennines and South Wales and their association with landslides. *Quarterly Journal of Engineering Geology and Hydrogeology*, **35**(part 1, Symposium in Print), 33–39. https://doi.org/10.1144/qjegh.35.1.33

DONNELLY, L.J., CULSHAW, M.G., BELL, F.G. & TRAGHEIM, D. 2006. Ground deformation caused by fault reactivation in towns and cities. *In*: CULSHAW, M.G., REEVES, H., SPINK, T. & JEFFERSON, I. (eds) *Engineering Geology for Tomorrow's Cities. 10th IAEG International Congress*, 6–10 September 2006, Nottingham, theme 4, paper 114, The Geological Society, London, 1–15.

DONNELLY, L.J., CULSHAW, M.G. & BELL, F.G. 2008. Longwall mining-induced fault reactivation and delayed subsidence ground movement in British Coalfields. *Quarterly Journal of Engineering Geology and Hydrogeology*, **41**(Subsidence-Collapse, Symposium-in-Print), 301–314. https://doi.org/10.1144/1470-9236/07-215

DONNELLY, L.J., CULSHAW, M.G. & BELL, F.G. 2010a. Geological and mining factors which may influence mining subsidence induced fault reactivation in the UK. *In*: WILLIAM, A.L. (ed.) *Geologically Active: Proceedings of the 11th IAEG Congress*, Auckland, New Zealand, 5–10 September 2010, Leiden, CRC Press, Balkena.

DONNELLY, L.J., SIDDLE, H.J. & NORTHMORE, K.N. 2010b. Geological and mining factors which may influence mining subsidence induced fault reactivation in the UK. Paper submitted for the 11th IAEG Congress 2010, Geologically Active, 5–10 September 2010, Auckland, New Zealand (pending).

DONNELLY, L.J., CULSHAW, M.G. & DENNEHY, J.P. 2019. Fault reactivation and fissures. *In*: PARRY, D. & CHIVERRELL, C. (eds) *Abandoned Mine Workings Manual*. C758D, CIRIA, London, UK, 162–172.

HELLEWELL, E.G. 1988. The influence of faulting on ground movements due to coal mining. The UK and European experience. *Mining Engineer*, **147**, 334–337.

INSTITUTE OF CIVIL ENGINEERS 1977. *Ground Subsidence*. ICE, London.

LEE, A.J. 1965. The effects of faulting on mining subsidence. *Mining Engineer*, **125**, 735–745.

LEE, A.J. 1966. *The effects of faulting on mining subsidence*. MSc thesis, University of Nottingham.

MARR, J.E. 1961. Subsidence observations in the South Lancashire Coalfield. *Sheffield University Mining Magazine*, 24–35.

PARRY, D. & CHIVERRELL, C. (eds) 2019. *Abandoned Mine Workings Manual*. C758D, CIRIA, London, UK.

PHILLIPS, K.A.S. 1991. *The influence of geological faulting on mining subsidence*. PhD thesis, University of Wales College of Cardiff.

PHILLIPS, K.A.S. & HELLEWELL, E.G. 1994. Three-dimensional ground movements in the vicinity of a mining activated geological fault. *Quarterly Journal of Engineering Geology*, **27**, 7–14, https://doi.org/10.1144/GSL.QJEGH.1994.027.P1.03

SMITH, J.A. & COLLS, J.J. 1996. Groundwater rebound in the Leicestershire Coalfield. *Journal of the Chartered Institution of Water and Environmental Management*, **10**, 280–289, https://doi.org/10.1111/j.1747-6593.1996.tb00046.x

THE COMMISSION ON ENERGY AND THE ENVIRONMENT 1981. *(Flowers Commission) Coal and the Environment*. HMSO, London.

WHITTAKER, B.N. & REDDISH, D.J. 1989. *Subsidence: Occurrence, Prediction and Control*. Elsevier, Amsterdam.

WIGHAM, D. 2000. Occurrence of mining-induced open fissures and shear walls in the Permian limestones of County Durham. *Transactions Institution Mining Metallurgy, Section A, Mining Technology*, **109**, A131–A244.

WILDE, P.M. & CROOK, J.M. 1984. The significance of abnormal ground movements due to deep coal mining and their effects on large scale surface developments at Warrington New Town. *In*: GEDDES, J. (ed.) *Proceedings of a Conference on Ground Movements and Structures*, Cardiff, UK. Pentech Press, London, 240–247.

YOUNG, B. & CULSHAW, M.G. 2001. *Fissuring and related ground movements in the Magnesian Limestone and Coal Measures of the Houghton-le-Spring Area, City of Sunderland*. British Geological Survey Technical Report **WA/01/04**.

YU, M-H., JEFFERSON, I. & CULSHAW, M.G. 2006. Geohazards caused by rising groundwater in the Durham Coalfield, United Kingdom. *In*: CULSHAW, M.G., REEVES, H., JEFFERSON, I. & SPINK, T. (eds) *The Engineering Geology for Tomorrows Cities. Proceedings of the 10th Congress of the International Association for Engineering Geology and the Environment*, 6–10 September 2006, Nottingham. On CD-Rom, Geological Society, London.

YU, M-H., JEFFERSON, I. & CULSHAW, M.G. 2007. Fault reactivation, an example of environmental impacts of groundwater rising on urban area due to previous mining activities. *In*: *Proceedings of the 11th Congress of the International Society for Rock Mechanics*, 9–13 July 2007, Lisbon.

Chapter 18   Radon gas hazard

J. D. Appleton[1]*, D. G. Jones[1], J. C. H. Miles[2] & C. Scivyer[3]

[1]British Geological Survey, Keyworth, Nottingham NG12 5GG, UK
[2]49 Nobles Close, Grove, Oxfordshire OX12 0NR, UK
[3]Building Research Establishment, Watford, WD25 9XX, UK

Correspondence: jda@bgs.ac.uk

Abstract: Radon (^{222}Rn) is a natural radioactive gas that occurs in rocks and soils and can only be detected with special equipment. Radon is a major cause of lung cancer. Therefore, early detection is essential. The British Geological Survey and Public Health England have produced a series of maps showing radon affected areas based on underlying geology and indoor radon measurements, which help to identify radon-affected buildings. Many factors influence how much radon accumulates in buildings. Remedial work can be undertaken to reduce its passage into homes and workplaces and new buildings can be built with radon preventative measures.

18.1 Introduction

Radon is a natural radioactive gas that you cannot see, smell or taste and that can only be detected with special equipment. It is produced by the radioactive decay of radium, which in turn is derived from the radioactive decay of uranium. Uranium is found in small quantities in all soils and rocks, although the amount varies from place to place. There are three naturally occurring radon (Rn) isotopes: ^{219}Rn (actinon), ^{220}Rn (thoron) and ^{222}Rn, which is commonly called radon. ^{222}Rn (radon) is the main radon isotope of concern to man. The production of radon by the radioactive decay of ^{238}U in rock, overburden and soil is controlled primarily by the amount of uranium within the rock-forming minerals and their weathering products. The ^{238}U decay chain may be divided into two sections separated by ^{226}Ra (radium), which has a half-life of 1600 years. Earlier isotopes mostly have long half-lives, while the later isotopes, including radon (^{222}Rn), have relatively short half-lives (^{222}Rn half-life 3.82 days). ^{220}Rn (thoron) is produced in the ^{232}Th decay series and has a half-life of 56 s. ^{220}Rn has been recorded in houses, and about 4% of the average total radiation dose for a member of the UK population is from this source. ^{219}Rn (actinon) has a very short half-life of about 4 s and this, together with its occurrence in the decay chain of ^{235}U, which is only present as 0.7% of natural uranium, restricts its abundance in gases from most geological sources.

There are a number of different ways to quantify radon. These include (1) the *radioactivity* of radon gas, (2) the *dose* to living tissue, e.g. to the lungs from solid decay products of radon gas and (3) the *exposure* caused by the presence of radon gas. In the UK and most countries apart from the USA, radioactivity is measured using the SI unit becquerel (Bq). One becquerel represents one atomic disintegration per second and the level of radioactivity in the air owing to radon is measured in becquerels per cubic metre (Bq m^{-3}) of air. The average radon concentration in houses in the UK is 20 Bq m^{-3}, which is 20 radon atoms disintegrating every second in every cubic metre of air. The population-weighted world average radon concentration is 40 Bq m^{-3} (HPA 2009; UNSCEAR 2009).

The dose equivalent indicates the potential risk of harm to particular human tissues by different radiations, irrespective of their type or energy. It is measured in sieverts (Sv), where 1 Sv represents 1 joule of energy per kilogram, but normally expressed in mSv as the Sv is a large unit. The average person in the UK receives an annual effective radiation dose, which is the sum of doses to body tissues weighted for tissue sensitivity and radiation weighting factors, of 2.8 mSv, of which about 85% is from natural sources: cosmic rays, terrestrial gamma rays, the decay products of ^{220}Rn and ^{222}Rn, and the natural radionuclides in the body ingested through food and drink (Fig. 18.1). Of this natural radiation, the major proportion is from geological sources. About 60% of the total natural radiation dose is from the decay products of radon isotopes (mostly owing to alpha particle activity) while about 15% is thought to be due to gamma radiation from the uranium, thorium and potassium in rocks and soils and from building products produced from geological raw materials. Exposure to radon in buildings provides about half the total radiation dose to the average person in the UK (Watson *et al.* 2005). X-rays and radioactive materials used to diagnose and treat disease are the largest source of artificial

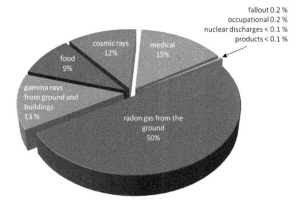

Fig. 18.1. Average annual radiation dose to the UK population (data from Watson *et al.* 2005).

exposure to people. The average dose owing to anthropogenic isotopes (radioactive fallout, nuclear fuel cycle, etc.) is less than 1% of the total annual dose (Fig. 18.1). The average annual dose to the UK population from radon is 1.2 mSv with a range from 0.3 to more than 100 mSv depending on area. In the most radon-prone area in Great Britain, the average person receives a total annual radiation dose of 7.8 mSv, of which 81% is from radon. On an individual basis, the dose would be dependent largely upon where one lived in the UK, the characteristics of the structure of one's home and one's lifestyle. Outdoors, radon normally disperses to low concentrations in the air, whereas in confined spaces such as buildings, mines and caves (Talbot *et al.* 1997; Gillmore *et al.* 2000; Gillmore *et al.* 2002), it may accumulate. Radon in indoor air comes principally from soil-gas derived from soils and rocks beneath a building with smaller amounts from the degassing of domestic water into the indoor air and from building materials. In very rare circumstances, radon may potentially be emitted from some anthropogenic sources, such as near-surface radioactive waste disposal sites (Appleton *et al.* 2011*a*). Building materials are the main source of thoron (^{220}Rn) in room air, although a minor contribution comes from soil-gas. Radon contributes by far the largest variation in the average dose from natural radiation sources.

Water in rivers and reservoirs usually contains very little radon, because it decays almost completely in a few weeks, so homes that use surface water do not have a radon problem from their water. Water processing in large municipal systems aerates the water, which allows radon to escape and also delays the use of water until most of the remaining radon has decayed. Mains water supplies pass through treatments which tend to remove radon gas but small public water works and private domestic wells often have closed systems and short transit times that do not remove radon from the water or permit it to decay. In such situations, radon from the domestic water released during showering and other household activities could add radon to the indoor air.

Areas most likely to have problems with radon from domestic water supplies include those with high levels of uranium in the underlying rocks.

In a study of private water supplies in southwestern England, a high proportion of water derived from granite areas exceeds the draft European Union action level of 1000 BqL^{-1} (Talbot *et al.* 2000). It was also found that radon concentrations varied significantly over the course of a week and between samples taken several months apart. For water from groundwater sources, mean values (by source type) at the tap were generally lower than those at the source. This is consistent with loss of radon owing to degassing as a result of water turbulence within the supply system and natural radioactive decay while the water is resident in the household supply system. All of the water sources sampled showed large variability in radon concentration over the summer sampling period, whereas less pronounced variability was observed during the winter sampling. Maximum values were observed during the summer.

Building materials generally contribute only a small percentage of the indoor air radon concentrations, although this may be 20–50% of the radon in an average UK dwelling (Gunby *et al.* 1993). However, Groves-Kirkby *et al.* (2008) noted that, in areas of generally low indoor radon concentrations, indoor radon may be mainly derived by emanation from building materials. The contribution of radon from the ground into homes varies over several orders of magnitude, in some cases giving substantial radiation doses to the occupants, while the contribution from building materials is much less variable. No cases have been identified in the UK where high indoor radon concentrations have been found to be caused by radon from conventional building materials alone, although it is reported that some buildings in SW England made from mining waste have high radon concentrations.

Radon levels in outdoor air, indoor air, soil-air and groundwater can be very different. Radon concentrations in outdoor air in the UK are generally low, on average 4 Bq m^{-3} whilst radon in indoor air in UK dwellings ranges from less than 10 to over 17 000 Bq m^{-3} (Rees *et al.* 2011) with a population-weighted average of 20 Bq m^{-3}. Radon in soil-air (the air that occupies the pores in soil) ranges from less than 1 to more than 2500 kBq m^{-3}.

18.2 Other natural sources of radiation

18.2.1 Gamma rays from the ground and buildings (terrestrial gamma rays)

Everyone is irradiated by gamma rays emitted by the radioactive materials in the Earth. Terrestrial gamma rays originate chiefly from the radioactive decay of natural potassium, uranium and thorium, which are widely distributed in terrestrial materials including rocks, soils and building materials extracted from the Earth. The average annual gamma radiation dose from all of these sources to the population in

Great Britain is about 350 µSv (Watson *et al.* 2005) with a range of 120–1200 µSv.

Within a masonry building, most of the gamma radiation is received from the building materials, whereas in wooden buildings a larger part of the dose is contributed from gamma radiation from the ground. The average person in the UK spends only 8% of their time outdoors so the contribution to total radiation dose from the ground is relatively small. The bulk of the radiation above the ground surface is derived from only the top 30 cm or so of soil or rock. Soils developed upon radioactive rocks generally have a much lower gamma radioactivity than the rock substrate. Whereas one can predict or identify areas of high geological gamma radioactivity, the resultant dose to the population depends on additional factors such as soil type, house construction and lifestyle.

Radioactive materials also occur in food. ^{40}K, in particular, is a major source of internal irradiation; natural radioactivity in the human diet gives an average annual dose for adults of around 250 µSv each year of which 165 µSv is from ^{40}K (Watson *et al.* 2005). The range for all internal radiation sources in Great Britain is 100–1000 µSv per annum. Shellfish concentrate radioactive materials so that, even when there is no man-made radioactivity, people who consume large quantities of mussels, cockles or winkles can receive a dose from natural radioactivity in food that is about 50% higher than average. Apart from restricting intake of shellfish, there is very little possibility of reducing the small exposure to natural radioactivity from food.

18.2.2 Cosmic rays

Little can be done about cosmic radiation because it readily penetrates ordinary buildings and aircraft. The average annual dose from cosmic rays in Great Britain is 330 µSv, with a range of 200–400 µSv at ground level (Watson *et al.* 2005), which equates to about 10% of the average annual radiation dose (Fig. 18.1). Aircrews and frequent air travellers receive higher doses because the dose increases with altitude.

18.3 Health effects of radiation and radon

Most of the radon that is inhaled is exhaled again before it has time to decay and irradiate tissues in the respiratory tract. Radon (^{222}Rn), however, decays to form very small solid radioactive particles, including ^{218}Po, that become attached to natural aerosol and dust particles, typically within minutes of formation. The attached and unattached decay products may remain suspended in the air or settle onto surfaces. When the decay products are inhaled, a large proportion of them are deposited in the respiratory tract and irradiate the lining of the bronchi in the lung with alpha particles (Fig. 18.2), increasing the risk of developing cancers of the respiratory tract, especially of the lungs. Only smoking causes more lung cancer deaths (Fig. 18.3).

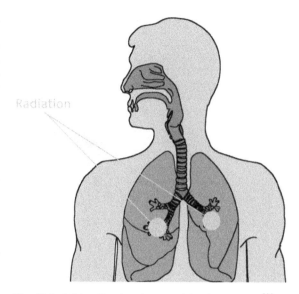

Fig. 18.2. Solid radioactive decay particles from radon (^{222}Rn), including polonium-218, attached to natural aerosol and dust particles are inhaled and irradiate the lining of the bronchi in the lung with alpha particles (image © Public Health England).

A study of lung cancer deaths from indoor radon gas, carried out for the Health Protection Agency (HPA 2009), estimated that in the UK in 2006:

(1) A total of 1100 lung cancer deaths were caused by radon, representing 3.3% of total lung cancer deaths (34 000).
(2) The dose–response relationship appears to be linear, in that the greater the concentration of indoor radon, the

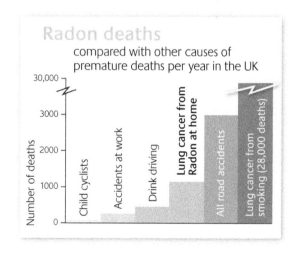

Fig. 18.3. Radon deaths compared with other causes of premature deaths per year in the UK (image © Public Health England).

greater the risk of developing lung cancer. Also, there is no evidence of any threshold below which there is no risk.

(3) Of the 3.3% of lung cancer deaths attributable to radon exposure, only 0.5% were due to radon acting alone; the remaining 2.8% were caused by a combination of radon and smoking, with nearly half the deaths likely to occur in people who had already given up smoking.

The Radon and Public Health report (HPA 2009) also highlighted that the vast majority of radon-induced lung cancer deaths in the UK occur in areas that are not currently designated as 'Radon Affected Areas' (areas in which over 1% of homes are estimated to exceed 200 Bq m^{-3}, the UK radon Action Level, AL). The overwhelming majority of the population live outside such areas, and around 95% of radon-attributable deaths are estimated to occur with residential concentrations below 200 Bq m^{-3}, and 70% at concentrations of less than 50 Bq m^{-3}.

Apart from lung cancer, there is no epidemiological proof of radon causing any other type of cancer (HPA 2009) and no consistent association has been observed between radon exposure and other types of cancer. No clear link between radon and childhood cancer (especially leukaemia) emerged from a number of ecological studies, and a review of ecological, miner and cohort studies did not find an association between radon and leukaemia (Laurier et al. 2001; Kendall et al. 2011). Radon ingested in drinking water may lead in some circumstances to organs of the gastrointestinal tract receiving the largest dose. Ingested radon is absorbed by the blood but most of the radon is lost quickly from the bloodstream through the lungs. However, the dose to the stomach from ingested radon can be significant and implies an increased risk of stomach cancer. Radon in soil under homes is the biggest source of radon in indoor air, and it presents a greater risk of cancer than radon in drinking water.

Estimation of the economic cost of radon-induced lung cancers is difficult. Using the quality adjusted life years approach (Gray et al. 2009), the total estimated cost of radon-induced lung cancers in the UK is £347 million per year.

18.4 Radon release and migration

Most of the radon atoms formed from the decay of radium remain in the mineral grains. In soils, normally 20–40% (in clays up to 70%) of the newly generated radon atoms are released to the pore space where they are mixed in with the gas (soil-air) or water that fill the pores. From the pore space, radon can be transported by diffusion or by flow in carrier fluids such as soil-air or water. The rate of release of radon from rocks and soils is largely controlled by their uranium and radium concentrations, grain size and the types of minerals in which the uranium and radium occur and their degree of alteration (Ball et al. 1991).

The most important factors controlling the migration of radon and its entry into buildings include:

(1) the characteristics of the bedrock and soils that affect fluid transport, including porosity and permeability;
(2) the nature of the carrier fluids, including carbon dioxide gas, surface water and groundwater;
(3) the construction of the building and its use, which includes the level of ventilation and heating;
(4) environmental factors such as temperature (because increased heating in the buildings during the colder months causes a chimney effect which draws soil gases including radon into the property), plus wind speed and direction which can increase the chimney effect.

The main mineralogical factors affecting the release of radon are the solubility, internal structure and specific surface area of uranium-bearing minerals. Most of the uranium in rocks can be attributed to discrete uranium-bearing minerals, even when there is only a few mg kg^{-1} of uranium present. Because radon is a gas with a limited half-life, its chances of escaping from the parent mineral are much greater if it is generated from grain margins. Other important controls are the openness of and imperfections in the internal structure of the mineral and the specific surface area of the mineral grains.

The fraction of radon mainly produced by radium decay that escapes from rock or soil (called the emanation coefficient) is dependent on the surface area of the source material. Emanation coefficients are greater for rocks than minerals, whereas soils usually have the highest values. When radium decays, the radon atom produced has a recoil energy that allows it to move a short distance through a rock grain, so it may escape into a pore, or may cross a pore space and become implanted into a neighbouring grain. If water is present in the pore space, however, the moving radon atom slows very quickly and is more likely to stay in the pore space. Differences in the uranium-bearing minerals, and especially in their solubility, control the amount of radon released. Uranium in the mineral uraninite (uranium oxide), commonly found in some granites, is easily weathered, especially near the surface. Uranium is more soluble in water so it is removed from the original mineral site, but the relatively insoluble radium, which is the immediate parent of radon gas, remains in a mixture of iron oxides and clay minerals. Uranium in other granites may occur in chemically resistant high-thorium uraninite, zircon, monazite and apatite, all of which liberate less radon. The mineral associations typically found in sedimentary rocks differ significantly from those in granites. In the Carboniferous limestone of northern England, for example, uranium is relatively uniformly distributed and associated with finely divided organic matter in the matrix of bioclastic limestones (usually <10 mg kg^{-1} U), although it may also be concentrated in stylolites, which typically contain 20–60 mg kg^{-1} U. Even though the overall concentration of uranium in the limestones is below 2 mg kg^{-1}, high radon emissions are probably derived from radium deposited on the surfaces of fractures and cavities. The high specific surface area of the radium permits efficient release of radon and high migration rates are promoted by the high permeability of the limestone. In addition, uranium and radium are concentrated in residual soil overlying limestone. In black shales in the UK, uranium is located mainly in the fine-grained mud matrix, where it may be present at levels up to 20 mg kg^{-1} U, and also in organic-rich bands at concentrations up to 40 mg kg^{-1} U. Uranium is rare in detrital phases and may also be remobilized and adsorbed on iron oxides. In sandstones, uranium is concentrated in primary detrital minerals, such as apatite and zircon, which can contain high concentrations of uranium (>100 mg kg^{-1}). Uranium may also be adsorbed onto Fe oxides in the matrix of sandstone or its weathering products. Emission of radon from sandstones is restricted by the relatively low specific surface area of the uranium minerals and appears to be more dependent upon fracturing of the rock.

Once the radon is released from the parent mineral into the space between mineral grains (the intergranular region), its migration is controlled mainly by (a) the fluid transmission characteristics of the rock including permeability, porosity, pore size distribution and the nature of any fractures and disaggregation features and (b) the degree of water retention (saturation) of the rocks. Faults and other fractures permit the efficient transmission of radon gas to the surface. The presence of faults with their enhanced fluid flow frequently results in high radon in soil-gases (Ball et al. 1991; Varley & Flowers 1993).

In highly permeable dry gravel, radon has decayed to 10% of its original concentration over a diffusion length of 5 m (UNSCEAR 2009). In more normal soils, which are generally moist, this distance would be substantially less. Diffusive ^{222}Rn in soil-gas can be determined from the specific ^{226}Ra activity, specific density, effective porosity and radon emanation coefficients of soils and rocks (Washington & Rose 1992). In caves, radon concentrations of c. 100 Bq m^{-3} would be expected if radon were generated by diffusion from solid limestone with 2.2 mg kg^{-1} U. However, the enhanced concentration of radon in caves suggests that structurally controlled convective (i.e. pressure driven) transport of radon in fluids along faults, shear zones, caverns or fractures is more significant than diffusive transport. Transportation of radon in this way may exceed 100 m.

Following radon release, migration in carrier fluids, such as carbon dioxide gas or water, is considered to be the dominant means of gas transmission to the surface. Water flow below the water table is generally relatively slow as is groundwater transport in the soil aquifer (<1–10 cm per day). Thus all hydraulically transported radon will have decayed over a distance of less than 1–2 m in most soil aquifers.

Radon dissolved in groundwater can migrate over long distances along fractures and caverns depending on the velocity of fluid flow. Radon is soluble in water and may thus be transported for distances of up to 5 km in streams flowing underground in limestone. Radon remains in solution in the water until a gas phase is introduced (e.g. by turbulence or by pressure release). If emitted directly into the gas phase, as may happen above the water table, the presence of a carrier gas, such as carbon dioxide, would tend to induce migration of the radon. This appears to be the case in certain limestone formations, where underground caves and fissures enable the rapid transfer of the gas phase.

The principal climatic factors affecting radon concentrations in buildings are barometric pressure, rainfall and wind velocity (Ball et al. 1991). Falling barometric pressure will draw soil-gas out of the ground, increasing radon concentrations in the near-surface horizons, whilst increasing barometric pressure forces atmospheric air into the ground and dilutes radon concentrations in the near-surface soil horizon. For permeable soils, radon concentrations are only affected during precipitation when saturation of small pore spaces with moisture effectively prevents the rapid out-gasing of radon from the soil. This causes the build-up of radon below the moisture-saturated surface layer and increases of an order of magnitude are sometimes observed.

A similar build-up of radon can be produced when the ground freezes and is also often observed during the night when dew forms on the surface, which can result in a two-fold increase in soil-gas alpha activity. Dry conditions cause clay-rich soils to dry out and to fracture, allowing easier egress for the soil-gases and hence an increase in radon activity at the soil surface. The radon concentration tends to be lower in the winter and higher in the summer, often varying by a factor of 3–10 (Rose et al. 1990). The variation is greater in the soil above 70 cm depth than below this depth, presumably owing to greater short-term fluctuations in soil moisture content. This suggests that radon in soil-gas measurements should be taken at depths greater than 70 cm in order to reduce the effects of temporal variations caused by rainfall and the effects of free exchange of soil-gas and atmospheric air, especially in permeable soils.

The principal soil properties that influence the concentration of radon in soil-gas, including the rate of release of radon and its transfer through soils, are soil permeability and soil moisture. Organically bound ^{226}Ra can be a principal source of ^{222}Rn in soil-gas (Greeman et al. 1990; Greeman & Rose 1996). In general, coarse gravelly soils will tend to have higher radon fluxes than impermeable clay soils. Soil permeability and rainfall (soil saturation) control radon concentrations in houses. Soil permeability generally closely reflects the permeability of the underlying rocks and superficial deposits such as glacial till, alluvium or gravel. Water-saturated soils impede the diffusion of radon enough for it to decay to harmless levels before it has diffused more than 5–10 cm. Consequently, radon from water-saturated soils is unlikely to enter buildings unless it is transported with other gases such as carbon dioxide or methane.

Radon migration may occur preferentially along natural planar discontinuities and openings including bedding planes, joints, shear zones and faults, although the precise location of such migration pathways is often difficult to establish, especially if the area is covered with soil or superficial deposits. In SW England, high radon is associated with uranium-enriched shear zones in granites, which are characterized by high radon in soil-gas and groundwater (Varley & Flowers 1993). Radon and other gases are known to concentrate and migrate upward along faults and through caves and other solution cavities. However, natural cavities such as potholes and swallow holes in limestone would also be difficult to locate precisely owing to their irregular and relatively unpredictable disposition. Whilst radon has been used to identify the location of faults and frequently reaches a maximum in the direct vicinity of faults (Barnet et al. 2008; Ielsch et al. 2010), in the UK a consistent decrease of indoor radon away from mapped faults is not observed (Appleton 2004) and it is likely that elevated radon concentrations will be associated mainly with active faults.

Artificial pathways for radon migration underground include mine workings and disused tunnels and shafts. Radon concentrations in old uranium and other mine workings are commonly 10–60 kBq m^{-3} and can be as high as 7100 kBq m^{-3} even when uranium is a very minor component of the metalliferous veins (Gillmore et al. 2001). High radon is known to be associated with gassy ground overlying coal-bearing rock strata. In addition, relatively randomly orientated and distributed blasting and subsidence fractures will affect areas underlain by mined strata. The sites and disposition of recent coal mine workings may be obtained from mine records. Other artificial pathways related to near-surface installations include electricity, gas, water, sewage and telecommunications services, the location of which may be obtained from the local service agencies. Land drains provide another potential migration pathway. The detection and prediction of migration pathways is difficult and may be imprecise, although a detailed geological and historical assessment together with appropriate radon gas monitoring and a detailed site investigation should provide a reasonable assessment of the source and potential radon gas migration pathways. Information on the local geology may be obtained from maps, memoirs, boreholes and site investigation records.

18.5 Factors affecting radon in buildings

The design, construction and ventilation of a building affect indoor radon levels, as do both the season and the weather. Radon from soils and rocks is transported into buildings through cracks in solid floors and walls below construction level; through gaps in suspended concrete and timber floors and around service pipes; and through crawl spaces, cavities in walls, construction joints and small cracks or pores in hollow-block walls (Fig. 18.4). Radon concentrations are generally highest in basements and ground floor rooms that are in contact with the soil or bedrock. Air released by well water during showering and other household activities may also contribute to indoor radon levels, although this generally makes a relatively small contribution to the total radon level.

In a typical masonry building in which radon occurs at the UK national average level of 20 Bq m^{-3}, c. 60% of radon comes from the ground on which the building stands, 25% from building materials, 12% from fresh air, 2% from the water supply and 1% from the gas supply. These figures apply to the average house in the UK, but can vary substantially, and the proportion of radon entering a home from the ground will normally be much higher in homes with high radon levels (Appleton & Ball 1995). The dominant mechanism of radon ingress is pressure-induced flow through cracks and holes in the floor, called the stack effect. Slightly negative pressure differences between indoor and outdoor atmospheres, caused by wind outside and heating inside the building, draw radon contaminated air into the building, especially through the floor. Energy-conserving measures such as double-glazing restrict the fresh supply of air and lessen the dilution of radon indoors. Conversely, such measures may also reduce the pressure difference between indoors and outdoors and thus reduce the influx of radon from the ground. Poor ventilation may increase radon concentrations, but it is not the fundamental cause of high indoor radon levels. Household energy efficiency improvements that decrease ventilation (e.g. better sealed windows and doors) could lead to an increase in exposure to radon.

Indoor radon concentrations are generally about 1000 times lower than radon in the soil underlying the house. Most houses draw less than 1% of their indoor air from the soil with the remainder from outdoors where, as noted earlier, the air is generally quite low in radon. In contrast, houses with low indoor air pressures, poorly sealed foundations and several entry points for soil-air may draw as much as 20% of their indoor air from the soil. Consequently, radon levels inside the house may be very high even in situations where the soil-air has only moderate amounts of radon.

18.6 Geological associations

Geology is usually one of the most important factors controlling the distribution and level of indoor radon and the radon hazard. Mapped bedrock geology explains on average 25% of the variation of indoor radon in England and Wales, whilst mapped superficial geology explains, on average, an additional 2% (Appleton & Miles 2010). The proportion of the total variation controlled by geology is higher (up to 37%) in areas where there is a strong contrast between the radon potential of sedimentary geological units and lower (14%) where the influence of confounding geological controls, such as uranium mineralization, cut across mapped geological boundaries.

In the UK, relatively high concentrations of radon are associated with particular types of bedrock and unconsolidated deposits, for example some granites, uranium-enriched phosphatic rocks and black shales, limestones, sedimentary ironstones, permeable sandstones and uraniferous metamorphic rocks. Permeable superficial deposits, especially those

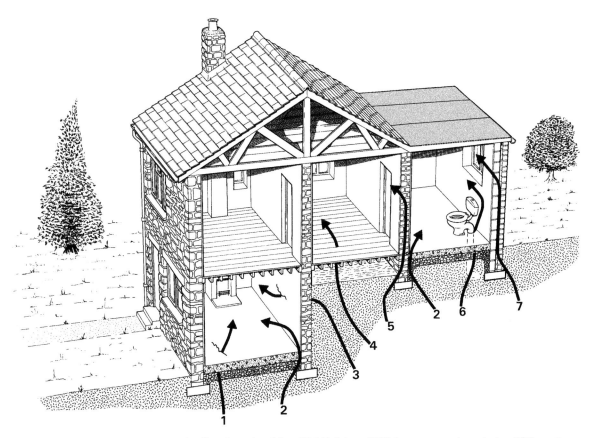

Fig. 18.4. Routes by which radon enters a dwelling. Reproduced from BR211, Scivyer 2007. Ingress routes: 1, cracks in solid floors; 2, construction joints; 3, cracks in walls below ground; 4, gaps in suspended floors; 5, cracks in walls; 6, gaps around service pipes; 7, cavities in walls.

derived from uranium-bearing rock, may also be radon prone. Geological units associated with the highest levels of naturally occurring radon (Figs 5–10) are: (1) granites in SW England, the Grampian and Helmsdale districts of Scotland and the Mourne Mountains in Northern Ireland; (2) Carboniferous limestones throughout the UK and some Carboniferous shales in Northern England and Wales; (3) sedimentary ironstone formations in the English Midlands; (4) some Ordovician and Silurian mudstones, siltstones and greywackes in Wales, Northern Ireland and the southern uplands of Scotland; (5) Middle Old Red Sandstone of NE Scotland; and (6) Neoproterozoic psammites, semipelites and meta-limestones in the western sector of Northern Ireland (Appleton & Ball 1995; Appleton & Miles 2005; Scheib et al. 2009; Appleton et al. 2011b, 2015; Scheib et al. 2013).

18.6.1 Granites

In SW England (Ball & Miles 1993; Scheib et al. 2013), the Grampian and Helmsdale areas of Scotland (Scheib et al. 2009) and the Late Caledonian and Paleogene acid intrusive rocks of the Mourne Mountains in the SE sector of Northern Ireland (County Down and County Armagh; Appleton et al. 2015), there is a correlation between areas where it is estimated that more than 20% of the house radon levels are above 200 Bq m^{-3} and the major granite intrusions. Granites in southwestern England are characterized by high uranium concentrations, a deep weathering profile and uranium in a mineral phase that is easily weathered. Although the uranium may be removed through weathering, radium generally remains *in situ* (Ball & Miles 1993). Radon is easily emanated from the host rock and high values of radon have been measured in groundwaters and surface waters (110–740 BqL^{-1}) and also in soil-gas (frequently >400 kBq m^{-3}). The highest radon potential values in Scotland are associated with Siluro-Devonian (late Caledonian) granite intrusions, notably those clustered within a zone to the west of Aberdeen and at Helmsdale (Scheib et al. 2009).

18.6.2 Black shales

The depositional and diagenetic environment of many black shales leads to enrichment of uranium. For example, some Lower Carboniferous shales in northern England and NE

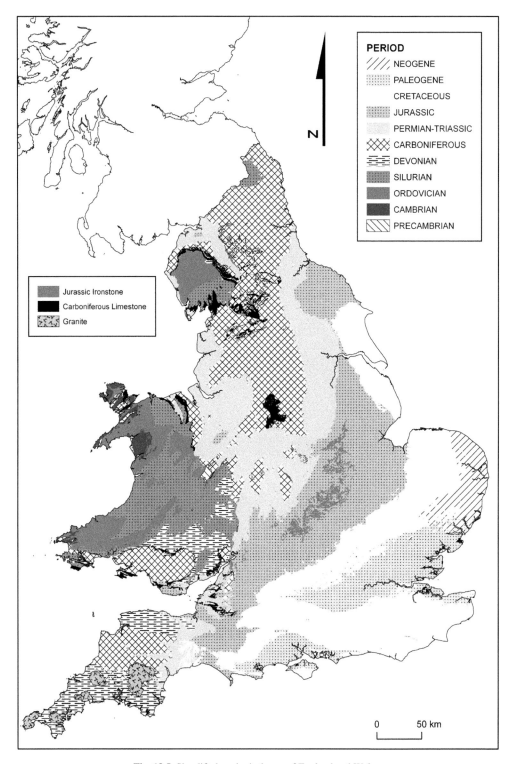

Fig. 18.5. Simplified geological map of England and Wales.

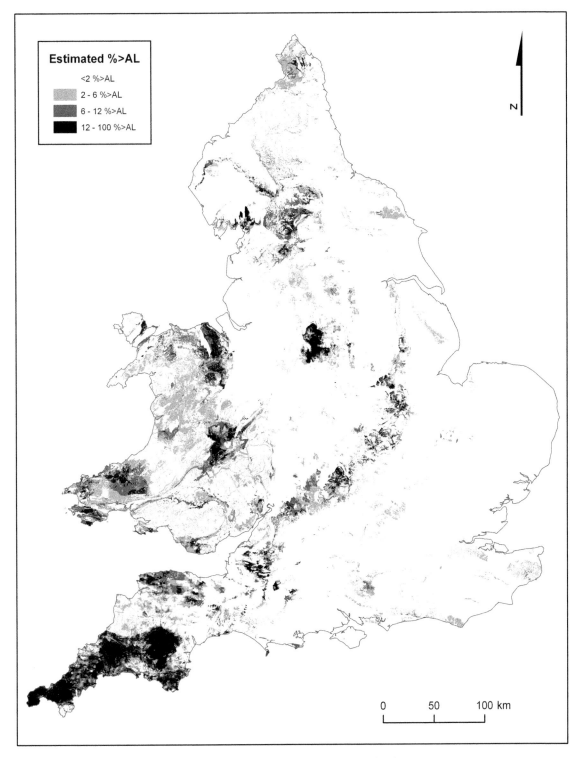

Fig. 18.6. Radon potential map of England and Wales.

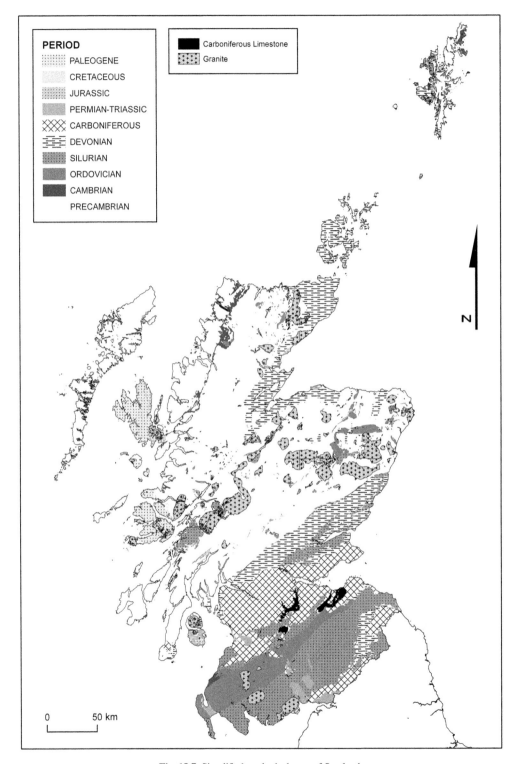

Fig. 18.7. Simplified geological map of Scotland.

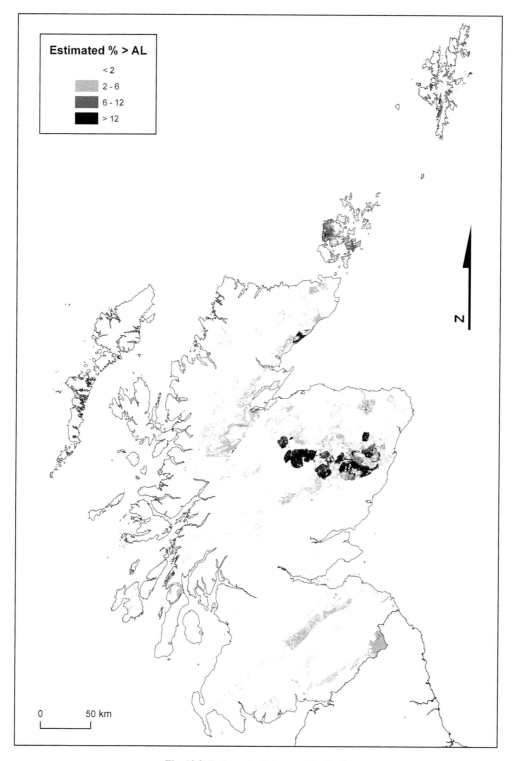

Fig. 18.8. Radon potential map of Scotland.

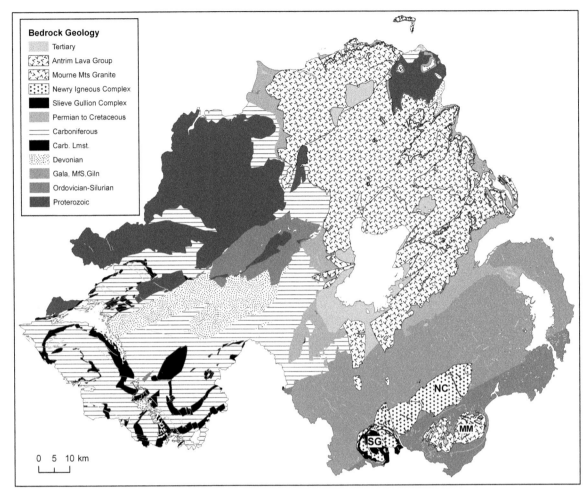

Fig. 18.9. Simplified bedrock geology of Northern Ireland. MM, Mourne Mountains Granite; SG, Slieve Gullion Complex; NC, Newry Igneous Complex.

Wales contain 5–60 mg kg^{-1} uranium, and weathering and secondary enrichment can substantially enhance uranium levels in soils derived from these shales. Some 15–20% of houses (rising to more than 65% in some areas) sited on uraniferous black shales with >60 mg kg^{-1} and high soil-gas radon (32 kBq m^{-3}; Ball et al. 1992) are above the UK radon AL.

18.6.3 Phosphatic rocks and ironstones

Uranium-enriched phosphatic horizons occur in the Carboniferous Limestone, the Jurassic oolitic limestones and the basal Cretaceous Chalk in the UK and these sometimes give rise to high radon in soil-gas and houses.

Many iron deposits are both phosphatic and slightly uraniferous and a large proportion (>20%) of houses underlain by the ironstones of the Northampton Sand Formation (NSF) and the Marlstone Rock Formation in England are affected by high levels of radon. Phosphatic pebbles from the Upper Jurassic and Lower and Upper Cretaceous phosphorite horizons in England contain 30–119 mg kg^{-1} U. Radon in dwellings is a significant problem in areas where these phosphatic rocks occur close to the surface, especially if the host rocks are relatively permeable. The NSF consists of ferruginous sandstones and oolitic ironstone with a basal layer up to 30 cm thick containing phosphatic pebbles. The ferruginous sandstones and ironstones mainly contain low concentrations of uranium (<3 mg kg^{-1}), whilst the phosphatic pebbles contain up to 55 mg kg^{-1}. However, it is probable that the mass of the NSF, which in many cases contains disseminated radium, may contribute more to the overall level of radon emissions than the thin uranium-enriched phosphate horizons. Laboratory investigations of Jurassic ironstones have demonstrated that the tightly cemented nature of the

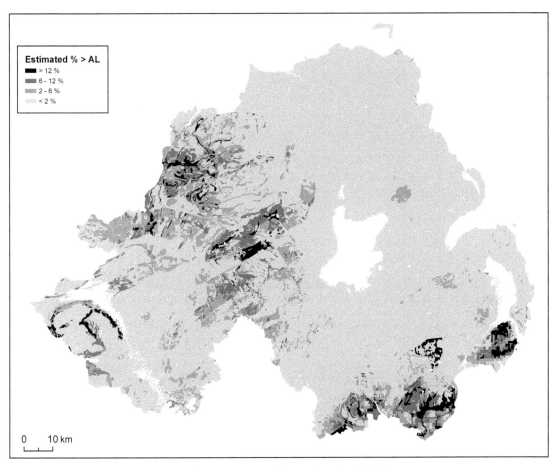

Fig. 18.10. Radon potential map of Northern Ireland.

Marlstone Rock Formation ironstone impedes radon emanation owing to low permeability, whilst the NSF has a fine-grained, more permeable and often more altered matrix. Near-surface weathering is likely to be very important in enhancing radon potential since weathered mineral phases emanate radon more efficiently. Bedrock fracturing and working of the ironstones both also have an important impact and increase the radon risk (Scheib et al. 2013). Worked ground on the Northampton Sand Formation exhibits increased radon potential relative to areas where the ironstone bedrock has not been extracted, possibly reflecting the phosphatic nature of the remaining basal beds, which were not worked.

18.6.4 Limestones and associated shales and cherts

High levels of radon occur in both soil-gas and houses underlain by Carboniferous and Jurassic limestone in the UK as well as in caves and mines in these bedrocks. There are 10% to more than 30% of houses built on the limestones in England with radon concentrations greater than the UK AL (Appleton et al. 2000a; Scheib et al. 2013). High radon in houses is associated with all areas underlain by Lower Carboniferous limestones throughout Wales and also with Visean shales, sandstones and cherts in NE Wales. Moderate to high radon levels occur in houses built on Namurian Pentre Chert Formation and Westphalian ('Coal Measures') strata in Flintshire and Wrexham (Appleton & Miles 2005). In the western sector of Northern Ireland, especially in County Fermanagh, Carboniferous limestone has >3% of dwellings above the AL (Appleton et al. 2015). In Scotland, elevated radon potential of limestones ranging in age from Dalradian to Carboniferous is likely to relate to the high specific surface area and permeability of the uranium minerals present that permit efficient release of radon and also to the high joint and fracture permeability of the limestone. Some of the radon is thought to emanate from uranium- and radium-enriched residual soils that overlie the highly permeable limestones. In south Wales, high radon is also associated with the Triassic basal Mercia Mudstone Group Marginal Facies and the limestone–mudstone sequences of the Jurassic Blue Lias Formation, a feature also seen in Somerset.

The radon emanation characteristics of chalk are different from the Carboniferous and Jurassic limestones. Chalk still retains its primary porosity, although most of the water and gas flow is through fissures. The proportion of dwellings on chalk with radon above the AL is much lower than over the Carboniferous Limestone.

18.6.5 Sands and sandstones

Thick, permeable Cretaceous sand formations in southwestern England, including the glauconitic Lower and Upper Greensand and the Upper Lias Midford Sands, all emanate high levels of soil-gas radon and are characterized by a high proportion of houses above the action level (13 and 22% for the Upper Greensand and Midford Sands, respectively). In Scotland, the geometric mean indoor radon for the Middle Old Red Sandstone bedrock of the Orcadian Basin is only slightly above average but there are high values probably related to uranium mineralization. Indeed, the locations of known uranium mineralization, in the Caithness area and on the Orkney Islands, show evidence of elevated radon potential. In Pembrokeshire, SW Wales, high radon potential characterizes the arenaceous rocks of the Cosheston Group (Devonian) and in SW England, the Devonian mudstones and sandstones in Devon and Cornwall.

18.6.6 Ordovician–Silurian greywackes and associated rocks

In Wales, the radon potential of the Cambrian is high in NW Wales and parts of Pembrokeshire whilst moderate to high radon levels occur in many areas underlain by Ordovician and Silurian mudstones, siltstones and greywackes throughout Wales. In Scotland, Ordovician greywackes of the Southern Uplands Terrain have a moderate radon potential but this decreases progressively from the NE to SW, suggesting a lateral variation in composition. The Silurian Riccarton Group greywackes derived from an evolved granitic sediment source also appear from limited data to have a moderate radon potential. In the SE of Northern Ireland, the Silurian Hawick Group greywackes also display a moderate radon potential (5.7% > AL); Appleton et al. 2015). Moderate radon levels are associated with some areas underlain by Ordovician and Silurian acid volcanic rocks in the Gwynedd, Conwy, Powys and Pembrokeshire districts of Wales.

18.6.7 Miscellaneous bedrock units

Other bedrock types that exhibit elevated radon potential include: Late Jurassic mudstones, siltstones and sandstones along the NE Scottish coast near Helmsdale, although it is likely that this is locally influenced by high uranium from the adjacent uranium-mineralized Helmsdale Granite; Devonian mafic lavas and tuffs, surrounding the Cheviot Granite on the England–Scotland border; and Dalradian metasediments on the Shetland Islands. In the western sector of Northern Ireland, the Neoproterozoic Argyll and Southern Highland Group psammites, semipelites and meta-limestones also have moderate radon potential 2–4% >AL (Appleton et al. 2015).

18.6.8 Superficial deposits

The source, composition and permeability of superficial deposits determine their radon potential. In areas of relatively uranium-rich bedrock, for example SW England and Scottish Grampian granites or the Carboniferous limestone in Derbyshire, transported material such as alluvium often exhibits a similar radon potential to the local bedrock. Relatively high radon potential characterizes Clay-with-Flints deposits which overlie parts of the Upper Greensand and the White Chalk Group in southern England. Clay-with-Flints is a heterogeneous, unbedded residual deposit formed partly by weathering and solifluction of the original Paleogene cover and earlier Quaternary deposits, and partly by dissolution of the underlying chalk. Soils developed over the Clay-with-Flints are characterized by enhanced zirconium and uranium concentrations, which may explain the slightly enhanced radon potential. Killip (2004) suggested that the moderate radon potential was caused by phosphate-rich components with high uranium contents derived from the chalky source.

In Scotland, increased radon potential of permeable glaciofluvial deposits relative to bedrock and more impermeable superficial deposits is observed on the biotite granite plutons of the Grampian Region and the Hawick Group greywackes of the Southern Uplands. Uranium-rich material influencing radon potential is evident on superficial deposits overlying the Argyll Group psammites, pelites and semipelites where large ranges of radon potential are exhibited, with the highest values found adjacent to the uranium-rich evolved biotite granite plutons of the Grampian Region. In Wales, where bedrock is overlain by unconsolidated deposits, such as glacial till or alluvium, the proportion of houses with high radon is usually reduced with respect to the underlying bedrock. In NE Wales, for example, c. 30% of dwellings situated directly on the Lower Carboniferous limestone exceed the AL, whereas the proportion is reduced to about 2% where the limestone is covered with glacial till. Whilst most other superficial deposits reduce radon potential relative to the underlying bedrock since they reduce permeability, the variation is often complex and locally controlled.

18.7 Measurement of radon

18.7.1 Radon testing in the home

The most common procedures for measuring radon make use of the fact that it is the only natural gas that emits alpha particles, so if a gas is separated from associated solid and liquid phases any measurements of its radioactive properties relate to radon or its daughter products.

Common short-term test devices are charcoal canisters, alpha track detectors, liquid scintillation detectors, electret ion chambers and continuous monitors. A short-term testing device remains in the home for 2–30 days, depending on the type of device. Because radon levels tend to vary from day to day and season to season, a long-term test is more likely than a short-term test to provide an accurate estimate of the home's year-round average radon level. If results are needed quickly, however, a short-term test followed by a second short-term test may be used to determine the approximate severity of the radon hazard. Long-term test devices, comparable in cost with devices for short-term testing, remain in the home for three months or more. Alpha track detectors and electret ion detectors are the most common long-term test devices. Ambiguous short-term measurements should be followed up by a long-term measurement.

The alpha (etched) track detectors recommended for radon testing in the UK consist of a container holding a small sheet of plastic. Radon can enter the container, and the alpha particles emitted by radon and its decay products damage the plastic as they strike it. The damaged areas can be removed by etching in a laboratory, leaving tracks in the surface where alpha particles have struck. The tracks are counted to determine the exposure of the detector to radon. Etched track detectors are relatively cheap and suitable for long-term measurement, and two detectors, placed in the living room and an occupied bedroom, are usually deployed for a period of three months.

Electret ion detectors contain an electrostatically charged PTFE disc. Radiation emitted by decay of radon and its decay products ionize the air inside the detector and reduce the surface voltage of the disc. By measuring the voltage reduction, the radon concentration can be calculated. Allowance must be made for ionization caused by natural background gamma radiation. Different types of electret are available for measurements over periods of a few days to a few months. The detectors must be handled carefully for accurate results.

Continuous monitors are active devices that need power to function. They require operation by trained testers and work by continuously measuring and recording the amount of radon in the home. These devices sample the air continuously and measure either radon or its decay products (NRPB 2000).

Charcoal detectors absorb radon from the air, and must be returned to the issuing laboratory quickly for assessment before the radon has decayed. They are not suitable for measurements longer than about 4 days. The standard procedure for deployment of such detectors is intended to give a 'worst case' result. If a high radon concentration is found, a follow-up measurement using long-term detectors can be carried out, but if the result is low then it is unlikely that a longer measurement would find a high concentration. The standard procedure requires that doors and windows must be closed 12 h prior to testing and throughout the testing period.

The test should not be conducted during unusually severe storms or periods of unusually high winds. The test kit is normally placed in the lowest lived-in level of the home, at least 50 cm above the floor, in a room that is used regularly, but not in the kitchen or bathroom where high humidity or the operation of an exhaust fan could affect the validity of the test. At the end of the test period, the kit is mailed to a laboratory for analysis; results are mailed back in a few weeks. If the result of the short-term test exceeds 100 Bq m^{-3} then a long-term test is normally recommended.

Remediation of the home is recommended if the radon concentration determined by a 3 month test exceeds 200 Bq m^{-3}. Radon levels are generally highest in winter so seasonal corrections are usually applied to estimate the average annual radon level (Miles 1998; Miles *et al.* 2011; Miles *et al.* 2012).

Bungalows and detached houses tend to have higher indoor radon than terraced houses or flats in the same area of the UK. Building material, double-glazing, draught-proofing, date of building and ownership also have a significant impact on indoor radon concentrations.

In workplaces, consideration needs to be taken of work practices and the building design and use. For small premises at least one measurement should made in the two most frequently occupied ground floor rooms. In larger buildings at least one measurement is required for every 100 m^2 floor area.

18.7.2 Measurement of radon in soil-gas and solid materials

Measurement of radon in soil-gas using pumped monitors is recommended as the most effective method for assessing the radon potential of underlying rocks, overburden and soil. Instruments for the determination of soil-gas radon are generally based upon either an extraction method, using a 'pump monitor' device for transferring a sample of the soil-gas to a detector, or simply emplacing the detector in the ground (passive methods). In the former method, a thin rigid tapered hollow tube is usually hammered into the ground to an appropriate depth (generally 70–100 cm), which causes minimum disturbance to the soil profile. Detection of radon is usually based upon the zinc sulphide scintillation method or the ionization chamber. Alpha particles produce pulses of light when they interact with zinc sulphide coated on the inside of a plastic or metal cup or a glass flask (Lucas cell). These may be counted using a photomultiplier and suitable counting circuitry. Because the radon isotopes are the only alpha-emitting gases, their concentration may be determined accurately using relatively simple equipment. The concentration of radon in soil-gases is usually sufficient that the level may be determined relatively quickly; a few minutes is generally enough. Determination of radon potential from soil-gas radon concentrations generally produces better results when soil permeability is also measured (Kemski *et al.* 2001; Barnet *et al.* 2008).

Radon release from disaggregated samples (soils, stream sediments and unconsolidated aquifer sands) may be determined by agitating a slurry of the material with distilled water in a sealed glass container, allowing a period of about 20–30 days for the generation of radon from radium, and then measuring the radon in the aqueous phase using a liquid scintillation counter. Emanation of radon from solid rock samples can be determined using a similar method. Alternatively a degassing kit can be used and the radon gas measured with similar equipment to that described for soil-gas measurement.

18.8 Radon hazard mapping and site investigation

18.8.1 Radon hazard mapping based on geology and indoor radon measurements

Requirements for mapping radon-prone areas using indoor radon data include (1) accurate radon measurements made using a reliable and consistent protocol, (2) centralized data holdings, (3) sufficient data evenly spread and (4) automatic conversion of addresses to geographical coordinates. It appears that the UK is the only country that currently meets all of these requirements for large areas (Miles & Appleton 2005)

Radon potential mapping is sometimes based on indoor radon data that have been normalized to a mix of houses typical of the housing stock as this removes possible distortion caused by construction characteristics. Maps based on results corrected for seasonal variations or temperature at the time of measurement but not normalized to a standard house mix reflect such factors as the greater prevalence of detached dwellings in rural areas, and hence the higher risk of high radon levels in rural areas compared with cities where flats are usually more prevalent. Radon potential estimates based on radon levels in the actual housing stock are more appropriate for the identification of existing dwellings with high radon.

In the UK, digital geological data compiled by the British Geological Survey (BGS) and indoor radon measurements by Public Health England (PHE, formerly the Health Protection Agency, HPA) are used to produce radon potential maps which indicate the spatial variation in radon hazard (Miles & Appleton 2005; Miles et al. 2007, 2011; Scheib et al. 2009, 2013; Appleton et al. 2015). The current radon potential maps indicate the probability that new or existing houses will exceed a radon reference level, which in the UK is termed the UK AL (of 200 Bq m^{-3}).

It is important to remember that a wide range of indoor radon levels is likely to be found in any particular area. This is because there is a long chain of factors that influence the radon level found in a building, such as radium content and permeability of the ground below it, and construction details of the building (Miles & Appleton 2000). Radon potential does not indicate whether a building constructed on a particular site will have a radon concentration that exceeds the AL. This can only be established through measuring radon in the building.

Approximately 25% of the total variation of indoor radon concentrations in England and Wales can be explained by the mapped bedrock and superficial geology. The proportion of the variation that can be attributed to mapped geological units increases with the level of detail of the digital geological data (Appleton & Miles 2010). As a consequence, the most accurate and detailed radon potential maps are generally those based on house radon data and geological boundaries provided that the indoor radon data can be grouped by sufficiently accurate geological boundaries.

The factors that influence radon concentrations in buildings are largely independent and multiplicative so the distribution of radon concentrations is usually lognormal. Therefore lognormal modelling was used to produce accurate estimates of the proportion of homes above the AL in the UK (Miles 1998; Miles et al. 2007). When indoor radon measurements are grouped by geology and 1 km squares of the UK national grid, the cumulative percentage of the variation between and within mapped geological units is shown to be 34–40% (Appleton & Miles 2010). This confirms the importance of radon maps that show the variation of indoor radon concentrations both between and within mapped geological boundaries. Combining the grid square and geological mapping methods gives more accurate maps than either method can provide separately (Appleton & Miles 2002; Miles & Appleton 2005).

Each geological unit within a map sheet or smaller area, such as a 1 km grid square, has a characteristic geological radon potential that is frequently very different from the average radon potential for the geological unit. The results of the integrated mapping method allowed significant variations in radon potential within bedrock geological units to be identified, such as in the radon-prone Jurassic NSF. Moderate radon potential (<4% > AL) occurs within and to the NW of the urban centre of Northampton with much higher potential (>12% > AL) to the north (Fig. 18.11). The NSF comprises a lower ironstone, often with a uraniferous and phosphatic nodular horizon at the base (Sutherland 1991), overlain by a massive yellow or brown calcareous sandstone (locally called the 'Variable Beds', Hains & Horton 1969). Variable Beds sandstones at Harlestone Quarry (Fig. 18.11) contain on average 11% Fe_2O_3T, 0.37% P_2O_5 and 1.1 mg kg^{-1} U (Hodgkinson et al. 2005) in an area with a radon potential of about 3% > AL whereas at Pitsford Quarry (Fig. 18.11), the ironstones contain 29% Fe_2O_3T, 1.0% P_2O_5 and 2.4 mg kg^{-1} U in an area with radon potentials in the range 8–21% > AL. Radon in soil-gas correlates positively with the radon potential of the NSF (Fig. 18.12), and there is also a close correlation ($r = 0.86$, $p < 0.0005$) between radon emanation and uranium in rock samples (Hodgkinson et al. 2005). Track etch alpha autoradiography studies indicate that uranium is associated predominantly with the goethite-rich clay matrix of the ironstones, and is also concentrated in occasional phosphatic nodules (Hodgkinson et al. 2005).

The reliability and spatial precision of the mapping method is, in general, proportional to the indoor radon measurement density and the accuracy of the geological boundaries. It is,

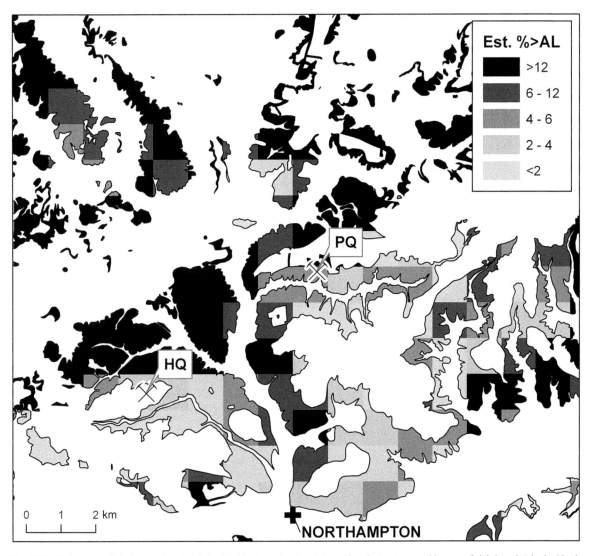

Fig. 18.11. Radon potential of ground underlain by the Northampton Sand Formation (but not covered by superficial deposits) in the Northampton area (HQ = Harlestone Quarry; PQ = Pitsford Quarry).

however, reassuring that even when the measurement density is as low as 0.2–0.4 per km^2, geological radon potential mapping discriminates between geological units in a logical way. These relationships can be explained on the basis of the petrology, chemistry and permeability of the rock units and are confirmed in adjoining areas with higher measurement densities (Miles & Appleton 2000). Other uncertainties in the mapping process relate to house-specific factors, proximity to geological boundaries and measurement error impact on the radon mapping process (Hunter *et al.* 2009, 2011).

Geological radon potential maps of the UK have been produced at 1:625 000 (Appleton & Ball 1995), 1:250 000, and most recently at 1:50 000 (Figs 6, 8 & 10) for the purpose of defining radon-affected areas in the UK.

18.8.2 Radon hazard mapping based on geology, gamma spectrometry and soil-gas radon data

In the absence of an adequate number of high-quality indoor radon measurements, proxy indicators such as information on uranium content or soil-gas radon data may be used to assess geological radon potential. The reliability of maps based on proxy data increases with the number of classes as well as the quantity and quality of available data. Uranium and

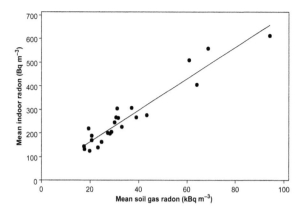

Fig. 18.12. Relationship between average soil-gas radon concentration (kBq m^{-3}) and the geological radon potential (estimated proportion of dwellings exceeding the UK radon AL, 200 Bq m^{-3}) for the NSF. Data grouped by 5 km grid square.

radium concentrations in surface rocks and soils are useful indicators of the potential for radon emissions from the ground. Uranium can be estimated by gamma spectrometry either in the laboratory or by ground, vehicle or airborne surveys (IAEA 2003).

The close correlation between airborne radiometric measurements and indoor radon concentrations has been demonstrated in parts of the UK (Appleton & Ball 2001; Scheib et al. 2006; Appleton et al. 2008, 2011b). Areas with high permeability tend to have significantly higher indoor radon levels than would be otherwise expected from the equivalent ^{226}Ra concentrations, reflecting an enhanced radon flux from permeable ground. In Northern Ireland linear regression analysis of airborne and soil geochemical parameters revealed that the most significant independent variables were eU (equivalent uranium), a parameter derived from gamma spectrometry measurements of radon decay products in the top layer of soil and exposed bedrock, and the permeability of the ground. The radon potential map generated from airborne gamma spectrometry data agrees in many respects with the map based on indoor radon data and geology but there are several areas where radon potential predicted from the airborne radiometric and permeability data is substantially lower than that found from radon measurements (Appleton et al. 2011b). This underprediction could be caused by the radon concentration being lower in the top 30 cm of the soil than at greater depth, because of the loss of radon from the surface rocks and soils to air.

It has been demonstrated in a number of countries, including the UK, Germany and the Czech Republic, that soil-gas radon measurements combined with an assessment of ground permeability can be used to map geological radon potential in the absence of sufficient indoor radon measurements. After uranium and radium concentration, the permeability and moisture content of rocks and soils are probably the next most significant factors influencing the concentration of radon in soil-gas and buildings. Radon diffuses farther in air than in water, so in unsaturated rocks and overburden with high fluid permeability, higher radon values are likely to result from a given concentration of uranium and radium than in less permeable or water-saturated materials. Weathering processes can also affect permeability. Enhanced radon in soil-gas is also associated with high-permeability features such as fractures, faults and joints. The fracturing of clays, resulting in enhanced permeability, combined with their relatively high radium content and their emanation efficiency, may also result in higher radon concentrations in dwellings. The permeability of glacial deposits exerts a particularly strong influence on the radon potential of underlying bedrock. Significant correlations between average indoor and soil-gas radon concentrations, grouped according to geological unit, have been recorded in the UK (Appleton et al. 2000a) as well as in the Czech Republic (Barnet et al. 2008), Germany (Kemski et al. 2009) and the USA (Gundersen et al. 1992). Other radon hazard mapping methods are discussed in Appleton & Ball (2001) and Appleton (2013).

18.8.3 Radon site investigation methods

Radon migrates into buildings as a trace component of soil-gas. Therefore the concentration of radon in soil-gas should provide a good indication of the potential risk of radon entering a building if its construction characteristics permit the entry of soil-gas. There is a growing body of evidence that supports the hypothesis that soil-gas radon is a relatively reliable indirect indicator of indoor radon levels at the local as well as the national scale (Ball et al. 1992; Barnet et al. 2008; Kemski et al. 2009). However, in some cases, soil-gas radon data may be difficult to interpret owing to the effects of large diurnal and seasonal variations in soil-gas radon close to the ground surface as well as on a scale of a few metres. The former problem may be overcome by sampling at a depth greater than 70 cm or by the use of passive detectors with relatively long integrating times, although this may not be a practical option if site investigation results are required rapidly. Small-scale variability in soil-gas radon may be overcome by taking 10–15 measurements on a 5- to 10 m grid to characterize a site. Radon in soil-gas varies with climatic changes including soil moisture, temperature and atmospheric pressure so weather conditions should be as stable as possible during the course of a soil-gas radon survey.

A range of methods such as controlled gas extraction, air injection procedures or water percolation tests can be used to estimate gas permeability at a specific site. In the absence of permeability measurements, more qualitative estimates of permeability can be based on visual examination of soil characteristics, published soil survey information or the relative ease with which a soil-gas sample is extracted.

In some areas and under some climatic conditions, site investigations using soil-gas radon cannot be carried out reliably, for example, when soil-gas cannot be obtained from waterlogged soils or when soil-gas radon concentrations are abnormally enhanced owing to the sealing effect of soil moisture. If soil-gas radon concentrations cannot be determined because of climatic factors, measurements of radon emanation in the laboratory or gamma spectrometric measurements of eU can be used as radon potential indicators in some geological environments. However, few data are available and the methods have not been fully tested.

Radon may also be a problem underground, such as in tunnels. Radon emanation from borehole core samples can be determined to derive values of radon emanation per unit of surface area, which is an effective and simple way of assessing radon hazard in tunnels at the site investigation stage although the effects of larger-scale features such as fractures may not be accounted for (Talbot *et al.* 1997).

18.9 Strategies for management: avoidance, prevention and mitigation

18.9.1 Introduction

Accurate mapping of radon-prone areas helps to ensure that the health of occupants of new and existing dwellings and workplaces is adequately protected. Radon potential maps have important applications, particularly in the control of radon through planning, building control and environmental health legislation. The overall aim of most countries that have identified a radon problem is to map radon-prone areas and then identify houses and workplaces with radon concentrations that exceed the radon reference level. An AL for radon in homes of 200 Bq m^{-3} has been established in the UK. Public Health England has recommended that parts of the UK with 1% probability or more of homes being at or above the AL should be designated as radon-affected areas. In 2010, PHE guidance on radon gas concentrations introduced a new 'target level' of 100 Bq m^{-3} above which remediation work should be seriously considered, especially by high-risk groups such as smokers and ex-smokers.

In the UK, testing for radon is recommended by government in radon-affected areas where more than 1% of dwellings exceed the AL of 200 Bq m^{-3}. Householders are encouraged to have radon measured in existing and new dwellings in affected areas and local authority environmental health departments are generally responsible for ensuring that radon in workplaces is monitored in such areas.

PHE and the BGS have produced a series of radon maps showing radon-affected areas, based on the underlying geology and indoor radon measurements. The maps are used (1) to assess whether radon protective (preventive) measures may be required in new buildings; (2) for the cost-effective targeting of radon monitoring in existing dwellings and workplaces to detect those buildings with radon above reference levels which need to be remediated; (3) to allow measurement campaigns and public awareness to be targeted on areas at greatest risk; and (4) to provide radon risk assessments for home buyers and sellers. In the UK, this is achieved by the online reports services developed by BGS and PHE: (1) for existing homes with a valid postcode, a Radon Risk Report can be ordered online from the UK Radon website (http://www.ukradon.org/); (2) for existing large homes and other large buildings and plots of land, a GeoReport can be obtained from BGS (http://shop.bgs.ac.uk/georeports/). The radon map data help local authorities to communicate quickly and effectively with existing home owners or developers planning to build new homes in high-radon areas

It is important, however, to realize that radon levels often vary widely between adjacent buildings owing to differences in the radon potential of the underlying ground as well as differences in construction style and use. Whereas a radon potential map can indicate the relative radon risk for a building in a particular locality, it cannot predict the radon risk for an individual building. Once identified as being in a high-risk area, homes and workplaces can have the radon levels measured and if found to be high, remedial work can be undertaken.

18.9.2 Environmental health regulations

There are environmental or safety thresholds of radioactivity such as *dose limit*, *AL* and *reference level*, which are used in legislation and advice. Governments set occupational *dose limits* in order to ensure that individuals are not exposed to an unacceptable degree of risk from artificial radiation. Historically, occupational exposure to radon decay products has been expressed in working level (WL) units. A WL is any combination of short-lived radon decay products (^{218}Po, ^{214}Pb, ^{214}Bi, and ^{214}Po) in 1 litre of air that will result in the emission of 1.3×10^5 MeV of potential alpha energy. Exposures may be measured in working level months (WLM). A WLM is the cumulative exposure equivalent to 1 WL for a working month (170 hours). In SI units, a WLM is defined as 3.54 mJ h m^{-3} (ICRP 1993). One WL is approximately equal to a radon exposure of 7500 Bq m^{-3} and 1 WLM to an average radon exposure of about 144 Bq m^{-3} yr (on the assumption that people spend most of their time indoors) (NRPB 2000).

The International Commission for Radiological Protection (ICRP) has recommended that all radiation exposures should be kept as low as reasonably achievable taking into account economic and social factors. In the UK, statutory regulations apply to any work carried out in an atmosphere containing ^{222}Rn gas at a concentration in air, averaged over any 24 h period, exceeding 400 Bq m^{-3}, except where the concentration of the short-lived daughters of ^{222}Rn in air averaged over any 8 h working period does not exceed 6.24.10^7 J m^{-3}. The limit on effective dose for any employee of 18 years of age or above is 20 mSv in any calendar year. This dose limit may be compared with the dose to the average person in the UK of 2.5 mSv, the dose of 7.5 mSv to the average person living in the high radon area of Cornwall, UK, and 0.4 mSv to the average nuclear industry worker in the UK (Watson *et al.* 2005).

A number of occupations have the potential for high exposure to ^{222}Rn progeny: mine workers, including uranium, hard rock and vanadium; workers remediating radioactive-contaminated sites, including uranium mill sites and mill tailings; workers at underground nuclear waste repositories; radon mitigation contractors and testers; employees and recreational visitors of natural caves (Gillmore *et al.* 2000; Gillmore *et al.* 2002); phosphate fertilizer plant workers; oil refinery workers; utility tunnel workers; subway tunnel workers; construction excavators; power plant workers, including geothermal power and coal; employees of radon 'health mines'; employees of radon balneotherapy spas (waterborne ^{222}Rn source); workers and visitors to hot spring hotels; water plant operators (waterborne ^{222}Rn source); fish hatchery attendants (waterborne ^{222}Rn source); employees who come into contact with technologically enhanced sources of naturally occurring radioactive materials; and incidental exposure in almost any occupation from local geological ^{222}Rn sources. Recreational and other visitors to abandoned metalliferous mines may also be exposed to high radon concentrations (Gillmore *et al.* 2001).

Environmental health responses include the provision of guidance for radon limitation including recommendations for dose limits and action levels, establishment of environmental health standards for houses and workplaces and enforcement of Ionizing Radiations Regulations to control exposure to radon in workplaces (Appleton *et al.* 2000*b*; WHO 2009). National authorities are recommended to set a reference level for radon that represents the maximum accepted radon concentration in a residential dwelling. Remedial actions may be recommended or required for homes in which radon exceeds the reference level. Various factors such as the distribution of radon concentrations, the number of existing homes with high radon concentrations, the arithmetic mean indoor radon level and the prevalence of smoking are usually considered when setting a national radon reference level. WHO (2009) proposed a reference level of 100 Bq m^{-3} to minimize health hazards owing to indoor radon exposure, although if this concentration cannot be reached, the reference level should not exceed 300 Bq m^{-3}. This represents *c.* 10 mSv per year according to recent calculations and recommendations by the International Commission on Radiation Protection (ICRP 2007, 2009), which also recommends 1000 Bq m^{-3} as the entry point for applying occupational radiological protection requirements. There are substantial variations in action levels (or their equivalents) in countries that recognize a radon problem. International and national recommendations for radon limitation in existing and future homes, given as the annual average of the gas concentration in Bq m^{-3}, range from 150 to 1000 for existing dwellings and from 150 to 250 for new dwellings (Åkerblom 1999). The majority of countries have adopted 400 and 200 Bq m^{-3}, respectively, for the two reference levels (HPA 2009).

The reasons for these different reference levels appear largely historical but are also due to a combination of environmental differences, different construction techniques and varying levels of political and environmental concern. There would be advantages in harmonizing standards because the existence of different levels may lead to confusion among the public.

In addition to variations in house radon standards between countries, there also appears to be some variation in standards applied within the field of radiological protection. However, international and national radiological protection authorities are united in acknowledging the need for a distinction in the ways radiation is approached in different situations, such as dwellings and nuclear installations.

The European Commission Recommendation (2001/928/EURATOM) on the protection of the public against exposure to radon in drinking water supplies recommends 1000 Bq/L as an action level for public and commercial water supplies above which remedial action is always justified on radiological protection grounds. Water supplies that support more than 50 people or distribute more than 10 m^3 per day, as well as all water that is used for food processing or commercial purposes, except mineral water, are covered by the European Commission Recommendation. The 1000 Bq/L action level also applies to drinking water distributed in hospitals, residential homes and schools and should be used for consideration of remedial action in private water supplies.

18.9.3 Radon and the building regulations: protecting new buildings

Provisions have been made in the building regulations to ensure that new dwellings are protected against radon where a significant risk of high radon concentrations in homes has been identified on the basis of house radon surveys.

The UK has regulations and guidelines for radon prevention in the planning stages of new development (e.g. where construction permits are applied for dwellings, offices, and factories). Cost-effectiveness analysis in the UK suggests that radon protective (preventive) measures, such as a sealed membrane under the ground floor in new homes, would be justified in all areas (HPA 2009).

Provision has been made in Requirement C1 (2) of Schedule 1 of the Building Regulations 2000 for the protection of buildings against the ingress of radon. The Approved Document C (Building Regulations 2010) refers to BRE Report, BR211 Radon: guidance on protective measures for new buildings (Scivyer 2007), for detailed guidance on where such protection is necessary and practical construction details. The guidance in the Approved Document C (Building Regulations 2010) applies to all new buildings including dwellings, extensions, conversions and refurbishment, whether they be for domestic or non-domestic use. BRE provides technical guidance on protective and remedial measures. Local Authority building control officers and Approved Inspectors enforce regulations and guidance for dealing with radon in new development through the Building Regulations.

In England and Wales, radon protective measures currently need to be installed in new dwellings, extensions to dwellings and buildings where there is a change of use to a residential or sleeping use where it is estimated that the radon concentration exceeds the AL in 3% or more of homes. In Scotland and Northern Ireland, radon protective measures have to be installed in all new dwellings where greater than 1% of dwellings are estimated to exceed the AL. Guidance is provided in BRE report BR376, *Radon: guidance on protective measures for new dwellings in Scotland*. Guidance for Northern Ireland is provided in BR413, *Radon guidance on protective measures for new dwellings*.

Supplementary guidance is contained in BRE Good Building Guide GG 73, *Radon protection for new domestic extensions and conservatories with solid concrete ground floors*, BRE Good Building Guide GG 74, *Radon protection for new dwellings* and BRE Good Building Guide GG 75, *Radon protection for new large buildings* (Scivyer 2015).

18.9.4 Radon and workplaces

Under the Health and Safety at Work etc. Act 1974 (HSW Act), employers must ensure the health and safety of employees and others who have access to that working environment. Protection from exposure to radon at work is specified in the Ionising Radiations Regulations 1999 (IRR 1999) made under the HSW Act. The concentration at which measures to protect employees should be taken in workplaces is 400 $Bq\ m^{-3}$ averaged over a 24 h period, as specified in the Health and Safety Executive Approved Code of Practice and guidance for IRR (1999) (*Working with ionising radiation*, L121).

Persons responsible for a workplace are required to assess the risks from radon in affected areas, and this usually requires a radon measurement. Enforcement of regulations is done through the Health and Safety Executive and Local Authority Environmental Health Departments. Further guidance is contained in Scivyer (2011).

As radon in affected areas contributes a higher dose to staff than work with ionizing radiations, it is important that all workplace affected rooms are located and remediated (Denman *et al.* 2002).

18.9.5 Radon and the planning system

Appleton *et al.* (2000*b*) recommended that the planning system should address the problem of radon in new development and that information on radon should be contained in development plans and in decision letters for individual planning applications. In 2004, ODPM issued Planning Policy Statement PPS23: Planning and Pollution Control, which complements the new pollution control framework under the Pollution Prevention and Control Act 1999 and the PPC Regulations 2000. Radon is mentioned in Annex 2 to PPS23 (Development on Land Affected by Contamination) on the basis that, since radon may pose a risk to human health, its presence is a material planning consideration. PPS23 indicates that Local Planning Authorities should include appropriate information on radon in the land condition and quality section of their Local Development Documents as well as in the determination of planning applications in relation to development on radon-affected land. This guidance applies only in England and no equivalent planning policy statement or technical guidance has yet been published for Northern Ireland, Scotland or Wales.

18.9.6 Remedial measures

Before considering how to reduce the radon level in a home, a reliable radon measurement must be taken. This will usually comprise a whole house average reading based upon a pair of detectors placed in the main living room and main bedroom. For most homes this average reading will be adequate for helping to select appropriate remedial measures. Larger houses, houses of unusual layout or construction and workplaces may need additional measurements to help target remedial measures. If the annual average radon level exceeds 200 $Bq\ m^{-3}$ in a home, householders or owners are advised to take action to reduce it. These measures should be designed to reduce the annual average as low as reasonably practicable, not just to get below the AL. Effective remedial measures to reduce domestic exposure to radon can be carried out to a typical house for around £1000.

In the case of high radon levels in workplaces, the law requires employers to avoid exposing their employees to excessive radiation doses, either by reducing radon levels or by other means. In the UK, owners of workplaces may be forced to carry out remedial measures whereas householders in dwellings with radon above the action level are generally only advised to take action to reduce the radon level. Guidance is provided in a number of BRE publications (BRE 1996; Pye 1993, Scivyer 1993; Scivyer & Jaggs 1998; Scivyer 1993, 2007, 2011, 2012, 2013*a*, *b*).

Guidance on reducing radon levels in dwellings is provided on the PHE Ukradon web site (http://www.ukradon.org/), but the cost of installing remedial measures in a dwelling is normally the householder's responsibility. Grant aid may be available

The principal ways of reducing the amount of radon entering a dwelling include; (1) active underfloor ventilation, i.e. drawing the air away from underneath the floor so that any air containing radon gas is dispersed outside the house; (2) subfloor depressurization (radon sump); (3) positive pressurization (i.e. pressurize the building in order to prevent the ingress of radon); and (4) ventilation (i.e. avoid drawing air through the floor by changing the way the dwelling is ventilated).

Radon gas may be easily removed from high-radon groundwaters by aeration and filter beds will remove daughter products. Various aeration technologies are available including static tank, cascade or forced aeration in a packed tower. Packed tower aeration is simple and cheap and is recommended for large drinking water supplies.

Monitoring and remediation in existing homes are not cost-effective at present in the UK but might become so in areas with mean indoor radon concentrations of 60 $Bq\ m^{-3}$ or above if the UK AL was reduced from 200 to 100 Bq m^{-3} (HPA 2009).

18.10 Scenarios for future events

Very high indoor radon concentrations (5000–17 000 Bq m^{-3}) have been measured in UK homes but it is possible that higher concentrations may be found in the future. Household energy efficiency drives that decrease ventilation (e.g. better sealed windows and doors) could lead to an increase in exposure to radon. Climate change and greater weather variability may influence annual indoor radon concentrations and also seasonal correction factors.

Acknowledgements This article is published with the permission of the Director of the British Geological Survey (Natural Environment Research Council). Two anonymous reviewers are thanked for suggesting improvements to an earlier version of this chapter.

Funding This research received no specific grant from any funding agency in the public, commercial, or not-for-profit sectors.

References

ÅKERBLOM, G. 1999. *Radon Legislation and National Guidelines*, **99:18**. Swedish Radiation Protection Institute.

APPLETON, J.D. 2004. Influence of faults on geological radon potential in England and Wales. *In*: BARNET, I., NEZNAL, M. & PACHEROVÁ, P. (eds) *Radon Investigations in the Czech Republic X and the Seventh International Workshop on the Geological Aspects of Radon Risk Mapping Czech*. Geological Survey & RADON corp, Prague.

APPLETON, J.D. 2013. Radon in air and water. *In*: SELINUS, O., ALLOWAY, B.J. *ET AL*. (eds) *Essentials of Medical Geology*. Springer, Dordrecht, 239–277.

APPLETON, J.D. & BALL, T.K. 1995. *Radon and background radioactivity from natural sources: characteristics, extent and the relevance to planning and development in Great Britain*. British Geological Survey Technical Report, **WP/95/2**.

APPLETON, J.D. & BALL, T.K. 2001. Geological Radon Potential Mapping. *In*: BOBROWSKY, P.T. (ed.) *Geoenvironmental Mapping: Methods, Theory and Practice*. Balkema, Rotterdam, 577–613.

APPLETON, J.D. & MILES, J.C.H. 2002. Mapping radon-prone areas using integrated geological and grid square approaches. *In*: BARNET, I., NEZNAL, M. & MIKSOVÁ, J. (eds) *Radon Investigations in the Czech Republic IX and the Sixth International Workshop on the Geological Aspects of Radon Risk Mapping Czech*. Geological Survey, Prague, 34–43.

APPLETON, J.D. & MILES, J.C.H. 2005. Radon in Wales. *In*: NICOL, D. & BASSETT, M.G. (eds) *Urban Geology of Wales*. National Museum of Wales, Cardiff, 117–130.

APPLETON, J.D. & MILES, J.C.H. 2010. A statistical evaluation of the geogenic controls on indoor radon concentrations and radon risk. *Journal of Environmental Radioactivity*, **101**, 799–803, https://doi.org/10.1016/j.jenvrad.2009.06.002

APPLETON, J.D., MILES, J.C.H. & TALBOT, D.K. 2000*a*. *Dealing with Radon Emissions in Respect of New Development: Evaluation of Mapping and Site Investigation Methods for Targeting Areas Where New Development May Require Radon Protective Measures*, British Geological Survey Research Report, **RR/00/12**.

APPLETON, J.D., MILES, J.C.H., SCIVYER, C.R. & SMITH, P.H. 2000*b*. *Dealing with Radon Emissions in Respect of New Development: Summary Report and Recommended Framework for Planning Guidance*, British Geological Survey Research Report, **RR/00/07**.

APPLETON, J.D., MILES, J.C.H., GREEN, B.M.R. & LARMOUR, R. 2008. Pilot study of the application of Tellus airborne radiometric and soil geochemical data for radon mapping. *Journal of Environmental Radioactivity*, **99**, 1687–1697, https://doi.org/10.1016/j.jenvrad.2008.03.011

APPLETON, J.D., CAVE, M.R., MILES, J.C.H. & SUMERLING, T.J. 2011*a*. Soil radium, soil-gas radon and indoor radon empirical relationships to assist in post-closure impact assessment related to near-surface radioactive waste disposal. *Journal of Environmental Radioactivity*, **102**, 221–234, https://doi.org/10.1016/j.jenvrad.2010.09.007

APPLETON, J.D., MILES, J.C.H. & YOUNG, M. 2011*b*. Comparison of Northern Ireland radon maps based on indoor radon measurements and geology with maps derived by predictive modelling of airborne radiometric and ground permeability data. *Science of The Total Environment*, **409**, 1572–1583 https://doi.org/10.1016/j.scitotenv.2011.01.023

APPLETON, J.D., DARAKTCHIEVA, Z. & YOUNG, M.E. 2015. Geological controls on radon potential in Northern Ireland. *Proceedings of the Geologists' Association*, **126**, 328–345, https://doi.org/10.1016/j.pgeola.2014.07.001

BALL, T.K. & MILES, J.C.H. 1993. Geological and geochemical factors affecting the radon concentration in homes in Cornwall and Devon, UK. *Environmental Geochemistry and Health*, **15**, 27–36, https://doi.org/10.1007/BF00146290

BALL, T.K., CAMERON, D.G., COLMAN, T.B. & ROBERTS, P.D. 1991. Behaviour of radon in the geological environment – a review. *Quarterly Journal of Engineering Geology*, **24**, 169–182, https://doi.org/10.1144/GSL.QJEG.1991.024.02.01

BALL, T.K., CAMERON, D.G. & COLMAN, T.B. 1992. Aspects of radon potential mapping in Britain. *Radiation Protection Dosimetry*, **45**, 211–214, https://doi.org/10.1093/rpd/45.1-4.211

BARNET, I., PACHEROVA, P., NEZNAL, M. & NEZNAL, M. 2008. *Radon in Geological Environment – Czech Experience*. Czech Geological Survey, Prague, Czech Republic, Special Papers, **70**.

BRE 1996. *Minimising noise from Domestic Fan Systems, and Fan-assisted Radon Mitigation Systems*. BRE, Good Building Guide 26, GBG26, GG26.

BUILDING REGULATIONS 2010. Approved Document C – Site preparation and resistance to contaminates and moisture (2004 Edition incorporating 2010 and 2013 amendments), https://assets.publishing.service.gov.uk/government/uploads/system/uploads/attachment_data/file/431943/BR_PDF_AD_C_2013.pdf

DENMAN, A.R., LEWIS, G.T.R. & BRENNEN, S.E. 2002. A study of radon levels in NHS premises in affected areas around the UK. *Journal of Environmental Radioactivity*, **63**, 221–230, https://doi.org/10.1016/S0265-931X(02)00029-2

GILLMORE, G.K., SPERRIN, M., PHILLIPS, P. & DENMAN, A. 2000. Radon hazards, geology, and exposure of cave users: a case study and some theoretical perspectives. *Ecotoxicology and Environmental Safety*, **46**, 279–288, https://doi.org/10.1006/eesa.2000.1922

GILLMORE, G.K., PHILLIPS, P., DENMAN, A., SPERRIN, M. & PEARCE, G. 2001. Radon levels in abandoned metalliferous mines, Devon, southwest England. *Ecotoxicology Environmental Safety*, **49**, 281–292, https://doi.org/10.1006/eesa.2001.2062

GILLMORE, G.K., PHILLIPS, P.S., DENMAN, A.R. & GILBERTSON, D.D. 2002. Radon in the Creswell Crags Permian limestone caves. *Journal of Environmental Radioactivity*, **62**, 165–179, https://doi.org/10.1016/S0265-931X(01)00159-X

GRAY, A., READ, S., McGALE, P. & DARBY, S. 2009. Lung cancer deaths from indoor radon and the cost effectiveness and potential of policies to reduce them. *British Medical Journal*, **338**, a3110, https://doi.org/10.1136/bmj.a3110

GREEMAN, D.J. & ROSE, A.W. 1996. Factors controlling the emanation of radon and thoron in soils of the eastern U.S.A. *Chemical Geology*, **129**, 1–14, https://doi.org/10.1016/0009-2541(95)00128-X

GREEMAN, D.J., ROSE, A.W. & JESTER, W.A. 1990. Form and behavior of radium, uranium, and thorium in central Pennsylvania soils derived from Dolomite. *Geophysics Research Letters*, **17**, 833–836, https://doi.org/10.1029/GL017i006p00833

GROVES-KIRKBY, C.J., DENMAN, A.R., PHILLIPS, P.S., TORNBERG, R., WOOLRIDGE, A.C. &CROCKETT, R.G.M. 2008. Domestic radon remediation of U.K. dwellings by sub-slab depressurisation: evidence for a baseline contribution from constructional materials. *Environment International*, **34**, 428–436, https://doi.org/10.1016/j.envint.2007.09.012

GUNBY, J.A., DARBY, S.C., MILES, J.C.H., GREEN, B.M.R. & COX, D.R. 1993. Factors affecting indoor radon concentrations in the United-Kingdom. *Health Physics*, **64**, 2–12, https://doi.org/10.1097/00004032-199301000-00001

GUNDERSEN, L.C.S., SCHUMANN, E.R., OTTON, J.K., DUBIEF, R.F., OWEN, D.E. & DICKENSON, K.E. 1992. Geology of radon in the United States. *In*: GATES, A.E. & GUNDERSEN, L.C.S. (eds) *Geologic Controls on Radon*. Geological Society of America, Special papers, **271**, Boulder, CO, 1–16.

HAINS, B.A. & HORTON, A. 1969. *British Regional Geology: Central England*. 3rd edn. BRG10. British Geological Survey.

HODGKINSON, E.S., JONES, D.G., EMERY, C. & DAVIS, J.R. 2005. *Petrography and geochemistry of two East Midlands Jurassic Ironstones and their relationship to radon potential*. British Geological Survey Research Report **RR/05/041**.

HPA 2009. *Radon and public health: Report prepared by the Subgroup on Radon Epidemiology of the Independent Advisory Group on Ionising Radiation, documents of the Health Protection Agency*. Health Protection Agency, UK.

HUNTER, N., MUIRHEAD, C.R., MILES, J.C.H. & APPLETON, J.D. 2009. Uncertainties in radon related to house-specific factors and proximity to geological boundaries in England. *Radiation Protection Dosimetry*, **136**, 17–22, https://doi.org/10.1093/rpd/ncp148

HUNTER, N., MUIRHEAD, C.R. & MILES, J.C.H. 2011. Two error components model for measurement error: application to radon in homes. *Journal of Environmental Radioactivity*, **102**, 799–805, https://doi.org/10.1016/j.jenvrad.2011.05.009

IAEA 2003. *Guidelines for Radioelement Mapping Using Gamma Ray Spectrometry Data*. IAEA-TECDOC-1363. International Atomic Energy Agency, Vienna.

ICRP (INTERNATIONAL COMMITTEE ON RADIOLOGICAL PROTECTION) 1993. *Protection against Radon-222 at Home and at Work*, Ann. ICRP, 23(2): 1–65, ICRP Publication, **65**.

ICRP 2007. *The 2007 recommendations of the International Commission on Radiological Protection*, Annals ICRP. ICRP, Oxford.

ICRP 2009. *International Commission on Radiological Protection Statement on Radon*. ICRP, **37**, 2–4.

IELSCH, G., CUSHING, M.E., COMBES, P. & CUNEY, M. 2010. Mapping of the geogenic radon potential in France to improve radon risk management: methodology and first application to region Bourgogne. *Journal of Environmental Radioactivity*, **101**, 813–820, https://doi.org/10.1016/j.jenvrad.2010.04.006

KEMSKI, J., SIEHL, A., STEGEMANN, R. & VALDIVIA-MANCHEGO, M. 2001. Mapping the geogenic radon potential in Germany. *Science of the Total Environment*, **272**, 217–230, https://doi.org/10.1016/S0048-9697(01)00696-9

KEMSKI, J., KLINGEL, R., SIEHL, A. & VALDIVIA-MANCHEGO, M. 2009. From radon hazard to risk prediction-based on geological maps, soil gas and indoor measurements in Germany. *Environmental Geology*, **56**, 1269–1279, https://doi.org/10.1007/s00254-008-1226-z

KENDALL, G., LITTLE, M.P. & WAKEFORD, R. 2011. Numbers and proportions of leukemias in young people and adults induced by radiation of natural origin. *Leukemia Research*, **35**, 1039–1043, https://doi.org/10.1016/j.leukres.2011.01.023

KILLIP, I.R. 2004. Radon hazard and risk in Sussex, England and the factors affecting radon levels in dwellings in chalk terrain. *Radiation Protection Dosimetry*, **113**, 99–107, https://doi.org/10.1093/rpd/nch436

LAURIER, D., VALENTY, M. & TIRMARCHE, M. 2001. Radon exposure and the risk of leukemia: a review of epidemiological studies. *Health Physics*, **81**, 272–288, https://doi.org/10.1097/00004032-200109000-00009

MILES, J.C.H. 1998. Mapping radon-prone areas by log-normal modelling of house radon data. *Health Physics*, **74**, 370–378, https://doi.org/10.1097/00004032-199803000-00010

MILES, J.C.H. & APPLETON, J.D. 2000. *Identification of localised areas of England where radon concentrations are most likely to have 5% probability of being above the Action Level*, Department of the Environment, Transport and the Regions Report, **DETR/RAS/00.001**.

MILES, J.C.H. & APPLETON, J.D. 2005. Mapping variation in radon potential both between and within geological units. *Journal of Radiological Protection*, **25**, 257–276, https://doi.org/10.1088/0952-4746/25/3/003

MILES, J.C., HOWARTH, C.B. & HUNTER, N. 2012. Seasonal variation of radon concentrations in UK homes. *Journal of Radiological Protection*, **32**, 275–287, https://doi.org/10.1088/0952-4746/32/3/275

MILES, J.C.H., APPLETON, J.D., REES, D.M., GREEN, B.M.R., ADLAM, K.A.M. & MYERS, A.H. 2007. *Indicative Atlas of Radon in England and Wales*. HPA-CRCE-023. HPA, Chilton.

MILES, J.C.H., APPLETON, J.D. *ET AL.* 2011. *Indicative Atlas of Radon in Scotland*. HPA-CRCE-023. HPA, Chilton.

NRPB 2000. *Health Risks from Radon*. National Radiological Protection Board, UK.

PYE, P.W. 1993. *Sealing Cracks in Solid Floors: A BRE Guide to Radon Remedial Measures in Existing Dwellings*. BRE Report 239, **BR239**.

REES, D.M., BRADLEY, E.J. & GREEN, B.M.R. 2011. *HPA-CRCE-015 – Radon in Homes in England and Wales: 2010 Data Review*. HPA, Chilton.

ROSE, A.W., HUTTER, A.R. & WASHINGTON, J.W. 1990. Sampling variability of radon in soil gases. *Journal of Geochemical Prospecting*, **38**, 173–191.

SCHEIB, C., APPLETON, J.D., JONES, D.J. & HODGKINSON, E. 2006. Airborne uranium data in support of radon potential mapping in Derbyshire, Central England. *In*: BARNET, I., NEZNAL, M. & PACHEROVÁ, P. (eds) *Radon Investigations in the Czech Republic XI and the Eighth International Workshop on the Geological*

Aspects of Radon Risk Mapping. Czech Geological Survey, Prague, 210–219.

SCHEIB, C., APPLETON, J.D., MILES, J.C.H., GREEN, B.M.R., BARLOW, T.S. & JONES, D.G. 2009. Geological controls on radon potential in Scotland. *Scottish Journal of Geology*, **45**, 147–160, https://doi.org/10.1144/0036-9276/01-401

SCHEIB, C., APPLETON, J.D., MILES, J.C.H. & HODGKINSON, E. 2013. Geological controls on radon potential in England. *Proceedings of the Geologists Association (in press)*.

SCIVYER, C. 2007. *Radon: Guidance on Protective Measures for New Buildings*. BRE Report 211, **BR211**, BRE, Watford.

SCIVYER, C. 2011. *Radon in the Workplace*. BRE Report **FB41**, BRE, Watford.

SCIVYER, C. 2012. *Radon Solutions in Homes: Improving Underfloor Ventilation*. BRE Good Repair Guide **GRG 37/1**, BRE, Watford.

SCIVYER, C. 2013a. *Radon Solutions in Homes: Positive Ventilation*. BRE Good Repair Guide **GRG 37/2**, BRE, Watford.

SCIVYER, C. 2013b. *Radon Solutions in Homes: Sump Systems*. BRE Good Repair Guide **GRG 37/3**, BRE, Watford.

SCIVYER, C.R. 1993. *Surveying Dwellings with High Indoor Radon Levels: A BRE Guide to Radon Remedial Measures in Existing Dwellings*. BRE Report 250, **BR250**, BRE, Watford.

SCIVYER, C 2015. Radon protection for new domestic extensions, dwellings and large buildings. *Good Building Guides 73 74 & 75*. BRE Report **AP 307**, BRE, Watford.

SCIVYER, C.R. & JAGGS, M.P.R. 1998. *Dwellings with Cellars and Basements: a BRE Guide to Radon Remedial Measures in Existing Dwellings*. BRE Report 343, **BR343**, BRE, Watford.

SUTHERLAND, D.S. 1991. Radon in Northamptonshire, England: geochemical investigation of some Jurassic sedimentary rocks. *Environmental Geochemistry and Health*, **13**, 143–145, https://doi.org/10.1007/BF01758547

TALBOT, D.K., HODKIN, D.L. & BALL, T.K. 1997. Radon investigations for tunnelling projects; a case study from St. *Helier, Jersey. Quarterly Journal of Engineering Geology*, **30**, 115–122, https://doi.org/10.1144/GSL.QJEGH.1997.030.P2.02

TALBOT, D.K., DAVIS, J.R. & RAINEY, M.P. 2000. *Natural Radioactivity in Private Water Supplies in Devon*. DETR/RAS/00.010, Department of Energy, Transport & The Regions, London.

UNSCEAR 2009. *United Nations Scientific Committee on the Effects of Atomic Radiation (UNSCEAR)*. UNSCEAR 2006 Report. Annex E. Sources-to-Effects Assessment for Radon in Homes and Workplaces. United Nations, New York.

VARLEY, N.R. & FLOWERS, A.G. 1993. Radon in soil-gas and its relationship with some major faults of Sw England. *Environmental Geochemistry and Health*, **15**, 145–151, https://doi.org/10.1007/BF02627832

WASHINGTON, J.W. & ROSE, A.W. 1992. Temporal variability of radon concentrations in the interstitial gas of soils in Pennsylvania. *Journal of Geophysical Research*, **97**, 9145–9159, https://doi.org/10.1029/92JB00479

WATSON, S.J., JONES, A.l., OATWAY, W.B. & HUGHES, J.S. 2005. *HPA-RPD-001 – Ionising Radiation Exposure of the UK Population: 2005 Review*. HPA, Chilton.

WHO 2009. *WHO Handbook on Indoor Radon: A Public Health Perspective*. World Health Organization, Geneva.

Chapter 19   Methane gas hazard

Steve Wilson* & Sarah Mortimer

The Environmental Protection Group Limited, Warrington Business Centre, Long Lane, Warrington WA2 8TX, UK

Correspondence: stevewilson@epg-ltd.co.uk

Abstract: This paper identifies potential sources, and the key chemical properties, of methane. Guidance is provided on deriving a conceptual site model for methane, utilizing various lines of evidence to inform a robust, scientific, reasoned and logical assessment of associated gas risk. Discussion is provided regarding the legislative context of permanent gas risk assessment for methane, including via qualitative, semi-quantitative and detailed quantitative (including finite element modelling) techniques. Strategies for mitigating risks associated with methane are also outlined, together with the legal context for consideration of methane both in relation to the planning regime and under Part 2A of the Environmental Protection Act 1990.

19.1 The source and chemical properties of methane

Methane (historically known as 'marsh gas') was discovered in 1776 by Alessandro Volta, who collected gas bubbles from disturbed sediments on Lake Maggiore. Methane is the most abundant organic compound in the Earth's atmosphere. Its occurrences in the Earth's crust are predominantly of biogenic origin (i.e. it is formed by bacterial decomposition of organic matter). Methane can also be formed by decomposition of organic matter as a result of geothermal heat and/or pressure, when it is known as thermogenic gas. Such gas is generated at great depth but even so the methane can migrate to the surface along faults or other features and accumulate in near-surface rocks. Abiogenic methane is thought to be formed by chemical reactions, for example during the cooling of magma or serpentinization of ultramafic rocks. Methane has been detected in many shallow drift deposits in soils in the UK (London Clay, River Terrace Deposits) where there is no apparent external source, such as landfill. It is thought that the methane occurs because of disturbance caused by installing monitoring wells and there is oxidation of small volumes of organic material in the soils to produce carbon dioxide that is subsequently reduced by methanogens. This low-level source of methane is not known to pose a hazard to developments.

Methane is a colourless, tasteless and odourless gas that is widespread within the subsurface environment. Its chemical symbol is CH_4 and it has a molecular weight of 16.04 g/mol. Methane is a gas at room temperature and standard pressure. It melts at −183°C and boils at −162°C. It is less dense than air (density, 0.716 kg m^{-3} at 0°C). It has a viscosity of 1.03×10^{-5} N s m^{-2} and a diffusion coefficient in air of 1.5×10^{-5} m^2 s^{-1} at standard temperature and pressure.

Methane is a tetrahedral molecule with four equivalent C–H bonds and is stable at standard temperature and pressure. It is flammable and explosive in air at certain concentrations and is also violently reactive with oxidizers and halogens. Methane is also a potent 'greenhouse gas'.

Methane is ubiquitous in the subsurface environment and is present in soils and rocks below many parts of the UK and other countries. With a few notable exceptions it is generally immobile and does not cause any particular problems unless a pathway is available that 'links' the gas source to the receptor, such as where enclosed spaces (e.g. basements and buildings) are constructed on top of the methane-bearing strata, there are excavations (e.g. mines and tunnels) through the methane bearing strata or methane-bearing water is pumped from depth such that changes in confining pressures occur to permit release of methane from solution. Potential sources of methane may typically be divided into two groups: anthropogenic (caused by humans) and natural. The principal potential sources of methane in the UK are summarized in Table 19.1.

Methane is produced (and consumed) by microbial processes, and in the oceans may be present as methane hydrates (crystals). In deeper sections of the Earth's crust, methane is a product of the conversion of organic matter under the influence of elevated temperatures and pressures. Deeper still methane is found in metamorphic rocks. Methane is also found emanating from geothermal waters on continents with hot water vents at oceanic spreading centres. However, in volcanic and geothermal environments carbon dioxide and other gases are typically more predominant than methane.

Engineering Group Working Party (main contact for this chapter: S. Wilson, The Environmental Protection Group Limited, Warrington Business Centre, Long Lane, Warrington WA2 8TX, UK, stevewilson@epg-ltd.co.uk)
From: GILES, D. P. & GRIFFITHS, J. S. (eds) 2020. *Geological Hazards in the UK: Their Occurrence, Monitoring and Mitigation – Engineering Group Working Party Report.* Geological Society, London, Engineering Geology Special Publications, **29**, 457–478,
https://doi.org/10.1144/EGSP29.19
© 2020 The Environmental Protection Group Limited. Published by The Geological Society of London. All rights reserved.
For permissions: http://www.geolsoc.org.uk/permissions. Publishing disclaimer: www.geolsoc.org.uk/pub_ethics

Table 19.1. *Principal Sources and Origins of Methane in the UK (adapted from Hooker & Bannon 1993)*

Source	Typical methane concentration (%v/v)	Derivation
Anthropogenic sources		
Landfill sites	20–65%	Microbial decay of organic materials derived from the disposal of putrescible materials. Wide variations in concentration can occur over the life of a landfill site. Methane concentrations around 65% are typical during the maximum gas generation phase but will reduce as gas generation reduces and oxygen diffuses back into the landfill. It also depends on the nature of the waste and can be affected by chemical reactions and solution of gases in groundwater or leachate (loss of carbon dioxide into water leads to enrichment of methane concentrations).
Made ground	0–20%	Microbial decay or organic materials contained in reworked natural ground containing demolition and other wastes.
Foundry sands	Up to 50%	Microbial decay of waste materials from the foundry process (phenolic binders, dextrin, coal dust, wood, rags and paper).
Sewage sludge, dung, cess pits/heaps	60–75%	Microbial decay of organic material.
Burial grounds (including cemeteries)	20–65%	Microbial decay or organic materials contained with human/animal remains.
Industrial/chemical/petroleum sites/ manufacturing	30–100%[a]	Degradation of organic chemicals derived from leaks or spills from storage, processing and disposal areas
Natural gas (supply pipes)	90–95%	Leakage from bulk pipeline transportation of natural gas.
Natural sources		
Soil	<2ppm	Physical, chemical and biological transformation of rock during weathering.
Oil and gas bearing strata	70–90%	Burial of vegetation under high temperatures and pressures leading to reservoirs of natural gas
Abiogenic gases	70–80%	Abiogenic methane formed by various processes including the reduction of carbon dioxide during magma cooling, water–rock interactions in hydrothermal systems and serpentinization of ultramafic rocks (Sherwood Lollar *et al.* 2002)
Coal measures strata	<1–99%	Burial of vegetation under high temperatures and pressures, liberating gases as a by-product of mining activities.
Peat/bog areas	10–90%	Gas formed by the microbial decay of accumulated plant debris under anaerobic conditions.
Alluvium (organic rich sediments)	0–5%	

Notes: Using infra-red bulk gas monitor reading as methane.

Methane is often present at elevated concentrations in uncontrolled and engineered fill materials in the unsaturated zone, especially where the soil is wet and an anaerobic zone exists below the groundwater (Eklund 2011). The probability of detecting methane tends to increase with increasing depth and even clean (inert) fill can generate methane. Although the volumes are low and do not pose a risk of surface emissions the generation can cause high concentrations in monitoring wells. A concentration of 30% methane was proposed as a rule of thumb to identify when gas production may pose a risk of significant emissions.

A study in the San Diego area determined that high concentrations of methane are normal in engineered fill containing silts or clays, and that the methane did not pose a hazard to structures (California Environmental Protection Agency 2012). The study involved methane gas testing on hundreds of mass grading projects located in San Diego County, over a two-year period (Sepich 2008). These studies found that the soils in the cut areas did not contain methane, but the material from the cuttings that was placed as engineered fill contained high concentrations of gases consistently exhibiting microbial characteristics (up to 40% methane). Soil containing organic materials as low as 0.4% by weight could produce elevated levels of methane. The risk associated with the methane in the fill was much lower than that associated with landfill sites or leaking oil wells because the methane is not under pressure and is in small volumes.

The greatest hazard posed by methane is that it is flammable and/or explosive. Therefore if an explosive mix occurs in a building, tunnel or mine, for example, there is a risk of explosion. When methane enters an enclosed space it can displace oxygen and cause asphyxiation. It can also displace air in soil pores and adversely affect plants by starving the roots of oxygen.

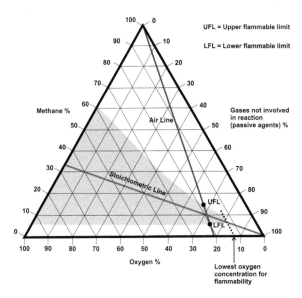

Fig. 19.1. Explosive range of methane.

Generally, the lower explosive limit (LEL) of methane is 5% by volume in air and the upper explosive limit is 15% by volume in air. The explosive limits of methane will change as the oxygen concentration reduces (**Fig. 19.1**). When carbon dioxide reaches 25% concentration, or nitrogen reduces to 36%, methane is not flammable (Kuchta 1985).

The stoichiometric line on Figure 19.1 is the composition at which the amounts of fuel and oxygen are in balance. This means that there is no excess fuel or oxygen left after the chemical reaction is complete. The grey-shaded area is the explosive range.

It is important to note that concentrations of methane above the upper explosive limit should not be taken to represent a state of safety, because dilution and dispersion of the gas in the air will occur by natural processes so that at some point in time the actual composition and concentration of a gaseous state will pass through the flammable range defined by the upper and lower explosive limits. Soil acts as a natural flame arrestor so methane in a typical soil matrix cannot explode (Eklund 2011). Therefore there is no LEL for methane in soil gas (unless methane is present in a large void in the soil).

Where carbon dioxide is also present, the range of methane concentrations over which it is flammable is reduced. There can be no guarantee that a gas mixture will remain constant as it will be subject to dilution and dispersion. Thus it is always prudent practice to adopt the LEL for methane of 5% by volume in air inside buildings or other structures as an upper threshold criterion for risk assessment. Above a methane concentration of 5% by volume in air it must be assumed that a risk of flammability can occur; all that is required is a point of ignition (e.g. a naked flame or electric spark). A flammable mixture of methane and air will burn rapidly when ignited. If it is confined the deflagration generates high overpressures so that the effects are explosive (Beresford 1989). It is highly conservative to apply the LEL as a limit to soil gas concentrations. Elevated concentrations of gas in the ground above these limits do not necessarily pose a risk. It is the rate at which gas is emitted from the surface that dictates the level of risk.

Sepich (2008) suggested that, even at 100% methane soil concentration, when flow is by diffusion it will not exceed 1% inside a building using the common attenuation factor of 100. Monitoring of unprotected buildings showed that the attenuation factor achieved in practice was actually 10 000 (i.e. the measured concentrations in the buildings were 10–100 times lower than predicted by modelling). Thus, there is no unsafe concentration of methane in the ground.

The size and intensity of an explosion can be large and are a function of the following:

(1) the properties of the confined space itself – a relatively rigid vessel or container (such as a building) will result in an intense explosion whilst a flexible or loose container (for example a balloon) can potentially dissipate the energy of an explosion by physically changing shape;
(2) the volume of methane between the lower and upper explosive limit range within the confined space.

There have been many instances around the world of fatal explosions associated with methane accumulation, typically associated with mining, oil and gas and landfill activities or leaks in gas supply networks. Pumping water with dissolved methane in it can also cause explosions if the methane accumulates in the pump house or chamber. Subsequent sections of this paper will provide more detailed information regarding some of these cases. However, some of the instances in the UK and USA where injury or death has been caused by explosions owing to methane gas from landfill sites are listed in Table 19.2.

Notably there are no recorded instances of explosions from sources of methane such as alluvium, general made ground (excluding domestic landfill) or coal measures where there are no workings present. The reason for this is that for a hazard to exist there must be enough volume or flow of methane in the available soil gas to result in gas concentrations greater than the LEL inside the structure. Only pressurized methane/soil gas can achieve explosive concentrations in an overlying building. The ASTM (2016) states that there are no known examples of methane incidents from diffusive transport alone. There is unlikely to be sustained gas pressure in alluvium, general made ground (excluding domestic landfill) or coal measures where workings are not present.

In addition to the explosive/flammable hazard posed by methane it is also important to note that at concentrations in excess of 33% v/v it may also act as an asphyxiant, by displacing oxygen. Typically physiological effects are observed when oxygen concentrations fall below 18% v/v. This begins with impairment of judgement (17% v/v), followed by anoxia and abnormal fatigue (10–16% v/v), nausea and

Table 19.2. Recorded incidents of injury from explosions caused by methane from landfill sites

Date	Location	Incident	Consequence	Geology and/or hydrogeology
Source: HSE TD5/030 Review of Landfill Gas Incidents 1970–1995				
1995	Bramfield, Hertfordshire, UK	Landfill gas ignited during installation of electro-fusion welded plastic joint.	One injured	Chalk[B = Bedrock]
1990	Ebbw Vale, Gwent, UK	Drilling in shaft used for venting on a landfill site. Spark from unprotected drill ignited gas.	Three injured	Mudstone, siltstone and sandstone[B]
1988	Appley Bridge, Lancs., UK	Partial blockage of passive venting trench was thought to have allowed excessive off-site migration of LFG. An accumulation of methane in the site offices some 50 m off site was ignited and exploded.	Structural damage; no staff in building at time of explosion.	Coal measures[B]
1987	Stone, Dartford, UK	Temporary throttling back of gas abstraction during Christmas vacation was thought to have allowed excessive off-site migration of LFG. An accumulation of methane, forming an explosive mixture within a house less than 50 m from the site boundary was ignited.	Occupant suffered mild burns and temporary deafness. Building was structurally damaged.	Chalk[B]
1986	Loscoe, Derbyshire, UK	Coinciding with a rapid drop in barometric pressure, methane migrating from an adjacent landfill site was drawn into a bungalow less than 20 m from the site boundary. An explosive mixture was ignited when the heating system was turned on.	Total destruction of the building, three occupants injured, structural damage to neighbouring properties.	Solid geology: Middle Coal Measures – sedimentary rocks comprising coals, mudstones, siltstones and sandstones. Superficial: topsoil over head deposits (silty clay).
1986	Northern Ireland	Explosion in factory.	One killed	–
1983	Nuttall, UK	Landfill gas explosion in manhole.	–	Mudstone, siltstone, sandstone[B]
1982	Morley, Leeds, UK	Explosive concentration of methane was ignited in a sewer beneath a landfill.	Two children injured	Sandstone[B]
1982	London, UK	Explosion in cable manhole.	Two injured	–
1980	Ormskirk, UK	Explosion in manhole.	Two injured	Sandstone[B]
Pre 1981	Unknown	Gas accumulating in buildings built on completed landfill caused explosion.	One injured	–
1979	Christchurch, New Zealand	Explosion in toilet.	One injured	–
1972	Airdrie, Lanarkshire, UK	Ignition in manhole.	One injured	–
Source: Air Emissions from Municipal Solid Waste Landfills – Background Information for Proposed Standards and Guidelines (USEPA 1991)				
1987	Pittsburgh, Pennsylvania	Suspected landfill gas explosion.	Structural damage	–
1984	Akron, Ohio	Landfill gas migrated to and destroyed a house.	Structural damage	–

Year	Location	Description	Casualties	Notes
1984	Comack, NJ	Gas migrated to on-site landfill weigh-station and caused explosion.	One killed; one injured	–
1984	Comack, NJ	Methane migrated to house on-site and exploded.	Structural damage	–
1983	Manchester, NJ	Spark from landfill pump probably ignited gas.	One injury	–
1983	Cincinnati, Ohio	Explosion destroyed residence close to landfill.	One house destroyed; minor injuries	–
1980	Cleveland, Ohio	Explosion at foundry adjacent to landfill.	One killed	–
1977	Commerce City, CO.	Explosion in tunnel being built under a railroad right-of-way.	Two killed; four injured	–
1975	Sheridan, Colorado	Gas migrated into drainage pipe under construction. Welding truck led to fire.	Two injured	–
1975	Sheridan, Colorado	Landfill gas accumulated in a storm drain pipe that ran through a landfill. Explosion occurred when children playing in the pipe lit a candle.	Several children seriously injured	–
1975	Richmond, VA.	Gas migrated from nearby landfill into apartment.	Two injured	–
1967	Atlanta, Georgia, USA	Methane from a nearby landfill entered the basement of a building and caused an explosion when someone lit a cigarette.	Two killed; six injured	–
1969	Winston-Salem, North Carolina	Landfill gas migrated from adjacent landfill into the basement of an armoury. A lit cigarette caused gas explosion.	Three killed; five seriously injured	–
–	Madison, WI.	Explosion destroyed sidewall of a townhouse.	Two seriously injured	–

Source: *Ground Gas handbook* (Wilson et al. 2009)

Year	Location	Description	Casualties	Notes
2000	Rochester, Michigan	Home exploded when landfill gas was ignited by a pilot light	–	–
1999	Atlanta, Georgia	Landfill gas exploded in a playground	Child burnt on arms and legs	–
1994	Charlotte, North Carolina	Landfill gas exploded in a park built over an old landfill	One person seriously burned	–
Circa 1987	–	House 50 m from waste-filled quarry damaged by gas explosion. Gas migrated along natural fissures in the limestone.	Structural damage	Limestone
1965	California	Boy playing in a cave dug in his backyard burned by flash fire when he attempted to light a candle. Gas attributed to the nearby Palos Verdes landfill.	One injury	–
1947	–	Fire in sewer manhole caused by landfill gas migrating from the overlying refuse.	Structural damage	–

unconsciousness (6–10%) and death (<6% v/v) (Card 1995). Displacement of oxygen at the root ball can also result in phytotoxic effects to plants and vegetation.

Methane is a potent 'greenhouse gas'. Infra-red radiation from the sun is absorbed and emitted by gases in the atmosphere, which results in the Earth's lower atmosphere and surface being warm. The effect was first suggested by Joseph Fourier in 1824 and it was first investigated by Svante Arrhenius in 1896. The 'greenhouse' effect is vital to life on Earth. However, as the amount of 'greenhouse gases' has increased, more of the radiation is trapped and cannot escape, resulting in rising global temperatures which affect many aspects of the environment. Effects of global warming include melting polar ice caps and the resulting rising sea-levels, and changing weather patterns. Extreme weather events are now occurring more frequently.

Methane has a high global warming potential, in excess of 20 times that of carbon dioxide averaged over 100 years. Historical methane concentrations in the atmosphere (estimated from ice cores) have varied from around 400 ppb during glacial periods to around 700 ppb during interglacial periods. More recently, in 2005, the atmospheric methane concentration was around 1770 ppb.

The majority of methane is removed from the atmosphere by reaction with the hydroxyl free radical (OH) which is present as a result of photochemical reactions between oxygen and water vapour in the atmosphere. The methane is converted to water and carbon dioxide. With a half-life of 7 years, methane lasts around 10 years in the atmosphere.

19.2 Guidance and best practice

19.2.1 Legislative background

There is no legislation in the UK that is specific to methane. However there is legislation relating to the various man-made sources of methane, e.g. landfills and landfill gas or oil and gas operations. Health and safety legislation also applies to methane in situations such as mining and tunnelling or other construction works. There is also some legislation with respect to preventing ground gases (including methane) entering developments and relating to the reduction of greenhouse gas emissions.

The UK probably has the most extensive set of guidance documents in the world on the investigation of methane in the ground and on providing protective measures to buildings to prevent ingress of methane from the ground. The list of good practice and guidance is extensive. Some of the key documents utilized within the UK are summarized below:

(1) European Landfill Directive 1999/31/EC, effective 16 July 2001.
(2) *Approved Document C* – Site preparation and resistance to contaminants and moisture, 2004 Edition incorporating 2010 and 2013 amendments. The Building Regulations 2010 (HM Government 2013).
(3) British Standard BS8485: 2015 + A1: 2019. Code of Practice for the design of protective measures for methane and carbon dioxide ground gases for new buildings (BSI 2019).
(4) *CIRIA Report C665* – Assessing Risks Posed by Hazardous Gases to Buildings (revised) (CIRIA 2007).
(5) The Local Authority Guide to Ground Gas. Chartered Institute of Environmental Health (CIEH 2008).
(6) British Standard BS8576: 2013. Guidance on investigations for ground gas. Permanent gases and Volatile Organic Compounds (VOCs) (BSI 2013).
(7) Statutory Instruments 2012 No. 3038. The Greenhouse Gas Emissions Trading Scheme Regulations 2012, enacted 1 January 2013.

Methane is often a contaminant present as a result of land contamination. In the UK the assessment of contaminated land, either via the planning process or as defined by Part 2A of the Environmental Protection Act, 1990, is based upon the principles of risk assessment (Environment Agency 2004 and DEFRA 2012). Underpinning the risk assessment is the concept that for there to be a risk of harm then there must exist a pollutant linkage. For a pollutant linkage to exist, there must be in place a 'source' of contamination, a 'receptor' and a means of transmitting the source to the receptor – known as a 'pathway'. Understanding potential sources, pathways and receptors is critical to identifying potential risks associated with methane gas.

The presence of high concentrations of methane in the ground does not necessarily mean that there is viable source that can move out of the ground and cause surface emissions. Pathways also need to be credible and to consider the permeability of soils and rocks. The fine particle content of granular soils is important and clayey silty sands or gravels are unlikely to have a sufficiently high permeability to allow large volumes of gas to migrate through them.

In all instances where land is being assessed it is important that the ground model is understood, with a robust conceptual site model determined such that all decisions are based on scientific, reasoned and logical judgements. Multiple lines of evidence should be sought when considering the potential impacts associated with methane migrating from the ground.

19.3 Developing the conceptual site model

19.3.1 Sources of methane

Methane is present in soils and rocks below many parts of the UK and other countries. In the UK the dominant natural geological sources of methane are coal deposits, hydrocarbon reservoirs, marshes, peat and tidal sediments. Other major sources are anthropogenic, notably landfills. Across the world other environments exist which can generate methane, including volcanic, hydrothermal and geothermal settings.

Soil gases containing methane with concentrations ranging from trace to more than 90% v/v have also been encountered in a wide range of igneous, metamorphic and sedimentary rock environments associated with various economic deposits including oil-shales, potash, salt, uranium, zinc, copper, lead, diamond, arsenic, apatite and gold. The deposits range in age from Recent to Precambrian.

Methane can be formed by the anaerobic biochemical reduction of organic matter by micro-organisms, or can be formed over geological periods following the burial, compression and heating of organic material. Thermogenic methane is a type of biogenic methane generated by thermal degradation of organic matter. Bacteriogenic methane is generated by microbes. Thermogenic methane can be distinguished from bacteriogenic methane by its heavier carbon isotope composition and by the presence of light hydrocarbon gases such as ethane, propane, butane, etc.

Coalification is the process whereby plant debris and other organic matter are turned into peat, then lignite and finally coal. The degree of coalification and the transformation of plant matter to carbon determines the nature of coal. The lowest grade is brown coal (which has least carbon) and the highest grade is anthracite (with the highest carbon content). Some evolution of methane occurs in the early stages of the accumulation of plant debris. However the major phase of methane generation is thought to occur at a later stage in the coalification process.

In coal measures methane is invariably adsorbed on coal, or may be trapped in gas pockets or dissolved in solution in groundwater within the whole of the coal measures deposition sequence, which is typified by mudstones, siltstones, sandstones, seatearths and coals. Methane is often a problem in areas where the coal measures have been worked either at the surface or at depth, owing to natural degassing of methane into the voids formed by the coal workings. Without the workings methane does not normally have a significant pathway to the ground surface and remains trapped in the rocks.

'Free' methane gas may be found in pores, fractures and cavities in soils, rocks and sediments of all types and also as bubbles in water or in modern marine sediments. Adsorbed methane is found in high concentrations on coals and other carbonaceous materials. Fine pores within coal provide a vast surface area on which gas molecules can be so tightly packed that the density of the gas approaches that of a liquid.

Typically background concentrations of methane in soil gas vary between 0.2m and 1.6 ppm, the latter being the mean concentration of methane in air. Methane concentrations above 0.1% v/v (1000 ppm) are rarely encountered in soil gas in the absence of an identifiable source (Butterworth 1991). In the proximity of landfill sites the distribution of methane concentrations within shallow soils (less than 2 m) may be used to give a general indication of the direction of movement of contaminated groundwater (Barber *et al.* 1990). Theoretical total landfill gas yields of around 400 m^3 per tonne are possible from fresh municipal solid waste, of which 50–60% would be methane. However, in practice yields ranging from 50 to 250 m^3 per tonne of refuse are typically reported for wastes over a 10-year period. UK fresh municipal solid waste landfills typically generate 10 m^3/tonne/year, which reduces over time. This reflects the less than optimal conditions within the landfill comparative with the theoretical yield. The composition of waste has a major influence on its degradation, as the waste influences the nutrients required for microbial growth and metabolism.

Methane has a very low solubility at atmospheric pressure but at elevated pressures significant volumes can be held in solution. It is known that normal shallow groundwater in the UK has very low background concentrations of methane, although trends indicate a general increase in dissolved methane concentrations with depth. This is probably due to a number of factors including degradation of indigenous traces of organics in the rock phase at depth and slower groundwater movements deep within the bedrock.

In the past decade advances in horizontal drilling have facilitated an increase in hydraulic fracturing (also known as 'hydro-fracturing' or 'fracking'). This process involves pumping fluid into an oil or gas well at high pressures to create fractures in the rock formation which allow oil or gas to flow from fractures in the wellbore (Jackson *et al.* 2011). The US Energy Information Administration estimates that the USA 2119 trillion cubic feet of recoverable natural gas, about 60% of which is 'unconventional gas' – i.e. stored in low-permeability formations such as shale, coal beds and tight sands at depth (US Energy Information Administration 2010). The US Energy Information Administration predicts that by 2035 shale gas production will increase to 340 billion cubic metres per year, amounting to 47% of the projected gas production in the USA.

Associated with the fracking process are the 'fugitive emissions' of methane to atmosphere, which are estimated to vary between 1 and 3% of total gas production per well (Kirchgessner *et al.* 1997). This is clearly an important consideration given the potency of methane as a 'greenhouse gas'.

Concerns have been raised regarding the potential for fracking to release dissolved methane into potable water supplies. Currently dissolved methane concentrations are not regulated in drinking water, on the basis that the compound does not alter the colour, taste or odour of water and it is not known to affect its potability. However, there is a secondary concern that, as groundwater is pumped to surface for use, the changes in pressure could result in methane coming out of solution and resulting in an asphyxiating and/or flammable/explosive atmosphere within the well head construction/pump room/dwelling. In reality, groundwater at the depth which fracking typically occurs is likely to already contain dissolved methane, reflecting the significant pressure at depth. Any water pumped from such depths would ordinarily contain dissolved methane. Fracking could allow this deeper groundwater to move upwards and contaminate shallower aquifers. It is important to manage all deep groundwater which is brought to surface in a way that prevents the risk of explosions occurring.

19.3.2 Pathways for migration

There are a number of mechanisms which can cause methane to move through the ground. These may be summarized as follows:

(1) *Diffusion* – gas migrates along a chemical gradient from points of high concentration to points of lower concentration. Methane will naturally diffuse from a region of high concentration to a region of low concentration. This transport mechanism is slower than advective flow (see below) unless driving pressures are very low.
(2) *Pressure driven flow (advection)* – gas migrates along a pressure gradient from a point of high pressure to a point of lower pressure.
(3) *Buoyant flow* – gas migrates through saturated ground or water along a density gradient in an upwards direction owing to the difference in density between the gas and water. Differences in gas temperature within a landfill is usually the reason for buoyancy effects.
(4) *Volume expansion* – gas migrates owing to volume expansion as a result of changes in pressure, although the distances over which this can occur will be limited to a few metres. This could be significant where a residential property is located very close to a reservoir of accumulate gas (for example if gas has accumulated in stone columns below a building). It was the main driving force at the site in Loscoe, where a bungalow exploded owing to methane migration from a landfill site. The public inquiry into the incident at Loscoe (Ryan *et al.* 1988) suggested that the following combination of events was the most probable cause of the explosion:
 (a) accumulation of an appreciable quantity of landfill gas (tens of cubic feet) in an old well or other void in the ground close to the bungalow;
 (b) expansion of the landfill gas in response to a large drop in barometric pressure;
 (c) flow of gas along the backfill to a drain into the floor space below the house; and
 (d) ignition from a central heating boiler.

 Landfilling was completed at the landfill only 4 years prior to the explosion. Therefore, the gas source was fresh domestic waste at the peak of gas generation. Contrary to common perception, it is likely that the gas migrated to the well or void by diffusion from the landfill via fractured rock or old coal workings (coal seams in and below the former quarry had been worked). It was not 'sucked out of the landfill' by the fall in pressure. The inquiry concluded that the chance of a repetition of the explosion was remote because the combination of circumstances which caused it was considered unusual.
(5) *Dissolved phase* – methane can become dissolved in leachate or groundwater under high pressure and migrate in the direction of groundwater or leachate flow. For the gas to return to a separate gaseous phase there has to be a reduction in pressure in the groundwater or leachate. It is rare that such a pressure change occurs in or around a landfill and it usually requires gas to migrate in solution from depths greater than around 30 m up to the surface where the pressure is equal to atmospheric pressure. Such phenomena occur widely in nature from natural gas eruptions owing to volcanic activity or active tectonic fault movement.

In practice advection (pressure driven flow) and expansion are typically the dominant gas transport mechanisms where gas migration problems have occurred. In a landfill gas pressures can quickly build up to above ambient atmospheric pressure as the gas is generated from degrading waste. This creates a pressure gradient and can cause gas to migrate rapidly from a region of high pressure to a region of low pressure. Advective gas migration can be caused by rapid differential pressure changes such as a fall in barometric pressure owing to a low-pressure weather system passing over the landfill. The UK Environment Agency LFTGN03 states that a large and rapid drop in atmospheric pressure, of say 3 kPa (30 mbar), can provide a significant driving force if the slight positive pressure within a landfill (compared with the atmosphere) is only a few kPa (mbar). This effect can be dramatic if the rate of change of pressure is fast, i.e. it takes place over a few hours, and there is a large reservoir of gas close to buildings and a highly permeable pathway. A number of explosions have occurred as a result of methane migration after such large drops in atmospheric pressure (Boltze & de Freitas 1996). A danger threshold for landfill gas migration was identified when the duration of the fall in pressure/drop in pressure ($\delta t/\delta P$) is less than 0.6 h/mbar. The definition of a worst case zone for gas monitoring has been developed in CLAIRE Technical Bulletin TB17 (Wilson *et al.* 2018). This identifies that the critical factor for collecting gas monitoring data is the rate of fall of atmospheric pressure. It suggests that arbitrary limits on atmospheric pressure, such as the requirement to collect data when pressure is less than 10 000 mb, are not appropriate.

The distance that methane gas will travel underground from a source depends on a number of factors:

(1) the nature of the source;
(2) the permeability of the source and the underlying/overlying/surrounding natural geology;
(3) groundwater/leachate conditions;
(4) the presence of open pathways in the ground such as fractures or coal workings;
(5) the gas pressure and diffusion gradients;
(6) any mitigation measures taken at source to prevent gas migration.

Where the source of methane is natural (e.g. peat bog, alluvium etc.), it is quite normal for high concentrations of methane to be encountered. In the case of older alluvium and peat deposits the methane is typically from historic generation, which has become trapped in the pore matrix of the soil by virtue of the small pore spaces and predominantly cohesive (low permeability) matrix. As such there is typically no driving pressure to promote migration of the methane from these sources over large distances or long timescales, with releases generally being of small volume and short lived (e.g. a short-term release of methane as a borehole is drilled into peat).

Where methane is from an anthropogenic source (e.g. landfill) the distance that it could potentially migrate will depend on a number of factors;

(1) The permeability of the landfill will influence the rate at which gases can migrate through the waste deposits. Where wastes are older and the degradation processes well under way it is not uncommon to see that the waste deposit has degraded to form a predominantly cohesive matrix. These latter degradation stages generally occur where the waste material is wet, thus the wet and cohesive nature of the waste starts to slow the gas migration potential within the landfill body as a whole.

(2) The permeability of any topsoil/capping material placed over the waste also plays an important role when considering the nature of gas generated by a landfill and the distance over which that gas can migrate. Where the capping is un-engineered, or of limited thickness and/or predominantly granular, it will permit flow of gas out of the landfill to atmosphere, whilst promoting atmospheric air to flow into the landfill during pressure changes. Thus, the overall effect is diluting the gas concentrations within the shallower depths of the landfill mass. Conversely, where the capping comprises thicker, more cohesive deposits, these can effectively 'seal' the surface of the landfill – promoting anaerobic degradation (i.e. increased methane generation) over time as the initial oxygen in the ground is used up by the bacterial reactions and resulting in gas pressures building up within the landfill. In some instances a cohesive cap can prevent water infiltration into the landfill material – resulting in a dry waste mass and slow degradation of the fill material, which inhibits gas generation. In these circumstance lower levels of methane may be anticipated – but generated over longer timescales.

(3) The thickness of the waste deposits are also an important consideration. Where landfills are shallow (i.e. <5 m) it is possible for the waste degradation cycle to be influenced by the ingress of atmospheric air during pressure changes, which can result in aerobic conditions and inhibit methane generation. However, deeper wastes are unlikely to be subject to such influences, with anaerobic degradation becoming dominant at depth over time as the initial oxygen in the ground is used up by bacterial reactions. The increasing depth of a landfill has a direct increase on the pressures existing at depth.

(4) Leachate conditions within a landfill can also influence the movement of landfill gas. If leachate levels rise within the landfill it can force landfill gas out of the soil pore space upwards, which can lead to an increase in landfill gas pressure. The effect is similar to a hydraulic ram. Landfill gas will then migrate along the path of least resistance from the region of high pressure to a region of low pressure. The significance of this will depend on the rate at which leachate levels rise. A slow rise may not pose any significant risk.

(5) Groundwater conditions within/surrounding a landfill will also influence movement of landfill gas. Saturated soils will minimize the distance that landfill gas will migrate (although it can migrate in unsaturated zones if they are present above or below the saturated zone). At a landfill in Australia gas migration has occurred along unsaturated sand layers confined below perched water tables. Indications are that landfill gas has migrated below a housing estate that is located over 400 m from the landfill site (Brouwer 2009). The presence of water in soil or rock will also reduce pressure-driven (advective) and diffusive flow as the water takes up pore space that is not readily available to gaseous phase flow. Effective permeability is the permeability to one fluid (in this case landfill gas) of a porous medium partially saturated with another immiscible fluid (in this case water). When soil or rock is fully saturated with water, the relative permeability to landfill gas flow will be zero. The effective air-filled porosity will be zero and diffusive flow through the soil pore water will be very low. This can limit horizontal migration where there is a high groundwater table or prevent gas escaping through the surface if it is saturated. The permeability of soils is often greater in the horizontal direction than the vertical direction owing to horizontal layering in the soil during placement or deposition. Gas will flow along the lines of greatest permeability (i.e. path of least resistance) if all other factors are constant.

(6) The nature/efficacy of any engineering measures installed as part of the landfill will also be an important consideration when assessing landfill gas migration potential. Modern landfills are typically designed with leachate collection and gas barriers/extraction systems as standard, so as to capture and control any leachate/gas prior to it migrating away from the landfill. However, older landfills are often unlined, with no engineering measures to capture leachate and/or gas.

In simplistic terms gas migration is generally impeded by soils with low permeability (i.e. silts, clays and fine silty/clayey-sands), although gas migration can still occur in heavily fissured, dry clays. In unconsolidated soils (e.g. clean sands and/or gravel) migration is much more likely as 'seepage' of gas occurs via pore spaces in the soil. Migration through an unconsolidated stratum occurs generally under lower pressures for short distances, the gas having significant pore spaces to accumulate within. Conversely in consolidated geological units (e.g. chalk and limestone), gas can migrate under higher pressures for longer distances, being 'channelled' along joints, bedding planes and fault lines/zones.

The presence of deep permeable layers of soil that are confined by less permeable layers will increase the distance over which landfill gas will migrate. Recent works in the Surrey area of the UK, where clay-with-flint quaternary drift geology overlies chalk bedrock, have indicated migration of methane at depth in the unsaturated chalk beneath the predominantly cohesive drift deposits. Methane gas has migrated towards existing residential dwellings located around 100 m from the landfill. With distance from the landfill the clay-with-flint thins and the chalk outcrops beneath the properties, thereby providing a preferential pathway (i.e. via the chalk) directly to the receptor (i.e. the residential dwellings). The principal mechanism for migration has been via diffusive flow, facilitated by the waste deposits locally within the area of landfilling being in direct contact with the underlying chalk (the clay-with-flint drift deposits having been eroded and/or deliberately removed to form daily cover layers as part of the landfilling processes).

Conversely, where permeable soils are shallow and not confined, the maximum distance gas will migrate is shorter as the gas can easily migrate vertically and be dispersed in the atmosphere. The presence of homogeneous clay soils around a landfill site will limit the distance that gas can migrate, if it occurs at all in such conditions.

Landfill gas can undergo changes in composition as it migrates through the ground. The most significant of these is oxidation of methane by bacteria in the ground. Methane is converted to carbon dioxide. This will also change the isotopic balance of the gases. Oxidation of the methane causes carbon dioxide to become increasingly enriched in ^{12}C with distance from a landfill and the residual methane is depleted in ^{12}C (Williams *et al.* 1999). Carbon dioxide is more soluble than methane and it is possible for gas enrichment to occur (i.e. higher methane concentrations) where carbon dioxide has dissolved in groundwater. Changes can also occur owing to differing rates of diffusion between gases.

Often there is a need to distinguish between landfill gas and thermogenic gas. This can be achieved by considering various isotope ratios, including δ^{13}C and δD for methane. A 'Bernard' plot of methane/(ethane + propane) against δ^{13}C for methane is another useful indictor of a gas source. Thermogenic gas can contain a significant proportion of ethane (between 2% and 10%). Biogenic gas on the other hand will have less than 100 ppm ethane. Therefore, the ratio of methane to ethane can be used as a reliable indicator to distinguish between biogenic and thermogenic methane in the ground. However, ethane and propane can be stripped from thermogenic gas as it migrates through the ground.

Pressure gradients can occur owing to a buildup of gas pressure inside the landfill and owing to variations in atmospheric pressure between the subsurface and land surface. Similarly pressure gradients can occur between the inside and outside of a building owing to thermal differences between indoor and outdoor air (affecting density and buoyancy), wind effects and imbalanced ventilation. Pressure gradients will also occur where suction is applied to the ground, for example with gas abstraction wells. Gas will flow along the line of greatest pressure gradient if all other factors are constant.

Gas will normally flow horizontally through or vertically upwards from the ground. It can flow downwards but this is rare and would only occur where the gas is trapped by a confining impermeable layer or there is a pressure gradient in the downward vertical direction creating flow. The rate of flow is dependent on the effective mass permeability of the soil and the magnitude of the pressure gradient. It is difficult to predict the direction that landfill gas migration will occur because of uncertainty in identifying all permeable pathways.

Open pathways such as faults, mine shafts, tunnels, abandoned oil wells and some service trenches can significantly increase the distance over which landfill gas can migrate. For service trenches to form a pathway they need to be backfilled with open gravel. Gas can also enter utility ducts and migrate inside the duct itself. In some areas of California abandoned oil wells have caused deep thermogenic gas to migrate to the surface and cause explosions in buildings.

19.3.3 Potential receptors to methane

Receptors may be defined as any people, animals, flora, buildings or structures that may be connected to a source of methane via a pathway. Examples of potential receptors include:

(1) building occupants/users;
(2) buildings/structures/infrastructure;
(3) construction/maintenance workers;
(4) flora.

Receptors may be located either immediately at the source of the methane, or at distance from the source, depending on the nature of the pathway which links the source to the receptor.

It is important to note that risks to a receptor can be altered via activities carried out in isolation of the source, pathway or receptor. Examples include:

(1) Decreasing the permeability of surfacing (e.g. by adding a clay cap to a landfill or increasing hard cover in the proximity of buildings), which may limit the potential for natural venting of gas to atmosphere and thus promote horizontal migration pathways. This has often been linked to problematic migration of landfill gas such as at Loscoe and Brookland Greens.
(2) Providing a new preferential pathway for gas flow (e.g. by constructing a new sewer which could facilitate flow of methane along service trench backfill from a source to a receptor).
(3) By adopting vibro stone columns for building foundations – which can provide a high-permeability conduit for migration and accumulation of gas immediately beneath a building.

19.4 Examples of methane impacts

The case studies presented in Table 19.3 demonstrate how 'complete pollutant linkages' resulted in hazards being realized in relation to methane gas such that an incident occurred.

Table 19.3. *Case studies demonstrating methane impacts*

Fairfax, California 1985	
Source	Methane gas from an oil field underlying an urbanized area
Pathway	Migration along faults and improperly abandoned historical oil wells
Receptor	People in and around the Ross Dress for Less Store, 23 people injured

In 1985, methane from deep geological deposits seeped into the Ross Dress for Less Store in Fairfax, California, and caused an explosion that injured 23 people (Council of the City of Los Angeles 2004). The property is located on the Salt Lake Oil Field, which is an urbanized oil field located beneath the city of Los Angeles, discovered in 1902 and still exploited today. Ground conditions comprise a layer of Quaternary sediments (alluvial and marine), underlain by several oil-bearing formations which include the Repetto Formation (a sandstone and conglomerate unit which is a prolific petroleum reservoir throughout the Los Angeles basin). All the rock units are faulted and folded, forming structural traps, with oil trapped in anticlinal folds and along fault blocks (University of Texas 2013). The source of the methane associated with the explosion was controversial, with initial studies indicating biogenic methane associated with decomposition of organic matter from an old swamp. However, more recent studies (which are widely accepted to be more likely) have indicated that the gas originated from the oil field itself, having migrated to the surface along a combination of faults and improperly abandoned historic boreholes. This incident resulted in the 400 blocks overlying the oil field to be designated as a 'High Methane Zone', with all structures required to have a methane detector to give warning of accumulation of the gas before it could attain explosive concentration. A later Fairfax incident in 1989 did not result in explosion, although pressures within a venting well built up such that 'a fountain of mud, water and methane gas' was emitted from the ground. As a result of these incidents, Los Angeles further upgraded their City Building Code to require new buildings to have adequate venting systems, and be underlain with an impermeable membrane to prevent methane from getting in beneath foundations (Perera 2001).

Abbeystead, Lancashire 1984	
Source	Methane gas coming out of solution
Pathway	Preferential migration along a tunnel into a valve house
Receptor	People in valve house, 16 people killed and 28 injured

An explosion occurred in a pumping station at Abbeystead in Lancashire in 1984 (Hooker & Bannon 1993). This was caused by methane accumulating in a tunnel constructed through a Carboniferous sequence of silty mudstones, siltstones and sandstones. Isotope testing of the methane following the explosion confirmed it to be $c.20\,000$ years old and to have most likely originated from fractures within the mudstone bedding. Water from the bedrock geology was able to flow into the tunnel via gaps in the concrete lining construction. It is believed that this water contained dissolved methane, which when it was exposed to a lower ambient pressure within the tunnel was able to come out of solution and accumulate. The tunnel provided a preferential pathway for methane migration, into the valve house. Prior to the explosion water flow within the tunnel had been stopped for a period of 17 days due to draught. When the pumping system of the tunnel was turned back on, as part of a demonstration for local people who had congregated below ground level within the valve house, an unknown ignition source triggered the explosion. Of the 44 people in the valve house for the demonstration, 16 were killed and all the others injured. It is of note that workmen constructing the tunnel complained of headaches and lightheadedness on three separate occasions. However, monitoring for gas during the tunnel construction was typically completed whilst the active ventilation systems were turned on and the three sets of results which did indicate gas to be present were not considered in any great detail. Following the explosion, the Health and Safety Executive monitored methane flow into the tunnel at $0.001\,\text{m}^3$ per second. Importantly, the designers of the tunnel system did not require any ventilation to be provided as part of the construction. In operating the system, the designers had required either the pumps to be working, or the tunnel to be full of water. However, this was not the case with a control valve typically left open, thus resulting in a void within the tunnel which enabled the methane to come out of solution. In 1987 Lancaster High Court apportioned blame for the disaster as being 55% due to design (Binnie & Partners), 30% due to operation (North West Water Authority) and 15% due to construction (Nuttalls Limited). An out of court total sum payable was agreed of £2.5 m.

Loscoe, Derbyshire 1986	
Source	Methane gas migrating from a landfill
Pathway	Preferential migration along fractures / bedding in sandstone
Receptor	People in bungalow at time of explosion, three injured

In 1986 an explosion occurred in a bungalow located approximately 60 m from an old brick pit. The brick pit had been used as a domestic refuse site between 1977 and 1982 and subsequently capped. Laboratory testing of samples removed from the site of the bungalow proved the migration to be from the landfill, having migrated via fractures and bedding joints within the solid bedrock sandstone geology underlying the bungalow and abutting the landfill. The effect of the migration via the bedrock geology had been amplified by many years of coal mining and rock blasting in the area, which had resulted in 'sagging' of the bedding joints. From meteorological records, it is clear that the few hours before the explosion had shown rapid decrease in atmospheric pressure in advance of a weather front crossing the area. Over a period of seven hours the atmospheric pressure had rapidly fallen by 29 mb, with hourly drops in pressure ranging between 3.3 mb and 4.8 mb. This significant decreasing atmospheric pressure had resulted in the volume of landfill gas that had accumulated in a soakaway pit close to the house expanding and migrating into the void below the bungalow floor. Here it accumulated beneath the property and gradually migrated into the lower portion of the dwelling via convection, with the spark from the boiler first thing in the morning providing an ignition source for the explosion. The three occupants of the bungalow were injured.

19.5 Managing risk

The presence of methane has an impact on tunnelling works and mining operations, where it can be a serious safety hazard during construction. This is normally managed by following well-defined health and safety procedures for working in these environments. One of the most common areas where methane has an impact is on developments constructed over or adjacent to methane sources. This aspect of methane in the ground is discussed in the following sections.

19.5.1 Site investigation for methane

Guidance on site investigation for methane (and other ground gases and vapours) is provided in British Standard BS8576 *Guidance on investigations for ground gas – Permanent gases and Volatile Organic Compounds (VOCs)* (BSI 2013). The standard, which is intended to be used in conjunction with BS 10175:2011+A2:2017, Investigation of potentially contaminated sites. Code of Practice. (BSI 2017). provides guidance on the monitoring and sampling of ground gases including volatile organic compounds and permanent gases such as methane.

It is particularly relevant to the assessment of development sites and the risks posed by gassing sites to neighbouring land and developments. It is also relevant to investigations being completed in order to assess potential liabilities as set out in Part 2A of the Environmental Protection Act 1990 and under the Environmental Damage Regulations (e.g. in respect of spills of oils). The standard sets out good practice for the desk study, installation of gas monitoring wells (where required) and gas monitoring procedures.

19.5.2 UK contamination practices

The risk associated with methane is often dealt with in the UK as a contaminated land issue, especially where migration is occurring from historic landfill sites that are not covered by the current landfill regulatory regime. The Government recognizes that the UK's industrial past has left the country with a legacy of contaminated land and has introduced various statutory instruments to deal with/prevent future contamination and to deal with past contamination. Individual Local Authorities have a responsibility to identify where contaminated land exists within their Borough and to deal with it in accordance with the Statutory Regulations.

The Government's objectives with regard to contaminated land are:

(1) to identify and remove unacceptable risks to human health and the environment;
(2) to seek to bring damaged land back into beneficial use; and
(3) to seek to ensure that the cost burdens faced by individuals, companies and society as a whole are proportionate, manageable and economically sustainable.

These three objectives underlie the Government's 'Suitable for Use' approach to the remediation of contaminated land.

The suitable for use approach means that any land where contamination is causing unacceptable risks to human health and the environment must be identified, assessed on the basis of the current use and circumstances of the land and if necessary remediated to ensure that it is reinstated to a position where such risks no longer exist. The suitable-for-use approach also means requirements for remediation must be limited to the necessary work to prevent unacceptable risks to human health and the environment in relation to the current use of the land. In other words, the mitigation works must not be over-engineered, over-remediated or indeed 'guesses' made as to what might constitute contaminated land.

The assessment of contamination is dealt with in one of two ways in the UK – either as part of the planning process (for new developments or alterations to existing buildings/infrastructure) or, if deemed appropriate, via legal determination of 'Contaminated Land' following the guidance set out in Part 2A of the Environmental Protection Act 1990 (as clarified by the Contaminated Land Statutory Guidance, DEFRA April 2012).

Methane gas, regardless of whether it is from an anthropogenic or natural source, is treated like all other forms of contamination as part of the assessment processes, requiring appropriate investigation, assessment, remediation and verification to ensure that it does not pose a risk of harm to identified receptors.

19.5.3 The planning process

When planning permission is granted in the UK, various 'planning conditions' will be attached to the consent. Where the Local Authority considers there is a risk of contamination being present, planning conditions relating to contamination will be included as part of the consent. The exact wording of these vary between Local Authorities, but typically necessitate a desk-based and intrusive assessment of contamination and ground conditions at the site to inform the development of a refined conceptual site model. The planning conditions will also require that, if found to be necessary, remedial works will be implemented as part of the development works – verified in accordance with current UK guidance and good practice. Again, methane gas, regardless whether it is from an anthropogenic or natural source, is treated like all other forms of contamination.

Such planning conditions do not only apply to residential housing schemes; similar conditions often apply to larger commercial/industrial developments as well as infrastructure projects. Indeed, for larger-scale developments an environmental impact assessment is generally required, considering all potential impacts of ground conditions, hydrology, hydrogeology and contamination both in relation to the scheme and cumulatively to receptors in the wider geographic area.

Planning conditions are not typically discharged until the Local Authority is satisfied that the information provided in relation to contamination at the development site is sufficiently robust and compliant with UK legislation/good practice. Undischarged planning conditions will typically be identified during legal searches associated with the purchase of a property, with mortgage funds/financing only being released where the lenders are sure that there are no unresolved contamination issues, or associated liabilities.

19.5.4 The definition of contaminated land

Outside of the planning process, the strategy for the legal determination of 'Contaminated Land' is as set out in Part 2A of the Environmental Protection Act 1990 (As Amended). This process is informed by the Contaminated Land Statutory Guidance document as issued by the Department for Environment, Food and Rural Affairs (DEFRA) in April 2012. Section 78A(2) defines 'Contaminated Land' as:

> any land which appears to the local authority in whose area it is situated to be in such a condition, by reason of substances in, on or under the land that – (a) significant harm is being caused or there is a significant possibility of such harm being caused; or (b) significant pollution of controlled waters is being caused, or there is a significant possibility of such pollution being caused.

The process of assessing Significant Possibility of Significant Harm (SPOSH) requires an estimation of the likelihood that significant harm might occur to an identified receptor, together with an estimation of the impact if the significant harm did occur. With regard to non-human receptors, Part 2A also applies to certain ecological receptors (in consultation with Natural England), property (e.g. crops, livestock, buildings, infrastructure) and controlled waters. It should be noted that Part 2A only relates to current land uses and existing migration pathways.

Methane gas risks are most likely to influence human receptors and buildings/infrastructure, although risk to crops should clearly not be overlooked.

19.6 The risk assessment process

The assessment of SPOSH requires an assessment of relevant information, notably scientifically based, authoritative and relevant sources. For planning purposes generic qualitative or semi-quantitative risk assessment models are typically utilized for gas risk assessment (e.g. Situation A (Modified Wilson and Card) or Situation B (NHBC Traffic Lights) of CIRIA Report C665 (CIRIA 2007)). However, when considering SPOSH these frameworks are too simplistic and conservative and should not be used. Instead, the assessment of SPOSH should be based on robust interpretation of data and quantitative mathematical modelling based on information obtained from multiple lines of evidence.

The guidance on SPOSH does not easily translate to methane gas risks. However, applying a systematic and consistent approach in line with the statutory guidance, it would be necessary to demonstrate the following:

(1) Site investigations must have identified with reasonable confidence that there are credible and realistic pathways for gas to reach a receptor.
(2) Once credible pathways have been identified an assessment of equilibrium gas concentrations inside the occupied parts of the building (or unoccupied parts where there may be an ignition source) should be completed. Gas flow into buildings should be based on analysis of gas migration through the ground and into the building/structure by advection or diffusion. This assessment is a function of the driving pressures of the gas source and the permeability/effective diffusion coefficients for the soils/rocks below a site.

Gas migration into a building/structure will be influenced by its construction and it is imperative that risk assessors understand the exact nature of floor slabs, foundations, service entries, etc., within an affected development. It is paramount that practitioners understand the intrinsic relationships of gas generation sources, ground hydraulics, building construction and ventilation principles. An example of a SPOSH decision framework that has been used successfully by the authors on several sites is provided in **Figure 19.2.**

Definitions of confidence/certainty may be summarized as follows:

(1) *Good* – the nature of source material is well characterized, potential pathways well characterized and understood, likely volume of gas generation is characterized, impact of ground gas relationships/variations is understood, building foundation and floor slab construction are known. Multiple lines of evidence are available.
(2) *Reasonable* – the nature of source material is well characterized, potential pathways are understood and reasonable parameters can be assigned, likely volume of gas generation can be estimated, building foundation and floor slab construction are known with reasonable certainty. Multiple lines of evidence are available.
(3) *Poor* – limited information/understanding of source material, pathways, volumes of gas generation, building foundations and floor slab construction. There are many assumptions and limited lines of evidence.

'Multiple lines of evidence' is using different information sources or assessment methods to determine the likely risk associated with ground gas. For methane the following are examples of separate lines of evidence:

(1) desk study identifying age and nature of gas source – whether it is likely to be producing large volumes of gas;
(2) desk study to identify how long receptors have been exposed to source and whether there have been any reported issues (odours, drowsiness etc.);
(3) site investigation to give visual assessment of gas source together with chemical test data and gas generation modelling;
(4) flow modelling of gas migration through the ground and into buildings;

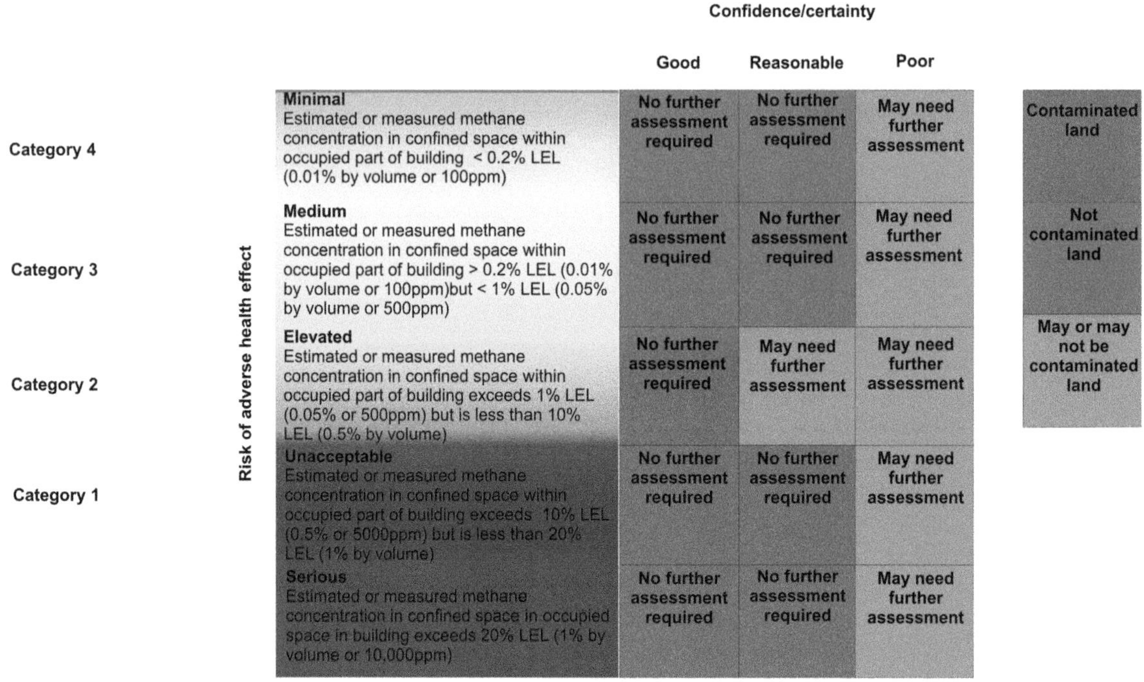

Fig. 19.2. Significant Possibility of Significant Harm (SPOSH) framework for methane.

(5) confirmation that gas migration is occurring by gas monitoring in the ground;
(6) flux chamber testing of surface emissions;
(7) confirmation of gas ingress into buildings by internal space gas monitoring.

19.6.1 Qualitative risk assessment

For many low-risk sites being considered as part of the planning process for a new development, qualitative or semi-quantitative assessment techniques will be sufficient for assessment of potential methane risks to identified receptors.

Applying qualitative risk assessment techniques to site assessment should not be seen as a 'soft' option. Multiple lines of evidence and robust consideration of the conceptual site model are required to demonstrate that ground gas risks are low.

However, where it can be robustly established that associated risks are low, it is not unreasonable to consider that ground gas monitoring is not required, or the number/period of monitoring events can be reduced.

A qualitative approach to site assessment is provided in CLAIRE Research Bulletin RB17 (CLAIRE 2012). The approach has been refined and included in British Standard BS8485: 2015. It provides a framework for the qualitative assessment of low-risk gassing sites, which are defined as sites where the gas source is one of the following:

(1) natural soils with a high carbonate content (e.g. chalk, some glacial till etc.);
(2) natural soils that are known to contain methane (e.g. alluvium, peat etc.);
(3) made ground with a low organic content (i.e. predominantly soils, ash or clinker with some occasional pieces of wood, etc.), where the maximum thickness of material does not exceed 5 m and the average thickness of material does not exceed 3 m;
(4) areas of flooded mine workings or mine workings that were abandoned by the early twentieth century – with the exception of where buildings are proposed within 20 m of a mine opening (shaft or adit) or where shallow workings are very close to the surface and/or connected to deeper unflooded mines.

The approach described in the paper could be considered as more robust than some semi-quantitative techniques as it requires a thorough understanding of the methane source, whereby it uses an assessment of the total organic carbon (TOC) of the source of methane to determine the scope of protective measures that are required. It removes the need for gas monitoring for methane for most low-risk sites, and identifies that gas monitoring is only required for:

(1) higher-risk sites where gas can be emitted from the ground in large volumes (domestic or industrial landfill sites with a high degradable content, made ground with a higher degradable content, mine workings where

there is still a large gas reservoir and a vent to the ground surface such as a shaft or fractured rock);
(2) sites with made ground where the maximum thickness exceeds 5 m and/or the average thickness exceeds 3 m;
(3) sites where migration from an off-site source with a credible migration pathway has been identified.

The qualitative process provides an indication of whether gas monitoring is required and/or if gas protection measures are necessary. The approach has also been used by the authors to control the reuse of excavated material from landfill sites as engineered fill below developments.

For natural soils, gas monitoring is not considered necessary where no viable source of anthropogenic ground gas (on- or off-site) has been identified. Alluvial soils and buried peat can give rise to high concentrations of methane (typically up to 90% v/v), although carbon dioxide concentrations are generally lower, having been dissolved out of the gas whilst trapped in the soil pore space. However, the rates of generation from alluvial soils and peat are generally low – reflecting that the gas is from historic generation and breakdown of the material which has become trapped in the pores. Given the extent of such material across the UK, and the experience gathered from many sites, it is not unreasonable to simply install gas protection measures equivalent to Characteristic Situation 2 (CS2) in accordance with BS8485:2015 +A1: 2019 instead of implementing a prolonged regime of gas monitoring. If gas monitoring is carried out it can, however, often be shown that, even when high concentrations of methane are present in the monitoring wells, the risk of surface emissions is negligible and gas protection measures are not necessary.

With regard to made ground soils where the maximum thickness of material does not exceed 5 m and the average thickness of material does not exceed 3 m, the assessment of associated ground gas risks has been calculated via gas generation modelling, which uses TOC content to determine limiting values for assessment of a site's Characteristic Situation designation. TOC data are easy and relatively inexpensive to obtain during intrusive site investigations, with each test costing a few pounds. This approach has already been incorporated into guidance published in Australia (EPA New South Wales 2012).

The limiting values for the TOC content of made ground from the CLAIRE bulletin are summarized in Table 19.4. The table also has to be used in conjunction with an assessment of the discrete biodegradable fractions of any made ground (e.g. the proportion of paper, wood, vegetable matter). Even if gas monitoring is carried out, where the source of gas is made ground it it useful to characterize it by taking samples for TOC testing at 0.5 m intervals over its full depth. This allows a much more robust assessment of the monitoring data.

19.6.2 Semi-quantitative risk assessment

A large proportion of sites being considered as part of the planning process will comply with the scenarios outlined as part of the approach to qualitative risk assessment in the preceding section, thus requiring no (or limited) actual gas monitoring works in order to confirm gas protection requirements. However, other sites being considered as part of the planning process may require semi-quantitative risk assessment.

Within the planning framework the most commonly utilized semi-quantitative methods of ground gas risk assessment are:

(1) the NHBC Traffic Lights approach (Boyle & Witherington 2007), for the assessment of low-rise residential houses which have passively ventilated sub-floor voids;

Table 19.4. *Limiting values of TOC for assessment of made ground*

Characteristic situation (Modified Wilson and Card, Situation A CIRIA C665, BS8485)	Depth of made ground	Maximum total organic carbon content of made ground (TOC, %)[1,2]		Comments
		New made ground <20 years old	Old made ground >20 years old	
CS1	Maximum thickness not exceeding 5 m Average thickness not exceeding 3 m	1.0	1.0	Limiting values based on reported soil organic material (SOM) content of natural soils up to about 1%
CS2		1.5	3.0	Limiting values based on gas generation modelling assuming slow degradation Equilibrium methane concentration in building <0.01%v/v
CS3		4.0	6.0	

Notes: (1) TOC = DOC × 1.33 (Hesse 1971). (2) TOC testing in accordance with the method described in the guidance on sampling and testing of wastes to meet landfill waste acceptance procedures (Environment Agency 2005).

(2) the Modified Wilson and Card framework (Wilson & Card 1999), for the assessment of all other building types.

These frameworks for the semi-quantitative assessment of ground gas risks are the foundation of the risk assessment approaches described in CIRIA Report C665 and British Standard BS8485:2015 +A1: 2019. In summary, both models require the derivation of a Gas Screening Value (GSV, litres of gas per hour), which is a function of the borehole flow rates and methane/carbon dioxide gas concentrations recorded on a site and consideration of the conceptual site model.

Careful consideration of the flow rates used in the assessment is necessary. A common scenario in the NW of England is where boreholes are installed deep into glacial till deposits and intercept sand layers at depth. The glacial till can cause slightly elevated methane concentrations, up to 5% v/v in some cases, as a result of inclusions of organic material. Because the borehole acts like a chimney it allows high air flow to occur from the sand layer, which combined with the methane concentration indicates a risk. In reality the gas is trapped in pore spaces within the sand and the high flow would not occur if the borehole was not present.

19.6.3 NHBC Traffic Lights

The NHBC Traffic Lights approach works by comparing the measured gas emission rates with traffic lights, i.e. green, amber-1, amber-2 and red. The Traffic Lights include 'typical maximum concentrations', which are provided for initial screening purposes, and risk-based 'gas screening values' for consideration of situations where the typical maximum concentrations are exceeded. The calculations are completed for both methane and carbon dioxide, with the worst-case designation adopted for site assessment and confirmation of gas protection requirements. The NHBC Traffic model has been defined utilizing a number of basic assumptions, which include:

(1) the dwelling being considered by the model is a 'standard' residential house, with a passively ventilated, suspended (open void) floor slab construction;
(2) the worst gas regime identified at the site, either methane or carbon dioxide, recorded from monitoring in the worst temporal conditions, will be the decider for which the traffic light GSV is allocated;
(3) the typical maximum concentrations can be exceeded in certain circumstances should the conceptual site model indicate it is safe to do so;
(4) the GSVs calculated as part of the model should not normally be exceeded without completion of a detailed gas risk assessment taking into account site-specific conditions;
(5) the model low-rise house utilized for calculation of the screening GSVs has a floor plan measuring 8.00 × 8.00 m, floor area 64 m^2;
(6) the model low-rise house utilized for calculation of the screening GSVs has an open sub-floor void of minimum height of 0.15 m, equivalent to a sub-floor void space of 9.60 m^3;
(7) a conservative ventilation rate of one air change per 24 h has been assumed within the sub-floor void (i.e. equating to $9.60/24 = 0.40$ m^3 h^{-1});
(8) for carbon dioxide the GSVs have been calculated via the assumption that there is a leak from the sub-floor void into a small downstairs room of dimensions 1.50 × 1.50 m × 2.50 m, total volume 5.63 m^3;
(9) a conservative ventilation rate of one change per 24 h has been assumed within the small downstairs room (i.e. equating to $5.63/24 = 0.24$ m^3 h^{-1});
(10) The leak from the sub-floor void has been assumed to account for 10% (0.024 m^3 h^{-1}) of the small room ventilation rate, which is considered to be representative of a significant leak.

For methane the above assumptions have been used to back-calculate screening GSVs which aim at maintaining an equilibrium concentration of gas within the sub-floor void at below 2.5% v/v (amber-2/red), 1% v/v (amber-2/amber-1) and 0.25% v/v (amber-1/green).

Where it is calculated that equilibrium concentrations will exceed the amber-2/red designation (2.5% v/v) within the sub-floor void, it may be concluded that a high gas regime has been identified and standard residential housing would not normally be acceptable without further gas risk assessment and/or possible mitigation measures to remove and/or reduce the source of the gas.

For carbon dioxide the above assumptions have been used to back-calculate screening GSVs which aim at maintaining an equilibrium concentration of gas within the small downstairs room at below 0.5% v/v (amber-2/red), 0.25% v/v (amber-2/amber-1) and 0.125% v/v (amber-1/green).

Where it is calculated that equilibrium concentrations will exceed the amber-2/red designation (0.5% v/v) within the small downstairs room, it may be concluded that a high gas regime has been identified and standard residential housing would not normally be acceptable without further gas risk assessment and/or possible mitigation measures to remove and/or reduce the source of the gas.

A summary of the NHBC Traffic Lights model is presented as Table 19.5. Once the site has been classified via the traffic lights system, the scope of protection required may then be determined utilizing Table 19.6.

19.6.4 British Standard BS8485: 2015

The BS8485 technique utilizes the same basic requirements as NHBC Traffic Lights, whereby it requires a thorough understanding of the conceptual site model in order to inform the semi-quantitative risk assessment. However, whereas the NHBC Traffic Lights models assumes that the development comprises a low-rise residential dwelling with passively vented, suspended floor slab, BS8485 permits the assessment

Table 19.5. *NHBC traffic light system for 150 mm void*

Traffic light designation	Methane		Carbon dioxide	
	Typical maximum concentration (%v/v)	Gas screening value (GSV, l/hr)	Typical maximum concentration (%v/v)	Gas screening value (GSV, l/hr)
Green	1	0.16	5	0.78
Amber-1				
	5	0.63	10	1.56
Amber-2				
	20	1.56	30	3.13
Red				

Notes:
See assumptions as detailed above prior to utilizing this model and refer to CIRIA C665 for detailed information

---

of various other building types, including residential dwellings with ground bearing floor slabs, commercial buildings and industrial buildings.

As with the NHBC Traffic Lights, Modified Wilson and Card requires calculation of a GSV for a site, utilizing parameters for derivation which are appropriately robust and justifiable based on a thorough understanding of the ground model.

The correct application of the guidance in British Standard BS8485:2015 +A1: 2019 is described in Clause 6.3.1. BS8485:2015 +A1: 2019 uses the concept of the borehole hazardous gas flow rate. This is obtained by multiplying the gas concentration by the flow rate of all gases from a well. It is calculated for each monitoring event in each well. Where flow is not detectable, the limit of detection of the instrument is used in the calculations.

The individual values of hazardous gas flow rate obtained from several monitoring locations over several visits are considered collectively to establish a GSV for the site as a whole. This requires consideration of the results in relation to the conceptual site model.

Table 19.6. *Gas protection measures for low-rise housing development based upon allocated NHBC traffic light (Boyle & Witherington 2007)*

Traffic light classification	Protection measures required
Green	Negligible gas regime identified and gas protection measures are not considered necessary.
Amber-1	Low to intermediate gas regime identified, which requires low-level gas protection measures, comprising a membrane and ventilated sub-floor void to create a permeability contract to limit the ingress of gas into buildings. Gas protection measure should be as prescribed in BRE Report 414. Ventilation of the subfloor void should facilitate a minimum of one complete volume change per 24 hours.
Amber-2	Intermediate to high gas regime identified, which requires high-level gas protection measures, comprising a membrane and ventilated sub-floor void to create a permeability contrast to prevent the ingress of gas into buildings. Gas protection measures should as prescribed in BRE 414. Membranes should always be fitted by a specialist contractor. As with Amber-1, ventilation of the sub-floor void should facilitate a minimum of one complete volume change per 24 hours. Certification that these passive protection measures have been installed correctly should be provided.
Red	High gas regime identified. It is considered that standard residential housing would not normally be acceptable without a further Gas Risk Assessment and/or possible remediation mitigation measures to reduce and/or remove the source of gas.

The process for developing a GSV for a site is summarized as follows:

- borehole hazardous gas values are calculated for each borehole standpipe for each monitoring event;
- the reliability of the measured gas flow rates and concentrations is assessed taking into account borehole construction (and geology);
- decisions are made as to whether to use peak gas flow rates or steady state rates in each calculation (note: BS8485 states that steady-state values are normally to be used, as does BS8576: 2013);
- decisions are made about how to deal with any temporal or spatial shortages in the data; and
- judgements are made about what GSV to use for design purposes taking all relevant information into account.

Using the GSV, the site is characterized as one of six characteristic situations with situation CS1 being the lowest risk and situation CS6 being the highest. In addition, there is a requirement to *consider* an increase from CS1 to CS2 if methane concentrations exceed 1% v/v or carbon dioxide concentrations exceed 5% v/v. A similar approach is taken in NHBC guidance on ground gas where the GSV is used to define the site traffic light colour. Typical maximum concentrations are provided but these are for guidance and consideration only. It states that 'The typical maximum concentration can be exceeded in certain circumstances should the conceptual site model indicate it is safe to do so'. It is not mandatory in either BS8485:2015 +A1: 2019 or the NHBC Traffic Lights guidance to automatically increase the classification of a site because of elevated gas concentrations. In practice this would only be necessary where a high-risk source of gas is present such as from domestic landfills or open mine workings. It is not normally appropriate to increase the classification when the source of gas is made ground with limited organic content or alluvium.

An assessment of worst-case conditions is then completed, if appropriate.

Once the CS designation is confirmed and justified within the conceptual site model, the gas protection measures which will be required as part of the proposed development can be defined.

19.6.5 Quantitative risk assessment

The qualitative and semi-quantitative techniques outlined in previous sections of this paper will prove appropriate for the vast majority of sites being considered as part of the

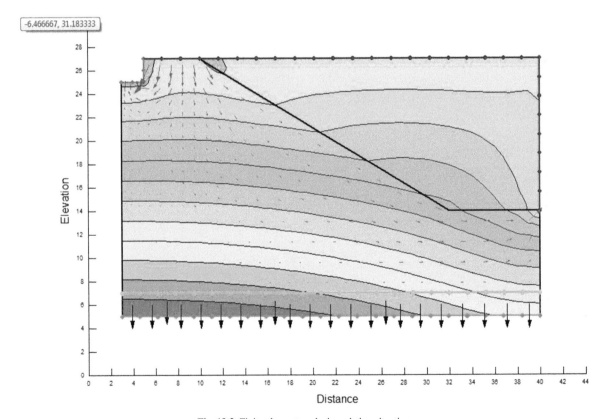

Fig. 19.3. Finite element analysis, existing situation.

planning process. However, there will be certain sites where these approaches are not appropriate and quantitative assessment techniques will be required. These scenarios are likely to include:

(1) sites being assessed for determination as Contaminated Land as defined by Part 2A of the Environmental Protection Act 1990;
(2) high-sensitivity developments (i.e. residential houses) where a high/very high gas generation potential source has been identified;
(3) complex underground structures (e.g. tunnels, deep basements).

With regard to complex underground structures the ventilation requirements outlined in Building Regulations 2010 Approved Document F will typically minimize risks associated with methane from most natural and anthropogenic sources, i.e. such structures are typically actively ventilated and constructed using reinforced concrete techniques. However, as demonstrated by the incident at Abbeystead (see discussion in Table 19.3), potential methane risks should never be overlooked and it is important that a detailed conceptual site model is developed to confirm that associated risks are being mitigated by design.

Quantitative risk assessment techniques typically involve mathematical modelling to determine the gas generation potential of a source and confirm the lateral/vertical rates of diffusion/advection of methane from an identified source towards an identified receptor. In addition, mathematical modelling will include calculation of anticipated equilibrium concentrations within a structure, together with any impact that other potential sources of methane (e.g. in the dissolved phase) may pose. Guidance on undertaking quantitative risk assessment modelling is provided in The Local Authority Guide to Ground Gas (Wilson *et al.* 2008). CIRIA Reports 152 (CIRIA 1995) and C665 also give guidance on estimating the probability of an explosive atmosphere occurring in a building (or the risk of an explosion or asphyxiating concentration of gas).

Practitioners should also consider the use of finite element models to assess the potential impact of gas migration through the ground. These are an extremely useful tool to aid understanding of the likely risk to identified receptors. An example of finite element modelling is provided in Figure 19.3. This relates to a site in Yorkshire, UK, where concerns had been raised regarding the potential for landfill gas to migrate into a residential dwelling located close to waste deposits via an underground, occupied cellar. The finite element modelling provided another line of evidence to practitioners when considering potential Part 2A liabilities associated with the landfill.

The finite element modelling also allowed the impact of varying landfill gas pressure, atmospheric pressure and waste permeability to be understood. This highlighted which properties were at greatest risk and allowed targeted intervention with remedial measures. It also allows a robust assessment of the effectiveness of potential remedial work, thus supporting a key principal of the Part 2A process, i.e. ensuring that remedial works are proportionate to the risks identified and cost effective (**Fig. 19.4**).

19.6.6 Acute situation

It is important to note that risks associated with methane gas are typically 'acute', i.e. reflecting exposure/contact with a substance that occurs once or over a short period of time. This is contrary to many other forms of contaminants, which typically result in chronic effects owing to exposure/contact over a longer, more sustained and prolonged period of time. Given the acute nature of methane gas risks it is important that the process of assessment is completed robustly and efficiently – such that all stakeholders are confident in the approach taken and conclusions reached.

For example, where methane migration from a landfill has been identified and residential properties are present in close proximity, long-term gas monitoring is not an appropriate risk management strategy. This is because the acute risk associated with the methane gas could be realized in a 'one-off' event – potentially as a result of temporal factors coming together to influence the gas generation/migration regime. Instead, the process of investigation should commence with a robust preliminary risk assessment which informs the development of a conceptual site model – where plausible pollutant linkages are identified and assessment works (be they intrusive, site monitoring, laboratory and/or theoretical) targeted efficiently.

The value of a robust conceptual site model cannot be underestimated. All too often poorly considered conceptual site models result in consultants being unwilling to make decisions on risk, resulting in recommendations for multiple phases of site investigation and/or prolonged periods of gas monitoring.

Ultimately the design of the assessment works should be sufficient such that assessment of risk can be undertaken in a robust, scientific manner, with minimal risk of additional assessment work being required.

19.7 Mitigating methane risks

In many situations the qualitative, semi-quantitative and quantitative modelling techniques outlined within this paper will demonstrate one of the following;

(1) gas protection measures are not required;
(2) generic gas protection measures (e.g. suspended, ventilated sub-floor voids and/or gas-resistant membranes) are adequate to manage risks associated with identified methane concentrations;
(3) the inherent construction of the proposed structure is sufficient to effectively mitigate risks associated with methane gas migration (e.g. owing to the adoption of active ventilation methods within the building and/or the nature of a reinforced concrete floor slab).

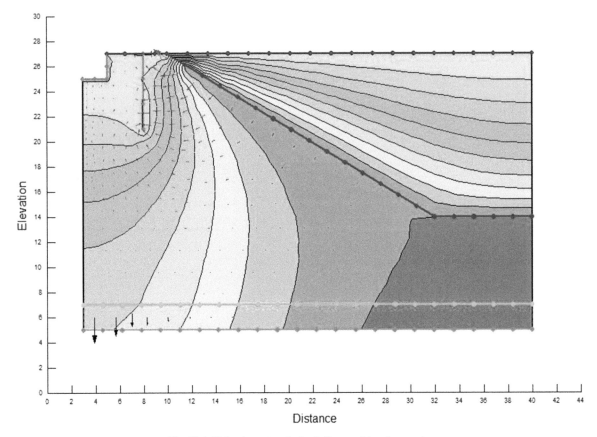

Fig. 19.4. Finite element analysis, shallow vent trench scenario.

However, inevitably there will be some sites where the risk to receptors is identified as being unacceptable and more specialized protection systems are required. Broadly these remedial works may be defined into three categories;

(1) source removal;
(2) barrier installation;
(3) dilution and dispersion.

In general terms 'source removal' of large gassing sources is impractical and unsustainable. However, there are instances where 'landfill mining' is used to remove the gassing source, treat/sort/stabilize it and redeposit the material such that gas risks are managed and the land made 'suitable for use'. An example of this is a site in Wokingham, where a former landfill is being excavated and sorted to minimize off-site disposal of waste, providing a development platform for the construction of 350 new private domestic properties.

'Barrier installation' incorporates both barriers to buildings (e.g. membranes and floor slabs/foundations) and in-ground barriers such as bentonite walls or vent trenches. The efficacy of these installations is highly variable depending on the specific conditions which exist at the subject site and the workmanship of the installations. Again the site conceptual model will drive the design of these systems (e.g. by ensuring that a vent trench intercepts all likely migration pathways).

'Dilution and dispersion' systems include both active and passive venting below the floor slabs of buildings. In the UK passive systems are preferred because they require less maintenance and do not require ongoing power. In tunnels or other excavations the only method of protecting against methane ingress is to provide adequate ventilation to dilute the gas, combined with detection and alarm systems (both area wide and on a personal level).

The overriding approach in the UK is to provide at least two different methods of protection against ground gas ingress so that if one fails or is damaged the building is still protected. As the level of risk associated with methane increases, the number of different protection levels required is increased, thereby increasing the available redundancy.

In some instances, the authors are aware of risks associated with methane gas ingress into buildings being managed via the installation of monitoring and alarm systems. Alarms do not prevent gas entry and should not be used as a means of gas control for buildings. In the short term, alarm systems may provide a useful tool for the localized management of methane risks. However, perception issues, erroneous alarms, accidental/deliberate interference and the need for continued calibration/maintenance should remind all involved with the specification of gas protection measures that an alarm within a gas protection system should be the last line of defence, appropriate only for low-sensitivity end uses (e.g. industrial units) as a final check of safety.

Fundamentally gas protection systems should be just that, 'systems' of defence whereby multiple layers of protection work together to mitigate methane gas risks to identified receptors.

19.8 Summary and conclusions

Methane is ubiquitous in the underground environment and can arise from a number of natural and manmade sources. Understanding the geological strata and processes in and around a site is a vital component of the risk management process for methane. Methane can pose a risk to buildings constructed over gas-bearing ground and also to the construction of tunnels and operation of mines.

In the UK there are several well-accepted approaches to assessing risk to new developments located over, or close to, methane sources (e.g. landfill sites). Most recently a new qualitative risk assessment tool has been introduced, which allows practitioners to carefully consider a methane gas source and determine the most appropriate risk mitigation strategy for adoption based on the gas generation potential of the source (as assessed via TOC laboratory testing).

On more sensitive, or highly gassing, sites increased understanding of landfill gas flow characteristics is informing more detailed quantitative risk assessment techniques/ modelling, with finite element analysis being utilized to aid the adoption of cost- and performance-effective mitigation solutions.

There are many different methods of preventing gas ingress into buildings but whatever the options the emphasis is on using a 'system' of measures to protect development. This provides redundancy such that if one element fails the buildings will still remain safe. 'Systems' of protection are essential when managing risks associated with methane – reflecting that it is an acute contaminant.

Funding This research received no specific grant from any funding agency in the public, commercial, or not-for-profit sectors.

References

ASTM 2016. Standard guide for evaluating potential hazard as a result of methane in the vadose zone. ASTM E2993- 16. ASTM International.

BARBER, C., DAVID, G.B., BRIEGEL, D. & WARD, J.K. 1990. Factors controlling the concentration of methane and other volatiles in groundwater and soil gas around a waste site. *Journal of Contaminant Hydrology*, **5**, 1990, 1555–1169, https://doi.org/10.1016/0169-7722(90)90003-Y

BERESFORD, J.J. 1989. *Permanent Works Design. Proceedings of Symposium on methane –facing the problems*. Paper 4.3, Nottingham, September.

BOLTZE, U. & DE FREITAS, M.H. 1996. Changes in atmospheric pressure associated with dangerous emissions from gas generating sites. The explosions risk threshold concept. *Proceedings of the Institution of Civil Engineers, Geotechnical Engineering*, **119**, July 177–181, https://doi.org/10.1680/igeng.1996.28509

BOYLE, R. & WITHERINGTON, P. 2007. *Guidance on Evaluation of Development Proposals on Sites where Methane and Carbon Dioxide are Present*. Report Edition No **4**, March 2007. NHBC.

BROUWER, G.E. 2009. *Brookland Greens Estate – Investigation Into Methane Gas Leaks*. Ombudsman Victoria, October 2009.

BSI 2013. Guidance on investigations for ground gas. Permanent gases and volatile organic compounds (VOCs). British Standard BS8576: 2013.

BSI 2017. Investigations of potentially contaminated sites – Code of practice. British Standard BS 10175:2011+A2:2017.

BSI 2019. Code of Practice for the design of protective measures for methane and carbon dioxide ground gases for new buildings. British Standard BS8485: 2015 + A1: 2019.

BUTTERWORTH, J.S. 1991. Methane, carbon dioxide and the development of contaminated sites. Paper 2.4, *Methane – Facing the Problems. Symposium*, Nottingham, March 1991.

CALIFORNIA ENVIRONMENTAL PROTECTION AGENCY 2012. *Evaluation of Biogenic Methane. A Guidance Prepared for the Evaluation of Biogenic Methane in Constructed Fills and Dairy Sites*. Department of Toxic Substances Control, 28 March.

CARD, G.B. 1995. *Protecting development from methane*. CIRIA Report **C149**. CIRIA, London.

CIEH 2008. *The Local Authority Guide to Ground Gas*. Chartered Institute of Environmental Health.

CIRIA 1995. *Risk assessment for methane and other gases from the ground*. CIRIA Report **C152**.

CIRIA 2007. *Assessing risks posed by hazardous gases to buildings*. CIRIA Report **C665**.

CLAIRE 2012. *A pragmatic approach to ground gas risk assessment*. Research Bulletin RB17. November 2012.

COUNCIL OF THE CITY OF LOS ANGELES 2004. Los Angeles Municipal Code, Ordinance no. 175790. *Meeting of Council*, 12 February.

DEFRA 2012. Contaminated Land Statutory Guidance. April 2012, https://assets.publishing.service.gov.uk/government/uploads/system/uploads/attachment_data/file/223705/pb13735cont-land-guidance.pdf

EKLUND, B. 2011. *Proposed regulatory framework for evaluating methane hazard due to vapour intrusion*. Air and Waste Management Association Magazine. February 2011.

EKLUND, B., SEPICH, J. & LEGRAND, R. 2014. Procedures for evaluating potential methane hazard at vapour intrusion sites. *Vapour Intrusion, Remediation and Site Closure Conference*, September, New Jersey.

ENVIRONMENT AGENCY 2004. *Model procedures for the management of contaminated land*. Contaminated Land Report **11**.

EPA NEW SOUTH WALES 2012. *Guidelines for the Assessment and Management of Sites Impacted by Hazardous Ground Gases*. State of NSW, Environment Protection Authority, Sydney, Australia.

HM Government 2013. *Approved Document C – Site preparation and resistance to contaminants and moisture. 2004 Edition incorporating 2010 and 2013 amendments.* The Building Regulations 2010.

Hooker, P.J. & Bannon, M.P. 1993. *Methane: its occurrence and hazards in construction.* CIRIA Report **130**.

Jackson, R.B., Pearson, B.R., Osborn, S.G., Warner, N.R. & Vengosh, A. 2011. *Research and Policy Recommendations for Hydraulic Fracturing and Shale-gas Extraction.* Centre on Global Change, Duke University, Durham NC.

Kirchgessner, D.A., Lott, R.A., Cowgill, R.M., Garrison, M.R. & Shires, T.W. 1997. Estimate of methane emissions from the US natural gas industry. *Chemosphere*, **35**, 1365–1390, https://doi.org/10.1016/S0045-6535(97)00236-1

Kuchta, J.M. 1985. Investigation of fire and explosion accidents in the chemical, mining and fuel-related industries – a manual. United States Department of the Interior, Bureau of Mines, Bulletin 680.

Martin, E. 2019. *Gas Explosions*, CBFT-US 2019, http://cfbt-us.com/wordpress/?p=421 (accessed 24 December 2019).

Perera, D. 2001. *Fresh Produce and Streets of Fire: Making Sense of the Methane Explosion in Fairfax*, 10 May, La Weekly, https://laweekly.com/fresh-produce-and-streets-of-fire/

Ryan, G., King, P.J. & Munday, G. 1988. *Report of the non-statutory public inquiry into the gas explosion at Loscoe, Derbyshire, 24 March 1986.* Submitted to Derbyshire County Council, February 1988.

Sepich, J. 2008. Hazard assessment by methane CVP (concentration, volume, pressure). Everything you knew about methane action levels was wrong. *Presented at the Sixth Annual Battelle Conference, Monterrey*, 19–22 May, California.

Sherwood Lollar, B., Westgate, T.D., Ward, J.A., Slater, G.F. & Lacrampe-Couloume, G. 2002. Abiogenic formation of alkanes in the Earth's crust as a minor source for global hydrocarbon reservoirs. *Nature*, **416**, 522–524.

University of Texas 2013. *Repetto Formation, at Bureau of Economic Geology.* University of Texas, Austin and Hamilton/Meechan, 151 http://www.beg.utexas.edu/environqlty/co2seq/co2dat/orepetto.htm. [accessed 5 October 2013].

US Energy Information Administration 2010. *Natural Gas Year-in-Review 2009*, July, http://www.eia.gov/pub/oil_gas/feature_articles/2010/ngyir2009/ngyir2009.html (accessed May 2011).

Williams, G.M., Ward, R.S. & Noy, D.J. 1999. Dynamics of landfill gas migration in unconsolidated sands. *Waste Management and Research*, **17**, 327–342, https://doi.org/10.1177/0734242X9901700502

Wilson, S. & Card, G.B. 1999. *Reliability and Risk in Gas Protection Design.* Ground Engineering, February.

Wilson, S., Card, G.B. & Haines, S. 2008. *The Local Authority Guide to Ground Gas.* Chartered Institute of Environmental Health. September.

Wilson, S., Card, G., Collins, F. & Lucas, J. 2018. Ground gas monitoring and 'worst-case' conditions. CL: AIRE Technical Bulletin TB17. August 2018.

Index

Page numbers in *italics* refer to Figures. Page numbers in **bold** refer to Tables.

Abbeystead (Lancashire) **467**, 475
Abbot's Cliff, Folkestone slope failure (Kent) 16, 69, 73
Aberfan (South Wales) 2, 3, *3*, 16, **18**, 81, 106, *108*
Abergorchi colliery 106, **109**
acid mine water 338
acrotelm 245
actinon 433
active-layer detachment slides 265
activity, quick clay 208
adit mining 291–292
Afen Slide *70*, **70**
Airdrie (Lanarkshire) methane gas explosion **460**
ALARP concept 149–150
Alaska, Lituya Bay tsunami 63
Allderman's Hill (Derbyshire) 102
alluvium, effect on radon distribution 446
Alport Castles (Derbyshire) 97
Alum Bay (IOW) **18**, *129*
Ampthill Clay Formation *227*
angle-of-draw, in coal mining subsidence 297
Anglian glaciation 259, 260, *262*
anhydrite *see* gypsum and anhydrite
Anna Maria Arsenic Works (Devon) 345, *345*, *346*
anthropogenic geohazards 10, 116–119, 132, 134
Antrim County 82, 115, 119, 121, 134
Appley Bridge (Lancashire) **460**
Arden Sandstone Formation *406*
Ardyne Point 213
area-of-influence, in coal mining subsidence 297
Argyll Group, effect on radon distribution 446
Ariake Bay Clay (Japan) 205
arsenic mining 328, 330
asteroid impact, role in tsunami 65
attenuation 49
Avachinsky, tsunamigenic activity **65**
avalanche *see* snow avalanche
Avonmouth (Bristol) 214

Bacton (Norfolk) 134
Ballycastle (N Ireland) 82
Barrow on Soar (Leicestershire) 404
Barrow-in-Furness (Lancashire) 47, 56
Barton Clay Formation 121, *227*
Barton-on-Sea (Hampshire) **83**, 108, *113*
bastard brining 373
Baswich (Staffordshire) 378
bat roosts 353–355
Bath 18, 82, 102, 129
Beachy Head (Sussex) **83**
Beacon Hill (Bath) 129
Beaconsfield (Buckinghamshire) *399*
Beaminster Tunnel (Dorset) 16, **18**, 82
Becton Sand Formation *113*
Bedford United Mine, Cornwall *338*

Bedfordshire 312, 317, **390**
Bedwellty colliery flow slide **109**
beidellite 224
Beinn Alligin rock avalanche 96, *100*
Belfast sleech 214
bell pits 292, 293, *294*, 314, *315*
Bellwin Scheme 147–148
Beltinge (Kent) **83**
Ben Attow 102
Ben Gorm (Mayo) *103*, *126*
Ben Lui Schist Formation *125*
Ben Nevis rock fall **18**
Benston, Loch of, tsunami deposits 67, *71*
bentonite clay 11, *263*
Berkshire 6–7, *8*, **18**, 312, 317, 318, **390**, *395*
Bilkely 371
Bindon (Devon) **83**, 121, 132
biplanar compound slide 105, *107*
Birches slide (Shropshire) 129
Birling Gap (East Sussex) *111*
Bishop Monkton (Yorkshire) 408
black shales 437, 439, 444
Black Ven (Dorset) 4, **83**, 97, 114
Blackbrook reservoir damage 47
Blackgang (IOW) **83**, 114, *114*, *117*
Blackhall (Durham) **83**
Blackwall Tunnel (London) 5, *6*
Blaencwm 99
Blue Anchor Bay (Somerset) **83**
Blue Anchor Formation *406*
bog burst 96, 102, **104**, 132
bog slide **104**
bogflow **104**, 132
Bolsover landslide (Derbyshire) 119
bord-and-pillar mining 292–293, *294*
Borrowdale Volcanics Group 264
Boscombe (Dorset) **18**
Bothkennar (Scotland) 215
boulders, subsurface 9
Bournville (Ebbw Fach, Blaina) 3, **99**, 119, *121*, *122*
Bovey Basin 327, 328
Bowland, Forest of 134
Bramfield (Hertfordshire) **460**
brammalite 224
Branscombe Mudstone Formation 404, *406*
brickearth 20, *20*, 188, 285
    characteristics 192–196, *194*, **195**
    defined 200
    maps *189*, *190*, *191*
    terminology **192**
Bridport Sands 82
Brierley Hill (West Midlands) 18, 84
Brighton Marina 135
brine extraction and subsidence hazards 29–30, *29*

brine springs 371
Bristol Channel 65, **68**
Brotherton Formation *406*, **407**
Buckinghamshire 268, *273*, **312**, 390, 395, 399
Buildwas (Shropshire) 3, 99, **99**, 101, 119, 129
bulking, in coal mining subsidence 297
buried periglacial terrains 285
Burnhope Seat (North Pennines) 252–253, *253*
Burnley (Lancashire), coal mining subsidence *294*
Burton Bradstock (Dorset) 16, **18**, 81, *82*
Bury Hill (West Midlands) 18, 84
Bwlch y Saethau rock fall **18**

Cadeby Formation *279*, *406*, **407**, 408
Cairndow (Scotland) 164, 168, *169*, *170*
California, Fairfax **467**
cambering 102, *105*, **266–267**, 276–280, *278*
Camborne School of Mines 331
Cambridgeshire **312**, **390**
Camden Crescent (Bath) 18, *82*
Came Clay 215, *216*
Campi Flegrei (Italy) 1
Canary Islands, as source for tsunami **67**
capping *305*, 306
carbonate rock dissolution (karst) 30–32, *31*, *32*, **268–269**, 284–285, 389
    features 392–395
        characteristics 392, 394, *394*, 395
        engineering management 395–400
    UK occurrence 389, **390**, *391*, 392, *393*
Carboniferous Limestone 119, *123*, 436
Caribbean, as source zone for tsunami **67**
Carmarthenshire, Cwmduad slide **18**
Carnon River (Cornwall) 338, *341*
Carrickfergus (NI) 373, 380
Carse Clay Formation 214, 215
Carsington Dam (Derbyshire) 5, *5*, 261
Carsington Pasture (Derbyshire) karstification 2
Castle Hill (Kent) 3
catotelm 245
caves 392, *394*, 408
Cefn Glas colliery flow slide **109**
celadonite **224**
cemented layers 9
Chale Cliff (IOW) **83**, *93*, 108, *114*
chalk 314
    dissolution *see* carbonate rock dissolution
    hazards 6–7, *8*
        engineering management 317–319
        mining and mine characteristics 27, 28, *28*, 311, 314–317
        UK distribution 311–313, **312**
Chalk Group 119, 389
chalkangles 316, *316*
chalkwells 315–316, *315*
Chama Sand Formation *113*
Channel Tunnel Terminal 134–135
channelized debris flows 19, *19*, 164, *164*
Channerwick (Shetland) 102
Chartley Moss 379
Charton Bay (Devon) **83**

Chellaston Quarries 411
Cheshire **390**
    salt strata 29, *29*, 369, 370, *370*, 371, 372, 377, *377*, *378*, *383*, 384
Cheshire Brine Subsidence Compensation Board 382
Chilton Chine (IOW) *93*
China, Zhouqu debris flow 178, *179*
china clay, Cornwall 321, 326, *327*, 331, *333*, 353
    opencast mining legacy 337, *340*
Cilfynydd colliery flow slide 106, **109**
clay 7, 9, 205, 274, 276
    see also quick clay
Clay Bank (Yorkshire) **99**
clay minerals **224**, 225, 230–231, *231*, 238
Clay-with-Flints, effect on radon distribution 446
Cleveland, dissolution features **390**
cliff behaviour units (CBU) 106–108, *110*
cliff recession 16, 81, **83**, 99
    development classification 120–121, 124
Cligga Mine, Cornwall *329*
climate change, impact on debris flows 179–183
Cliveden, solution pipe *395*
Clwyd Cap 336, *337*
Clyde Alluvium 214
Clyde Clay Formation 214
Clyde raised marine deposits *217*
coal mining
    hazards 7, *9*, 26, 27
        managing subsidence risks 301–304
            mitigation and remediation 304–306
                role of engineering properties 298–299
                subsidence classification
                    longwall 295–298
                    room-and-pillar 293, 295
                    shafts and bell pits 293
                subsidence prediction 299–301, *299*, *300*
    methods
        adits, drifts, shafts 291–292, *293*
        bell pits 292, *294*
        longwall 293
        room-and-pillar 292–293, *294*
    UK legacy 291
coalification 463
coastal flooding 12
coastal hazards 4–5
coastal landslides 106–108, *110*
    cliff behaviour 120–121, 124
coastal protection measures 146
coinage 330
Colchester (Essex) 5, 12, 45, 47, 56
collapse monocline *370*, 372
collapsible soils 19–21
    collapse potential 196–197
    defined 187, 200
    management 197–198
    non-engineered fills 196
    treating collapsibility **199**
    see also loess
colliery spoil flow slides 106, **109**
Colwich (Staffordshire) 378

Combermere Abbey 371
complex cliffs 107–108, *110*
complex rock block spreads 102
complex/composite movement, defined 85, *89*, **95**
Composite cliff systems 107, *110*, *113*
compressibility, peat, hazards studies 250–253
compressible ground
  defined 244–245
  hazard ratings 254, **255**
compressible ground potential 25
  map *23*, *244*
compression
  peat 246, 248
  subsidence hazard 260–261
compression index, peat 247, 248, *249*
Comrie (Perthshire) 12, 45, 53
conservation work 352–353, *361*
consolidation, peat 246, 247, 248
contaminated land, defined 469
copper mining *322*, 328, 330, 336–337
Cornish engine house 331, *332*
Cornubian Batholith 326
Cornwall
  china clay 321
  cliff recession **83**
  earthquake monitoring 42
  karst features **390**
  loessic deposits 188, 191
  Looe slide **18**
  mining and subsidence hazards 28–29
  Newquay **18**
  St Michael's Mount **68**
  tsunami events **68**
Cornwall and Devon mining industry
  environmental legacy
    contaminated water 338, 343
    flooding 343
    opencast 336–337
    silted estuaries 338
    tailings ponds 338
    underground 333–336.*334*, *335*, *336*
    waste tips 337
  geological setting 322, *323*
    plate tectonic setting *324*
    post-Variscan magmatism and mineralization 326–328, *327*, **327**
    regional structure 325–326
    Variscan basement 322–325
  history 328–333
  legacy of hazards
    assessment 344
      desk studies 344, **344**
      field mapping 346, **349**
      geophysics 345–346, **347**, **348**
      ground investigation 346, 349–350
      modelling 350, *350*
      monitoring 351–352
      remote sensing 344–345, *345*, *346*
      risk register *351*
    treatment 352–360

case studies 356–360
conservation 352–353, *361*
derelict land problems 355
planning framework 352
remediation 353–355
water contamination problems 355–356
cosmic rays 435
Cotswolds 102, 119, 127, 131
Coulomb Equation 108
Covehithe (Suffolk) **83**, 134
coversand, as a geohazard 265, **268–269**, 285
Craig y Ddelw 102
Craig-Duffryn colliery flow slide **109**
Crimdon (Durham) **83**
Cromer (Norfolk) **83**
crown hole 293, 295, 318, *318*, 319
cryogenic wedges 283–284
Cuilcagh Mountains (NI) 134
Cumbria **18**, 134, **390**
curved compound slide 105, *107*
Cwmduad slide **18**
Cwmtillary (South Wales) 125

Dalradian schist 119, 446
dams, earthquake damage 47
Darlington (Yorkshire), gypsum dissolution subsidence 409–411
Darren Ddu landslide 102, *430*
Darren Goch landslide 102, 119, *430*
DART 74
Dartford (Kent) **460**
Dawlish (Devon) **18**
debris avalanche *165*
debris floods 164, 166
debris flows 18–19
  channelized 19, *19*, 164, *164*
  defined 163, *165*
  hazard and risk assessment 172–173
  hillslope (open-slope) 19, *19*, 164, *164*
  impacts of 178–179
  risk reduction 173–178, *178*
  role of climate change 179–183
  Scotland 166, *167*, *169*
    A9 Dunkeld 168, *169*, *171*, *173*
    A83 Cairndow 168, *169*, *170*
    A83 Rest and be Thankful *169*, 171, 172, *176*, *177*, 179, *180*
    A85 Glen Ogle 82, 170, *174*, *175*
  travel distance *142*
deep weathering hazards 262–265, **266–267**
deglaciation
  isostatic uplift 45
  postglacial slope responses 128
deneholes 314–315, *315*
Derbyshire
  Allderman's Hill 102
  Alport Castles 97
  Bolsover 119
  Carsington Dam 5, *5*
  Carsington Pasture 2
  Gillot Hey 102
  gypsum mining 411

Derbyshire (*Continued*)
  karst features **390**
  Loscoe 2, *2*, **460**, 464, **467**
  Mam Tor 3–4, *4*, 16, 82, *86*, 97, **99**, 114
Derrybrien landslide (Galway) 251–252, *251*
Devensian glaciation 260, *262*, 328, 371
Devon
  chalk and flint mines **312**
  cliff recession **83**
  Dawlish **18**
  earthquake monitoring 53
  karst features **390**
  loessic deposits 188, 198, *199*
  mining and subsidence hazards 28–29
  tsunami events **68**
  see also Cornwall and Devon mining
Devon Great Consols 360, *360*, *361*
Devonian limestone 119
dickite **224**
dip, role in coal mining subsidence 297, *298*
Dirtwich (Shropshire) 371
dissolution hazard 2–3
    see carbonate dissolution *also* gypsum/anhydrite *also* salt
Dogger Bank (North Sea) 5
Doggerland 67
dolines *see* sinkholes
Dorset
  Beaminster Tunnel 16, **18**, 82
  Black Van 4, 97
  Boscombe **18**
  Burton Bradstock 16, **18**, 81, *82*
  chalk and flint mines **312**
  cliff recession **83**, 121, *130*
  Durdle Door 16, **18**, 81
  karst features **390**
  Kimmeridge Bay **18**
  Lulworth Cove 16, **18**, 81
  Lyme Regis **68**, 97, 116, 118
  Swanage Bay **18**
Dover, Guildford Battery **18**
Dover Harbour slope failure *73*
Dover Straits, earthquakes 14, 45, 55, 73
Downderry (Cornwall) **83**
Dragvoll quick clays (Norway) *209*, **210**
Drakelands Tungsten Mine (Devon) 337, *339*
drift mining 291–292
Droitwich (Worcestershire), salt strata *29*, 30, 379–380, *381*, 384
dropout sinkholes 396
dry rockhead 372
Dudley, The Crooked House *9*
Dungeness Nuclear Power Station 82
Dunkeld (Scotland) 168, *169*, *171*, *173*
Dunwich (Suffolk) 16, 81, **83**
Durdle Door (Dorset) 16, **18**, 81
Durham **83**, **390**
Durness Group limestone 389

Earl's Burn reservoir damage 47
earth flow *165*
earthquake intensity 44

earthquake magnitude 44
earthquakes
  isostatic uplift effects 128–129
  magnitude-depth relation *47*
  role of plate tectonics 44
  role in tsunami 43, 63, *64*
  UK distribution 12, *13*, 14, 44–47, *46*
    consequences of 47–48
    monitoring 53
    recognizing hazard 48–50
  wave motion 43
East Anglia, 1953 floods 12
East Leake subsidence 411
East Pentwyn landslide (South Wales) 3, *90*, **99**, 119, *122*
East Sussex, snow avalanche 11, *11*
Ebbw Fach Valley landslide *430*
Ebbw Vale (Gwent) **460**
Ebrington Hill (Warwickshire) 131
Eddystone gneiss 325
Eden, Vale of 407, **407**
  gypsum dissolution 410–411
Eden Project 337, *340*, 353
Eden Shales Formation 407
Edlington Formation 404, *406*, 407–408, **407**, 409
effective stress 108
elastic compression, peat 246
Elton Flashes *378*
emanation coefficient 436
engineering rockhead, defined 263
English Channel 12, 45, **68**
Essex
  chalk and flint mines **312**, 315
  chalk karst **390**
  cliff recession **83**
  Colchester 5, 12, 45, 47
  earthquakes 48, 56
  loess 20
estuary silting, Cornwall and Devon 338, *341*
Etna tsunamigenic activity **65**
European Macroseismic Scale (EMS) 44
Ewood Bridge 84
Exwick Farm (Exeter) 18, 84
Eyjafjallajökull (Iceland) 1, 6, *8*

Factor of Safety Equation 108, *115*
Fairfax (California) **467**
Fairlight Glen (Sussex) 4, **83**, 121, 124, *130*
Fal Estuary 338
falling, mechanism defined 85, *89*, **95**, *96*
Faroe Islands, Storegga slide deposits 67
faults, reactivation features *426*
  diagnostic features 426–427
  hazards 34–35, *36*
  mitigation 427
  occurrence 425
  rockhead anomalies 281
  role in earthquakes 43
  surface expression *428*, *429*, *430*
Fernhill colliery flow slide **109**
Fforchaman colliery flow slide **109**

Flamborough Head (N Yorkshire) **83**, 107, *113*, 120
flashes 372, 373, *373*, *374*, 377
Flint Hall Farm (Surrey) 3, *3*
flint mining 27, 28, 311
    mine characteristics 313–314
    UK distribution 311–313, **312**
flocculation in clays 206, *207*, **209**
flooding 11–12, 343
flow slides in colliery spoil 106, **109**
flowing, mechanism defined 89, *89*, **95**, **96**
fluvial scouring 281
Folkestone (Kent)
    Abbot's Cliff slope failure 69
    earthquake 55, 56
    Folkestone Warren landslide 4, *4*, 18, *73*, 97, 119, 121, *124*, 163
    tsunami event **68**
Foreland Slide Complex *70*, **70**
fracking, problems of methane 463
Franklands Village (West Sussex) 18, 84
freeze-thaw processes 230, 270
French House slide (Kent) 16
French Park Salt Mine (NI) 380
Frome Valley 102
frost creep 265
Fuller's Earth 270
Fyne, Loch 213

Galway, Derrybrien landslide 251–252, *251*
gamma rays 434–435
Gaping Gill (Yorkshire) 392
Garth Loch tsunami 15, 67, *71*
gas, role in peat 246, *246*
gas hazards *see* methane *also* radon
gas hydrates, role in tsunami 65, 66
gas screening values (GSV) 472, 474
gas storage 375
Gault Clay Formation *92*, 111, *117*, 119, 121, *124*, 126, *148*, *227*, 270
Geike escarpment 55, *55*
Geike Slide *70*, **70**
gelifluction 265
GEM Raft *70*, **70**
geohazard defined 1, 10, 10–11, **11**
Geological Society Engineering Group *see* Working Party on Geohazards
geology
    impact on landslides 119–124
    impact on radon distribution 439, *440*, *441*, *442*, *443*, 444–446, *444*, *445*
geotechnical hazards 7, 9
Gillot Hey (Derbyshire) 102
glacial till, effect on radon distribution 446
glacial/periglacial cycles, impact on landslides 126–127
glaciofluvial scouring 281
Glamorgan, Llantwit Major **18**
glauconite **224**
Glen Ogle (Stirlingshire) 3, 16, 82, *85*, *125*, *169*, 170–171, *174*, *175*
Glendevon dam 47

Glendun River 134
Glenrhondda colliery flow slide **109**
global warming and greenhouse gas 462
Gloucestershire, karst feautues **390**
Gogarth (Gwynedd) **18**
Gothenberg quick clays (Sweden) **209**, **210**
Graig Goch (Wales) 97
grain flow *165*
Gramscatho Basin 323–324, *323*, 325
Grand Banks (Newfoundland) 48
Grangemouth 214, *217*
granites *323*, 326, *327*, 436, 439
Gratwich (Staffordshire) 378
Great Glen Fault 50
Great Glen (Inverness) 12, 45
Great Oolite Group 264
Greater London **312**, 390
Greatham (Teesside) 375
greenhouse gas and global warming 462
Greensand Formation 119, *273*
greywackes, Ordovician/Silurian, effect on radon distribution 446
grillages/rafts 400
Grimes Graves (Norfolk) 312, 313
ground deformation, categories 230
ground ice 261, 283
ground investigation techniques **271**, **272**
ground motion characterization (GMC) 49
ground motion model 49
ground motion prediction equation (GMPE) 49
ground source heat pumps 417, *418*
groundwater levels 116
grouting *305*, 306, *319*, 353, 382, 384, 399
Guildford Battery, Dover **18**
gulls 102, *105*, 277, **277**, *279*, **279**, 280
Gunnislake 333, *336*
Gutenberg-Richter equation 14, *14*, 45
Gwash Valley (Leicestershire/Rutland) 126, 131
Gwynedd **18**, 135
gypsum and anhydrite 403
    dissolution 2–3, 404
        hazard management 32–34, *33*, *34*
            civil engineering techniques 413–416
            ground investigation 411–413
            relation to water abstraction 416–417
            subsidence, planning for 417, 419
    related subsidence
        Darlington 409
        Ripon 407–409
        Vale of Eden 410–411
    transition, expansion and heave 403–404
    UK distribution 404–405, *406*, 407
Gypsy Hill (London) 84

Haggerlythe (Whitby) 131
halite 29
    see also salt
halloysite **224**
Hambleton Hills (Yorkshire) 102
Hameldon Hill (Burnley) *294*

Hampshire 20, **83**, *113*, **312**, **390**, 397
Hanover Mine (Reading) *311*
Happisburgh (Norfolk) 4, *84*
Happisburgh Till Member 190
Harrison, Lake 127
Harthope Quarry (Pennines), peat flow 252, *252*
Hatfield chalk mining (Hertfordshire) 317
Hatfield Main Colliery (South Yorkshire) 3, 16, 82, *87*
Haverton Hill 375
Hawkesbury (Canada) quick clays **210**
Hawkins, Alfred Brian xvii–xviii
Hawkley (Hampshire) **99**, 129, 131
hazard, defined 135–136
head 268
heave
    anhydrite-gypsum 32–33
    defined 238
    see also shrink-swell soils
hectorite **224**
Hedgemead landslide 18, 82
Hell Kettles (Yorkshire) 409
Helland Hansen Slide *70*, **70**
Hemel Hempstead (Hertfordshire) 317, *318*, *319*
Hensroost Mine, Devon 336, *339*
Herefordshire 14, 45, **390**
Hermitage (Berkshire) *395*, *396*
Hertfordshire **83**, **312**, *316*, *318*, *319*, **390**, 399
Highcliff (Hampshire) **83**
Higher Wych 371
Hill Pot (Yorkshire) 392
hillslope (open-slope) debris flows 19, *19*, 164, *164*
Holbeck Hall Hotel (Scarborough) 4–5, *4*, 120, 132, 134, 135, 145, 147
Holderness coast (Yorkshire) 16, **83**
Holocene, climate change impact 129
Holocene Cold Event 66
Horndean (Hampshire) 397
Hornsea (Holderness) **83**
Hot Dry Rock Geothermal Project 53
house raising 385
Howgill Fells (Cumbria) 134
Humberside offshore earthquakes 12, 45
hydro-consolidation, defined 200
hydrocompaction, defined 200
hydrous mica **224**
Hythe-Lympne escarpment 16

ice-wedge pseudomorphs 283–284
Iceland 1, 6, *8*
illite clay **224**, 238, *263*
Indian Ocean tsunami 14, 43, 61, *62*
Indonesia, Aceh *62*
intrusive ice 261, 283
Inverness 12, 45, 48, 50
Ireland sesimicity 14, 45
Ironbridge Gorge landslides (Buildwas and Jackfield) 3, 18, 83, *88*, 118, 119, 127, 129
ironstones, effect on radon distribution 444–445
irregular compound slide 105, *107*
isostatic uplift 128–129, 212

    effect on quick clay 205, *216*
    role in earthquakes 45
Italy 1, **65**, 163, 178

Jackfield (Shropshire) 3, 18, 83, *88*, 119
Japan 61, **65**
Jersey, earthquakes 45

kandites **224**
kaolinite and kaolinitization 188, **224**, 326
karst and karstification 2–3, **268–269**, 284–285, 389, 394
Kellaways Formation *227*
Kent
    Abbot's Cliff 16
    chalk karst **390**
    cliff recession **83**
    Dover Harbour slope failure *69*
    earthquake 55, 56
    Folkestone tsunami event **68**
    Folkestone Warren 4, *4*, 18, *73*, 163
    Hythe-Lympne escarpment 16, 82
    loess 20, 190, 191, 192, 193, *194*
    methane explosion **460**
    Robertsbridge 116
    Roughs landslide 116, *118*
    Sevenoaks Bypass 5, *7*, 82, *100*, 126
    Warden Point **83**, 97, 116, *120*, *123*
Kent-Artois Shear Zone 73
Kessock Bridge (Inverness) 50
Keyworth subsidence 411
Kilnsea (Holderness) **83**
Kimmeridge Bay (Dorset) **18**
Kimmeridge Clay Formation *130*, 227
Kintail 53
Kintbury Mine (Berkshire) *312*, 317
Kirby Hill (Norfolk) **83**
Kirby Thore gypsum opencast 411
Knipe Point (Yorkshire) 135, 145
Knocknageeha bog (Kerry) 96, 132

Lake District, Threlkeld Knotts landslide 86
Laki Fissure eruption (1783-4; Iceland) 1, 6, *8*
Lambeth Group *227*
Lanarkshire, methane gas explosion **460**
Lancashire
    Abbeystead **467**, 475
    coal mining subsidence *294*
    dissolution features **390**
    Forest of Bowland 134
    Rampside earthquake 47, 56
    salt strata 29, *29*, 369, 370, *370*, 378
landfill site, problems of methane 463, 464, 465
landslides 3–5, *134*
    assessing risk 134–140
    causes of 108–124
        anthropogenic 116–119
        geology 119–121
        rainfall 114–116
    classification 84–86, *89*, 164
    Derrybrien (Galway) 251–252, *251*

hazard analysis 140–144
    impact on property 140, *141*, **141**
    influence of periglacial climates 284
    inventory for Great Britain and Ireland 91–92, 94, *97*, *98*
    main features
        coastal 106–108
            peat failures 101–102, *103*, **104**
            slope deformation 102, 105, *105*
    investigation methods 142–144
    phases of activity 124–134
        anthropogenic change 132, 134
        deglaciation adjustments 127
        glacial/periglacial 126–127
        Holocene climate change 129
        Little Ice Age climate change 129–132
        postglacial slopes 127–129
    reactivation damage curve 139, *139*, **140**
    retrogressive 211, *214*
    risk analysis 134–140
    risk management 144–146
        role of government 146–149
        tolerable risk 149–151
    role in tsunami 63, *65*
    susceptibility map *17*
    travel terminology *142*, **142**
Lapworth, Lake 127
Last Glacial Maximum 128
Late Glacial rockfalls 126
lateral spreading *see* spreading
Leach Slide Complex 6, *8*, 15, 48, *70*, **70**
leaching 207
lead mining 328
Leda Clay (Canada) 205, 207
Leeds (Yorkshire) *293*, **460**
Leicestershire 126, 127, **390**, 411
Leira quick clays (Norway) *209*, **210**
Lemieux (Ontario) 205
Levant Mine, Cornwall *330*
Lewes (East Sussex), snow avalanche 11, *11*
Lias Group 107, 119, 121, *227*, *233*, 270
Liassic Limestone, (Glamorgan) 124
limestone pavements 392, *394*
limestones 119
    effect on radon distribution 445–446
    see also carbonate rock dissolution
Lincoln, cathedral earthquake damage 47, 56
Lincolnshire 47, 56, **312**, **390**
liquefaction and earthquakes 43, 48
liquidity index, quick clay 208
Lisbon earthquake (1755) 5–6, *72*
    tsunami 15, 48, 65, 68–69
Little Ice Age 129–132, 260
Lituya Bay (Alaska) 63
Liverpool, earthquake damage 47
Lizard Complex 324–325
Lizard Peninsula (Cornwall) 188, 191, 264
Llantwit Major rock fall **18**
Lleyn Peninsula earthquake (North Wales) 47, 55
Loch Lomond Stadial (Readvance) 126, 128, 129, 260
Lochalsh, Kyle of 53

loess 20, *20*, 188, 328
    defined 200
    as a geohazard 265, **268–269**, 285
loessic brickearth 20, *20*
    characteristics 191–196, *194*, **195**
    defined 200
    terminology **192**
London 5, *6*
    chalk **312**, **390**
    methane gas explosion **460**
    rockhead anomalies 281, *282*
    Thames Barrier *12*
London Clay Formation 5, *6*, 11, 107, 111, 119, *120*, 121, *123*, *227*, 230, 232, 264, 270, 457
longwall mining 293, 295–298
Looe (Cornwall) **18**
Loscoe (Derbyshire) 2, *2*, **460**, 464, **467**
lower explosive limit, methane 459, *459*
LOWNET 50, 53
Luccombe (IOW) 116
Lulworth Cove (Dorset) 16, **18**, 81
Lyme Regis (Dorset) **68**, 96, 108, *115*, 116, 118, 145
Lympne (Kent) 16, 82, *87*

Maerdy colliery flow slide **109**
Maiden Mount Salt Mine (NI) 380, *382*
Mains Colliery (Wigan) 7
Mam Tor (Derbyshire) 3–4, *4*, 16, 82, *86*, 97, **99**, 114
Marcle landslide 129
marine erosion 16
marine isotope stages (MIS) 259, *260*
Market Rasen (Lincolnshire) 56
Marl Bluff (Norfolk) **83**
marsh gas *see* methane
Marston Big Hole 373
mass movement caves 102, *105*
Maypole Colliery (Wigan) 7
Mayuyama, tsunamigenic volcanism **65**
Mercia Mudstone Group *227*, 264, 265, 377, 378, 379
    gypsum beds 404, *406*
metastability, defined 200
meteorites, role in tsunami 65
meteotsunami 64–65
methane clathrates, role in tsunami 65, 66, *66*
methane gas 35, 38
    explosions **460**, **461**, 466, **467**
        Abbeystead **467**
        Loscoe 2, *2*, **467**
        risk assessment 469–475
        risk management 468–469
        risk mitigation 475–477
    greenhouse properties 462
    legislative framework 462
    migration pathways 464–466
    modelling behaviour of 463–466
    in peat 246, *246*
    properties 457, 459, *459*
    sources 457, **458**
        modelling behaviour of 462–463
Metropolitan Line closure 135

Mid-Atlantic Ridge, as source zone for tsunami **67**
Middlesborough 375
Middlewich (Cheshire) 380
Milford Haven, tsunami **68**
Miller Slide *70*, **70**
Millstone Grit Group 119
mining *see* coal; chalk; flint; salt; Cornwall and Devon mineral exploitation
mining hazards 6–7
mining subsidence
  effect of faults *427*
  see also faults, reactivation features
Mink Creek (Canada) *209*, **210**, **213**
mobilized shear strength 108
Moine schist 119
montmorillonite **224**, 239
Montserrat, tsunamigenic volcanism **65**
Morecambe Bay, sand fountains 48
Moston Flash 372, 373, *373*
Mundesley (Norfolk) **83**
Mynydd Corrwg Fechan colliery flow slide **109**
Mynydd Llangynidr (Wales) *396*
Myojin-sho, tsunamigenic volcanism **65**
Mythop Halite Formation 378

nacrite **224**
Nangiles Adit 338
Nantewlaeth colliery flow slide **109**
Natural Hazards Partnership (NHP) 149
needle-ice creep 265
Nefyn (Gwynedd) **18**, 135
Ness, Loch **18**
Netherwich (Worcestershire) 380
Nettlebed (Oxfordshire) 317
New Tredegar **99**
Newark Gypsum 404
Newbiggin (Northumberland) **18**, 410–411
Newfoundland 43, 48
Newquay (Cornwall) **18**
NHBC traffic light system 472–473, **473**
non-engineered fill, defined 196, 200
nontronite **224**
Norfolk
  chalk and flint mines **312**, 317, *317*
  chalk karst **390**, *399*
  cliff recession 4, 82, **83**, *84*, 107, *112*, 120, 132, 134, 135, *136*, *137*
  sand flows *92*
Normannian Nappe 325
North Faroes Slide *70*, **70**
North Foreland (Kent) **83**
North Sea
  Dogger Bank 5
  earthquakes *13*, 14, 45, 55
  Storegga slide 48
  submarine landslides *70*, **70**
Northamptonshire 2, **390**, 448, *449*
Northern Ireland 16, 82
  salt strata 29, *29*, 369, 370, *370*, 371, 373, 380, 384
Northumberland **18**, **390**

Northwich (Cheshire) 382, 385
Northwich Halite Formation 377
Norwegian Sea, submarine landslides *70*, **70**
Norwich Brickearth 190
Notre Dame de la Salette (Quebec) 205
Nottinghamshire **390**, 411
nuclear industry 50, 77, 82
Nunwick (Yorkshire) 408
Nyk Slide *70*, **70**

Ochil Hills 12, 45, 47
Ogle *see* Glen Ogle
Ogmore Valley landslide *430*
Ogwen, Llyn (Snowdonia) 82
open-slope debris flows 19, *19*, 164, *164*
opencast mining, Cornwall and Devon 336–337
Ormskirk (Lancashire) **460**
Ospringe (Kent) 190
Overstrand (Norfolk) 107, *112*, 132, 135, *136*, 137
Oxford Clay Formation *227*, *273*
Oxfordshire 47, **312**, 317, **390**

Pacific Ocean model for tsunami wave detection *74*
Pakefield (Suffolk) **83**
Palaeogene sands 121, *129*
Papua New Guinea, tsunamigenic activity **65**
paraglacial conditions, defined *262*
Parc colliery flow slide **109**
Peach Slide Complex *70*, **70**
peak ground acceleration (PGA) 44, 49, *52*
peat
  composition *246*
  compression 246–249
  defined 243
  engineering properties 245–249
  humification classification 245, **245**
  UK distribution *244*, 249–250, **250**
peat failures 101–102, *103*, **104**
peat fissures *106*
peat flow **104**
peat hazards 23, *23*, 25, 243–245
  mitigation 253–255
peat slide **104**, 134
  County Mayo *126*
  North Pennine 252, *252*
  Snowdonia 16
Pegwell Bay (Kent) loessic brickearth 190, *192*
Pembrokeshire **68**, **99**, 127
Penmaenbach talus slopes (North Wales) *132*
Pennine Hills 14, 45, 119
  Burnhope Seat peat slide 252–253, *253*
  Harthope Quarry peat flow 252, *252*
Penrith Sandstone Formation 407
Pentre colliery flow slide **109**, 119, *121*
perched water tables 9
periglacial, defined 25
periglacial environments 261
periglacial hazards 25–26, **266–269**
  cambering and superficial valley disturbances **266–267**, 276–280

deep weathering 262–265
shallow-slope movements 265, 268–270, 274–276
periglacial legacy 5
  impact on landslides 126–127
periglacial processes, impact on geotechnical properties *264*
permafrost 24, 261
Permian, halite and gypsum 30, 34, 369, 370, *370*, 375, *376*
Perthshire, Comrie 12, 45, 53
Peterhead (Aberdeenshire) **68**
Petrockstowe Basin 327, 328
phengite **224**
Phoenix Mine, Cornwall *334*
phosphatic rocks, effect on radon distribution 444–445
phreatic tubes 392, *394*, *396*
piles 399–400
pillar-and-stall mining
  chalk 316–317
  coal 292–293, *294*
  salt 373, 375
Pilling Moss **99**
pingos 5, 281
Pitlands Slip (IOW) **18**
planar translational slide 105, *107*
plastic limit 239
plasticity index 208, *209*, 229, 239
plate tectonics, role in earthquakes 44
plug-like deformation 265
Plumstead chalk mines *317*
Poldice Mine, Cornwall *332*
polyhalite 375
Porcupine Bank Mass Flow *70*, **70**
pore ice 261, 283
porewater pressure 108, 114, 269
Port Clarence 375
Port Glasgow eastern bypass 214
Port Mulgrave (N Yorkshire) **83**
Portavadie dry dock 213
Portland, Isle of 129, 131, 132
Portland Sand Formation *130*
Portugal *see* Lisbon earthquake
potassium, radioactive decay 434, *435*
pots/potholes 392
precipitation *see* rainfall
Preesall, salt strata *29*, 30
Preesall Halite 378
Purbeck (Dorset) **83**

Quaternary Period 259, *260*, 328
quick clay 21–22, *21*, 205
  failure mechanisms 208–212, **215**
  geotechnical character 208, **210**
  hazard management 216–218
  mode of formation 205–208, *206*, **209**
  UK examples 212–216
quick sand 43

Rabaul, tsunamigenic volcanism **65**
radiation dose equivalent 433, *434*
radioactivity 433

radium, in surface rocks 449–450
radon gas 35, *37*
  building build-up 438, *439*
  distribution, role of geology 438–439, *441*, *443*, 444, *445*
  hazard
    management strategies 451–454
    mapping and site investigation 448–451
    Northamptonshire 2, *449*
  health effects 435–436, *435*
  measurement 446–448
  potential hazard map *37*
  release and migration 436–438
  sources 433–435
rafts/grillages 400
rainfall effects
  carbonate dissolution 389
  debris flows 179, *181*, *182*
  landslides 96, 114–116, *117*, **139**
  shrink-swell soils **224**
Rampside (Lancashire) 47, 56
reactivation 111
Reading (Berkshire) 6–7, *8*, **18**, *311*, 317, *318*
Reading Formation 264
relict cliff systems 108, *110*
residual shear strength 274
residual subsidence 298
resonance 44
Rest and be Thankful (Argyll and Bute) 3, *169*, 171–172, *176*, *177*, 179, *180*
Rhondda Main colliery flow slide **109**
ribbed landslides 212
Richter Scale 44
ridge and trough topography 102, *105*
Ripon (Yorkshire) *409*, *418*
  gypsum dissolution 2–3, *3*, *405*, 407, 408, *409*, *409*, *410*, *413*, *414*, *418*
risk
  assessment 136, *137*
  defined 135–136
  economic **138**
  equations 136, 138, 139
  individual **138**
Rissa (Norway) 205
river flooding 11–12
Robin Hoods Bay (N Yorkshire) **83**, 131
rock glide 105, *107*
rock salt *see* salt
rock slides 128
rock slope failures 105–106, *107*
rock-cut stabilization measures *146*
Rockall Bank Mass Flow *70*, **70**
rockhead, wet and dry 372
rockhead anomalies **266–267**, 280–283
  mechanisms 281
rockhead (engineering), defined 263
room-and-pillar mining 292–293, *293*, *294*, 295, *295*
Roscommon County bog burst 96
Rossal Halite Formation 378
Rosseland quick clays (Norway) *209*, **210**

Rostherne Mere 371
rotational mudslide *93*
rough translational slide 105, *107*
Roughs landslide (Kent) 116, *118*
Roxby Formation 404, *406*, **407**, 408, 409
Royal Cornwall Polytechnic Society 331
Royal Geological Society of Cornwall 331
Royal Institution of Cornwall 331
running sands 9
Runswick Bay (N Yorkshire) **83**, 131
Runton (Norfolk) **83**
Russia, tsunamigenic activity **65**

sackung failure 102
St Bees Evaporite Formation 407
St Catherine's (IOW) 94
St Cleer Mine (Cornwall) *334*
St Dennis (Cornwall) *333*
St Dogmaels (Pembrokeshire) **99**, 127
St Jude (Quebec) 205
St Just (Cornwall) *329*, 357–358
Saint Liguori (Quebec) *215*
St Thuribe quick clays (Canada) *209*, **210**
Sakurajima, tsunamigenic volcanism **65**
salt (halite, rock salt, sodium chloride) 29
    karst and dissolution 369, *370*, 371–372
    mining
        brining 372–373
        modern mining 373, 375
        and subsidence hazards 29–30, *29*
            mitigation 380, 381, 384
    UK distribution 370–371
Saltwick Nab (N Yorkshire) **83**
Salwick (Lancashire) 378
sand fountains 48
sandstones and sands, uranium and radon 437, 446
Sandwith Anhydrite 407
saponite **224**
Sarno (Italy) 163, 178
sauconite **224**
Scarborough (Yorkshire) 4–5, *4*, 120, 131, 132, 134, 135, 145, 147
Scilly Isles, tsunami event **68**
Scotland
    carbonate dissolution **390**
    debris flows 163, 166–172
    earthquakes 12, 14, 53
    rock fall **18**
    snow avalanches 11
Scottish Highlands
    postglacial slope responses 128
    rock slope failures 105–106, *107*, 119, **133**
scouring 281
sea-level change 212, *216*
Seaford Chalk Formation *111*
segregated ice 261, *263*, 283
seismic events and hazards 5, 12–14
    isostatic uplift 128–129
    UK mapping 51, 53
    see also earthquakes

seismic risk, defined 43
seismic source characterization (SSC) 49
seismograph network *54*
Selborne (Hampshire) **99**, 129
sensitivity, quick clay 208, **211**, 214
Seven Sisters (Sussex) **83**
Sevenoaks Bypass (Kent) 5, *7*, 18, 82, *100*, 126, 268, *273*
shaft collapse 333, *334*, 335, *335*, 336, *336*, 337
shaft hedge 333, *335*, *358*
shaft mining 291–292, 293, 333
shallow-slope movements 265, 268–270, 274–276
Shanklin (IOW) **83**, 135
shear strength 108, *116*, 274
shearing 269
Sheffield, open cast mining site *295*
Shetland **68**, 102, 116
    Storegga slide deposits 67, *71*
Shih-Kang dam 43
Shirleywich (Staffordshire) 378–379
shoplifting 385
shrink-swell soils 22–23
    association damage costs 224–225, *228*
    behaviour 230–233
    characteristics 226, 228–230
    distribution 225–226, *226*, *227*
    engineering management 233–236, **234–235**
    formation 225
    mechanisms 230
    properties 223–224
shrinkage and shrinkage limit 239
shrinking soils *see* shrink-swell soils
Shropshire
    Buildwas 3, 99, **99**, 101
    earthquakes 14, 45
    Jackfield 3, 18, 82, *88*, **99**
    karst features **390**
    salt strata 371
Sicily, Etna, tsunamigenic volcanism **65**
silver mining 328
simple cliff systems 107, *110*, *111*
simple landslide systems 107
sinkholes 2–3, *3*, 392, *394*, *395*, *396*
    gypsum and anhydrite 32
    Ripon 409, *409*
Sklinnadjupet Slide *70*, **70**
Skye, Isle of *101*
sliding, mechanism defined 85, *89*, **95**, **96**
slope deformation 102, 105, *105*, 296
slope failure 3–5, 85
slope stability hazards 16–19
smectite **224**, 239
snow avalanches 11
Snowdonia (Gwynedd) 16, **18**, 82
soakaways 400, *417*
sodium chloride *see* salt
soil arches 392
soil erosion 11
soil gas, problems of methane 463
soil liquefaction 48
soil suction 229, *229*

soils
    collapsible *see* collapsible soils
    shrink-swell *see* shrink-swell soils
solifluction 265, **266–267**, 270, *274*
solution features *see* carbonate dissolution
solution pipes 392, *394*, *395*
solution sinkholes 392, *394*
Solway Moss 96, **99**, 102
Somerset **83**, **390**
Somerset Levels Formation 214
Sonning Cutting (Reading) **18**
Soufrière Hills, tsunamigenic volcanism **65**
South Crofty Mine (Cornwall) 343
South Oxhey (Hertforshire) *316*
South Wales coalfield 119
Southwold (Suffolk) **83**
spectral acceleration 44
SPOSH 469, *470*
spreading, mechanism defined 85, *89*, *91*, **95**, **96**
Staffordshire 378–379
    gypsum mining 411
    karst features **390**
    salt strata 29, *29*, 369, 370, *370*, 379
    Walton's Wood 5, *7*, 82, *88*
Staithes (Yorkshire) rock fall **18**
stalactites 392
stalagmites 392
Start Complex 325
stepping *297*
stepwise landslides 211
Stevenage (Hertfordshire) 317
Sticklepath Lustleigh Fault Zone 326
Stirlingshire, Glen Ogle 3
Stoke Hammond (Buckinghamshire) 268, *273*
Stonebarrow (Dorset) **83**
Storegga slide 6, *8*, 15, 48, *70*, **70**
    role in tsunami 63, 66–67, *69*
strain, in coal mining subsidence 296
Stromboli, tsunamigenic volcanism **65**
Studd Hill (Kent) **83**
sturzstrom 96, *199*
submerged periglacial terrains 285
subsidence *see* gypsum *also* mining *also* shrink-swell soils
subsidence factor in mining subsidence 297
subsidence in peat 250–251
    Derrybrien landslide 251–252, *251*
    Harthope Quarry 252, *252*
    peat slide 252–253, *253*
subsidence sinkholes 392, *394*
subsidence trough *292*, 295, 306
Suffolk 16, **18**, **83**, 134, **312**, **390**
suffosion sinkholes 396
Sula Sgeir Slide *70*, **70**
superficial valley disturbance (valley bulge) **266–267**, 276–280, *278*
Surrey **312**, 317, *317*, **390**
Sussex
    chalk and flint mines **312**
    chalk karst **390**
    cliff recession **83**, 121, *130*

Fairlight Glen 4, 124
    loess 20
Sutton Howgrave (Yorkshire) 408
swallow holes 392, *394*, *395*
Swanage Bay (Dorset) **18**
Swansea, earthquakes 14, 45
swelling index 239
swelling soils *see* shrink-swell soils
Syke, Isle of *90*

tailings ponds 338, *342*
Taiwan, Shih-Kang dam 43
talc **224**
Tamar Estuary 338, *343*
Tampen Slide *70*, **70**
Tangley (Hampshire) 397
Taren Taff Valley 3
tectonic hazards *see* seismic *also* tsunami
Teesside, salt strata 29, 30, 375, *376*, 384
Teifi, Lake (Wales) 127
Tennant Salt Mine (NI) 380
Thames Barrier (London) 12, *12*
Thames Valley rockhead anomalies *282*
thermogenic gas *see* methane
thorium, radioactive decay 434–435
thoron 433, 434
Thorpeness (Suffolk) rock fall **18**
Threlkeld Knotts landslide 86
tills, role in landslips 120, *125*
tilt, in coal mining subsidence 295–296
tin mining 328–330, 336–337
Tinley Park Mine (Sheffield) *295*
toe-buckling translational slide 105, *107*
Tohoku (Japan) tsunami 61
toppling, mechanism defined 85, *89*, *90*, **95**, **96**
toppling failures 105, *107*
Torbay (Devon) **83**, 92, 119, 188, 198, *199*
total organic carbon (TOC), significance for made ground 471, **471**
Totternhoe (Bedfordshire) 312, 317
Toys Hill (Kent) **99**
Traenadjupet Slide *70*, **70**
translational landslides 265
translational rockslide *93*, **142**
trees, impact on shrink-swell soils 233, *233*, 236–237
Treherbert (South Wales) 135
Trewavas Mine (Cornwall) 358, *359*, 360
Triassic
    anhydrite-gypsum 34
    salt strata 29–30, 369, 370, *370*, 372–375
    Sandstone 124
Trimmingham (Norfolk) **83**
Troedrhiwfuwch **99**
Trotternish Escarpment *101*
tsunami 2, 5–6
    defined 61
    generation processes 63–65, *64*
    hazards 14-16
    management and mitigation 74, *75*, *76*, 77
    mechanisms 15

tsunami (*Continued*)
    nomenclature 61, *62*
    source zones for UK 15, *15*, 65, **67**, *68*
    wave characteristics 62–63, *63*
    relation to earthquakes 48
tungsten mining, Cornwall and Devon 339
Tutbury Gypsum 404, *416*

Umboi, tsunamigenic activity **65**
Undercliff (IOW) 4, **83**, 119, 132
uninterupted landslides 211–212
uplift
    anhydrite-gypsum 32–33
    post-glacial 102
Upton Warren (Worcestershire) 380
Upwich (Worcestershire) 380
uranium 433, 434–435, 436, 449–450
USA, methane gas explosions **461**

valley bulging *see* superficial valley disturbances
Ventnor (IOW) 18, 163
vermicullite **224**
volcanic activity 6, *8*, 63–64, *65*
volcano flank collapse 48
vulnerability shadow 179, 183

Wadhurst Clay Formation 116, 124, *227*
Wales
    Aberfan and other coalmining landslips 2, 3, *3*, 16, **18**, 81, 106, *108*, **109**, 119
    carbonate dissolution features **390**, *396*
    earthquake monitoring 47, 48, 53, 55
    landfill methane **460**
    landslides 3, **18**, 90, 97, **99**, 119, *122*, *125*, 127, *132*, 135, *430*
    Snowdonia 16
    Wylfa Power Station 51
Walton-on-Naze (Essex) **83**
Walton's Wood (Staffordshire) 5, *7*, 18, 82, *88*
Warden Point (Kent) **83**, 97, 116, *120*, *123*
Warren Mine (Yorkshire) 409
Warwickshire, Ebrington Hill 131
waste disposal 375
waste tips and contaminated land, Cornwall and Devon 337, *340*, *341*
water
    impact on peat 245, *248*
    impact on shrink-swell soils 232–233, *232*
water consumption and use **118**
Wath (Yorkshire) 408
waves in tsunamis, characteristics 62–63, *63*
Weald Clay Formation *227*, 270
weathering, deep 262–265, *266–267*
wedge ice 261, 283
West Bay (Dorset) **83**
West Runton sand flows 92
wet rockhead 372
Wheal Anna Maria (Devon) *361*
Wheal Hermon (Cornwall) *358*
Wheal Jane (Cornwall) *342*

Wheal Maid (Cornwall) *342*
Wheal Owles (Cornwall) *358*
Wheal Peevor (Cornwall) 357, *357*
Whitby (N Yorkshire) **83**
White Chalk Subgroup 264
White Nothe (Dorset) **83**
Whitehaven (Cumbria) **18**
Whitestone Cliff (Yorkshire) 99, **99**, 101, 126
Whitlands Landslide 132
width-depth ratio, in coal mining subsidence 297
Wigan, collieries 7
Wight, Isle of
    Alum Bay **18**, *129*
    Blackgang *117*
    chalk and flint mines **312**
    chalk karst **390**
    cliff recession **83**, 114, 121
    landslides *148*
    Luccombe 116
    mud flows *92*
    Pitlands Slip **18**
    Shanklin 135
    Undercliff 4, 119, 132
    Ventor 18, 163
wild brining 372
Wilkesley Halite Formation 377
Wiltshire **312**, **390**
Windrush Valley 102
windypits 102, *105*
Winsford (Cheshire) 375
Withernsea (Holderness) **83**
Witton Flashes 373
Wolstonian Complex *262*
Wonder Landslide (Marcle) 129
Worcestershire 369, 370, *370*, 379–380
Working Party on Geohazards 9–12
World Heritage Site 321, 328, 337, 352
Wybunbury (Cheshire) 371
Wychbold (Worcestershire) 380
Wylfa Power Station 51

Yorkshire
    chalk and flint mines **312**
    chalk karst **390**, 392
    cliff recession 4–5, *4*, **83**, 120, 132, 134, 135, 145
    Darlington 409–411
    Kettleness 132
    Knipe Point 135, 145
    Ripon 2–3, *3*, *405*, 407–409
    Robin Hoods Bay 131
    Runswick 131
    salt strata *29*, *30*, 369, *370*
    Scarborough 4–5, *4*, 131, 134
    Staithes rock fall **18**
    Whitby 131
    Whitestone Cliff 99, **99**, 101, 126

Zechstein Group 370, 371, 375, 404
Zhouqu debris flow 178, *179*